HANDBOOK OF FRESHWATER FISHERY BIOLOGY

FISHERY BIOLOGY VOLUME ONE

*Life History Data on Freshwater Fishes of the United States
and Canada, Exclusive of the Perciformes*

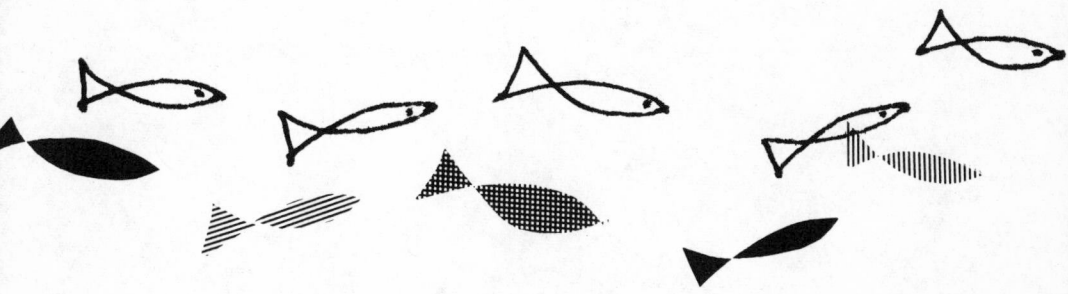

KENNETH D. CARLANDER

ROBERT L. PFEIFER

THE IOWA STATE UNIVERSITY PRESS, AMES, IOWA

KENNETH D. CARLANDER is Professor of Fisheries, Department of Zoology and Entomology, Iowa State University, where he teaches courses in fishery management, fishery research techniques, fishery resources, population dynamics, and herpetology. He has served as President of the Iowa Academy of Science and the American Fisheries Society, and has been United States Representative, International Association of Theoretical and Applied Limnology, attending Congresses in Madison, Wisconsin, in Poland, and in Israel. Besides this book his publications include chapters in books concerned with conservation, limnology, and fishing, as well as articles for the professional journals of his field and such magazines as *Field and Stream, Outdoor Life,* and *Successful Farming.*

© 1969 The Iowa State University Press
Ames, Iowa, U.S.A.
Printed in the U.S A.
Third edition, 1969
International Standard Book Number: 0-8138-0709-3
Library of Congress Catalog Card Number: 69-18736
Second printing, 1970

1950 edition. 1953 edition and Supplement © Wm. C. Brown Company

HANDBOOK OF FRESHWATER FISHERY BIOLOGY

VOLUME ONE

ROBERT L. PFEIFER

A publication of the Iowa Cooperative Fishery Unit, sponsored by the Iowa State Conservation Commission; the Bureau of Sport Fisheries and Wildlife (U.S. Department of the Interior); and the Iowa Agricultural and Home Economics Experiment Station, Iowa State University. Preparation of this revision was aided by a grant (G-9471) from the National Science Foundation, Office of Scientific Information.

Preface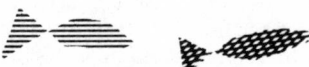

THE DECISION to publish in sections this revision of my 1953 *Handbook of Freshwater Fishery Biology* was forced by the realization that it will take at least two more years to complete the manuscript for the rest of the book and by the desire to make the compilation available for use as soon as possible. Furthermore, two volumes are better than one because of the bulkiness of the data.

The revision has taken much longer to prepare than was expected. The second edition and the first supplement published in 1953 by Wm. C. Brown Co., Inc., have been out of print for almost ten years. In 1955-56, when a second supplement was due, the National Research Council was planning a "Handbook of Biological Data for Aquatic Biology" for which many fishery biologists were asked to prepare sections. I felt that such a handbook might meet many of the needs for which my earlier handbook and supplement were designed. When the National Research Council decided in 1958 to drop plans for publication of its handbook, it was evident that a supplement to my 1953 handbook would not be adequate and that a more comprehensive revision was necessary.

The *Handbook of Freshwater Fishery Biology,* Vol. I, has a more limited scope and purpose than the abandoned "Handbook of Biological Data." In the latter it was proposed that existing knowledge in aquatic biology be tabulated in concise and parallel form, somewhat like the well-known *Handbook of Physics and Chemistry,* but it was found that the state of knowledge was not at a level where many constants or normals could be adequately expressed. This present handbook aims rather at serving as an index to the literature and at demonstrating the gaps in information on freshwater fishes of the United States and Canada.

This revision was begun in 1959 when the Office of Scientific Information, National Science Foundation, provided a grant to assist in preparation of the manuscript and to relieve the author of teaching assignments for a year. Tabulation of data from the literature had been continued since the appearance of the handbook and the first supplement, but revision and preparation of the manuscript obviously would take much more time than was available with my other duties. Unfortunately, conditions never permitted as much freedom from the other duties as was planned.

This revision includes data on several phases of the life histories of the fishes not previously covered. Some phases now included in the compilations are not as adequately covered as others. It was not possible to review all the literature already covered to abstract the data on the additional phases now included. It is hoped that the additional information will more than balance the delays and the inadequacies in some areas.

The data being published in the areas summarized in this handbook are too great to be conveniently handled without a computer. Much of the available data, however, are not presented in a fashion to be readily used in a computer program. Research for developing a suitable computer program and standard methods of presentation of data should be begun by someone in the near future.

The first edition, published in 1950, included data from 1,122 citations; the first supplement, published in 1953, brought the list to 1,593; the citations in the present volume and those already examined for the second volume number more than 4,400. The amount of time required to record the data from each citation is from 15 minutes to 15 hours and probably averages an hour, not counting the time needed in preparation of the tables and manuscript. Papers appearing after December, 1966, are not included in this volume.

It is not practical to list the many persons and organizations who have helped in the preparation of this Handbook. The Fish and Wildlife Service, United States Department of Interior, kindly loaned me its copies of the summaries prepared by many scientists for the proposed "Handbook of Biological Data in Aquatic Biology." Where these summaries have been used they are listed in the citations. The grant (G-9471) from the Office of Scientific Information, National Science Foundation, served as a stimulus to begin the revision and made much of it possible. The Sport Fishing Institute, Inc., Washington, D. C., provided a grant to aid in publication. Special recognition should also be made of the Iowa Cooperative Fisheries Research Unit, and those persons from the Iowa State Conservation Commission and the Department of Zoology and Entomology at Iowa State University of Science and Technology who have been associated with it. Special thanks, also to my wife, Harriet, for typing the bibliography and for helping with several other details.

KENNETH CARLANDER

Contents

HANDBOOK OF FRESHWATER FISHERY BIOLOGY

VOLUME ONE

Introduction

THIS HANDBOOK is first of all an index to the literature on certain aspects of the life histories of freshwater fishes of the United States and Canada. Much of the literature in fish biology is scattered and difficult to locate. Much of it is in separate leaflets, bulletins, or periodicals published by state conservation commissions, universities, research laboratories, and other agencies. *Biological Abstracts, Zoological Record,* and, in particular, *Sport Fishing Abstracts* summarize and provide help in locating some of this literature, but these do not provide as complete an indexing as is attempted in this handbook. Many of the life history items are in papers dealing with several species or with broader ecological studies, and thus the life history items for the various species often are not indexed in the abstracting journals.

Some of the information in the handbook comes from papers which cannot accurately be described as published: mimeographed or otherwise duplicated bulletins, theses, and typed manuscripts. Criteria for inclusion or exclusion of such data have not always been consistent. In general, availability of the manuscripts to the compiler has been the first criterion. Much other equally or more significant data are in the files of research laboratories around the country. The progress and final reports of federal aid projects have a great deal of life history data, only a small part of which is included in the handbook. These data are particularly difficult to summarize because many are provisional or "in progress" and because it is often difficult to determine which have also been included in other reports. I will attempt to provide Xerox or other duplicator type copies of papers not otherwise available, at the cost to me of about 10¢ per sheet. A sheet may often take two pages of a publication.

References to the sources of the data in the tables and text are indicated by code numbers consisting of the first letter of the author's (or first author's) last name and a number. These code numbers lead to the full citation in the Citations section. The code numbers are not always consecutive. The missing numbers refer to papers which are not cited in this volume but which will appear in a later volume of the handbook.

In addition to serving as an index to the literature, the handbook summarizes, for quick reference, the available data on several aspects of the life histories of each species. The student is advised to go to the original paper for further information and for verification. It is hoped that these summaries also will indicate the gaps in our knowledge and lead to further research and publication of the needed information. On some species, the general aspects of life

3

history and growth are so well documented that there is little need for further compilation.

I do not wish to imply that the lack of information in this handbook necessarily indicates a lack of available information. While I have tried to give as complete coverage as time and facilities permitted, I can make no claim that the handbook is complete. Many significant papers have undoubtedly been overlooked and I will appreciate having them called to my attention. Coverage is more complete on the length conversions, length-weight relationships, and age and growth sections than on other aspects of life history. In the early stages of preparation of the handbook these were the only data recorded and when it was decided to expand the areas of life history to be summarized it was not feasible to go back over the previously reviewed papers. In general, the data included on these other aspects were only the data uncovered in looking for the age and growth summaries and little effort was made to adequately search the literature in these other areas.

Age, growth, length-weight, and condition factor data are valuable in describing the general life history of a fish. They are more valuable from a management viewpoint, however, when they can be compared with similar data from other populations. The tabulations in this handbook provide such quick comparison and guide the biologist to the original papers for more detailed analyses.

The needs of fishery biologists and others working in fish conservation and management were of prime interest in selecting the life history areas for inclusion in the summaries. Numerical data which could be more readily included in tabular form were given precedence over material which would require extensive descriptive analysis.

Not all of the data in the tables are of equal value. Some represent careful, detailed studies and others are mere chance observations. Although it might be desirable to rule out certain data thought to be less accurate, it was almost impossible to draw any lines representing what might be considered sufficient accuracy. Inaccuracies may be present in even some of the most carefully executed studies.

As a measure of the variation in the relationships, ranges are given whereever possible. In some cases, the ranges as given are merely the ranges of means from various lakes, ponds, year classes, etc., and do not accurately depict the entire range shown in the samples. Range is not an entirely satisfactory measure of variation. Standard deviation (as used in S65, L30, E3, and others) and analysis of variance (as used in M77, M78, R74, S20, and others) give better measures of variation and also permit statistical treatment to determine the significance of differences between sets of data (S125 and other statistical texts).

Fish show great variations in growth, length-weight relationships, condition, and other measurements. Even within a given population or year class in a lake the variation may be large. For this reason, the application of data from one body of water to a population in another body of water, or to the same population at another time, should be made with great care and with full recognition of possible variation and probable errors.

Small samples may give erroneous pictures of growth or length-weight relationships. The size of sample needed depends upon the degree of accuracy desired and the variation within the sample. One study (C32) indicated that samples of about 240 fish were needed to be confident that in 19 of 20 times the calculated growth from the sample did not differ by more than 1 percent from the growth in the population and that about 70 fish were needed to keep the error to less than 2 percent. Selective sampling methods will cause errors in the estimate even though the number of fish in the sample is large.

The metric system of measurement is used throughout the tabulations unless otherwise indicated. Lengths are in millimeters, weights in grams, standing crops or production in kilograms per hectare, and temperatures in degrees centigrade. The manuscript had been prepared using the English system of measurement rather than the metric since the English system has been widely used in fishery biology and management in North America. In September 1967, the American Fisheries Society formally adopted "the metric system as its officially recognized system of weights and measures" and urged each member "to work for the universal adoption of the metric system in his personal contacts with others." Beginning with Volume 98 of the *Transactions of the American Fisheries Society* all measurements are to be in the metric system, although English equivalents may follow in parentheses.

Since most of the tables were originally compiled on the English system of measurement, the class ranges are a bit unnatural for the metric system. The class ranges were already somewhat unnatural since the ranges given in various publications differed and it was sometimes necessary to include data which did not really belong in the designated class. For example, data given for fish 45-64 mm, would be included in the class 51-75 mm since the mean value would be in this range and since it was not practical to try to get the original data and reassign it to the ranges used in these tabulations.

Tables for conversion from English to metric are given for ready reference.

The species are arranged according to the American Fisheries Society 1960 list of common and scientific names of fishes (A13) except in a few cases where more recent literature indicates the need for change.

Although the data are limited to freshwater species of United States and Canada, some references to these species in other parts of the world are included. Coverage of the foreign literature is not at all complete, however.

Data on the Pacific salmon, *Oncorhynchus* spp., have not been included, except for a small amount of data on the kokanee which have been introduced into areas outside of the Pacific Northwest.

Under each species the data are listed in the following order:

Range and habitat (These are usually given in general terms and are presented only for quick reference. They do not represent extensive literature review.)
Conversion factors for various length measurements
Length-weight relationships
Ponderal indexes or condition factors
Observed lengths and weights at various ages
Calculated lengths and weights at various annuli
Discussion of growth data
Age at maturity and reproduction data
Food habits
Sizes taken in various mesh sizes of gill nets

Occasionally, other data on mortality, blood counts, temperature tolerance, and management are briefly recorded. In some species where the total write-up is short, the above order may not be strictly adhered to. The following discussions will help in interpreting the various sections.

CONVERSION FACTORS FOR LENGTH MEASUREMENTS

Unfortunately, there is no generally accepted method of measuring fish, despite several papers on the subject (R74, R31, R41, C37, H68, S508, K132). At least eight standard lengths, one fork length, and two total lengths have been used in fishery work (H68). Total lengths are used throughout this text except as otherwise indicated.

Standard length (SL) has been variously defined, but is essentially the length of a fish from the tip of the snout to the end of the vertebral column. Standard lengths are extensively used in taxonomic studies, but now are rarely used in other aspects of fishery biology.

Fork length (FL) is from the tip of the snout to the end of the rays in the center of the caudal fin. Fork lengths are the usual lengths used in marine fisheries, and have been used extensively in fresh water fisheries in Canada, Europe, and some states in United States. Often they are reported as total lengths, "measured to the center of the fork of the tail."

Total length (TL) as defined by Hile (H68) is "the distance from the tip of the head (jaws closed) to the tip of the tail with the lobes compressed so as to give the maximum possible measurement." Total length to the tip of the tail in its "natural position" should be dropped from usage.

In many cases, no indication of the length measurement used is given in the original papers and it has been necessary to arbitrarily assign the data to one of the lengths. When an author, or group of workers in a given research unit, has used a given method of measurement in one paper, the same method is assumed to have been used in other papers which do not indicate the type of length used.

Factors for converting one type of measurement to another are given for each species when such data are available. Additional clues as to the relationship between various types of lengths sometimes can be obtained from the tabulated growth data. In most cases, the conversion factor has been determined by dividing the TL (or FL) by the SL. Where the ratio varies as the size of the fish increases, separate conversion factors often have been determined for fish in various size ranges. Better statistical methods of describing the relationships between SL and TL or FL are available, but present evidence indicates that differences in the ways in which various workers measure fish are more significant in causing errors of conversion than are the methods of calculating the conversion factors (C32).

LENGTH-WEIGHT RELATIONSHIPS

Since it is frequently necessary to estimate the weights of fish from their lengths, data have been compiled on the length-weight relationships of many of the species.

For several species the data from several populations are combined rather than listing data from each paper. In these compilations the mean values from each population or each paper are weighted as follows:

Weight	Data based on:
1	1 fish
2	2-5 fish, or number of fish not indicated
3	6-10 fish
4	11-99 fish
5	over 100 fish

While this weighting system is somewhat arbitrary, it is believed to be superior to an unweighted mean since small samples are not given equal weight with large samples and, at the same time, this weighting prevents the average from being unduly affected by a large sample from a single population. Furthermore, this system permits the inclusion of data for which the size of the sample is not indicated. The weighting system is used both in deriving the mean of the means and in determining the central 50 percent. The central 50 percent should not be considered as including 50 percent of the fish of the size range, but 50 percent of the means. In some cases the mean of the means may not be included in the central 50 percent, because the mean of the means is being affected by a widely divergent mean. In these cases the median might be a more representative average and this value can be approximated by taking the center of the central 50 percent range.

Several authors have determined the mathematical relationships between length and weight for various populations using the formula:

$$W = cL^n$$
$$W = \text{weight}$$
$$L = \text{length}$$
$$c \text{ and } n = \text{constants}$$

The value of the constant n will usually be near 3.0 since the weight of an object will vary as the cube of its length if shape and specific gravity remain the same. Since this formula must be calculated in the logarithmic form and is most usable in this form, the length-weight relationships in these compilations are given in the logarithmic form:

$$\log W = \log c + n \log L$$

With the conversion of all measurements to the metric system, it has also been necessary to convert the regressions based on English measurements. The value of the slope, n, remains the same but the value of log c has to be adjusted for the unit length and for the weight units.

For example, with W in .01 pound and L in inches: 1 mm = 0.03937 inch. Thus the new unit length is $\log L = -1.405$ instead of $\log L = 0.0$, and the intermediate $\log c = \log c + n(-1.405)$. But this intermediate log c must be adjusted for the new weight units: 0.01 pound = 4.536 g, which gives a log value of 0.6567. The new $\log c = $ intermediate $\log c + 0.6567$.

As an example, if $\log W_E = -1.2925 + 2.768 \log L_E$

then

$$\log W_M = -1.2925 + (2.768 - 1.405) + 0.6567 + 2.768 \log L_M$$
and

$$\log W_M = -4.5248 + 2.768 \log L_M$$

With W in pounds and L in inches, the correction factor for weight units is 2.6567.

With W in ounces and L in inches, the correction factor for weight units is 1.4524.

With W in grams and L in centimeters, the conversion to millimeters is

$$\log W = \log c - n + n \log L$$

because 1 mm = 0.1 cm, which gives the log value -1.0 which must be multiplied by n to make the adjustment for unit length.

It has been stated that the slope will usually be above 3.0 because most fish become plumper as they grow. If the entire life span were considered, this generalization would probably be true because fry and young fingerlings are usually

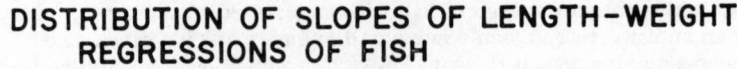

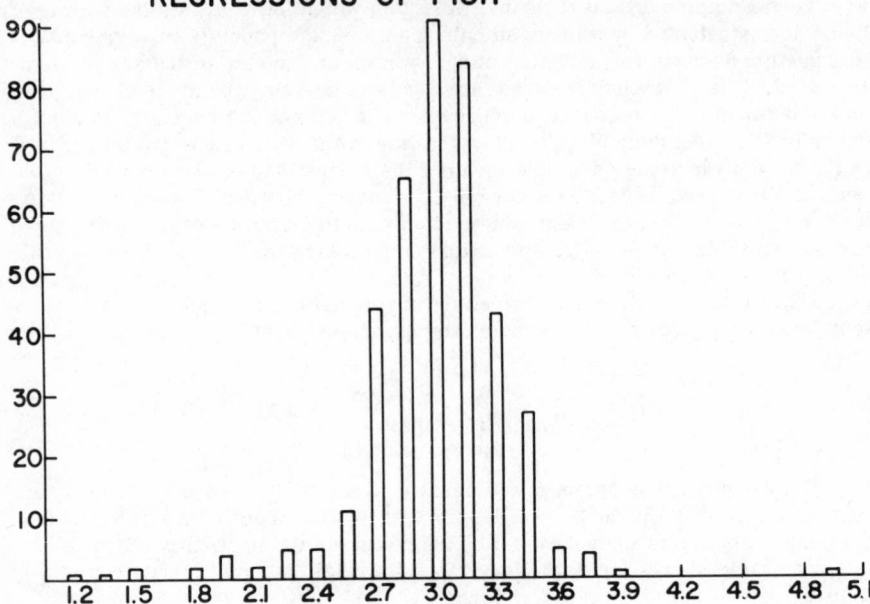

Fig. 1. Frequency distribution of slopes, n, of length-weight regressions of fish from 398 populations.

quite slender, but most length-weight relationships do not include these early stages of growth and changes in body form are less characteristic during most of the life span. A frequency chart (Figure 1) of the slopes of the 398 populations included in this volume indicates a slight tendency for the slopes to be above 3.0, but the mean is 2.993.

Some species of fish characteristically increase in plumpness with increase in length and thus usually have slopes of over 3.0. This is true of the larger

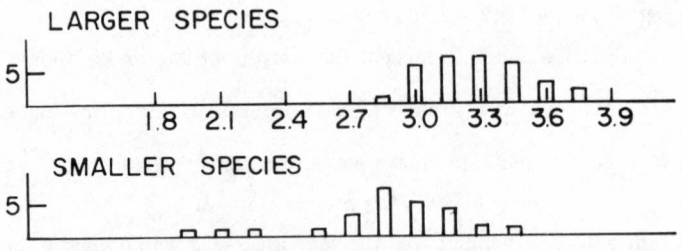

Fig. 2. Distributions of slopes of length-weight regressions of larger species of catfish (channel, blue, white, and flathead) and of smaller species (bullheads, stonecats, and madtoms).

species of catfishes (channel, blue, white, and flathead) (Figure 2). On the other hand, most of the length-weight regressions for the smaller species of catfish (bullheads, stone cats, etc.) have slopes below 3.0. Many of the populations of bullheads which have been studied were overcrowded and stunted with the longer fish in poorer condition than the smaller individuals in the same population.

Slopes of all 7 populations of gar (*Lepisosteus* spp.) were above 3.1. Slopes of most of the *Salvelinus* populations were above 3.0 but the *Salmo* populations

DISTRIBUTION   OF   SLOPES   OF   LENGTH-WEIGHT REGRESSIONS   OF   SALMONIDAE

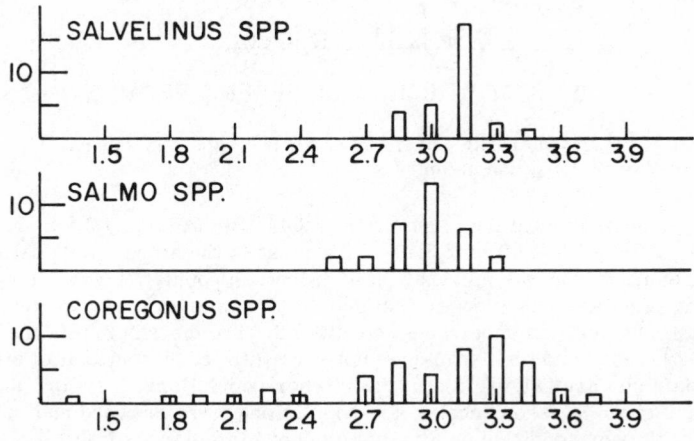

Fig. 3. Distributions of slopes of length-weight regressions of trout and whitefish.

usually had slopes below 3.0 (Figure 3). The *Coregonus* populations showed a wide variation in slope with the mode at 3.3. *Esox* populations usually had slopes above 3.0 (Figure 4). The three values below 2.0 represent prolarval length-weight relationships which will be discussed later.

Length-weight regressions are available (S356) for 24 populations of carp harvested from ponds in Japan (Figure 5). These fish were mostly 250-500 mm long. In some cases the populations represent the same ponds in different years and there seems to be a tendency for the slopes to be characteristic of a pond. For example, the three highest slopes, 3.21, 3.32, and 3.52, were from one pond; the two lowest slopes, 2.42 and 2.45, were from another pond which had 2.77 a

DISTRIBUTION OF SLOPES OF LENGTH–WEIGHT REGRESSIONS OF <u>ESOX</u> SPP.

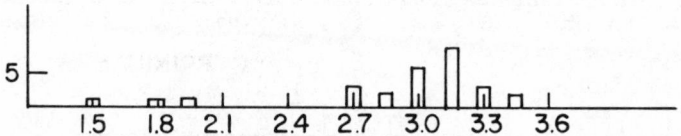

Fig. 4. Distribution of slopes of length-weight regressions of *Esox* spp.

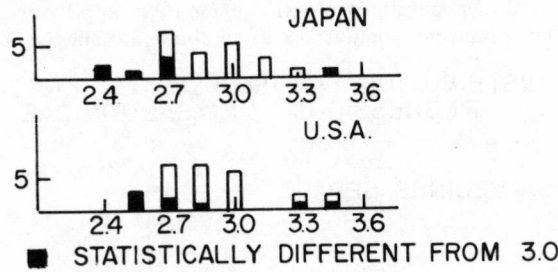

Fig. 5. Distribution of slopes of length-weight regressions
of carp in Japan and in United States.

third year; in another pond the slopes were 2.64, 2.70, 2.73, and 2.98, and in a
fourth pond, 2.89, 2.94, 3.00, and 3.10. That most of the slopes were below 3.0
may have resulted from the fact that the fish in each population were harvested
at about the same age. The longer fish may not be quite as heavy for their
length as the shorter fish under these conditions. The distribution of the slopes
of carp from these ponds in Japan does not seem different from that of other
available data on carp (Figure 5). In these other populations, the slope may be
less than 3.0 in some cases because the populations were crowded and in poor
condition, with poor condition being most evident in the larger fish.

Shimadate *et al*. (S356) tested whether the various slopes differed from 3.0,
at the 95 percent confidence level, and those which did differ are so indicated
on the graph. Significant deviations from 3.0 are also indicated on the U.S.
carp data, but in several of these populations no test of significance was made
and thus some of the others might also be significantly different. Variance
within the sample is an important factor in tests of significance (see S125, or

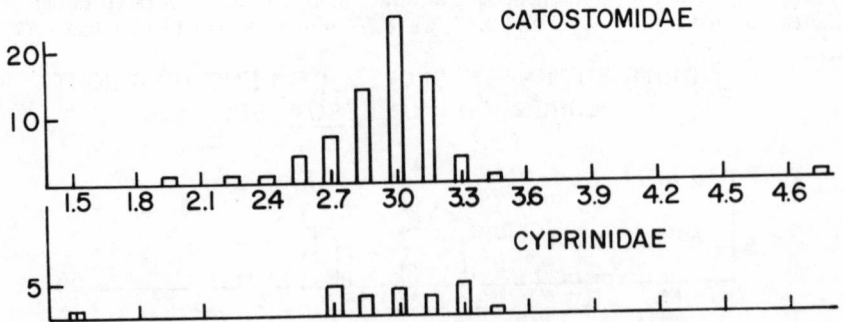

Fig. 6. Distributions of slopes of length-weight regressions of suckers and of minnows.

other texts for suitable tests, and H200 and M279 for examples), and most authors have not analyzed the variance.

Data are available on only a few other cyprinid fishes, which show slopes about evenly spaced above and below 3.0 (Figure 6).

Regression slopes outside the range of 2.5 to 3.5 cannot apply over a wide range of lengths without profound changes in body form of the fish. As an illustration, five regressions were computed with the 200 mm fish in each population weighing 120 g and with slopes of 2.0, 2.5, 3.0, 3.5, and 4.0. With regression slopes of 2.0 and 4.0, the body depths at 100 mm and at 400 mm are obviously outside the usual range (Figure 7). At 2.5 and 3.5 the body form is near the usual limits and fish farther from 200 mm would show more extreme body forms even with these slopes. At a slope of 2.0 the condition index is halved each time the fish length doubles and at a slope of 4.0 the condition index doubles with a doubling of length. At slopes of 2.5 and 3.5 the same change in condition index occurs with a quadrupling of length.

What was the situation in each of the regressions which were reported with slopes less than 2.5 or over 3.5? The tabulated slopes (Figure 1) did not include data on 21 species from one paper (S472). These eliminated slopes ranged from 0.48 to 2.59, and all involved fish from 25-152 mm long (in 18 species the maximum was 102 mm). In each case, the tabulated weights and condition indexes for the 25 mm fish (or the 51 mm fish in 4 species with no 25 mm fish) were obviously too high, indicating errors in weighing the fish in the field (see data on silver chub for an example). It is highly unlikely that the 25 mm fish could have an average condition factor 2 to 3 times as high as the longer fish. The reported slope was 2.44 but the condition factors of the 51 to 152 mm fish indicate that the slope should be about 3.0 for this range in lengths. Data on one other regression, with a slope of 4.12, was also eliminated since it included only 51 and 76 mm fish. Regressions from 47 populations in S472 are included in the tabulations because in these populations the size range was much greater and the effect of errors in weighing the smallest fish was believed to be small. In some of these, the reported slope is probably somewhat lower than it should be to correctly represent the population. Errors in weighing or measuring the fish may have affected some of the slopes in other papers reported in the tabulation.

Errors in computation may also have affected some of the tabulated slopes. One paper (H156) reported a slope of 2.80, although the condition indexes obviously increased with increase in length of these black buffalo fish. Many papers do not give adequate information to judge the accuracy of the regression slope.

One paper (H64) on ciscos, *Coregonus artedi*, included 3 populations with regression slopes of 3.62 to 3.69, and 5 populations with slopes of 1.38, 1.77, 2.00, 2.05, and 2.31, as well as 5 with slopes between 2.67 and 3.51. The slopes of over 3.6 were based upon samples in which the length ranges were 200-230 mm. These three populations represented different years from the same lake — a lake with a deep-bodied cisco which apparently gets more than proportionately deeper bodied with increase in length. The slopes of less than 2.32 were based on samples in which the range in lengths was not over 49 mm. The maximum length in these samples was 164-179 mm. These fish were taken with gill nets and mostly in one mesh size. It is likely that the shortest fish in the gill nets were usually among the plumpest at that length and the longest fish were among the thinnest and that the regression slopes were biased downward by this gill net selection. Such a bias was suggested by J6 when the slope of 2.73 was reported for bloaters in Lake Michigan; by J5 when a slope of 2.47 was reported for shortnose cisco; and by M364 when a slope of 1.27 was reported for male

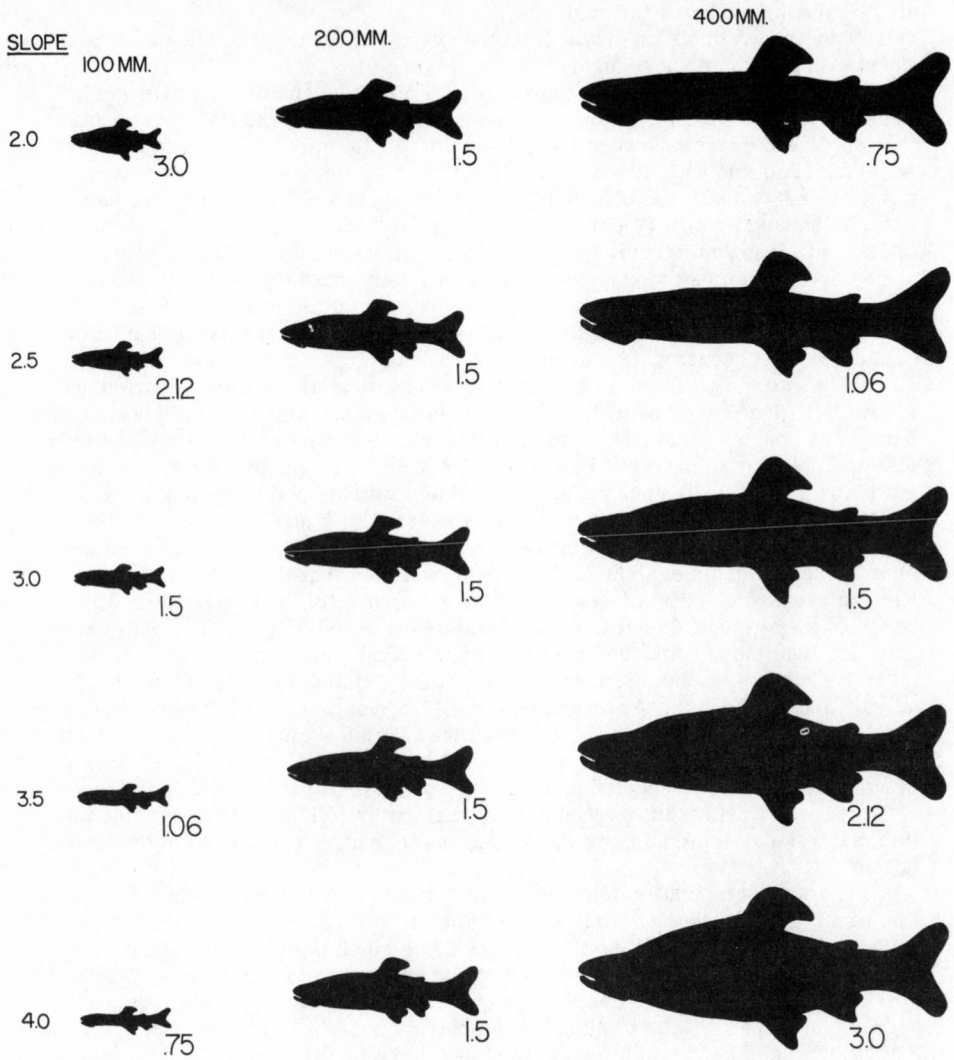

Fig. 7. Body form of trout at 100, 200, and 400 mm in theoretical populations with length-weight regression slopes of 2.0, 2.5, 3.0, 3.5, and 4.0. The condition factor, k, is given for each individual.

shad in linen gill nets although the slope was near 3.0 for nylon nets. K134 concluded that length-weight regressions could be satisfactorily completed with gill net samples provided the sample included a wide range of lengths, but that usually gill net catches from one mesh size do not give reliable estimates of the population length-weight regression.

Some of the regressions which are outside the 2.5-3.5 range refer only to short spans in the early life of the fish. For example, F114 report slopes of 1.54, 1.74, and 1.96 for prolarval stages of pike while the slopes for the alevin-juvenile stages were 2.7-3.0. M303 reported a slope of 4.91 for suckers 12-20 mm long.

Length-weight relationships of fish collected during spawning runs may be affected by the relatively short range of lengths included and by differences in stages of gonad development associated with lengths of fish. R14 reported slopes of 2.23 and 2.42 for female and male suckers migrating upstream to spawn, and 2.51 and 2.59 for females and males going downstream. Each sample included 25 fish and the range in lengths was 102, 83, 76, and 64 mm respectively. O1 reported a slope of 2.28 for 180 adult female alewife, 110-145 mm, and of 2.41 for 211 adult male alewife, 97-138 mm, in Seneca Lake.

The slopes of 1.97, 2.16, and 2.29 reported for bullheads (F65) illustrate another situation. The 1.97 and 2.16 slopes represent 48 females and 38 males of the 1946 year class collected in Ventura Marsh in the summer of 1951. The data cover a fairly narrow range of lengths, from 178 to 221 mm, except for one male at 257 mm. The average condition factor K(SL) dropped from 3.12 to 2.60 for females and 3.28 to 2.67 for males from mid-July to the end of August. The longer fish in the sample were collected later in the year when the condition was poor and thus the slopes were low. The 2.29 slope represents 257 bullheads collected in Clear Lake in 1952, almost entirely from the 1946 year class (age VI) which was showing definite signs of slowing growth and poorer condition, particularly of the longer individuals.

P122 reported a slope of 1.57 for golden shiners but this was based upon only 8 fish presumably all of one age group. The slope of 1.98 reported for male river carpsuckers (H420) appears to be biased by 2 fish which are unusually heavy at 400 mm, the lower limit of size range, and by 1 fish which is unusually thin at the upper limit of the range. The other 33 fish in the scatter diagram obviously have a slope near 3.0.

Five slopes between 3.55 and 3.74 for channel and flathead catfish appear to be valid and related to the tendency for larger catfish to be obviously heavier bodied as they grow. A slope of 3.74 was also reported for 33 male blackfish (B260).

All slopes outside 2.5-3.5 have been explained above except: The slope of 2.20 for whitefish from Waswanipi River and Lake (M431) which was based upon whitefish from 310 to 470 mm FL but data for checking the accuracy of the regression are not given.

PONDERAL INDEXES OR CONDITION FACTORS

Since the weight of a fish varies with the cube of its length provided the shape and specific gravity remain the same, any change in the shape or relative plumpness of a fish will cause a change in the value of c in the formula:

$$W = cL^3$$

Fishery biologists have used this fact in describing the condition, plumpness, or well-being of a fish. The coefficient of condition, K, has been widely
used:

$$K = \frac{W \, 10^5}{L^3}$$

where W = weight in grams

L = length in millimeters

and 10^5 is a factor to bring the value of K near unity.

The method of length measurement is indicated as K(SL), K(FL), or K(TL).
The same formula, using the English rather than the metric system, can be
used to calculate C. The value of C is usually calculated from lengths in inches and tenths and weights in hundredths of pounds and is usually adjusted so
that there are two digits in front of the decimal point, instead of one as in K.
A condition factor, R, has also been used where weight has been recorded in
grams and length in inches (C132). In this handbook all ponderal indexes are
given as K, but conversions can be made as follows:

$$K = 0.277C = 0.610R$$

The K values are reported according to the type of length measurement used,
SL, FL, or TL. If the ratio, r, of SL to FL or TL is known, conversion can be
made as follows (K16, B30):

$$C(TL) = 3.61 \, r^3 \, K(SL)$$

To eliminate the trends in condition associated with increase in length,
LeCren (L213) proposed a relative condition factor K_n where

$$K_n = W/aL^n$$

or where the observed weight is divided by the expected weight as computed
from the length-weight regression for the population. While the relative condition factor is useful in certain studies, it is not suitable for comparisons among
populations and it assumes that the length-weight relationship remains constant
over the period of study.

Condition may also be studied by comparisons of length-weight regressions
using analysis of covariance (M77, M78, H200, B344).

Length-weight relationships and condition factors vary with season, sex,
sexual maturity, age, and various other factors. These relationships are reported for each species when they have been discussed in the reviewed papers
or were evident from the data. In the compilation of the data, however, these
trends are often not discernible and one should refer to the original papers for
more detailed analysis.

OBSERVED LENGTHS AND WEIGHTS AT VARIOUS AGES

Throughout the compilations, the ages are indicated by Roman numerals. A
numeral III indicates that the fish has passed through three winters. In some of
the reviewed literature, such a fish would be referred to as being in its fourth
summer. In accordance with a suggestion by Hile (H68), the first of January is
taken as the birthday of all fish, and fish taken in the early part of the year are
therefore credited with an annulus at the edge of the scale even though such

annulus was not yet completed. For fall spawning fish it is assumed that the eggs do not hatch until after the first of the year — even though they may do so, particularly in hatcheries. These fish are listed as age 0 until January of the next year, when the fish are just over one year old.

The ages of the fish were determined by one of the following methods (L135):

Direct observation: This method applies, when the age of the fish is known because the fish have been under observation, as in hatcheries, and rearing ponds; or when fish of known age have been tagged or otherwise marked so that they can be recognized upon recapture; or when fish are introduced at a known age into waters where the species was not previously present and as long as the introduced fish can be distinguished from their progeny.

Length-frequency or Petersen method: Age groups can often be separated and identified as peaks in frequencies when the numbers of fish are plotted against their lengths. In later years of life, the sizes may overlap so much that the peaks cannot be detected, but the method is usually useful during the first year or two. Where growth is fairly uniform, peaks may be followed several years.

Growth rings: Scales, otoliths, vertebrae, spines, fin rays, opercles, and other bones, frequently show annual growth rings by which the age of the fish can be determined. Examination of such annual rings or annuli is the source of most of the growth data past the second year of life.

It must be recognized that the estimation of age from length frequencies and from growth rings may involve some errors, particularly among older fish. In the compilations, where mean of means and central 50 percent ranges are given, the data are weighted as described under length-weight relationships. Also, comparisons of lengths and weights at a given age are somewhat approximate since some of the fish may have been measured during the early part of the age range and others toward the end. For example, some fish listed as age I may have been measured in March before the season's growth started and others in November after an additional season's growth.

CALCULATED GROWTH

From measurements on the scales, bone, or spines, the size of a fish at certain times earlier in life can be estimated. Measurements to the growth rings are usually assumed to be directly proportional to the length of the fish at the time that the growth ring was formed. The lengths at each annulus are then computed by the Dahl-Lea method:

$$\frac{Sn}{Sc} = \frac{Ln}{Lc}$$

where

    Sn is scale (or bone) measurement to a given annulus, n
    Sc is scale measurement to edge
    Ln is length of fish at the time of annulus, n
    Lc is length of fish at capture

It has been found that the growth of body length and of scales or bones is not directly proportional and that more accurate calculations can be made using other formulae (L135). The commonest correction is the Fraser (V5, p303; R282, p324) or Lee Method (L135, p152) which assumes a straight-line body-scale regression with an intercept at some place other than zero on the ordinate.

$$L = a = cS$$

$$\text{or } Ln = a = \frac{Sn}{Sc} (L_c - a)$$

The value of a has frequently been interpreted as the length of the fish at the time of scale formation (R282, p324) but this interpretation is not necessarily correct. The value of a is the intercept that will give the best straight-line relationship. Early growth of the scale usually is not proportional to that of the body even though scale growth may be proportional through much of life.

More accurate estimation of past growth may come from more precise description of the body-scale relationship (the Sherriff, Monastyrsky, etc. methods in L135).

The estimated past growth histories by these various methods are usually comparable except for the first year or two of life and the data are thus tabulated together. The method of computation, if not the Dahl-Lea method, is usually mentioned in the discussion which follows the tabulation.

The increments given in the tables are often not the increments as reported in the original paper but were computed from data given in the paper. The increments are computed to include data only from the fish completing the year's growth. Frequently, increments are recorded by subtracting the mean length at the end of one year of life from the mean length at the end of the next year of life even though the first mean may include several fish not included in the second mean. These growth-curve increments can be readily computed from the average growth data given in the tables and there is no reason to repeat them. More importantly, it is recognized that growth potential is determined to some extent by size at the beginning of the period (L91, P123) and therefore the average increment should be computed using at the beginning of the year only the fish that also complete the year. The size of the fish to which the increments apply can be determined by subtracting the average increments from the average length of the fish at the end of the year of growth. This length may differ somewhat from the mean length reported for the end of the previous year because the latter mean includes fish which did not complete the increment under consideration. Summation of annual increments (as suggested by L1, page 156, and some other papers) is not recommended since it applies the increment to fish at a different length than the fish which actually made the increment.

For some species, increment data are given as percentages of length at beginning of year (relative growth as defined in R282, page 314). Since these data were derived by dividing the mean increment by the mean length at the beginning of the year they are only approximate measures of relative growth since the sizes may have varied considerably around the mean.

Calculated growth in weight is given for some populations. In these cases the calculated lengths are converted to weights through use of a length-weight regression for the given population. If necessary, growth in weight of other populations may be estimated from the calculated lengths and an acceptable length-weight regression.

DISCUSSION OF GROWTH DATA

The discussions which follow the growth tabulations cover a variety of topics. First, they may describe the methods of age and growth determination, the characteristics of annuli, and evidence that the scale method is valid. In general, scale measurements are along the anterior radius unless otherwise indicated in this discussion.

Lee's phenomenon of apparent change in growth rate is characteristic of much growth data. The calculated lengths of the older fish in the earlier years of life are systematically lower than those of younger fish at the same age. Several causes of this phenomenon have been proposed (see particularly V5, pages 302-6, 328-44) but usually it is the result of (1) failure to use a corrected body-scale relationship in computing the lengths, (2) selective sampling, or (3) differential mortalities of the fish in relation to their growth rates. Many sampling techniques tend to select only the larger fish in younger age groups. Faster-growing fish are frequently shorter-lived than the slow-growing fish which, thus, constitute most of the fish in the older age groups.

Growth compensation is the tendency for slow-growing fish to grow more rapidly in later years than the fish which showed faster early growth.

Instantaneous growth rates and mortality and survival rates (see R283) are given for some species but no special effort has been made to include all such data. Instantaneous growth rates, particularly those pertaining to length rather than to weight, are greatly affected by the size of the fish. In a few cases, the growth curves have been represented by a Walford line (see R282, R283, W218) or by the von Bertalanffy formula (see B361, T128).

Wherever possible, the regional, genetic, and environmental factors affecting growth rates are discussed. There is, however, a basic problem in evaluating the effects of environmental factors upon the average growth rates of fish in a body of water (C348). Most environmental factors control the growth of the biomass of the fish population rather than the average growth rate. Where the population numbers can be kept constant or are known, the effects of environmental factors might be readily detected if only one factor at a time is modified. But in most natural conditions, several factors fluctuate independently or in various interdependent relationships and the changes in population numbers are not known.

The potential change in biomass growth brought about by any factor can be met by change in population numbers or by change in the average growth rates of the individual fish or by combination of the two. In many situations the population changes, which are more difficult to document and measure, are much greater than the changes in average growth rates. Population density is so often the most significant factor affecting growth that fishery biologists consider slow growth and low condition factors as indications of overpopulation. The problem is further complicated by the fact that it is not absolute population density of the fish which has such a dominant effect on growth, but it is the population density in relation to the carrying capacity of the body of water. Growth is a function of the degree to which biomass is below carrying capacity. Carrying capacity is in turn difficult to measure.

AGE AT MATURITY AND REPRODUCTION DATA

Age at maturity is fairly uniform in some species, but is subject to considerable variation in others. In several species, maturity is controlled by both age and size, and slow-growing individuals may not mature until somewhat older than fast-growing individuals. The slow-growing individuals may nevertheless mature at a smaller size than the fast-growing fish. One source of variation in maturity data is that some authors record a fish as mature in the year that the gonads begin more rapid development and other authors not until the fish is actually in spawning condition. Usually these two stages are not far apart and the variance from this source is small.

Courtship and spawning behavior have usually not been described in these summaries except very briefly but references to such descriptions are recorded where known.

Fecundity data, ova or egg counts, are given where available. Variation in ova counts may be partly a function of the degree of ripeness. In early stages of gonad development it may be difficult to distinguish which ova are to be spawned in the coming season. Even after the ova are quite advanced in development, a significant number may become atretic and not develop into mature eggs (V61). The determination of fecundity is more difficult in species which spawn intermittently over a prolonged spawning season than those which spawn within a single period. The numbers of eggs produced is related to the amount of care given the eggs by the parents. Species which build nests usually produce less than 2,000 eggs a season compared to the 50,000 to a million by less discriminating spawners.

A comprehensive survey of fish reproduction (B358) has appeared as the present handbook reached final stages. No attempt has been made to include the references given in B358, nor to eliminate duplication. In general, the sections on reproduction in this handbook are suggestive of available data rather than comprehensive.

FOOD HABITS

Data on food of fishes from various studies are difficult to tabulate or to summarize because of differences in the method of reporting (e.g., frequency of occurrence, percentage of volume, number, or fullness) and in the categories of food reported. The summaries provided here merely suggest some of the findings and lead to some of the original papers. Many other references are available and can often be located in the bibliographies of the papers listed.

OTHER DATA

Some references to standing crops and harvest are given but it is planned that such population data will be in a future volume and therefore no attempt has been made to cover the literature on populations in this volume.

Sex ratios have usually not been reported in these summaries since the sex ratio is often a function of the method of capture of the fish.

ABBREVIATIONS

In the tables, P., L., R., and C. refer to pond, lake, river, and creek. Reservoirs are usually referred to as lakes but sometimes Res. is used as the abbreviation. The two-letter abbreviations for states and U.S. possessions are those recommended by the American Fisheries Society in their computorized membership list, and abbreviations for Canada have been made up on the same model.

United States and Possessions

Alabama	AL	Kentucky	KY	Ohio	OH
Alaska	AK	Louisiana	LA	Oklahoma	OK
Arizona	AZ	Maine	ME	Oregon	OR
Arkansas	AR	Maryland	MD	Pennsylvania	PA
California	CA	Massachusetts	MA	Puerto Rico	PR
Colorado	CO	Michigan	MI	Rhode Island	RI
Connecticut	CT	Minnesota	MN	South Carolina	SC
Delaware	DE	Mississippi	MS	South Dakota	SD
District of Columbia	DC	Missouri	MO	Tennessee	TN
Florida	FL	Montana	MT	Texas	TX
Georgia	GA	Nebraska	NB	Utah	UT
Guam	GU	Nevada	NV	Vermont	VT
Hawaii	HI	New Hampshire	NH	Virginia	VA
Idaho	ID	New Jersey	NJ	Virgin Islands	VI
Illinois	IL	New Mexico	NM	Washington	WA
Indiana	IN	New York	NY	West Virginia	WV
Iowa	IA	North Carolina	NC	Wisconsin	WI
Kansas	KS	North Dakota	ND	Wyoming	WY

Canadian Provinces and Territories

Alberta	AT	Northwest Territories	NT	Ontario	ON
British Columbia	BC	Dist. of Franklin Bay	FB	Prince Edward I.	PE
Manitoba	MB	Dist. of Keewatin	KE	Quebec	QU
New Brunswick	NK	Dist. of Mackenzie	MC	Saskatchewan	SA
Newfoundland	NF	Nova Scotia	NS	Yukon Territory	YU

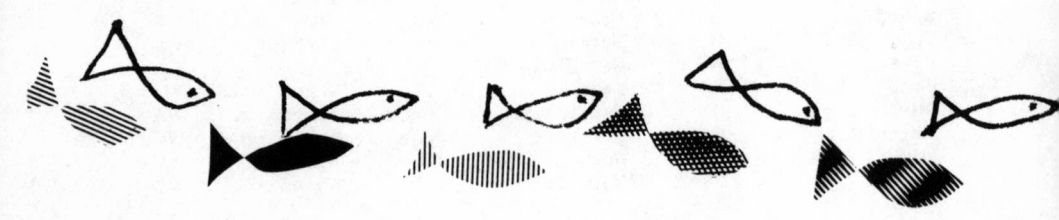

LIFE HISTORIES BY SPECIES

LAMPREYS, *Petromyzontidae*

Young lampreys, or ammocoetes, live in streams, feeding on minute organic particles for several years prior to metamorphosing into adults (E119, C254). The adults usually migrate to lakes, oceans, or large rivers, and most species feed on fish by attaching themselves with a suctorial disc and rasping a hole with their horny-plated tongue. A few species are nonparasitic and do not feed as adults. All species build nests in gravel areas of streams and die shortly after spawning. Ammocoetes of all species are particularly vulnerable to siltation and to mine wastes and such pollutants, since silting over of beds of sand and organic debris destroys their habitat (T113).

Most of the growth data on the ammocoetes were determined by the length frequency method, the accuracy of which beyond age I was questioned by Schultz (S24).

CHESTNUT LAMPREY, *Ichthyomyzon castaneus* Girard

Western Manitoba and northern Wisconsin to eastern Oklahoma and Georgia. Also in the Lake Michigan drainage (E119, H382). (K120 reports it in Texas.) Maximum length about 380 mm (S471). A 284 mm female had 42,000 eggs (H154). In the Manistee River, Michigan, lampreys lived about 18 months in the parasitic stage but fed actively only 5 months (May to October) near the middle of the adult life span (H394). In May, lampreys averaged 104 mm and 1.8 g; in October, 201 mm and 9.1 g (H394).

NORTHERN BROOK LAMPREY, *Ichthyomyzon fossor* Reighard and Cummins

Great Lakes drainage (except Lake Ontario) and the Mississippi River drainage in Wisconsin and northern Indiana, in creeks and small rivers (H382). A few records in Ohio River drainage in Ohio (T113). After several years as ammocoetes, these lampreys metamorphose in late summer or fall and spawn the next spring without feeding (H382). They spawn in May and June at water temperatures of 13-17C (S471). In Indiana they spawn in late May (L21). Adults are 127-150 mm (K118) or 127-254 mm long and weigh up to 5.5 g (S471). Growth of ammocoetes is as follows:

23

			No.	Mean TL	Range
Age 0, 3 months					
O12	MI		-	20	10-30
Age 0, July					
C48	WI	Brule R.	-	28	-
Age 0, Sept.					
C48	WI	Brule R.	-	-	38-43
Age 0, Nov.					
C48	WI	Brule R.	-	-	36-48
Age 0					
H89	MI	Gilchrist C.	-	20	-
H89	MI	Thunder Bay R.	-	25	-
H89	MI	Rifle R.	-	25	-
Age I					
L21	IN	Tippecanoe R.	12	38	36-43
H89	MI	Thunder Bay R.	-	41	-
H89	MI	Rifle R.	-	46	-
O12	MI	Thunder Bay R.	-	-	46-51
H89, O12	MI	Gilchrist C.	-	51	36-61
Age II					
H89	MI	Thunder Bay R.	-	56	-
L21	IN	Tippecanoe R.	10	56	53-61
H89	MI	Rifle R.	-	61	-
O12	MI	Thunder Bay R.	-	61	-
H89, O12	MI	Gilchrist C.	-	66	-
Age III					
H89, O12	MI	Thunder Bay R.	-	69	-
L21	IN	Tippecanoe R.	13	74	69-79
H89, O12	MI	Gilchrist C.	-	76	-
H89	MI	Rifle R.	-	76	-
Age IV					
H89, O12	MI	Thunder Bay R.	-	81	-
H90, O12	MI	Gilchrist C.	-	86	-
L21	IN	Tippecanoe R.	22	104	102-135
Age V					
H89, O12	MI	Thunder Bay R.	-	89	-
H89, O12	MI	Gilchrist C.	-	94	-
L21	IN	Tippecanoe R.	16	124	122-135
Age VI					
O12	MI	Thunder Bay R.	-	104	-
O12	MI	Gilchrist C.	-	109	-
H89	MI		-	114	-

The numbers of eggs per female are reported as follows:
V43, QU 9 females, 127-150 mm TL, 3-5 g, 1115-11979 eggs, Mean 1524
L21, IN 2 females, 112-122 mm 1050-1340 eggs
  The 9 females from Quebec had an average K(SL) of 0.16 and a range of 0.13-0.18 (V43).

  I. fossor nested in southern Michigan May 10-27 at water temperatures 20-22C (R265, H360) and in Quebec in late May (V77).

SOUTHERN BROOK LAMPREY, *Ichthyomyzon gagei* Hubbs and Trautman
Lower Mississippi River drainage (E119) and Gulf Stream drainage from Florida to Texas (D63). Nine adult males ranged 89-104 mm TL and 7 females, 107-124 mm in Alabama and Georgia (R264). In the western part of the range they may be larger, 5 from Oklahoma and Texas averaging 135 mm, range 107-165 mm (K120, R264). Nesting groups were observed March 24 in Alabama at water temperature 15C (R264), and late March to early May in Georgia (D63). Young ammocoetes were 8-33 mm in June (D63). Ten females, 102-152 mm TL averaged 1787 eggs, range 1000-3264 (D63) and a 150 mm female had 1300 eggs (H154).

ALLEGHENY BROOK LAMPREY, *Ichthyomyzon greeleyi* Hubbs and Trautman
Upper Ohio River drainage (E119, T113)
R2 PA Age 0, fall 28-33 mm
The maximum length of Allegheny brook lamprey ammocoetes is 178 mm and of adults 198 mm, with most adults being 127-178 mm (T113). Adults do not feed. Spawning in western Pennsylvania was observed May 19-26 at water temperatures of 19C (R2).

MOUNTAIN BROOK LAMPREY, *Ichthyomyzon hubbsi* Raney
Upper Tennessee River system in North Carolina and Georgia (R264). Nonparasitic, adults being 102-150 mm TL. Males average smaller than the females (R264). Spawning groups were found March 25 to April 11 with water temperatures of 10-12C.

SILVER LAMPREY, *Ichthyomyzon unicuspis* Hubbs and Trautman
Ohio and upper Mississippi drainage, western tributaries of Hudson Bay, Great Lakes, St. Lawrence River, and Lake Champlain (H382). Adult life mostly in larger lakes and rivers.
At 1 month, ammocoetes in New York were 12-15 mm long (G1). Ammocoetes may reach 178 mm and newly transformed individuals are about 102-152 mm long when they drift downstream usually in late April or May (T113) or in May and June in Michigan (A75). The next spring they enter the streams to nest after water temperatures reach 10C (T113), males being up to 279 mm and females up to 381 mm and 85 g (S471). In Quebec the largest males were 287 mm and females, 318 mm (V78). The number of eggs per female is reported as 11000 (K118), or 65000 (G1) or 12000-29412 eggs averaging 19012 for 10 females, 201-312 mm TL, weighing 13-75 g from Quebec (V43). The average K of these 10 females was 0.26, range 0.17-0.41.

LEAST BROOK LAMPREY, *Lampetra aepyptera* (Abbott)
Small streams of upper Ohio drainage and from Potomac to Neuse Rivers and in some Gulf coastal streams (E119). Maximum size for ammocoetes is 178 mm and adults are 122-178 mm TL (T113). Ten adults, collected January 28, 1937, in Mississippi ranged 89-104 mm (C330). Adults are nonparasitic.

			No.	Mean TL	Range
S219	MD	0- Oct.	15	33	28-43
		I- Feb.	67	48	33-66

		No.	Mean TL	Range
S219 MD (continued)	I-Aug.-Sept.	30	66	56-84
	II-Mar.	12	89	
	II-Oct.	--	109	
Maturing larvae	III-Feb.	28	107	89-127
Adults	III	22	104	89-119

In Ohio this species spawns in late March or early April, earlier than other lampreys, because the small streams in which they live warm to 10C earlier than other streams (T113). In Maryland they spawned from April 10 to May 10 at water temperatures of 15.5C (S219). One female had 1164 eggs (S219). First-year ammocoetes live in fine organic silt beds, moving to fine sandy drifts in the second year and coarser sand in the third year (S219).

RIVER LAMPREY, *Lampetra ayresi* (Günther)
Anadromous in streams of California to British Columbia. Length to 305 mm. They probably spawn from late April through May in California (C254).

V62 CA 175 mm, 20.2 g female 37288 eggs
         229 mm, 24 g female 11398 eggs.

ARCTIC LAMPREY, *Lampetra japonica* (Martens)
This species is anadromous in Alaska and northern Asia. In the Naknek River of southwest Alaska most of these lampreys complete their life in freshwater (H395).

	No.	Mean TL	Range
Probable anadromous adults	9	254	218-312
Freshwater adults	74	155	117-188
Immatures	69	168	114-226

Nesting and spawning occurred May 28-July 2 at water temperatures 12-15C. Spawning is not monogamous (as reported for *Petromyzon marinus*).

AMERICAN BROOK LAMPREY, *Lampetra lamottei* (LeSueur)
A nonparasitic lamprey in small streams from Missouri and Minnesota east to Maryland and Connecticut (E119, M420). The same, or a virtually indistinguishable form, is in the Yukon River system in Alaska and in northeastern Asia (H382). Ammocoetes may reach 203 mm TL, and adults are 135-208 mm (T113) or 114-203 mm, and up to 8 g (S471).

S338   NH   107-112 mm   6   2.6 g   2.3-2.8
            117-122 mm   6   3.1 g   2.5-3.5
            132 mm   -   4.1 g
V43   QU   10 females 117-157 mm   avg. K   0.175   range 0.14-0.23

			Mean TL	Range
	Age 0, Aug.			
O11	MI		20	-
H89	MI	Honey C.	25	-
H89	MI	Maple C.	36	-
	Age I			
O11	MI		41	-

			Mean TL	Range
Age I (continued)				
H89	MI	Honey C.	-	46-51
H89		Maple C.	64	-
S1			-	30-61
Age II				
O11	MI		89	-
H89	MI	Honey C.	-	76-81
H89	MI	Maple C.	94	-
S1	MI		-	89-109
Age III				
H89	MI	Maple C.	109	-
H89	MI	Honey C.	-	104-119
O11	MI		119	-
S1	MI		-	170-198
Age IV				
O11	MI		-	145-160
H89	MI	Honey C.	-	147-180
Age V				
O11	MI		170	-
H89	MI	Honey C.	-	147-180

Metamorphosis occurs at about 5 years (O11, H89). Spawning is in May (K118). Adults were taken in Michigan from mid-April to the first week of June (A75).

V43 QU 10 females, 117-157 mm TL, 3.1-7.7 g, 1085-3648 eggs. Mean 2339.
G2   NY   1 female   3276 eggs
D36 NY   820 eggs per female

WESTERN BROOK LAMPREY, *Lampetra planeri* (Bloch)
Europe, Asia, and western North American coast south to Alameda County, California (M420). Nonparasitic. The maximum length is about 203 mm (S471).

Length-weight relationship, Great Britain (H11)

Larvae			Larvae			Adults		
Length	No.	Weight	Length	No.	Weight	Length	No.	Weight
21-30	28	0.042	101-110	32	1.52	106-115	2	2.45
31-40	55	0.085	111-120	33	1.95	116-125	3	3.01
41-50	80	0.17	121-130	18	2.46	126-135	18	4.24
51-60	190	0.26	131-140	13	3.19	136-145	42	5.14
61-70	182	0.41	141-150	12	4.26	146-155	49	6.03
71-80	134	0.58	151-160	3	5.22	156-165	29	7.30
81-90	81	0.83	171-180	1	8.40	166-175	9	8.50
91-100	59	1.16				186-195	2	11.39

H11 Great Britain. In larvae, weight increases as 2.6 power of length up to 75 mm, as 2.9 power of length from 76-120 mm, and as 3.7 power of length over 120 mm. In adults, weight increases as 2.8 power of length. For larvae, K(TL) is 0.12-0.20; for adults, 0.16-0.19.

		Mean TL	Range
	Age 0		
S24	WA	-	5-20
H89	CA May	-	10-20
H89	Holland	-	10-28
H11	Great Britain	33	-
H89	CA Oct.	-	33-74
	Age I		
S24	WA Newauken C.	-	25-46
S24	WA Evans C.	-	30-51
S471	-	-	41-51
H11	Great Britain	-	36-76
H89	Holland	-	51-99
	Age II		
H89	Holland	79	-
H11	Great Britain	-	44-104
S471	-	-	81-86
	Age III		
S471	-	-	102-107
H89	Holland	112	-
H11	Great Britain	145	-
	Age IV		
S471	-	-	117-122

I1   Finland, additional growth data.

Western brook lampreys mature at 3-5 years at 127-155 mm. They produce up to 1500 eggs per female and incubation is 3-4 days (S471). They spawn March to May (S471).

PACIFIC LAMPREY, *Lampetra tridentata* (Gairdner)
Anadromous in streams of the Pacific from southern California to Gulf of Alaska (C254). Adults are 305-686 mm long when they migrate upstream in the spring. Landlocked lampreys of this species, 193-218 mm long in the parasitic stage, were reported in California (C328) and in Oregon (H89). Metamorphosis was reported to take place after 3 to 4 years in the stream and the lampreys were about 152 mm long at downstream migration (G137).

			No.	Mean TL	Range
H89	CA	Age 0, May	158	13	10-20
		0, Oct.	90	38	20-51
		I	1	76	-
		II	1	140	-

In Oregon, Pacific lamprey spawn in June and July (M434).
Marks from this species of lamprey were found on several whales (P170).

SEA LAMPREY, *Petromyzon marinus* Linnaeus
This species has been present along the Atlantic seaboard from northern Florida to the St. Lawrence River and in Europe and in Lake Ontario and Finger Lakes, N.Y. for an indefinite period but was first reported in Lake Erie in 1921, Lake St. Clair 1934, Lake Michigan 1936, Lake Huron 1937, and Lake Superior 1946 (A33, H397). It has had pronounced effects upon the Great Lakes

fishes and fisheries (A73, A74, H396, M327). Little is known of the life history
of the anadromous populations (H396, B100).

W109 NY Cayuga L. Ammocoetes $W = 0.00163 - 0.000139\ L + 0.000062\ L^2$

TL	No.	Mean W	Range	TL	No.	Mean W	Range
42-50	2	0.21	0.17-0.25	102-114	27	1.84	1.1-2.2
52-61	11	0.37	0.25-0.48	116-126	22	2.58	2.1-3.0
68-75	4	0.57	0.51-0.65	127-139	26	3.12	2.6-4.0
78-86	9	0.91	0.79-1.08	140-151	11	4.36	3.7-5.4
91-101	11	1.42	1.08-1.67	152-156	2	4.90	4.5-5.2

W109 NY Cayuga L. Transforming lampreys.
$W = 0.00505 + 0.000656\ L - 0.000046\ L^2$

TL	No.	Mean W	Range
119-124	2	3.12	2.9-3.3
127-139	17	4.17	3.5-4.9
140-151	20	5.07	4.2-6.2
152-164	8	6.49	5.7-7.4

A33 Great Lakes

TL	No.	Weight	TL	No.	Weight	TL	No.	Weight
302-328	3	68	405-429	484	153	504-531	74	269
329-353	27	96	430-455	539	176	532-556	14	329
354-378	124	111	456-480	376	210	557-581	8	360
379-404	321	133	461-503	210	241			

B100 Atlantic Ocean TL 838 mm, 1049 g.

Females with developing eggs

TL	No.	Mean W	Range	Mean K(TL)	Range	Citation
290	1	65	-	0.26	-	V43 ON
297	1	51	-	0.19	-	W109 NY
310-325	3	74	60-147	0.23	0.20-0.29	V43 ON
330-355	2	91	82-99	0.23	-	V43 WI
330-343	6	74	65-94	0.19	0.17-0.24	W109 NY
343	1	96	-	0.24	-	V43 ON
361-376	5	108	99-119	0.22	0.19-0.24	W109 NY
358-378	3	116	108-130	0.23	0.22-0.24	V43 WI
366-373	2	167	-	0.33	0.33-0.34	V43 ON
381-401	7	142	108-176	0.24	0.19-0.31	W109 NY
391-396	2	133	130-136	0.22	0.22-0.23	V43 WI
406-427	3	153	133-176	0.21	0.20-0.23	W109 NY
409-419	2	164	153-176	0.23	0.22-0.24	V43 WI
411-424	2	196	184-207	0.27	0.27-0.28	V43 ON
439	1	159	-	0.19	-	V43 ON
434	1	221	-	0.27	-	V43 WI
445-450	2	230	224-235	0.25	0.25-0.26	W109 NY
457-478	2	227	216-241	0.22	0.22-0.22	W109 NY

TL	No.	Mean W	Range	Mean K(TL)	Range	Citation
490	2	262	241-284	0.22	0.20-0.24	W109 NY
511	1	332	-	0.25	-	W109 NY
665-688	2	582	564-601	0.18	0.18-0.19	V43 QU anadromous
721-754	6	975	680-1000	0.21	0.18-0.23	V43 QU
767	1	860	-	0.19	-	V43 QU
789	1	1145	-	0.19	-	V43 QU

L103. In aquaria, ammocoetes average 10 mm SL at 1 month and 11 at 3 months.

	No.	Mean TL	Range
A33 MI Leaving nest, 0-June	-	8	5-10
0-July	107	10	8-15
0-Aug.	129	18	10-28
0-Oct.	252	23	15-33
0-Dec.	50	46	30-58

Transformation at 41-47 months.

W109 NY Cayuga L. length of ammocoetes, from length frequencies; July-August.

0	12	IV	107
I	37	V	127
II	60	VI	142
III	82	VII	137

In the period August 24 to March 13, transforming sea lampreys gained 7 mm, but lost 0.5 g.

		No.	Mean TL	Range	No.	Mean W	Range
A33 MI							
Parasitic life	0-April	21	190	124-231	21	17	6-28
	0-May	39	206	163-262	39	20	8-45
	0-June	85	211	140-302	85	23	6-65
	0-July	23	259	170-394	23	45	8-102
	0-Aug.	58	320	241-427	58	79	37-193
	0-Sept.	32	345	251-518	32	105	37-306
	0-Oct.	24	406	345-490	24	153	65-261
	0-Nov.	4	424	373-470	4	170	96-230
	0-Dec.	51	439	325-536	21	170	79-363

P64 in aquaria parasitic life 22 lampreys.

	TL	Weight		TL	Weight
0-May	140	3	0-Dec.	371	79
0-July	183	11	I-Feb.	371	79
0-Aug.	241	26	I-Mar.	363	77
0-Sept.	333	60	I-May	343	62
0-Nov.	366	77	I-June	320	51

G66 Mean sizes in spawning runs show a decline in Great Lakes.

	Number	Mean TL	Number	Mean W
L. Superior				
1954	3939	460	2474	227
1955	6174	437	6168	196

L. Superior (cont)	Number	Mean TL	Number	Mean W
1956	9593	452	9593	204
1957	11015	432	11015	176
1958	12985	427	12985	164
L. Michigan, west and north				
1954	572	450	500	173
1955	4972	437	4972	173
1956	2222	445	2222	170
1957	14435	424	14435	130
1958	7373	404	7373	136
L. Michigan, east shore				
1957	2647	404	2647	116
1958	3049	399	3049	116

Using the data from A33, a Gompertz-type curve was fitted by R284. Sea lampreys mature at 3-6 years, yet some at least have 7 years of larval life (K118). In Carp Lake River, Michigan, 663 larvae, 124-183 mm long, averaging 160 mm, were taken 7 years after the last spawning (S417).

Downstream migration takes place in late autumn or early winter for the anadromous fish (B100) and late October to early May in the Great Lakes region (A75, A76). Downstream migrants ranged from 94-190 mm and from 2.3 to 8.4 g (A75). Increase in water level increased downstream movement (A75).

In Ontario streams, the size of larvae increased in proportion to the distance downstream from the spawning grounds, indicating a downstream movement (T97). Larval lampreys were found along shore in Lake Michigan, probably coming from streams (H312).

Adults feed on a variety of fish by sucking blood and other body fluids (L194, B100). In the laboratory an average lamprey killed 8394 g of fish during its parasitic life (P64). Since these lampreys grew only to half the weight of lampreys in the lake, normal food consumption may be 16-18 kg (H396). For growth to sizes attained by anadromous fish, even greater food consumption is indicated.

Newly metamorphosed lampreys were found feeding on menhaden, in Chesapeake Bay in January and February (M382). Parasitic life apparently lasts 12 to 20 months (H396).

A body temperature of 29.5C was reported for sea lamprey (T102) which must be near the maximum sustainable.

Spawning runs take place from April to July in the Great Lakes (H396), March to June on the Atlantic Coast (B100), late April to July in Quebec (V77) and mid-April to mid-May in Cayuga Lake, N.Y. (W109). Nests are built in gravel areas (C329, W109). Eggs hatched in 13-14 days at 14-18C (W109). Successful hatching was reported only within 15.5-21C (P142). At 18.4C, a hatch of 78% was secured compared to 12% at 15.5C and 5% at 21.2C. The embryology was described (P142).

	Size of females		No. of	No. of eggs per female	
	TL	Wt.	females	Mean	Range
W109 NY Cayuga L.	297	51	1	19107	
	330-343	65-94	6	21723	13974- 26214
	361-376	99-119	5	31371	21692- 40139
	381-401	113-176	7	52955	18747- 71324
	406-427	133-176	3	53059	35429- 63905

| | Size of females | | No. of | No. of eggs per female | |
	TL	Wt.	females	Mean	Range
W109 (cont.)	445-450	224-235	2	64938	58279-71597
	457-478	216-241	2	65412	63974-66850
	490	241-284	2	70144	68861-71427
	511	332	1	85162	
A34 MI	320	60-71	2	27614	24021-31206
	340-353	79-102	3	41614	36388-46438
	361-380	94-139	8	41925	21000-55594
	381-405	111-179	8	50716	38302-74972
	406-431	136-181	8	54154	43808-69327
	432-456	133-247	14	61605	44010-82389
	457-472	179-255	11	73504	47718-91394
	483-503	196-303	11	77510	53147-97288
	513-521	221-329	4	87947	76235-94526
	536	318	1	107,138	
V43 WI	290-439	60-210	10	62870	38678-85712
	526	-	1	78762	
ON	330-434	79-221	10	55913	28891-74023
QU, sea-going	665-844	559-1145	10	171,589	123,873-258,874

A15 MI 527 mm female with 78762 eggs; NY 236000 eggs per female.
L1   62400 eggs per female; one 312 g female with 107,138 eggs.

Bibliographies and summaries of life history of sea lampreys are given by B100 and H396.

=====

STURGEONS, *Acipenser* spp.
   In a review of growth of sturgeons (M423), it was stated that they grow throughout life, but that the growth rate decreases after first maturity. The growth of males and females is similar. Growth is less rapid in fresh water than in the sea.

=====

SHORTNOSE STURGEON, *Acipenser brevirostrum* LeSueur
   Anadromous from Florida to New Brunswick on the Atlantic Coast (M438, V79). The record for Maryland is 838 mm (S386), but in the Connecticut River they have been taken at 1000 mm TL, or 899 mm FL (V79).

M411, M438 NK Gulf of Maine, St. John R.
Ages as determined by finrays, one fish per age group.

XIV	XV	XVI	XVII	XVIII	XIX	XX	XXII	XXIII
729	800	691	769	789	1002	820	769	882

G22 NY Hudson R.

		No.	Mean TL	Range	No.	Mean W	Range
Age	III	3	480	437-556	3	766	426-1333
	IV	5	536	480-592	5	807	626-880
	V	19	564	434-665	19	1129	766-1814
	VI	12	615	546-714	12	1469	993-2182
	VII	14	615	495-798	14	1460	740-2409
	VIII	8	653	574-752	8	1660	1248-2214
	IX	4	795	706-853	4	3098	2268-3461

		No.	Mean TL	Range	No.	Mean W	Range
Age	X	3	732	620-798	3	2150	1474-2495
	XI	4	678	640-701	4	1955	1615-2327
	XII	3	787	676-851	3	3093	1729-3828
	XIII	4	665	564-884	4	1941	1021-4055
	XIV	2	711	648-775	2	2622	1955-3289

In the Hudson River males mature at age V, 475-541 mm and females at age VI, 541-584 mm (B308). Period of hatching 4 to 6 days (B308) and 13 days at 8-12C (M441).

Young sturgeon to 3 months of age have minute conical teeth and probably feed on algae, protozoa, crustacea and small insects (B308). Adults feed on bottom organisms, small plants and animals intermingled with mud (V79, B308).

LAKE STURGEON, *Acipenser fulvescens* Rafinesque

A freshwater species, occasionally in brackish water. In large lakes and rivers from Hudson Bay to the Tennessee River in Alabama (V79, H382). The largest female reported was approximately 125 kg from Lake Winnipeg, Manitoba, and the largest male was 100 kg from Lake Erie (V79).

C136, C234 QU FL = 0.9166 TL
M423 FL = 0.960 TL - 30 mm;   r = 0.984

QU (C234, C135, V60)

FL	No.	Mean Kg	Range	FL	No.	Mean Kg	Range
406-429	3	0.32	0.11-0.41	1016-1066	16	8.71	6.89-10.02
434	3	0.50	-	1067-1117	17	10.14	8.16-12.25
457	1	0.68	-	1118-1167	9	10.41	7.71-12.25
486	3	0.82	-	1168-1218	10	13.20	10.66-15.20
546	7	1.13	-	1219-1269	7	14.92	11.79-17.69
564	15	1.27	-	1270-1320	5	16.78	15.42-17.69
587	17	1.59	-	1321-1371	8	19.28	15.74-21.77
612	24	1.86	-	1372-1421	5	22.53	15.88-26.40
635	20	1.95	-	1422-1472	2	25.85	20.87-30.84
663	24	2.00	-	1473-1523	3	24.90	22.54-26.31
686-709	19	2.40	2.09-2.45	1524-1574	2	29.12	22.45-35.83
711	32	2.54	-	1575-1625	1	29.48	-
739	18	2.81	-	1575-1625	+	34.0	-
762-812	45	3.36	3.22-3.81	1702	+	45.4	-
813-863	38	4.04	3.54-4.49	1753	1	48.5	-
864-913	20	4.49	4.81-5.44	1854	1	52.6	-
914-964	31	5.94	5.58-7.98	2616	1	96.2	-
965-1015	15	6.91	5.22-7.71				

C234 QU

28 immature males	log W = -3.1135 + 2.732 log FL
36 mature males	log W = -5.0799 + 3.314 log FL
43 immature females	log W = -3.1772 + 2.749 log FL
14 mature females	log W = -3.6723 + 3.100 log FL
9 spent females	log W = -2.3689 + 2.288 log FL

M423 QU

TL	200	300	500	600	800	1000	1200	1500
Wt	33	86	561	1013	2750	5952	11350	23250

TL	No.	Mean Kg	Range		
1020	3	9.21	8.6-10.0	S9	WI
1140	11	13.29	8.6-20.4	S9	WI
1260	12	16.06	9.1-20.4	S9	WI
1400	7	19.0	13.6-22.7	S9	WI
1520	8	30.6	24.5-37.2	S9	WI
1650	3	34.8	33.6-35.4	S9	WI
1980	1	69.2	-	W67	MI
2007	1	81.6	-	W115	WI
2440	1	107.0	-	E4	MN

P111 WI L. Winnebago 981 sturgeon $\log W = -6.156 + 3.304 \log TL$
M423 QU St. Lawrence R. $\log W = -6.190 + 3.327 \log TL$
M439, M431 QU Waswanipi R. and L. $\log W = -5.302 + 2.993 \log TL$

Weight of dressed sturgeon

C234	QU	12 males	60.9% of live weight	57.4-67.4%
		7 mature females	53.9% of live weight	48.2-58.9%
		2 spent females	66.9% of live weight	

C234	QU		Mean K(FL)	Range
		28 immature males	2.81	2.04-3.37
		36 mature males	2.79	2.35-3.32
		43 immature females	2.79	2.36-3.23
		14 mature females	3.18	2.73-3.85
		9 spent females	2.44	2.13-2.88

M423 QU K(TL)   increased from 0.36 at 203 mm to 0.83 at 2540 mm
C136 QU 337 sturgeon mean K(FL) 0.74 range 0.60-0.95
P111 WI Winnebago L. 981 sturgeon mean K(TL) 0.60

Age group	No.	Mean FL	Mean TL	Range	No.	Mean W	Range
Age 0-Aug.							
V79 QU St. Lawrence	4	102	112	86-122	4	5	2-8
Age 0-Sept.							
V79 QU St. Lawrence	4	94	104	64-122	4	4	1-6
Age 0-Oct.							
V60 QU	-	-	-	58-114	-	-	-
V79 QU St. Lawrence	2	91	97	84-107	2	3	2-5
P111 WI L.							
Winnebago	2	-	196	180-208	2	30	27-32
Age 0-Nov.							
V79 QU St. Lawrence	1	89	99	-	1	3	-
Age I							
H13 ON L. Nipigon	4	241	-	196-282	4	68	45-91

Age group	No.	FL Mean	TL Mean	Range	No.	W Mean	Range
M411, M423							
St. Lawrence	7	234	251	201-300	7	54	32-68
Age II							
H13 ON L. Nipigon	3	300	-	-	3	136	-
P111 WI L. Winnipeg	13	-	366	318-445	13	213	127-408
Age III							
M440 QU Waswanipi	5	297	340	320-371	5	195	150-277
H13 ON L. Nipigon	4	345	-	330-366	4	172	136-181
M423 QU Ottawa R.	1	345	366	-	1	240	-
C136 QU St. Francis							
L.	2	419	-	-	2	454	-
P111 WI L.							
Winnebago	10	-	480	429-513	10	499	318-635
Age IV							
M423 QU Ottawa R.	2	361	399	371-429	2	286	249-349
M440 QU Waswanipi	9	371	445	391-450	9	354	299-477
H13 ON	5	450	-	404-470	5	390	272-499
M411,M423							
St. Lawrence	3	409	455	450-460	3	390	381-531
C136 QU St. Francis							
L.	9	511	-	-	9	953	-
P111 WI L.							
Winnebago	9	-	579	528-653	9	953	680-1361
Age V							
M423 QU Ottawa R.	1	376	419	-	1	349	-
M440 QU Waswanipi	6	421	478	450-599	6	508	200-1098
H13 ON	3	503	-	483-521	3	544	499-590
C136, V60 QU							
St. Francis L.	16	569	-	-	16	1225	-
P111 WI L.							
Winnebago	8	-	663	605-698	7	1393	1179-1588
H260 L. Huron	-	889	-	-	-	-	-
Age VI							
M440 QU Waswanipi	5	429	486	429-531	5	494	376-726
M411,M423							
St. Lawrence	8	480	531	521-541	8	608	600-617
H13 ON	3	544	-	538-551	3	717	680-771
C136 QU St. Francis							
L.	27	605	-	-	27	1588	-
M423 QU Ottawa R.	1	541	610	-	1	998	-
P111 WI L.							
Winnebago	6	-	800	699-889	6	2722	1497-4309
Age VII							
M440 QU Waswanipi	10	478	559	480-610	10	726	500-1052
M423 QU Ottawa R.	2	564	625	599-650	2	1540	1230-1850
H13 ON	5	579	-	546-605	5	890	726-1043
C136 QU St. Francis							
L.	34	627	-	-	34	1724	-
P111 WI L.							
Winnebago	11	-	783	732-864	11	2449	1950-4309

Age group	No.	FL Mean	TL Mean	Range	No.	W Mean	Range
Age VIII							
M440 QU Waswanipi	19	503	564	490-640	19	843	499-1652
M423 QU Ottawa R.	3	589	640	561-739	3	1415	853-2160
M411,M423							
St. Lawrence	2	579	635	579-691	2	1200	798-1602
H13 ON	2	604	-	602-605	2	1021	953-1089
R180 QU Ottawa R.							
males	2	-	648	584-711	-	-	-
C136 QU St. Francis							
L.	36	671	-	-	36	2041	-
P111 WI L.							
Winnebago	14	-	840	749-914	13	3357	2041-5443
Age IX							
M440 QU Waswanipi	4	551	615	541-671	4	1170	798-1530
H13 ON	4	599	-	561-630	4	989	771-1225
R180 QU Ottawa R.							
males	4	-	648	559-711	-	-	-
M411,M423							
St. Lawrence	1	-	681	-	1	1400	-
M423 Ottawa R.	3	660	721	671-800	3	2223	1406-2404
C136 QU St. Francis							
L.	31	696	-	-	31	2449	-
P111,P110 WI L.							
Winnebago	33	-	874	775-991	30	3538	2268-6804
Age X							
M440 QU Waswanipi	2	536	610	599-620	2	890	780-1000
H13 ON	5	665	-	607-747	5	1390	1225-1860
R180 QU Ontario R.							
males	2	-	686	660-711	-	-	-
M411,M423							
St. Lawrence	5	630	691	660-732	4	1497	1202-1900
M423 QU Ottawa R.	5	729	780	760-840	5	2980	2313-3719
C136,V60 QU							
St. Francis L.	25	742	-	-	-	2858	-
C93 QU	-	762	-	-	-	3629	-
P111 WI L.							
Winnebago	47	-	907	813-1118	47	4082	2268-8165
H260 L. Huron	-	1041	-	-	-	-	-
Age XI							
M440 QU Waswanipi	3	579	653	589-711	3	1361	998-1678
R180 QU Ontario R.							
males	4	-	698	635-787	-	-	-
M411,M423							
St. Lawrence	3	681	729	711-770	2	1800	1400-2200
M423 QU Ottawa R.	4	744	789	508-820	4	3021	2798-3220
C136 QU St. Francis							
L.	25	764	-	-	25	3220	-
P111 WI L.							
Winnebago	74	-	940	838-1067	71	4580	2858-8845

Age group	No.	FL Mean	TL Mean	Range	No.	W Mean	Range
Age XII							
M440 QU Waswanipi	6	612	688	600-780	6	1406	816-1996
M423 QU Ottawa R.	8	759	820	789-871	8	3340	2812-3720
R180 QU Ottawa R.							
males	2	-	737	711-762	-	-	-
H13 ON	1	773	-	-	1	2313	-
C136 QU St. Francis							
L.	23	787	-	-	23	3447	-
P111 WI L.							
Winnebago	51	-	985	850-1143	50	5262	3175-9526
Age XIII							
M440 QU Waswanipi	6	630	696	569-840	6	1497	816-2223
M423 QU Ottawa R.	3	789	851	780-910	3	3970	3220-4990
R180 QU Ontario R.							
males	5	-	749	686-864	-	-	-
H13 ON	1	742	-	-	1	1905	-
C136 QU St. Francis							
L.	21	833	-	-	21	4264	-
M411,M423							
St. Lawrence	1	962	940	864-1295	1	4990	-
P111 WI L.							
Winnebago	40	-	1031	-	38	6623	3629-14062
Age XIV							
M440 QU Waswanipi	3	676	752	749-759	3	1724	1588-1814
M423 QU Ottawa R.	2	864	940	889-991	2	4672	4400-4944
R180 QU Ottawa R.							
males	5	-	762	635-991	-	-	-
M411,M423							
St. Lawrence	2	851	929	882-980	2	4100	3500-4700
C136 QU St. Francis							
L.	17	874	-	-	17	5262	-
P111 WI L.							
Winnebago	59	-	1046	940-1143	58	6760	4536-15876
S9 WI L. Winnebago	1	-	1194	-	1	12700	-
Age XV							
M440 QU Waswanipi	2	711	780	719-840	2	2450	1680-3220
M423 QU Ottawa R.	3	775	836	660-1001	3	3900	1860-6305
H13 ON	3	780	-	764-798	3	2223	1814-2585
R180 QU Ottawa R.							
males	4	-	838	737-965	-	-	-
C136,V60 QU							
St. Francis R.	14	889	-	-	-	5670	-
P111 WI L.							
Winnebago	62	-	1130	940-1422	62	8573	3175-17237
S9 WI L. Winnebago	2	-	1232	1168-1295	2	12700	11794-13608
H260 L. Huron	-	1168	-	-	-	-	-
Age XVI							
M440 QU Waswanipi	11	742	815	630-960	11	2812	1179-6804
R180 QU Ottawa R.							
males	6	-	864	660-1041	-	-	-

Age group	No.	FL Mean	TL Mean	Range	No.	W Mean	Range
M411,M423							
St. Lawrence	3	840	929	869-971	3	4100	3200-4900
M423 QU Ottawa R.	1	1000	1050	-	1	7800	-
C136 QU St. Francis							
R.	19	885	-	-	19	5080	-
P111 WI L.							
Winnebago	41	-	1130	864-1448	41	8300	3630-15422
S9 WI L. Winnebago	4	-	1189	1067-1422	4	12474	8618-18144
Age XVII							
M440 QU Waswanipi	5	754	833	769-940	5	2722	2087-3583
C135 QU St. Peter	1	800	-	-	1	3946	-
R180 QU Ottawa R.							
m&f	9	-	840	711-1040	-	-	-
H13 ON	1	878	-	-	1	2948	-
M411,M423							
St. Lawrence	2	900	980	960-1000	2	5200	5100-5300
C136 QU St. Francis	10	955	-	-	10	7121	-
P111 WI L.							
Winnebago	30	-	1160	1041-1295	29	9888	6260-16330
Age XVIII							
R180 QU Ottawa R.							
m&f	10	-	838	711-1016	-	-	-
M440 QU Waswanipi	9	762	844	760-920	9	2676	2087-3583
H13 ON	1	882	-	-	1	2903	-
M411,M423							
St. Lawrence	6	929	1000	940-1030	5	6300	5900-7050
M423 Ottawa R.	1	840	910	-	1	4536	-
C136 QU St. Francis	3	965	-	-	3	6214	-
P111 WI L.							
Winnebago	12	-	1240	1092-1422	12	11657	6350-18144
S9 WI L. Winnebago	6	-	1308	1194-1473	7	16783	10433-20866
Age XIX							
R180 QU Ottawa R.							
m&f	9	-	836	711-991	-	-	-
M440 QU Waswanipi	7	793	876	770-960	7	3493	1996-5900
H13 ON	1	907	-	-	1	4445	-
M411,M423							
St. Lawrence	7	960	1030	970-1062	7	6400	5400-6700
C136 QU L.							
St. Francis	2	1067	-	-	2	9072	-
S9 WI L. Winnebago	4	-	1245	1143-1321	4	13517	8618-17237
P111 WI L.							
Winnebago	13	-	1278	1118-1448	13	12655	7938-19505
Age XX							
R180 QU Ottawa R.							
m&f	9	-	889	762-1041	-	-	-
M440 QU Waswanipi	14	805	891	780-991	14	3717	1996-5307
M411,M423							
St. Lawrence	7	1000	1080	930-1030	7	7600	5806-9685
M423 Ottawa R.	2	1062	1120	920-1320	2	12383	4536-20230

Age group	No.	FL Mean	TL Mean	Range	No.	W Mean	Range
C136,C93 QU L.							
St. Francis	7	1016	-	-	7	8573	-
B12 MB L. Winnipeg	1	1016	-	-	-	-	-
S9, P111 WI L.							
Winnebago	17	-	1265	1118-1422	17	13245	6800-19050
H260 L. Huron	1	1295	-	-	-	-	-
Age XXI							
R180 QU Ottawa R.							
males	8	-	889	838-940	-	-	-
M440 QU Waswanipi	9	869	947	810-1041	9	4990	2812-7711
M423 L. Nipissing	1	1179	1250	-	1	14515	-
C136 QU L.							
St. Francis	3	1016	-	-	3	7847	-
P111, S9 WI L.							
Winnebago	21	-	1300	1168-1500	20	14561	9072-22680
Age XXII							
M440 QU Waswanipi	5	767	846	671-960	5	3357	1406-6305
R180 QU Ottawa R							
m&f	8	-	900	787-1041	-	-	-
H13 ON	1	890	-	-	1	3810	-
C136 QU L.							
St. Francis	2	953	-	-	2	6580	-
M411,M423							
St. Lawrence	2	1107	1204	1163-1245	2	10297	-
C93 QU	1	1156	-	-	1	11340	-
P111, S9 WI L.							
Winnebago	27	-	1328	1148-1676	27	14923	7258-35380
Age XXIII							
R180 QU Ottawa R.							
males	6	-	914	737-1041	-	-	-
M440 QU Waswanipi	3	907	983	960-1020	3	5035	3992-5897
M441,M423							
St. Lawrence	4	1082	1163	1067-1214	4	9516	6800-11700
C136 QU L.							
St. Francis	3	1252	-	-	3	17100	-
P111, S9 WI L.							
Winnebago	30	-	1275	965-1450	30	14424	4082-21773
M423 L. Nipissing	1	1323	1410	-	1	21000	-
Age XXIV							
R180 QU Ottawa R.							
m&f	11	-	914	737-1041	-	-	-
M440 QU Waswanipi	5	853	991	831-1052	5	4264	2313-6800
M411,M423							
St. Lawrence	7	1090	1166	1100-1219	6	10569	7983-13608
P111, S9 WI L.							
Winnebago	46	-	1356	1054-1651	45	16828	7711-33566
M423 L. Nipissing	1	1372	1471	-	1	21320	-
Age XXV							
R180 QU Ottawa R.	6	-	965	787-1118	-	-	-
M440 QU Waswanipi	10	810	983	900-1082	10	5900	3538-7711

Age group	No.	FL Mean	TL Mean	Range	No.	W Mean	Range
M411,M423							
St. Lawrence	2	1087	1153	1120-1200	2	10523	9027-12020
M423 Ottawa R.							
males	1	1260	1321	-	1	18144	-
H13 ON	1	1001	-	-	1	5443	-
C234 QU St. Peter	1	1219	-	-	1	17690	-
P111, S9 WI L.							
Winnebago	34	-	1369	1118-1588	34	17191	10886-29483
H260 L. Huron	1	1322	-	-	-	-	-
S88 IA	-	-	-	-	1	27215	-
Age XXVI							
R180 QU Ottawa R.							
m&f	11	-	958	787-1118	-	-	-
M440 QU Waswanipi	6	914	991	828-1128	6	6123	2495-9117
M411,M423							
St. Lawrence	1	1138	1209	-	1	12383	-
C135 QU L. St. Peter	1	1067	-	-	1	10433	-
P111, S9 WI L.							
Winnebago	45	-	1445	1316-1651	45	18642	9525-41277
Age XXVII							
R180 QU Ottawa R							
m&f	6	-	1013	864-1194	-	-	-
M440 QU Waswanipi	6	947	1021	800-1118	6	6849	3220-9117
P111, S9 WI L.							
Winnebago	25	-	1422	1316-1651	25	18597	12247-27669
M423 L. Nipissing	1	1372	1494	-	1	22000	-
Age XXVIII							
R180 QU Ottawa R.							
m&f	10	-	1029	864-1194	-	-	-
M440 QU Waswanipi	6	1013	1067	869-1097	6	7303	2903-9117
M411,M423							
St. Lawrence	2	1200	1300	1240-1360	2	14696	13109-16284
P111 WI L.							
Winnebago	39	-	1458	1435-1651	39	19822	12701-31751
Age XXIX							
R180 QU Ottawa R.							
males	9	-	1016	914-1118	-	-	-
M440 QU Waswanipi	11	935	1024	861-1097	11	5670	2722-9117
M411,M423							
St. Lawrence	2	1229	1310	1290-1330	2	12882	12020-13789
C234 QU L. St. Peter	1	1270	-	-	1	17690	-
P111 WI L.							
Winnebago	31	-	1433	1219-1600	31	20457	13608-32658
Age XXX							
R180 QU Ottawa R.							
m&f	10	-	1041	813-1168	-	-	-
M440 QU Waswanipi	9	988	1072	902-1199	9	6985	2994-11793
M411,M423							
St. Lawrence	1	1330	1420	-	1	19096	-
C93 QU	-	1270	-	-	-	15422	-

Age group	No.	FL Mean	TL Mean	Range	No.	W Mean	Range
P111 WI L.							
Winnebago	22	-	1540	1417-1803	22	24900	12701-44906
H260 L. Huron	-	1524	-	-	-	-	-
M424 WI L. Mendota	1	-	1567	-	-	-	-
Age XXXI							
R180 QU Ottawa R.							
m&f	12	-	1062	838-1194	-	-	-
M440 QU Waswanipi	9	1011	1092	930-1200	9	7167	4128-10433
P111 WI L.							
Winnebago	27	-	1552	1295-1829	27	25128	14061-37422
B12 MB L.							
Winnipeg	2	1562	-	1524-1600	2	30844	24494-37194
Age XXXII							
R180 QU Ottawa R.							
m&f	12	-	1082	889-1194	-	-	-
M440 QU Waswanipi	4	1067	1143	1000-1210	11	8618	4400-11793
P111 WI L.							
Winnebago	24	-	1567	1422-1930	23	26762	19278-42637
M423 L. Nipissing	1	1570	1658	-	1	33293	-
R180 QU Ottawa R.							
m&f	11	-	1176	1067-1270	-	-	-
H13 ON	1	1367	-	-	1	19051	-
C234 QU L. St. Peter	1	1491	-	-	1	26308	-
P111 WI L.							
Winnebago	8	-	1679	1524-1753	8	29664	22680-36740
Age XXXIII							
R180 QU Ottawa R.							
m&f	10	-	1092	940-1219	-	-	-
M440 QU Waswanipi	4	1021	1115	1008-1179	4	8210	4899-9616
P111 WI L.							
Winnebago	13	-	1590	1379-1765	13	26263	21319-30844
Age XXXIV							
R180 QU Ottawa R.							
m&f	6	-	1092	940-1448	-	-	-
M440 QU Waswanipi	3	1097	1207	1179-1229	3	9389	8890-10206
M411,M423							
St. Lawrence	1	1280	1339	-	1	14968	-
P111 WI L.							
Winnebago	14	-	1585	1422-1689	13	27624	20412-36288
Age XXXV							
R180 QU Ottawa R.							
m&f	17	-	1123	914-1295	-	-	-
M440 QU Waswanipi	12	1062	1143	960-1199	12	8029	4899-9480
M411,M423							
St. Lawrence	1	1280	1361	-	1	16783	-
C234 QU L. St. Peter	1	1207	-	-	1	14061	-
P111 WI L.							
Winnebago	5	-	1656	1473-1778	5	31117	25855-40824
H260 L. Huron	-	1638	-	-	-	-	-
Age XXXVI							
M440 QU Waswanipi	4	1067	1148	1080-1240	4	7847	6396-11294

Age group	No.	FL Mean	TL Mean	Range	No.	W Mean	Range
M411,M423							
St. Lawrence	1	1361	1440	-	1	26308	-
R180 QU Ottawa R.							
m&f	11	-	1158	864-1372	-	-	-
C234 QU L. St. Peter	2	1427	-	1372-1501	2	23814	21772-25855
P111 WI L.							
Winnebago	11	-	1670	1500-1778	11	30844	25401-43545
Age XXXVII							
M440 QU Waswanipi	2	991	1077	930-1219	2	7938	4128-11793
Age XXXVIII							
R180 QU Ottawa R.							
m&f	8	-	1176	1016-1295	-	-	-
M440 QU Waswanipi	5	1095	1179	1087-1250	5	9344	6804-12292
H13 ON	1	1486	-	-	1	20412	-
P111 WI L.							
Winnebago	2	-	1486	1448-1524	2	31071	30844-31298
Age XXXIX							
R180 QU Ottawa R.							
m&f	8	-	1201	1067-1397	-	-	-
M440 QU Waswanipi	3	1194	1275	1240-1341	3	11340	9480-12701
C234 QU L. St. Peter	1	1427	-	-	1	30844	-
P111 WI L.							
Winnebago	3	-	1651	1549-1702	3	35380	28123-39462
Age XL							
M440 QU Waswanipi	7	1090	1191	1006-1290	7	9707	5897-15876
R180 QU Ottawa R.							
m&f	14	-	1209	1041-1397	-	-	-
H13 ON	1	1328	-	-	1	13154	-
C234 QU L. St. Peter	1	1389	-	-	1	25855	-
C93 QU	-	1473	-	-	-	28123	-
B12 MB L. Winnipeg	1	1575	-	-	1	36740	-
H260 L. Huron	-	1727	-	-	-	-	-
Age XLI							
M440 QU Waswanipi	4	1128	1209	1029-1326	4	10070	5897-11793
R180 QU Ottawa R.							
m&f	13	-	1229	1067-1422	-	-	-
P111 WI L.							
Winnebago	3	-	1720	1549-1829	3	34609	29483-43544
S88 IA	-	-	-	-	1	40822	-
Age XLII							
M440 QU Waswanipi	4	1143	1224	1179-1250	4	11793	9480-14606
R180 QU Ottawa R.							
m&f	7	-	1285	1219-1397	-	-	-
Age XLIII							
M440 QU Waswanipi	5	1151	1234	1199-1260	5	10342	4491-13109
R180 QU Ottawa R.							
m&f	6	-	1308	1219-1372	-	-	-
P111 WI L.							
Winnebago	1	-	1778	-	1	31751	-
M423 L. Nipissing	1	1623	1730	-	1	39462	-

Age group	No.	FL Mean	TL Mean	Range	No.	W Mean	Range
Age XLIV							
M440 QU Waswanipi	3	1138	1222	1189-1260	3	10115	6577-12701
R180 QU Ottawa R.							
m&f	6	-	1326	1219-1397	-	-	-
Age XLV							
M440 QU Waswanipi	8	1123	1247	1140-1311	8	11385	8210-16284
R180 QU Ottawa R.							
m&f	11	-	1306	1118-1524	-	-	-
H260 L. Huron	-	1803	-	-	-	-	-
Age XLVI							
M440 QU Waswanipi	5	1189	1275	1229-1321	5	11113	9480-11793
S88 IA	-	-	-	-	1	49895	-
Age XLVII							
M440 QU Waswanipi	5	1204	1285	1107-1349	5	12564	6396-18189
R180 QU Ottawa R.							
females	12	-	1321	1118-1473	-	-	-
H13 ON	1	1605	-	-	1	29030	-
Age XLVIII							
M440 QU Waswanipi	2	1214	1306	1300-1312	2	12701	12474-12927
R180 QU Ottawa R.							
females	8	-	1346	1143-1448	-	-	-
Age XLIX							
M440 QU Waswanipi	4	1179	1267	1138-1341	4	11249	7711-14106
R180 QU Ottawa R.							
females	4	-	1372	1321-1448	-	-	-
P111 WI L.							
Winnebago	1	-	1829	-	1	48081	-
Age L							
M440 QU Waswanipi	5	1199	1278	1199-1349	5	11430	7484-13608
M411,M423							
St. Lawrence	2	1580	1684	1671-1697	2	33293	29619-37013
R180 QU Ottawa R.							
females	5	-	1372	1245-1499	-	-	-
H13 ON	1	1562	-	-	1	22680	-
M411 QU							
St. Lawrence	2	-	1684	1666-1702	-	-	-
C93 QU	-	1676	-	-	-	45360	-
C234 QU L. St. Peter	1	1854	-	-	1	52617	-
Age LI							
R180 QU Ottawa R.							
female	1	-	1372	-	-	-	-
C234 QU L. St. Peter	1	1765	-	-	1	48535	-
Age LIII							
R180 QU Ottawa R.							
female	1	-	1397	-	-	-	-
Age LIV							
M440 QU Waswanipi	1	1260	1349	-	1	13290	-
Age LV							
M440 QU Waswanipi	2	1240	1316	1300-1331	2	14106	-
R180 QU Ottawa R.							
female	1	-	1448	-	-	-	-

Age group	No.	FL Mean	TL Mean	Range	No.	W Mean	Range
Age LVII							
M440 QU Waswanipi	1	1349	1430	-	-	-	-
R180 QU Ottawa R.							
female	1	-	1600	-	-	-	-
Age LVIII							
R180 QU Ottawa R.							
female	1	-	1676	-	-	-	-
Age LX							
M440 QU Waswanipi	3	1341	1837	1382-1458	3	17826	12882-21092
M411,M423							
St. Lawrence	1	1699	1808	-	1	35879	
Age LXII							
M440 QU Waswanipi	1	1290	1400	-	1	13200	-
Age LXIII							
W129 WI L.							
Winnebago	1	-	2159	-	-	-	-
Age LXX							
M440 QU Waswanipi	4	1367	1458	1400-1499	-	-	-
Age LXXX							
M440 QU Waswanipi	2	1412	1486	1471-1501	2	21228	18008-24449
Age LXXXII							
P111 WI L.							
Winnebago	1	-	2007	-	-	-	-
Age LXXXV							
M440 QU Waswanipi	1	1331	1422	-	1	20003	-
Age CLII							
C206 ON L. of Woods	1	-	2057	-	1	97524	-

Age was determined in the above studies from rings on the cross section of fin rays. Tagging gave verification of age determination by fin ray rings (M423). In Lake Michigan a tagged sturgeon 533 mm TL (907 g) grew to 610 mm (2268 g) in 3 years and 5 months and another, tagged at 6350 g, increased to 34019 g in 28 years (S76). Three of 4 tagged sturgeon were recaptured, one having traveled at least 154 miles in 8 months (S76).

Sturgeon from Lake Winnebago, Wisconsin, showed above-average growth, but growth varied considerably in the various regions so that no geographical or ecological trends are evident. Males and females apparently grow at about the same rate, but females may live longer (P111, R180). In Nottaway River, the oldest male was 55 years and the oldest female 80 years (M514).

Males mature at 760-1350 mm (C234), 889 mm (H13, B12), 762 mm (C93), 838-940 mm (M514), 1016 mm (C136) or 1219-1270 mm (W129), at ages XV (C136), XIV (C93), XIV-XXXIX (C234), XII-XIX (R180), XVIII-XX (M514), XXII (H13), XXV (B12), or XVI-XVIII (W129). The smallest mature male reported by V60 and V79 was 965 mm and weighed 5103 g.

Females mature at 889 mm (H13), 889-991 mm (M514), 1270 mm (C136), 914-1753 mm (C234), 1372-1397 mm (W129), or 1397 mm (C93), and at ages XX-XXIII (M514), XXII (H13), XXIII (C136), XXV (C93, B12), XXIII-LVI (C234), XIV-XXIII (R180), or XXIV-XXVI (W129). The smallest mature female reported by V60 and V79 was 1143 mm and weighed 7711 g.

Lake sturgeon spawn at intervals of 4 to 7 years after first reaching maturity, and the slower growth 2 to 3 years prior to spawning results in "belts" of good growth and poor growth (R180). Tagging indicates a periodicity of spawning in females, but some males spawn in consecutive years (W129). Periods between spawnings for females were 4 years in Minnesota, 4 to 5 years in the St. Lawrence River, and 6 years in Lake Nipigon and the Nottaway basin (M516).

Spawning occurs April 15-May 15 in central Wisconsin at water temperatures 12-15C. Spawning occurs just under the surface near shore, or at depths of 2.4m or more where currents are pronounced. Tagging indicates that males at least return to the same spawning grounds in different years (W129). In Ontario spawning is at the end of May and in early June, at water temperatures 12-19C (H260). In Quebec, spawning is from early May to the end of June at water temperatures about 18.4C (M514, V60, V79). In many cases they run to rivers for spawning, frequently spawning at the foot of falls in 0.6-4.6 m of water (V60).

Ripe eggs, dark brown or blackish in color, are 2.5-3.3 mm in diameter (V60). Females of 90 kilo are reported to lay 3 million eggs (V60, S471) but the numbers of eggs reported from counts are as follows:

	No.	Weight	FL	Eggs per female Mean	Range
C234	3	13608-17690	1207-1270	184,913	181,720-188,890
	1	21772	1372	277,560	-
	3	25855-26308	1389-1504	323,817	273,940-412-420
	1	30844	1427	485,500	-
	1	35834	1549	370,912	-
	2	48535-52617	1765-1854	667,772	652,904-682,640
V60	1	18144	-	650,275	-
D12	-	-	-	200,000	-
V67	-	-	-	up to 500,000	-

After the early fingerling stage, lake sturgeon appear to be strictly bottom feeders, with insect larvae, leeches, molluscs, isopods and crustacea as principal foods. Food items are sensed with the barbels, picked up with the protractile lips, and rapidly strained by washing soft bottom materials through the gill openings. Contents of the anterior portion of the stomach are clean and entire (W129). Feeding may continue throughout the winter (W129) or may be largely discontinued in winter and during the spawning migration (R199). The fish spend the winter in deep well-aerated holes in relative inactivity (R199). Tagged sturgeon travelled up to 48 kilometers in the St. Lawrence River, but most were recaptured less than 16 kilometers from point of release, even when they had been out 3 years (M515).

Population

The population of Lake Winnebago, Wisconsin, has been estimated at one 22680 g adult per 4.05 hectares and the harvest to spearing was estimated at 5-15% annually (R199).

Summaries of life history and extensive bibliographies are given by V79, M423.

ATLANTIC STURGEON, *Acipenser oxyrhynchus* Mitchill

Anadromous on the Atlantic coast of North America. These sturgeon enter streams for spawning from Hamilton Inlet in Labrador to St. John's River,

Florida and a subspecies, *A. o. desotoi* Vladykov is found in the Gulf of Mexico and Caribbean Sea to the north coast of South America (V79).

M423   FL = 0.867 TL + 10 mm   r = 0.989

V60	FL	2502 mm	TL 2667 mm	lb. 160 kg
C205		2769 mm		lb. 170 kg
V60	TL	4270 mm		lb. 369 kg

M423, St. Lawrence R.

TL	200	300	400	500	600	800	1000	1200	1500
Wt.	24	89	191	416	741	1872	4113	7618	15840

log W = -5.943 + 3.180 log TL

M423 K(TL) increased from 0.30 at 203 mm to 0.425 at 1524 mm

	No.	Mean TL	Range	No.	Mean Wt.	Range
Age 0-Aug.						
V79 St. Lawrence	2	79	64-97	2	2	1-4
Age 0-Sept.						
V79 St. Lawrence	3	104	102-109	3	4	-
V79 St. Lawrence	108	183	124-246	108	23	5-50
Age I						
M411, M423 St. Lawrence	28	221	195-251	12	27	18-50
Age II						
M411, M423 St. Lawrence	17	279	231-310	6	68	50-91
G22 NY Hudson R.	3	300	287-305	3	141	86-227
Age III						
M411, M423 St. Lawrence	26	351	300-442	15	136	91-191
G22 NY Hudson R.	13	460	404-531	13	408	269-567
Age IV						
M441, M423 St. Lawrence	54	419	351-531	7	313	159-440
G22 NY Hudson R.	4	607	580-620	4	1025	907-1135
Age V						
M411, M423 St. Lawrence	75	490	380-600	27	426	181-608
G22 NY Hudson R.	4	670	620-720	4	1310	1134-1700
Age VI						
M411, M423 St. Lawrence	60	580	450-730	23	698	400-1300
G22 NY Hudson R.	4	650	565-730	4	1347	1021-1928
Age VII						
M411, M423 St. Lawrence	48	660	580-843	16	1052	798-2110
G22 NY Hudson R.	14	716	648-780	14	1534	1066-1928
Age VIII						
M411, M423 St. Lawrence	25	749	620-910	11	1751	998-2700
G22 NY Hudson R	3	830	815-850	3	3003	2214-3856
Age IX						
M411, M423 St. Lawrence	34	870	720-1010	26	2600	1112-4110
Age X						
M411, M423 St. Lawrence	49	900	760-1015	45	3102	1450-4110
G22 NY Hudson R.	1	2146	-	1	36288	-
Age XI						
M411, M423 St. Lawrence	58	980	830-1200	56	4500	2400-6800
G22 NY Hudson R.	1	2154	-	1	39916	-

	No.	Mean TL	Range	No.	Mean Wt.	Range
Age XII						
M411, M423 St. Lawrence	33	1052	930-1180	33	5400	3400-7484
G22 NY Hudson R.	2	2388	2235-2540	3	85277	79834-90720
Age XIII						
M411, M423 St. Lawrence	29	1090	920-1250	28	5806	3720-9117
Age XIV	14	1148	1010-1250	14	6895	3400-9525
Age XV	9	1200	1000-1300	9	7575	6532-11612
Age XVI	11	1220	1100-1390	9	8936	6440-13608
Age XVII	3	1330	1232-1420	3	10296	7983-12927
Age XVIII	6	1392	1200-1540	6	13109	6804-19504
Age XIX	1	1525	-	2	18915	14288-23587
Age XX	1	1625	-	1	21772	-
Age XXI	1	1760	-	1	30480	-
Age XXIV	1	2017	-	1	31800	-
Age XLVI	1	2600	-	1	152,863	-
Age LX	1	2670	-	1	160,000	-
V79 NY Hudson R.						
Age XVIII	1	2655	-	1	102,060	-

Age was determined from otoliths (G22) and from fin rays (M411,M423). Females are larger than males at maturity (V60).

The smallest mature males were 1750 mm TL weighing 31750 g (V79), and 1892 mm (G22). The smallest mature females were about 2540 mm (G22) and 68040 g (V79). The spawning migration begins in February in Georgia, in April in Chesapeake Bay, in late April and May in the Hudson River, and May-June in the Gulf of Maine (V79). Spawning took place at water temperatures of 13-18C (B63). The eggs are light to dark brown when laid and are adhesive to weeds, stones, etc. (V79). The number of eggs per female is from 500,000 to 2.5 million (B63, B73, A61). One 160 kg female had ovaries of 41.3 kg and an estimated 3,755,745 eggs (V60, V79). The fry averaged 11 mm at hatching (R75).

Young sturgeon in fresh water feed on aquatic insects, amphipods, oligochaetes, and *Pisidium*. In marine habitats the sturgeon feed on gastropods, shrimp, amphipods, and other bottom organisms. They also eat small fish, particularly launce *(Ammodytes)* (V79). Sturgeon were larger as samples were taken downstream in the St. Lawrence River (M515). One sturgeon travelled 1056 kilometers downstream in 8 years and another, 1008 kilometers in 2 years (M515).

Summaries of life history and extensive bibliographies are given by V79, M423.

WHITE STURGEON, *Acipenser transmontanus* Richardson

Anadromous on the Pacific coast from Monterey northward (E119). Lengths up to 61 m and weights to 816 kg have been reported (S471) or 680 kg (G33) and 907 kg (C254).

D100 Columbia R.

FL	No.	Weight	FL	No.	Weight
0-74	2	5	1448-1523	30	30755
150-228	3	42	1524-1599	39	33203
229-304	4	136	1600-1675	23	39010
305-380	6	227	1676-1752	16	43092

FL	No.	Weight	FL	No.	Weight
381-456	14	454	1753-1828	20	51257
457-532	11	680	1829-1904	9	54432
533-609	5	907	1905-1980	11	68947
610-685	4	1724	1981-2056	4	73030
686-761	1	2268	2057-2133	3	79834
762-837	7	3266	2134-2209	3	106,596
838-913	5	4400	2210-2285	1	88906
914-990	3	6169	2286-2361	2	98431
991-1066	7	8255	2362-2437	3	128,822
1067-1142	11	11521	2438-2514	1	129,276
1143-1218	7	12927	2743-2818	2	157,853
1219-1294	17	18642	2819-2895	2	272,160
1295-1371	22	19913	3200-3276	1	395,993
1372-1447	15	23269			

M337 OR 1372 mm FL 20412 g

P115 CA

FL	grams	FL	grams
508	454	1524	25855
762	2268	1778	35834
1016	6350	2032	49896
1270	17236	3048	125,647

	TL	No.	Mean Weight	Range
J13	890	1	3402	-
G33	3200	1	460,404	-
G33	3505	2	376,715	374,674-378,756
D100	3505	1	395,993	-
G33	3658	1	362,880	-
G33	3810	1	582,876	-

Age	Location	No.	Mean FL (except as marked)	Range
0	D100 OR Columbia R.	10	190	-
	S388 CA Old R.	30	203	-
	P115 CA San Pablo R.	30	264	178-302
I	D100 OR	14	272	-
	P115 CA	2	457	406-508
II	D100 OR	28	373	-
	P115 CA	6	538	457-584
III	D100 OR	52	450	-
	I7 ID Snake R.	-	381 TL	-
IV	D100 OR	25	564	-
V	D100 OR	33	655	-
	P115 CA	4	800	635-940
VI	D100 OR	36	742	-
	P115 CA	54	1011	838-1194
VII	D100 OR	44	826	-
	P115 CA	12	1092	914-1270
VIII	D100 OR	52	866	-
	P115 CA	20	1133	940-1321
IX	D100 OR	29	1036	-

Age	Location	No.	Mean FL (except as marked)	Range
	P115 CA	4	1087	991-1194
X	D100 OR	31	1214	-
	P115 CA	6	1156	1016-1270
XI	B90 OR Columbia R.	1	1041	8845
	I7 ID	2	1156 TL	1143-1168
	D100 OR	29	1316	-
	P115 CA	1	1410	-
XII	P115 CA	2	1295	1245-1321
	D100 OR	34	1402	-
	I7 ID female	1	1397 TL	-
XIII	P115 CA	2	1334	1168-1473
	D100 OR	17	1488	-
XIV	P115 CA	18	1389	1143-1651
	D100 OR	21	1509	-
XV	P115 CA	35	1438	1219-1575
	D100 OR	26	1623	-
XVI	P115 CA	125	1529	1295-1854
	D100 OR	11	1676	-
XVII	P115 CA	40	1582	1295-1930
	D100 OR	11	1694	-
XVIII	P115 CA	15	1636	1448-1803
	D100 OR	17	1727	-
XIX	D100 OR	7	1712	-
	P115 CA	6	1740	1549-2007
XX	D100 OR	2	1765	-
	P115 CA	5	1775	1626-1880
XXI	D100 OR	1	1524	-
	P115 CA	2	1753	1626-1930
XXII	D100 OR	6	1844	-
	P115 CA	2	2019	1930-2108
XXIII	D100 OR	4	1798	-
	I7 ID female	1	2032 TL	-
XXIV	D100 OR	6	1875	-
XXV	D100 OR	5	1915	-
	P115 CA	1	2019	-
XXVI	D100 OR	2	2090	-
	P115 CA	1	2096	-
XXVII	D100 OR	1	1930	-
XXVIII	D100 OR	1	1778	-
	I7 ID male	1	1905 TL	-
XXIX	P115 CA	1	2400	-
XXX	D100 OR	2	2329	-
	I7 ID female	1	2489 TL	-
	P115 CA	1	2400	-
XXXI	D100 OR	1	2464	-
XXXIII	D100 OR	1	2743	-
XXXV	D100 OR	1	2426	-
XXXVI	D100 OR	1	2286	-
XXXIX	O15 OR	1	2591	-
XL	D100 OR	1	2159	-

Age	Location	No.	Mean FL (except as marked)	Range
XLII	D100 OR	2	2852	-
	O15 OR	1	2362	-
XLVII	S347 CA	1	2819	209,563
XLVI-L	O15 OR	1	3048	-
LIII	D100 OR	1	2819	-
LXXXII	D100 OR	1	3226	-

Ages were determined from fin ray counts. Most of the growth takes place in the ocean. Males mature at age IX-X (D100), females at XI-XII (P111), XII-XIV (D100), XV (B90) or XV-XX (S239) and at 22.7-34 kg and 1270-1651 mm (S347). In the Columbia River they spawn May-July at temperatures 9-17C (D100). A 210 kg female, age XLVII, had 1,700,000 eggs (S347).

As small fingerlings, white sturgeon feed on plankton and minute bottom organisms (D100, S347). Amphipods, shrimp, and tendipedid larvae were the food of 30 young sturgeon, average 203 mm FL, taken from August to October (S388). As adults they are omnivorous and scavengers. They will eat onions, wheat, lampreys, molluscs, insect larvae, crayfish, frogs, frog eggs, small adult salmon, trout, squawfish, suckers, carp, anchovies, smelt, or mud and the contained organisms (D100, S347). Pacific lampreys and eulachon seem to be important foods. A house cat was found in one sturgeon stomach (D100). A 20.4 kg sturgeon had 14 salmonids, 100-290 mm long, in its stomach (M337).

Small sturgeon were successfully shipped to Japan, at 4.5C when various anesthetics failed (R266).

PALLID STURGEON, *Scaphirhynchus albus* (Forbes and Richardson)
Rare in the upper Mississippi River and its larger tributaries (E119) but fairly common in the Missouri River and its new reservoirs. Largest specimen reported, 30.8 kg in North Dakota (W192).

B127   15 sturgeon   167-637 mm SL.   FL = 1.08SL

B127   1473 mm TL   14288 g
B172 MT 1676 mm TL   17236 g

F100,124,125 SD Oahe Reservoir 71 sturgeon

TL	470	495	521	546	572	648	673	724	749	800	826	851
g	249	299	354	417	499	749	834	1089	1216	1588	1742	1964

TL	876	902	927	978	1003	1029	1054	1080	1156
g	2195	2440	2745	3266	3547	3932	4196	4427	6174

	No.	TL	g	K(TL)
F100,F105,F124,F125, SD Oahe	76	457-1143	-	0.27-0.35
N115,S276,S229 SD Gavins Point	12	660-1295	1588-8505	0.27-0-42

The K(TL) values generally increased with increase in length.

F125 SD Oahe L. Calculated growth of pallid sturgeon, from pectoral rays

Year	1	2	3	4	5	6	7	8	9	10
TL	279	378	470	574	638	673	732	790	838	881
Incr.	279	99	92	104	64	56	58	79	76	86
No.	6	6	6	6	6	5	5	4	3	2

F100 SD Oahe L. Males mature at 533-584 mm TL

SHOVELNOSE STURGEON, *Scaphirhynchus platorynchus* (Rafinesque)
In the Mississippi River and its larger tributaries (E119). Largest specimen: 813 mm, 4536 g (T113).

B127   15 sturgeon   136-730 mm SL   FL = 1.080 SL

| M132 Mississippi R. | | | E54 Ohio R. | | |
FL	No.	Mean W	No.	Mean W	Range
406-431	3	281	1	454	-
432-456	12	295	-	-	-
457-482	17	376	2	567	-
483-507	28	499	3	567	-
508-532	57	562	6	662	567-680
533-558	52	644	12	789	680-907
559-583	96	762	7	857	680-907
584-609	87	902	2	1188	1134-1248
610-634	96	1038	9	1197	1021-1361
635-659	113	1239	6	1252	1021-1474
660-685	126	1465	9	1511	1361-1814
686-710	90	1647	3	1665	1361-1814
711-736	57	1896	1	1928	-
737-761	27	2186	1	2155	-
762-786	12	2540	-	-	-
787-812	4	2681	-	-	-

TL	SD Oahe F100,105,124, 125 (1313 fish)	SD Ft. Randall & Gavins Point N94, N115, S226,227, 229,276 (587 fish)	Upper Mississippi R. U8
		Weight in grams	
279-304	50	-	-
305-329	82	64	86
330-355	104	86	127
356-380	122	104	172
381-405	145	136	213
406-431	181	168	249
432-456	222	213	318
457-482	277	249	372
483-507	318	290	454
508-532	386	336	522
533-558	454	390	654
559-583	513	468	766
584-609	622	567	867
610-634	671	635	998
635-659	839	699	1143
660-685	993	780	1420
686-710	-	843	-
711-736	-	921	1928 (J13)
737-761	-	1206	2019
762-786	-	1297	-
787-812	-	1397	-

N94,115, S224,226,227,228,229,276,393. SD Ft. Randall and Gavins Point. 635
sturgeon, 305-787 mm, K(TL) = 0.25, range 0.22-0.27
     The K value decreases slightly June-September (S229, S276). No sex dif-
ference nor trend with size (S227, S229).

F100,105,124,125 SD Oahe. 1322 sturgeon, 279-660 mm, K(TL) 0.25-0.34
F125 SD Oahe L. calculated TL, from pectoral rays.

Year	1	2	3	4	5	6	7	8	9	10
TL	213	274	325	366	399	437	470	483	457	503
Incr.	213	61	51	51	51	48	48	36	25	51
No.	35	35	35	31	23	17	9	6	2	1

     One summary reports spawning from April to June, with males maturing at
551 mm and females at 645 mm (K118). In Oahe, South Dakota, sturgeon over
508 mm were considered mature (F100, F105). They spawn in South Dakota
prior to June 20 (S227, W192) or in late June and early July (S229, S276). Ripe
males and females were found April 20 in Kansas (M215).

Smallest mature males   356 mm TL (S226) 495 mm FL (M132) 495 mm SL (B94)
Smallest mature females 356 mm TL (S226) 521 mm FL (M132) 635 mm SL (B94)

Food of 74 shovelnose sturgeon, 445-610 mm TL, was 75% *Trichoptera*, 17%
Hexagenia spp., and 8% other insects (H243, Mississippi R.).

===

PADDLEFISH, *Polyodon spathula* (Walbaum)
     The paddlefish is found in the larger streams of the Mississippi River
drainage (E119, T113). A relict population was in Lake Erie until 1903 (T113),
and there are a few records from Lakes Huron and Ontario (S471). The
largest specimen was 1524 mm long and weighed 83.5 kg (T113). Another
weighed 74 kg (F10).
     The length of the paddle, from the medial edge of the lower jaw to the tip
was described by the following formula based on 240 specimens from 17 to
2159 mm (M305):

$$P = 0.2366 \text{ TL} + 37.6 \text{ mm, with } r = 0.988$$

For fish under 254 mm a better fit was

$$P = 0.365 \text{ TL} - 4.6 \text{ mm, with } r = 0.986$$

M305 Missouri R. 101 paddlefish, 521-1295 mm, FL TL = 1.103 FL (no consis-
     tent trend)
H399 OK 5 paddlefish, TL = 1.182 SL

A3 IL Standard Length-Weight Relationship					
SL	No.	Weight	SL	No.	Weight
178-202	1	54	991-1015	4	5534
305-329	1	100	1016-1040	1	6260
533-558	2	930	1040-1066	1	6895
610-634	1	1905	1067-1091	1	8936
635-659	2	1451	1092-1117	2	6758
660-685	1	1134	1118-1142	1	6940
686-710	1	1633	1143-1167	1	9979

SL	No.	Weight	SL	No.	Weight
737-761	2	2313	1168-1193	1	7257
761-786	3	1950	1194-1218	1	10433
838-863	2	3175	1270-1294	1	11340
940-964	1	3311	1880-1904	2	70988

TL	No.	Mean W	Range	Citation, Locale
76-101	3	5	-	S472 AL
152-177	1	7	-	S472
178-202	8	36	14-36	S472
208-228	6	18	14-18	S472
	2	26	-	P155 MO
229-253	14	27	23-36	S472
254-278	25	36	32-45	S472
	5	68	59-86	P155
279-304	19	50	45-59	S472
	7	100	86-127	P155
305-329	11	59	50-73	S472
	7	127	113-141	P155
330-355	5	100	73-136	S472
	2	154	-	P155
356-380	2	200	-	P155
381-405	1	136	-	S472
	1	227	-	P155
406-431	1	177	-	S472
	1	254	-	P155
432-457	3	200	181-213	S472
458-482	2	249	227-272	S472
	2	322	281-363	M305 IA
483-507	2	240	227-254	M305, S472
508-532	10	281	227-363	S472
	1	399	-	P155
533-558	6	426	-	S472
	3	553	481-590	P155, M305
559-583	5	345	272-454	S472
	2	726	680-771	M305
584-609	19	508	454-613	S472
	3	735	481-907	M305
610-634	12	568	454-680	S472, M305
635-659	15	671	318-1089	S472, M305
660-685	15	726	567-907	S472
	4	998	821-1179	M305
686-710	7	857	680-1043	S472
	8	1103	853-1497	M305
711-736	13	1052	680-1406	S472, M305
737-761	8	1125	953-1497	S472
	6	1511	1161-1874	M305
762-786	30	1107	907-1361	S472
	10	1715	1220-2631	M305
	1	1950	-	F132 TN
787-812	4	1243	1021-1451	S472
	7	1669	1361-2096	M305
813-837	4	1451	1225-1905	S472
	15	1887	1420-2767	M305, L59

TL	No.	Mean W	Range	Citation, Locale
838-863	3	1474	1248-1814	S472
	12	1950	1565-2676	M305
864-888	3	1987	1588-2223	S472
	16	2291	1729-2722	P155, M305
889-913	6	1850	1474-2041	S472
	22	2318	1814-3606	L59, M305, P155
914-939	31	2571	2041-3343	M305, S472
940-964	31	2849	2354-4264	M305
965-1015	95	3066	1928-5049	L59, M305, S472
1016-1066	57	4042	2948-6350	L59, M305, S472
1069-1117	42	4904	3402-7938	L59, M305, S472
1118-1167	29	5475	4082-7484	L59, M305, S472
1168-1218	20	6858	5216-8800	L59, M305, S472
1219-1269	13	8128	6350-11140	M305
1270-1320	8	9707	8391-11567	M305, S472
1321-1359	5	11998	8165-16216	J13, B106, M305, S472
1372-1410	4	12673	9979-18597	M305
1422-1453	3	16974	14061-18597	B106, L59, M305
1778	1	35834	-	J13
1880	2	62052	56020-68040	F10, J13
2210	1	31751	-	O21
2235	1	56020	-	E55

M305 IA Mississippi R.	533-1270 mm TL	$\log W = -4.975 + 2.842 \log TL$
M305 SD Missouri R.	406-1524 mm TL	$\log W = -6.815 + 3.464 \log TL$
H399 OK Ft. Gibson	432-1194 mm TL	$\log W = -6.248 + 3.284 \log TL$
S472 AL	610-1321 mm TL	$\log W = -7.896 + 3.81 \ \log TL$

S226 SD 47 paddlefish, 457-1041 mm, K(TL) = 0.32

    K value increased with increase in length of males but not females.

M305 IA Mississippi R. 245 paddlefish, 406-1346 mm, K(TL) = 0.34, range 0.25-0.72. Missouri R. 204 paddlefish, 406-1778 mm, K(TL) = 0.37, range 0.19-0.67. Some increase in K value with increase in length in the Missouri River but not in the Mississippi River.

S472 AL	203-900 mm	K(TL)	0.21-0.31
	914-1332 mm	K(TL)	0.29-0.52

The code in the following table indicates that age was determined by Petersen's length frequency method (P), fin ray sections (F), dentary bones (B), or known age because of rearing (K). Otoliths were not found satisfactory for age determination (A3).

	Code	No.	SL	TL	Range	No.	Mean W	Range
0-May								
T11 IL	P	7	x	-	17-20	-	-	-
B215 MO	P	1	-	36	-	-	-	-
0-July								
M305 IA	P	200	-	x	64-81	-	-	-
D101 MO	P	3	-	x	74-91	-	-	-
0-Aug.								
P155 MO	K	29	-	297	213-406	29	113	36-254
H221,H399 OK								
Ft. Gibson L.	F	9	-	495	401-551	2	318	268-367

	Code	No.	SL	TL	Range	No.	Mean W	Range
0-Sept.								
M305 IA	P	5	-	x	104-178	-	-	-
T11 MO	P	-	x	-	37-93	-	-	-
H221,H399 OK								
Ft. Gibson L.	F	3	-	589	561-622	3	735	595-880
0-Aug.-Oct.								
W192 SD Lewis								
& Clark L.	P	6	-	-	150-274	-	-	-
0-Oct.								
H221,H399 OK								
Ft. Gibson L.	F	4	-	688	635-721	4	1125	880-1248
0-Nov.-Dec.								
H399 OK								
Ft. Gibson L.	F	6	-	724	698-749	-	-	-
I-Feb.								
H399 OK								
Ft. Gibson L.	F	2	-	775	749-800	-	-	-
I-April								
P155 MO	K	2	-	538	526-551	2	440	399-481
I-June								
H399 OK								
Ft. Gibson L.	F	7	-	879	826-909	-	-	-
I-July								
A52 MO (as inter-								
preted by M305)	P	25	-	x	102-152	-	-	-
M305	P	7	-	x	152-305	-	-	-
H221,H399 OK								
Ft. Gibson L.	F	14	-	879	838-953	-	-	-
I-Sept.								
D101 Mo (as inter-								
preted by M305)	P	-	-	x	152-305	-	-	-
H221,H399 OK								
Ft. Gibson L.	F	3	-	958	927-1003	-	-	-
I-Oct.								
P155 MO	K	+	-	-	864-902	+	-	2495-2722
I-Oct.-Nov.								
H399 OK								
Ft. Gibson L.	F	12	-	1029	927-1156	-	-	-
Age I								
S393 SD								
Ft. Randall L.	B	+	-	-	152-203	-	-	-
M305 IA	B	1	-	206	-	-	-	-
A3 IL	B	2	250	-	197-305	2	77	54-100
T11 IL	P	-	254	-	-	-	-	-
Age II								
S393 SD								
Ft. Randall L.	B	+	-	-	305-330	-	-	-
M305 IA	B	7	-	541	483-597	6	726	454-907
S330 SD								
Ft. Randall L.	B	1	-	572	-	1	454	-
A3 IL	B	8	604	-	533-686	7	1347	907-1905

56 PADDLEFISH

	Code	No.	SL	TL	Range	No.	Mean W	Range
T11 IL	P	-	686	-	-	-	-	-
H221,H399 OK								
Ft. Gibson L.	F	51	-	1095	1001-1232	-	-	-
Age III								
S330,S393 SD								
Ft. Randall L.	B	-	-	-	457-546	1	595	-
M305 IA	B	15	-	724	622-800	15	1451	907-2495
A3 IL	B	6	762	-	737-812	5	2105	1633-2359
T11 IL	P	-	1220	-	-	-	-	-
Age IV								
S393 SD								
Ft. Randall L.	B	+	-	-	686-737	-	-	-
S330 SD	B	9	-	798	-	9	1588	-
M305 IA	B	16	-	869	775-889	16	2132	1588-2722
A3 IL	B	2	838	-	-	2	3175	-
Age V								
S330 SD	B	27	-	834	-	27	2014	-
M305 IA	B	26	-	914	864-940	26	2540	2041-3175
A2,A3 IL	B	2	851	-	762-940	2	2767	2268-3311
Age VI								
S330 SD	B	17	-	970	-	17	3202	-
M305 IA	B	74	-	968	864-991	74	2858	2268-4309
A3 IL	B	1	966	-	-	-	-	-
Age VII								
S330 SD	B	22	-	965	-	22	3655	-
M305 IA	B	24	-	1006	965-1041	24	3357	2722-4082
A3 IL	B	3	998	-	990-1016	3	5443	4989-6260
Age VIII								
S330 SD	B	17	-	968	-	17	3765	-
M305 IA	B	19	-	1024	991-1067	19	3719	3175-4989
Age IX								
M305 IA	B	12	-	1067	1016-1118	12	4128	3629-4763
S330 SD	B	11	-	1105	-	11	5942	-
A3 IL	B	1	1066	-	-	1	8936	-
Age X								
S330 SD	B	11	-	1085	-	11	5715	-
M305 IA	B	9	-	1087	1041-1118	9	5670	4082-6350
A3 IL	B	4	1118	-	1092-1168	4	6918	6123-7348
Age XI								
M305 IA	B	10	-	1107	1016-1270	10	5216	4309-6350
S330 SD	B	9	-	1158	-	9	7303	-
Age XII								
M305 IA	B	9	-	1148	1118-1194	9	5398	4989-5897
S330 SD	B	6	-	1167	-	6	8800	-
A3 IL	B	1	1194	-	-	1	10433	-
Age XIII								
M305 IA	B	6	-	1167	1143-1194	6	5579	5216-5897
S330 SD	B	6	-	1214	-	6	9571	-
Age XIV								
S330 SD	B	7	-	1125	-	7	6532	-
M305 IA	B	5	-	1196	1168-1219	5	6940	5897-8165
A3 IL	B	1	1283	-	-	1	11340	-

	Code	No.	SL	TL	Range	No.	Mean W	Range
Age XV								
S330 SD	B	5	-	1212	-	5	8709	-
M305 IA	B	5	-	1224	1143-1257	5	7394	7257-7484
Age XVI								
M305 IA	B	1	-	1219	-	1	8845	-
S330 SD	B	3	-	1252	-	3	8437	-
Age XVII								
S330 SD	B	1	-	1194	-	1	5897	-
P171 MO	B	26	-	1478	-	26	13880	-
Age XIX								
M305 IA	B	2	-	1295	-	2	9435	8845-9979
Age XX								
M305 IA	B	2	-	1308	1295-1321	2	10070	8845-11340
Age XXII								
M305 IA	B	2	-	1378	1372-1384	2	14968	11340-18597
S330 SD female	B	1	-	1443	-	1	12428	-
Age XXIV								
S330 SD female	B	1	-	1420	-	1	18235	-
Age XXIX								
P171 MO	B	1	-	1308	-	1	12247	-
Age XXX								
P171 MO	B	1	-	1308	-	1	13154	-

		Calculated TL at each annulus						
	No.	1	2	3	4	5	6	7
L195 OK Arkansas R.								
and Cimarron R.	19	292	424	521	612	701	732	795
L195 OK Ft. Gibson L.	28	363	533	706	1006	1148	1257	1331
J42 OK Ft. Gibson L.	16	206	432	-	-	-	-	-

Growth during the first 2 years may be rapid (H221, H399, P155) or relatively slow (M305, A3). Rapid growth in Fort Gibson was associated with changes brought on by impoundment (L195).

In an Iowa pond, 26 paddlefish, 356-864 mm, averaging 460 mm September, 1916, grew to average 762 mm by December, 1918 (F34). In Alabama ponds (S473) where paddlefish were fed Auburn No. 2 feed, growth from July 1 to October 22 was:

						Percent weight increase
2 fish	787 mm	1021-1134 g	to	991 mm	3402 g	220-233
3 fish	889-965	1701-2155	to	1016-1118	4309-5715	110-153
2 fish	1092	3402-4082	to	1168-1194	7257-8346	103-113

and from June 19 to November 10 was:

1 fish	724 mm	680 g	to	965 mm	3148 g	363
1 fish	991	2495	to	1176	6918	177
2 fish	1118-1143	5103	to	1295-1455	11567-12020	127-136

At Fort Randall, South Dakota, 3 paddlefish tagged at 1016 mm grew 64 mm in 11 months (S393); and a tagged 18824 g paddlefish grew 907 g in 16 months, in Center Hill Lake, Tennessee (R182).

Female paddlefish averaged larger than males (P171) but no sex difference in growth rate was noted by S330.

Males can be distinguished from females by the papillae which encircle the urinogenital opening (M377). Gravid females were collected in September and from January through June (M305) and males with sperm were found at all seasons. They spawn April to June in Missouri (P130, B309), and from March to June (K118). They spawn over sand and pebbles in schools (K118), or over a large gravel bar at 16C in strong current (P130). Eggs taken in Missouri on April 29 hatched May 10-22 (P155). Fertilized eggs are 2.7-4.0 mm in diameter and are adhesive to gravel (P130). They hatched in 7 days at 18.4-21C and the fry were 8-10 mm TL at hatching, and 48 mm at 29 days (P130). At 14C they hatched in 10 days (B309) and the embryological stages are described. At 70 days, a young fish was 61 mm long and resembled an adult in form (B309). Successes with artificial propagation are reported (M377, P155).

The smallest mature males weighed 2270 g and females 2950 g (M305). Males first mature at 998 mm or 7 years and females at 1067 mm or 9-10 years (K118). Females apparently spawn at intervals of 4-7 years, judging from crowding of annuli (S393, M305). Sperm or eggs begin developing at age V and take about 2 years to mature (S393). Two females 1374 mm long had 137,247 and 141,531 eggs (M305).

In Fort Randall Lake, South Dakota, snagging by sport fishermen brought a return of 9.8% of 61 tagged and 12.6% of 198 tagged (N93, S393), and 12.4% of 306 tagged (S330).

Food of paddlefish is largely plankton (E127) but may include insects (F133). In the Mississippi River mayfly naiads comprised 46-95% of the volume of the food of adult paddlefish (H243, M305). One stomach contained a 64 mm river darter, *Percina shumardi* (M305). Another contained 13 threadfin shad, 50-100 mm long (F132). In 155 adult paddlefish collected in the winter in South Dakota, *Daphnia* and *Diaptomus* comprised most of the food with phytoplankton less abundant proportionately in the food than in the water (S330). Paddlefish fry chased *Daphnia* but not copepods (B309).

L135 (p. 133) reports that Lagler and Obrecht were unable to satisfactorily age gar otoliths.

SPOTTED GAR, *Lepisosteus oculatus* (Winchell)

The spotted gar is found in the Lake Erie drainage in Ohio and southern Michigan, and in the Mississippi River to Iowa, and in the Gulf of Mexico streams from Florida to Mexico (T113, E119). They are found in clear waters with abundant vegetation (T113).

TL	No.	Mean W	Range	Citation
152-177	2	8	-	S472 AL
178-202	2	6	-	S472
203-228	10	23	17-45	S472
229-253	24	37	14-85	S472
254-278	9	43	23-45	S472

TL	No.	Mean W	Range	Citation
279-304	6	68	45-94	S472
305-329	11	82	45-142	S472
330-355	5	99	85-105	S472
356-380	3	105	74-173	S472
	1	213	-	H156 OK
381-405	2	181	-	S472
	3	267	201-400	G39 IN
406-431	10	250	218-289	G39, S472
432-456	18	292	227-351	G39, S472
457-482	14	321	227-386	G39, S472
483-507	15	414	272-680	G39, H156, S472
508-532	15	428	318-500	G39, S472
533-558	16	641	445-1715	G39, H156, S472
559-583	9	615	454-779	G39, S472
584-609	10	689	545-770	G39, S472
	4	1100	981-1191	H156
610-634	6	819	680-907	G39, S472
635-659	8	998	726-1225	G39, S472
	1	1424	-	H156
660-685	1	907	-	S472
	1	1180	-	G39
	3	1537	1452-1638	H156
686-710	3	1157	1134-1208	S472
	1	1602	-	H156
711	1	1323	-	G39
737	2	1980	-	S472
1092	1	2722	-	T113 OH

S472 AL	74 fish	152-420 mm, TL log W = -6.33 + 3.31 log TL
	89 fish	421-737 mm, TL log W = -6.54 + 3.40 log TL

H156 OK Canton L. 14 gar, 378-688 mm, log W = -5.8921 + 3.4530 log TL mean K(TL) = 0.47, range 0.39-0.55, increasing with increased TL

S472 AL	71 fish	150-370 mm TL	K(TL)	0.12-0.31
	92 fish	380-490 mm TL	K(TL)	0.28-0.49

0-July	6 fish	188-224 mm TL		R263 OK
0-Aug.	6 fish	234-264	TL	R263 OK
0-Oct.	-	178-254	TL	T113 OH

Young gar grew 1.7 mm (1.4 to 2.1 mm) or 1.0 g (0.7 to 1.3 g) per day in July-August in Oklahoma (R263). Scalation is complete at 140-150 mm (S474).

After a review of papers on the food of spotted gar D57 says that fish are the principal food and that feeding is mostly in early morning. Spotted gar were more piscivorous than shortnose gar taken at the same time (L196). Blue crab (S474), crayfish, amphipods and insects are eaten in addition to fish (B310). In Louisiana brackish waters (L137, L140), spotted gar, 254-1000 mm, fed about equally on blue crabs and sunfishes.

L216 LA Sizes in gill net meshes

Bar measure	No.	TL	Range
25 mm	10	762	445-838
38	36	584	483-737
51	22	688	445-775
64	4	777	546-914

LONGNOSE GAR, *Lepisosteus osseus* (Linnaeus)
Montana to Quebec and south to the Gulf of Mexico and the Rio Grande River (E119, T113, S474). Frequently in brackish water (S474).
C30 IA   TL = 1.15 SL, in 6 fish.

TL	No.	Mean W	Range	Citation
51	2	1	-	S472 AL
127	1	4	-	S472
178	2	8	6-9	S472
203	10	14	10-15	S472
229	7	48	14-113	S472
254	9	27	23-45	S472
279	3	59	45-91	S472
305	11	73	41-181	S472
330	17	77	45-367	S472
356	6	91	77-413	S472
381	2	91	-	S472
406	1	172	-	S472
432	4	122	118-145	S472
508-558	8	286	227-454	G39 IN, S472
559-609	5	408	318-454	S472
610-659	5	504	454-530	G39, W132 TN
	4	726	680-907	M242 FL
	3	816	680-998	S472
660-710	3	649	544-703	W132 TN
	3	771	703-843	G39
	19	907	726-1134	F74, M242 FL
	1	1179	-	S472
711-761	7	839	630-953	G39, W132
	1	924	-	C30 IA
	4	920	816-967	S472
	47	1089	771-1225	F74, M242 FL
762-812	1	912	-	G39
	10	1043	816-1165	S472
	3	1257	1134-1446	C30
	97	1270	907-1406	F74, M242 FL
813-863	1	1170	-	G39
	6	1438	1165-1710	S472
	1	1474	-	C30
	63	1588	1361-1950	F74, M242
864-913	1	1451	-	G39
	3	1860	1497-2495	S472
	36	1905	1451-2177	F74, M242
	1	2078	-	C30

TL	No.	Mean W	Range	Citation
914-964	4	2268	1588-2948	S472
	26	2449	1905-3311	F74, M242
965-1015	2	3062	2948-3175	S472, M319 MD
	9	3447	2495-5125	F74, M242
1016-1066	2	2785	2580-2994	G39, W132
	19	3901	3266-11793	F74, M242
1067-1117	13	4627	4264-5443	F74, M242
	1	5216	-	S472
1118-1167	8	6668	4627-12065	F74, M242
1168-1218	1	4853	-	S472
	15	6260	4309-8573	F74, M242
1219-1269	1	4989	-	N52 WI
	9	7257	6214-9072	F74, M242
1270-1320	9	9026	7530-9934	F74, M242
1321-1371	13	9979	9072-12247	F74, M242
1372-1421	1	6350	-	T113 OH
	8	11430	10569-12746	F74, M242
1422-1472	17	12020	8981-14061	F74, M242
1473-1523	6	14696	12701-17010	M242
1524	3	14606	12247-16103	M242
1600	1	18371	-	M242

N102 MO 139 gar 94-1369 mm TL, $\log W = -7.067 + 3.507$ TL
S472 AL 120 gar 51-1194 mm TL, $\log W = -5.97 + 3.12 \log$ TL

C36 IA 6 gar 630-757 mm SL, average K(SL) 0.40 range 0.35-0.48
D179 TX 461 gar 456-1290 mm SL, average K(SL) 0.41 range 0.22-0.90
S472 AL 118 gar 127-1194 mm TL K(TL) range 0.13-0.41
Spawning may result in 50% loss in weight (N102).

F74 FL tanks. Growth of young longnose gar

Days age	No.	Mean TL	Range
2	10	13	-
28	10	30	28-33
49	10	46	41-56
64	10	53	43-58
85	1	155	-
88	1	180	-
95	1	201	-
102	1	231	-
115	7	109	94-127
183	1	335	-

A5 VA Potomac R. At 2 weeks gar averaged 19 mm.
H86 MI Portage Lake 2.33 mm growth per day for a month in early first
    summer.
R263 OK 4 young gar grew 2.7 mm per day in July-August.
        5 young gar grew 3.2 mm (2.3 to 4.5 mm) per day in July-August.
        5 young gar grew 1.8 g (1.3 to 2.3 g) per day in July-August.

N102 MO 20 gar kept in the laboratory showed the following growth.

Days of experiment	Mean TL	Range	Increase per day	Mean W	% of weight eaten per day
0 (July 12)	137	124-160	-	3	-
14	173	132-206	0.11	-	10.1
28	229	188-264	0.15	-	11.3
42	264	231-302	0.10	-	7.3
52	290	262-325	0.10	41	8.0

The average increase in weight was 0.72 g per day, and the food conversion ranged from 1.57 to 2.94.

	No.	Mean TL	Range
Age 0-June			
M215 KS Big Blue R.	-	20	13-30
D166 KS	43	53	-
T33 NY	6	56	41-71
Age 0-July			
F5 NY L. Erie	1	41	-
T113 OH	-	-	76-127
R263 OK	-	-	224-305
Age 0-Aug.			
T113 OH	-	-	102-203
R263 OK	11	345	226-368
Age 0-Sept.			
T33 NY	5	234	203-279
Age 0-Oct.			
T113 OH	-	-	254-381

N102 MO Ages determined by branchiostegal ray rings.

Age	I	II	III	IV	V	VI	VII	VIII	IX	X	XI
TL	475	607	640	747	714	767	823	853	826	765	843
No.	6	4	6	6	6	10	10	8	7	4	5
Age	XII	XIII	XIV	XV	XVI	XVII	XVIII	XIX	XX	XXII	
TL	1054	960	1130	1181	1153	1120	1166	1135	1199	1257	
No.	4	2	4	1	2	5	3	1	2	2	

All longnose gar in the above (N102) older than XVII and longer than 1029 mm were females.

R63 OH	Age	I	II	III
	TL	165	318	470

A2 IL 1041 mm gar age XI (from otoliths).
In captivity longnose gar have lived 20 years (F6) and 30 years (B69).

N102, MO calculated TL (from branchiostegal rays, with a Fraser-type correction of 130 mm):

Year	1	2	3	4	5	6	7	8	9	10
Males	495	610	668	706	732	762	782	800	792	815
Incre.	495	109	58	36	33	28	20	13	15	10
No.	54	51	47	43	38	33	24	18	12	9
Females	559	655	706	765	810	856	894	922	958	998
Incre.	559	94	51	53	43	38	38	28	23	28
No.	45	42	42	40	39	37	36	32	30	26

Year	11	12	13	14	15	16	17	18	19	20	21	22
Males	856	968	978	993	1003	1016	1029					
Incre.	10	10	10	15	10	13	13					
No.	6	2	2	1	1	1	1					
Females	1034	1059	1087	1115	1128	1143	1153	1179	1194	1222	1250	1257
Incre.	25	23	20	23	13	18	10	15	5	8	3	8
No.	25	24	20	19	15	14	12	8	5	4	2	2

The growth of various individuals of the same age was quite variable.

A caudal filament at the termination of the upturned vertebral column atrophies as the gar ages, but has been found on individuals up to 323 mm TL (S474). Several studies have been made of the embryology and early development of young (S474).

Male longnose gar mature at 3-4 years, and females at 6 years (N102) in Missouri, and at about 500 mm in Kansas (D166). In Florida they spawn March-August with the peak in April (M242), in Kansas in June (D166), and in Missouri in May when they run upstream (N102). The left ovary tends to be larger than the right and the eggs are green, toxic, and 2.1-3.2 mm in diameter (N102).

N102 MO log number of eggs = -7.1866 + 3.818 log TL
    26 females 767-1370 mm, average 27830 eggs range 4273-59422
M426   36500 eggs/female

H213 FL	TL	No.	Eggs per female
	680-750	6	6210
	760-870	42	11920
	880-1000	37	13038
	1010-1130	14	26378
	1140-1260	15	38751
	1270-1390	11	43380
	1400-1500	2	34750-77156

A26 FL 1270 mm TL female 46803 eggs; 1422 mm TL female 14515 g 77156 eggs

---

D57, summarizing papers on longnose gar food, says that gar up to 50 mm long feed on entomastraca and insect larvae, but when longer they eat fish primarily.

L216 LA Sizes in gill net meshes

Bar measure	No.	TL	Range
25 mm	45	869	635-1295
38	59	993	686-1321
51	28	1031	864-1575
64	10	1006	686-1168
76	5	925	737-1092
89	1	960	
114	2	1181	1143-1219

SHORTNOSE GAR, *Lepisosteus platostomus* Rafinesque
Mississippi and Ohio River drainage from southern Minnesota and Ohio to northern Louisiana and Texas (T113) in open silty rivers (H382).

C30 IA   TL = 1.15 SL, based upon 2 fish.
B330 WI   TL = 1.153 SL, based upon 5 fish.
H7 FL   TL = 1.4884 mm + 1.1245 SL, based upon 225 fish, 350-1160 mm SL.

In the following length-weight table, the South Dakota data is based upon 1,477 shortnose gar but the numbers at each length were not given (S224,226,227, 228,229,276, N115,94, F105,100,124,125).

TL	No.	Mean W	Range	Citation
279	1	91	-	S472 AL
305	7	91	82-91	S472
318	-	132	-	SD
330	2	109	18-136	S472
343	-	141	-	SD
368	-	168	-	SD
381	1	172	-	S472
394	-	204	-	SD
419	-	245	-	SD
445	-	295	-	SD
470	-	345	-	SD
483	1	494	-	S472
483	1	635	-	F74 FL
495	-	395	-	SD
508	1	590	-	F74
521	-	477	-	SD
533	2	771	-	F74
546	-	562	-	SD
559	7	821	762-	S472
559	1	1089	-	F74
572	-	644	-	SD
574	1	867	-	C30 IA
584	3	844	771-880	S472
584	3	998	907-1043	F74
597	-	757	-	SD
610	1	1134	-	F74
622	-	871	-	SD
635	2	1239	1225-1252	S472
648	-	1030	-	SD
660	2	1347	1306-1388	S472
673	-	1170	-	SD
698	-	1366	-	SD
711	1	2041	-	F74
724	-	1651	-	SD
726	1	571	-	C30
737	1	2404	-	F74
749	-	1747	-	SD
775	-	1469	-	SD
800	-	1506	-	SD
826	-	1547	-	SD

S224 SD Ft. Randall 292 gar 483-686 mm log W = -7.533 + 3.122 log TL
S472 AL 26 gar 279-660 mm log W = -6.53 + 3.43 log TL

B330 WI 5 gar K(SL) 0.55 range 0.44-0.70
C36 IA 2 gar 500-632 mm SL, average K(SL) 0.66 range 0.62-0.69.
S472 AL 26 gar 279-660 mm TL, average K(TL) 0.40 range 0.30-0.48
F100,105,124,125, N94,115, S224,226,227,228,229,276,393 SD 1559 gar 305-
  813 mm, K(TL) 0.34 range 0.31-0.40
No change with sex or size except S224 reports some increase in K with in-
crease in TL.

	Age	No.	Mean TL	Range
R27 IL	2 weeks	1	19	-
R27 IL Illinois R.	0-May	-	-	15-64
R27 IL Illinois R.	0-June	1	44	-
S226 SD Ft. Randall	0-July	1	69	-
R27 IL Illinois R.	0-July	1	89	-
P13 IA L. Okoboji	0-Aug.	-	-	64-127
P13 IA L. Okoboji	0-Oct.	-	-	102-152
T113 OH	0-Oct.	-	-	178-254
W192 SD Lewis and Clark L.	I	31	417	-
	II	31	486	-
	III	60	536	-
	IV	93	587	-
	V	26	605	-
	VI	16	671	-
	VII	2	734	-

A shortnose gar was kept 20 years in captivity (F6).

Shortnose gar spawned May-June in 300-900 mm of water in Illinois (R27) and May-July in Iowa (P13). In South Dakota shortnose gar spawned May 30-June 15 at water temperatures 19-23.5C in 1956 (S228) and in late June and early July in 1957-8 (S229, 276). A few males mature at 457 mm and females at 483 mm (S229).

The bright green eggs (E4) are deposited in small masses held together by a clear gelatinous substance which attaches to weeds (P13). A 4082 g female had 36460 eggs (P13). Eggs hatch in 8 days (P13).

Young gar start feeding at 16 days after hatching (R27) and feed on entomostraca and mosquito larvae (P13). Shortnose gar in Lake Okoboji, Iowa (P133) fed on crayfish, perch, sunfish, and bluegills, feeding mostly in the morning. On clear warm days they lined up along shore in 100-250 mm of water, with their tails toward shore, remaining until prey were in easy reach. In May and June, adult gar were observed coming to the surface to seize emerging gnats and mayflies in Illinois (R27). Shortnose gar fed heavily on carp until the latter were over 125 mm long (S227).

L216 LA Sizes in gill net meshes

Bar measure	No.	TL	Range
25 mm	6	627	381-991
38	7	605	483-889
51	5	805	737-889
64	4	1003	914-1067
76	2	879	737-991

FLORIDA GAR, *Lepisosteus platyrhincus* DeKay
   In Florida and southern Georgia (S474). The largest reported was
1331 mm TL.

TL	No.	Mean W	Range	Citation
254-278	1	68	-	H233
279-304	3	86	68-113	H233
305-329	5	113	86-154	H233
330-355	13	168	113-236	H233
356-380	13	213	177-249	H233
381-405	18	263	159-313	H233
406-431	16	340	272-413	H233
432-456	5	404	372-449	H233
457-482	3	440	372-504	H233
483-507	6	513	454-652	H233,M242
508-532	14	649	513-816	H233,M242
533-558	14	721	454-1089	H233,M242
559-583	4	898	735-1043	H233,M242
584-609	3	921	862-1007	H233,M242
610-634	2	1075	967-1188	H233
660	1	1547	-	M242

Florida gar spawn mostly in April and May, but spawning may continue into
October (M242).

H213 TL	No.	Eggs per female	TL	No.	Eggs per female	TL	No.	Eggs per female
330	1	523	483	118	4079	635	10	7340
356	1	1887	508	104	5605	660	4	9495
381	6	1934	533	87	5897	686	3	11151
406	18	2133	559	47	6471	711	1	10526
432	31	3096	584	29	7038	737	2	7962
456	95	3407	610	15	7988	864	1	15450

A26 FL 610 mm TL female 10,705 eggs; 864 mm TL female, 2948 g, 15,540
eggs.
   Fish constitute most of the food, but various crustaceans and insects are
eaten in significant quantities (H213, H233). H244 determined that Florida gar
take 36 days to consume their body weight in food, compared to 15 days for
largemouth bass.

ALLIGATOR GAR, *Lepisosteus spatula* Lacépède
   Gulf drainage and Mississippi-Ohio system to southwestern Ohio and
northern Missouri (T113, E119), frequently in brackish and marine waters
(S474).

TL	Weight in kg	Citation	TL	Weight in kg	Citation
1651	24.86	B94	2337	80.7	G33
1651	24.95	W57	2692	54.9	L66
2235	69.9	G33	2794	105.2	G33
2261	73.9	G33	2959	137.0	G33
2286	84.4	G33	3048	104.3	M114

Alligator gar spawn April-June in Louisiana (S474).

D102 in a review of literature on food of alligator gar referred to them as primarily scavengers and predators on fish and crabs. In brackish water in Louisiana (L137,140,144) blue crabs constituted most of the food of alligator gar 838-1854 mm long. Four gar examined in Texas had all eaten birds: anhinga, ducks and grackles (R228). One 1981 mm alligator gar had a 660 mm shortnose gar in its stomach (H400).

L216 LA Sizes in gill net meshes

Bar measure	No.	TL	Range
38 mm	2	711	698-724
64	1	737	

BOWFIN, *Amia calva* Linnaeus

Bowfin occur in sluggish rivers and lakes of the Mississippi and Ohio River drainage and in parts of the Great Lakes drainage into Quebec, south to the Gulf of Mexico and up the Atlantic coast to Connecticut (E119, H382). Generally in clear water with abundant vegetation (T113).

C24 MN   TL = 1.162 SL 100-599 mm
C30 IA   TL = 1.17 SL 588 mm

TL	No.	Mean W	Range	
356	1	510	-	G39 IN
381	2	567	533-601	G39
406	1	595	-	G39
457	1	910	-	G39
508	3	1228	1171-1267	G39
533	3	1588	1100-2132	G39
559	2	1640	1500-1752	G39
584	1	1500	-	G39
610	1	2016	-	G39
737	1	3912	-	N52 WI
749	1	4508	-	R152 IN

C23 MI Average K(SL) 1.38 16 fish   300-650 mm
C36 IA   K(SL) 1.36 1 fish   688 mm

	Age	No.	Mean TL	Range	Weight
R27 IL Illinois R.	0-May	-	-	10-58	-
G133 MI	0-June	-	-	89-99	2
S8 WI	0-July	-	-	43-71	-
G141 AL	0-Aug.	24	406	-	680
T113 OH	0-Oct.	-	-	127-229	-
C215 MI Whitmore L.	III	18	475	-	-
C215 MI Whitmore L.	IV	12	566	-	-
C215 MI Whitmore L.	V	3	632	-	-
C215 MI Whitmore L.	VI	5	607	-	-
C215 MI Whitmore L.	VII	1	648	-	-
R72 OH	VII	1	711	-	3969

A bowfin was kept in captivity 20 years (F6), 24 years (F7), and 30 years (B69).

Females tend to be larger than males (G133). The largest male was 686 mm and 2268 g while the largest female was 787 mm and 3856 g (T113).

Males build and guard the nest and young (A1, G133, K118). They usually nest in colonies (K118) in sheltered areas clearing out the vegetation for the nest (G133). Bowfin are most active at twilight and dawn (G133) and they spawn at night (K118). They spawn in April in Illinois (R27), in June in Ontario (D180), or in April to July over the entire range (K118). The eggs hatch in 9 days and the young remain another 7-9 days in the nest. They are 8 mm at hatching and 10-13 mm at the time they leave the nest (G133).

	TL of female	Weight	No. of eggs
L14 OH	489	-	30170
V21, E4 MI	483	2268	23600
	533	-	64000
R72 OH	711	3969	55244

Bowfin feed mainly on fish and crayfish (G133,P132,L188,L197,S475). Insects, molluscs, earthworms, frogs and leeches were eaten in small numbers (L188,F133,L197). Bowfin often feed at night (E4).

L216 LA Sizes in gill net meshes

Bar measure	No.	TL	Range
38 mm	1	579	-
64	2	650	559-737

BLUEBACK HERRING, *Alosa aestivalis* (Mitchill)
Anadromous from Cape Breton, Nova Scotia, to St. Johns River, Florida (H401). H401 lists it as *Pomolobus aestivalis* (Mitchill).

	Age	No.	Mean FL	Range	Mean W	Range
H61 MD Chesapeake B.	0-June	27	-	20-38	-	-
H61 MD Chesapeake B.	0-July	31	-	30-58	-	-
H61 MD Chesapeake B.	0-Aug.	93	-	36-64	-	-
H61 MD Chesapeake B.	0-fall	514	-	41-74	-	-
H61 MD Chesapeake B.	I-spring	1340	-	64-119	-	-
H61 MD Chesapeake B.	II	30	-	140-208	-	-
G22 NY Hudson R.	III	4	267	259-274	207	198-213
G22 NY Hudson R.	IV	16	292	274-302	261	213-340
G22 NY Hudson R.	V	15	297	277-325	275	213-340
G22 NY Hudson R.	VI	5	318	310-330	346	255-340

The maximum reported in Maryland (S386) was 279 mm, 312 g, but they reach 381 mm and 369 g (H401). The eggs are 0.87-1.11 mm in diameter and the fry at hatching are 3.1-4.2 mm (M393).

Many individuals become sexually mature and perform their first spawning migration at 4 years of age (H401). The inshore migration is about a month later than that of alewives, at water temperatures of about 21C. In the lower Chesapeake Bay the peak runs are early April, in the Potomac River late April to early May, and in Massachusetts in June and July (H401).

Food consists of plankton (H401). An extensive bibliography is given in H401.

ALABAMA SHAD, *Alosa alabamae* Jordan and Evermann
I include here the data on the Ohio shad, *Alosa ohiensis* Evermann, following H401. As thus defined the species is anadromous, entering the principal streams tributary to the Gulf of Mexico from near Pensacola, Florida, to the Mississippi River and ascending the Mississippi River system to Hot Springs, Arkansas; Fairport, Iowa; and Montgomery, West Virginia (H401). Not now found in the Mississippi drainage, except for recent reports in Oklahoma (E119) and Kentucky (H401).

E54 KY Ohio R.

TL	No.	Mean W	Range
381	7	811	680-907
406	22	920	794-1134
432	17	1094	907-1361
457	2	1361	-

The females in this collection averaged heavier at each length than the males.
Probably the adult shad eat little while in fresh water on their spawning migration (H401, C88). An intensive bibliography is given in H401.

SKIPJACK HERRING, *Alosa chrysochloris* (Rafinesque)
In the Gulf of Mexico and in streams, and occasionally lakes and borrow pits, of the Gulf drainage. Probably anadromous in many streams. As far north as Lake Pepin, Minnesota, and Pittsburgh, Pennsylvania (H401). Listed as *Pomolobus chrysochloris* Rafinesque by H401.

S472 AL

TL	No.	Mean W	Range	K(TL)
25	6	0.5	0.2-0.54	2.95
51	212	1.7	0.9-2.3	1.14
76	923	6.4	1.3-33.6	1.45
102	2288	10	3-50	0.94
127	556	15	5-40	0.76
152	783	26	20-50	0.73
178	383	45	23-91	0.80
203	311	68	36-200	0.82
229	258	104	45-172	0.88
254	369	154	68-227	0.95
279	530	227	136-318	1.05
305	214	281	136-318	0.99
330	99	354	249-454	0.99
356	158	449	318-499	0.99
381	156	517	376-726	0.93
406	78	585	399-811	0.87
432	9	749	590-853	0.93
457	4	843	590-1080	0.88

S472 4762 herring 51-152 mm TL, log W = -4.43 + 2.69 log TL
2519 herring 178-457 mm TL, log W = -5.48 + 3.17 log TL

At Keokuk, Iowa, spawning occurred from early May to early July but little else is known about spawning of skipjack herring (C88, H401).

Skipjack are carnivorous; the young feed on insects and the adults feed on fish (F10,C88,H401).

An extensive bibliography is given by H401.

HICKORY SHAD, *Alosa mediocris* (Mitchill)

Anadromous from Florida to the Bay of Fundy, Canada (M393), less common north of New York and Cape Cod (H401). They enter brackish or fresh water in the spring and the fall. H401 lists the hickory shad as *Pomolobus mediocris* (Mitchill). Maximum length is 610 mm (H401), but 381 mm and 907 g was reported as a record for Maryland (S386).

M393 MD	No.	mm	TL/SL
	32	5-8	1.033
	13	9-15	1.101
	42	16-26	1.190
	37	27-90	1.23
	5	100-150	1.29

A graph is given for TL/FL conversion (M393)

Age	No.	TL	Citation
0-May	34	8-33	M393 SC, MD
0-June	158	18-114	M393 NC, VA, MD
0-July	1	130	M393 MD
0-Aug.	9	51-170	M393 GA, VA, MD
0-Sept.	3	102-185	M393 MD, NJ
0-Oct.	11	107-193	M393 MD, NC
0-Nov.	3	132-173	M393 GA
I-Jan.-Mar.	6	142-193	M393 GA, VA, MD
I-June	2	183-198	M393 VA
III	2	358-391, 851-1219 g	M393 MD

There was no evidence of faster growth to the south of Chesapeake Bay than in the Bay (M393). The wide spread of lengths suggests an extended spawning season. Few young hickory shad have been collected, however.

Although it has been questioned that hickory shad are anadromous and spawn in fresh water (H401), eggs were successfully taken and hatched in fresh water in the Chesapeake (M393). Spawning fish were age III. The first ripe females were seen in early May and all were spent in early June (M393). Eggs hatched in 2-3 days at 18-21C and the prolarvae were 5-6.6 mm TL at hatching (M393). The embryology was described (M393).

Hickory shad feed chiefly on small fish (H401).

Bibliographies are given by H401 and M393.

ALEWIFE, *Alosa pseudoharengus* (Wilson)

While primarily an anadromous fish on the Atlantic Coast from the St. Lawrence to North Carolina (H401), alewives became established in the Finger Lakes, New York, about 1868; in Lake Ontario, 1883; Lake Erie, 1931; Lake Huron, 1935; Lake Michigan, 1952; Lake Superior, 1954 (T75). M446 gives

1873 for the first specimen in Lake Ontario; 1931 for Lake Erie; 1933, Lake
Huron; 1949, Lake Michigan; and 1954, Lake Superior. As the alewife population increased in South Bay, Lake Huron, the cisco population declined proportionately (S440). Alewives are referred to as *Pomolobus pseudoharengus*
(Wilson) in H401.

R128 ME	SL	Mean W	Range
	225	168	146-191
	250	219	197-241
	275	276	254-299
	300	333	301-364
	325	393	349-437

TL	No.	Weight	
216	1	77	E105 MD
338	1	275	S398 MD
381	1	408	S386 MD

O1 NY Seneca L. 225 immature alewives log W = -4.550 + 2.78 log SL
      211 adult males   log W = -3.604 + 2.41 log SL
      180 adult females log W = -3.312 + 2.28 log SL
    combined, 616 alewives log W = -3.770 + 2.51 log SL

The Massachusetts alewives are reported to be heavier for their lengths than
those from Maine (R128).

Age 0-June	No.	SL	FL	TL	Range	No.	Mean W	Range
H401 MD	-	-	-		15-30	-	-	-
H61 MD Chesapeake	26	-	-	X	30-58	-	-	-
Age 0-July								
H61 MD Chesapeake	79	-	-	X	46-74	-	-	-
G64 Atlantic N.S.	84	45	-	-	-	-	-	-
Age 0-Aug.								
H61 MD Chesapeake	80	-	-	X	51-81	-	-	-
R276 NY Cayuga L.	507	-	-	36	23-58	-	-	-
V80 NY Pond F								
(at 40 days)	-	-	-	43	-	-	-	-
H280 ME Mt. Desert	50	-	-	58	-	-	-	-
G97 NJ Hopatcong L.	371	-	-	58	41-74	-	-	-
Age 0-Sept.								
O1 NY Seneca L.	108	24	-	-	21-31	-	-	-
H61 MD Chesapeake	134	-	-	X	51-79	-	-	-
G97 NY Hopatcong L.	316	-	-	66	51-81	-	-	-
V80 NY Black P.								
late spawn	67	-	-	33	28-43	67	0.3	-
V80 NY Black P.								
early spawn	65	-	-	76	69-89	65	4.8	-
R276 NY Cayuga L.	302	-	-	76	66-79	-	-	-
Age 0-Oct.								
G97 NJ Hopatcong L.	370	-	-	64	51-86	-	-	-
R276 NY Cayuga L.	6	-	-	81	-	-	-	-
C275 L. Erie	-	-	-	112	-	-	-	-

Age 0-Fall	No.	SL	FL	TL	Range	No.	Mean W	Range
B71 NY Kensico L.	364	40	-	-	26-63	-	-	-
S57 NC	-	-	-	X	51-76	-	-	-
H61 MD Chesapeake	352	-	-	X	56-94	-	-	-
B32 MA	-	-	-	X	51-102	-	-	-
H401 ME	-	-	-	X	51-102	-	-	-
Age I								
O1 NY Seneca L.	113	55	-	-	30-70	-	-	-
B71 NY Kensico L.	-	55	-	-	34-82	-	-	-
V80 NY Pond F, May	600	-	-	74	-	-	-	-
G97 NJ L. Hopatcong	1839	-	-	74	56-112	-	-	-
G64 L. Ontario	37	62	-	-	-	-	-	-
P19 ON L. Ontario	36	-	84	-	-	25	6	-
G25 NY L. Ontario	5	-	-	89	69-109	-	-	-
R267 NY Cayuga L.	580	-	-	102	91-107	-	-	-
G97 NJ Kemah L.	+	-	-	127	-	-	-	-
B32 MA	-	-	-	127	-	-	-	-
H61 MD Chesapeake	-	-	-	X	66-224	-	-	-
Age II								
G97 NJ L. Hopatcong	273	-	-	117	102-142	-	-	-
R267 NY Cayuga L.	576	-	-	122	-	-	-	-
G64 L. Ontario	27	108	-	-	-	-	-	-
O1 NY Seneca L.	89	116	-	-	100-130	-	-	-
P19 ON L. Ontario	6	-	130	-	-	5	17	-
G25 NY L. Ontario	4	-	-	135	132-137	-	-	-
G97 NJ Kemah L.	-	-	-	203	254	-	-	-
Age III								
R267 NY Cayuga L.	404	-	-	130	-	-	-	-
G22 NY Kensico L.	48	-	-	142	127-163	48	26	20-40
P19 L. Ontario	13	-	142	-	-	11	37	-
O1 NY Seneca L.	284	123	-	-	104-148	-	-	-
G64 L. Ontario	28	126	-	-	-	-	-	-
G25 NY Ontario R.	4	-	-	163	155-170	-	-	-
G64 Atlantic N.S.	6	147	-	-	-	-	-	-
G22 NY Hudson R.	2	-	-	274	269-279	2	193	184-227
R128 MA Nemasket R.	83	263	-	-	-	83	264	-
Age IV								
R267 NY Cayuga L.	62	-	-	137	-	-	-	-
P19 L. Ontario	48	-	152	-	-	48	40	-
G22 NY L. Kensico	24	-	-	163	147-241	24	34	26-79
O1 NY L. Seneca	49	137	-	-	119-158	-	-	-
G25 L. Ontario	14	-	-	170	157-190	-	-	-
G64 L. Ontario	6	141	-	-	-	-	-	-
G64 Atlantic N.S.	27	171	-	-	-	-	-	-
G22 NY Hudson R.	10	-	-	292	277-305	8	216	199-255
H280 ME Mt. Desert	45	-	-	312	-	-	-	-
R128 Atlantic ME	1068	274	-	-	-	1068	261	-
Age V								
R267 NY Cayuga L.	6	-	-	145	-	-	-	-
P19 L. Ontario	52	-	160	-	-	52	43	-
G22 NY Kensico L.	14	-	-	165	160-175	14	37	31-45
O1 NY Seneca L.	25	139	-	-	129-150	-	-	-
G64 L. Ontario	12	149	-	-	-	-	-	-

	No.	SL	FL	TL	Range	No.	Mean W	Range
G25 L. Ontario	7	-	-	183	168-203	-	-	-
G64 Atlantic N.S.	9	209	-	-	-	-	-	-
G22 NY Hudson R.	7	+	-	295	279-312	7	233	184-340
Age VI								
G22 NY Kensico L.	2	+	-	168	165-170	2	41	40-42
G64 L. Ontario	5	157	-	-	-	-	-	-
P12 L. Ontario	12	+	180	-	-	12	62	-
G25 L. Ontario	3	+	-	216	173-259	-	-	-
G64 Atlantic N.S.	4	231	-	-	-	-	-	-
G22 NY Hudson R.	5	+	-	300	277-325	4	224	213-241
Age VII								
G22 NY Kensico L.	1	+	-	160	-	1	34	-
G64 L. Ontario	1	164	-	-	-	-	-	-
P19 L. Ontario	1	+	188	-	-	1	85	-
V6 L. Huron	1	188	-	216	-	-	-	-
Age VIII								
G64 L. Ontario	1	172	-	-	-	-	-	-
Age IX								
G64 Atlantic N.S.	1	270	-	-	-	-	-	-

				Calculated SL at each annulus				
	No.	1	2	3	4	5	6	7
B71 NY Kensico L.	8	44	68	80	91	107	111	120
O1 NY Seneca L.	668	55	111	126	137	144		

		Calculated TL at each annulus				
H297 ME Long P.	No.	1	2	3	4	5
males	74	132	221	272	300	310
females	70	135	221	274	305	318

Scales taken from above the anal fin insertion and near the midline were uniform and clear (R235). Dry scales were better than impressions. Annuli appear as fine lines cutting across ridges and transverse grooves. They are continuous in the anterior field and evident in the posterior as bands of increased opacity. Annuli were formed at spawning time in immature and mature fish. The annuli were verified by consistency of distance beyond the last annulus and by agreement with length-frequencies (R235). Body-scale relationship was found to be relatively constant after 75 mm (H297).

Alewives grow faster and larger in marine (to 381 mm) than freshwater environments (to 216 mm). G64 points out that alewives usually mature one year earlier in fresh water than in the ocean and that this is one reason for the smaller size in the freshwater populations. He refers to Hoar's (H261) statement that the thyroid may not function properly in landlocked populations.

Growth is faster in young of the year in Lake Ontario than in Nova Scotia, probably because of higher summer temperatures in Ontario (G64). Young from shallow water averaged 66 mm TL compared to 61 mm from deep water (colder water) in October in Lake Hopatcong, New Jersey (G97). Seasonal growth of young stops in the first half of October at water temperatures of 18.4C in Lake Erie (C275). No relationship was noted between year class abundance and growth rate in Long Pond, Maine (H297). Females may grow a bit faster and live longer than males (R128, T75, H297, H280). Ultimate lengths as determined by Walford lines were 333 mm for males and 356 mm for females in Long Pond, Maine (H297).

A stunted population with few individuals reaching 127 mm TL was found in Cayuga Lake, New York (W93), usually in deep water where they were important food for lake trout. They move inshore to spawn in late July. They sometimes show high mortality in late spring. Adults were transferred from this lake to 4 ponds and reproduced successfully in 2 (V80). When stocked at 127 mm, 14 g each, they grew to 150 mm, 43 g from June 25 to August 19 and 160 mm, 45 g by October 3.

In the Atlantic, maturity is reached in 3 years off Massachusetts but 4 off Maine (R128). In Long Pond, Maine, some matured at age III and some not until V, but some females spawned as many as 3 times (H297). In Lake Hopatcong, New Jersey, a landlocked race mostly spawns at age II with some not maturing until age III and only one spawning per life cycle is known (G97). Alewives stocked in a pond in New York spawned at age II (F135). In Lake Ontario, males mature at age II, 109 mm SL, and females at III, 125 mm SL (T75). Two spawnings were evident from the two sizes of offspring after adult alewives were placed in a pond in New York on June 25 (V80). In Lake Michigan, mature alewives were as small as 132 mm TL (E118).

Marine alewives enter streams and lakes to spawn. South of Cape Cod runs start in late March, in Maine usually not until May (R128). They were observed spawning in the Hudson River in mid-May when water temperatures were 11C (H401). Probably few anadromous alewives spawn more than once (R128). Freshwater populations of alewives spawn in streams or in shallow water areas along shore on sand or gravel bottoms (T75, G114) often in areas with some vegetation (G97). They spawn at night in 150-300 mm of water (G64, G97). Spawning occurs at water temperatures of 13-21C (R128), 13-16C (T75), 17-19C (G97), or 22C (C275). In Lake Ontario spawning is in late May to early August (P19) or begins in April with the peak mid-June to early July (G64). Spawning occurs in late June in Lake Erie (C275) and in Lake Michigan (E118), late May and June in New Jersey (G97), late June and early July in Cayuga Lake, New York (G114), early June in a New York pond (F135), and early June to late August in the Finger Lakes, New York (O1).

Annual mortalities between ages V and VI, and VI and VII were estimated at 78.6% and 74.7% (H297) in Long Pond, Maine. Little relationship could be shown between brood stock and progeny (H297).

Incubation time ranges from 48-96 hours at 22C to 6 days at 15.5C (R128), or from 81 to 132 hours at 13-23C (O1), or from 2 to 3 days at 15.5C (M442). Eggs were 0.95-1.27 mm in diameter (M393). At hatching the fry are 4-5.6 mm TL (M393, M442). The fry are positively phototropic and pelagic (O1).

H280 marine	60000-100,000 eggs per female.
B73, H61, S57 marine	average 102,800 eggs per female, 644 females stripped.
O1 NY Seneca L.	10000-12000 eggs per 178 mm female.

That the young return to the "parent stream" is evident from runs started by stocking gravid adults (H401).

Zooplankton constitutes most of the food of alewives (T75, G97, P19). The young while in fresh water feed on diatoms, copepods and ostracods (H401). Shrimp and small fish may be the chief diet in salt water, but diatoms and other plankton organisms are also important. As is true of most anadromous fish, the adults feed little while in fresh water to spawn. Adults in freshwater populations may eat insects, crustacea and small fish (O1). During their spawning season alewives may eat a number of their own eggs (E118, G97) though most of the eggs were eaten by immature alewives. As many as 440 eggs were found in the stomach of a 100 mm alewife (E118). Many of the stomachs also contained shiner scales, probably mistaken for eggs (E118).

Alewives provide forage for many other fishes. They are the sole host of the glochidia of the mussel, *Anodonta implicata,* which has disappeared where alewives have disappeared (D181). In Cayuga Lake, N.Y. (G114) alewives are generally distributed at various depths but tend toward deeper water in October and November.

Alewives in 27-28C water died when scared into 33.3C water, though some still survived this temperature change (T102).

Additional life history data are in a thesis (R267) and in papers listed in the bibliography of H401.

AMERICAN SHAD, *Alosa sapidissima* (Wilson)

Anadromous from the St. Lawrence and the Nova Scotian banks to Florida (H401). Also introduced in the Pacific and established from southeastern Alaska to San Pedro, California. On the Atlantic the maximum length is 762 mm and weight 5443 g, but females of 6350 g are reported from the Pacific (H401). Early records on the Atlantic went to 6804 g (T119).

P35 NC

FL	No.	Mean W	Range	FL	No.	Mean W	Range
31-40	2	0.3	-	81-90	9	5.0	3.5-6.2
41-50	4	0.7	0.5-1.0	91-100	7	6.9	6.0-8.2
51-60	7	1.4	1.2-1.7	101-110	2	10.7	9.0-11.5
61-70	10	2.1	1.7-2.5	111-120	2	13.2	11.4-15.0
71-80	4	2.8	2.4-3.1	121-130	1	15.8	-

H61 MD Chesapeake Bay

FL	No.	Wt.	FL	No.	Wt.	FL	No.	Wt.
356-380	2	1021	457-482	13	1388	559-583	4	2495
381-405	2	1021	483-507	8	1502	584-609	14	2631
406-431	13	993	508-532	4	2068	610-634	5	2975
432-456	12	1048	533-558	8	2155	635-659	3	3574

W141 FL mature females

FL	No.	Weight		FL	No.	Weight	
13.9-353	1	794		417-429	5	1406	1247-1531
356-378	3	814	595-992	432-455	4	1588	1389-1814
391-404	4	1177	1021-1304	457-460	2	1828	1814-1843

M364 MD males from nylon net $\log W = -2.187 + 2.959 \log TL$
males from linen net $\log W = -0.080 + 1.275 \log TL$

The correlation coefficients, $r = 0.88$ and 0.94 respectively

The best explanation of the 1.275 slope with the linen gill nets is that the nets are taking fish of a very specific girth, thus taking only the heaviest short fish and the lightest long fish. The nylon net, being more elastic, was not so selective.

N83 An age IV male which remained at least 120 days in the stream after the spawning season weighed only 454 g at 437 mm FL when average weight is 1451 g — a 69% loss of body weight.

M185 MD ponds Growth of young shad

Age (in days)	No.	FL		Age (in days)	No.	FL
2	9	8.6		30	296	24
3	4	9.7		58-65	155	30
4	3	10.4		91	43	36
5	35	10.9		121	104	45
10	27	13.7		127	500	46
15	9	17.0				

	No.	SL	FL	Range	No.	Mean W	Range
Age 0-May							
S57 NC	-	-	10	-			
B73 —	-	-	X	10-15			
H37 CA	461	-	36	-			
P35 NC ponds	542	-	36	25-43			
Age 0-June							
L23 Canada	90	-	X	8-15			
P35 NC ponds	24	-	38	33-46			
H37 CA	112	-	41	-			
H61 MD Chesapeake	128	-	X	20-66			
Age 0-July							
L23 Canada	399	-	X	8-58			
H61 MD Chesapeake	138	-	X	30-74			
H37 CA	105	-	46	-			
S56 VA Potomac R.	-	51	-	-			
B9, B21 DC	-	X	-	50-75			
Age 0-Aug.							
L23 Canada	2	-	X	41-51			
W79 NC Neuse R.	98	-	51	36-84			
B73 --	-	-	X	51-58			
W79 FL St. Johns R.	112	-	53	38-74			
W79 NY Hudson R.	110	-	53	46-64			
W79 Connecticut R.	217	-	53	36-84			
W79 GA Ogeechee R.	130	-	56	41-76			
P35 NC ponds	49	-	56	48-66			
H61 MD Chesapeake	67	-	X	46-79			
W79 VA Pamunkey R.	223	-	64	46-86			
H37 CA	110	-	84	-			
Age 0-Sept.							
C45 VA Potomac R.	-	X	-	25-76			
P35 NC ponds	169	-	61	48-74			
M408 VA York R. system	2384	-	69	46-94			
H61 MD Chesapeake	84	-	X	51-91			
B73 --	84	-	X	76-102			
H37 CA	24	-	91	-			
Age 0-Oct.							
P35 NC ponds	62	-	76	69-86			
H6 MD ponds	-	X	-	40-108			

	No.	SL	FL	Range	No.	Mean W	Range
H61 MD Chesapeake	225	-	X	46-112			
B17 CT	-	-	X	69-91			
H37 CA	84	-	89	-			
M7, 8, 9, 10, B9 DC	-	X	-	60-180			
C331 MD Chesapeake	25	-	107	94-114			
T25 --	-	95	-	-			
B73 --	-	-	X	102-114			
Age 0-Nov.-Dec.							
H61 MD Chesapeake	139	-	X	61-119			
B73 VA Potomac R.	-	-	X	76-102			
S56 VA Potomac R.	-	X	-	85-100			
P35 NC ponds	134	-	91	64-124			
H37 CA	6	-	99	-			
B21 DC	-	X	-	100-150			
B23 DC	-	-	152	-			
S57, B73 NC	-	-	X	178-229			
Age 0							
B64 CT	26	X	-	58-113			
Age I							
P35 NC Mar.-May	49	-	91	81-124			
H37 CA	5	-	119	-			
H61 MD Chesapeake	23	-	X	86-152			
B17 CT	1	-	142	-			
B64 CT	4	-	X	145-180			
L82 MD Susquehanna R.	6	-	282	-			
Age II							
V81 CA Millerton L.	2	-	178	160-196			
L23 Canada	192	-	229	-			
B64 CT	7	X	-	210-290			
L82 MD Susquehanna R. m.	4	-	318	-			
L82 FL St. Johns R. m.	36	-	348	-			
L82 NC Neuse R. m.	34	-	361	-			
L82 FL St. Johns R. f.	14	-	366	-			
L82 NC Neuse R. f.	2	-	401	-			
S55 Pacific Coast	1	-	432	-			
Age III							
B17 CT	1	-	295	-	-	-	-
L23 Canada	29	-	305	-	-	-	-
B64 CT	10	X	-	300-350	-	-	-
W141 FL St. Johns R. m.	37	-	348	-	-	-	-
W141 FL St. Johns R. f.	5	-	363	-	-	-	-
L82 MD Susquehanna R. m.	19	-	386	-	-	-	-
G22 NY Hudson R.	23	-	399	269-526	22	821	227-2100
L82 FL St. Johns R. m.	31	-	394	-	-	-	-
L82 NC Neuse R. m.	44	-	401	-	-	-	-
L82 FL St. Johns R. f.	33	-	401	-	-	-	-
L82 NC Neuse R. f.	19	-	434	-	-	-	-
Age IV							
B17 CT	2	-	323	315-330	-	-	-
B64 CT	22	X	-	360-390	-	-	-
W141 FL St. Johns R. m.	60	-	388	-	-	-	-
W141 FL St. Johns R. f.	66	-	411	-	11	1107	539-1534

	No.	SL	FL	Range	No.	Mean W	Range
C167 NY Hudson R. m.	1	-	411	-	1	1134	-
L82 FL St. Johns R. m.	7	-	419	-	-	-	-
L82 MD Susquehanna R. m.	17	-	421	-	-	-	-
L23 Canada	14	-	432	-	-	-	-
L82, D58 FL St. Johns R. f.	25	-	437	381-	1	967	-
D58 VA Potomac R. f.	2	-	X	419-488	2	1375	1193-1565
L82 MD Susquehanna R. f.	22	-	455	-	-	-	-
D58 GA Ogeechee R. f.	2	-	457	-	2	1755	1700-1810
L82 NC Neuse R. m.	9	-	457	-	-	-	-
L82, D58 NC Neuse R. f.	26	-	465	447-	2	1800	1787-1814
D58 SC Edisto R. f.	1	-	465	-	1	1615	-
C167 NY Hudson R. f.	1	-	450	-	1	1814	-
G22 NY Hudson R.	44	-	483	356-564	43	1841	454-2382
Age V							
B17 CT	6	-	351	325-391	-	-	-
W141 FL St. Johns R. m.	26	-	424	-	-	-	-
W141 FL St. Johns R. f.	37	-	450	-	6	1107	539-1533
L82 MD Susquehanna R. m.	4	-	432	-	-	-	-
D58 VA Potomac R. f.	1	-	460	-	1	1446	-
L82, D58 FL St. Johns R. f.	22	-	465	368-	4	-	907-1561
C167, D58 NY Hudson R. f.	4	-	472	442-486	4	1633	1361-1814
D58 GA Ogeechee R. f.	1	-	478	-	1	1928	-
L23 Canada	46	-	488	-	-	-	-
D58 SC Edisto R. f.	3	-	X	478-498	3	-	1474-1700
L82 MD Susquehanna R. f.	12	-	490	-	-	-	-
L82, D58 NC Neuse R. f.	28	-	495	-	1	2182	-
G22 NY Hudson R.	61	-	511	401-584	61	1715	680-2522
Age VI							
B17 CT	3	-	381	376-388	-	-	-
W141 FL St. Johns R. m.	2	-	437	-	-	-	-
W141 FL St. Johns R. f.	7	-	462	-	2	1828	1814-1842
D58 GA Ogeechee R. f.	2	-	478	475-480	2	2041	1928-2155
D58 VA Potomac R. f.	2	-	495	483-505	2	2232	2127-2354
D58 SC Edisto R.	1	-	503	-	1	2068	-
L82, D58 NC Neuse R. f.	7	-	513	483-	2	2410	2100-2720
L23 Canada	17	-	536	-	-	-	-
G22, C167, D58 NY Hudson R.	24	-	533	450-570	22	1932	1134-2495
Age VII							
W141 FL St. Johns R. f.	2	-	467	-	-	-	-
B17 CT	25	-	470	450-516	-	-	-
G22, D58 NY Hudson R.	10	-	503	414-597	10	1923	967-2608
L23 Canada	8	-	536	-620	-	-	-
Age VIII							
B17 CT males	7	-	478	460-516	-	-	-
B17 CT females	4	-	511	500-521	-	-	-
L23 Canada	1	-	569	-	-	-	-
G22 NY Hudson R.	4	-	582	470-658	4	2513	1275-3461
Age IX							
B17 CT males	3	-	470	450-500	-	-	-
B17 CT females	4	-	516	480-550	-	-	-
Age XI							
C167 NY Hudson R. f.	2	-	574	572-577	2	3189	2948-3430

	No.	1	2	3	4	5	6	7	8
			Calculated FL at each annulus						
L82 MD Susquehanna R. m.	50	165	267	330	376	404			
L82 N.S. Scotsmans C. m.	-	180	262	328	388				
L23 Canada	-	130	236	330	404	465	516	554	584
L82 FL St. Johns R. m.	74	175	277	345	394				
L82 FL St. Johns R. f.	89	180	284	353	409	447			
L82 N.S. Scotsmans C. f.	-	193	284	353	417				
L82 MD Susquehanna R. f.	54	175	284	361	404	445			
L82 N.C. Neuse R. m.	87	178	287	368	421				
L82 N.C. Neuse R. f.	77	183	295	376	429	472	513		

	1	2	3	4	5	6	7
		Calculated FL and annual increments					
W141 FL St. Johns R.							
Males	173	290	348	388	424	437	
incre.	173	117	58	43	33	25	
number	125	125	125	88	28	2	
Females	178	300	358	409	445	457	467
incre.	178	122	58	51	38	33	25
number	117	117	117	112	46	9	2

Scale readings were validated by fin clipping, although there was much fin regeneration (J86). Good scale photos were given (J86). Considerable difficulty was reported in the early studies of growth from scales (L23, B17, B64, B311). On the basis of spawning marks, mortality rates were estimated at about 70% (M447).

Female shad apparently grow faster than males. There is no obvious trend in growth from north to south. W79 found no trend in size of young the first summer even though shad spawning in the St. Johns River, Florida, was in March while that in the Connecticut River was in May. The sizes in Canadian waters (L23) were generally shorter during the first summer than those from farther south, however.

Shad enter streams above brackish water to spawn, always in freshwater, but within the limits of tidal influence in the Virginia areas studied (M444). They reportedly spawned at least 240 kilometers (150 miles) up the Hudson River (B308). In the Connecticut Rever they were stopped by a dam 84 miles upstream but are passed over with a mechanical fish-lift and spawn farther upstream (N83). In the St. Lawrence River they spawn as far upstream as Montreal (V82). One tagged in Quebec was caught off Cape Cod — at least 1900 kilometers (1200 miles) by sea. There are no landlocked populations but young shad apparently from shad which never went to sea were found in Millerton Lake, California, in 1963 (V81). S465 reports that shad enter the rivers earlier in the south (January in Georgia and South Carolina) than in the north (May-June in Canada) with the run starting at 10-12C and peaking at 13-15.5C. In Florida the run begins in November but peaks in January (T119). The run in the York River, Virginia, peaks in March to mid-April but may extend from January through May, with few fish running at temperatures below 4.5C (N113). In Maryland the run starts in February but peaks in April (M444); in the Hudson River the run is in March and early April (B308); in Connecticut from early April to mid-June (N83); and in the St. Lawrence mid-May to June (V83). In California ripe males are found from February to June and eggs have been taken from females as early as April or as late as October (G137).

Shad spawn over a fairly wide temperature range in Virginia but most spawn at 15.5-17C (M445). Other temperatures noted are 11-14C (S465), 12C in Connecticut (L23), and 13-19C in Maryland (M444). Spawning is mostly in the evening (B308).

Some male shad mature at age II and females at age III in Florida (W141) but generally not until a year or two later elsewhere (K118). In Maryland some spawn at age III but most at IV and V (M444). In Georgia the spawning run was of age III to age VI fish, all first spawns (S476). In St. Johns River, Florida (W141), and in Ogeechee River, Georgia (S476), there was no evidence of spawning more than once, but shad in York River, Virginia, may spawn as many as 4 times in their life span (N113) and in the Connecticut River as many as 5 times (M447).

L19, B73, J13, H61   30,000 eggs per female, maximum 156,000 eggs per female

B23   25,000 eggs per 1800-2300 g female, 60,000-100,000 eggs per 2700 g female

W43   27,800 eggs per female

M30 NY Delaware R.   28,000 to 98,000 eggs per female

B22   33,000 to 49,000 eggs per female

H5 MD   150,000 eggs per female

Y1   100,000 to 150,000 eggs per female

M173 MD   Average 30,000 eggs per female stripped. One female 156,000 eggs

---

L100 Hudson R. Volumetric measure

FL	Weight	No.	Thousands of eggs/female Mean	Range
353-399	794-1247	5	149	116-219
419-465	1360-1701	6	207	138-237
486-526	1871-2608	7	319	241-359
536-587	2722-3657	4	448	436-468

Thousands of eggs = -462.7 + 1.58   FL;   r = 0.930

Thousands of eggs =   18.8 + 0.1307 W;   r = 0.939

Thousands of eggs =   -91   + 62.182 Age in years;   r = 0.980

N113 VA York R.

FL	Weight	No.	Thousands of eggs/female
378	964	1	252
384-401	1049-1219	4	169-226
406-427	1191-1503	4	194-304
432-455	1474-1814	4	246-421
462-475	1758-2070	4	360-436
483	2268	1	358

Thousands of eggs = -574.2   + 2.003 FL;   r = 0.80

Thousands of eggs =   17.174 + 0.180 W;   r = 0.82

Thousands of eggs =   -35.6   + 71.85 Age in years;   r = 0.74

W141 FL St. Johns R.

Thousands of ova = 65 + 0.140 W;   r = 0.90

D58 found the number of eggs per female increased to the south even though the average size of the female decreased. The fish were captured while ova were early in maturation, however.

	No.	FL	Weight	Thousands of eggs per female Mean	Range
Hudson R.	5	442-495	1361-1843	253	209-307
Potomac R.	5	419-505	1191-2353	364	295-525
Neuse R.	5	447-498	1786-2722	484	423-547

	No.	FL	Weight	Thousands of eggs per female Mean	Range
Edisto R.	5	465-503	1446-2155	413	360-480
Ogeechee R.	5	457-480	1701-2155	430	359-501
St. Johns R.	5	368-445	907-1559	473	320-616

The eggs are demersal, laid in small batches (S465) and are 3.0-3.74 mm in diameter when water-hardened (M393).

Incubation periods	
3-8 days	B308
8-12 days at 11-15C	L23
17 days at 12C	S465
12-15 days at 12C	H401
6 days at 14C	S465, M444
52 hours at 14C	H401
109 hours at 18C	S465
70 hours at 23.5C	S465
36 hours at 23.5C	H401
3 days at 24C	M444

Development of embryos almost stopped at 7C (L23). At 17C eggs developed normally with salinities of 1.5% but not at 12C where maximum suitable salinities were 0.7% (L23).

The fry at hatching are 6.7 to 10 mm long (M393).

Young usually remain in the rivers until temperatures drop in the fall when they proceed to the ocean (B308). A few young overwinter in the brackish water of Chesapeake Bay but most proceed to the ocean (M444).

In the Connecticut River, W176 reported that the brood stock was proportionate to the number of recruits, but R215 reported that re-analysis of Fredin's 1954 data indicated little evidence of correlation between brood stock and progeny.

Of 842 shad tagged with streamer tags in the York River, Virginia, 118 were recaptured outside the river (N113) and of 313 tagged on spawning grounds none were taken at other spawning areas (N110). Most shad probably spawn in their "home streams" (M444).

M364

Stretch measure and type of net	Males No.	Mean TL	Range	Females No.	Mean TL	Range
89 mm linen	2	462	437-488	2	511	503-523
89 mm nylon	13	419	366-493	9	503	498-579
102 mm linen	20	414	381-488	5	503	478-544
102 mm nylon	60	419	371-523	19	493	411-554
127 mm linen	29	457	417-518	31	490	432-559
127 mm nylon	65	462	406-523	74	500	432-594

The nylon gill nets caught more fish than the linen, and the males at least were heavier for their lengths.

In the York River, Virginia, young shad feed most actively at dusk with little feeding in midday (M408). The food is mostly insects, with terrestrial insects predominating upstream and aquatics downstream. The young may also eat small fish, copepods, and other crustacea (G137, M185). Most growth occurs at sea where the shad feed upon copepods and mysids and other plankton organisms (L23, B308). Shad usually do not eat while in freshwater for spawning and may lose considerable weight (N83). Adult shad were observed to eat in a freshwater pond, however, and A77 suggested that the usual lack of feeding might indicate a shortage of available food.

GIZZARD SHAD, *Dorosoma cepedianum* (LeSueur)

Minnesota to St. Lawrence River and New Jersey, south to the Gulf and into Mexico (E119, T113), in large rivers, reservoirs, lakes, swamps, floodwater pools, estuaries, and brackish water bays (M443). Gizzard shad were probably native to Lake Erie, but were first reported from Lake Huron about 1919 and from Lake Michigan in 1953 (M446).

L2 IN   FL = 1.097 SL

	No.	SL	TL =
M302 IA	61	119-200	1.30 SL
	185	201-426	1.25 SL
W105 IL	11	85-134	1.247 SL
	31	135-184	1.265 SL
	4	185-234	1.287 SL
M230 IA Black Hawk L.	602	63-375	1.274 SL
B146 FL Newman L.			1.31 SL
B158, 312 OH	611	20-380	4.9 mm + 1.23 SL

In the following tables, data given in SL or FL were converted to TL using the B158 formula, unless a conversion factor was given in the paper. For fish under 51 mm TL was assumed to equal 1.25 SL.

Total length-weight

There was some indication that the average weights of gizzard shad from the southern part of the range (FL F74, H227, M242; AL A39, L97, S472; and LA V44) were below those in the north (SD N94, S226,228,229,330; IA M230,302; OH T113, B312; QU B65) in the size range 127-355 mm TL but heavier in the size range 356-532 mm TL. The average weights for the central part of the range (MO P74, P113; KY T84; TN W132, K71; MD H61, S398,386; and OK H156) were slightly above or below those of the southern range. It is not thought that the regional differences in length-weight are significant but rather represent the population sampling and therefore the following table combines the data from all the above references.

TL	Mean of means	Weight Central 50 percent	Range	No. of fish	Lowest and highest mean
25	0.3	-	-	379+	B312; S472
48	1.1	-	-	2	T84
51-63	1.6	-	0.6-20	11+	T84; L97
64-75	2.8	-	2.6-3.1	31	T84; H227
76-101	4.8	4.0-5.1	3.7-15.3	198+	H227; M230
102-126	11.3	8.5-12.5	8.5-18.7	180+	H227; B312
127-151	22.4	19.8-24.1	8.8-42.5	387+	H61; B312
152-177	39.7	36.9-45.4	17-62	128+	W132; M230

TL	Mean of mean	Weight Central 50 percent	Range	No. of fish	Lowest and highest mean
178-202	61.2	57-71	23-91	370+	W132; B312
203-228	91	79-96	45-145	389+	W132; H61
229-253	144	122-145	76-216	372+	W132; F74
254-278	185	170-198	105-284	601+	W132; M302
279-304	241	224-269	145-326	324+	V44; M242
305-329	318	295-349	181-545	462+	S472; S386
330-355	406	354-448	227-590	838+	N94; M242
356-380	516	485-559	250-624	408+	T84; S228
381-405	678	652-720	508-794	342+	W132; F74
406-431	824	757-921	454-1021	343+	P74; M242
432-456	1041	986-1134	743-1191	208+	W132; F74
457-482	1267	1205-1443	817-1443	133	W132; F74
483-507	1641	-	1446-1690	24	M302; F74
508-532	1852	-	1673-2041	4	M302; F74

	No.		
B312 L. Erie	-	20-380 mm	$\log W = -4.8177 + 3.071 \log SL$
M320 IA			
Black Hawk L.	602	63-375 mm	$\log W = -4.7812 + 3.042 \log SL$
L107 IL			
Crab Orchard L.	888	51-216 mm	$\log W = -4.1365 + 2.711 \log SL$
W105 IL			
Crab Orchard L.	46	-	$\log W = -5.175 + 3.198 \log SL$
			$r = 0.99$
L57 IL			
Beaver Dam L.	-	-	$\log W = -5.090 + 3.034 \log TL$
J28* OK			
Lower Spavinaw L.	274	-	$\log W = -7.5567 + 3.025 \log TL$
Upper Spavinaw L.	244	-	$\log W = -7.7115 + 3.086 \log TL$
T84 KY			
Herrington L.	237	48-356 mm	$\log W = -4.970 + 2.970 \log TL$
J37* OK			
Rod and Gun L.	243	-	$\log W = -3.889 + 2.685 \log TL$
P74, P113 MO			
Black R.	231	-	$\log W = -3.739 + 2.981 \log TL$
H156 OK Canton L.	761	173-335 mm	$\log W = -6.254 + 3.066 \log TL$
S472 AL	192,307	51-165 mm	$\log W = -4.41 + 2.73 \log TL$
	74892	166-432 mm	$\log W = -4.74 + 2.89 \log TL$

*The *a* values for J28 and J37 are as given in the original and not converted to metric system because they do not seem consistent with the measurements used in these papers and are believed to be in error.

	SL	No.	Mean K(SL)	Range
C23 MN	150-309	11	1.95	-
E28 TN	-	197	2.2	-
W105 IL Crab Orchard L.	85-234	46	1.78	-
L107 IL Crab Orchard L.	51-216	888	1.82	-
D179 TX	103-362	7023	1.85	0.99-3.93
TX (after thinning population)	120-350	378	2.22	1.53-2.89
M302 IA	TL120-415	246	2.11	1.41-3.00

	TL	No.	Mean K(TL)	Range
R64 OH	-	-	1.11	-
J28 OK Lower Spavinaw L.	-	274	0.91	-
Upper Spavinaw L.	-	244	0.96	-
H156 OK Canton L.	173-335	761	0.94	0.69-1.14
S224,226,227,228,229,330				
N94, N115, SD	127-381	838	1.02	0.94-1.18
S472 AL	76-432	-	-	0.82-1.19

No tendency for K values to increase or decrease with length of the fish was reported, but the M302 data suggest a trend. The slopes of the log L-W regressions are mostly close to 3.0 suggesting no trend in K values, except those from Oklahoma (J37) and Alabama (S472) where condition decreases with increased length. S226 reported that females had higher K(TL) values in 1955, but not in the other years (S225,227,228,229,330). Weights of males and females at various sizes were similar in Lake Erie except in June and July when females were heavier than males (B312). The differences were more than the gonadal differences and suggest greater loss of weight by males from greater activity during the spawning season. Condition was highest August-October in Lake Erie, declined during the winter and reached a low at spawning May-June (B312). Condition was highest November-February and lowest in June in Texas (D179). Reduction in population density improved condition (D179).

Scale and annuli characteristics were described by Berry (B146), who found the annuli forming February-April, in Florida. He also reported a tendency toward false annuli in the first year. Length at hatching was reported as 3.25 mm (M326), 3.5 mm (B158), and 5 mm (B146), and at 10 days as 6.2 mm (B158). During the first year of life shad in South Dakota grew 25 mm per 13 days (S227, S393) or 25 mm per 10-19 days (S228) and age I fish 25 mm per 17 days (S229) or 18 days (S228). Growth in South Dakota reservoirs began 3rd week of May at temperatures of 15.5C (S228) and the first week of June (S229). In Lake Erie (C275) growth of young gizzard shad ends the first half of October at water temperatures of 18.5C. During the season they grew 7.1 mm per week. Young gizzard shad grew most rapidly July-September with peak growth in August (B312).

		Total Length			Weight		
	No.	Mean of means	Central 50%	Range	No.	Mean of means	Range
Age 0-April							
FL B146	1457	28	-	18-56	-	-	-
Age 0-May							
FL B146	227	30	-	23-43	-	-	-
Age 0-June							
IN L2 L. Erie B158, OH C284, B312	71+	36	-	23-61	+	0.1	-
OK B183, R206, C271, G140, TN E28	2838	41	-	13-97	136	1.0	-
FL B146	75	66	-	51-104	-	-	-
Combined	2984	43	30-53	13-104	-	-	-

	Total Length				Weight		
	No.	Mean of means	Central 50%	Range	No.	Mean of means	Range
Age 0-July							
IN L2, IL W105, OH B312, C284, B158, T113, IA M195	1119+	43	41-51	13-76	+	0.6	-
OK H95, B183, C134, C271, G140, TN E28	3671	56	25-81	18-109	-	-	-
SD S224,229	70+	56	-	30-124	-	-	-
FL B146	228	69	-	38-114	-	-	-
MN L. Pepin 174	178	71	-	38-84	178	4.8	0.9-6.8
Combined	5266+	54	41-69	13-124	-	-	-
Age 0-Aug.							
South: B183, E28, J34, S117, R206, C271, G140	1126	85	71-91	36-119	2493+	4.3	1.4-57
North: B158,312, L2,107, S224,226, S227, C284	226+	88	69-91	30-203	+	28	-
MN L. Pepin P174	981	109	-	91-147	991	15	10-35
Age 0-Sept.							
South: B146,183, E28, C271, G140	7400	100	84-107	51-147	-	-	-
SD S224,229,393	+	132	-	38-165	-	-	-
OH T113, B312	+	152	-	25-152	+	79	-
MN P174	110	130	-	127-132	110	24	22-26
Age 0-Oct.							
TN, OK, LA E28, J25, V44, C271, G140	1175	115	86-178	61-226	17	47	23-94
OH, MO, IA K74, P74, C275	+	99	-	76-117	-	-	-
L. Erie B312	+	193	-	-	+	85	-
MN P174	40	155	-	-	40	37	-
Age 0-Nov., Dec.							
AL S117, OK J34	+	81	-	76-86	+	105	-
OH C284, T113	+	81	-	56-152	-	-	-
IN L2, IL L57	70+	107	-	81-188	-	-	-
L. Erie B312	+	185	-	-	+	77	-
Age 0							
MD H61, KY T84, OK T44	44+	81	53-130	13-160	1	6	-
OH R64	+	89	-	76-102	-	-	-
Age I							
FL, LA B146, V44	133	201	178-239	137-274	8	108	54-170
OK B183, H127, C271, O37, J25,34,41,20, T44	453	169	145-190	117-282	217	54	26-82
TN, KY, MO E28, P74, T84	60+	203	-	160-254	-	-	-

		Total Length				Weight	
	No.	Mean of mean	Central 50%	Range	No.	Mean of means	Range
OH, IL, IN, IA K74, L2,57,107, M230, R64, W105	554+	193	147-244	99-315	-	-	-
L. Erie B158,312	599	198	-	104-251	+	-	43-499
Combined	1799+	185	147-211	99-315	225+	-	26-499
Age II							
OK H127,156, J28, J25,41, T44, O37	198+	238	208-262	175-386	149	119	45-340
OH, IN, IL, IA L2, L57,107, M230, R64, W105, M448	735+	251	190-282	183-343	27	201	122-241
KY, TN, MO E28, P74, T84	111+	264	244-279	188-318	-	-	-
LA, FL B146, V44	153	282	-	234-335	3	145	116-196
L. Erie B158,312	1019	348	-	264-411	+	-	383-995
Combined	2216+	257	231-292	183-411	179+	-	45-995
Age III							
IN, IL, IA, SD L2, L57,107, M230, S231, W105, M448	306	279	196-361	185-394	23	306	184-369
OK H127,156, J25,41 T44, W7, O37	233+	292	257-315	198-429	217	184	57-496
KY, MO P74, T84	159	313	287-333	216-345	-	-	-
FL B146	12	320	-	-	-	-	-
L. Erie B158,312	135	406	-	-	+	-	539-1168
Combined	845	302	267-333	185-429	240+	-	57-1168
Age IV							
IN, IL, IA L2,57, L107, M230, W105	77	333	287-381	216-411	-	-	-
MO, KY P74, T84	51	339	330-353	290-368	-	-	-
FL B146	3	345	-	315-368	-	-	-
OK H127,156, J25, T44	14	352	330-353	330-421	9	397	320-652
L. Erie B158,312	25	437	-	399-472	+	-	624-1228
Combined	170	349	330-353	216-472	9+	-	320-1228
Age V							
IN, IL, IA L2, 107, M230, W105	44	338	325-432	190-432	-	-	-
MO, KY P74, T84	6	357	-	330-391	-	-	-
OK H127, J25	5+	432	-	363-452	2	652	-
L. Erie B158,312	5	470	-	437-486	+	-	808-1350
Combined	60+	381	356-445	190-486	-	-	652-1350
Age VI							
IL Crab Orchard L. W105	1	267	-	-	-	-	-
IA Black Hawk L. M230	52	437	-	-	-	-	-
L. Erie B158,312	3	450	-	399-475	+	-	728-1115
Age VII							
IA Black Hawk L. M230	15	447	-	-	-	-	-
Age IX							
IA Black Hawk L. M230	4	447	-	-	-	-	-

Median average calculated TL at each annulus.

	No.	1	2	3	4	5	6	7	8	9	10
KY, TN, IL, MO, NC:											
T84, P112,113,73, 74, K71, L57,107, W105, L157	3129	130	218	279	315	338	272	310	373	381	399
KS, OK											
M215, E61, J25, 27,28,31,34,37,41, 45,101, H150, T44, H303, L195, S391, O37	6948	130	208	262	292	333	328	333	292		
IA, SD											
M230, S225,226, 228,229,330	554	140	282	333	381	399	409	421	421	432	
FL L. Newman B146	349	244	302	323							
L. Erie B158	714	175	348	401	442	447	399				
Combined	8794	130	224	284	318	345	328	333	373	406	399

Highest average caclulated TL

	1	2	3	4	5	6	7	8	9	10
KY, TN, MO, IL										
K71, P113, L57	213	269	328	361	406	421	419	429	437	442
KS, OK										
H150, J25,28,31	196	307	363	411	439	429	401			
SD, IA										
M230, S226,330	160	345	345	381	399	409	409	421	432	
Combined										
B146,158, H150, M230, P113	244	348	401	442	447	429	421	429	437	442

	Mean increments					Mean increments		
Year of life	No. of fish	Median	Range	Mean TL at beginning	No. of fish	Median	Range
2	1,995+	94	33–168	102–126	1134	89	33–157
3	925+	61	23–175	127–151	262	89	15–175
4	444+	41	15–66	152–177	54+	38	15–132
5	196+	25	10–30	178–202	149	117	69–168
6	81+	18	5–25	203–228	173+	61	23–76
7	20	20	13–25	229–253	67+	66	18–76
8	4	18	-	254–278	68+	48	5–66
9	4	8	-	279–304	217	56	30–61
10	-	-	-	305–329	6	30	25–33
11	-	-	-	330–355	152	30	28–38
12	-	-	-	356–380	109	25	25–28
13	-	-	-	381–405	91	18	13–25
14	-	-	-	406–431	8	13	8–18

from: B146, J31, L107, M230, P71, P74, S226, S228, S330, T84, W105, H303, N94, O37, T44

In general, the growth of gizzard shad appears to be more rapid in the southern part than in the northern part of the range, except that growth in Lake

Erie appears to be faster than elsewhere. Growth appeared to be better in the middle stretch of the Black River, Missouri than in the lower portion of the same river (P74) and in the headwaters of streams than in middle or lower sections (P113). Females were larger than males at ages II-IV in Newman Lake, Florida (B146), but no sex differences were noted in Beaver Dam Lake, Illinois (L57). Growth of females was more rapid than that of males in Lake Erie after June of the second year of life (B312). The calculated TL of males and females were as follows:

	No.	1	2	3	4	5
Males	212	178	340	388	427	432
Females	267	178	356	417	455	483

In ages IV-VI females were more abundant than males, indicating a lower mortality rate. Gizzard shad showed growth compensation in Lake Erie (B312). Growth was faster in new reservoirs than in the stream before impoundment (S226, P71), or in old reservoirs (J28, P71), but growth in streams in general did not differ from that in reservoirs or lakes in general. Reduction of population density resulted in increased growth rate and reproduction in Lake Beulah, Florida (H230). Gizzard shad are known to reach 521 mm TL (M326).

Males mature at age I (B158, B146, S229) at a length of 239-290 mm (B158), but not usually until age II in Lake Erie (B312) and South Dakota (S228, S226) or even age III in South Dakota (N94, W192). Females usually do not mature until age II (S228, B146, S226, B158) but a few may spawn at age I (B158, B312, S229) and at 152 mm (V66). The mean length in June of mature age II females in Lake Erie was 361 mm TL compared to 348 mm for age II females which were not mature (B312). In Florida (B146) it was noted that the age I females did not spawn until late May. Females in Lake Erie which matured at age I spawned in July at the end of the spawning season and were large for their age (B312). Peak of spawning in Florida is late March to early April (B146, M242). In the United States spawning occurs from mid-March to late August with most spawning in April to June at temperatures of 10-21C (M326). In Norris Reservoir, Tennessee spawning occurred from May 15 to June 8 at surface temperatures of 23-29C (D163). In Lake Erie gizzard shad spawned June 15-30 at 22C, and the fry were 6.4 mm at hatching (C275). In Lake Erie spawning was from early June to early July, at temperatures mostly above 19.5C (B312). Elsewhere in Ohio spawning has been observed in May (B312, L80). In South Dakota spawning occurs from June 10-August 10 at 21-25.5C (S228). Spawning observed in Iowa May 8-June 21 at temperatures of 21-27C occurred only in shallow water in early morning or evening (M196). Heavy post-spawning dieoff may occur (B146).

B158, B312 L. Erie, Fecundity

Age	No.	TL	Weight	Eggs per female Mean	Range
I	2	282-297	261-303	59,482	22,405-96,560
II	5	353-381	524-592	378,958	211,378-543,912
III	3	404-429	700-848	344,784	260,509-406,174
IV	2	434-452	882-896	308,749	267,216-350,283
VI	1	442	1015	215,331	

H281 Mississippi R. 315 mm female 50,000 eggs.
Embryology was described by B312.

Gizzard shad can withstand rapid transfer from fresh water to 3.37% salt (G86) and may be common in brackish water, size often increasing with increased salinity. They can be acclimatized at 35C and withstand temperatures to 36.5C (S333) but sudden temperature changes seem to be a cause of mortality (M326). No overwintering of young shad was observed when ice cover was over 103 days in South Dakota, but survival was noted when ice cover lasted 88, 101, 103 days (W192).

Young of the year school, but schooling behavior is little developed after the first year in Norris Reservoir (D163). In a mild winter gizzard shad migrated into small streams in Minnesota (S477).

During the first few weeks of life, gizzard shad eat mainly protozoa, rotifers, and entomostraca (M326, B312, D57, D163, K87, P134, T120, T121, P173). Throughout the rest of life most of the food is phytoplankton and zooplankton. In a Kentucky stream bottom fauna, tendipedids, oligochaetes, diatoms and spirogyra were listed as the principal foods (M404). Tendipedids were also reported by B312, and D163 reports small shad being eaten by adults. Some algae were still viable after passing through the intestinal tract (V83).

While gizzard shad are not esteemed by man and have not been widely used for animal food (M326), they at times are important forage for game fish (K74, M326).

Reduction in the shad population in Lake Beardshear, Georgia (W213) resulted in a better size distribution of shad and of largemouth bass. White bass and threadfin shad stocked after the gizzard shad population was thinned grew well. Threadfin shad were eliminated when the gizzard shad were reduced in Lake Beulah, Georgia (H230). In Kentucky shad reduction resulted in improved angler success and in increased bass spawning and survival (S342). The gizzard shad was classed as a poor bioassay species because it is difficult to handle and is very susceptible to injury and disease (B349).

L217 LA Sizes in gill net meshes

Bar measure	No.	TL	Range
25 mm	10	257	165–384
38	111	345	241–419
51	433	363	305–483
64	236	394	330–495
76	42	396	343–483
89	5	409	368–445

M443 and M326 present good summaries of the biology and systematics of gizzard shad. Hybrids between *D. cepedianum* and *D. petenense* were found in the Ohio River, Kentucky (M449).

THREADFIN SHAD, *Dorosoma petenense* (Günther)

The threadfin shad is found in the Gulf of Mexico drainage from northern Guatemala to Florida (M443, M448). Until 1950 the northern edge of the range was probably in Oklahoma and Tennessee but the range has been extended by stocking, as a prolific forage fish (M449). It is now in the Ohio River between Kentucky and Ohio (M449), in the Atlantic drainage in Virginia (P68), in Arizona (H248, G142), in California (K121, P175, I14), and in Hawaii (H321). Threadfins show a preference in Texas for brackish water with salinities of 10-20 pp thousand (M443). In the northern part of the range the adults do not exceed 125-152 mm; in the southern states they may be 178 mm; in Guatemala they may reach nearly 220 mm (M443). In Alabama, however, threadfins are reported up to 330 mm (S472).

TL	AL A39 No.	Mean W	S472 No.	Mean W	Range
25	10294	0.7	83973	0.4	0.1-1.05
51	31764	1.5	103,391	1.3	0.5-5.1
76	1240	4.0	133,115	4.5	1.7-29.5
102	974	12.2	41458	9.9	0.8-22.7
127	13103	18.1	25891	21.8	7.4-45.4
152	382	23.5	4267	31.8	-54.0
178	3	22.7	535	41.7	22.7-63.5
203	-	-	119	54.5	45-85
229	-	-	38	95	91-122
254	-	-	25	141	113-181
279	-	-	24	190	181-204
305	-	-	9	236	221-244
330	-	-	2	277	264-286

H248 AZ lakes (D.p. atchafalayae) 120 shad, 102 mm FL, range 69-127 mm; mean W 15.6 g, range 4.3-32 g

S472 AL 303,855 shad   51-127 mm TL log W = -4.63 + 2.80 log TL
5019 shad 150-330 mm TL log W = -5.46 + 3.16 log TL

Threadfin shad have a short life span, few individuals reaching 2 years of age (M443).

G142 AZ Pena Blanca L.	Age 0-July	20-38 mm TL
	0-Aug.	36-46
	0-Nov.	119-130
Introduced stock	I March-April	61-71
	I May-June	64-84

These introduced fish apparently died shortly after producing the young stock, which reached 119-130 mm by November.

Age I threadfin shad in July in Lake Martin, Alabama were 102 mm (A78).

R206 OK L. Texoma	Age 0-Aug.	4829	1.3 g
	I-June	51	12.5
	I-Aug.	903	15.9

Threadfin shad may spawn when less than a year old. There may be two spawning peaks — spring and fall (M443). In Arizona threadfin shad spawned from late May through July with the peak in mid-June (G142).

In California, threadfin shad spawned April to early July (M422, R238) and in Hawaii from June throughout the summer (H321). Although previous reports had indicated spawning only at about 21C, shad spawned in April at 14-17C, and in May at 18C in another lake (R238). They spawn in schools under brush and floating logs, the eggs adhering to the brush (R238). They may also spawn in open water (B318, M443). A 102 mm female may have 6700-12400 eggs (B318).

Mortality of spawned and spawning threadfin shad was noted in Florida in April (B313).

Plankton is the principal food (D102, B318, G142, H248, H290, M354). Some *Chaoborus* and Tendipedids are also eaten (M354). In Arkansas they were feeding during the spawning season in contrast to some earlier reports (M354). Plankton is apparently stored in the pharyngeal organs prior to swallowing (M450).

Introduction of threadfin shad to Lake Martin, Alabama, had no measurable depressing effect on other species (A78). Bluegills, other sunfish, and minnows showed some increase in weight. Improved condition of black crappie in Roosevelt Lake, Arizona, was associated with threadfin shad introduction (B315).

Although mortality usually occurs when temperatures drop to 7C, threadfin shad survived in California ponds at 1.1C (B318), and survived the winter in lakes that do not go below 9C (S515).

LONGJAW CISCO, *Coregonus alpenae* (Koelz)

Lakes Michigan and Huron at 5.5-183 m but usually in moderately deep water (H382, B209). Also in Lake Erie (S425).

J7 L. Michigan	1028 fish	140-229 mm SL	TL = 1.1976 SL
	5105 fish	230-329 mm SL	TL = 1.1891 SL
	38 fish	330-469 mm SL	TL = 1.1792 SL

J7 L. Michigan TL-Weight relationship

TL	No.	Weight	TL	No.	Weight
173-179	1	35	305-329	1656	234
180-202	6	52	330-355	913	290
210-226	17	84	356-380	183	388
227-250	183	120	381-405	58	484
251-278	908	156	406-431	11	625
279-304	1374	194	432-456	4	763

J7 L. Michigan   log W = -5.0163 + 3.0606 log SL
J7 L. Michigan   5314 fish   average K(SL) 1.31

J7 L. Michigan

		No.	Mean SL	Range
	Age II	20	233	205-265
	III	99	246	215-285
	IV	190	257	205-315
	V	35	274	245-305
	VI	20	300	275-385
	VII	9	317	285-375
	VIII	4	357	335-385
	IX	1	372	-

S425 L. Erie

		No.	Mean TL	Range
	Age III	31	300	259-345
	IV	2	323	297-348

Calculated SL at each annulus

	No.	1	2	3	4	5	6	7	8	9
J7 N.E. L. Michigan	320	113	169	209	238	262	289	312	344	360
J7 South L. Michigan	58	123	179	215	235	245				

	No.	1	2	3	4	5	6	7	8
				Calculated TL at each annulus					
J7 North L. Michigan	320	135	203	251	282	312	371	406	421
J7 South L. Michigan	58	147	213	257	279	292			
S425 L. Erie	25	130	218	269					

The percentage of females increased with age of the fish, indicating a differential mortality, but the oldest male was IX and the oldest female VIII (J7). Maturity is attained in the third or fourth year (M327, S425). They spawn in deep water in November (K88). Food is mostly *Mysis relicta* (K88, B209).

CISCO or LAKE HERRING, *Coregonus artedii* LeSueur
    The cisco is in all of the Great Lakes and in the deeper, cold, inland waters of the St. Lawrence River, Hudson Bay and Mississippi River drainages (H382). Several subspecies, differing in body form and growth and designated as species by some authors, are recognized. The averages given in this section cover many populations and thus probably do not apply to any particular subspecies or growth form. Neither is sampling such as to refer to the average of the species as a whole. More important are the averages by groups or regions and the variation which is indicated.

C37	L. Superior	142 fish	200-299 mm SL	FL = 1.049 SL
C35	MN L. Vermilion	–	up to 250 mm SL	FL = 1.080 SL
		–	over 250 mm SL	FL = 1.085 SL
H64	WI Silver L.,			
	Muskellunge L.	–	–	TL = 1.18 SL
H64	WI Trout L.,			
	Allequash L.	–	–	TL = 1.19 SL
C59	MI Blind L.	–	–	TL = 1.197 SL
C35	MN L. Vermilion		up to 250 mm SL	TL = 1.197 SL
			over 250 mm SL	TL = 1.201 SL

G92	Saginaw Bay	27 fish under 200 mm SL	TL = 1.2104 SL
		600 fish 201-300 mm SL	TL = 1.2032 SL
		23 fish over 300 mm SL	TL = 1.1985 SL

C28	MN L. of the Woods (tullibee)	FL = 1.076 SL	TL = 1.175 SL
C35	MN L. Vermilion (tullibee)	FL = -2.6 mm + 1.094 SL	TL = 1.3 mm + 1.205 SL
C37	MN 2537 tullibee (100-349 mm SL)	FL = 1.071 SL	1073 fish TL = 1.186 SL

All data are converted to TL, with FL = 1.075 SL and TL = 1.185 SL, where conversion factors were not given in the original paper.
    Lake herring of smaller lakes are usually deeper bodied than those from the Great Lakes and are frequently called tullibee. Examination of the available length-weight data indicates that the Muskellunge, Silver and Trout Lakes, Wisconsin, populations studied by Hile (H64) are closer to the Great Lakes populations (C20, W41, V5, S265) whereas the Clear Lake population (H64) was

similar to tullibee populations in Minnesota (C20, C28, C35, S62). Data taken from growth data indicate that ciscoes of Pine Lake, Wisconsin (C1) are of the slender variety while those of Oconomowoc Lake, Wisconsin (C1), Irondequoit Lake, New York (S93), Swain Lake, Michigan (B79) and lakes in Indiana (H62), Alberta (M113) and Northwest Territories (S165) are more of the tullibee type. The Great Bear Lake population (K32) appears to be intermediate.

TL	Weight Mean of means	Range	No. of fish	Populations, from lightest to heaviest
102-126	40	28-45	1+	C20, S62
127-151	24	23-26	13	S265 Great Lakes; D182
	40	28-45	11+	H64, C1 ciscoes
152-177	31	28-39	66	D182, S265 Great Lakes
	36	34-37	+	H64 ciscoes
	74	40-105	34+	S93, M113, C1 tullibee
178-202	50	42-61	74	D182, S265 Great Lakes
	52	47-60	41+	H64, C1 ciscoes
	95	28-142	132+	C1,28,35,20, S165 tullibee
203-228	67	54-79	+	H64 ciscoes
	73	62-81	108	D182, S265 Great Lakes
	120	57-170	213+	H64, S62, C28, C35, C20, C1 tullibee
229-253	95	85-105	54+	C1, H64 ciscoes
	98	88-121	317	D182, S265 Great Lakes
	152	113-227	41+	H64, S62, C35, C28 tullibee
254-278	145	108-181	2944	S265, V5, C20, D182 Great Lakes
	235	113-340	392+	H64, S93, C28,20,35, C1, S62 tullibee
279-304	173	142-227	4674	W41, V5, S265, D182, C20 Great Lakes
	119	-	108	C1 ciscoes
	306	227-397	438+	K32, H64, S93, C28, C35, S93, C20, S62 tullibee
305-329	218	204-281	3586	V5, C20, D182, W41, S265 Great Lakes
	372	255-680	595+	K32, H62, H64, C28, C35, S93, S62, C20 tullibee
330-355	190	-	172	C1 ciscoes
	287	230-397	1014	C20, V5, S265, W41, D182 Great Lakes
	464	284-794	1388+	C1, K32, H64, C28, C35, S93, S62, C20 tullibee
356-380	261	-	95	C1 ciscoes
	385	281-471	226	D182, S265, V41 Great Lakes
	548	445-964	898+	C1, K32, H62, H64, C35, C28, S93, M113, S62, C20 tullibee
381-405	392	-	53	C1 ciscoes
	464	408-510	120	S265, D182 Great Lakes
	643	510-822	261+	K32, C1, C35, B79, H62, H64, S62, S93, M113 tullibee
406-431	436	-	3	C1 ciscoes
	584	422-666	60	S265, D182, P21 Great Lakes
	817	454-1077	339+	C1, B79, C20, S93, S62, H64, S93 tullibee

TL	Weight Mean of means	Range	No. of fish	Populations, from lightest to heaviest
432-456	731	-808	13	D182 L. Superior
	851	624-1211	83+	C1, H62, S62, C20, S165, S93, H64 tullibee
457-482	898	679-1256	19+	C1, S62, C20, S93 tullibee
483-507	1247	-	+	M113 tullibee

		No.	Length	
H64	Muskellunge L. 1928	263	145-179	log W = -1.3412 + 1.3771 log SL
H64	Muskellunge L. 1930	244	140-179	log W = -2.2462 + 1.7654 log SL
H64	Muskellunge L. 1932	93	145-179	log W = -2.7378 + 2.0039 log SL
H64	Muskellunge L. 1931	505	130-179	log W = -2.8480 + 2.0507 log SL
H64	Trout L. 1931	338	125-164	log W = -3.4116 + 2.3076 log SL
H64	Trout L. 1928	138	125-154	log W = -4.1856 + 2.6726 log SL
H64	Trout L. 1930	483	125-169	log W = -4.3113 + 2.7103 log SL
G92	Green Bay	349	180-299	log W = -4.3622 + 2.7626 log SL
G92	Saginaw Bay	684	160-329	log W = -4.6399 + 2.8906 log SL
S265	Green Bay	3849	147-424	log W = -5.3040 + 3.0729 log TL
C17	MN L. of Woods 1939-41 tullibee	1267	80-370	log W = -5.0564 + 3.168 log SL
C28	MN L. of Woods 1942-43 tullibee	293	158-320	log W = -5.2173 + 3.2134 log SL
G92	Grand Traverse	59	150-319	log W = -5.4936 + 3.2666 log SL
H64	Silver L. male 1930-31	110	130-199	log W = -5.5963 + 3.3237 log SL
H64	Silver L. female 1930-31	160	135-204	log W = -5.9093 + 3.4537 log SL
H64	Clear L. male tullibee 1931	85	160-359	log W = -5.9955 + 3.5053 log SL
H64	Clear L. male tullibee 1932	100	160-379	log W = -6.3050 + 3.6248 log SL
H64	Clear L. female tullibee 1931	105	170-379	log W = -6.3415 + 3.6499 log SL
H64	Clear L. female tullibee 1932	85	150-389	log W = -6.4375 + 3.6849 log SL

Ponderal indices of "cisco" populations

		Size of fish	No.	Mean K(SL)
V5	L. Huron		609	1.13
H64	WI Muskellunge L.		1032	1.17
E3	MN	216-292 mm	143	1.19
H64	WI Trout L.		1032	1.20
H64	WI Silver L.		405	1.31
H64	WI Allequash L.		113	1.45
H64	WI Tomahawk L.		68	1.58

Ponderal indices of "tullibee" populations

			Size of fish	No.	Mean K(SL)	Range
H62	IN		Age II-VII	45	1.66	–
H64	WI	Clear L. males		186	1.67	–

			Size of fish	No.	Mean K(SL)	Range
C34	MN	Leech L.		101	1.75	–
H64	WI	Clear L. females		198	1.84	–
C35	MN	L. Vermilion		499	2.03	–
C28	MN	L. of Woods 1942-43	158-320 mm	293	1.99	1.30–2.60
C17	MN	L. of Woods 1939-41		1261	2.29	–
E3	MN		76-405 mm	1841	2.28	–
C23	MN	1941-43	100-399 mm	929	1.77	–

						K(TL)
C23	MN	standards:	Excellent		over 2.1	over 1.19
			Good		1.9-2.0	1.13-1.18
			Average		1.5-1.9	0.89-1.12
			Fair		1.2-1.5	0.72-0.88
			Poor		under 1.2	under 0.72

In some populations condition improves with increase in length (that is, the slope of the length-weight relationship is over 3.0) and in others it declines. The low values of the length-weight slopes, 1.377 to 2.051, of the Muskellunge Lake populations (H64) are far below those usually found. These slopes are derived from samples covering a fairly narrow range, 130-179 mm or less. The average K(SL) for 147 mm fish in the 1928 sample is 1.40 and drops to 0.98 by 176 mm. By 200 mm (236 mm TL) the average K would be 0.84 and the average weight 68 g. Hile (H64, page 246) pointed out that the calculated weights at 100 mm would indicate very robust body form for these Muskellunge Lake ciscoes.

Selectivity of gillnetting might greatly affect the slope of the length-weight curve if the gillnetting selected the individuals with the greatest girth (and thus weight) at the shortest lengths in the sample and those with the least girth at the longer lengths. Although gill nets of several mesh sizes were used in Muskellunge Lake, ciscoes were taken only in the 36 mm mesh in 1930 and 1931. Hile (H64, page 301) stated that selectivity according to length was greater in Muskellunge Lake cisco samples than in those from Trout Lake.

There was no consistent sexual difference in the ponderal indices in Lake Huron (V5), in Muskellunge and Trout Lakes (H64) and in Lake of the Woods (C28). Males had higher K values than females of the same size in Silver Lake, but in Clear Lake females had the higher values (H64). In spawning runs, the weight differences between ripe, partially spent, and spent females were so great that good comparisons could not be made from year to year (D182). Males, however, were obviously heavier from 1956-61 than from 1950-54 in spawning runs in Lake Superior.

Scale and annuli characteristics were described by Van Oosten (V5). Annuli form in March-April (V4), in May (F20), and May-June (S265). Annulus formation in Lake Superior may spread over 8 weeks, May through July (D182). Younger fish start growth earlier than older ones (D182).

		TL				Weight	
	No.	Mean of means	Central 50%	Range	No.	Mean of means	Range
Age 0-First wk.							
L. Erie F5, L. Huron P20	-	-	-	10-13	-	-	-
Age 0-May							
Great Lakes S265, P20	95+	-	-	10-18	-	-	-
Age 0-June							
St. Lawrence G28	233	30	-	18-46	-	-	-
Age 0-July							
ON L. Nipissing F20	111	61	-	-	-	-	-
Age 0-Aug.							
ON L. Nipissing F20	82	71	-	-	-	-	-
Age 0-Sept.							
WI C1, ON F20	10	81	-	74-86	3	23	-
Age 0-Dec.							
MI C29, S265	3	152	-	147-155	-	-	-
Age 0							
Canada M98, K32	81	71	-	56-84	-	-	-
WI H64	17	79	-	-	-	-	-
NY S93	8	168	-	-	8	40	-
tullibee MN C17	208	79	-	61-94	1	28	-
Age I							
ciscoes Canada K32, M98, F20	300	145	-	58-170	-	-	-
ciscoes WI MI H64, N37, C59	197	168	160-168	145-218	168	34	26-34
Great Lakes C52, P21, V17, G92	10+	190	152-236	76-287	+	31	-
tullibee M113, E3, C17, C25, H64, S93	180+	211	208-216	152-254	164+	91	28-167
Combined	687+	183	163-211	76-287	332+	60	26-167
Age II							
tullibee Canada M113	+	160	-	-	+	57	-
cisco WI MI C1, H64, N37, C59	1061	190	160-211	147-279	866	54	31-99
cisco Canada K32, D33, F20, M98	419	213	-	112-234	-	-	-
Great Lakes C52, S265, D182, V17, P21, V5, S137, G92	293+	272	236-302	132-366	98+	224	57-680
tullibee C1,35,17, 25, E3, S93, H64, H62, J81	615	264	254-302	160-351	384	235	51-454
Combined	2388+	244	203-297	112-366	1338+	176	31-680

	No.	TL Mean of means	Central 50%	Range	No.	Weight Mean of means	Range
Age III							
cisco WI MI H64, C59, N37	1252	213	193-236	150-297	970	72	37-125
cisco Canada K32, D33, F20, M98	929	257	-	224-272	-	-	-
Great Lakes C52, S265, V17, P21, V5, S137, G92, D182	2690	290	267-300	190-421	1363+	?18	71-1000
tullibee Canada M57, R67, M113	22+	318	-	290-373	+	581	567-595
tullibee C1,35,17, 25, E3, S93, H64, H62, J81	1110	315	305-361	203-432	893	380	113-709
Combined	6003+	279	254-305	150-432	3226+	247	37-1000
Age IV							
cisco WI MI H64, C1, C59, N37	729 / 729	229	208-249	155-307	428	85	43-145
Great Lakes S265, C52, V17, P21, V5, G92, D182	5601+	295	284-305	221-386	3496+	198	62-369
cisco Canada K32, F20, M98, D33	1059	292	-	224-345	18	227	-
tullibee Canada M57, M113	37+	330	-	297-399	+	822	-
tullibee WI MN NY MI C1,35, 17,25, E3, S93, H62, H64, B79, J81	1254	345	340-373	239-421	806	477	170-879
Combined	8680+	300	272-325	155-421	4748	272	43-879
Age V							
cisco WI MI H64, C59, N37, C1	351	249	224-290	168-315	200	96	48-156
tullibee Canada M57	31	297	-	-	-	-	-
Great Lakes S265, C52, V17, P21, V5, G92, D182	1731	307	297-323	246-447	1021+	213	105-468
cisco Canada K32, F20, M92, D33	581	318	-	279-366	31	255	-
tullibee MN WI NY MI C35, C17, C1, C25,							

		TL				Weight	
tullibee (cont.)	No.	Mean of means	Central 50%	Range	No.	Mean of means	Range
E3, S93, H64,							
H62, B79, J81	1366	368	340-399	272-462	908	564	241-879
Combined	4066+	318	297-340	168-462	2160+	321	48-879
Age VI							
cisco WI MI							
H64, C59, C1	218	267	-	175-343	185	133	51-190
cisco Canada							
F20, K32, D33	402	323	-	221-391	75	298	227-369
Great Lakes							
S265, C52, V5,							
P21, V17, G92,							
D182	276	325	307-328	246-503	85	247	133-403
tullibee Canada							
M57, M113	88+	381	-	320-500	+	1247	-
tullibee MN WI							
NY MI IN C35,							
C1, C25, S93,							
H64, B79, H62,							
E3, J81	809	388	376-401	229-483	536	655	312-1049
Combined	1793	340	310-386	175-503	871	417	51-1049
Age VII							
cisco WI MI H64,							
C59, C1	101	323	-	190-373	97	210	111-261
tullibee Canada							
M57	74	333	-	-	-	-	-
cisco Canada							
M98, F20, K32,							
D33	277	348	-	307-419	106	397	-
Great Lakes C52,							
V5, P21, V17,							
G92, D182	43+	345	325-353	284-455	4+	346	238-488
tullibee MN WI							
NY MI C35,							
S93, C25, J81,							
H64, C1, B79,							
E3, H62	457	404	396-417	267-475	370	723	340-1106
Combined	952+	368	333-404	190-475	577+	567	111-1106
Age VIII							
cisco WI H64, C1	41	335	-	216-391	36	343	-
Great Lakes C52,							
V17, P21, V5,							
G92	12+	353	-	312-432	+	406	354-454
cisco Canada							
F20, K32, D33	115	361	-	307-419	36	454	-
tullibee MN WI							
MI NY C35,							
C25, H64, B79,							
S93, C1, E3	186	411	404-427	318-483	167	778	312-967
Combined	354+	384	351-421	216-483	239+	656	312-967

	No.	TL Mean of means	Central 50%	Range	No.	Weight Mean of means	Range
Age IX							
cisco WI H64, C1	19	345	-	234-401	17	406	-
cisco Canada K32, F20	39	386	-	335-419	20	510	-
Great Lakes C52, V17, P21	+	411	-	338-460	+	751	-
tullibee MN MI WI NY C35, B79, H64, S93, C1, E3	44	424	386-442	333-516	42	723	425-1086
Combined	102+	401	386-437	234-516	79+	638	406-1086
Age X							
cisco Canada F20, K32	20	384	-	363-475	13	510	-
cisco WI C1, Pine L.	3	409	-	-	3	436	-
Great Lakes P21, Ontario	1	409	-	-	1	666	-
tullibee MN MI WI NY C35, B79, H64, C1, E3, S93, C25	27	442	439-470	333-541	23	973	454-2240
Combined	51	422	391-457	333-541	40	811	436-2240
Age XI							
cisco WI Canada H64, F20, K32	10	345	-	226-419	2	119	-
tullibee MN WI C35, H64	3	391	-	345-447	2	893	539-1191
Combined	13	361	-	226-447	4	493	119-1191
Age XII							
H64, K32, S165	4	356	-	267-406	2	539	173-907
Age XIII							
K32 NT Great Bear L.	1	391	-	-	-	-	-

	Mean calculated TL and increments at each annulus											
	1	2	3	4	5	6	7	8	9	10	11	12
cisco-type												
H64 WI Trout L.	97	140	163	175	188	198	213	224	234	251	259	267
Increment	97	43	23	13	13	13	13	13	10	18	8	8
Number	1281	1279	1080	499	114	26	14	10	5	3	3	1
H64 WI Silver L.	94	152	183	201	213	224	231					
Increment	94	48	33	20	18	15	13					
Number	564	564	529	397	168	25	1					
H64 WI Allequash L.	86	137	185	208								
Increment	86	56	51	43								
Number	113	44	16	1								
H64 WI Muskellunge L.	114	170	190	206								
Increment	114	56	20	15								
Number	1097	1097	373	2								

	Mean calculated TL and increments at each annulus										
	1	2	3	4	5	6	7	8	9	10	11
H64 WI Tomahawk L.	91	170	213	234	249						
Increment	91	79	48	28	20						
Number	78	77	66	17	1						
C59 MI Blind L.	124	185	231	241	257	272	290	320			
Increment	124	58	33	25	18	18	18	8			
Number	152	151	148	125	59	21	11	1			
Great Lakes											
S265 Green Bay	135	201	244	269	282	333	371				
Increment	135	66	43	28	23	25	25				
Number	4193	4175	4082	2772	279	12	2				
G92 L. Michigan	145	213	251	277	295	312	323	345			
Increment	145	69	43	30	23	20	20	18			
Number	1159	1149	968	609	281	85	22	7			
V5 L. Huron	150	218	259	279	290	302	325	345			
Increment	150	69	41	25	20	18	18	18			
Number	2310	2310	2294	1722	590	120	17	3			
D182 L. Superior	119	188	236	279	300	310	323	340			
Increment	119	69	48	41	28	23	20	15			
Number	3777	3777	3777	3699	2757	721	55	4			
tullibee-type											
C28 MN Lake of Woods	117	188	244	290	323	345	373	394	447	472	
Increment	117	71	58	53	43	38	30	20	18	15	
Number	1561	1453	1230	836	389	100	21	7	2	2	
S93 NY Irondequoit L.	155	236	279	312	340	363	381	404	427	442	
Increment	155	81	46	33	30	30	23	18	18	20	
Number	500	497	433	214	201	158	90	14	3	2	
H64 WI Clear L. males	130	231	302	340	366	384	396	409	419		
Increment	130	107	76	41	25	18	18	13	10		
Number	216	176	122	73	52	41	28	11	1		
females	130	236	307	345	371	386	404	417	427	437	445
Increment	130	107	76	46	33	20	18	13	13	10	8
Number	224	195	147	101	83	74	61	28	7	4	1

	Mean of mean calculated TL at each annulus												
	Number of fish	1	2	3	4	5	6	7	8	9	10	11	12
cisco H64, C59, F20	3572	102	160	198	216	234	231	244	272	234	251	259	267
Great Lakes S425, G92, D182, V5, S265	10498	127	203	249	277	292	315	335	343				
tullibee C35, K33, R136, C28, B12, S93, H64	6105+	122	201	259	302	335	361	381	399	419	429	421	
Combined	20085+	117	188	234	264	295	323	338	363	391	399	381	267

	Highest mean calculated TL at each annulus									
	1	2	3	4	5	6	7	8	9	10
cisco C59, F20	124	185	231	254	269	272	290	320		
Great Lakes G92, D182, S265	152	218	259	297	318	333	371	356		
tullibee S93, H64, B12, C28	155	236	307	345	371	386	404	419	447	472

	Lowest mean calculated TL at each annulus									
	1	2	3	4	5	6	7	8	9	10
cisco H64	86	137	163	175	188	198	213	224	234	251
Great Lakes G92, S425, D182	69	173	224	264	274	290	290	325		
tullibee C35	91	142	198	249	279	305	325	335	345	345

		Calculated weight at each annulus							
	No.	1	2	3	4	5	6	7	8
S265 Green Bay	3132	17	60	108	145	164			
R136 Peter Pond L.	-	11	51	122	255	392	630		
C28 Lake of Woods	1561	20	74	176	329	504	718	910	1044
G92 Grand Traverse	59	11	45	96	187	235			
G92 Green Bay	415	23	105	125	139	147	176	190	
G92 Saginaw Bay	685	40	94	153	193	221	247	298	307
D182 L. Superior	3677	14	45	99	162	204	255	310	354

		Average increments in grams						
		1	2	3	4	5	6	7	8
G92 Grand Traverse		11	43	51	54	60			
	Number	59	59	48	19	8			
G92 Green Bay		23	79	43	37	28	26	23	
	Number	415	415	379	235	87	16	2	
G92 Saginaw Bay		57	57	48	45	43	43	43	43
	Number	685	675	541	355	186	69	20	7

Median annual increment in length as percentage of length at beginning of year. (Percentage increments were computed for each age group or year class given in each paper.)

TL	WI Trout, Silver, Allequash, Muskellunge, and Tomahawk Lakes H64	L. Huron V5	Green Bay S265	L. Superior D182	MI Blind L. C59	WI Clear L. H64	NY Irondequoit L. S93	MN L. of Woods C28	Range
51-75	100	-	-	-	-	-	-	-	-
76-101	62	-	-	63	-	-	-	-	45-81
102-126	46	-	53	57	-	-	-	61	33-78
127-151	22	47	48	55	46	82	-	-	16-85
152-177	12	43	59	26	-	-	52	-	7-61
178-202	8	19	22	26	17	-	-	30	7-34
203-228	6	17	20	19	12	34	-	-	4-34
229-253	8	13	13	13	7	33	20	23	3-33
254-278	3	8	10	11	7	-	-	-	3-18
279-304		6	11	8	2	15	12	15	2-15
305-329		6	8	8	-		10	12	4-12
330-355			9	5		9	9	9	5-11
356-380			10			5	6	5	5-10
381-405						3.6	5	-	3-5
406-431						2.5	4.5	4	2.4-5
432-456						1.7		-	-
457-482								3	-

Calculated lengths were computed from scale measurements by direct proportion in most of the papers. Van Oosten (V5) found the longitudinal diameter of the scale grew in more direct proportion to the body length than did the anterior radius, and Smith (S265) found that the scale diameter-body length straight-line regression gave an intercept of only 0.016 mm and therefore used direct proportion. Stone (S93) found that lengths calculated with a correction factor of 35 mm corresponding to the size at scale formation (V5) compared more closely with lengths at capture than did the direct proportion calculations, but he used the latter for his study. Carlander (C28, C35) used anterior radius measurements and a curvilinear body-scale relationship in his computations. G92 used correction factors of 17 mm, 32 mm, and 40 mm in three populations.

It is obvious from the above tabulations that the shallow-bodied "cisco" forms grow more slowly (particularly in weight) than the deep-bodied "tullibee." Whether this is a matter of genetics or environment is not known. There does not seem to be any latitudinal or regional trend in growth except that the Canadian ciscoes grow faster than those from Wisconsin. Smith (S265) found that growth for the first year was less in northern Green Bay than in the southern part, but that these differences disappeared by the fourth year. In Lake Superior growth was more rapid in the eastern end of the lake (D182) and was more rapid in 1956-1959 than in 1950-1955. The Great Lakes ciscoes appear to grow more rapidly than the other shallow-bodied populations. Hile (H64) found that female ciscoes grow more rapidly and possibly live longer in Clear Lake, but not in three other Wisconsin lakes. In three Lake Michigan populations, no sex difference in growth rate was noted, but females tended to be longer-lived than males. No sexual differences in growth were detected in ciscoes in Lake Huron (V5), Green Bay (S265), Lake Nipissing (F20), Blind Lake (C59), Lake Irondequoit (S93), or Lake of the Woods (C28).

Hile (H64) found the growth of ciscoes in four Wisconsin lakes to be in inverse order to the bound carbon dioxide, with fastest growth in the least fertile lake. Some of the other lakes listed above with faster growth have higher carbonate contents than any listed by Hile.

The growth of ciscoes in the four Wisconsin lakes was also in reverse order to the population density of ciscoes and the latter are therefore in the same order as the carbonate contents of the lake. The growth in the lakes was also slower for a period of years in which populations were high than when population densities were lower. In Lake Superior increased growth was associated with decreased population density (D182).

Average water temperatures appeared to show no correlation with rate of growth in the four Wisconsin lakes (H64) either from lake to lake or from year to year. No correlations were found between annual fluctuations in air temperatures and first year growth of ciscoes in Saginaw Bay (V5) nor between temperatures and annual growth in Green Bay (S265), Lake of the Woods (C28) or Lake Superior (D182).

The growing season for ciscoes in Clear Lake was longer than in Trout Lake and the annual growth was greater in Clear Lake (H64). In Trout Lake the season's growth was completed in July, and in Clear Lake in October. Muskellunge and Silver Lakes had intermediate lengths of growing season and intermediate growth. Hile suggested that the differences in the growing seasons for these lakes in the same region of Wisconsin might be related to differences in plankton cycles or to differences in the time needed for maturation of gonads. Development of gonads in small spawners may begin earlier in the season than in large spawners. Then differences in rate of growth would be the cause and not necessarily the result of differences in length of growing season. In Lake Superior most rapid growth was in June and 90% took place in May to September (D182). In Green Bay, greatest growth was in July but there was some growth from May to October (S265).

Fry (F20) indicated that most of the growth takes place in the spring and early summer while the ciscoes are in the epilimnion, however, a proportion of the fish continued to feed and grow after migration to the hypolimnion. The growth increments in Lake Nipissing increased about the eighth year, correlated with an increase in the percentage of the ciscoes feeding in the hypolimnion. Pritchard (P21) noted a similar increase in the annual increments at about the eighth year.

Hile (H64) noted that while Clear Lake ciscoes had the fastest growth and the lowest rate of parasite infestation of four Wisconsin lakes, the relationship between parasites and growth was not clear. Tullibee heavily parasitized with *Triaenorphorus* showed slower growth than those not so heavily parasitized (M57).

Lee's phenomenon was noted to some extent in all growth studies on ciscoes where data have been examined for it. It was not evident in the data from the Grand Traverse, where there was no commercial fishing, but was evident in Saginaw and Green Bays which are commercially fished (G92). Selective sampling by gill nets was thought to be the principal cause in some populations (H64, C28, E3, D182). Sampling from spawning runs with faster-growing individuals spawning earlier in life was a factor (V5, C59). Differential mortality of fast- and slow-growing fish may be coupled with the sampling (H64, C28, C59, D182). Fry (F20) believed that the growth changes associated with migration to the hypolimnion explained Lee's phenomenon at Lake Nipissing, and Smith (S265) believed that segregation of ciscoes on the basis of size and the resulting differences in fishing mortality were significant factors in Green Bay.

Growth compensation was demonstrated for several populations of ciscoes (V5, H64, C28, E3, S265). Both within and between four populations in Wisconsin (H63) more rapid growth in length and weight was associated with relatively shorter heads, maxillaries, paired and dorsal fins, wider bodies, and smaller eyes.

Both male and female ciscoes spawn at age I, the fall of their second year, in Lake Huron (V5), in four Wisconsin lakes (H64), in Lake of the Woods (C28) and in Green Bay (S265), but these may be only the faster-growing individuals. The youngest mature fish in Blind Lake (C59), Irondequoit Lake (S93), Lake Erie (C52), Lake Ontario (P21), and Lake Superior (D182) were age II. In Lake Oconomowoc, Wisconsin (C1) and in Hudson Bay (D33) the youngest mature ciscoes were age III and in Manitoba lakes (B12) tullibee were mature at age IV. The smallest mature males in Lake Superior were 221 mm and females 246 mm (D182), but ciscoes may mature in some lakes at 152 mm or 113 g (S471). Males have tubercles on head and body at the breeding season (P20).

The proportion of females may increase with age (+), decrease with age (-) or show little change (0):

% females	Total number	Age groups	Change with age	Gear	Locale and citation
29	152	0-VII	-	gill net	Blind L., MI C59
51	2950	II-VIII	-	pound net	Saginaw Bay, MI V5
51	440	I-X	+	gill net	Clear L.,WI H64
53	1524	II-X	0	gill net	L. Nipissing, ON F20
54	421	I-X	+	gill net	Lake of Woods, MN C28
54	3213	I-VII	-	pound net	Green Bay, WI S265
55	496	I-VII	0	gill net	Silver L., WI H64
57	494	II-X	0	pound net	Irondequoit L., NY S93
58	861	I-IV	+	gill net	Muskellunge L., WI H64
63	658	II-VII	+	gill net	Gull L., MN E3
67	84	IV-X	+	gill net	Swains L., MI B79
68	762	I-VIII	+	gill net	Lake Superior D182
73	1101	I-X	+	gill net	Trout L., WI H64

		Fecundity				
		Size of female		No. of	No. of eggs/female	
	Age	TL	Wt.	females	Mean	Range
C1 WI Oconomowoc L.			454	1	15238	-
B23, J13 Canada			1134	1	23700	-
L13 OH		343		1	30328	-
B79 MI Swain L.		386-411	680-860	9	30328	23272-37272
M65 Great Lakes		-	-	1	16040	-
S93 NY Irondequoit L.	II	277	337	8	13723	-
	III	335	457	78	21824	-
	V	376	581	2	27846	-
	VI	384	734	9	38606	-
	VII	391	779	4	35928	-
	VIII	421	939	3	48999	-
B12 MB		-		-	-	15000-20000
L80 L. Erie		421		-	30000	-
S137 L. Erie	II			12	29225	16000-42500
	III			6	23017	14000-38600
S265 L. Michigan		216-221		1	3748	-
		254-274		21	6397	3968-11212
		277-305		45	6154	3471-10250
		307-351		5	7642	5304-10442
D182 L. Superior		269-274		2	4314	3834-4794
		277-305		19	5966	3728-9417
		307-328		8	6549	5305-8685
		353-358		1	10250	-

While the number of eggs per female increased with length of the female, the number per gram of female decreased (S265, D182). Fish of different ages but of the same length showed no difference in fecundity. The average number of eggs per gram of female was 34 in Green Bay and 31 in Lake Superior (D182).

Shallow water ciscoes spawn in late fall, usually about the time of ice formation. In Hudson Bay and James Bay spawning may start as early as September 10 (D33), in the Great Lakes usually mid-November to mid-December (P20, S265, D182). Water temperatures at spawning are from 1.1-5.0C (C1, S93, B79, P20, S265, D182). While there is a general movement to shoal areas for spawning, spawning may take place at depths of 43-46m (S265, K88) over almost any type of bottom. Spawning apparently often takes place some feet off the bottom, even at 14m from the surface over bottoms of 160m (D182).

The peak of spawning in Lake Mendota, Wisconsin (J81) is December 2-11, and incubation takes 120-132 days in the lake. At 2.4C incubation takes 125 days; at 2.6C, 120 days; at 2.8C, 116 days; at 3.3C, 111 days (J81). Fry appeared about May 9 in Lake Ontario (P20).

Mortality rates of 93-95% were reported from age II to IV in Lesser Slave Lake, 80% from IV to VI and from VI to VIII; and 45% from VI to VII and 77% from VII to VIII in Baptiste Pond; and 71% from V to VII in Square Lake (M300).

Shallow water ciscoes avoided water of over 15.5C (D182), 17C (C1, H64) to 20C (F20) or water having less than 3-4 ppm dissolved oxygen (S265). As surface waters warmed up, the ciscoes moved into the hypolimnion (H64, F20, S265). When the surface water dropped again to 13C, cisco catches increased at depths of 0-40m in Lake Superior (D182). Ciscoes were not taken in bottom trawls and their pelagic distribution was influenced by temperature, abundance of plankton and spawning activities (D182).

In Cayuga Lake, *Coregonus artedi* were not found deeper than 200 feet, but were in 35 to 50m depths in the summer (G114). A tagged cisco travelled at least 100 kilometers in 32 days (S76); at least 5.4% of those tagged were recaptured.

Schooling, often by size classes, is typical, but is less strong in the warm summer months (S269).

Ciscoes are predominantly plankton feeders throughout life but sometimes feed on aquatic insect larvae and terrestrial insects that fall on the water (S369, D182, D183, F133, L198, K88, B79, C332). Ciscoes over 12 inches sometimes eat minnows (S369). They eat their own eggs (S93, D182) and eggs of other fish when available.

Fry can stand about 18 days starvation after hatching. If kept 22 days some will survive but most will die even if fed at that time (J81). Algae and small copepods and cladocerans constitute the first food (P20).

Occasionally during large emergences of *Hexagenia* in July in Lake Simcoe, Ontario, considerable numbers of *Coregonus artedi* are caught by flycasting (W144). In fall, when *Notropis atherinoides* are abundant, ciscoes are caught on bait. Schools of ciscoes (and also gulls and terns) follow the schools of shiners.

In Lake Nipissing, Ontario, larger ciscoes which migrated early to the hypolimnion fed almost continuously from spring to middle September on plankton crustacea with high fat content (L198). Those remaining longer in shallow water ate large numbers of emerging mayfly for a short time but then

fasted until some weeks after migration into the hypolimnion. A marked increase in growth of older ciscoes was probably the result of their feeding upon
abundant *Diaptomus oregonensis* in deep water all summer.

Sizes in gill nets, mesh size, stretch measure	No.	FL Mean	Range	Central 50%	Mean weight
38 mm					
C17 MN L. of Woods, tullibee	124	168	-	152-196	-
R87 NT, ciscoes	1233	178	99-376	-	68
B251 Georgian Bay, Huron, nylon net	321	198	165-290	-	-
B251 South Bay, Huron, nylon net	280	213	178-284	-	-
J81 WI L. Mendota	1	241	-	-	-
45 mm					
B251 Georgian Bay	641	216	165-328	-	-
B251 South Bay	385	234	190-315	-	-
51 mm					
C17 L. of Woods	415	180	-	160-185	-
R97 NT	1030	231	99-450	-	159
J81 L. Mendota	4	251	221-351	-	-
B251 South Bay	345	254	178-315	-	-
B251 Georgian Bay	394	259	170-315	-	-
57 mm					
C17 L. of Woods	725	229	-	216-241	-
B251 South Bay	208	269	178-315	-	-
B251 Georgian Bay	294	272	165-335	-	-
64 mm					
J81 L. Mendota	63	249	201-381	-	-
B251 South Bay	80	277	206-335	-	-
B251 Georgian Bay	131	282	178-340	-	-
70 mm					
B251 Georgian Bay	46	272	165-348	-	-
B251 South Bay	32	282	203-361	-	-
73 mm					
C17 L. of Woods	586	264	-	236-272	-
76 mm					
C17 L. of Woods	886	246	-	229-254	-
B251 L. Huron	28	277	183-351	-	-
R97 NT	173	295	130-439	-	367
J81 L. Mendota	24	320	229-401	-	-
83 mm					
C17 L. of Woods	2418	264	-	236-272	-
86 mm					
C17 L. of Woods	679	287	-	262-297	-

Sizes in gill nets, mesh size, stretch measure	No.	FL Mean	Range	Central 50%	Mean weight
89 mm					
B251 L. Huron	22	259	196-323	-	-
J81 L. Mendota	61	356	300-419	-	-
95 mm					
C17 L. of Woods	515	292	-	274-302	-
102 mm					
B251 L. Huron	8	231	196-297	-	-
C17 L. of Woods	15484	302	-	282-318	-
J81 L. Mendota	59	368	340-429	-	-
R97 NT	123	373	180-450	-	816
105 mm					
J81 L. Mendota	38	371	310-411	-	-
114 mm					
B251 L. Huron	9	241	196-297	-	-
J81 L. Mendota	122	378	320-429	-	-
127 mm					
B251 L. Huron	4	183	165-201	-	-
R97 NT	15	391	165-450	-	953
140 mm					
R97 NT	15	305	180-460	-	635

The sizes in the gill nets are subject to considerable variation in this species because of the great difference in body form and girth in various populations, which was evident in the length-weight relationships.

B251 Girth-FL relationship

South Bay   780 ciscoes   log girth = 0.2841 + .0399 FL

Georgian Bay 182 ciscoes   log girth = 0.3109 + .0376 FL

LAKE WHITEFISH, *Coregonus clupeaformis* (Mitchill)

The lake whitefish is found in lakes of the Arctic drainage of Alaska and Canada south to northern New England, New York, the Great Lakes, northern Minnesota, and northern Manitoba (T113, H382, B314). Although whitefish enter Hudson Bay regularly, there are few records from the Atlantic Ocean (B314).

	No.	SL	
C22 MN L. of Woods	22	250-349	FL = 1.062 SL
	10	350-530	FL = 1.071 SL
C33 MN	69	250-350	FL = 1.067 SL
C35 MN L. Vermilion	-	-	FL = 1.073 SL
Q1 SA Hunter Bay	53	-	FL = 1.075 SL
Q1 SA L. La Ronge	50	-	FL = 1.089 SL
V30 L. Erie	42	under 350	TL = 1.186 SL
	1052	350-449	TL = 1.175 SL
	36	450-499	TL = 1.166 SL
	17	over 500	TL = 1.155 SL

	No.	SL	
C22 MN L. of Woods	5	250-299	TL = 1.156 SL
	10	300-530	TL = 1.175 SL
C37 MN	69	250-350	TL = 1.180 SL
C35 MN L. Vermilion	-	-	TL = 1.186 SL
V10 L. Huron	1495	250-549	TL = 1.175 SL
	37	under 250	TL = 1.182 SL
	50	over 550	TL = 1.163 SL
V19 NY L. Champlain	257	300-499	TL = 1.19 SL
	36	500-599	TL = 1.18 SL
C255 L. Michigan	55	0-349	TL = 1.217 SL
	524	350-449	TL = 1.209 SL
	312	450-499	TL = 1.201 SL
	94	over 500	TL = 1.196 SL

	No.	TL	
N91 NY	-	432-483	TL = 2.44" + 0.98 FL
Q1 SA Hunter Bay	39	-	TL = 1.118 FL
SA L. La Ronge	68	-	TL = 1.128 FL

For the purposes of this summary TL = 1.18 SL = 1.12 FL

Q1 Round weight = 1.129 dressed weight

There appeared to be no consistent difference in the length-weight relationships of whitefish from the Great Lakes (E100, V10, V30, W41, D174, M451) and from smaller lakes (C20, C126, C236, L83, R136).

TL	Mean of means	Weight Central 50%	Range	No. of fish	Population with smallest & largest
150	27	-	-	1	D174
152-177	32	-	27-54	37+	R136-V30
178-202	50	41-54	27-54	60+	C236-R136
203-228	77	68-86	41-113	108+	C236-C20
229-253	109	91-118	59-186	125	C236-V30
254-278	154	141-159	100-195	203	C236-V30
279-304	204	186-213	141-240	344+	C236-V30
305-329	272	254-299	200-336	264+	C236-C20
330-355	367	345-372	304-495	266	E100-C20
356-380	445	431-481	363-540	327	E100-R137
381-405	576	531-580	463-907	470	E100-R137
406-431	694	675-731	481-880	1248	C236-R137
432-456	834	794-907	481-1070	1585	C236-R136
457-482	989	948-1034	707-1193	1162	C236-C20
483-507	1161	1070-1243	794-1306	687	C236-C236
508-532	1384	1320-1401	1220-1561	511	V10-R136
533-558	1642	1606-1692	1134-1955	309	W41-R136
559-583	1955	1850-1964	1819-2467	122	D174-R136
584-609	2164	2127-2228	1941-2295	87	V10-V30
610-634	2567	2522-2662	2218-2921	51	V10-C20
635-659	2953	-	1987-3107	15	V10-V30

TL	Mean of means	Weight Central 50%	Range	No. of fish	Population with smallest & largest
660-685	3125	-	2745-3742	9	D174-M451
686-710	3951	-	3629-4114	3	D174-V30
711-736	4876	-	-	1	V30
737-760	4037	-	-	1	D174

K93 reports a maximum weight of 10 kilo in Great Slave Lake and questions the record 19 kilo whitefish caught in 1918 and reported by Van Oosten, since 11.8 kilo is the next largest.

	No.		
V30 L. Erie	-	-	$\log W = -5.1772 + 3.1523 \log SL$
C255 Big Bay de Noc	848	340-519 mm	$\log W = -5.4271 + 3.2544 \log SL$
High Island L. Mich.	174	370-509	$\log W = -4.2567 + 2.8166 \log SL$
Gull Island L. Mich.	254	350-529	$\log W = -4.7239 + 2.9886 \log SL$
B227 Flathead L. MT (no	189	-	$\log W = -4.456 + 2.84 \log SL$
difference in Yellow Bay and Polson Bay)			

H25 ON Shakespeare Is. L.	-	-	Weight increases as 3.3 power of FL
K55 Great Slave L.	-	-	$\log W = -5.706 + 3.333 \log FL$
K56 MB L. Winnipeg	-	-	$\log W = -4.651 + 2.96 \log FL$
Q1 SA L. La Ronge	480	-	$\log W = -6.16 + 3.48 \log FL$
Hunter Bay	134	-	$\log W = -5.58 + 3.26 \log FL$
C303 L. Ontario females	-	-	$\log W = -5.973 + 3.426 \log FL$
C333 Georgian Bay, spring,	-	-	$\log W = -4.912 + 3.019 \log FL$
fall	-	-	$\log W = -5.472 + 3.209 \log FL$
M431 QU	-	310-470 mm	$\log W = -2.732 + 2.196 \log FL$

M451 L. Michigan	-	-	$\log W = -5.979 + 3.359 \log TL$
Green Bay	-	-	$\log W = -6.032 + 3.386 \log TL$
D174 L. Superior	-	-	$\log W = -5.677 + 3.241 \log TL$
E100 L. Superior, in June	379	-	$\log W = -5.563 + 3.171 \log TL$

				K(SL)		K(TL)
	Size range	No.	Mean	Range		Range
V30 L. Erie	-	-	1.7	1.25-2.07		-
C22 MN L. of Woods	300-514 mm SL	6	2.05	-		-
C23 MN	150-550	25	1.80	-		-
C23 MN standards:	Excellent	-	-	over 2.2		over 1.33
	Average	-	-	1.6-2.0		0.97-1.22
	Poor	-	-	under 1.4		under 0.86
V10 L. Huron	250-550 mm	515	1.44	-		-
C34 MN Leech L.	-	3	1.77	-		-
V19 NY north L. Champlain	-	120	1.65	-		-
V19 NY south L. Champlain	-	175	1.77	-		-
V20 MN Red Lake	268-409	2	2.11	-		-
H69 WI Trout L.	-	48	1.30	-		-
C35 MN L. Vermilion	-	24	1.84	-		-

	Size range	No.	K(SL) Mean	K(SL) Range	K(TL) Range
B227 MT Flathead L.	170-370	409	1.45	.99-1.98	K(FL)
Q1 SA L. La Ronge	Age II-up	480	-	-	0.53-0.86
Hunter Bay	Age II-up	134	-	-	0.69-1.03
N91 NY Little Moose L.	17-19"	1116	-	-	1.16-1.58

N91	K(TL) = 0.7202 K(FL)

Condition was lowest in February and March, particularly in spent fish, and highest in summer, in Flathead Lake (B227). In Lake Erie (V30) and in Lake Huron (V10) also, condition was higher in summer than just prior to spawning. This was true of both mature and immature whitefish. The smaller fish tended to have higher condition perhaps due to gill net selection in Flathead Lake (B227). In Little Moose Lake (N91) and Lake La Ronge and Hunter Bay (Q1) condition increased with length and age. In the breeding season, females were heavier than males (N91, E100) but no difference in length-weight relationship was noted between males and females at other seasons in Lake Superior (E100) or in Flathead Lake (B227). Females had higher condition values than males in summer and fall in Lake Huron (V10). In Lake Huron there appeared to be no change in condition with age or with length (V10).

Whitefish data on sizes at various ages are reported in five groups: United States, Eastern Canada, Western Canada, Northern Canada, and Great Lakes.

	No.	TL Mean of means	TL Central 50%	Range	No.	Weight Mean of means	Weight Central 50%	Range
Age 0-May								
N. Can. R21	+	13	-	-	-	-	-	-
Great Lakes F5, P20	+	15	-	13-18	-	-	-	-
Age 0-June								
Great Lakes F5, P20	+	23	-	-	-	-	-	-
W. Can. R21, B11	17+	23	13-30	-	-	-	-	-
E. Can. B11, H25, V69	7	48	-	41-53	-	-	-	-
Combined	24+	30	23-25	13-53	-	-	-	-
Age 0-July								
Great Lakes F5	-	33	-	-	-	-	-	-
Age 0-Sept.								
ON hatchery V69	11	81	-	64-119	-	-	-	-
Age 0-Oct.-Dec.								
Can. M15, N41	4+	157	-	145-170	-	-	-	-
hatcheries M65, B73	+	178	-	127-229	-	-	-	-
U.S. B23, B73	864+	183	-	127-203	-	-	-	-
Combined	868	175	152-196	127-229	-	-	-	-
Age 0								
N. Can. K32	3	58	-	-	-	-	-	-
Age I								
N. Can. K55	-	-	-	-	5	23	-	23-113
E. Can. H24, H25	16	147	-	-160	9	43	-	-
U.S. H69, K5, G21, B227, C35, F121	29	180	157-218	152-257	8	54	-	26-122
W. Can. R138, G62, Q1, R203, M113, M15, R21	37	203	170-236	135-259	13	122	-	45-198
Great Lakes F3, C67, V30, C255, M251, D174	202	241	216-305	84-358	185	159	91-250	57-250
Combined	284	203	163-236	84-358	215	108	43-170	23-250

	No.	TL Mean of means	Central 50%	Range	No.	Weight Mean of means	Central Range	50%
Age II								
E. Can. H24, H25, M65	32	190	-	168-221	22+	164	57-71	57-567
N. Can. D33, K55	2	211	-	206-213	42	71	-	
W. Can. C236, R139, Q1, R138, G62, R203, M15, R21, M209, M57	131+	264	211-310	163-429	39+	295	54-638	43-813
U.S. H69, C35, E3, B227, B73, V19, F121	191	300	193-368	183-457	5	269	-	54-454
Great Lakes E100, H24, D174, M356, C67, V30, C255, V10	521	333	328-358	183-464	325	335	323-477	85-595
Combined	877+	287	211-356	163-464	404+	272	71-454	43-813
Age III								
E. Can. H24, S155	25	208	-	193-251	11	102	-	85-170
N. Can. R26, D33, K55, K32	26	239	-	152-284	57	99	-	-
W. Can. R139, C236, Q1, R138, K56, G62, M115, M113, M209, M57, R67	260+	310	244-345	173-470	186+	496	96-813	45-1134
U.S. H69, G21, C35, E3, B227, V10, L17, F121	203	307	254-361	185-419	16+	691	386-728	79-2041
Great Lakes E100, H24, C67, D174, M356, L83, V30, M451, C333, C255, V10	4340+	409	404-450	183-597	1992+	685	643-912	227-947
Combined	4854	338	241-427	173-597	2762+	548	96-813	45-2041
Age IV								
E. Can. S155, H24	60	244	226-259	224-378	13	167	-	142-193
N. Can. K55, R76, D33, K32, M59, K122	14+	302	-	246-371	123+	420	-	113-1588
W. Can. C236, R139, Q1, R138, G62, M15, H207, K56, M113, M209, M57, R18	500+	368	279-465	218-526	923+	779	210-1134	45-1814
U.S. H69, C35, E3, G21, B227, F121, V19, N91	218	348	279-439	188-498	44	658	403-998	113-1106
Great Lakes E100, H24, D174, M356, M451, C67, L83, C333, V30, V10, C255	1711+	450	437-498	221-607	1295+	907	737-1094	482-1191
Combined	2503+	378	284-475	188-607	2398+	714	210-1091	45-1814
Age V								
E. Can. S155, H24	48	269	-	267-274	35	224	-	221-227
N. Can. K55, C67, R76, D33, K32, K122	17	305	-	262-471	153	669	-	147-2268
W. Can. Q1, R139, R138, C236, G62, B15, R203, H207, K56, M113, M57, M209, R18	270+	404	330-480	218-541	1976+	930	354-1298	91-2041

	No.	Mean of means	TL Central 50%	Range	No.	Mean of means	Weight Central Range	50%
U.S. H69, G21, C35, D174, E100, H24, L83, M356, M451, V10, V30	982	483	467-528	241-655	962+	1160	936-1361	737-1531
Combined	1558+	414	328-513	218-655	2921+	879	354-1162	145-2268
Age VI								
E. Can. H24	26	287	-	282-292	18	284	-	-
N. Can. K55, D33, R76, K32, K122	17	376	-	302-457	228	912	-	113-2892
W. Can. Q1, R139, R138, G62, C236, M138, M15, M209, R203, H207, K56, M113, M57	297+	414	368-472	290-559	2226+	887	510-1247	227-2381
U.S. H69, E3, C35, G21, B227, F121, V19, N91	256	414	388-500	264-503	174	893	521-1191	176-1361
Great Lakes E100, C67, C333, D174, H24, V30, L83, M451, C255, V10	1273+	495	465-559	234-691	372+	1282	1134-1542	680-1956
Combined	1869+	432	368-505	234-691	3018+	959	510-1295	113-2892
Age VII								
E. Can. H24	13	320	-	318-323	12	397	-	-
N. Can. K55, D33, R76, K32, K122	24	419	-	333-513	231	1256	307-2466	113-3176
W. Can. Q1, R139, G62, R138, C236, K55, M209, M57, M113	306+	434	388-470	320-579	1456+	986	624-1383	318-2495
U.S. V4, H69, C64, B227, G21, F121, E3, C35, V19, N91	290	434	328-513	269-516	234	1063	598-1361	241-1542
Great Lakes E100, C67, D174, C333, H24, R203, L83, M451, C255, V30, V10	483+	505	467-566	246-665	184	1531	1219-1871	1049-2268
Combined	1116+	452	388-513	246-665	2117+	1145	598-1540	113-3176
Age VIII								
E. Can. H24	21	351	-	338-361	21	519	-	454-567
U.S. H69, V4, G21, C35, N91, V19	273	442	363-523	246-528	235	1239	1239-1542	287-1379
N. Can. K55, R19, D33, R76, K32	66+	457	424-508	399-569	245	1205	397-1304	113-3628
W. Can. Q1, R139, G62, R138, C236, R203, K56, M113, M57	349+	467	411-541	333-579	674+	1106	782-1364	618-3062
Great Lakes E100, C67, H24, D174, C333, C255, L83, V30, M451, V10	188+	523	470-592	267-701	82+	1608	1219-2126	1219-2608
Combined	897+	467	404-528	246-701	1257+	1202	655-1542	113-3628
Age IX								
E. Can. H24	25	376	-	353-401	20	671	-	510-851
N. Can. D45, K55, D33, R76, K32, K122	52	478	432-533	381-572	358	1347	697-1531	227-4082

	No.	TL Mean of means	Central 50%	Range	No.	Weight Mean of means	Central Range	50%
W. Can. Q1, R139, R138, C236, G62, R203, K56	314+	475	442-495	427-559	310+	1094	703-1556	500-3176
U.S. H69, V5, C64, C35, G21, N91, V19, F121	223	475	447-541	363-544	212	1323	1276-1579	408-1579
Great Lakes E100, C67, H24, D174, C333, V30, C255, V10	87+	546	528-612	282-726	20+	2000	-	1191-2807
Combined	701+	486	432-541	282-726	920+	1320	760-1588	227-4082
Age X								
E. Can. H24	26	417	-	388-432	16	851	-	737-907
W. Can. C236, Q1, R139, R138, G62, R203, K56	248+	495	460-528	452-569	167+	1157	893-1619	680-3176
N. Can. C67, K55, R19, R76, D45, K32, M57, K122	17+	498	470-554	330-569	495+	1585	1021-1729	340-4082
U.S. C64, N91, V19, C35, F121	210	500	513-549	371-561	203	1580	1285-1950	652-1950
Great Lakes E100, D174, C333, C67, C30, L83, C255, V10	59+	566	569-625	305-714	30+	2427	2367-2552	992-3458
Combined	560+	511	460-569	305-714	911+	1585	958-1950	340-4082
Age XI								
E. Can. H24	30	424	-	406-439	15	851	-	794-907
N. Can. D45, R76, K55, K32, K122	10+	505	-	457-521	887	1522	1097-1758	454-2268
W. Can. C236, Q1, R139, R138, K56, R203	212+	526	488-599	445-655	113+	1352	1060-1797	770-3176
U.S. C64, N91, V19, F121	135	511	434-574	386-574	133	1471	1301-2041	397-2041
Great Lakes E100, H24, V30, V10, C255, L83, C333	29+	574	605-671	310-749	17	2602	-	1361-3572
Combined	416+	518	455-599	310-749	1165+	1540	1097-1899	397-3572
Age XII								
E. Can. H24	23	409	-	401-417	15	1012	-	964-1077
U.S. C64, N91, V19, F121	95	528	465-577	414-584	93	1650	1180-2132	624-2132
N. Can. D45, R19, K55, K32, K122	10+	521	-	388-597	1240+	1542	1361-1843	567-2268
W. Can. Q1, R139, R138, K55	167+	533	513-569	490-597	79+	1551	1217-1950	819-3402
Great Lakes E100, V30, C333, V10, C255, L83	19+	602	597-620	348-716	11+	2716	-	1531-3118
Combined	314+	513	465-592	348-716	1438+	1627	1205-1950	567-3402
Age XIII								
E. Can. H24	31	452	-	417-480	18	1120	-	992-1247
U.S. C64, N91, V19	52	546	516-592	414-605	51	1740	1318-2132	754-2495
N. Can. D45, K55, K32, K122	14	549	-	447-605	1153	1698	1315-1895	680-2835
W. Can. Q1, R139, R138, R203, K55	93+	564	538-589	490-668	42+	1831	1412-2770	865-4536
Great Lakes V30, C255, C333, E100	12	607	-	386-729	8+	2866	-	1588-3110
Combined	202+	541	480-605	386-729	1272+	1712	1389-1758	680-4536
Age XIV								
E. Can. H24	13	465	-	455-488	8	1264	-	1162-1418

		TL					Weight		
	No.	Mean of means	Central 50%	Range	No.	Mean of means	Central Range	50%
U.S. C64, N91, V19	34	528	-	419-597	34	1500	-	595-2268
N. Can. D45, R19,								
K55, K32, K122	19+	579	-	450-711	932+	1882	1389-2353	680-3402
W. Can. Q1, R139,								
R138, R203, K55	68+	584	559-592	533-648	28+	1899	1597-2041	1225-2835
Great Lakes V30	5	640	-	630-739	5+	2931	-	1871-3130
Combined	139+	556	521-597	419-739	1007+	1874	1389-2353	595-3402
Age XV								
E. Can. H24	10	481	-	455-495	5	1332	-	1106-1559
U.S. N91, V19	17	546	-	508-599	17	1738	-	1304-2313
W. Can Q1, R139,								
R138, K56	25+	589	584-589	559-620	9+	2546	-	1497-4536
N. Can. D45, K32,								
K55, K122	15	607	599-650	516-711	628	1967	1480-2098	907-3176
Georgian Bay, L.								
Huron C333	-	-	-	-	+	2835	-	2126-3346
Combined	67+	561	508-599	455-711	659+	2013	1559-2449	907-4536
Age XVI								
E. Can. H24, M37	14	490	-	455-607	13	1398	-	1361-1418
U.S. N91, V19	8	569	-	513-625	8	1880	-	1282-2480
Great Lakes E100,								
V30, C333	3	592	-	424-749	2+	3458	-	2438-4000
N. Can. D45, K32,								
K55, K122	3+	650	-	597-711	433+	1919	1585-2089	794-3176
W. Can. K56, Q1	3	703	-	-	1	1361	-	-
Combined	31+	572	505-653	424-749	457+	2002	1418-2410	794-4000
Age XVII								
E. Can. H24	3	475	-	467-488	2	1361	-	-
NY L. Champlain								
V19	3	617	-	-	3	2855	-	-
W. Can. Q1, K56	4	673	-	-	1	3742	-	-
N. Can. K32, K55	3	711	-	683-739	271	1976	-	1021-3628
Georgian Bay, L.								
Huron C333	-	-	-	-	+	3090	-	2807-3374
Age XVIII								
E. Can. H24	6	470	..	-	4	1512	-	-
N. Can. K32, K55	1	769	-	-	101	2135	-	1021-3516
Georgian Bay C333	-	-	-	-	+	3176	-	3118-3260
Age XIX								
E. Can. H24	7	518	-	472-587	7	1899	-	1559-2410
W. Can. Q1	1	716	-	-	-	-	-	-
N. Can. K32, K55	1	769	-	-	43	2364	-	1021-3288
Age XX								
K32 NT Great Bear	1	798	-	-	-	-	-	-
K55 Great Slave S.	-	-	-	-	10	2293	-	1247-4762
K55 Great Slave N.E.	-	-	-	-	9	2798	-	2041-3288
Age XXI								
H24 ON Shakespeare	6	483	-	-	6	1616	-	-
K55 Great Slave S.	-	-	-	-	4	2013	-	1928-2155
Age XXII								
H24 ON Shakespeare	3	488	-	-	3	1701	-	-
K55 Great Slave S.	-	-	-	-	1	1588	-	-
Age XXIII								
H24 ON Nipigon	1	688	-	-	1	3204	-	-
K55 Great Slave S.	-	-	-	-	1	4310	-	-
Age XXIV								
H24 ON Shakespeare	1	508	-	-	1	1758	-	-
Age XXV								
K55 Great Slave N.E.	-	-	-	-	1	4762	-	-
Age XXVI								
H24 ON Shakespeare	1	559	-	-	1	2523	-	-
Age XXVIII								
K55 Great Slave S.	-	-	-	-	1	578	-	-

Van Oosten (V4) demonstrated the applicability of the scale method to whitefish with known-age whitefish in aquaria, and Fry (F90) used lead versenate to show that the scale and body ratio remained the same over a year. Young whitefish developed false annuli in July due to a period of starvation (V69). Independent readings of 668 whitefish collected in 1952 showed 77% agreement, and of 200 fish collected in 1953 showed 85% agreement (N91).

C303 questioned the scale readings on the Lake Ontario fish in L83 and therefore reread the scales considering "cutting over" as essential for annuli and eliminating any annuli designated merely by crowding. "Good" agreement (78%) was reached by 3 scale readers and year class abundance appeared consistent. In this sample 9% of the scales were regenerative: Consistency in abundance of year classes in successive collections indicated the validity of age determination in C333.

Most papers assume direct proportion in the back calculation of growth, and D174 recorded an intercept of only 1 mm with 694 whitefish from Lake Superior. The following correction factors were used in other studies:

L. Superior E100	38 mm TL
ME F121	33 mm TL
L. La Ronge Q1	24 mm FL
Hunter Bay Q1	22 mm FL

The coefficients of correlation in the last two cases were 0.84 and 0.97. Annuli form from mid-June to mid-July in Flathead Lake, Montana (B227), with little scale growth before water temperatures reach 10C.

Lee's phenomenon was noted in V30, D174, and E3 but not in K13 or M451. It was believed to be associated with the selectivity of gill nets in most of the E3 samples and with year selectivity and shorter life span of faster-growing fish in D174. Growth compensation was shown in V10, V30, and E3, but not in Lake Opeongo (K13).

In the following table, mean calculated lengths at various annuli are averaged (unweighted) from populations which show somewhat similar growth and are from somewhat similar areas. (Populations of dwarf whitefish are not considered here but are discussed later.)

Average TL at each annulus

	No.	1	2	3	4	5	6	7	8	9	10	11	12	13	14	15	16	17
E. Can. and U.S. K13, F121, V5, H69	1438	104	155	193	229	259	284	315	335	366	388	409	442	465	432			
MN C35, E2, E3, M258, S62	228	104	168	254	310	353	376	417	437	486	538							
ME E93	+	-	262	318	330	340	358	384	399	421	414							
MT streams P159	6	99	198	290	333	371												
MT lakes P159	54	178	277	356														
SA lakes R136, R138	89+	165	213	259	297	330	368	401	434	472	513	551	584	594	620	648		
AB lakes R136, M88	827	122	239	338	396	434	465	488	521	584	605	630	643	655	663	673	686	698
MB lakes B11	+	137	196	378	434	470	503	538	566									
BC Altin L. W117	22	97	178	262	371	475	584	620										
NY L. Champlain V19 north	120	140	246	330	399	442	475	500	521	541	554	572	584	594	607	615	625	635
south	175	135	267	363	429	465	493	513	528	541	559	572	582	589	599	607	615	612
L. Superior E100	178	140	183	213	239	257	274	292	307	328	345	366	391	399	406	414	424	
L. Superior D174	1569	142	231	310	368	401	414	437	465	549	602							

Average TL at each annulus (cont.)

	No.	1	2	3	4	5	6	7	8	9	10	11	12	13	14	15	16	17
L. Huron, L. Michigan R162, V5, V10, V17, C255, M451	2722+	135	239	330	417	480	528	569	605	630	638	660	686	719	734	744		
L. Erie, Green Bay V30, M451	3553	173	312	406	460	503	538	566	597	605	602	615	625	635	643	650		
Unweighted Mean	10981+	130	216	300	351	386	421	455	470	500	511	549	569	577	597	627	607	660

Calculated weight at each annulus

	No.	1	2	3	4	5	6	7	8	9	10	11	12	13	14	15	16	17
E100 L. Superior M60 AB	178	18	41	68	91	122	159	191	222	263	299	349	395	417	463	495	535	
Pigeon L. R138 SA	300	32	141	449	807	1216	1184	1220										
Wollaston L. R138 SA	89	41	86	168	254	313	454	567	735	967	1188	1451	1814	2132	2449	2858		
Cree L. W117 BC	+	-	86	159	240	313	363	499	635	907	1179	1497	1905					
Atlin L. B11 MB L.	22	-	-	318	590	1678	2994	3039										
Winnipeg	+	23	64	590	771	1043	1134	1361	1497	1678	1905	2177	2495	2631	2948	3175	3629	4536
V30 L. Erie M451 L.	2766	45	313	658	989	1265	1506	1715	1910	2101	2282	2440	2590	2745	2894	2994	3116	
Michigan M451 Green	201	23	118	372	776	1206	1574	1865	2155	2382								
Bay	787	27	222	635	1116	1520	1870	2110	2475	2727								

Annual increments are given for certain populations. For some of these the increments have been computed as percentages of the lengths at the beginning of the year.

Calculated TL and annual increments at various annuli

	1	2	3	4	5	6	7	8	9	10	11	12	13	14	15	16
H69 WI Trout L.	97	137	178	213	244	272	297	325	358	373	394	404	421	432		
Incr.	97	43	41	36	33	28	28	30	28	23	15	15	18	11		
No.	55	35	55	55	43	34	17	10	6	3	3	2	2	2		
E3 MN	107	185	257	302	330	358	399	417								
Incr.	107	78	72	58	51	48	41	33								
No.	203	203	202	122	56	16	6	2								
M15 BC Okanagan L.	155	264	306	358	396	439										
Incr.	155	112	71	51	43	32										
No.	40	33	5	3	2	1										
E100 L. Superior	140	183	213	239	257	274	292	307	328	345	366	391	399	406	414	424
Incr.	140	43	30	25	20	20	18	15	15	15	18	13	8	13	8	10
No.	178	178	176	171	154	135	104	67	45	29	14	7	3	1	1	1
D174 L. Superior	142	231	310	368	401	414	437	465	549	599						
Incr.	142	89	84	81	74	64	53	43	43	36						
No.	1569	1565	1460	993	576	283	173	74	9	5						
V10, V17 L. Huron	127	226	312	409	488	544	582	607	630	643	658	676				
Incr.	127	99	86	97	79	56	38	28	23	18	18	15				
No.	982	982	982	954	616	497	154	65	48	17	5	2				
C255 L. Michigan	132	229	330	406	470	513	533	546	599	627	665	706	719	734	744	
Incr.	132	97	122	94	86	58	43	36	28	20	23	13	18	18	13	
No.	1062	1056	1034	870	707	541	249	87	25	9	6	5	1	1	1	
M451 L. Michigan	142	249	351	434	495	538	594	640	660							
Incr.	142	107	102	86	61	41	28	25	20							
No.	201	201	197	157	130	110	4	1	1							
M451 Green Bay	168	300	401	462	508	551	592	612	622							
Incr.	168	132	107	74	43	36	23	25	18							
No.	787	785	685	243	153	120	6	4	2							

Annual increments in length as percentage of length at beginning of the year. The data given are median values from the various year classes.

TL at start	ON L. Opeongo K13 Dwarfs	Normal	WI Trout L. H69	L. Superior E100	BC Okanagan R. M15	Big Bay Noc R162	AB lakes M88	Range
76-101	-	-	45	-	-	-	-	39-61
102-186	19	33	37	64	64	-	98	5-103
127-151	16	31	29	29	74	78	94	10-114
152-177	-	18	22	22	73	79	-	13-84
178-202	-	14	19	16	-	-	-	8-37
203-228	-	13	14	11	35	-	-	7-38
229-253	-	10	11	8	31	-	41	5-50
254-278	-	10	10	7	20	48	42	5-48
279-304	-	8	10	6	23	38	23	4-40
305-329	-	7	7	5	13	-	24	3-24
330-355	-	7	7	5	11	-	19	3-20
356-380	-	7	5	4	11	-	18	3-20
381-405	-	5	4	3	-	-	11	2-13
406-431	-	5	2	2	7	-	10	2-11
432-456	-	6	-	-	-	-	7	4-12
457-482	-	4	-	-	-	-	6	3-10
483-507	-	7	-	-	-	-	5	3-7
508-532	-	1	-	-	-	-	4	1-6

In general, whitefish grew faster in the Great Lakes than in the other areas studied. Growth in Lake Superior (E100), the coldest of the Great Lakes (normal maximum 7C), was, however, slower than most other waters. Growth was faster in Lakes Huron, Michigan, and Erie than in Ontario (V30). In the first years of life whitefish grew faster in Lake Erie (V30) and in Green Bay (M451) than in Lake Michigan. Whitefish showed lower average growth in eastern Canada than in the other regions. Average growth in northern Canada was similar to that in western Canada and in the United States, but it may be significant that most of the northern Canadian lakes studied were larger than those in the other two regions. Whitefish grew more rapidly in Lower Watertown Lake, which is shallower and warmer, than in Upper Watertown (C236). Growth was also faster in southern than northern Lake Champlain (V19). Growth seemed more rapid in central Minnesota (the southern edge of the range of whitefish) than in northern Minnesota (E3).

Two separate populations of whitefish were found in Lake Opeongo with the "dwarf" race having a shorter growing season, slower growth, and shorter life. The "dwarf" population spawned at about 140 mm TL and the other race at 279 mm (K13).

Dwarf and normal populations were also found in six Maine lakes (F121). The dwarfed forms differed from the normal in gill raker counts and blood antigens, but showed the same body-scale relationship. The dwarfed forms showed earlier maturity and shortened life span and are not analogous to "stunted" populations where maturity is late. In Webster Lake the dwarf whitefish actually grew more rapidly than the normal, similar to situations reported in Sweden by Svärdson (S478, S479). In Cliff Lake, the dwarfs spawn after freezeup, whereas the normals spawn before. In Second Musquacook

Lake the dwarfs spawn in November about 150 m up a tributary stream on gravel and rubble, and the normals spawn later. The dwarfs mature at 152-178 mm and at age I, compared to 229 mm for normals in Cliff Lake; at 152-203 mm and age II, compared to 305 mm for normals in Second Masquacook Lake; and at 178-305 mm and age II in Rowe Lake (F121). Several populations of whitefish were recognized in Lake Superior (D174), and in Georgian Bay (C333).

Growth data for "dwarf" populations compared with those of associated "normal" populations are given in the following table.

Calculated TL and increments at each annulus

	1	2	3	4	5	6	7	8	9	10	11	12	13
K13 ON L.													
Opeongo dwarf	107	127	135	140	142								
incr.	107	23	10	5	5								
no.	167	160	82	22	1								
normals	117	160	190	218	244	259	282	305	358	404	421	470	508
incr.	117	41	33	33	30	25	28	25	28	20	18	18	8
no.	1042	757	611	505	415	215	153	88	33	15	6	3	1
F121 ME													
Webster L.													
34 dwarfs	122	163	193	224									
66 normals	99	135	180	226	259	302	323	333	358	399			
Musquacook L.													
78 dwarfs	122	168	183	198	211								
62 normals	122	173	213	254	282	312	371	391	396	401			
Clear L.													
dwarfs	114	155	168	178	185								
incr.	114	43	18	15	8								
no.	182	174	84	23	1								
normals	114	173	211	241	269	292	315	340	353	368	409	452	
incr.	114	58	38	30	28	23	20	20	20	18	20	20	
no.	177	175	175	165	154	114	58	25	14	10	4	4	

In Lake Erie no correlation could be shown between growth rate and population density, wind, or sunshine, but there was a tendency for inverse correlation between summer temperature and growth: $r = 0.6$ (V30). Furthermore, low turbidity in May, June, and October and high rainfall in July and August seemed to be related to good growth of whitefish. Van Oosten (V30) also suggested a cyclic fluctuation in growth rate. In Trout Lake, Wisconsin, slow growth was believed to be due to the high density of the fish population, particularly ciscoes, in the strata inhabited by whitefish. Whitefish heavily parasitized with *Triaenorphorus crassus* were slightly shorter and considerably lighter than nonparasitized whitefish of the same age (M57). Increased growth in the fourth year in Lake Huron was believed to be due to movement from shallow to deep water (V10).

No whitefish older than XX were reported except from eastern and northern Canada. B227 noted that few old whitefish were taken from Flathead Lake, Montana, even though there was little exploitation.

The abundant 1943 year class of whitefish in Lake Huron grew more slowly than the less abundant year classes (C333). Growth rate increased in Pigeon and Waubonum Lakes, Alberta, when the populations were overexploited (M88). The average age and size of the whitefish in the catch also markedly declined.

It was suggested that if the age composition of the catch is such that over 35% are age IV or younger, the fishery cannot maintain itself. There was also evidence of a decrease in growth of the young fish as the population density increased (M88). Thinning of a population in Little Moose Lake, New York, resulted in some increased growth (N91). There was little increase in size noted between ages IV and XVI, though the sample may have been somewhat biased because all fish were taken from spawning runs. Thus the younger age groups may have been represented only by the faster-growing, earlier-maturing individuals.

In Lake La Ronge most growth occurred from June to mid-August with growth ceasing in mid-September. Growth of older whitefish slowed earlier in the season than that of young fish (Q1) perhaps because the older fish became sexually mature. In Lake Erie growth was from May to October but the limits are uncertain (V30). In Lake Opeongo, Ontario, the oldest fish stopped growing by the end of July while younger whitefish continued to grow into September (K13).

Tagging slowed the growth of some of the tagged fish (D184). Tagged fish had increased 41 mm and 71 mm TL when recaptured 1 and 2 years after tagging.

Males were longer than the females in most age groups (N91). No sex difference in growth was noted in Lake La Ronge, Northwest Territories (Q1), in Lake Superior (D174), or in Lake Michigan (M451). Females grew faster than males in Lake Erie (V30), and in Lake Huron (V10) they grew faster in weight but not in length than males.

Male whitefish first mature at age I (dwarf whitefish in Cliff Lake, Maine, F121); at age II in Lake Michigan (M451), in Montana (B227), and in some Maine lakes (F121); at age III in Maine lakes (E93) and Georgian Bay (C333); at age IV in Great Slave Lake (K55), and Bay of Quinte, Ontario (H23), and not until VII in Lake Superior (E100). The minimum total lengths at maturity were 140 mm for the dwarf race in Lake Opeongo (K13), 152-178 mm for dwarfs in Cliff Lake (F121), 208 mm in Lake La Ronge (Q1), 297 mm in Lake Superior (E100), 373 mm in Lake Michigan (M451), and 394 mm in Georgian Bay (C333).

Dwarf females may also mature at age I (F121). Females first mature at age III in Montana (B227), Georgian Bay (C333), and Lake Michigan (M451), at age V in Maine (E93), at age VIII in Great Slave Lake (K55) and at age X in Lake Superior (E100). The smallest mature females, except the dwarfs, are 183 mm in Lake La Ronge (Q1), 178-203 mm in Maine (E93), 307 mm or 395 g in Great Slave Lake (K55), 310 mm in Lake Superior (E100), and 424 mm in Lake Michigan (M451) and Georgian Bay (C333). Increased growth as the result of overexploitation was followed with the whitefish maturing at age II whereas they had first spawned at age IV (M88). Females apparently live a little longer than males (Q1) in Lake La Ronge, but the older fish taken in Lake Huron were males (V10). Females were more abundant than males at age V and in all older age groups in Lake Ontario (C303).

Females probably spawn only in alternate years in Lake La Ronge (Q1) and in Great Slave Lake (K55), but in most lakes they spawn each year after reaching maturity.

In the northern part of Great Slave Lake, whitefish spawn September 16 to October 10, but do not spawn before October 1 in the southern part of the lake (R19). The spawning migration occurs in October in Bay of Quinte, Ontario (H23), but spawning does not occur until the first week of November. They spawn in November and December in Yukon Territory (L175), in September

and October in the far north (V68), late October to December in Montana
(B227), Maine (E93) and New York (B308), from November 8 to December in
Lake Erie (W161, V30).

Spawning occurs at temperatures of 0.5-1.7C (E4, K118), 5.5C (Q1), 4.5-
10C (H23), 4.5C (B308), and 0.5-4.5C (S471). They may spawn over humus,
rock, gravel, or hard bottom in depths of 1-30 m in Montana (B227), or over
rocky bottom in water not more than 3 m deep in Lake La Ronge (Q1), over
gravel at 2-2.5 m depths in Red Lake or at 2-23 m in Lake Superior (E4), over
gravel or sandy shoals at 2-3 m (S471). In Lake Winnipeg one population
spawns in the lake and another enters the rivers for this purpose (B11). Whitefish have also been noted spawning in rivers elsewhere.

Males usually precede females to the spawning grounds (E93, H23). The
males have pronounced breeding tubercles at this time, the females weaker
tubercles (H23). Eggs are broadcast with promiscuous spawning, pairs releasing sex products near the surface after display and upward movement
(E93, B308) mostly at night (H23, S471, B308). The eggs are released in small
batches over a period of about 10 days (B11). Eggs are semi-buoyant, nonadhesive, 2-3 mm in diameter, 53000 per fluid quart (V68). In Lake Superior
eggs averaged 34000 per quart compared to 49500 to 50000 per quart in Red
Lake (E4). Fry hatch in March-April (E93). The fry are about 13 mm at 3
weeks (F108). In Lake Erie 11% of 1709 whitefish had visible deformities
compared to almost no deformities in Heming Lake whitefish (L145). Water
temperature probably remains at 0.5C in Heming Lake during development of
whitefish embryos, but fluctuates somewhat in Lake Erie. In nature whitefish
hatch in March-April, with an incubation period of 120-140 days at 0.5-1.7C
(V68). Natural fertilization averages 80% (30-100%) but mortality to fry stage
usually exceeds 87% (V68). Low air and water temperatures were correlated
with high fertility of eggs taken at Lake Erie (W161). When water temperatures rose from 5.5C to 8C in mid-spawning season, fertility of eggs dropped
significantly. Descriptions of various stages of embryos and fry are given in
F5, H23, and H308. Rate of growth is slow at first, but fast from the end of
May to the end of July (H23). In the laboratory (P140) no eggs survived at 0C
or at 12C but at other temperatures:

Constant temperature	0.5C	2C	4C	6C	8C	10C
Length at hatching, mm	12-14	11-13	-	11-12	-	89.5
% mortality through hatch	27	42	41	42	81	99
% abnormal at hatching	0	0	1	10	25	50
Days incubation	141	121	80	59	40	30

	Size of females	No. of females	No. of eggs per female Mean	Range
D25, B11, L1	-	-	35000	-
D25 Great Lakes	4.9 kg	1	150,000	-
B11 MB	V-year olds	-	22000	-
	X-year olds	-	51000	-
	XV-year olds	-	91000	-
N25 NY	-	-	-	10000-75000
L13 L. Erie	508 mm	1	34760	-
L199 L. Erie	467-533 mm	14	47000	32169-59906
	617 mm	1	121,700	-
B23 NY	3.4 kg	1	66606	-
L17, B73	-	-	-	10000-75000
N39 Great Lakes	1.8 kg	1	50000	-

	Size of females	No. of females	No. of eggs per female	
			Mean	Range
D12 Great Lakes	-	-	-	25000-28000
M65 Great Lakes	0.9-1.4 kg	2	24856	21229-28500
	1.8 kg	1	48000	-
	3.4 kg	1	66606	-
K55 Great Slave L.	907-1360 g	6	-	7000-25000
	1361-1810 g	7	-	16000-37000
	1815-2267 g	2	-	31000-36000
	2268-2721 g	3	-	35000-66000
	2.7 kg	1	73000	-
	4.5 kg	1	89000	-
B227 Flathead L.	age III 408-590 g	3	10318	6285-15400
	age V 680 g	1	8844	-
Q1 L. La Ronge	590-730 g	6	9569	5958-14025
	1 kg	2	19892	17850-21933
Q1 Hunter Bay	1-1.4 kg	4	23634	17850-29120

Eggs per kg of female

B73, N25	22050	
L17	22050-26460	
M65	24250-26460	
H23 ON	8800-14300	by stripping
B227 MT	13000-25930	
C303 L. Ontario	21795	average for 29 females
L199 L. Erie	35495	
C333 Georgian Bay	18042	95% confidence interval 17310-20225
Q1 L. La Ronge	15775	
Hunter Bay	19880	

The differences in egg numbers per kg of female in Georgian Bay (C333), Lake Erie (L199), and Lake Ontario (C303) were believed to be due to environment rather than genetics since there has been so much hatchery stocking (C333).

In Lake Erie, female whitefish lose approximately 11% of their weight at spawning, but no weight loss was demonstrated for males (V30).

Annual mortality rates were estimated at 52% for Lake La Ronge and 55% for Hunter Bay (Q1), at 57%, 74%, and 81% for three Northwest Territory lakes (K122), at 61% for Great Slave Lake whitefish age XVII-XXII (K55), at 64% for Lake Winnipeg whitefish age VII-XIV (K56), at 41-59% for Lake Opeongo whitefish age VI-XII, at 8-17% for Shakespeare Island whitefish age XI-XX and 21-45% for age XXI-XXVII (R196). In Big Bay du Noc, annual mortality from III to IV was estimated at 94% (R162).

Natural mortality coefficients averaged 0.41 for whitefish beyond age III in Georgian Bay (C333) and fishing took 40% annually of the fish in those age groups which were vulnerable. Maximum sustained yield per recruit would require taking the fish at an average of age IV in Georgian Bay (C333). In Lake Superior the annual exploitation rates were a little over 20% (D184).

Fry or egg stocking in lakes with established whitefish populations had no detectable effect on the abundance of whitefish in the lakes studied (V54, V71, M381, M88, C303, C333). Wind was a possible factor in the strength of year classes in Alberta lakes (M381) but in Lake Erie no correlations could be found between year class strength and various meteorological conditions (V30).

Warm Novembers and warm winters seemed to result in poor year classes in Lake Ontario (C303), but could not be demonstrated for Georgian Bay (C333). The decline in population in Lake Ontario was believed due to overfishing and to some decline in habitat conditions, and not to lamprey effects (C303). Reproduction seemed hindered by high population density in Little Moose Lake, New York (N91) but did not increase after population reduction.

Fry begin feeding on entomostraca at 13 mm or age 2 weeks (H23, F108). The fry have teeth on the lower jaw for capturing their prey (F108). At about 6 weeks and 25 mm, the mouth becomes inferior and fingerlings start feeding on the bottom organisms (V68).

Whitefish are primarily bottom feeders, with *Ponteporeia, Hyallela, Gammarus,* sphaeriids and other molluscs, and insect larvae, particularly chironomids, as the principal foods (C236, R138, Q1, J53, B236, B11, K88, H309, R19, V19, Q3, M15, W194, and B227). Plankton constituted an important part of the food of whitefish until their fifth year of life in Shakespeare Island Lake, Ontario (H309). Lake whitefish fed mostly on pelagic and surface foods in contrast to the bottom feeding of *Coregonus pidschian* in Yukon Territory lakes (L175). A few sticklebacks were found in whitefish stomachs in Cree and Wollaston Lakes (R138), a few sculpins in larger whitefish in Okanagan Lake, British Columbia (M15), and some fish remains were found in Lake La Ronge whitefish (Q1). In Flathead Lake, chironomid larvae were the principal food January to June, and microcrustacea July to December (B227). In Lake La Ronge whitefish stomachs were usually empty at spawning season and in the winter (Q1, Q3), and in Lake Champlain 86% were empty prior to the spawning season (V19). In Watertown Lakes greatest feeding was noted in morning and evening (C236).

Sizes in gill net meshes	No.	Mean FL	Range	Mean weight	Range
25 mm					
R268 South Bay	51	124	109-135(173)	-	-
38 mm					
R268 South Bay	215	196	137-216(523)	-	-
R97 NT	261	213	102-528	227	-
Q1 L. La Ronge	+	312	155-483	422	27-1315
Q1 Hunter Bay	+	239	155-424	367	45-953
B227 Flathead L.	2	168	152-180	-	-
44 mm					
R268 South Bay	690	206	165-290	-	-
M294 South Bay	90	218	190-310	-	-
51 mm					
R268 South Bay	739	239	(170)196-307(467)	-	-
R97 NT	701	267	130-521	367	-
C22 L. of Woods	1	190	-	-	-
Q1 L. La Ronge	+	295	183-470	386	45-4309
Q1 Hunter Bay	+	274	208-493	354	91-1633
B227 Flathead L.	25	-	201-371	-	-
M294 South Bay	184	249	203-462	-	-
57 mm					
R268 South Bay	962	259	(178)216-318(406)	-	-
M294 South Bay	309	282	221-399	-	-

Sizes in gill net meshes	No.	Mean FL	Range	Mean weight	Range
64 mm					
R268 South Bay	1030	282	(178)234-338(394)	-	-
B227 Flathead L.	152	-	206-396	-	-
M294 South Bay	563	295	241-475	-	-
70 mm					
R268 South Bay	748	305	(239)259-368(488)	-	-
M294 South Bay	521	312	234-401	-	-
76 mm					
R268 South Bay	723	323	(229)282-373(465)	-	-
R97 NT	773	318	190-500	590	-
Q1 L. La Ronge	+	340	251-488	535	136-1678
Q1 Hunter Bay	+	328	254-437	499	114-1547
B227 Flathead L.	164	-	264-419	-	-
M294 South Bay	449	335	267-450	-	-
89 mm					
R268 South Bay	526	373	(282)325-427(488)	-	-
C22 L. of Woods	6	325	279-381	-	-
B227 Flathead L.	89	-	305-411	-	-
M294 South Bay	257	371	284-475	-	-
102 mm					
R268 South Bay	452	411	(328)351-462(518)	-	-
R97 NT	496	391	221-579	1043	-
C22 L. of Woods	43	358	279-584	-	-
Q1 L. La Ronge	+	381	272-523	721	227-2268
Q1 Hunter Bay	+	396	325-523	916	454-2041
B227 Flathead L.	81	-	295-445	-	-
M294 South Bay	225	429	284-513	-	-
114 mm					
R268 South Bay	281	439	(345)388-490(612)	-	-
M294 South Bay	231	450	381-610	-	-
127 mm					
R268 South Bay	109	465	(371)437-498(584)	-	-
R97 NT	367	421	269-589	1451	-
Q1 L. La Ronge	+	414	295-493	1034	318-1547
Q1 Hunter Bay	+	442	356-518	1279	590-2268
140 mm					
R97 NT	216	429	320-589	1547	-
Q1 L. La Ronge	+	437	414-508	1184	794-2268
Q1 Hunter Bay	+	465	419-503	1460	953-2041

Selectivity curves were developed (M294, R268) for use in correcting data on size distributions and catch per effort (C333). Relative efficiency of various gill net mesh sizes was also computed for Lake Huron whitefish (M356) using the relationship: girth = 0.68 FL - 367 mm. No difference in size selection was found between nylon and cotton gill nets, though the nylon caught more fish (M294).

Bathymetric distribution and movements (summarized by V68)
  Larvae: Within 2 weeks after hatching, they concentrate in less than

450 mm; after 4 weeks move into 0.9-1.2 m but remain near surface; in mid-June or when 33 mm go to bottom in 3-15 m.

Yearlings: School in less than 15 m; when about 550 g move with adults into deeper water.

Adults: Concentrate late spring in less than 18 m; midsummer 18-60 m with stragglers in 105+ m. Early fall move inshore to spawn and return to deeper water by the time ice breaks. In Lake Erie they migrate about 280 kilometers between the deep eastern waters to the western spawning grounds. In Lake Michigan 4% of tagged fish travelled beyond 40 kilometers; maximum distance 115 kilometers in 12 months. In Great Slave Lake average movement about 12.8 kilometers during a year.

Of whitefish tagged in Lake Michigan 22% were recaptured (S76).

A list of parasites was compiled by (V68):

TREMATODA
Diplostomulum
Crepidostomum

CESTODA
Bothriocephalus
Cyathocephalus
Diphyllobothrium
Eubothrium
Proteocephalus
Schistocephalus
Triaenophorus

NEMATODA
Cystidicola
Philometra
Rhabdochona
Spinitectus

ACANTHOCEPHALA
Echinorhynchus
Leptorhynochoides

COPEPODA
Achtheres
Ergasilus
Salmincola

PROTOZOA
Lymphocystis

PETROMYZONIDAE
Ichthyomyzon
Petromyzon

Although copepods were found in only about 1% of the stomachs in Heming Lake (W194), there was a high incidence of *Triaenophorus* in the whitefish.

BLOATER, *Coregonus hoyi* (Gill)

Except for a single specimen from Eva Lake, Ontario (L33), the bloater has been taken only from Lakes Ontario, Huron, Michigan, Superior, and Nipigon (S405). It is mostly an open-water species at depths of 27-30 m, with greatest abundance (J6) associated with temperatures of 3.8-7C. Extremes (J6) were 1.5 and 11.4C.

Increases in abundance of bloaters in Lake Michigan since 1950 resulted in some disruption of depth segregation and possible hybridization. *C. artedi* and *C. hoyi* have been crossed (S434).

J6 L. Michigan TL-weight relationship

TL	No.	Weight	TL	No.	Weight
102-126	1	8	254-278	1526	142
127-151	4	17	279-304	1850	179
152-177	247	31	305-329	372	227
178-202	1876	45	330-355	45	272
203-228	958	57	356-380	2	351
229-253	1061	102			

An inflection in the curve at about 200 mm suggested that it would be better to analyze two bloater groups separately (J6):

under SL 200 mm $\quad \log W = -5.5626 + 3.286 \log SL$

over SL 200 mm $\quad \log W = -4.2283 + 2.733 \log SL$

The curve for those over 200 mm was biased (J6) by gill net selectivity while those under 200 mm were entangled and not gilled in the nets.

J6 L. Michigan 8109 bloaters average K(SL) 1.25

The K value increased with increase in length from 90 to 220 mm and then showed a gradual decline, suggested by the length-weight regression given above. Older fish tended to have higher K values than younger fish at the same length, but some evidence was contradictory. No sex difference in K was evident. Both sexes were in poorest condition in May and June and improved to October or November.

	No.	SL	FL	TL	Range	No.	Mean W	Range W
Age I								
P21 L. Ontario	+	144	155	-	-	+	31	-
Age II								
J6 L. Michigan	35	145	-	175	125-195	-	-	-
P21 L. Ontario	+	161	175	-	-	+	51	-
Age III								
J6 L. Michigan-Mar.	28	140	-	170	155-185	-	-	-
J6 L. Michigan-Nov.	6	159	-	193	145-195	-	-	-
S94 L. Ontario, females	2	-	-	211	201-218	1	60	-
P21 L. Ontario	+	188	208	-	-	+	88	-
Age IV								
J6 L. Michigan-Mar.	113	151	-	183	125-185	-	-	-
J6 L. Michigan-Nov.	32	178	-	213	135-235	-	-	-
S94 L. Ontario, females	30	-	-	221	193-267	26	71	-
P21 L. Ontario	+	214	231	-	-	+	116	-
Age V								
J6 L. Michigan-Mar.	77	164	-	198	135-215	-	-	-
S94 L. Ontario, females	60	-	-	236	206-282	55	102	-
J6 L. Michigan-Nov.	29	206	-	249	175-235	-	-	-
P21 L. Ontario	+	222	246	-	-	+	139	-
Age VI								
J6 L. Michigan-Mar.	37	178	-	213	155-225	-	-	-
S94 L. Ontario, females	74	-	-	249	216-297	67	116	-
J6 L. Michigan-Nov.	27	212	-	254	185-245	+	218	-
P21 L. Ontario	+	255	282	-	-	+	218	-
Age VII								
J6 L. Michigan-Mar.	13	204	-	246	175-235	-	-	-
S94 L. Ontario	122	-	-	259	221-312	92	139	-
J6 L. Michigan-Nov.	14	227	-	274	215-255	-	-	-
Age VIII								
J6 L. Michigan-Mar.	6	209	-	251	195-225	-	-	-
S94 L. Ontario	89	-	-	269	231-307	65	162	-
Age IX								
J6 L. Michigan	1	231	-	277	-	-	-	-
S94 L. Ontario	50	-	-	282	246-323	42	184	-
Age X								
S94 L. Ontario	11	-	-	290	251-318	7	216	-
Age XI								
S94 L. Ontario	2	-	-	315	295-338	2	312	232-389

	No.	Average calculated SL at each annulus										
		1	2	3	4	5	6	7	8	9	10	11
J6 L. Michigan	442	72	119	141	158	174	189	206	211	231	-	-

	No.	Average calculated TL at each annulus										
S94 L. Ontario, females	360	94	142	168	190	211	229	246	262	277	290	315

	No.	Average calculated weight										
J6 L. Michigan	442	3	20	34	48	65	82	111	119	156	-	-

In Lake Ontario, scale growth did not resume until about mid-August (S94).
Growth calculations were made on the assumption of direct proportional growth of scale. Lee's phenomenon was evident in Lake Ontario (S94) and in some Lake Michigan samples (J6). Growth compensation was demonstrated in Lake Michigan (J6). In Lake Ontario many of the bloaters showed stunted condition after the third year. In Lake Michigan females were larger than males in most age groups (J6). In Lake Ontario, while no sex difference in length was noted the first 6 years, females averaged heavier than the males and attained a greater maximum length.

		Average annual increments in total length										
		1	2	3	4	5	6	7	8	9	10	11
J6 L. Michigan	males	86	56	25	18	15	15	15	13	-	-	-
	no.	223	223	210	183	104	46	13	4	-	-	-
	females	89	56	28	23	20	20	15	13	13	-	-
	no.	219	219	197	166	100	52	21	3	1	-	-
S94 L. Ontario	females	94	48	28	23	23	23	23	25	23	25	30
	no.	360	360	360	358	335	282	222	126	57	13	2

Annual increase in length as percentage of length at beginning of year (median value, with total number of specimens in parentheses and extreme values with the number of specimens on which they are based):

TL	Lake Ontario females S94	Lake Michigan J6
51-75		90(9), 90(8) - 93(1)
76-101	47(336), 27(1) - 66(14)	67(399), 52(25) - 80(8)
102-126	40(54), 11(1) - 80(2)	47(34), 46(21) - 49(13)
127-151	19(282), 13(1) - 21(31)	20(371), 13(7) - 25(29)
152-177	13(362), 7(1) - 19(39)	14(297), 9(113) - 22(21)
178-202	11(405), 9(43) - 16(8)	11(278), 7(81) - 15(5)
203-228	10(430), 7(1) - 15(14)	8(118), 4.8(4) - 10(26)
229-253	11(167), 9(51) - 12(110)	6(31), 4.6(2) - 8(6)
254-278	9(60), 7(7) - 11(24)	5(1)
279-304	13(1)	

Bloaters spawn November-January in Lake Ontario (P121) and in Lakes Michigan and Huron in February and March (J6, K88). Newly hatched fry with or without yolk sacs and 9-13 mm long were collected from April 9 to August 14 in southern Lake Michigan, suggesting a prolonged spawning season (W195). Adults may spawn several times and no care is given eggs or young (S405). The fry were in water temperatures below 4.7C (W195). They spawn at depths of 36-54m (K88). Maturity is reached at the end of the second year (P21).

Pontoporeia and *Pisidium* were principal foods in shallower water and *Mysis* in deeper water in Lake Huron (K88) but enough incidental items were also taken for Koelz (page 466) to state "the bloater will eat whatever occurs in his environment." *Mysis* and *Pontoporeia* were principal foods in Lake Ontario (P21, S94) and in Lake Michigan (B209). Alewives may compete with bloaters for food (S94).

Lengths of bloaters caught in gill nets of various mesh sizes

Stretch mesh size	No.		Mean	Range		Citation
25 mm nylon	308	FL	180	140-213	B251	L. Michigan
38 mm nylon	12986	FL	185	140-302	B251	L. Michigan
44 mm nylon	10665	FL	198	152-302	B251	L. Michigan
51 mm nylon	2865	FL	211	140-310	B251	L. Michigan
57 mm nylon	906	FL	213	152-335	B251	L. Michigan
60 mm	766	TL	254	152-318	S94	L. Ontario
-	1621	TL	234	152-302	J6	lower L. Michigan
64 mm	1067	TL	216	140-315	J6	lower L. Michigan
-	2030	TL	274	163-351	J6	upper L. Michigan
-	564	TL	257	183-333	S94	L. Ontario
- nylon	498	FL	218	145-361	B251	L. Michigan
67 mm	1106	TL	206	152-315	J6	lower L. Michigan
-	1298	TL	274	140-351	J6	upper L. Michigan
-	416	TL	259	196-323	S94	L. Ontario
70 mm	980	TL	203	152-302	J6	lower L. Michigan
-	625	TL	269	175-363	J6	upper L. Michigan
-	264	TL	251	193-318	S94	L. Ontario
- nylon	447	FL	203	152-373	B251	L. Michigan
73 mm	219	TL	249	157-353	S94	L. Ontario
76 mm	1174	TL	201	152-282	J6	lower L. Michigan
- nylon	186	FL	201	140-386	B251	L. Michigan
89 mm nylon	169	FL	193	140-348	B251	L. Michigan
102 mm nylon	138	FL	183	142-262	B251	L. Michigan
114 mm nylon	150	FL	183	142-246	B251	L. Michigan
127 mm nylon	76	FL	183	152-236	B251	L. Michigan

Above 64 mm, stretch measure, the nets do not appear to be particularly selective in any of these populations of bloaters. B251 worked out gill net selectivity measures using the relationship (from 147 L. Huron bloaters): girth = 22 mm + 0.0179 FL

DEEPWATER CISCO, *Coregonus johannae* (Wagner)
The deepwater cisco is known only from Lakes Michigan and Huron at depths of 54 to 180 m, with a few specimens in 29 m (K88). They spawn in late August and September and some individuals may spawn only biennially. The principal food consists of *Mysis* and *Pontoporeia*.

KIYI, *Coregonus kiyi* (Koelz)
The kiyi is found only in Lakes Ontario, Michigan and Superior at depths greater than 54 m (H382).

D8 L. Michigan up to 220 mm SL: TL = 1.217 SL; over 220 mm: TL = 1.209 SL

D8 L. Michigan TL — Weight Relationship

TL	No.	Weight	TL	No.	Weight
178-202	3	51	254-278	2626	153
203-228	21	74	279-304	1060	181
229-253	900	122	305-329	83	224
			330-355	3	312

D8 L. Michigan. 4696 fish, average K(SL) 1.43

During May through September, females were slightly heavier than males at corresponding lengths with a tendency for K to increase during this period. The average condition was lower in the island region of northwestern Lake Michigan than in the central basin of the lake. The length-weight regressions were therefore computed separately (D8):

$$\text{central basin:}\quad \log W = -5.0343 + 3.077 \log SL$$
$$\text{island region:}\quad \log W = -5.2732 + 3.167 \log SL$$

Females lost 6.2-14.4% of their weight at spawning, mean 11.8%, but males showed 0-7.4% loss, mean 1.6%. Kiyi of the same length but taken in increasing mesh sizes of gill nets showed increasing K values, as would be expected (D8).

		No.	SL	FL	TL	Range	No.	Mean weight	Range
	Age II								
P21	L. Ontario	+	178	190	-	-	+	82	-
D8	L. Michigan	32	212	-	257	236-272	32	139	-
	Age III								
P21	L. Ontario	+	203	216	-	-	+	125	-
D8	L. Michigan	184	211	-	254	188-307	184	142	-
	Age IV								
D8	L. Michigan	596	218	-	264	211-307	596	156	-
P21	L. Ontario	+	222	246	-	-	+	167	-
	Age V								
D8	L. Michigan	397	224	-	272	221-307	397	164	-
P21	L. Ontario	+	241	269	-	-	+	207	-
	Age VI								
D8	L. Michigan	136	230	-	277	236-307	136	179	-
P21	L. Ontario	+	274	297	-	-	+	303	-
	Age VII								
D8	L. Michigan	32	233	-	284	236-307	32	184	-
	Age VIII								
D8	L. Michigan	5	241	-	290	272-320	5	213	-
	Age IX								
D8	L. Michigan	1	228	-	269	-	1	162	-

Calculated TL and annual increments at each year

	1	2	3	4	5	6	7	8	9	10
D8 L. Michigan										
males	122	173	208	234	244	257	267			
incr.	122	51	36	25	18	15	13			
no.	172	172	165	128	57	13	2			

Calculated TL and annual increments at each year

	1	2	3	4	5	6	7	8	9	10
D8 L. Michigan (cont.)										
females	124	173	218	241	259	267	272	274	259	269
incr.	124	49	46	25	20	18	13	13	13	10
no.	1477	1477	1451	1297	728	242	50	7	1	1

Calculated weight at each annulus

	1	2	3	4	5	6	7	8	9	10
D8 central L. Michigan										
males	13	39	70	98	113	131	151			
females	14	40	79	109	133	149	160	164	137	153
Fox Islands area										
males	12	35	64	91	105	122	142			
females	13	36	73	102	124	140	150	154	128	144

Annual increment as percentage of length at first of year
(Median with total number in parentheses and range)

TL	D8 L. Michigan
76-101	47(1)
102-126	42(1080), 36(192) - 45(39)
127-151	34(569), 18(1) - 55(26)
152-177	22(1463), 13(1) - 25(761)
178-202	13(259), 7(1) - 29(154)
203-228	9(1426), 6(1) - 13(265)
229-253	6(942), 3.9(2) - 9(486)
254-278	5(52), 3.7(1) - 5(51)

Females averaged larger and heavier than males of the same age (D8). Lee's phenomenon was evident in the data from Lake Michigan (D8) and faster-growing individuals seemed to be shorter-lived. Growth compensation was also demonstrated. By July 9, 80% of the season's growth was completed and growth was terminated about the end of August.

Spawning peaks in the last half of October and early November in Lake Michigan but some spawning occurs in late September (H310) and may extend into January (P21). Kiyi spawn at depths of 100-150 m. Both males and females mature at age II and at length 173 mm (D8).

Kiyi are usually at depths of 108 m or more, but occasionally may be taken at only 36 m (K88). The temperature range is about 3.7-4.6C but other factors are more important since water of the same temperature at a lesser depth is not utilized. They feed mostly on *Pontoporeia* and *Mysis* (P2, K88, B209).

Sizes taken in gill nets, Lake Michigan D8

Mesh, stretch measure	60	64	67	70	76
Number of fish	2556	2225	1226	506	112
Mean TL	262	269	269	272	251
Range	201-307	188-343	175-320	201-333	175-333

BROAD WHITEFISH, *Coregonus nasus* (Pallas)
Broad whitefish are found in fresh and brackish waters of Siberia, Alaska

and arctic Canada (B314, S471) and weigh up to 6.35 kg. A 627 mm broad
whitefish from Beaver Lake, Alberta, weighed 4.3 kg (B11).

C201   SL = 0.924 FL   TL = 1.091 FL

W122 AK Ikrouvik L.

Age group	No.	TL	Range	Weight	Range
VIII	1	368	-	479	-
IX	1	498	-	1389	-
X	3	437	427-452	921	822-1041
XI	1	551	-	2064	-
XII	1	498	-	1375	-
XIV	1	572	-	-	-
XIII	1	549	-	-	-
XV	1	554	-	1991	-
XVI	1	610	-	2977	-

Calculated TL at each year

C201 AK Ikrouvik L.

No. 1	2	3	4	5	6	7	8	9	10	11	12	13	14	15
73	-	104	150	190	216	267	295	328	373	421	432	460	488	526	551

The lengths were calculated from scale measurements using the regression FL = 57.2 mm + 1.925 scale radius (x43). Since the young fish do not have scales the first winter, no annulus is formed.

Annual increment as percentage of length at first of year
(Median with total number in parentheses and range)

TL	C201 AK Ikrouvik L.
76-101	88(1)
102-126	95(73), 57(1) - 126(16)
127-151	47(53), 19(5) - 76(15)
152-177	34(16), 23(1) - 44(5)
178-202	30(27), 13(5) - 39(1)
203-228	26(41), 21(13) - 52(7)
229-253	24(20), 23(14) - 28(5)
254-278	17(14), 13(1) - 25(7)
279-304	18(35), 6(7) - 30(1)
305-329	19(16), 10(1) - 22(1)
330-354	16(7), 14(5) - 19(1)
355-380	19(1)
381-405	15(21), 9(5) - 20(1)
406-431	11(8), 10(6) - 22(1)
432-456	10(5)
457-482	8(3), 36(1) - 12(1)
483-507	10(5)
508-532	10(1)

In rivers, the broad whitefish spawns in August (W178), but in Ikrouvik Lake they spawn in July or even earlier (C201). Eggs hatch in 30-60 days. A

4.3 kg whitefish was feeding on mosquito larvae (B10). Instantaneous death rate per day was 0.0079 in Ikrouvik Lake (W121).

BLACKFIN CISCO, *Coregonus nigripinnis* (Gill)
J. J. Keleher, in a letter, indicates that the data reported for this species in Lake Erie by Clemens (C52) undoubtedly referred to *Coregonus artedi*. The blackfin is found in Lakes Huron, Michigan, and Superior, north to Hudson Bay and west to Alberta (S406), usually at depths of 27 to 180 m. Food of 56 black-fin ciscoes from Lake Huron consisted entirely of *Mysis relicata* (K88). Males and females mature at age IV (K37). They spawn from September to January (K88).

K37 MB L. Winnipeg

		No.	Mean	SL in each age group Range	Standard error
Age	II	1	140	-	-
	III	14	138	115-170	3.6
	IV	30	183	125-270	6.61
	V	41	225	140-285	5.21
	VI	32	256	195-315	6.02
	VII	16	287	240-335	7.7
	VIII	16	313	260-350	4.95
	IX	11	313	275-350	7.4
	X	2	330	310-350	-
	XI	1	330	-	-

NIPIGON CISCO, *Coregonus nipigon* (Koelz)
The nipigon cisco is found in shallow waters of a few lakes in Ontario and Manitoba (and possibly Quebec) in the Lake Superior and Hudson Bay drainages (H382). The maximum length is 495 mm and weight 907 g (S471).

M431 QU   log W = -1.499 + 2.811 log FL

K37 MB L. Winnipeg

		No.	Mean SL	Range	Standard error
Age	IV	12	230	210-270	4.9
	V	62	252	200-310	2.1
	VI	57	262	220-310	2.1
	VII	42	267	200-330	2.7
	VIII	3	286	275-290	-
	IX	1	250	-	-

HUMPBACK WHITEFISH, *Coregonus pidschian* (Gmelin)
These whitefish are more definitely bottom feeders than are *Coregonus clupeaformis*, when the two are found together. They spawn in June in the Yukon Territory (L175).

SHORTNOSE CISCO, *Coregonus reighardi* (Koelz)
Shortnose cisco are found in Lakes Michigan, Ontario, Superior, and Nipi-gon, usually in shallow water but sometimes to depths of 164 m (H382).

	No.		
J5 L. Michigan	384	160-209 mm SL	TL = 1.196 SL
	6423	210-249 mm SL	TL = 1.182 SL
	764	250-319 mm SL	TL = 1.178 SL

Length-weight relationship Lake Michigan, J5

Mean TL	Mean SL	No.	Mean W	Range of means
201	167	1	51	-
218	182	12	79	68-85
249	708	570	130	105-136
267	226	2633	159	145-170
287	244	2294	193	179-216
310	265	153	238	230-267
340	288	7	281	272-315
368	312	1	445	-

log W = -3.6071 + 2.468 log SL

The sample, however, probably was selective in that the gill nets favored heavier individuals at the shorter lengths. In general the K(SL) value decreased with increase in length and in age but the average for 5671 ciscoes 165-314 mm SL was 1.36, range of means 1.06-1.52. There was a suggestion of correlation between growth and condition with older fish at the same lengths having lower K values. K values were higher in May than June but increased each month thereafter until November. Males were lighter than mature and ripe females but heavier than spent females. Females lost 8% of their weight on spawning (J5).

	No.	SL	FL	TL	Range	No.	Mean weight	Range
Age I								
P21 L. Ontario	+	175	-	-	-	+	48	-
Age II								
J5 L. Michigan	14	224	-	264	212-237	14	162	-
Age III								
P21 L. Ontario	+	227	246	-	-	+	221	-
J5 L. Michigan	61	231	-	272	202-257	61	173	-
Age IV								
P21 L. Ontario	+	233	251	-	-	+	221	-
J5 L. Michigan	166	234	-	277	202-262	166	173	-
Age V								
J5 L. Michigan	71	241	-	284	217-262	71	187	-
P21 L. Ontario	+	250	269	-	-	+	258	-
Age VI								
P21 L. Ontario	+	241	254	-	-	+	201	-
J5 L. Michigan	18	245	-	290	222-262	18	201	-
Age VII								
P21 L. Ontario	+	270	290	-	-	+	357	-
J5 L. Michigan	1	234	-	277	-	1	156	-
Age VIII								
P21 L. Ontario	+	272	282	-	-	+	416	-

Annual increment as percentage of length

TL	L. Michigan J5
102-126	22(1)
127-151	24(19), 15(1) - 32(18)
152-177	30(313), 20(1) - 39(14)
178-202	18(255), 16(18) - 20(71)
203-228	18(62), 16(1) - 20(61)
229-253	10(275), 7(1) - 12(18)
254-278	6(89), 6(71) - 7(18)

The data (J5) show Lee's phenomenon, probably because of gill net selectivity. They also show growth compensation. No sex differences in growth were noted.

Shortnose cisco occur only in Lakes Michigan, Ontario, Superior, and Nipigon (K88), usually at depths of 30-150 m where temperatures range from 3.8-4.8C (J5). Spawning is in May and June at 36-142 m over sand, silt or clay bottoms in Lake Michigan though a few may spawn as late as September (J5). In Lake Ontario spawning may start in April (P21) and K88 suggests that they spawn in November in Lakes Superior and Nipigon. In Ontario one mature female was in the fall of its second year (age I) but most were 4-5 years old (P21). The smallest mature male was 198 mm TL in Lake Michigan (J5). Normally, shortnose cisco mature in the third or fourth year (M327). *Mysis* and *Ponteporeia* are the principal foods (B209, P21).

LEAST CISCO, *Coregonus sardinella* Valenciennes

Least ciscoes are anadromous in Siberia, Alaska, and northwest Canada (S471). There are freshwater populations in British Columbia.

W120 estuary forms	183 fish log W = -4.615 + 2.830 log FL
AK Ikrouvik L.	724 fish log W = -5.962 + 3.428 log FL

The lake fish were heavier for their lengths than the estuarine fish and the difference increased with increased length. A few unusually heavy-bodied fish in each population had spinal deformities.

	No.	FL	Weight
Age I			
C201 AK Ikrouvik L.	93	41-94	-
Age VI			
C126 BC Teslin L.	1	305	3402

Calculated FL at each year

	No.	1	2	3	4	5	6	7	8	9	10	11	12
C201 Ikrouvik L.													
Marine type	153	-	86	117	142	170	201	226	246	267	295	305	295
Stream type	116	-	109	132	157	180	201	218	224	224	231		
Lake type	1019	-	89	140	170	211	234	254	279	287	302	348	

Average annual increments in FL

		1	2	3	4	5	6	7	8	9	10	11	12
Marine type		-	86	30	25	28	30	28	25	20	23	23	18
	No.	-	153	153	153	151	146	127	90	40	12	6	1
Stream type		-	109	23	25	23	23	28	20	30	18		
	No.	-	116	116	115	106	90	54	20	3	1		
Lake type		-	89	53	43	43	36	30	28	23	25	28	
	No.	-	1019	988	421	356	210	113	56	14	4	2	

In Ikroavik Lake, three types of growth were recognized, representing fish that had spent their lives in the lake, in estuaries or ocean, and in the stream. The body scale relationships in the three populations were described as:

marine type　　　　FL = 52 mm + 1.68 scale radius X43
stream type　　　　FL = 78 mm + 1.44 scale radius X43
lake type　　　　　log FL = 1.077 + 0.636 log (scale radius X43)

There was also a difference in the length-weight relationships of these fish, with the lake type in best condition (W120). No annulus was formed during the first winter.

Annual increment as percent of length (number of fish in parentheses)

C201 AK Ikroavik Lake

	Marine type		Stream type		Lake type	
FL	Average	Range	Average	Range	Average	Range
76-101	85(153)	72(5)-89(19)	100(2)		50(988)	40(10)-73(567)
102-126	44(153)	24(28)-77(5)	74(118)	28(1)-92(9)	31(210)	28(99)-33(97)
127-151	30(151)	22(1)-48(28)	38(132)	27(17)-54(36)	19(356)	14(2)-30(146)
152-177	25(104)	21(28)-28(6)	22(141)	17(1)-54(36)	40(130)	35(65)-45(65)
178-202	18(128)	10(1)-24(42)	23(91)	9(1)-35(2)	17(113)	13(2)-19(42)
203-228	16(127)	15(51)-21(6)	19(18)	15(1)-22(17)	14(212)	9(2)-19(97)
229-253	11(34)	8(1)-14(5)	-		10(14)	6(2)-10(12)
254-278	8(12)	8(6)-12(6)	-		10(56)	6(2)-11(44)
279-304	10(11)	9(5)-10(6)	-		12(2)	
305-329	-		-		8(2)	

Annual mortalities of least ciscoes from Ikroavik Lake were estimated at 18%, 36%, 44%, 75%, and 67% at ages V-X respectively (W121). Spawning occurs in September at the time of freeze-over and the eggs hatch out under the ice (C201).

SHORTJAW CISCO, *Coregonus zenithicus* (Jordan and Evermann)
Shortjaw ciscoes are found in Lakes Michigan, Superior, Nipigon, Winnipeg and Athabasca, and from the Northwest Territories of Canada (H382, S406). They are usually found at less than 54 m.

V8 L. Superior　　　112 fish TL = 1.214 SL

Length-weight relationship, L. Superior V8

Males				Females			
SL	TL	No.	Wt.	SL	TL	No.	Wt.
195	239	3	85	182	221	1	64
223	269	6	110	200	244	4	83

Males				Females			
SL	TL	No.	Wt.	SL	TL	No.	Wt.
242	295	101	135	222	269	12	121
257	312	132	170	242	295	86	148
274	333	19	200	258	315	163	177
				277	335	47	219
				299	363	7	270

The average K(SL) of 261 males was 0.99
of 390 females   1.04

K decreased with increase in length but increased slightly with age. Within each length group, older fish were definitely heavier (V8).

	No.	Mean SL	Range	TL	Wt.
Age II					
K37 L. Winnipeg	12	138	115-220	-	-
Age III					
K37 L. Winnipeg	40	135	105-210	-	-
Age IV					
K37 L. Winnipeg	56	182	115-260	-	-
V8 L. Superior	5	193	182-202	234	77
Age V					
K37 L. Winnipeg	141	221	110-270	-	-
V8 L. Superior	35	229	202-252	277	125
Age VI					
K37 L. Winnipeg	145	240	180-272	-	-
V8 L. Superior	196	245	192-373	297	150
Age VII					
K37 L. Winnipeg	86	250	180-300	-	-
V8 L. Superior	243	257	237-287	312	170
Age VIII					
K37 L. Winnipeg	18	258	220-310	-	-
V8 L. Superior	95	268	247-292	325	198
Age IX					
K37 L. Winnipeg	31	277	210-340	-	-
V8 L. Superior	13	289	272-302	351	250
Age X					
K37 L. Winnipeg	1	270	-	-	-
V8 L. Superior	2	284	272-297	345	250

	Average calculated SL and increments at each annulus									
		1	2	3	4	5	6	7	8	9	10
V8 L. Superior	SL	80	107	136	165	191	215	232	247	265	268
	Incr.	80	27	29	29	26	25	22	21	20	20
	No.	589	589	589	589	584	549	353	110	15	2

	Average calculated FL at each annulus									
	1	2	3	4	5	6	7	8	9	10
B12 L. Winnipeg	94	150	193	229	264	295	315	325	328	330

Growth was computed from scale measurements on a direct proportion basis. In general, data indicate Lee's phenomenon and growth compensation (V8).

Annual increment as percentage of length at beginning of year

SL	Lake Superior V8	SL	Lake Superior V8
60-79	38(7)	160-179	16(489), 11(13)-16(474)
80-99	34(384), 23(2)-37(35)	180-199	13(549), 9(2)-14(196)
100-119	26(589), 22(13)-34(5)	200-219	10(353), 10(256)-12(2)
120-139	22(547), 19(95)-23(196)	220-239	9(110)
140-159	18(137), 14(2)-22(35)	240-259	8(15)

In Lake Huron spawning occurred in early October, in Lake Nipigon in early November, and in Lake Superior in late November and early December (K88). Both males and females mature at about 150-200 mm SL (V8). Food is mostly *Mysis* and *Pontoporeia* (K88, B209) but may include many aquatic insects from the bottom (B12, E4).

KOKANEE, *Oncorhynchus nerka kennerlyi* (Walbaum)

I have arbitrarily excluded the extensive data on the various Pacific salmon, *Oncorhynchus* spp., but since the landlocked form of the sockeye, the kokanee, is being introduced into a number of lakes outside the normal range of the species, I feel it necessary to mention some of the literature on this species.

B156 ID Priest L.   95 kokanee 190-254 mm TL   TL = 1.202 SL = 1.093 FL
     FL = 1.100 SL

	B156 Priest L.		R269 Flathead L.	
TL	No.	Wt.	No.	Wt.
178-202	10	62	-	-
203-228	53	91	-	-
229-253	175	122	16	136
254-278	238	159	56	184
279-304	187	210	264	233
305-329	36	238	443	286
330-355	-	-	59	335
358	-	-	3	378

K101 CA Donner L.   a battered male 483 mm FL, 1270 g.

	No.	Mean TL	Range	No.	Mean Weight	Range
Age 0-June						
C334 OR Elk L.	6	25	18-33	-	-	-
Age 0-July						
C334 OR Odell L.	26	33	25-43	-	-	-
Age 0-Aug.						
C334 OR Odell L.	47	46	38-53	-	-	-
C334 OR Elk L.	14	64	58-71	-	-	-
Age 0-Sept.						
C334 OR Odell L.	6	51	43-61	-	-	-
C334 OR Elk L.	21	84	76-91	-	-	-
Age I						
C334 OR Elk L.	161	147	135-165	-	-	-
C334 OR Odell L.	35	178	165-190	-	-	-

	No.	Mean TL	Range	No.	Mean Weight	Range
M300 CO Horsetooth L.	14	190	170-206	14	68	-
M384 CO	+	-	178-229	-	-	-
W163 NY Black Pond	+	211	-	+	96	-
R225 BC Cultus L. "residuals"	74	224	178-287	-	-	-
S480 CA Arrowhead L.	253	244	-	-	-	-
R225 BC Cultus L. "kokanee"	1	244	-	-	-	-
S480 CA Twin L.	23	267	224-290	-	-	-
F109 CA Tahoe L.	+	406	-	-	-	-
Age II						
W196 ID L. Pend Oreille	75	175	-	-	-	-
C334 OR Elk L.	60	196	183-221	-	-	-
V72 BC Kootenay L. south end	143	203	175-229	143	57	45-85
B156 ID Priest L.	81	211	198-221	-	-	-
C334 OR Odell L.	10	259	218-264	-	-	-
V72 BC Kootenay L. west arm	104	267	234-305	104	159	113-213
S480 CA	+	-	224-340	-	-	-
H300 CO Horsetooth L.	7	257	262-279	-	-	-
C293 CA Salt Springs	+	-	279-305	-	-	-
W163 NY Black Pond	85	295	231-361	85	269	108-408
R225 BC Cultus L. "residuals"	110	292	221-363	-	-	-
R225 BC Cultus L. "kokanee"	13	381	330-419	13	500	-
Age III						
W196 ID L. Pend Oreille	217	221	-	-	-	-
B156 ID Priest L.	110	234	229-246	-	-	-
V72 BC Kootenay L. north arm	104	239	213-262	104	105	68-128
C334 OR Elk L.	46	249	234-264	-	-	-
R225 BC Cultus L. "residuals" females	17	323	292-353	-	-	-
H300 CO Horsetooth L.	7	325	305-345	7	343	-
R225 BC Cultus L. "residuals" males	48	333	292-421	-	-	-
W163 NY Black Pond	200	330	-	200	334	-
C334 OR Odell L.	61	356	323-388	-	-	-
S480 CA Arrowhead L.	+	388	-	-	-	-
R224 BC Cultus L. "kokanee"	2	442	-	-	-	-
W163 NY Third Bishop L.	11	439	427-	11	819	-
Age IV						
W196 ID L. Pend Oreille	254	231	-	-	-	-
B156 ID Priest L.	296	259	234-277	262	150	113-196
S480 CA Arrowhead L.	13	381	-	-	-	-
Age V						
W196 ID L. Pend Oreille	97	231	-	-	-	-
B156 ID Priest L.	1427	269	246-300	918	181	122-221
Age VI						
W196 ID L. Pend Oreille	2	229	-	-	-	-

		No.	1	2	3	4	5
		Average calculated TL and increments at each year					
R225 Cultus L.	females	31	74	163	224		
	Incr.		74	89	89		
	No.		31	31	16		
	males	198	84	163	249		
	Incr.		84	86	109		
	No.		198	148	44		
B156 ID Priest L.		205	79	180	216	239	251
	Incr.		79	102	36	28	20
	No.		205	181	158	111	53
Upper Priest L.		96	89	193	264	297	325
	Incr.		89	104	71	36	33
	No.		96	94	71	51	33
F137 CO Granby L.		+	130	224	264		
P159 MT lakes		124	109	249	307	384	467

Lengths were calculated for R225 and B156 using the Fraser modification. The correction used in the Priest Lakes was 30 mm.

Annual increments as percentage of length at beginning of year			
TL	R225 Cultus L.	B156 Priest L.	B156 Upper Priest L.
51-75	107(109), 96(1)-142(12)	-	-
76-101	110(70), 89(6)-156(4)	125(181), 112(53)-133(23)	122(94), 106(23)-131(10)
102-126	68(52), 67(16)-86(30)	-	-
127-151	-	-	-
152-177	64(8)	-	-
178-202	-	21(158), 17(53)-72(47)	38(53), 38(30)-39(23)
203-228	-	13(111), 11(53)-14(58)	33(18)
229-253	-	9(53)	-
254-278	-	-	14(51), 10(10)-15(23)
279-304	-	-	13(23), 10(23)-16(10)

Population density was believed to be the most important factor affecting growth of kokanee in Priest and Upper Priest Lakes in various years. The faster growth in Upper Priest Lake was associated with a lower population density (B156). Most of the growth is from July to October, the period of maximum zooplankton abundance. In Kootenay Lake (V72) three separate populations were found differing in growth, age at maturity, fecundity, and spawning time. Ricker (R225, R226) distinguished two groups in Cultus Lake, probably an introduced stock which he referred to as kokanee and a "residual" or landlocked stock developed in Cultus Lake. He thought the residuals were an intermediate stage in the development of the kokanee (R227) but changed his mind when further information became available (R224). Males generally grow faster than females (B265, B156, S408, B264). A sea run has been developed from stocked kokanee indicating some variation in migrating behavior (F110).

Generally kokanee mature at 200-533 mm (much smaller than the sea run sockeye salmon) with a record weight of about 1814 g (B265). In California mature kokanee range from 203 to 610 mm FL (S480). In northern British Columbia lakes the average size is 203-229 mm while in southern British Columbia they average 305-381 mm. In Maine they have not exceeded 254 mm, but reached 533 mm in a Vermont lake (B265). In Lake Wenatchee, Washington, kokanee are 178-203 mm at maturity (R224). In Echo Lake, California, a high-altitude, very oligotrophic lake, they matured at 203-229 mm while at nearby

Donner Lake they were 305-406 mm (C293). Some kokanee spawn at the end of the second year (I+) (R225, V72, F109, S480), but most at II+ or III+ and in Priest Lake, Idaho, some were found at V+ (B156). In northern British Columbia most do not spawn until 5 years old and even 7 years (S480).

They spawn in tributary streams or in lakes on gravel. Wave action seems essential (C334). Courtship and nesting behavior have been described (K100, S410, S411). Kokanee may spawn from August to February (B265) peaking in December in Lake Pend Oreille, Idaho; in November in Flathead Lake, Montana; in October and November in Oregon (C334). In California some kokanee spawn from August to early October, and another group from late October to February, or even mid-April (S480). The early spawners were from British Columbia lineage. Spawning is usually at temperatures of 7-12C (S480). Almost all die shortly after spawning but occasionally a few males and females may survive and resume feeding after spawning (R224, B156). Nests with eggs were left exposed when the water level dropped in Donner Lake and although the surface of the redds froze, some of the eggs survived and hatched (K101). Some frozen and shrunken eggs were developed when thawed out in water.

Female kokanee lay from 300 to 1500 eggs depending on their size (B265).

		Eggs per female	
		Mean	Range
S408 WA Bear Creek	23 females, 224-268 mm SL	451	319-592
S408 BC Kootenay L.	2 females, 203-213 mm SL	-	360-375
B264 MT Flathead L.	121 females, 269-353 mm TL	701	309-963
S480 CO Granby L.	7 females, 388 mm TL	1676	-
C293 CA Salt Springs	126 females stripped	480	-
S480 ID L. Pend Oreille	254-272 mm TL	-	297-499
S480 CA Twin L. and Arrowhead L.	267-318 mm FL stripped	-	400-534
S480 CA Donner L. and Bucks L.	381-406 mm FL stripped	-	1174-1764

In Kootenay Lake, British Columbia, 26 females from the South End produced 3.44 ova per gram of body weight; which was significantly greater than did 23 females from the North Arm which produced 2.33 ova per gram and 10 females from the West Arm which produced 2.06 ova per gram (V72).

Food of the kokanee is almost entirely plankton with copepods and cladocera most important (F137, R226, C334, S480). In Skaguay Lake, Colorado, kokanee continued to feed on plankton when rainbow trout switched to aquatic insects in the summer (M384). In Cultus Lake they ate some insects and a few fry of Cottus and Oncorhynchus (R226). In Nicola Lake, British Columbia, crustacean plankton formed the bulk of the food in spring and autumn but chironomid pupae formed the bulk in summer (N121). A decrease in food intake was noted in summer, perhaps related to approaching maturity. Chironomid pupae were taken largely at night (N121) but feeding on other organisms is greater at dusk and early morning (N121, F137).

Kokanee were most abundant at water temperatures of 10.5-13C and their depth distribution increased as the summer progressed in a Colorado reservoir (H338). Water temperatures near 10C are preferred and severe mortality occurs at temperatures about 15.5C (S480).

R225 BC Cultus L.

Gill net mesh	38	44	51	57	64	70	76	89
Mean FL	173	190	239	274	274	292	295	310
No.	7	37	106	71	175	469	49	17

After a review of the literature Buss (B265) listed the advantages and dis-
advantages of stocking kokanee in Pennsylvania, indicating lack of competition
with shoal or bottom feeders as a major advantage. A similar analysis has
been prepared for California (S480).

BEAR LAKE WHITEFISH, *Prosopium abyssicola* (Snyder)
This species in the Bear Lake drainage, Idaho and Utah, may be a subspecies of *P. coulteri* (E119).

	No.	Mean calculated TL at each annulus

	No.	1	2	3	4	5	6	7	8
M152, S235 UT Bear L.	72	33	76	112	132	150	165	178	190
M152 203 mm female		2000 eggs							

Bear Lake whitefish are usually in water over 23m deep. They spawn
January to March. Ostracods, midge larvae and oligochaetes constitute their
food (M152).

PYGMY WHITEFISH, *Prosopium coulteri* (Eigenmann and Eigenmann)
This species, known from Alaska, British Columbia, Washington, and Montana, was found in Lake Superior in 1952-53 when bottom trawling was first introduced (E90). The maximum TL reported is about 206 mm for Alaska (R270)
and 150 mm for Lake Superior (E90).

H402   under 100 mm   TL = 1.0777 FL;   over 100 mm   TL = 1.0845 FL

	No.	Mean TL	Range	No.	Mean W
Age 0-June					
H402 AK Brooks L.	9	18	15-20	-	-
Age 0-July					
H402 AK Naknek area	475	-	15-33	-	-
Age 0-Aug.					
H402 AK Naknek area	413	-	25-56	-	-
Age 0-Sept.					
H402 AK Naknek area	117	-	41-66	-	-
E90 L. Superior	43	36	30-41	-	-
Age 0-Oct.					
E90 L. Superior	149	43	36-53	-	-
Age I					
H402 AK Brooks L.	150	61	-	-	-
E90 L. Superior	145	74	-	118	2.3
H402 AK South Bay	336	81	-	-	-
R270 AK Wood R.	9	102	-	9	8.5
W95 MT Bull L.	8	91	81-97 SL	-	-
Age II					
H402 AK Brooks L.	166	74	-	-	-
E90 L. Superior	56	89	-	29	5.4
H402 AK South Bay	395	119	-	-	-
R270 AK Wood R.	28	150	-	28	28.4
W95 MT Bull L.	15	105	91-129 SL	-	-

	No.	Mean TL	Range	No.	Mean W
Age III					
H402 AK Brooks L.	12	79	-	-	-
E90 L. Superior	44	91	-	15	7.0
H402 AK South Bay	30	132	-	-	-
R270 AK Wood R.	51	173	-	51	44.0
Age IV					
E90 L. Superior	62	112	-	51	10.4
H402 AK South Bay	16	147	-	-	-
R270 AK Wood R.	10	180	-	10	56.5
Age V					
E90 L. Superior	53	122	-	53	13.0
H402 AK South Bay	2	168	-	-	-
R270 AK Wood R.	3	193	-	3	61.3
Age VI					
E90 L. Superior	3	127	-	3	16.1
R270 AK Wood R.	1	206	-	1	70.6
Age VII					
E90 L. Superior	2	137	-	2	20.4

		Calculated TL and increments at each annulus						
		1	2	3	4	5	6	7
E90 L. Superior								
Isle Royale	males	41	69	81	89			
	incr.	41	28	13	8			
	no.	58	41	19	1			
	females	38	64	81	97			
	incr.	38	28	20	10			
	no.	36	26	21	10			
E90 L. Superior								
Keweenaw	males	48	71	86	97	107		
	incr.	48	28	18	13	8		
	no.	123	62	48	43	18		
	females	46	69	89	107	117	124	130
	incr.	46	30	20	18	15	10	8
	no.	148	91	76	66	40	5	2
H402 AK Brooks L.	males	46	66	74				
	incr.	46	23	3				
	no.	114	34	2				
	females	46	71	76				
	incr.	46	25	8				
	no.	216	146	12				
H402 AK South Bay	males	66	109	124	130			
	incr.	66	41	15	3			
	no.	393	232	18	4			
	females	69	112	127	140	147		
	incr.	69	41	18	13	10		
	no.	386	211	30	14	2		

		Calculated TL at each annulus						
	No.	1	2	3	4	5	6	7
E90 L. Superior								
Laughing Fish Pt. males	22	48	79	91	102	107		
females	15	46	69	91	107	119	124	

		No.	Calculated TL at each annulus						
			1	2	3	4	5	6	7
E90 L. Superior									
Apostle Islands	males	6	51	76	94	107			
	females	26	46	71	94	114	124	132	140
E90 MT									
L. McDonald	males	26	64	89	102	109			
	females	6	61	91	109				
E90 MT Bull L.	males	20	58	114	130				
	females	2	61	119	152				

	Annual increments as percentages of TL at start				
	E90	E90	H402	H402	
TL	Isle Royale	Keweenaw	Brooks L.	South Bay	Range
25-50	76	65	61	-	48-111
51-75	42	32	8	56	4-62
76-101	11	16	-	56	9-43
102-126	-	9	-	15	7-17
127-151	-	-	-	7	3-9

Growth in two British Columbia lakes (M452) was somewhat faster than in South Bay (H402) and two lakes had slightly slower growth.

Growth was computed on the assumption of direct proportion (E90) or on the basis of quartic body-scale equations (H402). In most cases females grew a little faster and lived longer than males. Growth rate was correlated with the dominance of insects over zooplankton in the food of several Alaskan populations (H402). Brooks Lake, with a slow-growing plankton-feeding population, had fewer associated whitefish species, less littoral area, and lower productivity than the lakes with faster-growing pygmy whitefish. No difference in diet with different growth rates were noted in British Columbia lakes (M452). Possible effects of interspecific competition on growth are discussed by H402.

In Brooks Lake 10% of the females and 36% of the males reach sexual maturity at the end of their second growing season (age I+). The smallest mature males were 58 mm and females 61 mm (H402). In South Bay no females and only 2% of the males matured at age I+ but most were mature at age II+. The smallest mature males were 86 mm and females 89 mm. No age I fish were mature in Wood River (R270). In Lake Superior, a few males matured at age I, 76 mm, and a few females at age II, 97 mm, but most matured a year later (E90).

Pygmy whitefish spawn in mid-November and December (E90, H402, M452) probably only at night in tributary streams (H402). The number of eggs per female increases sigmoidally with the total length (E90).

		Number	Eggs per female		Eggs per gram of body weight	
TL	Weight	of females	Mean	Range	Mean	Range
86-101	5.9	4	121	93-156	26	-
102-113	8.9	9	235	146-285	27	21-30
114-126	13.5	26	344	248-440	26	20-31
127-139	17.0	20	457	385-586	26	20-31
140-151	21.9	4	531	489-597	26	-

In the Naknek system the females are somewhat more fecund (H402) and
log (number of eggs) = -2.955 + 2.751 log FL

TL	FL	eggs per females
64-83	60-79	100-200
84-108	80-99	120-300
109-129	100-119	340-750
130-163	120-149	380-1150

Ostracods and *Pontoporeia* were the principal foods of 112 pygmy whitefish examined from Lake Superior (E90). Chironomids and sphaeriids were found in a number of stomachs. Fish eggs, probably coregonine, were taken frequently enough to suggest that they may be important as food at certain seasons. In some populations pygmy whitefish may feed on zooplankton and in others on insects (H402, R270).

H313 AK Average FL in gill nets

Stretch measure, gill net	No.	FL	Range
10 mm	15	64	58-69
13	337	69	58-79
19	7	71	66-74

ROUND WHITEFISH, *Prosopium cylindraceum* (Pallas)

Round whitefish are found in streams and lakes from New England and the Great Lakes across Canada and Siberia (H382). They are not a deepwater species, seldom being found at depths over 12m in Lake Nipigon or 54m in Lake Huron (S407). They go into salt or brackish water in the Churchill area, but not in Labrador.

The following ratios were used for converting FL to TL and SL to TL:

N91 NY Little Moose L. $TL = 1.3 mm + 1.085 FL$
C64 ME Moosehead L. under 270 mm SL $TL = 1.160 SL$; over, $TL = 1.147 SL$

Mean TL	No.	Weight	Location
117	3	11	B283 L. Superior
140	5	20	B283 L. Superior
170	38	31	B283 L. Superior
188	62	43	B283 L. Superior
216	64	68	B283 L. Superior
218	64	85	R97 Great Slave L.
234	4	85	M406 L. Michigan
241	74	94	B283 L. Superior
264	74	128	B283 L. Superior
274	+	199	R97 Great Slave L.
292	74	179	B283 L. Superior
297	5	179	M406 L. Michigan
318	15	216	M406 L. Michigan
318	74	235	B283 L. Superior
325	+	340	R97 Great Slave L.
343	41	278	M406 L. Michigan
343	74	315	B283 L. Superior
368	92	337	M406 L. Michigan
368	74	400	B283 L. Superior
381	+	567	R97 Great Slave L.
391	68	519	B283 L. Superior

Mean TL	No.	Weight	Location
394	43	437	M406 L. Michigan
409	+	879	R97 Great Slave L.
417	40	635	B283 L. Superior
419	7	527	M406 L. Michigan
437	2	590	M406 L. Michigan
442	24	754	B283 L. Superior
457	1	680	C126 BC
460	7	822	B283 L. Superior
470	3	808	M406 L. Michigan
488	+	1389	R97 Great Slave L.
495	4	956	M406 L. Michigan
516	1	1814	R97 Great Slave L.
521	2	1077-1134	M406 L. Michigan
551	1	2041	K93 Great Slave L.

Females were slightly heavier than males at most lengths (M406).

B283 L. Superior 755 fish 117-460 mm log W = -5.276 + 3.223 log TL
M406 L. Michigan 216 fish 234-521 mm log W = -4.695 + 3.294 log TL

K88 lists 2268 g as the maximum size in Lake Superior.

N91 NY Little Moose L. 408 whitefish average K(FL) = 1.05 range 0.82-1.13
No change in K was noted with age or sex.
$$K(TL) = 0.7682 \ K(FL)$$

	No.	Mean TL	Range	No.	Mean Weight	Range
Age 0-July						
R97 NT Great Slave L.	33	38	-	-	-	-
Age 0-Aug.						
H402 AK Brooks L.	65	51	33-58	-	-	-
Age 0-Oct.-Dec.						
N91 NY Little Moose L.	3	165	-	-	-	-
Age I						
W117 BC 3 lakes	+	74	61-81	-	-	-
R97 NT Great Slave L.	+	89	81-99	-	-	-
B283 L. Superior	39	109	76-137	-	-	-
N91 NY Little Moose L.	28	259	-	28	136	-
Age II						
W117 BC 3 lakes	+	140	117-155	+	102	-
R97 NT Great Slave L.	7	178	127-211	-	-	-
B283 L. Superior	152	180	152-226	-	-	-
E93 ME	+	234	-	-	-	-
N91 NY Little Moose L.	46	305	-	46	244	-
M406 L. Michigan	15	310	292-335	-	-	-
Age III						
W117 BC 3 lakes	+	180	145-203	+	164	-
B283 L. Superior	202	229	178-290	-	-	-
C64 ME Moosehead L.	20	259	241-284	20	139	113-187
E93 ME	+	272	-	-	-	-
R97 NT Great Slave L.	1	284	-	-	-	-
N91 NY Little Moose L.	148	328	-	148	300	-
M406 L. Michigan	138	361	310-391	-	-	-

	No.	Mean TL	Range	No.	Mean Weight	Range
Age IV						
W117 BC 3 lakes	+	236	190-262	+	187	-
B283 L. Superior	294	272	216-340	-	-	-
R97 NT Great Slave L.	1	267	-	-	-	-
C64 ME Moosehead L.	9	274	246-310	9	167	116-250
E93 ME	+	277	-	-	-	-
N91 NY Little Moose L.	151	345	-	151	340	-
M406 L. Michigan	43	394	371-417	-	-	-
Age V						
W117 BC South Tagish L.	+	234	-	-	-	-
E93 ME	+	290	-	-	-	-
C64 ME Moosehead L.	12	292	241-335	12	207	99-309
W117 BC 2 lakes	+	300	297-302	+	267	-
B283 L. Superior	143	305	267-366	-	-	-
R97 NT Great Slave L.	8	340	290-366	-	-	-
N91 NY Little Moose L.	35	348	-	35	363	-
K32 NT Great Bear L.	6	353	302-411	6	397	-
M406 L. Michigan	5	442	421-475	-	-	-
Age VI						
W117 BC South Tagish L.	+	251	-	-	-	-
C64 ME Moosehead L.	9	312	279-363	9	250	167-363
E93 ME	+	320	-	-	-	-
B283 L. Superior	127	333	267-391	-	-	-
W117 BC 2 lakes	+	351	345-356	+	451	-
N91 NY Little Moose L.	1	358	-	1	363	-
K32 NT Great Bear L.	11	363	274-439	11	397	-
R97 NT Great Slave L.	5	371	333-401	-	-	-
W113 CT	1	394	-	1	765	-
M406 L. Michigan	1	465	-	-	-	-
Age VII						
W117 BC South Tagish L.	+	274	-	-	-	-
C64 ME Moosehead L.	3	312	305-330	3	241	221-261
E93 ME	+	330	-	-	-	-
B283 L. Superior	75	356	318-404	-	-	-
W117 BC Atlin L.	+	371	-	+	491	-
K32 NT Great Bear L.	22	381	330-467	22	454	-
R97 NT Great Slave L.	3	401	388-411	-	-	-
M406 L. Michigan	6	498	486-526	-	-	-
Age VIII						
W117 BC South Tagish L.	+	279	-	-	-	-
S165 NT Tern Point	1	348	-	1	312	-
C64 ME Moosehead L.	6	363	330-399	6	383	272-476
B283 L. Superior	37	381	356-404	-	-	-
K32 NT Great Bear L.	24	396	330-521	24	539	-
R97 NT Great Slave L.	4	442	429-455	-	-	-
Age IX						
C64 ME Moosehead L.	3	376	353-404	3	485	394-598
B283 L. Superior	21	399	356-417	-	-	-
C126 BC Teslin L.	3	439	-	-	-	-
K32 NT Great Bear L.	6	457	384-551	6	907	-
R97 NT Great Slave L.	5	467	434-483	-	-	-

	No.	Mean TL	Range	No.	Mean Weight	Range
Age X						
S165 NT Tern Point	9	381	-	9	416	-
C64, E93 ME Moosehead L.	2	404	399-409	2	595	584-606
B283 L. Superior	6	406	381-429	-	-	-
K32 NT Great Bear L.	3	447	411-493	-	-	-
R97 NT Great Slave L.	1	523	-	-	-	-
Age XI						
B283 L. Superior	1	452	-	-	-	-
S165 NT Tern Point	1	467	-	1	907	-
Age XII						
B283 L. Superior	1	439	-	-	-	-
R97 NT Great Slave L.	2	511	508-513	-	-	-
K32 NT Great Bear L.	1	521	-	-	-	-
Age XIII						
R97 NT Great Slave L.	1	561	-	-	-	-
Age XIV						
R97 NT Great Slave L.	1	513	-	-	-	-

Average calculated TL and increments

	1	2	3	4	5	6	7	8	9	10	11	12
B283 Isle Royale	86	152	213	262	307	338	358	381	399	411	437	439
incre.	86	66	58	51	46	33	25	23	18	20	13	18
number	103	103	103	91	67	54	38	29	19	8	2	1
B283 Apostle Is.	117	183	231	274	307	333	358	378	394			
incre.	117	66	48	43	33	25	23	20	18			
number	995	956	804	614	344	214	103	37	10			
M406 L. Michigan	117	229	312	361	399	447	472	498				
incre.	117	112	83	51	33	30	28	25				
number	208	208	208	193	55	12	7	6				

Average calculated weight at each annulus

	1	2	3	4	5	6	7	8	9	10	11	12
B283 Isle Royale	3	23	65	133	224	309	391	474	548	641	703	800
B283 Apostle Is.	11	43	88	153	224	286	357	425	493			
M406 L. Michigan	11	74	210	340	457	581	718	860				

The calculated lengths in B283 and M406 were computed with Fraser-type corrections of 28 mm. Lee's phenomenon in the Lake Superior data (B283) was believed to be due to size selection in the gill nets. Annulus formation is usually in June or early July in Lake Superior but one sample from Isle Royale in August of a cold summer had only 32% with the annulus yet formed (B283).

Growth appears to be fastest in the big lakes of the Northwest Territories and in Lake Michigan. Females were longer than males in all adequately represented age groups in Little Moose Lake (N91).

In Lakes Huron and Superior, round whitefish mature at about 250 mm SL (K88). In Lake Superior, however, B283 reports females mature at 140-175 mm TL and males at 178-188 mm, and at age II. Females were more abundant than males at ages V and up (B283). In Lake Michigan the smallest mature male was 310 mm, age II, and the smallest mature female was 335 mm, age III (M406). In Great Bear Lake most mature at ages V and VI (K32).

Round whitefish spawn in October and November over rock or gravel at 4-14 m (S407). They were spawning December 14-23, 1923, in Lake Superior (E4). Spawning was completed on December 13 in Lake Michigan (M406). In Maine round whitefish spawned earlier than common whitefish, at the mouth of streams or in the lake proper (F121).

B73 Average of 3500 eggs per female. One 794 g female produced 12000 eggs.

B283 L. Superior 12028 eggs per kg of female

TL of female	No.	Eggs per female	Range
267	1	1076	–
292	1	2688	–
318	7	2751	1906-4595
343	3	3623	2986-3951
368	6	4288	–
394	10	5768	–
419	8	8718	-11888
437	1	10187	–

Of a group tagged in Lake Michigan, 20% were recaptured (S76).

When round whitefish first start feeding they eat plankton but they soon change over to bottom organisms: chironomids, caddis larvae, *Hyalella*, sphaeriids, snails, mayfly larvae (S407, R138, W117). Fish were eaten in moderate numbers by 279-330 mm round whitefish in Atlin Lake (W117). The round whitefish has been called a spawn eater but the evidence is not conclusive (S407).

Stretch measure of gill net	No. fish	B251 Mean FL	Range	No.	NT R97 Mean FL	grams	No.	AK H313 Mean FL	Range
19	-	-	-	-	-	-	58	137	97-178
22	-	-	-	-	-	-	2	147	117-175
25	-	-	-	-	-	-	4	150	132-170
38	31	231	165-417	22	221	154	-	-	-
44	85	257	165-450	-	-	-	-	-	-
51	76	310	241-462	73	279	295	-	-	-
57	78	345	241-462	-	-	-	-	-	-
64	68	376	284-488	-	-	-	-	-	-
70	126	386	272-462	-	-	-	-	-	-
76	91	394	335-475	74	348	590	-	-	-
89	74	411	348-475	-	-	-	-	-	-
102	39	417	356-488	18	414	953	-	-	-
114	4	373	361-437	-	-	-	-	-	-
127	6	437	399-475	-	-	-	-	-	-

B251 determined the relative efficiency of various gill net sizes using the following formula based upon 235 fish:

$$girth = -.254 + .0207 \ FL$$

BONNEVILLE CISCO, *Prosopium gemmiferum* (Snyder)
Bonneville cisco are found only in Bear Lake, Utah (S463).

TL at each annulus

	No.	1	2	3	4	5	6	7	8	9	10
M152, S235 UT											
Bear Lake 1938-41	1215	53	104	145	165	178	183	188	193	193	196
1952	35	56	104	140	165	178	183				

No sex difference in growth was noted (M152). Most growth is in June and July. Spawning is in late January and early February and the fish first mature at ages I and II. Bonneville ciscoes feed primarily on copepods and cladocerans (M152). They usually are in the upper part of the hypolimnion, but seek temperatures below 15C.

BONNEVILLE WHITEFISH, *Prosopium spilonotus* (Snyder)
Bonneville whitefish are also found only in Bear Lake, Utah (S463). Their growth is very similar to that of mountain whitefish in the same watershed.

TL at each annulus

	No.	1	2	3	4	5	6	7	8
M152 UT Bear L.									
1951-54	245	81	145	190	208	272	323	371	417

These whitefish reach spawning size, 203 mm, at age III. They spawn in November-January on rocky shoals (M152).

203 mm females had 600-900 eggs
229 mm female had 1200 eggs

Food consisted mostly of chironomid larvae and pupae.

MOUNTAIN WHITEFISH, *Prosopium williamsoni* (Girard)
Mountain whitefish are in lakes and streams on the west slope of the Rocky Mountains from Fraser River to Truckee River, in the Lahonton Basin, and east of the Rockies in the headwaters of the Missouri and Saskatchewan Rivers (E119).

S144 UT Logan River   313 fish   58-413 mm SL   FL = 1.0990 SL
                                                 TL = 1.1648 SL = 1.0599 FL
H263 WY   FL = 1.098 SL
             TL = 1.1518 SL (75 fish) = 1.084 FL (500 fish)

females FL = .9365 TL -3.8 mm SL = .854 TL -3.8 mm
males   FL = .931 TL -2.2 mm SL = .841 TL -0.18 mm

Where data were given as SL or FL in the original papers, without conversion factors, the data have been converted to TL as follows:
TL = 1.16 SL = 1.1 FL

TL	UT Logan R. S144 No.	Mean W	WY H263 No.	Mean W	MT B107 No.	Mean W	Combined No.	Mean W	Range
51-75	6	3	-	-	-	-	6	3	-
76-101	24	5	-	-	-	-	24	5	5-6
127-151	2	24	-	-	-	-	2	24	-
152-177	-	-	9	43	-	-	9	43	34-48
178-202	2	57	45	62	-	-	*51	65	45-88
203-228	7	113	82	96	-	-	*90	99	79-119
229-253	31	147	145	145	-	-	176	128	99-147
254-278	31	198	165	170	5	193	201	176	144-241
279-304	27	264	127	230	6	247	160	235	193-275
305-329	35	326	67	289	2	312	104	303	241-349
330-355	50	389	12	372	1	340	63	383	292-428
356-380	42	468	-	-	1	369	43	465	369-527
381-405	13	606	-	-	3	481	16	584	454-606
406-431	9	686	-	-	1	567	10	674	567-686
432-456	11	774	-	-	1	680	12	765	680-774
457-482	6	986	-	-	-	-	6	986	927-1058
483-507	-	-	-	-	1	1361	1	1361	-

*Includes a few fish from Carter Creek, Utah M340.

S144 UT Logan R.	296 fish	58-413 mm	log W = -4.7578 + 2.9803 log SL
H263 WY males			log W = -5.1976 + 3.1178 log FL
females			log W = -5.2919 + 3.1574 log FL
S144 UT Logan R.	296 fish	58-413 mm	average K(SL) = 1.572
M340 UT Carter C.	5 fish	177-200 mm	average K(SL) = 1.52

	No.	TL Mean of means	Central 50%	Range	No.	Weight Mean of means	Range
Age 0-March							
B107 MT	20	13	-	10-15	-	-	-
Age 0-April							
B107 MT	10	15	-	15-18	-	-	-
Age 0-May							
B107 MT	10	28	-	25-30	-	-	-
Age 0-June							
L51 MT *P. w. cismontanus*	+	38	-	-	-	-	-
B107 MT	20	43	-	23-58	-	-	-
Age 0-July							
L51 MT *P. w. cismontanus*	+	51	-	-	-	-	-
B107 MT	20	58	-	38-86	-	-	-
Age 0-Aug.							
H263 WY Phelps L.	4	56	-	-	-	-	-
M325 CA L. Tahoe	+	-	-	66-102	-	-	-
Age 0-Oct.							
B107 MT	20	109	-	86-127	-	-	-
Age 0							
Can. lakes M16, M15, E99, G62	5+	66	-	41-99	-	-	-
U.S. lakes H263, M16	30+	68	-	41-99	-	-	-
U.S. rivers M16	+	69	-	48-91	-	-	-
Can. rivers M16	+	75	-	56-94	-	-	-
Combined	35+	68	51-91	41-99	-	-	-

		TL Mean of	Central			Weight Mean of	
	No.	means	50%	Range	No.	means	Range
Age I							
Can. lakes R98, E99, M16, M15, G62	26+	142	86-180	69-196	-	-	-
Can. rivers M16	+	146	-	145-147	-	-	-
U.S. lakes M16, H263	104+	146	-	127-229	-	-	-
U.S. rivers S144, M16	2+	168	-	135-188	-	-	-
M16 Ghost R. Res.	+	170	-	-	-	-	-
Combined	132+	147	130-180	69-229	-	-	-
Age II							
Can. lakes R98, E99, M16, G62, M15	51+	197	168-234	89-254	-	-	-
Can. rivers M16, B10	4+	207	-	198-216	-	-	-
Can. reservoirs M16	+	215	-	198-231	-	-	-
U.S. lakes M16, W95, H263	192+	217	-	180-257	-	-	-
U.S. rivers M16, S144	4+	235	-	190-267	-	-	-
Combined	251+	209	193-234	89-267	-	-	-
Age III							
Can. lakes M16, E99, R98, G62	40+	229	185-272	135-277	+	28	-
Can. rivers M16, B10	2+	237	-	236-239	-	-	-
U.S. lakes M16, H263	183+	241	-	231-269	-	-	-
Can. reservoirs M16	+	262	-	244-279	-	-	-
U.S. rivers S144, M16	43+	264	-	208-297	-	-	-
Combined	268+	241	221-272	135-297	+	28	-
Age IV							
Can. lakes M16, R98, E99, G62	67+	259	229-295	185-340	+	79	-
Can. rivers M16	+	265	-	264-267	-	-	-
U.S. lakes H263, M16	141+	271	-	254-287	-	-	-
U.S. rivers M16, S144	79+	275	-	229-300	-	-	-
M16 Ghost R. Res.	+	330	-	-	-	-	-
Combined	287+	267	244-295	185-340	+	79	-
Age V							
Can. rivers M16	+	268	-	251-284	-	-	-
H263 WY Phelps L.	202	284	-	269-297	-	-	-
Can. lakes M16, E99, R98, G62	60+	291	259-310	165-353	+	142	-
U.S. rivers M16, S144	75+	296	-	254-318	-	-	-
M16 Kananaskia L.	+	310	-	-	-	-	-
Combined	337+	288	259-310	165-353	+	142	-
Age VI							
M16 Kananaskia L.	+	284	-	-	-	-	-
H263 WY Phelps L.	151	297	-	282-312	-	-	-
Can. lakes M16, E99, R98, G62, M113	49+	310	287-343	180-394	1+	295	216-454
S144 UT Logan R.	39	366	-	-	-	-	-
Combined	239+	311	287-343	180-394	1+	295	216-454
Age VII							
H263 WY Phelps L.	99	305	-	292-325	-	-	-
M16 Bow R.	+	330	-	-	-	-	-
Can. lakes M16, E99, R98, G62	37+	355	351-373	216-409	+	312	-
Can. reservoirs M16	+	358	-	330-386	-	-	-
S144 UT Logan R.	22	414	-	-	-	-	-
Combined	158+	337	355-386	216-414	+	312	-

	No.	Mean of means	TL Central 50%	Range	No.	Weight Mean of means	Range
Age VIII							
U.S. lakes O22, H263	67	311	-	297-330	-	-	-
Can. lakes M16, E99, R98, G62	16+	374	351-411	229-434	+	400	-
Can. reservoirs M16	+	376	-	340-411	-	-	-
S144 UT Logan R.	8	427	-	-	-	-	-
Combined	91+	370	315-424	229-434	+	400	-
Age IX							
H263 WY Phelps L.	20	335	-	318-345	-	-	-
Can. lakes M16, G62, R98, E99, C135	4+	367	363-404	254-411	+	439	340-539
M16 Bow R.	+	371	-	-	-	-	-
S144 UT Logan R.	9	452	-	-	-	-	-
Combined	33+	374	335-411	254-452	+	439	340-539
Age X							
H263 WY Phelps L.	4	353	-	345-361	-	-	-
E99, R98 AT Pyramid L.	1	432	-	-	1	737	-
Age XI							
M16 Bow R.	+	318	-	-	-	-	-
H263 WY Phelps L.	2	358	-	353-366	-	-	-
Age XII							
H263 WY Phelps L.	3	361	-	353-368	-	-	-
M16 Bow L.	+	371	-	-	-	-	-
M16 Waterton L.	+	384	-	-	-	-	-
Age XIV							
M16 Waterton L.	+	429	-	-	-	-	-
Age XVI							
H263 WY Phelps L.	1	429	-	-	-	-	-
Age XVII							
M16 Bow L.	1	434	-	-	-	-	-

Average TL and annual increments at each annulus

		1	2	3	4	5	6	7	8	9
S144 UT Logan R.		117	206	259	295	325	358	391	417	442
	incr.	117	89	53	33	28	33	33	33	28
	no.	281	279	275	232	153	98	79	17	9
M15 BC Okanagan L.		135	221	292	323					
	incr.	135	89	71	31					
	no.	33	17	1	1					
B145 WY Madison R.		130	226	305	348	388	429			
	incr.	130	96	76	48	30	23			
	no.	36	36	28	17	8	3			

Average calculated TL at each annulus

	No.	1	2	3	4	5	6	7	8	9	10	11
S144 CA Upper Twin L.	+	127	157	180	196	221						
H263 WY Phelps L.	915	160	193	229	251	269	290	307	318	335	340	
P78 WY	+	99	206	262	290	318	338					
M453, P159 MT lakes	587	91	190	241	272	300	325	356				
P159 MT Doctor L.	7	46	84	119	157	193	226	254	282	323	343	345
P159 MT reservoirs	232	86	183	246	290	312	335	351	371			
M453, P159 MT rivers	1212	86	180	246	292	328	353	368	419	442		

Average calculated weight at each annulus

	No.	1	2	3	4	5	6	7	8	9
S144 CA Upper Twin L.	+	-	51	57	68	113	-	-	-	-
S144 UT Logan R.	281	17	88	170	250	337	451	589	706	845

Annual increment as percentage of length at beginning of year
(average and range, with number of fish in parentheses)

TL	WY Madison R. B145	UT Logan R. S144	BC Okanagan L. M15
102-126	92(28), 81(5) - 109(3)	75(279), 64(4) - 78(75)	-
127-151	-	-	69(17), 68(16) - 71(1)
152-177	26(8)	-	-
178-202	-	24(43)	-
203-228	34(14), 33(9) - 36(5)	24(232), 22(9) - 26(75)	32(1)
229-253	34(14), 33(3) - 35(11)	13(9)	-
254-278	18(9)	13(215), 13(136) - 14(79)	-
279-304	14(5)	10(153), 9(114) - 10(39)	11(1)
305-329	-	10(56), 9(9) - 10(47)	-
330-355	13(8), 11(5) - 16(3)	10(31)	-
356-380	-	9(39), 8(22) - 11(9)	-
381-404	6(3)	7(8)	-
405-431	6(3)	7(9)	-

Growth was calculated from scales using direct proportion (B145, P78, M15) or from a curvilinear body-scale relationship (H263, S144). Lee's phenomenon was overcome by using the curvilinear relationship (M16). No growth compensation could be demonstrated in the populations studied by McHugh (M16); the faster-growing fishes continued to be faster growing throughout life. In paired reading of scales, second readings agreed with the first reading 52.6% of the time and 89.8% agreement was reached with a 3rd reading (H263) indicating fairly difficult scales. Annuli were assumed to form in May a couple of weeks after the ice goes off Phelps Lake, Wyoming (H263). Scales first appeared at 35 mm TL and 48 mm fish were fully scaled (H263). The following constants were computed for the von Bertalanffy formula (H263): $k = 0.1612$; $t_o = 2.103$ and $l_\infty = 360.7$ mm FL. No sex differences were noted in growth (M16, S144, H263).

In California streams larger young of the year whitefish were taken from the richer streams (M325). Slower growth of Pyramid Lake whitefish was believed to be due to a population density greater than in Lake Minnewanka (E99). McHugh (M16) found faster growth in the warmer lakes and streams in most cases. Growth seems to be faster in U.S. lakes and streams than in the more northern lakes and streams. In Wyoming (H263) growth of immature fish was found to continue into November but mature fish stop growing earlier in the season.

Mountain whitefish feed primarily on aquatic insects (S407, J53, M385). A greater variety of food may be found in stream fish than in lake fish with terrestrial insects taken by stream fish when bottom fauna is sparse (M385). Cladocera comprised the principal food of mountain whitefish in Phelps Lake, Wyoming (H263) in Pyramid Lake, (R98, E99) and in Lake Okanagan, B.C. (M15) perhaps indicating a scarcity of bottom fauna. In Pyramid Lake, mayflies comprised a major food item when they were emerging (E99). Mountain whitefish sometimes eat fish eggs (S412).

In northern lakes, mountain whitefish are usually at depths of less than 10 meters (G62) and in Phelps Lake, Wyoming, most were found at depths of 20-45 feet with the average size of the fish increasing roughly with depth (H263).

Some males and females mature in the second year (B107) in Montana but in Phelps Lake, Wyoming, the youngest mature fish were age II, minimum 198 mm (H263), and in Logan River, Utah, the youngest age was III (S144). Spawning started as early as September 6 in Wyoming (H263) and continued to November, initiating at temperatures of 12C. In Montana, mountain whitefish spawn in October and early November at temperatures of 5.5C or below (B107). In Logan River, Utah, they spawn in November and early December and show little upstream prespawning movement (S144). At Lake Tahoe, Nevada, they move into tributaries to spawn in October (S412) but some spawn in the lake (M458). Spawning is at night (K118).

	Size of female	No. of females	No. of eggs per female Mean	No. of eggs per female Range
B107 MT	254-278 mm TL	5	1803	1426-2160
	279-304 mm TL	6	2875	1470-4321
	305-380 mm TL	4	3584	3242-4131
	381-430 mm TL	4	5106	4494-5377
	432 mm TL	1	7271	-
	495 mm TL	1	24143	-
H263 WY Phelps L.	249 mm TL, 198 g	1	2955	-
	269-295 mm FL, 252-332 g	5	4831	3578-5727
	310-323 mm FL, 363-431 g	5	5970	5132-7466
	340 mm FL, 408 g	1	6148	-
S144 UT Logan R.	312-680 g	5	-	5500-14000

B107 MT   11780 eggs per kg of female

S146 WY   15787 eggs per kg of female

The embryological development is given (B107) with hatching time as 36 days at 11C. Hemoglobin determinations, hematocrits, and blood cell counts are given, with size, sex, and seasonal differences in M454.

Mountain whitefish in the Snake River, Idaho (I7) fed on Diptera, amphipods, trichopters, ephemerids, beetles, and snails. In Jasper Park (N13) they were found to be feeding on the same foods as rainbow trout, with less surface feeding. In British Columbia lakes mountain whitefish were mostly bottom feeders, chironomids and mayflies being predominant in the food (M385). Feeding was somewhat restricted as the spawning season approached. Summaries and bibliographies: S407, S144, M458.

GOLDEN TROUT, *Salmo aguabonita* Jordon

Golden trout were originally found only in the High Sierras of California, but have been introduced elsewhere (E119).

N36 CA 28 fish average K(SL) 1.315 range 0.839-1.991

C67 CA average K(FL) 1.07 $\log W = -1.770 + 3 \log FL$

G36 WY A 711 mm TL golden trout weighed 4989 g (world's record).

	No.	Mean FL	Range	No.	Weight Mean	Range
Age 0-Aug.						
C79 CA Cottonwood L.	-	25	23-30	-	-	-
Age 0-Sept.						
N36 CA High Sierras	-	30	-	-	-	-
C79 CA Cottonwood L.	-	38	33-43	-	-	-
Age I						
N36 CA High Sierras	12	127	109-137	12	26	17-48
Age II						
C149 OR introduced	45	160	130-211	-	-	-
N36 CA High Sierras	35	211	135-269	16	113	28-233
Age VI						
M461 CA Third Darwin L.	+	-	381-432	-	-	-

	Code	No.	FL at each annulus 1	2	3	4	5
C79 CA Lake 3	F	123	51	127	188	208	221
C79 CA Lake 4	F	147	41	147	229	284	295

	No.	TL at each annulus 1	2	3	4	5
M453, P159 MT lakes	28	114	206	287	338	

Scales begin to form at a FL of 46 mm, about the same time as the fingerlings migrate to the lakes (C79). Growth was calculated (C79) using Fraser's modification with 46 mm as the correction value. Some fish did not form annuli the first winter but these could usually be recognized by counting circuli. Lake 3 had good spawning areas and plenty of food for young, but because of the high population density showed slower growth in later years than Lake 4, with poor spawning conditions. Growth is greater in lakes than streams (M461).

Spawning is in June and early July as water temperatures reach 7-10C in stream gravels. Only 1 of 233 females matured at age II, most at III or IV (C79). Females matured at 165 mm FL in Oregon (S380).

C80 CA   200 mm female, 305 eggs stripped; 300 mm female, 715 eggs; 375 mm female, 1030 eggs

C79 CA   Number of eggs equals 4.11 FL - 508 based on 450 females, stripped.

S146 WY   2 females, 450-860 g, 1380 eggs per female
             2 females, 1360-1770 g, 2193 eggs per female, 2105-2280

L101 CA   382 eggs per female stripped

Food during the summer was primarily caddis and midge larvae and pupae and cladocera (M461, C79).

CUTTHROAT TROUT, *Salmo clarki* Richardson
     Cutthroat trout were originally found in streams of the Pacific coast from

the Eel River in northern California into Alaska (N101). Many, but not all, went to the ocean or at least into estuaries for a short part of their life cycle. Cutthroat trout were also found in the intermountain and Rocky Mountain lakes and streams from Arizona and New Mexico to southern Canada. Many subspecies were recognized which have now been lost through stocking and other changes. Hybridization with rainbows has occurred in many places where the latter species has been stocked (M455, N101, M460).

O22 OR

No.	FL	FL-TL normal	FL-TL compressed	TL normal-TL
35	76-126	1.0315	1.0644	1.0322
74	127-177	1.0270	1.0601	1.0422
80	178-228	1.0224	1.0558	1.0322
28	229-278	1.0179	1.0515	1.0322
40	279-329	1.0133	1.0472	1.0322
37	330-380	1.0088	1.0429	1.0322

	No.	Size	TL/SL	TL/FL	FL/SL
F50 UT Logan R.	241	-	1.1950	-	-
R65 WY	-	-	1.159	-	-
I8 ID Henry's L.	317	58-909 mm SL	1.1532	1.0306	-
C217 WY Yellowstone	-	239-450 mm TL	1.1461	1.0539	1.0874
males	-	-	1.1480	1.0524	1.0907
females	-	-	1.1413	1.0550	1.0817
P119 UT	-	-	1.139	1.121	-
A54 AB George C.	101	145-498 mm TL	1.131	-	-
M348 ID	-	-	-	1.055	-

I8 reports no trend in length conversion with size, age, or sex.

Much of the cutthroat trout growth data have been published as FL but all data, unless otherwise noted, are given as TL here. Where conversion factors are not given in the specific paper it has been assumed that TL = 1.15 SL = 1.05 FL.

S363

TL	2.3	2.5	2.8	3.2	3.4	3.7	4.0	4.3	4.6	5.3
Wt.	0.1	0.11	0.16	0.29	0.32	0.47	0.61	0.77	0.91	1.29
No.	62	76	11	7	9	8	6	4	2	1

TL	Mean of means	Weight Range	No. of fish	Citations
25-50	-	0.2-1.7	44	C265, W145
51-75	3.1	1.4-4	108	W145, I8, C335
76-101	8.5	4-12	415	W145, I8, C355
102-126	16.4	5-24	616	W145, I8, C335
127-151	25	15-28	527	W145, A54, C335
152-177	43	25-60	385+	C265, W145, A54, I8, C335, M348
178-202	66	43-99	290+	C263, B156, S368, A54, I8, C335, M348
203-228	88	60-120	209+	C263, B156, S368, A54, I8, C335, M348
229-253	133	79-201	78+	M390, S368, A54, B156, C265, M165, I8, C335, M348

TL	Mean of means	Range	Weight No. of fish	Citations
254-278	174	142-210	302+	S368, C265, B156, C335, M348
279-304	236	198-278	1090+	S368, B156, I8, C335, M348
305-329	270	184-340	755	S368, C265, B156, C335
330-355	333	221-482	334+	C265, S368, B156, I8, C335, M348
356-380	414	213-550	301+	C265, S368, B156, C294, I8, A54, C335, M348
381-405	482	261-697	187+	C265, S368, I8, M348
406-431	584	298-848	141+	C265, S368, M165, I8, M348
432-456	791	502-1060	53	C265, S368, I8
458-482	1156	765-1253	60	C265, I8
483-507	1321	765-1514	46	C265, A54, I8
508-532	1747	-	32	I8
533-558	1945	1403-2115	14	H258, I8
559-583	1679	-	1	H258
584-609	1937	1049-2234	9	C265, I8
698	4990	-	1	S146 WY
991	18597	-	1	F40, *S. c. henshawi*

C5 CA	$\log W = -5.272 + 3.086 \log FL$
F50 UT Logan R. 241 trout	$\log W = -4.3620 + 2.8253 \log SL$
I8 ID Henry L. 351 trout	$\log W = -4.635 + 2.9529 \log SL$
L200 OR streams summer	$\log W = -4.8594 + 2.948 \log FL$
winter	$\log W = -4.6303 + 2.827 \log FL$
C335 CO Trapper's L. 426 juveniles	$\log W = -5.188 + 3.10 \log TL$
H258 WY Pathfinder L.	$\log W = -5.978 + 3.00 \log TL$
W145 Yellowstone L. Arnica C.	$\log W = -5.063 + 3.00 \log TL$

		No.	Mean K(SL)	Range
S10 OR		214	1.30	-
M145 AB	May	+	0.89	-
	August	+	1.00	-
F50 UT Logan R.		341	1.819	-
P104 ID	spawners	+	1.35	-
B178 MT Flathead L.		32	1.36	1.16-1.56
I8 ID Henry's L.		351	1.79	-
M390 UT Carter C.	157-205 mm	3	1.43	0.93-2.07
P119 UT	ages 0-V	+	-	1.30-1.80
	ages VI-VII males	+	-	2.0-2.1
	females	+	-	2.15-2.55
C69 OR streams	190-343 mm	+	0.76	0.62-0.92
C69 OR sea-run	211-394 mm	+	1.29	1.14-1.52
G31 OR lakes		35	1.45	1.15-1.93

Males had higher condition factors than females and K decreased with length in the Logan River (F50), but size, age, and sex were not related to K in Henry's Lake (I8).

A79 ID Simons C. and St. Joe R.   15 trout   average K(FL) 1.09

M348 ID Salmon R. 815 trout average K(FL) 1.30

S387 OR Sand C. Sea-run over 279 mm average K(FL) 0.99

			No.			
C335	CO	Trapper's L.	2509	51-381 mm TL average K(TL) 1.52 range 1.0-3.5		
R65	WY	Upper No Name L.	63	K(TL) 1.05		
		Lower No Name L.	44	K(TL) 0.79		
P55	MT	W. Gallatin R.	111	K(TL) 0.99	range 0.72-1.05	
P78	WY		1	K(TL) 1.00		
H258	WY	Pathfinder R.	4 538-559 mm	K(TL) 1.06	range 0.97-1.19	

The coefficients of condition were higher in June and July and decreased as the summer progressed in the Salmon River (M348), and were higher in June than in July and August in Trapper's Lake (C335). Condition decreased as length increased from 102 to 381 mm in Trapper's Lake (C335), but the length-weight regression slope for immature trout from the same population was 3.10. Analysis of variance indicated seasonal differences in length-weight relationships but not differences between streams in three Oregon coastal streams (L200).

I8 ID at hatching 15 mm

M56 9-17 days 0.03-0.06 g

I8 ID 30 days 23 mm

			TL				Weight	
		Mean of	Central				Mean of	
	No.	means	50%	Range	No.	means	Range
Age 0-April							
L200 OR streams	10	38	-	-	-	-	-
Age 0-June							
S77 CA Convict C.	2021	24	-	18-30	-	-	-
G14 WY L. Irene	-	38	-	36-41	+	-	0.2-0.4
L200 OR streams	197	48	-	-53	-	-	-
Age 0-July							
S77 CA Convict C.	135	30	-	24-53	-	-	-
W145, B193 Yellowstone			-				
Arnica C	3027	30	-	20-48	-	-	-
G14 WY L. Irene	-	58	-	48-71	+	-	0.7-2.7
B62 WA hatchery	615	64	-	-	615	2.9	-
Age 0-Aug.							
W145, B193 Yellowstone							
Arnica C.	1172	36	-	20-71	-	-	-
CA streams C265, S77	55	36	-	23-66	-	-	-
OR streams L200	67	53	-	-	-	-	-
Hatcheries I8, R65, B62,							
D83	1731	71	61-79	41-79	1620	-	3.0-5.5
I8 ID streams	88	86	-	-	-	-	-
Lakes G14, P89, I6	9+	135	-	81-201	+	-	3.6-7.1
Age 0-Sept.							
W145, B193 Yellowstone							
Arnica C.	1132	48	-	20-89	-	-	-
L200 OR	299	56	-	-	-	-	-
S77 CA Convict C.	769	53	-	36-94	-	-	-
B62 WA, S10 OR	633	94	-	91-99	633	-	8.6-10.0
G14 WY	+	107	-	104-109	+	-	7.5-8.7
Age 0-Oct.							
C265 CA hatchery	+	-	-	28-43	+	-	0.3-0.7
W145 Yellowstone							
Arnica C.	11	58	-	41-76	-	-	-

	No.	TL Mean of means	TL Central 50%	Range	No.	Weight Mean of means	Range
Age 0-Nov.							
L200 OR streams	53	61	-	-79	-	-	-
V57 BC	88	89	-	53-122	-	-	-
D83 WA 3 stocks	493	163	-	145-173	493	50	30-65
P89 OR Devil's L.	7	201	-	-	-	-	-
Age 0							
P119 UT Strawberry L.	9	104	-	-	9	14	-
M113 AT N. Pork R.	+	107	-	-	-	-	-
Age I							
C265 CA Blue L.							
(*S. c. henshawi*)	10	64	-	51-76	-	-	-
C223 CO *S. c. stomias*	+	107	-	-	-	-	-
Hatcheries B154, S126,							
L18, M207	+	127	-	76-160	+	26	8-37
Other U.S. streams P119,							
F50, B156, I8, M348, A79	371	132	124-147	89-152	11	20	-
OR streams S387, S10,							
S126, O24, O22, L200,							
L201	1973+	152	107-221	79-246	261	113	102-125
Can. streams A54(5), M113	31+	183	163-188	140-254	25+	60	43-133
Other U.S. lakes P119,							
S368, E58, I6, C335, D185	152	206	142-254	43-312	95	91	1-235
OR lakes B236(2), J48, L34,							
P89, M456	164+	216	198-254	160-315	84	65	34-108
BC lakes M165 (4)	10	224	-	206-246	10	122	105-145
D83 WA 6 stocks	940	226	211-249	198-259	940	116	65-187
Combined	3641+	180	127-226	43-315	1331+	88	1-235
Age II							
C223 CO *S. c. stomias*	-	150	-	-	-	-	-
OR streams S126, O22,							
O23, O24, S387, B154,							
L200, L201	1118+	168	140-206	79-424	42	111	57-255
Other U.S. streams I8,							
P119, D66, O23, O24,							
S387, L200, L201, B154	390	180	140-221	102-221	16	26	-
Hatcheries M207, L18,							
L39, C69	6+	208	-	127-274	201	74	-
C265 CA Blue L.	17	211	-	157-264	-	-	-
Can. streams M145, N49,							
M113	351+	211	173-229	132-312	325+	102	57-255
Other U.S. lakes B154,							
B178, E58, L69, B156,							
S368, B249(3), P119,							
C335, D185, M456	706	274	264-340	135-358	308	198	31-369
Can. lakes M165(4), N49,							
G95	26+	302	272-343	267-455	22	275	201-354
D83 WA 4 stocks	97	328	315-328	302-391	97	304	290-321
Combined	2711+	236	190-277	79-455	1011+	176	26-369
Age III							
C223 CO *S. c. stomias*	-	201	-	-	-	-	-
OR streams S387, O22,							
O23, O24, L200, L201	594	201	183-196	107-414	-	-	-
Other U.S. streams P119,							
D66, F50, B156, M348,							
A79	415	239	201-269	140-368	12	74	-
Hatcheries M207, L39, L18	-	241	-	203-305	589	130	-
Can. streams M113, M145,							
M207, A54	280+	297	201-361	175-401	330+	145	48-482
Other U.S. lakes I6, L69(2),							
D83, B156, P119, C335,							
D185, M456	1600+	338	302-351	221-460	1382	372	119-689
Can. lakes M165(4), G95	22	391	343-406	325-508	19	783	383-1775
Combined	2911+	292	229-343	107-508	2332+	360	48-1775

	No.	TL Mean of means	TL Central 50%	Range	No.	Weight Mean of means	Range
Age IV							
C223 CO *S. c. stomias*	-	236	-	-	-	-	-
OR streams O22, O23, O24, S387, L200, L201	262	264	224-343	132-493	-	-	-
Other U.S. streams B156, P119, D66, A79, M348	162	282	216-325	175-386	2	116	-
Hatcheries M207, L39, L18	-	305	-	279-381	102	170	-
U.S. lakes B156, S368, E58, L69(2), I6, P119, C335, D185, M456	1172+	366	305-414	257-503	589	548	250-1018
Can. lakes M165, G95	6	411	-	386-462	2	581	-
Combined	1602	320	246-363	132-503	695	445	116-1018
Age V							
ID streams B156, M348	32	353	-	267-373	-	-	-
OR streams S387, O23, O24, S126, L201	201	338	279-361	185-439	-	-	-
Can. streams M113, A54	1+	378	-	254-498	1+	592	170-1106
M165 BC Battle L.	1	429	-	-	1	601	-
U.S. lakes B156, L69(2), S368, I6, P119, C335, D185	157+	457	384-551	345-587	66	1253	527-2197
Combined	392+	406	345-429	185-587	68+	1066	170-2197
Age VI							
Streams O22, O23, D66, M114	81+	404	-	279-493	+	609	-
Lakes B156, L69, I6, P119	16+	480	-	384-663	8	2364	-4020
Age VII							
O22, O23 OR Sand C.	10	421	-	361-467	-	-	-
M113 AT Spray R.	-	434	-	-	-	851	-
P119 UT Strawberry L.	1	734	-	-	-	6810	-
Age VIII							
O24 OR Sand C.	1	442	-	-	-	-	-
Age IX							
O24 OR Sand C.	1	442	-	-	-	-	-
Age X							
O24 OR Sand C.	1	450	-	-	-	-	-
Age 1-6 years freshwater plus 1 summer sea life							
O22, S387 OR	288	380	-	323-424	-	-	-
Age 2-5 years freshwater plus 2 years sea life							
O22, S387 OR	112	399	-	386-414	-	-	-
Age 2-5 years freshwater plus 3 years sea life							
O22, S387 OR	19	442	-	429-460	-	-	-
Age 2-6 years freshwater plus 4 years sea life							
O22, S287 OR	3	505	-	442-627	-	-	-

Average calculated TL at each annulus							
Streams	No.	1	2	3	4	5	6
B156 ID tributaries to Priest L.							
poorest	-	64	102	114	165	-	-
mean	294	86	127	170	201	254	-
best	-	104	183	262	259	-	-
S275 MT Flint C.	22	66	119	175	-	-	-
D66 CA Prairie C.	9	69	117	175	229	297	358
S387, O24 OR Sand C.*	376	69	112	173	211	249	251
A79 OR upper tributaries	118	53	102	152	224	-	-
lower tributaries	195	71	135	226	292	-	-
adfluvial	90	71	140	216	-	-	-

Average calculated TL at each annulus (cont.)

Streams	No.	1	2	3	4	5	6
P119 UT tributaries to Strawberry	41	89	135	183	216	-	-
P159 MT streams	2323	74	132	198	279	330	302
F50 UT Logan R.	234	99	170	231	-	-	-
M348 ID Salmon R.	464	107	150	213	279	330	394
E58 MT Thompson R.	41	130	198	262	318	-	-
P55 MT W. Gallatin R. at 6,000'	91	104	175	257	274	-	-
at 5,000' elev.	7	119	185	274	-	-	-
at 4,000' elev.	15	104	185	310	-	-	-
Mean of Means	4320	86	145	211	251	292	323

*These Sand Creek data do not include any sea life fish, which are given below:

year of sea life	1	2	3	4
mean TL	343	394	432	455

Average calculated TL at each annulus

Lakes	No.	1	2	3	4	5	6	7
R65 WY Lower No Name L.	64	102	145	190	221	231	-	-
B156, B238 ID Priest L.	90	81	135	211	287	348	371	-
Upper Priest L.	92	94	142	216	292	338	391	-
S237 UT White Rock and Ted's L.	22	130	185	201	221	-	-	-
F137 CO Granby Res.	+	109	196	251	290	-	-	-
B262 Yellowstone L.	5057	46	130	224	312	394	442	486
M453, P159 MT	2158	76	163	241	307	384	363	368
S237 UT Island L.	61	157	211	249	300	343	-	-
M152 UT Bear L. 1951-1952	108	-	150	257	363	541	635	729
1953-1955	39	38	163	264	358	434	498	513
E58 MT Thompson L.	41	130	198	262	318	-	-	-
R65 WY Upper No Name L.	75	112	178	274	381	421	478	-
C265 CA Blue L.	419	66	180	307	378	361	-	-
C265 CA Heenan L.	117	97	216	330	445	-	-	-
I6 ID Henry's L.	356	170	325	437	503	551	594	-
P119 UT Strawberry Res.	166	132	241	378	478	572	650	719
Mean of means	8843	102	185	269	340	406	490	564

Average calculated weight at each annulus

	1	2	3	4	5	6
I6 ID Henry's L.	62	408	958	1455	1902	2356
increment	62	346	578	454	380	196
no.	356	331	237	71	7	2

Average annual increment, TL

	1	2	3	4	5	6	7
B156, B238 ID Priest tributaries	81	53	48	48	-	-	-
No.	90	90	56	5	-	-	-
Priest L.	-	-	124	94	56	43	-
No.	-	-	33	53	17	2	-
U. Priest tributaries	94	46	43	-	-	-	-
No.	92	86	51	-	-	-	-

Average annual increment, TL (cont.)

		1	2	3	4	5	6	7
U. Priest L.		-	117	124	94	58	23	-
	No.	-	6	36	62	15	2	-
B156 ID streams		86	43	41	36	36	-	-
	No.	294	273	205	67	2	-	-
I6 ID Henry's L.		170	168	135	74	38	18	-
	No.	356	331	237	71	7	2	-
B262 Yellowstone L.		46	69	94	84	79	53	-
	No.	6417	6417	6226	4695	2088	248	-
E58 MT Middle Thompson L.		130	71	66	56	-	-	-
	No.	41	35	17	7	-	-	-
P119 UT tributaries		89	51	51	43	-	-	-
	No.	41	30	14	2	-	-	-
P119 UT Strawberry Res.		132	109	109	94	84	69	64
	No.	157	142	119	59	15	4	1
C265 CA Blue L.		66	114	124	66	28	-	-
	No.	419	419	419	402	32	-	-
C265 CA Hennan L.		97	127	114	114	-	-	-
	No.	55	117	117	117	-	-	-
F50 UT Logan R.		99	81	81	-	-	-	-
	No.	234	82	9	-	-	-	-
M348 ID Salmon R.		107	43	66	64	64	58	-
	No.	464	459	294	133	33	2	-

Annual increment as percentage of length at beginning of year

TL	UT Logan R. F50	UT streams P119	ID streams B156	MT lake E58	UT P119	Yellow-stone R. B262	Priest tributaries B156	Priest lakes B156	Range
25-50	-	-	-	-	-	134	-	-	95-324
51-75	-	-	49	-	-	172	-	-	40-226
76-101	90	64	46	-	-	106	55	114	31-130
102-126	-	40	28	57	114	89	40	109	20-124
127-151	54	37	33	49	117	64	35	86	16-122
152-177	-	25	22	-	61	44	28	63	15-77
178-202	-	-	22	33	-	42	-	57	18-61
203-228	-	-	28	-	-	42	-	44	27-51
229-253	-	-	-	-	49	35	-	25	25-43
254-278	-	-	3	21	42	30	-	22	3-42
279-304	-	-	-	-	47	25	-	21	18-56
305-329	-	-	-	-	30	17	-	15	13-32
330-355	-	-	-	-	-	15	-	-	13-16
356-380	-	-	-	-	-	15	-	6	6-18
381-405	-	-	-	-	25	-	-	-	-
406-431	-	-	-	-	-	-	-	-	-
432-456	-	-	-	-	18	-	-	-	-
457-482	-	-	-	-	21	-	-	-	-
483-507	-	-	-	-	14	-	-	-	-
508-532	-	-	-	-	15	-	-	-	-
533-558	-	-	-	-	16	-	-	-	-
584-609	-	-	-	-	8	-	-	-	-
635-659	-	-	-	-	9	-	-	-	-

While many of the papers computed growth by direct proportion, several used the Fraser modification, a straight-line relationship with intercepts of -16.7 mm SL (Strawberry Reservoir, P119), 1.4 mm SL (tributaries, P119), 50 mm SL (M348), 36 mm TL (Priest L., B156) and 43 mm TL (Wyoming, R65) and curvilinear body-scale relationships were found by I6, A79, B262, and F50.

In Yellowstone Lake, the different body-scale regression slopes found for various groups of juvenile cutthroats suggested different populations (L192). Scale measurements were made along the radius at 20° ventral to horizontal in the anterior direction (A79) but at the horizontal in most studies.

Scales first form at 43 mm TL on the caudal peduncle along the lateral line (A79, C265) and trout were fully scaled at 64 mm TL (C265). Scales formed at 25 mm TL in Henry's Lake, Idaho (I8), first at 41-43 mm in Yellowstone (L69) and at 43-66 mm TL in Montana (B173).

Annuli form in May (A79, I8) or in the first 2 weeks of October (F50). In Forest Canyon, Colorado, *Salmo clarki stomias* were found to form no scales until after the first winter (C223). This was also true of cutthroat trout in Montana (B173). In Logan River, Utah, all formed scales by the first winter but some did not form annuli (F50). Failure of some cutthroat trout to form annuli the first winter was also noted by L69, B156, A79, M348, D185, and R65. In Priest Lake (B156) the number not forming annuli was believed to be low because of an early spawning season. In mountain lakes in Utah (L191), scales with 8 or more circuli before the first annuli were designated as having failed to form the first annulus. The number of fish forming first annuli increased with the average growth rate index (correlation coefficient 0.79) (L191). Over 11 circuli before the first annulus was an indication of a missing annulus in Idaho (M348).

Tagged age IV fish grew 38.3 mm in a year compared to 35.3 mm computed from their scales and age V trout grew 95 mm compared to 107 mm as computed from their scales (I8).

Lee's phenomenon and growth compensation were found in data from Yellowstone Lake (B262).

Size selectivity of electrofishing was believed to have biased upward the estimates of mean lengths at each age in L200.

No sex differences in growth were detected by B262, S368, D185, but males grew faster than females in Henry's Lake (I8). Males appeared to be longer-lived (B262).

Growth in streams is generally slower than in lakes, as shown in the above tabulations. In Alberta streams, larger streams had larger, faster-growing fish (A54) and growth was faster in warm water than colder streams.

Stream growth of fish caught in lakes could be detected on the scales because of the slower growth in streams. In Priest and Upper Priest Lakes, cutthroat trout remained 2 to 4 years in streams and size at later years was inversely related to the years of stream life (B156, B238). A similar situation was noted in Strawberry Reservoir, Utah (P119). The faster-growing fish leave the streams for the lake at a younger age than slower-growing individuals (B156). The longer the trout remained in the stream, the smaller was the increment the first season in the lake (B156).

In Kiakho L., British Columbia, differences in stream and lake growth were shown as 2 stanzas of growth when the Parker equation was applied (L150). Sea run trout also show the difference between freshwater and sea growth. The sea trout are not necessarily larger than the lake varieties, however, since they usually spend rather short periods in the sea.

Racial differences in growth rates were demonstrated by D83 with hybrids

between the races showing intermediate growth. The Yellowstone River cut-
throat, *Salmo clarki lewisi*, was found to be faster growing than the Colorado
River cutthroat, *S. c. pleuriticus* in Trapper's Lake, Colorado (S368). At age
IV the former averaged 358 mm TL compared to 318 mm TL.

Slow growth in some streams was related to sparse bottom fauna (A54).
More bottom fauna was associated with the better growth of cutthroat trout in
Upper No Name Lake than in Lower (R65) but in this case the Lower Lake also
had the greater population density of trout, about 49 kg per hectare, due probably to better spawning facilities. This population difference occurred even
though migration between the lakes was possible at most water levels. The
Upper Lake had a bit more basic fertility which may have contributed to the
greater bottom fauna, but fish foraging on the fauna probably was also important. The Lower Lake cutthroats were in poorer condition in addition to being
slow-growing.

Slow growth after cutthroat trout reached 279 mm in Trapper's Lake was
believed to be associated with the fact that the trout continued to feed largely
on plankton (C335).

No correlation could be shown between summer air temperatures and
growth increments of Yellowstone Lake cutthroats (B262). Also it was noted
that there were little differences in growth in various areas of the lake despite
some size differences in the runs to different streams. Growth in the first 3
years apparently was controlled by different factors than growth in later years
of life (B262).

In Blue Lake, California, growth of age III cutthroats was greatest in
spring (C265), slowing in midsummer. In Trapper's Lake, Colorado, maximum growth was in midsummer (C335). Immature cutthroats grew in length
more rapidly than in weight from June through August, resulting in a decline in
condition coefficients (C335). In Oregon streams growth was maximum in
early April and minimum in early December (L200). Fingerlings showed little
growth before water temperatures rose in early April (C336). Growth was
slower in the upper tributaries than in the lower tributaries in Idaho probably
because of the lower water temperatures at high altitudes (A79). Also in Montana growth was more rapid at lower elevations (P55). In hatcheries growth
may continue through the winter as indicated by the average sizes in August,
November, and the next April of 5, 65, and 189 g (D83). Hatchery growth, with
a good diet, was also better than the same stock in Whatcom Lake, Washington
(D83). In Idaho an average size of 76-102 mm at 16 months was cited as evidence of slower growth in hatcheries than in nature (B156). After spawning the
first time, cutthroat trout showed no further growth in Blue Lake (C265) and
the larger spawners appeared to have higher mortality rates. In Priest Lake
few live to spawn a second time (B156).

In Yellowstone Lake, growth rate changes were caused primarily by differences in population density (B316). The growth rate increased as the fishing pressure and harvest increased (B262). The percentage of older fish in
the catch decreased. Increases in growth were evident only in ages IV-VI
which were most heavily exploited (B316). In Priest Lake, cutthroat averaged
282 mm TL in the creel compared to an average of 338 mm TL in Yellowstone
Lake (before fishing pressure became heavy) where the fish mature at the
same size (B156). Fishing pressure probably was a factor in causing the
smaller size caught in Priest Lake.

Competition with introduced brook trout and possibly with other species is
believed to be the most important factor in the decline of cutthroats in Priest
Lake (B156) but spawn taking, poaching, and fishing pressure may also have

had an effect. Stocking of fingerlings seemed to have had little effect. Cope
(C218) also listed competition, plus hybridization with rainbow trout, as factors threatening cutthroats in Utah. In Yellowstone Lake, the greatest year
classes resulted when the largest downstream migrations occurred in July and
August. Such migrations result when water levels are low in the streams
(B193). No relation of year class strength to brood stock or to predation could
be found (B193). B271 however reported no correlation with time of run, but
found that strong year classes resulted during years with low water levels in
Yellowstone Lake:

   log recruitment = 5.8083 - 0.4299 water level in feet
      (in thousands)

No relation of year class strength was found with brood stock (B193, B271,
D185), with predation (B193), or with summer air temperatures (B271).

   In Trapper's Lake (D185, D186) the number of fry recruited to the lake increased with stream flow up to a point after which it decreased with increased
stream flow. Snow-water content could be used as a measure of predicting
stream flow and showed a significant inverse correlation with fry recruitment
above the point where recruitment was highest. In general, fry recruitment
was correlated with strength of the year class at spawning and at harvest size.

   Cutthroat trout spawn from May to August in Arnica Creek, Yellowstone
(*S. clarki lewisi*) (W145, B259), from June 1 to mid-July in Sheep River, Alberta (A54); from April 14-August 16 in Logan River, Utah (F50); from April
25-May 24 in Nevada (F50); late April and early May in Thompson Lake, Montana (E58), and in Idaho (A79); from April to mid-June in Priest Lakes, Idaho
(B156); at the time ice cover goes off the lake at Henry's Lake, Idaho (I8); July
1-15 in Forest Canyon, Colorado (*S. c. stomias*) (C223); April to early July,
California (*S. clarki henshawi*) (C294, M459). Sea run cutthroats migrate from
the ocean October 1-February 1 and spawn in January-February and February-
May in British Columbia (C294). Cope (C220) reports night upstream migration (unusual in salmonids) of cutthroats in Arnica Creek, and explained it on
the basis of freshets in late evening due to snow melting. Negative phototro-
pism may explain the usual nighttime downstream migration of fry. In Colorado spawning runs were also in late afternoon and evening, in response to
slackening of the freshets (S368).

   Males spent 12-35 days upstream to spawn (average 17 days) while females remained upstream 6-21 days (average 7 days) in Arnica Creek, Yellowstone (B259). Artificially stripped fish remained in the streams 6-9 days.
In an Oregon stream, a female trapped going up and downstream had spawned
within one week (W101). Females lost 94 g and males lost 79 g in the spawning
and migration in Arnica Creek (B259). A strong tendency for homing was
demonstrated by Platts (P104). Cope (C219) reported that early spawners
spawned in the headwaters and later spawners downstream, but he believed
that this was due to heredity and not due to crowding.

   Apparently cutthroat trout spawn only in streams, on gravel. The male is
attentive and drives away other males while the female builds the nest (C294).

   Males mature at age II and most females at age V but some at III-VI, at
305-406 mm TL, while in the tributary streams most matured at age IV, 152-
254 mm TL (B156). In Oregon males as small as 102 mm FL and females
190 mm FL were mature (W101). In the St. Joe River system, Idaho, some age
II females of 163 mm FL were maturing and would spawn the next spring (A79),
but in the Salmon River they first matured at age V (M348).

   A bisexual cutthroat was reported (T109) and several were found in another population (V84) where fecundity was low.

While mortality following spawning is heavy in many populations (P104, B156) there may be considerable second and third spawning (C5, C265, B259). Some sea run cutthroat spawn as many as 3 times in their lives (S387). In the spawning runs of Arnica Creek, Yellowstone, 10 to 26% were alternate year spawners, 46% skipped two years and 1-15% spawned each year after reaching maturity (B259). In this stream in 1950 and 1951, 28% died at the spawning sites and about 1% more before leaving the stream (W145). Males and females, tagged and untagged, showed similar mortality. At Trapper's Lake, Colorado, the river cutthroat, *S. c. pleuriticus,* seemed to spawn every year in contrast to the Yellowstone cutthroat (S368). Mortality during migration and spawning varied from 12.8 to 62.7% at Blue Lake, California (C265). Spawning mortality in Yellowstone Lake was estimated at 48%, fishing mortality at 11.6%, and natural mortality of mature fish at 61.4% (B259). Eggs in gravel in Arnica Creek showed a 60-75% mortality to hatching at 36-40 days. About 0.4% of the eggs produce fingerlings to leave the creek and by September there are 4.54 young trout per spawning female (B259). To maintain the stock 1.67 of these must survive to spawn.

In 3 other tributaries to Yellowstone Lake, eggs in natural redds had 12, 13, 17, 29, and 42% mortality while eggs in hatching boxes showed 44, 45, and 46% mortality (M457). Most mortality occurred in the pre-eyed stage. Failure of fertilization was 1.3%.

A 98% mortality from egg to September fingerling was found in Convict Creek, California (S77). In Trapper's Lake, survival from potential eggs to downstream fry was 0.5 to 12.2% (S368), and in Convict Creek, 0-6% (S77). Mortality from time of fertilization to hatching was estimated at 46% and failure of fertilization at 1% at Trapper's Lake (S368).

	TL	Wt.	Age	No. of females	No. of eggs Mean	Range
C69 OR II year hatchery fish	203-254	-	-	6	373	226-467
Wild fish	239-305	-	-	5	480	380-605
Sea-run fish	320-404	-	-	2	1170	1098-1242
L18 III year fish	-	-	-	-	-	500-800
IV year fish	-	-	-	-	-	800-1200
older fish	-	-	-	-	-	950-2000
S146 WY	-	2041	-	3	3727	3120-4420
	-	2495	-	1	5355	-
	-	3402	-	1	6500	-
A79 ID St. Joe R. system	163-165	-	II	3	550	408-661
	190	-	III	1	204	-
	295-315	-	IV	2	801	751-852
	373-394	-	IV	3	1180	936-3539
I8 ID Henry's L. stripping	-	-	-	4692	2703	-
I6, I8 ID Henry's L.	366	567	-	10	1577	-
	472	1162	-	10	1914	-
	594	2410	-	10	2930	-
W101 OR	201	-	-	1	80	-
	241	-	-	1	157	-
S368 CO Trapper's L.	249	179	-	1	546	-
	259-274	173-252	-	4	627	551-700
	284-302	233-269	-	6	698	500-873
	307-323	255-323	-	5	686	490-885
W145 Arnica C. Yellowstone	284-302	-	-	5	918	745-1103
S. c. lewisi	305-329	-	-	6	1051	910-1241
	330-355	-	-	9	1071	910-1250

W145 Arnica C. Yellowstone (cont.)	TL	Wt.	Age	No. of females	No. of eggs Mean	Range
	356-380	-	-	6	1136	990-1342
	381-405	-	-	8	1148	846-1278
	406-431	-	-	6	1231	966-1344
C265 CA Blue L.	305-315	1st spawners		5	739	470-990
S. c. henshawi	325-340	1st spawners		17	904	415-1220
	325-340	2nd spawners		2	936	920-952
	345-371	1st spawners		12	1010	780-1240
	345-371	2nd spawners		10	811	560-1080
	376-394	1st spawners		10	1291	1120-1537
	376-394	2nd spawners		12	862	570-1370
	401-421	1st spawners		9	1523	1190-1670
	401-421	2nd spawners		11	955	550-1280
	432	1st spawners		2	1893	1690-2095
	432	2nd spawners		3	1036	678-1220
L162 Yellowstone R. Stripping	-	-		104	1113	-
S. c. lewisi	292-368	1st spawners		29	-	400-1200
	333-368	2nd spawners		8	-	550-1050
	373-434	1st spawners		26	-	800-2100
	373-434	2nd spawners		30	-	540-1400
S77 CA Heenan L. stripping S. c. henshawi	-	-		844	2806	-

R179 46.9 eggs per cm of FL, or 1,548 eggs per Kg of female
mean egg diameter 4.3-5.1 mm

The estimate of eggs per female based just on stripping appears to be too high (S77, I8) unless only large females were stripped. In 50 females examined after natural spawning, there were 3 eggs per female (range 0-17) (W145). The number of eggs produced during second and later spawnings is apparently lower than that produced the first season (C265, C5). Eggs hatch in 28-40 days (C294), or 30 days at 10C with 2 days longer for each 0.55 degree below 10C (S368).

In streams food is mostly insects (F50, C294, B156, I8, L200). In the Idaho streams (B156) flying ants comprised a high percentage of the food. Snails were important as food in the Salmon River (M348). In Arnica Creek, Yellowstone, trout eggs and fry comprised an important part of the food. At 80 mm some ate fry 25 mm long. Of 50 trout 76-254 mm long, 39 had eaten trout fry of 30 mm, and 36 had eaten trout eggs (up to 36 in one stomach) (W145).

Fry began feeding 14 days after hatching and all were feeding by 23 days (B78). In British Columbia streams (I11) cutthroat trout fed mostly on insects principally Trichoptera, up to 127 mm FL, on insects and sticklebacks, Gasterosteus, up to 305 mm, and then mostly fish. Frogs, earthworms, small fish, and insects were the major foods of cutthroat trout in Deer Creek, Oregon (L200).

In Heenan Lake, California (C265), cutthroat trout ate mostly planktonic crustacea, chironomids, and a few scuds. In Blue Lake, chironomids and other insects comprised a greater share of the food and a few minnows and small trout were eaten (C265). In the No Name Lakes, Wyoming, insects comprised the major food and the average volume of food in 203-254 mm TL trout was 0.98 cc in the Upper Lake compared to 0.3 cc in the Lower Lake (R65). Terrestrial insects comprised 33% of the food in Lower Lake compared to 19% in the Upper. As mentioned earlier, growth was faster in Upper Lake. Insects, particularly zygoptera, and amphipods were the major foods in Henry's Lake, Idaho (I8). In Thompson Lake, Montana, 40% of the trout had eaten small perch (average 7.5 perch per stomach). The smallest trout to have eaten perch was 165 mm TL (E58). Fish were the most important food of cutthroats in Bear Lake, Utah (M152). Insects, adult and immature, *Daphnia* and *Gammarus* were the principal foods. Chironomids, *Daphnia* and *Gammarus* were the important foods in Fish Lake, Utah (H287), in Idaho lakes (J53) and in Yellowstone Lake (B261). The larger trout ate bottom fauna and a few trout in Yellowstone while the trout under 305 mm TL included plankton in their food. The bottom fauna was more abundant in areas of heavy fishing pressure suggesting that the trout are caught off before depleting the bottom fauna and that other trout then move into the areas with greater bottom fauna (B196, B261). Trout 152-203 mm TL were found in deeper water in Yellowstone Lake than the larger trout, but cutthroats were rarely deeper than 20 m (B261).

Plankton was the principal food in Munsel Lake, Oregon (C336) except in the early summer when insects were eaten most frequently. Aquatic insects were the major food in Mann Lake, Oregon (B317). *Daphnia* and amphipods were the most important foods in Trapper's Lake, Colorado, but terrestrial and aquatic insects were also taken (C335). In British Columbia lakes (I11) cutthroats ate few fish until they were over 432 mm FL.

Increased fishing pressure in Trapper's Lake resulted in increased total harvest but decreases in the numbers of trout over 356 mm and over age IV, decreases in catch per man hour, and reduction in spawning potential (D185). The annual yield from Yellowstone Lake almost doubled from 1950 to 1959, catch per man hour varied but did not show a trend, growth rate of the catchable-size fish increased, and spawning runs were reduced particularly in areas of heavy fishing (B316). A maximum equilibrium yield was estimated.

Tagged trout, of spawning size, showed 170% higher mortality than did fin-clipped trout (B316). In Idaho, 189 of 253 tagged cutthroat trout were later recaptured over a mile away, averaging 30.5 kilometers (B284), and in another study tagged cutthroats were recaptured at distances up to 130 kilometers (M348).

Biomass growth was determined on three Oregon streams (L200).

An extensive bibliography on cutthroat trout has been published (C337). Summaries and more extensive papers on cutthroat trout include C222, B262.

RAINBOW TROUT, *Salmo gairdneri* Richardson

Although several subspecies have been described, all are included here. In some cases, but not all, subspecies are indicated. Few subspecific criteria are now valid because of mixing of stocks (N101, M455).

Rainbow trout which migrate to sea are usually referred to as steelhead trout. These usually grow larger than the freshwater forms and may differ in other life history details, but are included in this general summary. Originally rainbow trout were largely restricted to the Pacific drainage from latitude 24° N in Mexico to the Kuskokwim River, just south of the Yukon River, in Alaska (N101). They have been introduced in many waters throughout the United States and Canada and in Australia, New Zealand, Chile, Peru, South Africa, Japan, south Asia, and Europe. The origins of some stocks so introduced are traced (D187). Some were also in the Great Basin and in the upper Athabasca River in British Columbia (M462).

Length conversions

N46 NY	-	-	TL = 1.149 SL	-
S234 UT	396 fish	175-516 mm SL	TL = 1.1626 SL	-
	203 fish	-	TL = 1.0466 FL	FL = 1.114 SL
R134 AT	-	120-480 mm SL	-	FL = 1.105 SL
K16 VA	-	-	TL = 1.071 FL	FL = 1.083 SL
E130 NY	138 fish	-	TL = 1.059 FL -0.85 mm	

022 OR	No.	FL	FL-TL normal	FL-TL compressed	TL normal TL compressed
	55	76-126	1.0377	1.0684	1.0296
	100	127-177	1.0391	1.0730	1.0326

For those data which are published in SL or FL, without conversion factors in the paper, the following conversion factors were used
TL = 1.145   SL = 1.071 FL

LENGTH-WEIGHT SUMMARY FOR RAINBOW TROUT

The length-weight table contains data from the following references: B171,249,279,319; C20,25,66; E105; F40,42,46; G25,36,104,114; H55,116,161, 166,238,258; K15; L101,126; M61,67,78,98,108,113,136,340; N27,46,49; O24; P2,31,39-41,45,46,49-51,53,54,92-99,128; R66,84,87,98,134,159,181; S28,62, 80,82,178,234; T27; V55; W2,52,132,134.

The weight from fish over 508 mm may be somewhat biased by the inclusion of mature females from a selected stock of hatchery fish (B279). Specimens up to 23.59 kg are reported from Jewel Lake, British Columbia (M462).

The lengths given are not class centers but the lower limits of each class.

SL	FL	TL	Mean of means	Weight central 50%	Range	No. of fish	References for extremes	No. of references weighted
16	18	19	0.082	-	0.079-0.085	+	F46, S28	4
22	23	25	0.26	0.14-.34	0.14-0.43	638+	N46, L101	15
33	36	38	0.60	0.57-0.62	0.43-1.4	422+	L101	17
44	48	51	1.45	1.1-1.7	0.8-2.3	264+	M98, L101	19
55	59	64	2.33	1.87-2.98	1.08-3.7	40+	T27, L101	17
66	71	76	4.0	3.06-4.5	3.06-6.0	198+	W2, B319	11
77	84	89	8.0	-	7.3-9.4	+	N46, B319	8
89	94	102	11.6	-	9.1-13.0	+	B319	6
99	107	114	15.9	-	12.5-18.1	+	L101, B319	10
110	119	127	29	-	20-43	2+	B319, P125	12
132	142	152	47	43-51	28-68	38+	K15	16
155	165	178	79	71-91	43-136	49+	K15, P125	26
177	190	203	99	94-113	45-198	59+	W132, P125	32
199	213	229	149	142-156	85-227	101+	B181, C20	35
221	236	254	204	184-216	113-312	137+	K15, P125	40
243	262	279	243	224-252	142-454	171+	P99, P125	41
265	284	305	309	298-329	162-482	213+	S80, P125	40
287	307	330	394	386-451	213-680	202+	S80, P99	40
309	333	356	493	454-536	284-794	172+	S80, M180	44
331	356	381	604	542-643	411-1106	141+	M78, H161	37
353	378	406	760	709-803	454-1021	121+	B279, P92	38
375	404	432	930	865-1001	510-1474	124+	K14, P93	33
398	427	457	1060	907-1137	510-1451	92+	B279, H166	33
420	450	483	1321	1264-1298	794-1956	87+	B279, P92	27

SL	FL	TL	Means of means	Weight central 50%	Range	No. of fish	References for extremes	No. of references weighted
442	475	508	1483	1335-1497	907-2608	73+	B279, P97	24
464	498	533	1650	1537-1899	1021-2240	38+	O24, G114	20
486	521	559	2079	1814-2586	1531-3628	37+	P99, B279	20
508	546	584	2328	2126-2523	1474-3671 1021*	43+	B279, P98	19
530	569	610	2447	2268-2331	1814-3671	40+	B279, P94	17
552	592	635	2849	2662-3410	1814-3856 1474*	25+	B279, O24	13
574	617	660	3308	2835-3890	2155-4990 1588-1814*	23+	O24, B279	15
597	640	686	3827	3447-4315	2523-5585	17+	P41, B279	11
619	663	711	3830	3076-4842	2722-5670	16+	O24, B279	11
641	688	737	4354	3164-5525	1588-6336	26+	O24, S234	14
663	711	762	4868	3493-6265	3175-7768	24	O24, B279	11
685	734	787	5001	-	2410-7711	9	O24, B279	6
707	759	813	5279	-	4082-8165	8	O24, S82, W52	5
729	782	838	6021	-	4536-8250	4	O24, B279	4
-	914	-	10659	-	-	1	R84 OR	-
-	-	1029	16782	-	-	1	G36 WY	-
-	1219	-	21772	-	-	1	R87 B.C.	-

*Note: R159 listed 4 "sick" fish, very low in weight. These fish while listed here were not put in the averages.

N67 HI (with FL in cm) $W = 6.052 + (1.1403)$ FL

S234 UT $\log W = -5.608 + 3.340 \log SL$

S418 BC Paul L.   Weight of 349 females increased as 3.021 power of FL ($r = 0.970$) This varied during the season from 2.716 to 3.197 power of FL

D84 WA Selected stock   males in January $\log W = -3.341 + 2.589 \log FL$

mature females $\log W = -4.068 + 2.828 \log FL$

V55, H287 CA Sacramento R.

484 steelhead 325-691 mm FL $\log W = -5.057 + 3.063 \log FL$

M462 CA Beardsley Res. $\log W = -6.344 + 3.081 \log FL$

Castle L. $\log W = -5.204 + 2.604 \log FL$

Pine Flat L. $\log W = -6.700 + 3.233 \log FL$

H258 WY Pathfinder L. 121 trout, to 795 mm TL $\log W = -6.303 + 2.764 \log TL$

No sex difference was noted (H258) but the 2 fish larger than 673 mm were both females.

Condition factors

	SL	No.	Mean K(SL)	Range
H43 UT Fish L.	-	20	1.41	-
S10 OR	-	251	1.46	-
N46 NY	25-179	-	-	1.22-1.60
M370 CO hatchery	188-231	10	1.28	1.17-1.40
M370 CO wild	168-282	10	1.51	1.1 -1.87
M340 UT Green R.	135-309	10	1.48	1.19-1.68
R151 CA Convict C.	81-326	124	1.51	-
C23 MN	128-275	9	1.57	-
E3 MN	203-532	48	1.60	-
W134 OR Crater L. streams	121-264	10	1.63	1.43-1.84

Condition factors (cont.)		SL	No.	Mean K(SL)	Range
S234 UT Fish L.		175-647	249	1.65	1.46-2.44
M67 CA Colorado R.		-	-	1.66	-
K15 VA		-	-	1.75	1.5 -2.8
P68 TN Dale Hollow		152-330	2025	1.76	-

		FL	No.	Mean K(FL)	Range
W1 CA Castle L.		130-370	273	1.05	-
W2 CA Castle L.		-	400	1.07	-
S122 Germany		-	-	1.1	-
N23 CA Convict C.		-	-	1.12	-
K16 VA Big Spring		-	708	1.18	-
V22 CA June L.		-	201	1.32	-
R159 New Zealand "sick" fish		559-622 mm	4	0.66	0.58-0.75
K16 VA Big Spring		Over 356 mm	19	0.98	-
K16 VA St. Mary's R.		-	99	1.05	-
K16 VA North C.		-	245	1.11	-
N27 CA Arrowhead L.		-	28	1.12	0.91-1.39
H2 BC Vancouver I.		-	+	1.13	1.05-1.39
K16 VA Big Spring		-	708	1.18	-
L206 New South Wales lake		-	-	1.05	-
After fertilizing		-	-	1.33	-
S105 VA Big Spring		-	427	1.18	-
P176 N.Z. L. Lynden 1948-59		254-304	-	-	1.13-1.30
		305-355	-	-	1.08-1.25
		356-405	-	-	0.91-1.19
		over 406	-	-	0.80-1.14
	1931-38	over 406	12	1.38	0.97-2.02
N.Z. L. Georgina		254-483	141	1.28	1.05-1.64
L. Coleridge		305-381	33	1.30	-
		457-558	112	1.11	-
		559-635	38	0.94	-
F139 N.Z. Ngapouri L.		318 g mean	512	1.16	-
Okaro L.		726 g mean	146	1.22	-
Okataina L.		1678 g mean	401	1.36	-
S104 VA Big Spring		-	92	1.23	0.94-1.50
P128 S. Afr.		190-253	10	1.28	0.86-1.94
P38,41,43,53-4,92-9,128 H166 S. Afr.		254-380	213	1.39	0.75-2.49
P39,41,43,45,53-4,92-9,128, H166, M108 S. Afr.		381-507	329	1.30	0.85-2.02
P39-41,43,45,53-4,92-9,128, H166, M108 S. Afr.		508-737	122	1.26	0.87-1.86
Y13 N.Z.	April	Age I	196	1.16	-
	June	Age I	180	0.98	-

		No.	Mean K(FL)	Range
R149 CA Convict C.	Starved trout	-		.646-.924
R149 CA Convict C.	Trout dying from starvation	-		.480-.702
N78 CA Convict C.	127-203 mm TL at stocking	-		.939-1.351
N78 CA Convict C.	After several months in stream	-		.666-.99

		No.	Mean K(FL)	Range
R233 CA Convict C.	Aug. after stocking	45	0.98	0.90-1.09
	September	90	0.82	-
	October	90	0.76	-
	November	90	0.74	-
	May	-	0.68	0.60-.78
	Normal Convict C. trout	-	-	0.80-.90
	Lean and vigorous	-	-	0.70-.80
	Marginal, tiring easily	-	-	0.60-.70
M122 CA Convict C.	July, at stocking 221 mm	50	1.15	-
	Nov., 229 mm	25	0.89	-
	Nov., from enclosure	-	0.82	-
B319 ID Big Springs	64-229 mm	-	-	0.76-1.17
R151 CA Convict L.		-	1.02	-
R151 CA Dorothy L.		-	0.88	-
R151 CA Mildred L.		-	1.15	-
S481 ON Heart L.	1960 and 1961 means	-	1.15-1.16	-
M67 CA Colorado R.	-	15	1.04	-
R151 CA Convict C.	81-326 mm SL	124	1.05	-
R149 CA Convict C.	Several groups	-	-	0.94-1.136
C132 MI	173-371 mm	7369	0.95	0.68-1.20
K34 MT	-	478	1.03	-
P55 MT W. Gallatin R.	-	346	1.03	0.80-1.14
H258 WY Pathfinder L.	335-795 mm	121	1.05	-
F138 MT Cliff L.	nonparasitized	15	1.05	-
	178-429 mm parasitized	54	0.93	0.86-1.22
Meadow L.	nonparasitized	5	0.93	-
	parasitized	25	0.98	-
B155 MT Prickly Pear C.	Age I-IV	501	1.10	-
S276 SD Gavins Pt.	-	1	1.26	-
N75, N73 NH Swift R.	173-257 mm at stocking	2010	1.16	-
	188-259 mm at recapture	802	0.99	-
S275 MT hatchery	155-373 mm at stocking	2688	1.09	0.69-1.63
	163-386 mm 1-2 months later	1303	0.98	0.66-1.36
	155-353 mm Wild trout	448	1.04	0.80-1.28
	157-353 mm 1 month after tagging	472	0.93	0.70-1.18
P78 WY	At tagging	103	1.08	-
	Recaptured tagged fish	103	1.02	-
C132 MI	173-371 mm	7369	1.23	0.89-1.56
S283 WI hatchery	Normal	+	-	1.31-1.59

		No.	Mean K(FL)	Range
S282 WI hatchery,				
No yeast	127-249 mm	+	1.38	1.07-1.74
Yeast added	124-267 mm	+	1.46	1.07-2.04
M437 WI				
Fall-hatched	152-292 mm	3242	1.34	1.17-1.52
Spring-hatched	145-236 mm	671	1.31	1.14-1.42
Fall, at stocking	152-274 mm	1400	1.42	1.32-1.52
Later	152-292 mm	1842	1.27	1.17-1.35
Spring, at stocking	145-218 mm	400	1.40	1.38-1.42
Later	160-236 mm	271	1.16	1.14-1.20

An increase in the ponderal indexes with increase in length or age of the rainbow trout was noted by S234; while decreases were noted by M340, K16, K55, K34, P68, P176, and in the South African data given above; and H258, B155 reported no change with size or age. Ponderal indexes decreased with increase in length from 76 to 229 mm in steelhead trout in Idaho in September, but not in April, May, August and October (B319). Condition increased during the summer growth period and declined in winter (B319). In Tennessee yearling rainbows taken by electric shocker had higher K values than those taken by angling in each of the 6 months of study (P68). Male and immature female trout had similar K values, October to February (P68). Stocked trout had lower ponderal indexes than native trout (K16) which Klak laid to the hatchery fish not perfecting natural foraging ability. Planted rainbow lost condition, while native fish retained theirs (K16, N75, N78, S276). Stocked trout lost condition (M437, R233, N122). In California (N34) loss of condition after stocking was noted, but it was reported that wild trout also showed a decrease in condition in May to July. Low condition of rainbow trout, Kennedy Creek, Virginia, was believed to be due to wide fluctuation of water levels and to exhaustion of food when the trout are confined to small pools. Parasitized trout showed lower condition than nonparasitized fish in Cliff Lake, Montana (F138). Condition showed a positive correlation with growth rate (P176). Condition factors were in reverse order to primary production in 3 New Zealand lakes (F139).

In Paul Lake, British Columbia, length-weight relationships differ so much with sex, season, and length that growth analysis is complicated (L150).

	No.	Total length Mean of means	Central 50%	Range	No.	Weight Mean of means	Range
Size at hatching							
M98 MI	18	-	-	13-18	-	-	-
Age 0-March							
B78, H155, S28, D187, K123	+	-	-	20-30	+	-	0.17-0.5
Age 0-April (or 2 months)							
H155, S326, T27, W35, D187, M464, K123	+	-	-	25-69	+	-	0.3-2.4
Age 0-May (or 3 months)							
F75 steelhead, downstream	1	30	-	-	-	-	-
M76 BC streams							
S. g. kamloops	+	38	-	-	-	-	-
S326 N.Z.	+	-	-	23-76	-	-	-
B183 WI Weber L.	+	-	-	30-58	-	-	-
S79 CA Waddell C.	+	-	-	23-56	-	-	-
M98, D3, D4, H155, H391, K123, D187 hatcheries	+	-	-	33-66	+	-	0.2-5.2

	No.	Total length			No.	Weight	
		Mean of means	Central 50%	Range		Mean of means	Range
S396 CO Shadow Mt.	100	53	-	-	100	1.4	-
N29 *S. g. nelsoni*	-	-	-	-	+	-	1.3-3.3
M76 BC lakes *S. g. kamloops*	+	86	-	-	-	-	-
Combined	101+	51	-	23-86	100+	-	0.2-5.2
Age 0-June							
(or 4 months)							
F75 steelhead, downstream	4	30	-	20-36	-	-	-
S37 MI streams	156	41	-	38-43	-	-	-
W1, S79 CA	+	-	-	28-76	-	-	-
B184 WI Weber L.	+	51	-	-	-	-	-
W5, M98, D4, W35, F68,							
D187, H155 hatcheries	-	-	-	-	+	1.6	0.1-10.0
S326 N.Z.	+	-	-	23-99	-	-	-
O39 Japan	100	64	-	-	100	2.3	-
B266, B213 PA selected fish	+	127	-	-	+	-	11-17
Combined	260+	53	-	20-127	100+	-	0.1-17.0
Age 0-July							
(5 months)							
F75 steelhead, downstream	54	30	-	18-43	-	-	-
L61 BC Lordeau R.							
S. g. kamloops	+	-	-	23-43	-	-	-
H208 BC hatchery							
S. g. kamloops	-	-	-	-	+	-	0.2-0.4
B319, S37, S79 streams	80+	48	-	33-84	-	-	-
W5, M98, D4, N31, D3, T27,							
W35, B104, F68, H155, D187,							
C265 U. S. hatcheries	414+	56	46-64	33-81	+	3.7	0.2-18.0
B184, N98 lakes	+	58	-	41-66	-	-	-
W125 G. Brit. Shasta trout	91	94	-	61-135	91	11.4	-
Combined	639+	56	43-64	18-135	91+	4.6	0.2-18.0
Age 0-August							
(sixth month)							
F75 steelhead, downstream	183	33	-	20-84	-	-	-
L61 Lordeau R.							
S. g. kamloops	+	-	-	30-53	-	-	-
M61, M464 Can. hatcheries	+	53	-	41-71	+	-	0.5-1.8
B105, H208 BC hatcheries							
S. g. kamloops	-	-	-	-	+	-	0.4-0.7
M98, W5, D4, D3, S28, W2,							
N31, W35, G34, D187,							
C265 U. S. hatcheries	200+	61	-	41-69	200+	5.3	0.4-32.0
H215, B10, W2, B155 U.S.							
streams, West	528+	61	-	38-84	198	3.4	-
G18, S37, S107, H246 U.S.							
streams, East	703+	74	66-74	43-97	-	-	-
B105, F68 selected stock,							
hatcheries	2070	74	-	-	6188+	-	4.7-10.0
N98, P89, B141, W1,							
H296 lakes	37+	89	-	36-132	9	17.0	-
T34 Argentina	+	-	-	150-201	-	-	-
Combined	3721+	69	53-76	20-201	6595+	4.2	0.4-32.0
Age 0-September							
M61 AT hatchery	+	48	-	-	+	0.7	-
H208 BC hatchery							
S. g. kamloops	-	-	-	-	+	-	1.1-1.4
F75 steelhead, downstream	100	51	-	30-86	-	-	-
M98, D4, E8, D24, D187,							
E130, N31, W35, G34, P11,							
T27 U.S. hatcheries	1800+	94	-	74-112	1800+	9.5	0.6-50.0
S37, B155, B319, L105							
streams	257+	89	-	74-173	-	-	-
W1, B184, B141, N98,							
S396, M180 lakes	47+	112	76-142	36-163	39+	28.0	20-40
Combined	4684+	102	79-112	30-173	6789+	8.9	0.6-50.0

	No.	Total length				Weight	
	No.	Mean of means	Central 50%	Range	No.	Mean of means	Range
Age 0-October							
M61 AT hatchery	+	53	-	-	+	0.9	-
B105, H208 BC hatcheries							
S. g. kamloops	-	-	-	-	+	-	1.6-1.9
B319 ID streams	717+	-	-	66-130	-	-	-
A19, M275, T27, D187, N26, L41, S33, E130, T27 U.S. hatcheries	197+	132	104-165	79-216	+	18.6	1.4-90.0
S107 VA St. Mary's R.	+	145	-	-	-	-	-
N98, B184, B141, M181, M179(2), B236 lakes	102+	157	124-178	43-264	84	82.1	17-207
F68, L26 select stock, hatcheries	-	-	-	-	+	-	30-227
Combined	1016+	135	99-165	43-264	84+	44.0	1.4-227.0
Age 0-November (or nine months)							
H208 BC hatchery							
S. g. kamloops	-	-	-	-	+	-	2.0-2.6
M61 AT hatchery	+	51	-	-	-	-	-
L159, B155, B319 streams	93+	91	-	58-119	-	-	-
V57 BC steelhead, hatchery	96	91	-	69-173	-	-	-
B73 hatchery troughs	-	-	-	-	+	8.8	-
B73 hatchery ponds	-	-	-	-	41	31.2	-
B23, S33, D187, E130 hatcheries	468+	-	-	102-163	+	-	45-115
B154, B184 lakes	+	155	-	109-203	+	5.1	-
F68 select stock, X-rayed	-	-	-	-	+	41.0	-
F68 select stock	-	-	-	-	+	31.0	-
T34 Argentina, steelhead	+	-	-	239-269	-	-	-
Combined	657+	119	89-122	51-269	41+	35.1	2-115
Age 0-December							
H208 BC hatchery,							
S. g. kamloops	-	-	-	-	+	-	2.2-3.4
H246(3) NY streams	1521	81	-	74-89	-	-	-
B319 ID streams	+	89	-	84-94	-	-	-
F68 select stock, X-rayed, hatchery	-	-	-	-	+	47.0	-
F68 select stock, hatchery	-	-	-	-	+	60.0	-
D187 UT hatcheries	-	-	-	-	+	-	70-170
H287 CA steelhead, hatchery	1877	180	-	81-290	-	-	-
P98 OR Devils L.	45	234	-	-	-	-	-
B266 PA hatchery	-	267	-	-	-	-	-
Combined	3443+	137	81-180	74-290	+	-	2.2-170.0
Age 0							
N23 AT Jasper L.	10	41	-	28-53	-	-	-
S220 CA steelhead	+	-	-	28-117	-	-	-
A61 Sweden	12	104	-	-	12	82	-
W100, N33, K15, A46, R158, S82, J56 other	2+	119	71-190	38-279	+	-	0.3-11.0
P99, P128 S. Afr.	14	211	-	147-259	-	-	-
Combined	38+	130	71-211	28-279	12+	-	0.3-82.0
Age I							
L158 Russia, parasitized by Triaenophorus	35	89	-	-	35	6	-
Not parasitized	58	97	-	-	58	9	-
V53 Kenya	274	140	-	53-282	16	82	-
B105 BC, S. g. kamloops, hatchery, Jan	-	-	-	-	+	3	-
H246, B160, E13, S482 U.S. NE streams	3239+	157	130-178	119-234	4	68	-
B10, N49, N23, M113 W. Can. streams	16+	152	-	107-211	+	85	71-99
W125, G11, B48, G17 G. Brit.	115+	157	142-175	76-221	114	31	-

	No.	Total length Mean of means	Central 50%	Range	No.	Weight Mean of means	Range
S37, G18, S211, S221, W100 Midwest streams	375+	160	147-168	104-231	+	102	-
F75, S220, C251, G93, D29, S100 steelhead, downstream	227+	160	152-203	43-229	39	-	17-45
N29 CA, *S. g. nelsoni*	+	-	-	109-244	-	-	-
R98, R134, M113, E99, B187, B10, M210 W. Can. lakes	73+	170	66-241	48-284	4+	218	23-383
B155, S79, H215, A46, B319, K34, O24, I7, P78, U.S. West, streams	10122+	173	140-224	81-267	-	-	-
M76 BC *S. g. kamloops* streams	+	180	-	-	-	-	-
N67 HI	+	-	-	137-244	-	-	-
S104, L105, L160, L159, K15, S107, U.S. SE streams	179+	185	163-198	89-269	-	-	-
B152, D24, B73, K80, E11, M273, N31, M103, W2, K8, P1, K3, S82, L18, A62, Y13, B23, E103 hatcheries	3015+	188	155-226	53-305	648+	65	1.7-227
S481 ON lakes	56	193	-	-	-	-	-
G25, R66, H50, H285(2), B124, B184, P68, W198, W199, W200, J66(3) U.S. eastern lakes	2664+	234	183-282	97-546	1942+	190	8-510
C66, M166, P176 Australia N.Z.	1462+	231	-	119-470	8+	-	23-1361
J48, C197, B152, C198, J89, S396, P89, R189, H258,391,295,296,35, B249, B154, B236, G98, V22, B141, W1, W2, W150, L126, O34, C265, S234, O33, U.S. West, lakes	6165+	236	196-267	66-437	963+	176	3-1106
M463 CA *S. g. aquilarum*	+	-	-	203-254	-	-	-
A61 Sweden	7	-	-	216-290	7	216	-
R68, M165(2), R78, L62, M462, L90 CA, BC, *S. g. kamloops*, lakes	374+	249	244-292	188-366	11	-	79-150
M179(11), M180, M181, AK lakes	30	287	244-318	203-445	+	303	-
A21 QU streams, introduced	21	340	-	-	21	491	-
B154 OR kamloops, reservoir	24	378	-	353-470	24	463	346-765
P39,41,43,46,49,53,54,92, 94,98,99,128 S. Afr.	192	384	-	218-503	182	655	71-1503
H287 steelhead, returning	965	411	-	307-559	-	-	-
F68, B196, B266, B279, D84 hatcheries, selected stock	53+	465	-	419-531	38+	-	907-1860
Combined	29741+	216	155-257	53-559	4113+	218	1.7-1860
Age II							
C151, G93, P6, O22, O24, S220 steelhead, down-stream	1365+	198	178-190	114-333	396	-	34-51
A51, A61 Sweden	125	206	-	147-386	5	539	-
H246, B106, S482, B68 U.S. NE streams	904+	208	168-246	168-368	19	147	-
B10, N49, N23, M113 W. Can. streams	2+	206	-	150-292	1+	-	71-227
S221, S37, G18, E3, W100, S211 U.S. Midwest, streams	74+	213	173-249	132-338	+	122	-
S107, K15, S194, L105(2) U.S. SE streams	452	229	201-236	114-323	-	-	-

	No.	Total length Mean of means	Central 50%	Range	No.	Weight Mean of means	Range
S79, L39, S80, K34, I7(2), B155, P78, A46, B319, H215 U.S. Western streams	1098+	231	211-269	155-320	-	-	-
N67 HI	+	-	-	203-287	-	-	-
V53 Kenya	1322	251	-	102-526	1108	204	-
R98, E99, R134, C230, B10, M113, B187, B263, M210 W. Can. lakes	846+	254	170-302	112-490	+	-	43-1418
S481 ON Heart L.	149	259	-	-	-	-	-
R66, G25, H246, W193, W197, W200, A62, H285, B124, B184, P68 U.S. eastern lakes	1154+	297	290-333	142-483	968+	267	28-808
M181 AK Boleo L.	+	-	-	-490	-	-	184-851
W125, G17, G11 G. Brit.	6+	292	241-305	178-414	+	-	-
M165,76,78,463, L62, L90 CA, BC, kamloops, lakes	919+	312	244-381	231-460	100	295	102-1298
B141,84,249,152, W1, W2, H35,391,32,258, J89, V22, S234, L126, O34, M462, R158, J48, C198(2), M456, G98, S380, W150, C265 Western lakes	4997+	325	282-396	160-653	312+	315	77-1219
M166, P2, P176, C66 Australia, N.Z.	915+	338	287-462	234-546	2+	-	227-6123
B152, L18, B73, S263, M273, E103, M103, K80 hatcheries	22+	340	325-366	203-437	+	-	130-397
M76 BC, kamloops, streams	+	368	-	-	-	-	454-907
R76 PE streams	-	-	-	-	+	-	-
T61, S10,80,220,103, M167, H287 steelhead, returning	1403	437	358-526	300-711	268	334	-
P39,41,43,46,48,49,53,54, 92,94,95,96,99,128, M108, H57 S. Afr.	177	442	-	244-665	177	1071	227-5443
M463 CA *S. g. aquilarum*	+	-	-	432-457	-	-	-
B266, B195, B279, D84 selected hatchery	109+	584	-	498-686	109+	2700	1497-4536
Combined	16039+	292	216-340	102-711	3430+	581	34-6123
Age III							
A51 Sweden	101	224	-	188-277	-	-	-
G93, C151, S220, P6, S80 steelhead, downstream	164+	226	196-262	155-295	121	-	45-77
S107, L105, K15, S104 U.S. SE streams	228	257	213-269	165-406	-	-	-
M113, B10, N23, N49 W. Can. streams	+	264	-	119-381	+	269	14-482
R134, E99, R98, M113, B10, C230, M210, W. Can. lakes	557+	277	193-388	142-417	+	921	85-2156
S39, L39, I7, A46, B155, O22, O24, P78, K34 U.S. Western streams	478+	287	254-353	188-353	-	-	-
S482 NH stream	15	290	-	246-424	15	269	-
V53 Kenya	713	310	-	135-653	713	300	-
B160 ME Kennebec R.	+	318	-	-	-	-	-
S481 ON Heart L.	286	323	-	-	-	-	-
M76,78,165,462, L62, L90, CA, BC, Kamloops, lakes	540+	343	284-363	262-503	156+	-	133-4082
N67 HI	+	-	-	287-381	-	-	-
S37, E3, W100, G18 U.S. Midwest streams	64+	343	-	155-605	-	-	-
R66, S228, B184, W200, H246, S226, H285 U.S. eastern lakes	569+	366	333-386	221-635	75	507	113-709

	Total length				Weight		
	No.	Mean of means	Central 50%	Range	No.	Mean of means	Range
M213, W125, G11, G. Brit.	4+	368	-	323-462	1+	-	680-2041
M6, E103, K80, L18, M273, B73 hatcheries	+	394	-	178-511	+	652	454-907
L126, S234, H35, H32, N27, J49, C265, R158, W150, J48, G98, H258 U.S. western lakes	1160+	388	328-465	251-640	62	661	474-1777
C66, P2, M166, P176, W125 Australia, N.Z.	64+	421	315-523	259-627	103+	-	1247-9072
M76 BC, kamloops, streams	+	452	-	-	-	-	-
M436 CA *S. g. aquilarum*	+	-	-	457-559	-	-	-
P39,41,43,46,48,53,54,92,94,95,96,99,128, M108 S. Afr.	107	511	-	361-721	108	1398	454-3856
D84, B195, B279, B266 selected hatchery stock	71+	645	638-683	574-805	71+	3575	2087-7756
O34, M167, S220 steelhead 1 year sea	1021	505	-	500-531	1	1134	-
L87, O34, M167 steelhead 2 years sea	146	721	-	691-759	4+	3374	3175-3572
O34, M167, S220, L87, H287, S10, L93, D29, S103, T61 steelhead 1-3 years sea	1421+	605	500-696	394-759	77+	1919	516-3572
Combined Age IV	6541+	378	290-462	119-805	1489+	1556	14-9072
M113, B10, N23 W. Can. streams	+	262	-	140-363	+	278	28-482
A1 Sweden	90	300	-	274-348	-	-	-
R134, E99, R98, B10, R76 W. Can. lakes	33+	297	-	193-544	+	986	159-1814
S107, S104, K15, L105 U.S. SE streams	64	307	262-348	190-432	-	-	-
S481 ON Heart L.	1	343	-	-	-	-	-
P6, S220, O22, O24, S80 steelhead, downstream	14+	351	-	262-551	-	-	-
D22, K34, I7, B155, A46 U.S. western streams	53+	361	315-445	208-478	-	-	-
B73 hatcheries	-	-	-	-	+	-	907-1361
D59, W53 Kenya, Natal	179	388	-	198-577	137	663	592-907
B160, S482 NE streams	1+	399	-	318-439	1	340	-
M76, L62, M165 BC, kamloops	8+	419	-	272-549	8+	-	235-10886
W150, B233, L126, M462, M456, S234, H35, H32, H258, G143, G98 Western lakes	320+	434	340-505	254-706	58	737	340-822
G25, H246, W200 U.S. NE lakes	1052	439	394-490	305-660	27	1329	609-945
G11, W125, N213 G. Brit.	4+	495	-	338-627	2+	2231	1814-2722
P2, P176 Australia, N.Z.	39	437	-	330-610	8	-	907-2268
R67 E. Can. streams	+	-	-	544-599	+	1890	1361-2268
E3, G18 U.S. Midwest streams	37	564	-	378-769	-	-	-
B184, W100 U.S. Midwest lakes	+	599	-	533-602	-	-	-
P39,41,50,54,92,94,95,96,99,128, M108 S. Afr.	43	612	-	483-836	46	2152	964-5545
M167, S220, T61 steelhead, 1 year sea	341	582	-	462-681	-	-	-
B266, B279 hatcheries, selected stock	27+	721	-	584-849	27	5035 ·	2994-6760
O23, M167, S220, L87, T61 steelhead, 2 year sea	1231	726	-	584-754	83	3595	3544-3617
O23, M167, S220, L87 steelhead, 3 year sea	12	844	-	831-896	4	5216	5103-5443

	No.	Total length Mean of means	Central 50%	Range	No.	Weight Mean of means	Range
O23, M167, S220, L87, S103, H287 steelhead,							
1-3 years sea	2279	711	660-754	462-1012	87	4026	3544-5443
Combined	4244+	486	330-612	140-1012	352+	1562	28-10886
Age V							
M113, B10, N23 W. Can. streams	+	282	-	157-404	+	298	28-779
R98, E99, R67, B10, R134 W. Can. lakes	11+	356	-	277-503	+	955	321-1588
P176 N.Z. lakes	9	391	-	376-457	-	-	-
O24, K34, B155, I7 Western U.S. streams	8	460	-	368-508	-	-	-
V53 Kenya	44	475	-	305-681	35	1150	-
K15, E3, G18, B160, L105 Eastern U.S. streams	39+	508	411-658	343-787	-	-	-
H246, B184, W200, W100 Eastern U.S. lakes	751+	541	480-602	356-737	20	1003	-
S234, H32, G143, M462, H258 Western U.S. lakes	33+	599	541-615	541-844	9	1021	-
P41,50,51,94,95,96,99,128 S. Afr.	18	627	-	503-789	18	2515	893-3856
M213 G. Brit.	2	681	-	653-709	2	3232	2608-3856
B266, B279 hatchery, selected stock	6+	734	-	650-800	6	5701	2994-7711
S220, T61 steelhead, 1 year sea	20	610	-	597-627	-	-	-
T61, L87, O23, S220, M167 steelhead, 2 year sea	270	759	739-767	734-851	10	4567	4196-5126
L87, O23, S220, M167 steelhead, 3 year sea	282	841	-	795-909	40	8004	5542-6583
L87, O23, S220, M167, S80, S103, L93 steelhead,							
1-4 year sea	999	762	737-805	597-909	50+	5307	2495-8165
Combined	1920+	605	480-747	157-909	121+	2912	28-8165
Age VI							
M113, B10, N23 W. Can. streams	+	277	-	178-343	+	62	57-65
A51 Sweden	57	394	-	373-434	-	-	-
P176 N.Z. lakes	7	381	-	345-566	-	-	-
E99, R98, R134 W. Can. lakes	11+	414	-	373-483	+	613	-
V53 Kenya	10	541	-	363-653	7	1421	-
K34, K15, B160, G18 U.S. streams	12+	569	-	432-787	-	-	-
G25, H246, W200, W100 Eastern U.S. lakes	202+	592	511-693	406-787	2	1100	1063-1134
H35, H258, G143, S234 Western U.S. lakes	17	663	-	508-828	-	-	-
P50, P99, P128 S. Afr.	3	660	-	544-734	3	2920	1162-4224
B266, B279 hatcheries, selected stock	4+	737	-	650-767	4	4564	2750-5897
T61, S220 steelhead, 2 years sea	8	780	-	749-795	-	-	-
T61, M167, L87 steelhead, 3 years sea	141	856	-	836-876	14	6433	-
T61, M167, L87 steelhead, 4 years sea	4	874	-	762-1034	2	4196	-
T61, M167, L87, S80, B249, S103, L93 steelhead,							
2-4 years sea	236+	818	770-836	681-1034	17+	5559	2495-9072
Combined	569+	703	511-795	178-1034	49+	3311	57-9072
Age VII							
M113, B10 W. Can. streams	+	274	-	196-366	+	57	-
P176 N.Z. L. Lynden	3	394	-	-	-	-	-
E99, R98, R134 W. Can. lakes	8	495	-	445-549	1	1077	-

| | | Total length | | | | Weight | |
	No.	Mean of means	Central 50%	Range	No.	Mean of means	Range
K34 MT Missouri R.	1	500	-	-	-	-	-
V53 Kenya	1	513	-	-	1	1446	-
G25, H32, H246, W100							
U.S. lakes	47+	671	528-780	495-869	-	-	-
G18 MI streams	2	744	-	698-787	-	-	-
N28 BC Paul's L. kamloops	-	-	-	-	1	3175	-
C80, S103, S220, M167, T61							
steelhead	19	853	-	737-953	-	-	-
Combined	81+	653	500-795	196-953	3+	1180	57-3175
Age VIII							
B134 AT Amethyst L.	1	486	-	-	-	-	-
H246, W100 East U.S.	8+	681	-	483-813	-	-	-
G143 OR Klamath L.	2	798	-	719-884	-	-	-
S103 OR steelhead	1	884	-	-	-	-	-
Combined	12+	701	-	483-884	-	-	-
Age IX							
S103 OR steelhead	1	925	-	-	-	-	-
G143 OR Klamath L.	1	904	-	-	-	-	-

Average calculated TL and annual increments at each annulus

		1	2	3	4	5	6	7
Steelhead, stream life								
B217 OR north		99	152	198				
	Incr.	99	53	48				
	No.	410	387	89				
B217 OR south		97	163	216	279			
	Incr.	97	66	64	76			
	No.	216	215	113	5			
R155, B235 OR Rouge R.		84	132	175				
	Incr.	84	79	71				
G93 WA Minter C.		94	137	163				
	Incr.	94	46	41				
	No.	330	321	36				
Steelhead, sea life								
R155 OR Rouge R.	Incr.	-	147	224	241	140		
B217 OR north	Incr.	551	388	371	234	140		
	No.	6	29	313	335	83		
OR south	Incr.	-	421	404	315	190	74	
	No.	-	1	102	146	31	1	
Streams — not steelheads								
B155 MT Prickly Pear C.		89	168	239	300	424		
	Incr.	89	79	69	53	41		
	No.	860	249	61	10	1		
K34 MT Missouri R.		81	201	282	343	404	421	470
	Incr.	81	119	86	64	51	58	64
	No.	478	371	169	40	4	2	1
B145 WY Firehole R.		135	234	328	396			
	Incr.	135	109	104	81			
	No.	198	109	23	3			
Madison R.		127	244	356	417			
	Incr.	127	124	117	86			
	No.	125	78	42	9			
I7 ID Snake R.		130	257	353	462	495		
	Incr.	130	130	102	58	38		
	No.	80	-	-	-	-		

Average calculated TL and annual increments at each annulus (cont.)

		1	2	3	4	5	6	7
V53 Kenya		112	216	290	353	442	495	478
	Incr.	112	117	76	61	53	51	79
	No.	2550	2256	955	232	58	11	1
P93, 95 S. Africa		140	340	483	561	612	681	
	Incr.	140	198	114	99	81	38	
	No.	100	54	23	10	4	1	
Lakes								
H35 OR Crater L.		84	190	292	391	442	452	
	Incr.	84	89	127	117	102	48	
	No.	318	311	272	189	41	5	
S234 UT Fish L.		163	259	333	417	523	617	
	Incr.	163	104	81	81	81	66	
	No.	530	387	182	57	14	5	
H258 WY Pathfinder L.		122	307	447	523	597	640	
	Incr.	122	185	142	102	66	61	
C230 BC Paul L.	Incr.	81	109	109				
	No.	1264	1206	518				
Stocked as fingerlings	Incr.	76	109	145				
	No.	60	56	6				

			Calculated TL at each annulus					
	No.	1	2	3	4	5	6	7
Streams								
K104 Czech. Vrica C.	32	79	145	188	218			
S80, S81 CA, Waddell C.	34	69	137					
S81 Baja California								
S. g. *nelsoni*	20	91	157	198	221			
S275 MT Flint C.	64	66	135	183	229			
T2, T3 NM	65	86	155	206	257			
P55 MT Budge's Stream	98	97	168	218	284			
Elevation 6000'	133	79	155	239	305	335		
5000'	76	81	165	226	307	396		
4000'	142	86	188	292	368	437		
P159, M453 MT streams	5144	84	170	251	323	363		
I7 ID hatchery	40	127	244	333	445			
W125, F76 G. Br. Lough								
Share	78	109	231	277	320			
M213, W124 G. Br. Blogdon	35	119	284	396	505			
C53 BC kamloops	6	66	122	300	465	546	635	759
Steelhead, stream life								
C174 OR Alsea R.	853	122	163	173				
M167 BC Chilliwack R.	770	112	165	201	229			
H287 CA Sacramento R.	83	117	203					
Steelhead, sea life only								
B217 OR	632	549	739	912				
B325, R155 OR	+	421	592	698				
Steelhead, stream and sea life								
L87 BC Veddar R.	110	124	198	251	467			
O34 OR south	36	135	231	554	747	864	965	

Additional life history data on steelheads in Puget Sound (M35,36).

Lakes	No.	Calculated TL at each annulus						
		1	2	3	4	5	6	7
R151 CA Mildred L.	70	91	145	196	236	269		
Dorothy L.	22	86	147	208	259			
Convict L.	22	97	163	282				
M371 ON pond	+	107	190	244	292	318		
S263 BC Corbett L.	88	81	188					
F137 CO Granby L.	+	132	213	269	307			
L206 N.S. Wales, lake	128	51	114	140	160			
after fertilizing	476	76	132	152	168			
L90, C230, BC Paul L.	1656	89	206	287				
L91 BC slowest	-	58	145	198	206			
28 lakes	3080	86	211	282	356			
fastest	-	112	340	500	589			
H43 UT Fish L.	25	74	190	315	391	447		
P156 MT Willow Creek L.	427	86	216	351	447	498	503	
Canyon Ferry L.	13	86	183	351	437	483		
E112 NY pond	+	206	282	343	361			
P159, M453 MT lakes	2905	89	206	323	406	465	495	521
P159 MT ponds	215	107	218	330	467	528	549	
C53 BC kamloops	34	107	290	406	528	663		
H32 OR Crater L.	124	109	213	340	445	546	671	757
M462 CA Crowley L.	+	257	401	505				
M152 UT Bear L.	33	107	206	277	376	356		
S483 OR Cascade L.	4	132	251	328	295			
H179 NY Skaneateles L.	+	76	147	361	445	493	513	
Seneca L.	+	102	216	345	483	607	714	729
Reuka L.	+	86	155	376	472	526	559	577
S326 N.Z. 6 lakes	+	300	442					
Mixed lakes and streams								
K80 OR central	16	124	234	244	272			
west	41	109	264	432	508			
E2, E3, M136,258, S262 MN	+	124	231	325	409	521		
G18 MI	251	84	198	439	556	638	703	
P2 Australia	+	130	244	452	452			
Unweighted averages								
Streams, 21 means	10358	99	196	282	353	445	523	569
Steelhead, stream, 7 means	2662+	104	160	183	254			
Lakes, 26 means	9687+	117	224	325	396	486	572	645
Combined, 59 means	23015+	109	208	300	378	478	564	612

(handwritten annotation across the Unweighted averages rows: $G = 2.05$ 1.1 0.7 0.7 0.5)

	No.	Average calculated weight at each annulus					
		1	2	3	4	5	6
E112 NY ponds	+	113	281	481	595		
H258 WY Pathfinder L.	121	27	327	984	1515	2046	2449

Annual increments as a percentage of length at beginning of year
(number of specimens in parentheses)

TL	MT K55, K34 B155	ID Snake R. I7	Kenya V53	Crater L. H35	steelhead G93	Range (single variants in parentheses)
51-75	-	-	147(6)	108(75)	-	96-242
76-101	123(632)	91(2)	129(539)	110(215)	48(291)	45-225(433)

TL	MT K55, K34 B155	ID Snake R. 17	Kenya V53	Crater L. H35	steelhead G93	Range (single variants in parentheses)
102-126	84(11)	107(41)	102(1696)	104(26)	42(66)	25-187
127-151	43(17)	98(66)	94(148)	85(162)	-	17-160
152-177	39(11)	82(6)	50(48)	71(22)	-	31-125
178-202	45(196)	-	36(248)	67(44)	-	30-88(157)
203-228	30(7)	-	34(512)	65(54)	-	6-65(125)
229-253	25(2)	42(41)	30(162)	-	-	8-85
254-278	20(7)	-	34(103)	44(107)	-	10-88
279-304	23(40)	39(6)	25(50)	41(69)	-	6-57
305-329	24(2)	27(2)	23(46)	-	-	6-36
330-355	-	28(4)	26(24)	21(3)	-	5-42
356-380	-	-	22(28)	30(9)	-	5-41
381-405	11(5)	16(4)	18(35)	24(19)	-	5-27
406-431	16(1)	-	17(37)	8(3)	-	7-24
432-456	-	13(3)	17(15)	-	-	12-19
457-482	-	8(2)	16(4)	-	-	8-31
483-507	-	5(2)	14(30)	-	-	5-21
508-532	-	-	10(2)	-	-	-
533-558	-	-	-	-	-	-
559-583	-	-	5(2)	-	-	-
584-609	-	-	6(1)	-	-	-

Fingerlings stocked in Dale Hollow, Tennessee, grew 15 mm per month for 13 months while yearlings grew 13 mm per month (P67). A 3583 g tagged female weighed 4808 g when caught the next year in the Finger Lakes (H179).

Von Bertalanffy equations were computed for 3 New Zealand lakes (F139):

	ultimate length	K (slope)
L. Okataina	389	-0.094
L. Okaro	310	-0.150
L. Ngapouri	232	-0.094

In most cases the lengths were calculated on a direct proportion basis from the scale measurements. The Fraser modification was used by B155, B217, G93, H287 and K80 with corrections ranging from 28-43 mm TL. Body-scale relationship of Fish Lake, Utah (S234) rainbows was described as $SL = 0.01 mm + 4.61 R - 0.00007 R^2$, but this curvilinear relationship was not much superior to the straight line, $SL = 0.4 mm + 4.6 R$. Sex did not appear to affect the body-scale relationship. Size at scale formation in Washington was reported as 32 mm (L87, P6) and 34-37 mm for kamloops trout (S262). At Paul Lake, BC (L90, S262), lengths of kamloops trout, 45 mm and up, were back calculated with a 45° slope on log-log paper. Lengths of British Columbia steelheads (M167) were computed from scale measurements at annuli and a log-log regression of FL and scale diameter based upon presmolt fish.

M167 reports that scales from a large elliptical area on the side of steelhead trout from somewhat in front of to behind the dorsal fin and above and below the lateral line are about the same size and most suitable for age and growth studies. Freshwater annuli showed distinct cutting off of incomplete circuli by a complete circulus, but ocean annuli showed no crossing over and were distinguished by crowding of the circuli. Spawning marks were left through resorption of the scale edges. Such resorption was also noted on Lake Michigan (G18) and Finger Lakes (H246) rainbows. The first spawning mark was not always clear and spawning marks were less evident on females than males (G18). In addition to the erosion markings, large clear streaks on the anterior and lateral fields were reported as characteristic of spawning marks (H246). Age determination of trout over age III was not considered satisfactory in a New Zealand lake, because of erosion of the scale edge (P176). Scale pictures are given by B145, H246, S234, R151.

In Loon Lake, British Columbia, age determination was considered unreliable because of complications in stream and lake life (H306). In Paul Lake (L62) difficulties were also reported in separating stream from lake growth. Scale characteristics were utilized for selecting hatchery from naturally spawned trout in Fish Lake, Utah (S234) and Yellowstone Lake (B145) where the recovery rates determined in this fashion were believed to be more reliable than those from fin-clipping.

Second readings agreed with the first in 85-97% of the rainbow trout (H246). Scales showed the proper number of additional annuli on 48 of 49 tagged fish caught 1 to 3 years later. Furthermore, modes in length frequencies agreed with scale readings (H246).

In West Virginia, 30% of the 1933-year class rainbows and 20% of the 1934-year class went through the first winter without forming annuli, probably because of abundant food and relatively uniform water temperatures (S105). When rainbows were fed throughout the year, no annuli were formed and circuli were widely spaced even in the winter (G17). There was no evidence of an inherent growth rhythm. In Pennsylvania 152-254 mm rainbows stocked in December grew 0.15-0.33 mm per day by April (T110).

The annuli of steelhead in British Columbia were reported as formed in March (M167). In Montana, annuli form from April 28 to June 3, earlier at the lower altitudes than at the higher, and earlier than brook trout (B175). In Michigan, younger rainbows had formed annuli by April 11-18, but the older ones had not, perhaps indicating a shorter growing season associated with maturity (G18). In trout migrating to Lake Michigan in May, a false annulus forming near the scale margin and just outside the true annulus was reported (G18). Some summer- and fall-caught steelhead from the Columbia River failed to show any summer growth on their scales (L163).

Lee's phenomenon was observed on data from Paul's Lake, British Columbia, and was believed to be associated with selection of faster-growing young fish by angling (L90). Lee's phenomenon was also reported by H35.

The maximum age reported was 11 years for Eagle Lake rainbows, *S. g. aquilarum* (M463), 9 years for steelhead (S103, G143, M462), and 8 years for others (B134, H246, W100). Poor food conditions may result in poor survival after the first spawning (M79). Heavy angling may also shorten life span (M462).

Females appeared to grow faster and live longer in Kenya (V53), Finger Lakes (H246), Pathfinder Lake (H258), as did steelhead females in Waddell Creek (S220). At age VI, the remaining few males were as large as the females, however, in the Finger Lakes (H246). Males were reported to be the longer-lived by G18. The sex differences in growth were not statistically significant in Pathfinder Lake (H258) and G93 reported no sex difference in growth of steelheads. G18 reports females growing at least as fast as males in Michigan. M167 reports that female steelheads average larger than the males but suggests that this may be because a higher percentage of the females return to spawn more than once. No difference in the mean number of years of freshwater life was noted for male and female steelheads (M167). In Kenya (V53) females were reported as living longer than males, presumably due to earlier maturity and more intensive activity of the males at spawning times. In Michigan (G18) it appeared that males outlived females.

In general, the growth is slower in streams than lakes and fastest in the ocean. Growth in Klamath Lake, Oregon, was reported as similar to that in the ocean, because of abundance of food (G143). Stream growth can usually be distinguished on the scales from lake and sea-run trout by the closer spacing of the circuli and of annuli. Few stream trout exceed 508 or even 406 mm. Larger trout have been reported from many streams, but these larger trout have usually had some growth in lakes or the ocean. G18 mentions that in Michigan trout migrating from lakes are larger than the stream fish. Sea-run steelheads of over 1016 mm were reported in the tabulations above, but no non-sea-run rainbow of over 813 mm was noted. Length of steelheads at maturity was reported as being a function of sea life rather than total life (M167, L163). Steelheads may spend 1-4 years in freshwater (L87). No correlation was detected between number of years spent in fresh and in salt water (M167). Slower-growing smolts tended to remain longer in freshwater than the faster-growing (M167). In the Finger Lakes, trout migrating to the lakes earlier in their lives showed greater size at the same age (H246).

Growth was more rapid in waters where the trout were introduced and therefore had little competition (M79, A21, D171, R26, V53, W125, M210, B152, R158, M179, B187, O33, M210). In Malheur Reservoir, Oregon, fingerlings stocked in February reached 328 mm in July, in Beulah Reservoir 135 mm yearlings stocked in March were 414 mm in October, and in Warm Springs Reservoir 51 mm fingerlings reached 236 mm in 5 months (D75). In British Columbia, trout in lakes with poor spawning areas usually showed faster growth than those in lakes where spawning areas were good and the numbers of trout therefore high (M79). In Chickaka Lake, Alberta, the first rainbows stocked averaged 383 g at age I, those stocked the next year 284 g, and those stocked the third year 142 g (M210). At age II the first group averaged 1418 g and the second 992 g. Rainbows stocked in barren waters in Australia averaged 6577 g at age III, and the range of 103 fish was 3856-9072 g. Slower growth in Ngapouri Lake than in Okaro Lake was associated with the higher population (F139).

Rainbow trout transferred from Lake Lynden to three other lakes grew more rapidly because of lower population density, lower altitude and thus shorter winters, and abundance of forage fish in the three lakes (P176). In California lakes at higher elevations, rainbow trout often grew more rapidly than at lower elevations because of lower population densities than in the more heavily stocked lakes (M462). Growth of trout stocked in a California lake was twice as fast in 1941 as in 1942, due primarily to the relative paucity of wild trout in 1941 (N34). Rapid growth was associated with small numbers stocked in New York (W199). In an Oregon hatchery, increased population density reduced growth rate and increased mortality rate (M378). In Montana (B155), growth per day increased from 1949 through 1951 as the population density decreased. In Paul Lake, British Columbia (L62), no correlation between growth rate and year class strength was evident, but increased population density apparently resulted in slow growth of all year classes. H35 reported that abundant year classes showed more rapid growth than the less abundant. In Big Springs Creek, Idaho Springs, however, the most abundant year class had the smallest yearlings (B319). Yearling rainbow trout in Big Springs Creek grew more slowly after July, probably due to increased competition for food from the new year class (B319).

Slow growth in an Ontario pond (M371) was believed to be due to a high sucker population.

Selective breeding may result in faster growth (D2, D84, L26, B195, B266). Treatment of fry and fingerlings with X-ray reduced their subsequent growth rate (F68). In Corbett Lake, British Columbia, hatchery stock grew more rapidly than the wild stock in the 1954 year class but more slowly in the 1955 year class (S263). Trout stocked as fingerlings were larger at age II than those stocked as yearlings (C230). Rainbows planted earlier in the season in Castle Lake, California, grew faster than those planted late in the season (W1). The greater variation in growth of rainbow than in brook and brown trout was considered an indication of greater genetic variability (W3).

Although wood fiber in water did not affect hatching or survival of eggs, it did reduce growth and survival of fingerlings. Fingerlings maintained in water with 250 ppm wood fiber had instantaneous growth rates of 0.0061-0.0062 compared to 0.0213-0.0345 in the control fish (K123).

Growth was faster in hatchery units which recirculated the water (M464). M396 reported that rainbow fingerlings grow best at 13C. M378 reports that fry and fingerlings grow more rapidly and have lower mortality in ponds heated to 13.4C than at 10C. Under favorable conditions growth to fingerling stage is logarithmic (B105). When temperatures drop, metabolic rates slow, growth slows, and food conversion goes down (B105). Faster growth at lower elevations in West Gallatin River, Montana (P55) was probably related to warmer temperatures and longer growing season. Slower growth in two Wyoming lakes (B214) in 1957 was attributed to lower temperatures; growth was slower in the cooler lake despite its greater fertility than the other because of its lower temperature. Faster growth in Catherine Creek than in two other New York streams (H246) was believed due to its warmer temperatures and lower population density. Growth during the first season was related to earliness of emergence of the fry from gravel (B319). Size of steelhead trout was not correlated with the size or latitude of the home stream (M462, S220).

Growth was found to continue throughout the winter in a Washington hatchery (F68).

Differences in feeding levels in hatcheries can compensate for the differences in growth rate from 10 to 15.5C (D187).

Many papers not reviewed here have dealt with hatchery diets and growth and survival of trout. Brewer's yeast was reported to be a growth stimulating supplement and mixed diets were found to result in better growth than straight beef liver (B105). Trout fed on *Gammarus* grew more slowly than those on a meat and meal diet but showed greater food conversion (E111). Growth rate increased as protein levels in food increased (N123, H403).

Growth in lakes where the primary food is zooplankton and insect larvae is usually rapid up to 305 to 356 mm but slow thereafter (M462). Where forage fish are abundant and used by rainbow trout, growth may continue to be rapid. For example, in Shasta Lake, rainbows feed on threadfin shad and in Eagle Lake on tui chub and grow much larger than in many other lakes (M462). Drastic decline in Lahontan redside shiners in Lake Almanor was followed by a decline in large rainbow trout (M462).

Introduction of the shiner, *Richardsonius balteatus*, as food for trout in Paul Lake, British Columbia, had a number of effects but, in general, the increased growth of trout over 203 mm FL more than made up for earlier losses from competition with the shiners (C230, C228). It took some years for the trout to fully utilize the shiners as food.

Slower growth of older rainbows in Lake Skaneateles than in Lakes Seneca and Keuka was believed to be due to the scarcity of alewives and other forage fishes (H246).

In some California lakes growth of rainbows appeared to be correlated with total dissolved solids, but not with the food supply (R151). In Dorothy Lake, growth was slow perhaps due to scarcity of food. In Mildred Lake, food was abundant but the trout were too abundant for rapid growth. In Convict Lake, with low food supply and heavy population density, faster growth was believed to be related to the longer growing season associated with its lower altitude. Faster growth in Lake Okataina than in two other New Zealand lakes was believed to be related to the lesser eutrophication (F139).

Slow growth of rainbows compared to cutthroat trout in Bear Lake, Utah (M152), was believed to be associated with the rainbow's preference for shallow water. Slow growth in Green Lake, Alaska, was believed to be related to copper sulfate treatment and low amount of bottom fauna (M179).

High productivity of upwelling water off southern Oregon was believed to be associated with faster growth and earlier maturity of steelheads there than in northern Oregon (B217).

Growth is slower after maturity is reached (M79, P176, S326), particularly where large forage organisms are lacking (M462). In the Sacramento River, steelheads were noted to continue to grow during their spawning migration (M287). However, second spawners were significantly smaller than first spawners of the same age in Waddell Creek (S220), indicating a definite slowing of growth during the spawning migration.

Tagging had no obvious adverse effect upon growth (F139).

Year classes with faster growth were found to have less variation in growth rate than those with slower growth (L62). Mean length and standard deviation were inversely correlated.

A few general statements may be made upon geographical variation in growth rates, but variation due to other factors is often more significant. In general, growth in streams in northeastern United States and in western Canada (except for kamloops trout) is slower than that in midwestern and western United States. Growth in western Canadian lakes, except for kamloops, is also slower than growth in more southern lakes. Growth is more rapid in the large tributaries of the Sacramento River and in rich streams on the east side of the Sierras than in streams on the west slope and in the Klamath and Coast Ranges (M462).

Faster growth of rainbow trout than brook trout in the same water was believed to indicate more successful competition and more efficient use of plankton (W198). At the end of the first summer, brook trout were larger than rainbows because of earlier fry emergence (B319). Under identical conditions, growth of brook and rainbow was about the same, and faster than that of brown trout (W3).

Male rainbows have been reported mature at 9 months (V53 Kenya), at 12 months (M391), or first at age II (G18, E130, H246, H51). The smallest mature male listed was 170 mm (H51), and in the Finger Lakes the smallest was 388 mm (H246).

Female rainbows were mature at 22 months (M391), or first at age III (H51, H246, E130, B160) with most at age IV (H51, H246). Some are not mature until age V (G18), or even VI (H246). The smallest mature female listed was 239 mm (H51) and in the Finger Lakes the smallest was 480 mm (H246). Starving during the second year reduced the percentage mature in their third year

RAINBOW TROUT 189

(S418). In the Finger Lakes, fish with more than one parr year usually matured later than those remaining in streams only one year (H246).

Steelheads are usually age IV at first spawning (L93, B217, M462, S103), with a few males and females maturing at age II (B217, L93, S220). In the Columbia River, most mature at age V, with about 13% at age IV and with a few not maturing until age VIII (L163). About 41% of the male steelheads matured after one year at sea compared to 10% of the females (B217). About 85% of the steelheads off southern Oregon mature after one year at sea compared to 25% off northern Oregon (B217). Longer smolt life was usually related to shorter sea life to maturity (B217).

The number of repeat spawners in spawning runs have been reported as 18-33% in the Finger Lakes (H247), 5.1% for Washington steelhead (P6), 10.5% for British Columbia (W201), 11.4 to 26.2% for Oregon (S103, B321), 28% for northern Oregon (B217), 53% for southern Oregon (B217), and 17.2% for California (S220, H289). In general, the number of repeaters increased to the south (B217, W201). The high incidence, 31.3%, of repeat spawning in Cheakamus River, British Columbia, is associated with late runs and short residence in fresh water (W201). Fewer repeaters were reported from hatchery-reared than wild steelhead (H289). Large females were less likely to survive to spawn again than the smaller ones (H306), and males less likely than females (H306, H246, W201). As many as 5 spawning seasons are reported (H246). No fourth-spawning steelhead were reported north of California (W201). In a hatchery, spawn has been taken from an age XIII female (A63). Survival of spawners was inversely related to the number of trout entering the stream (H306). Higher survival of fin-clipped than of tagged trout was observed and unmarked fish has a higher survival than fin-clipped (H306).

Location	Size of female	No. of females	Number of eggs per female Mean	Range
S122 Germany	726 g	1	1900	-
S122 Germany	907 g	1	2400	-
S122 Germany	1678 g	1	2900	-
L26 CA	II year olds	13	-	723-1693
L26 CA	III year olds	9	-	2112-3072
N28 -	-	-	-	200-9000
S57 -	-	-	-	500-3000
L15 -	II year olds	-	-	500-800
L15 -	III year olds	-	-	1000-1200
M77 BC	907 g	17	2500	-
B23 NY	-	-	900	-
B22 MI	-	-	968	-
E31 MI	-	-	900	-
L18 -	III-VI year olds	-	-	1200-1500
L18 -	907 g	-	-	2500-3000
B73 -	227-680 g	-	-	500-800
B73 -	907-1814 g	-	-	2500-3000
M463 CA S. g. aquilarum	-	-	-	2500-3000
M110 WA	I age group	-	-	600-900
O24 OR Diamond L.	ave. 442 mm FL	9820	1562	1337-2050
S146 WY	1814 g	1	3000	-
F38 MT	3290 g	1	3960	-
F40 WY	2948 g	1	5280	-
F38 MT	5444 g	1	5805	-
N105 CA steelhead	635-660 mm	-	-	4800-6400
P128 S. Afr., sea run	1134 g, age III	-	-	650-700
W134 OR Crater L.	178-287 mm	2	670	620-720
A55 CA	706-1440 g	48	2455	1590-3161
B195, B270 PA	Age II, 1270 g	38	2615	1618-4300
	Age III, 2812 g	38	4520	1240-7279
	Age III, 1950 g	44	2692	1505-4386
	Age IV, 3493 g	44	4550	2348-6738

Location	Size of female	No. of females	Number of eggs per female Mean	Range
H289 Sacramento R. steelhead	larger females, stripped	-	2808	-
D84 WA selected stock	Age III, 680 g	-	-	800-1200
	Age I, 9752 g	28	3894	-5123
	Age II, 11430 g	28	5029	-9950
L101 CA	Age II, spring spawners	-	1553	-
	Age II, fall spawners	-	2600	-
	Age III, spring	-	2210	-
	steelhead	-	4304	-
N124 Tasmania	907-2268 g	135	2379	720-4310
S220 CA Scott C. steelhead	406-520 mm	-	-	1500-2900
	521-634 mm	-	-	2700-4100
	635-748 mm	-	-	4700-6100
	749-864 mm	-	-	6700-8100
	865-965 mm	-	-	8600-11000
V53 Kenya	304-397 g	4	909	658-1218
O33 OR Diamond L.	stripped, averages	-	-	1337-2050
B279 PA selected stock	Age I, 427-508 mm; 907-1860 g	15	2803	1270-4040
	Age II, 498-686 mm; 1497-4536 g	27	4324	2109-8637
	Age III, 579-805 mm; 2495-7756 g	27	5290	3389-10282
	Age IV, 584-850 mm; 2948-6758 g	27	5487	2084-10589
	Age V, 650-800 mm; 2994-7711 g	6	8406	2974-12749
	Age VI, 650-767 mm; 2767-5897 g	4	6229	2074-8659

N124 Tasmania   Number of eggs = 7.12 (FL - 164 mm)
                        Number of eggs = 1.6 (W - 82 g)
S220 Steelhead, logarithm of number of eggs per female = -2.093 + 2.1169
     log FL.
R179 reviews fecundity of rainbow trout and other salmonid fishes.

A55 reported that 13 of 17 female rainbows had more eggs in the left than the right ovary. A bisexual, mature sea-run steelhead has been reported (G117). Starvation in the second year of life reduced the number of maturing eggs the next year (S418). Egg size does not seem to be related to female size (S418). Selective breeding increased egg production (L26, B195, B270, B279).

	Size of female Age	W	No.	Eggs per kg of female Mean	Range
M79 BC Kamloops	-	-	-	-	1764-2205
S418 BC Paul Lake	III	-	-	4083	-
	IV	-	-	4422	-
B195, B270 PA	II	1270 g	38	2059	1274-2542
	III	2812 g	38	1607	441-2588
	III	1950 g	44	1380	772-2248
	IV	3493 g	44	1332	672-1929

Spawning seasons cover a wide range, though individual populations spawn over a relatively short time. In California where they are native they spawn February-April (A63). In Mexico the *S. g. nelsoni* subspecies spawns in January-February (N29). A few females were observed to spawn in July, though no males were ripe, in a California hatchery (H319) where water temperatures were high, 12-17C. Fertility at this hatchery during the usual January-February spawning is low and the incidence of abnormal fry is high.

Other spawning times listed by A63 are November through February for
Neosho, Missouri, and Virginia; November 15-March 15, Manchester, Iowa;
November to mid-June, New Hampshire; November-December and April-June,
Pittsford, Vermont; April 15-June 1, Montana; and May-July, Colorado. In
Maine they spawn February to June (B160). April 9 to May 3 was listed for
Michigan (G116). Although females were ripe only in April in New York ponds,
males were ripe in November and April (E130). In the Finger Lakes, New
York, the spawning run occurs February to late May, usually starting before
ice breakup (R229). The peak is in April, however, with temperatures of
5.5-13C. Some females may stay in the streams only 5 days. Migration
seems to be fairly uniform night and day, but thunderstorms caused a drop in
the migration. In Loon Lake, British Columbia (H306) spawners seem to enter
the outlet stream about the same time each year despite differences in ice
breakup. Spawners to inlet streams are usually 3-5 weeks later and appear to
be controlled by temperature and breakup. Most migration occurs in the daytime and when the temperatures are at the maximum.

In Kenya (V53), rainbow trout may spawn any month, but most spawn June-
August, the coldest season, with a secondary peak in November when rains
come.

Upstream migration of steelheads in the Sacramento River extends from
July to mid-March with a peak in September-October. Spawning occurs from
December to April (H289). In Oregon, the spawning peak is in March-April
(B217). Males migrate and mature a little earlier in the season than females.
In British Columbia they spawn late winter or early spring (L87), peaking in
March (M167), but adults may be in the Fraser River at any time. Winter-run
steelheads were observed to show little color or sexual characteristics when
they spawn compared to summer-runs (S353). In British Columbia (W201)
summer-run fish enter rivers from April to October to spawn the next spring,
and winter-run fish enter from December to April with the peak usually in
January to spawn in March and April. Summer and winter steelheads appear
to be genetically different and thus separate stocks (W201). Natural barriers
which are negotiable only in summer, at higher water than winter, may explain
the evolution of summer-run steelheads which must remain in the streams
several months before spawning.

Summer or spring-run salmon are those which enter streams while still
green. They usually enter streams on dropping water levels in April or May
and summer-over in pools to spawn in November and December. They do not
usually feed in fresh water, but remain fat and in good condition. Spring-run
salmon are not common in California, probably because of scarcity of deep,
snow-fed pools (S220). Fall-run or winter salmon are those which usually
enter streams on rising water levels, with sex organs in various stages of development but spawning within the same season. They may enter from salt
water throughout the year from August (early fish) through July (late fish).
Late fish usually spawn within a month (S220).

Spawning is almost always in streams. In June Lake some spawning on
shore areas was noted but there appeared to be no survival (V22). Courtship
and spawning behavior have been described (N105, V53).

Artificial propagation of rainbows has been extensive and no attempt is
made here to review this phase. The time to hatching has been listed as (E110).

101 days at 3.2C	27 days at 11.5C
75 days at 4.8C	25 days at 12.0C
44 days at 7.5C	21 days at 14.5C
29 days at 10.3C	18 days at 15.5C

Eggs hatched at 23 days at 10C and the embryology was figured and described (K103). Difficulty was encountered in securing eggs from rainbows held at 15.5C, but these fish produced satisfactorily when held at 13.4C (L26).

Embryo survival in nature was found to be positively correlated with water flow and dissolved oxygen through the gravel (C282). Of eggs planted in gravel, 42 to 92% emerged as fry (B319) and 4 to 11% survived to downstream emigration. In Japanese streams, floods washed out 75-96% of the eggs, a smaller percentage than of brook trout eggs (O39). Pre-eyed egg losses in redds averaged 8% with nonfertilization loss equalling 2.3% in New York streams (H317, H318). Survival of steelhead eggs in natural redds for 29-36 days ranged from 0-99% and for 41-63 days from 3-99% (B171). H317 gives a good review of egg losses. Sex differentiation occurs at an age of 95-104 days (M391) on fish reared at 13.4-14.5C. Number of pyloric caeca increases until the trout are about 50 mm long, but is established by then (N107).

Fry emerged 51-55 days after spawning, with 639 temperature units (degree days above 0C) (B319).

Fry on emerging from gravel were found to show a net downstream movement if the water temperatures were below 13C, because in the dark the fry and fingerlings do not maintain contact with the bottom (N98). At higher temperatures the fish retain frequent contact with the bottom and may show upstream movements on temperature rises. Stocked fingerlings moved downstream if waters were below 10C, but showed little drop downstream in warmer waters (C210, N106, C297). Water level fluctuations did not seem to affect this relationship (N106).

Downstream migration of steelheads at ages I-IV usually are greatest in the spring, with some in the fall (M167, B319, H289). Most migration seems to be at night (F75, B319, N98). At the time of migration, increased thyroid activity appears to be associated with the silvery color, the deposition of guanine on the internal surface of the scales and the disintegration of melanophores (R66, R232). Hatchery-reared steelheads stocked at 190 mm apparently go to sea shortly after release (B217). Most steelheads return to the home tributary to spawn with little straying (T108). A 203 mm trout, not known as a steelhead, tagged in the North Umpqwa River, Oregon, was recovered 296 days later as a 368 mm fish in the Garcia River, a trip of at least 160 river kilometers plus 480 kilometers in the ocean (S413).

Stocked trout in Montana were usually recaptured less than 45m from point of release, with more downstream than upstream (L171). In winter they were usually in riffle areas under surface ice but at other seasons mostly in pools. In Idaho, stocked rainbows showed less movement than wild cutthroats, Dolly Varden or rainbows (B284). Of the hatchery trout, 90% of those recovered the first year had moved less than 3.2 kilometers and 90% of those recovered after the first year had moved less than 8 kilometers (B284).

Extensive reviews of the effectiveness of trout stocking are given (C300, H47, N78, B322, S22, W202).

Planting of migrant-sized steelheads in Washington rivers resulted in returns of 6 and 12.7% as adults and showed definite buildup of the population (L93). Steelheads released at 3629 g or larger brought a 4% return as adults in the Sacramento River (H289), and costs per creeled steelhead were estimated at $10.00 (H287).

Big differences in survival, believed to be related to conditions at stocking, were reported when rainbow trout were stocked in various Quebec lakes (A21). In Crater Lake, Oregon, no correlation between fry stocking and year class abundance could be demonstrated (H35). In Quabbin Lake, Massachusetts,

50.5% of the 229-305 mm rainbows introduced were caught by anglers in 2 years (M296) but only 10.1% of a second stocking were caught the first year.

Survival of trout stocked at 51 mm in New York ponds was estimated at 20% the first year, and for 127 mm trout at 40% to the first year and 20% to the second (E103). In Paul Lake, British Columbia (M90) about 5% of 25 mm fingerlings survived to the creel. Harvest of 46 to 57% of yearlings stocked was reported for Green Lake, New York (W199). In Nevada ponds, survival of catchable-sized rainbows for 100 days was 90-100% and for 272 days was 3.3-90%, compared to survival in streams for 100 days of 59-94% and for 272 days, 0.6-33.3% (N84).

In the Peshtigo River, Wisconsin (B208) less than 1% of the stocked fingerlings, 67% of the yearlings, 33% of age II rainbows, and 40% of 305 mm stocked rainbows were caught by anglers in 4 years, by fly fishing only. Spring and open season stocked rainbows have been taken by anglers in streams at rates of 18% (S34), 21-25% (S324), 36% (N75), 45% (C143), 50% (N75), 54-60% (C143), 59% (W168), 60% (T110), and 73% (W169). Fall-stocked rainbows have been taken at rates of 7% (S34) and 46% (T110). Percentage returns were about the same, 30%, for fall- and spring-stocked "catchable" trout in Mad River, Ohio (W203).

One tagged 254 mm rainbow moved downstream 338 kilometers in Montana in one year (S378). Tag returns from rainbows tagged during spawning runs in the Finger Lakes (H246) showed that males were more heavily exploited by anglers while still in the streams, while females were more heavily exploited in the lakes. Exploitation rates in the streams were 13-17% and in the lakes the first year were 5-12%. Maximum first year exploitation was 28% with the average from 19.5-26.3%. Averages for the 2nd year were 0.8-6.1%, for the 3rd year 0.6-2.1%, and the 4th year 0.3-1.1%. A tagged rainbow in Lake Michigan travelled at least 190 kilometers in 9 months (S76) and 6.1% of those tagged were recaptured.

The instantaneous mortality rates for spawning-sized Finger Lakes rainbows were reported as (H246)

	i	p	q
Grant C.	1.45	.42	1.03
Catherine C.	1.43	.37	1.06
Cold C.	1.46	.38	1.08
Naples C.	1.69	.55	1.14

In East Fish Lake, Michigan, the instantaneous mortality rate per day was reported as .00045-.01821 (A62).

	Red blood cells per cubic mm	Range
W74 CA normal hatchery		
rainbows	1,122,000-1,346,000	-
anemic rainbow	508,000	-
T57 NY healthy hatchery		
rainbows	1,210,000-1,260,000	-
M370 CO wild rainbows	1,470,000	1,050,000-2,100,000
hatchery	1,400,000	970,000-1,810,000
M370 CO wild rainbows	9435 white b.c./cu. mm	2500-26850
hatchery rainbows	7195 white b.c./cu. mm	5000-9550

Young kamloops trout acclimated at 11C died at 24C (B200). In general
the range for rainbow trout is 0 to 28C (M462) with an optimum below 21C.
Lethal temperature of rainbows in Utah was found to be 24C which was believed to be a little lower than normal because the water was low in calcium
and magnesium (A59). High temperature also increased sensitivity and mortality to fluoride concentrations. Rainbow trout were most abundant at 19-21C
in a Colorado reservoir and they moved downward as the summer progressed
(H338). Rainbows were found to acclimate to concentrations of 0-3.5% salt
(B204), and to pH from 5.8 to 9.5 (M462).

Slow-growing trout had higher stamina than faster-growing trout (H404)
but stamina increased with average length. Stamina was better at 18.5-21C
than at lower or higher temperatures and was better within 70 hours of feeding
than later. Fin-clipping did not reduce stamina except for clipping of caudal
fin. Spaghetti and dart tags reduced stamina. The average number of vertebrae decreased by 0.45 for each degree centigrade from 2.5 to 10C during time
of development (G144).

The food of rainbow trout in streams seems to consist mostly of bottom-
living and terrestrial insects (O40, B317, L168, M392, N67, C295, M390, H405,
I7, N13, N122, H300, W134, S105, I11, T122). Amphipods are also often important (O40, I7, S105) and oligochaetes (N122). Frogs were taken in some numbers by trout 178-279 mm long in the Brule River, Wisconsin, and fish were
taken by those over 279 mm (O40). Crayfish and fish, mostly *Cottus* species,
were fairly important foods in Michigan streams for trout 178-406 mm, and
the major foods of those 430-711 mm (M390). Fish were the only food of two
rainbows 610-635 mm long in Horsetooth River, Colorado, and fish were eaten
by trout as small as 165 mm (H300). In the Gunnison River, Colorado, Cladophora was the principal food from late July through September; in the upper
river where Cladophora was missing most stomachs were empty (P125). In
June and July the trout fed on stoneflies, caddisflies and mayflies. Miscella-
neous items found in stream rainbow trout stomachs included blackberries,
hemlock needles, chicken feathers (L168), young birds, and lizards (P128).
Rainbows were found to feed actively all winter, even in frazil conditions
(M392), mostly on aquatic insects. Only 1 of 53 stomachs was empty. Rainbows in beaver ponds continued to feed primarily on stream organisms, in
contrast to brook and brown trout (G106). Compared to brook trout, rainbows
ate few Diptera larvae (L168). Rainbows fed slightly less on surface insects
than did brook or brown trout in North Carolina streams and yearling rainbows
less than older trout (T122). Species which were taken in greater frequency
than their abundance on the bottom were active crawlers or swimmers, fairly
large species, or species on exposed positions on rocks (T122). N67 suggests
stocking aquatic insects in Hawaiian streams to supplement trout food there.
In South Africa (P128, V53) rainbows seem to feed mostly on aquatic insects,
plus some terrestrial insects, snails, freshwater crabs and earthworms. The
larger trout eat quite a number of fish. Ant bites killed fry and some adults in
rearing ponds (V53). Some salmon eggs were eaten by rainbows in winter in
British Columbia (I11).

Aquatic and terrestrial insects are also the primary foods of rainbow
trout in most lakes (M152, R151, B320, W1, H35, N13, A21, H43, L165, J84,
J53, S481). Fish are eaten, particularly by the larger trout in many lakes
(M152, W1, M325, G97, M79, S326, S481, J53). Seventeen species of fish,
mostly cyprinids, but none of the abundant *Coregonus*, were eaten in Lake
Tahoe (M325). Cladocera and other plankton are a considerable part of the
food in some lakes, (E99, W1, J83). Amphipods are also important in many

(E99, R151, A21, H43, J83). Algae were found in 16.6% of the stomachs of
rainbows from an Ontario lake (S481) and algae are more frequent in rainbow
than in other trout (M462).

Rainbow trout depended less on bottom foods than did brown and brook
trout in Castle Lake, California (W1). Fish, mostly Utah chub and redside
shiner, were the dominant foods of rainbow trout in Fish Lake, Utah, in the
summers of 1935, 1938, and 1940, but crustacea, *Gammarus* and *Daphnia*
were dominant in 1922, 1935, and 1948 (S234). In New Zealand lakes, the trout
over 356 mm fed mostly on molluscs and crayfish, while the smaller trout ate
smelt (S326). In Paul Lake, British Columbia, *Gammarus* were the principal
trout foods in early summer, with Cladocera becoming the most important in
mid- and late summer (L62). Aquatic insects, terrestrial insects, molluscs,
leeches and algae were also eaten. The volume of food in December was only
slightly less than in summer (L62). The introduction of redside shiner in Paul
Lake resulted in trout over 330 mm feeding to a great extent on these shiners,
particularly July-September. The volume of food in the stomachs was larger
and the growth of the trout, after age II, was greater after the redside shiner
became abundant (L90). It took several years for the trout to utilize the shiners effectively. Growth of the younger trout was slowed somewhat by the introduction of the shiners. A correlation between size of shiners and the trout
eating them was noted. Large trout apparently do not try to catch the smaller
shiners (C230). Hartman (H324) demonstrated a correlation between mouth
size of young rainbow trout and the sizes of stonefly nymphs, caddis larvae,
and trout fry eaten.

Steelhead rainbows at sea seem to feed mostly on fish, with some squid
and amphipods (T61, L202).

In Paul Lake a sharp decrease in stomach contents was noted with spring
warmup from mid-May to mid-June and coinciding with ovarian maturation
(S148). No evidence of less feeding during the spawning season was evident in
Kenya rainbows, however (V53).

Rainbow trout would feed readily even when water temperatures were reduced to the freezing point, but digestion took 2 to 3 times as long at 1.6C as
at 10C. Trout gorging themselves with fresh trout eggs were unable to digest
their meals in 10 days at 0-3.4C (R149).

Feeding of trout has received much attention in hatcheries, but is not
reviewed here (see P177).

		Sizes of rainbow trout in gill nets	
H313 Alaska	19 mm stretch	13 trout 117 TL	91-198
	25 mm stretch	7 trout 140 TL	109-178
S326	51 mm stretch	most were	193-251
	76 mm stretch	most were	236-546
	102 mm stretch	most were	419-546

A greater catch in gill nets at night than day was noted (S326).

Comprehensive summaries of life history of steelhead trout include S220,
and of inland rainbows, M462.

GILA TROUT, *Salmo gilae* Miller

The Gila trout, more closely related to the rainbow than the cutthroat, is
now found only in 3 headwater streams of the Gila River in New Mexico (R271)
but was once more widely distributed in that river system (M455).

R271 NM   log W = -5.278 + 3.140 log TL

Analysis of covariance with 50 male and 48 female Gila trout showed no significant difference in the length-weight relationship. The K(TL) values increased with age and length (R271) and were higher in June and August than in November. The average for 337 trout, 51-203 mm TL, was 1.04 with a range of 0.71 to 1.27.

R271 NM Main Diamond C.

		No.	Mean TL	Range
Age	I	46	81	51-109
	II	159	117	91-137
	III	87	137	122-157
	IV	28	155	140-168
	V	13	170	160-185
	VI	4	193	180-206

Hatchery fish (13 fish, 185-269 mm TL) produced an average of 150 eggs per female and 2 fish from the creek had 96 and 196 ova (B271). Insects were the principal food items.

ATLANTIC SALMON, *Salmo salar* Linnaeus

The Atlantic salmon is anadromous in streams from New England to Hudson Strait and from Portugal to the Tschernaja River which flows into the White Sea (D188). A few are found on the southwestern coast of Greenland. Data on populations which spend their lives entirely in fresh water are separately considered in this handbook, even though the anadromous and freshwater populations were treated together in the steelhead-rainbow and brown trout-sea trout situations. Much of the European literature has not been reviewed here.

The record Atlantic salmon taken on rod and line was 35880 g, from Norway, but there are reports up to 46764 g from Europe (D188). The largest from American waters was 24947 g.

N46 NY   TL = 1.152 SL   B37 NK   FL = 0.9173 TL - 25 g

Not enough data are available on length-conversion to list all the data in total lengths as has been done for most other species.

N46 NY hatcheries

TL	25	38	51	64
Mean weight	0.13	0.48	1.11	2.40

H325 British Isles   4621 salmon

FL	W	FL	W	FL	W
508	907	762	4581	1016	11158
533	1406	787	4944	1041	11884
559	1588	813	5987	1067	12837
584	1950	838	6033	1092	13700
610	2223	864	6758	1118	14333
635	2585	889	7439	1143	15920
660	2950	914	8210	1168	15650
686	3220	940	8890	1194	16920
711	3629	965	9480	1219	18550
737	4082	991	10342	1245	18950

D188 From a curve based on American and European records

FL	508	635	762	889	1016	1270	1397
W	1134	2722	4536	7938	12474	25400	34019

P136 Scotland $\log W = -5.038 + 3.0 \log FL$

Nall (N57) defines $K = \dfrac{\text{Weight in pounds} \times 100,000}{L^3 \times 36}$. I believed that all here as
K(FL) were computed in this way. However, the formula in M465 does not
have the 36 in the divisor and is thus the usual ponderal index and this may be
true of other data which do not indicate the formula used.

			K(SL)
N46 NY hatcheries	25-64 mm fish		0.95-1.43
W204 Sweden	97 fish, age I, II		1.11-1.26

		K(FL)	Range
H74 Canada	1433 smolt	0.97	0.57-1.44
M11 Canada Moisie R.		1.06	-
C224, H108, J11,23, N55, W59, N57, O19, Great Britain	4737+	1.09	0.75-1.39
H74 Canada	210 parr	1.09	0.75-1.45
B37 NK	1473 fish	1.12	K(TL) 0.91
B35 NF	137 parr, age II	1.12	K(TL) 0.89
B35 NF	57 parr, age III	1.14	K(TL) 0.91
A12 England	896 freshwater fish	1.25	-
V41 France	423 Maiden salmon	1.009	-
	330 kelt	0.98	-
M465 Scotland, River Bran	531 fingerlings	1.10	1.01-1.22
B224 Canada	3 parr, age II	1.04	0.99-1.13
	same fish starved to death	.60	0.58-0.64
A60 NS hatchery	50 yearlings	.60	0.4-1.0
J75 British Isles	380 grilse	1.00	-
	986 salmon	1.07	-
J68 Czechoslovakia	90 adult salmon	1.14	-
E109 Latvia	smolt	-	0.9-1.07
	parr	-	0.92-1.25
	ripe male smolt	-	1.03-1.17
H320 NS	salmon	1.08	-
	kelt	-	0.46-0.60
	flattened kelt	-	0.69-0.75
P152 Ireland	92 parr	1.17	-
H241 British Isles	2442 grilse	0.97	-
	5575 2-summer salmon	1.02	-
	12751 2-summer salmon	1.00	-
	12057 3-summer salmon	1.03	-
	574 3-summer salmon	1.04	-
	758 4-summer salmon	1.04	-
	1988 spawned salmon	1.03	-
W96 British Isles	10 small spring salmon	0.94	0.73-1.06
N56 Scotland	30 grilse	0.95	-
	268 2 and 2+ summers	1.04	-
	16 3 and 3+ summers	1.06	-
	2 4 and 4+ summers	1.13	1.07-1.19

		K(FL)	Range
J71, J72, J73 Taw and Torridge		0.96	some increase with size
S92 N.Z. Te Anau L.	70 immature	0.86	-
	41 spawned	0.87	-
H239 British Isles	20 females over 18 kg	1.16	0.91-1.30
	53 males over 22.5 kg	1.15	0.91-1.51

V73 reports that condition increased from May through August and then declined, in the Petchora River, Russia. Condition of fingerlings was higher March to August than October to January in Scotland (M465). Condition of age I and age II fish was higher in August than February, in Sweden (W204).

							Weight	
	No.	SL	FL	TL	Range	No.	Mean	Range
M396 MA hatchery at 10-17C								
Age 0-Feb.	-	-	-	-	-	+	0.18	-
Age 0-Mar.	-	-	-	-	-	+	0.30	-
Age 0-April	-	-	-	-	-	+	0.46	-
M465 Scotland	100	-	28	-	-	+	-	-
Age 0-May (or first month)								
H38, B38 Can. hatcheries	+	-	28	-	25-30	+	0.025	-
I12 Russia Baltic hatchery	+	-	25	-	20-28	+	0.138	.08-.18
M396 MA hatcheries	-	-	-	-	-	+	0.18	-
M396 MA heated water	-	-	-	-	-	+	0.8	-
A11 NS	+	-	18	-	15-20	-	-	-
M465 Scotland	328	-	28	-	-	-	-	-
H108, H325 G. Brit.	3+	-	36	-	33-38	-	-	-
Age 0-June								
S485 PE	+	-	23	-	-	-	-	-
A11 NS	+	-	23	-	20-25	-	-	-
M465 Scotland	30	-	30	-	-	-	-	-
B23 NY	+	-	-	38	-	-	-	-
B38, H38 Can. hatcheries	+	-	38	-	36-41	+	0.049	-
M396 MA hatcheries	-	-	-	-	-	+	0.44	-
M396 MA heated hatcheries	-	-	-	-	-	+	1.8	-
H109, H325 G. Brit.	6+	-	46	-	38-48	+	0.136	-
S328 ME Sheepscot R.	+	-	-	43	-	-	-	-
I12 Russia, Baltic hatcheries	+	-	51	-	43-58	+	1.7	0.8-2.4
Age 0-July								
H38, B38 Can. hatcheries	+	-	51	-	46-53	+	0.10	-
A11 NS	+	-	25	-	-	+	0.133	-
M465 Scotland	58	-	36	-	-	38	0.45	-
B1 NH	28	42	-	51	41-61	-	-	-

	No.	SL	FL	TL	Range	No.	Weight Mean	Weight Range
R120 ME	-	-	-	-	-	+	0.7	-
V73 Russia Petchora R.	17	-	x	-	33-46	+	-	0.3-1.2
H108, H325 G. Brit.	23+	-	56	-	41-86	-	-	-
M396 MA hatchery	-	-	-	-	-	+	0.84	-
M396 MA hatchery, heated	-	-	-	-	-	+	4.4	-
H331 NS Cape Breton	4	-	56	-	53-58	-	-	-
Age 0-August								
R120 ME	-	-	-	-	-	+	0.8	-
M465 Scotland	280	-	46	-	41-48	146	0.9	-
V73 Russia Petchora R.	7	-	-	-	46-56	+	-	1.0-2.0
H314 NS Mosers R.	5	-	-	-	56-64	-	-	-
G19 NY Raquette R.	1	-	-	56	-	-	-	-
S328 ME Sheepscot R.	+	-	-	-	64-66	-	-	-
H108, H325 G. Brit.	7+	-	64	-	48-89	-	-	-
M396 MA hatcheries	-	-	-	-	-	+	1.88	-
M396 MA hatcheries, heated	-	-	-	-	-	+	9.7	-
Age 0-September								
R120 ME	-	-	-	-	-	+	0.8	-
M465 Scotland	1007	-	48	-	-	-	-	-
V73 Russia Petchora R.	3	-	-	-	46-56	+	-	1.2-2.0
I12 Russia Baltic hatchery	+	-	69	-	58-89	+	4.0	2.3-8.6
H108, H325 G. Brit.	7+	-	79	-	53-99	-	-	-
M396 MA hatchery	-	-	-	-	-	+	4.2	-
M396 MA hatchery, heated	-	-	-	-	-	+	16.2	-
Age 0-October								
M465 Scotland	39	-	46	-	-	17	0.76	-
A18 ME hatchery	-	-	-	-	-	+	-	1.8-2.4
V73 Russia Petchora R.	2	-	x	-	56-58	-	-	2.0-2.2
H108, H325 G. Brit.	19+	-	79	-	51-117	-	-	-
P178 Scotland L. Kinardochy	1636	-	97	-	38-132	-	-	-
M396 MA hatchery	-	-	-	-	-	+	6.6	-
M396 MA hatchery, heated	-	-	-	-	-	+	27.0	-
Age 0-November								
M465 Scotland	280	-	48	-	-	236	1.21	-
Age 0-December								
M465 Scotland	35	-	51	-	-	-	-	-
S484 Ireland, Furnace	+	-	137	-	-	-	-	-
Age 0-Fall								
R120, R121 ME	-	-	-	-	-	+	-	2.8-5.3
A17, B73 NS hatcheries	+	-	-	-	64-76	+	-	0.4-14.5
A45 Sweden hatchery	13760	-	46	-	43-51	-	-	-
S336, A12 G. Brit.	558+	-	56	-	48-61	559	2.0	-
E107 NK	+	-	x	-	33-43	-	-	-
B36 NS Moisie R.	130	-	38	41	-	-	-	-
B37 NS Margarie R.	1094	-	53	58	-	+	1.8	-

	No.	SL	FL	TL	Range	No.	Weight Mean	Range
I12 Russia Baltic hatchery	29	-	86	-	-	+	7.3	-
S328 ME hatchery	+	-	-	-	84-140	-	-	-
A17 NK Miramichi R.	+	-	-	127	-	-	-	-
Age I								
M465 Scotland, Jan.-Mar.	98	-	51	-	46-56	94	1.4	-
V73 Russia Petchora R.	68	-	61	-	46-74	68	2.8	1.1-5.7
A45 Sweden, hatchery, spring	40	-	56	-	-	-	-	-
R120 ME	-	-	-	-	-	+	11.9	-
K2 NY Tuxedo	+	-	-	76	-	-	-	-
S328 Greenland	+	-	-	76	-	-	-	-
I12 Russia Gulf of Riga	155	-	102	-	86-130	155	13.3	5.7-25.5
E109 Latvia pond	+	-	-	-	109-203	+	-	11.3-96
B35, B36, B37, B224, D169, E77, H314, H331, S354 eastern Canada	1181+	-	97	-	71-130	363+	11.3	2.8-17
A12, H108, H375, J76, I9, O35, S336, W23, W59 G. Brit.	2652+	-	112	-	51-163	340	8.5	-
S328 ME Sheepscot R.	+	-	-	-	117-160	-	-	-
J68 Czechoslovakia	67	-	130	-	97-175	-	-	-
G19 NY Raquette R.	3	-	119	-	114-127	-	-	-
B1 NH	10	99	-	119	104-137	-	-	-
S484 Ireland, April	+	-	152	-	-	-	-	-
W204 Sweden hatchery	42	130	-	-	110-167	42	28	14-57
B22 NY Long Island	-	-	180	-	-	-	-	-
Age II								
R120 ME	-	-	-	-	-	+	20	-
V73 Russia Petchora R.	233	-	97	-	79-119	233	10.5	5.7-18.4
I12 Russia Baltic hatchery	430	-	122	-	-	430	24	-
B35, B36, B37, E107, H314, S354 eastern Canada	177+	-	124	-	99-193	150	14	-
A12, I9, J76, M367, M465, O35, S336, W23, S484, W59 G. Brit.	4474+	-	140	-	64-274	158	17	-
J68 Czechoslovakia	26	-	165	-	102-226	-	-	-
C66 Australia	+	-	175	-	-	-	-	-
K2 NY Tuxedo	+	-	-	x	152-229	-	-	-
S328 ME Sheepscot R.	+	-	-	x	183-203	-	-	-
W204 Sweden hatchery	55	152	-	-	142-173	56	40	34-57
B1 NH	1	168	-	206	-	-	-	-
Age III								
V73 Russia Petchora R.	42	-	127	-	112-157	42	26	14-60
B35, B36, E107, S354 eastern Canada	60+	-	140	-	114-190	57	20	-
A12, M367, W23, M465, W59 G. Brit.	568	-	157	-	109-236	3	14	-
G19 NY Raquette R.	6	-	310	-	180-549	-	-	-

	No.	SL	FL	TL	Range	No.	Weight Mean	Range
D1 Norway	+	-	320	-	-	-	-	-
C66 Australia	+	-	-	-	297-450	-	-	-
Age IV								
V73 Russia Petchora R.	6	-	160	-	130-188	6	34	28-85
W23, M465 Scotland	10	-	157	-	132-178	-	-	-
B35 NF	2	-	147	157	-	2	37	-
G19 NY Raquette R.*	4	-	-	498	470-526	-	-	-
C66 Australia*	-	-	493	-	-	-	-	-
D1 Norway*	1	-	980	-	-	-	-	-
G19 NY Raquette R.*								
Age V	2	-	-	528	505-549	-	-	-

*probably include some sea life

	No.	FL Mean of means	Central 50%	Range	No.	Weight Mean of means	Central 50%	Range
Sea life 1+, "grilse"								
M204, M283 NF	290	523	-	432-610	290	1724	-	1134-2858
D52, D48 Norway	+	549	-	-	293+	1633	-	1089-3719
J2 Baltic	4366	531	-	-	4362	1315	-	-
M229, H110, H234, eastern Canada	450	577	508-612	500-686	449	1814	1497-2449	816-3311
C205, H104,108,110,238,245, J11,23,69,70,73,75, N55,56,58, O19, S272, S287, W23,59,76 G. Brit.	5621+	627	617-643	472-773	5055	2676	2495-2858	907-5897
V41 France	28	650	-	-	-	-	-	-
Sea life 2, "small spring fish"								
K90 NK Miramichi R.	+	734	-	-	+	4491	-	-
D48, D52 Norway	+	726	-	-	165+	4853	-	2722-9072
M204, M283 NF	95	724	-	610-813	95	4445	-	3266-7031
M109, M249 CT	2	726	-	711-744	2	4082	-	3583-4581
M11 Canada Moisie R.	54	742	-	-	54	4128	-	-
J2 Baltic	7494	752	-	-	7435	4808	-	-
C243 ME	+	-	-	737-813	+	4536	-	-
S354 PE	8	-	-	696-782	-	-	-	-
B229 NK Miramichi R.	133	769	-	640-851	133	4445	-	2631-6713
C207, H104,108,110,238, 241, J11,23,69,70,71,72, 73,75, M212, N55,56,58, O19, S272,190,423,287, W23, W59, W96 G. Brit.	10966+	765	759-792	495-965	11166+	4808	4400-5216	907-12701
J68 Czechoslovakia	3	869	-	-	3	7575	-	-
Sea life 2.5 years, "small summer fish"								
H237 Canada	362	734	-	-	362	4218	-	-
M204 NF	40	737	-	615-787	40	5080	-	2359-6350
B229 NK Miramichi R.	1047	787	-	681-1000	1047	4627	-	2404-9435
H104,110,238,241, J69,70, 72,73, N55,56,58, S272, S423 G. Brit.	13596	813	800-831	660-953	13620+	5761	-	1134-13517
Sea life 3 years, "large spring fish"								
D48, D52 Norway	+	833	-	-	63+	8210	-	6350-17236
M238 NF	23	831	-	-	23	7167	-	-
K90 NK Miramichi R.	+	925	-	-	+	8936	-	-
B229 NK Miramichi R.	21	947	-	889-1031	21	8845	-	6260-12610
C207, H104,108,110,238, 239,241, J11,23,69,70,71, 72,73,75, N55,56,57,58, O19, S190,272,287, W76 G. Brit.	14603+	942	912-970	813-1372	14594+	9344	-	3856-29030
J68 Czechoslovakia	157	988	-	-	76	10886	-	-

		FL				Weight		
	No.	Mean of means	Central 50%	Range	No.	Mean of means	Central 50%	Range
V41 France	689	968	-	-	-	-	-	-
J2 Baltic	9423	950	-	-	9423	10024	-	-
M11 Canada Moisie R.	+	960	-	-	+	9979	-	-
Sea life 3.5 years, "large summer fish"								
B229 NK Miramichi R.	8	892	-	800-960	8	7938	-	5579-9661
H237 Canada	4	897	-	-	4	8391	-	-
H110,238,241, J69,72,73, N55,56,58 G. Brit	588	1013	-	927-1092	606	10750	-	3629-19504
Sea life 4 years, "very large spring and summer fish"								
D52 Norway	+	894	-	-	-	-	-	-
M283 NF	6	953	-	-	6	9344	-	8890-11249
J2 Baltic	2818	1067	-	-	2818	13517	-	-
M11 Canada Moisie R.	2	1113	-	-	2	16392	-	-
H104,108,239,241, J11,71, 73, 75, N55,56,58, S190, S287 G. Brit.	852	1130	-	1026-1422	849	16194	-	8845-33566
J68 Czechoslovakia	20	1092	-	-	9	13381	-	-
H108 Norway	-	-	-	-	1	33566	-	-
V41 France Adau R.	4	1069	-	-	-	-	-	-
Sea life 5 years								
J2 Baltic	100	1217	-	-	100	19958	-	-
M283 NF	1	993	-	-	1	11113	-	-
J68 Czechoslovakia	3	1133	-	-	2	14877	-	-
D53 Norway	1	1219	-	-	1	12701	-	-
H239 G. Brit.	5	1354	-	1321-1372	5	26082	-	23587-30980
Sea life 6 years								
S245 G. Brit.	-	-	-	-	1	23587	-	-
Sea life 7 years								
S190 G. Brit.	-	-	-	-	1	11793	-	-

The oldest salmon was age XIII from Scotland, weighing 13381 g (F7).

		Life in fresh water				
	No.	Calculated SL at each annulus				
		1	2	3	4	5
B33 NK Upsalquitch R.	402	47	77			
D1 Baltic Bornholm	-	48	114			
D1 Norway Bygland	-	88	170	266		

	No.	Mean calculated FL at each annulus				
B36 NS Moisie R.	-	38	64	91		
D18 Polish Baltic	-	38	81	124	160	201
D17 Polish Baltic	-	41	86	127	150	
D17 Scotch rivers	-	41	84	130	152	
N13 G. Brit. Ewe R.	510	41	97	135	152	
N13 G. Brit. average	-	36	84	137		
V41 France Adour R.	1022	114	178	193		
M465 Scotland	-	43	89	114	142	
W23, W59, O19 Scotland	1525	51	114	127	157	
K90 NK Miramichi R.	-	53	91	109	130	
M204 NF	143	36	86	142	165	183
N56, N58, N59 Scotland	734	48	107	132	160	
J68 Czechoslovakia	93	122	163			
S415 Meerbasen	44	99	160			

		No.	Mean calculated TL at each annulus				
B34	Gaspe St. John R.	-	41	64	86		
B36	Can. Moisie R.	-	41	69	99		
B35	NF	148	38	79	127		
B35	NF Codry R.	-	46	79	112		
B37	Can. Margaree R.	-	56	86			
B35	Can. 21 rivers average	-	51	91	114		
B35	NF 6 rivers	196	51	89	124		
B35	NF Barachois R.	-	53	97	140		
B35	NF	-	38	117	130	152	160
B34	Can. Margaree R.	-	66	140			
C243	ME	-	64	114	178		

As an indication that, for most purposes, growth in fresh water can be disregarded in considering the growth in the ocean and that age can be listed as years of life at sea, the following data are recorded from D18 on the Polish Baltic salmon. Growth after the * is growth in the sea.

	Calculated FL at each annulus							
Age Group	1	2	3	4	5	6	7	8
Freshwater, sea								
II, 5	51	114	*404	775	973	1074	1138	
III, 5	43	89	147	*434	782	970	1072	1171
IV, 4	38	74	114	157	*460	810	975	1044
V, 2	25	56	86	132	178	*361	742	

		Calculated FL at each year of sea life				
		No.	1	2	3	4	5
D18	Polish Baltic	-	399	752	973	1064	1100
D17	Polish Baltic	-	434	747	970	1062	
D1	Norway	-	427	749			
D17	Scotch rivers	-	442				
N13	G. Brit. average	-	475	754	942	1080	
J11	Scotland Dee R.	1033	488	765	927	980	
V41	France Adour R.	1022	503	780	973	1100	
W23, W59, O19, N56, N58, N59 Scotland		2049	475	757	902	1077	
J68	Czechoslovakia	93	450	724	937	1080	1200
K90	NK Miramichi R.	-	452	683			
S415	Meerbasen	44	450	790	1029	1200	

R120 described growth in a Maine hatchery by the following formula:
   Y = 3.4933 - 0.0844X where Y is the log of the number to the pound and
   X is number of months after hatching.
Additional data on the growth of Atlantic salmon are given in the following papers:

Canada B59, B363, C3, C4, M44, M54
British Isles B82, F18, H28,29,99,100,101,102,104,105,106,107,109,110,111,112, 113,114,115,116,235, M12,13,39,40,41,42,43,46,47,48,49,50,51,52,53,55, N21, P10,151, S84, W15,17,18,19,20,21,22,23
Baltic D16, J12
New Zealand G15, P5, S92

Average annual increments FL

		1	2	3	1	Ocean 2	3	4	5
V41	France Adour R.	114	84	56	363	277	193		
	No.	1,022	303	11	(1,022)				
J68	Czechoslovakia	122	69		310	269	213	170	132
	No.	93	24		(93)	(93)	(93)	(6)	(1)

Fourteen years is the greatest age reported for Atlantic salmon (D188).
    In salmon parr of the same lengths, older fish could be separated by their
larger eye size (W170). Scale studies on salmon date from the work of H. W.
Johnston, 1905 (P153). With 142 known-age hatchery-reared salmon, 80% were
accurately aged from scales (H188). Many studies calculate growth from scale
measurements on a direct-proportion basis, but others used a correction factor for body-scale intercept (B34, B35, B36, B37). A curvilinear body-scale
relationship, FL = 24 mm + 12.95 + 1.22 S^2, was used by K90. On parr scales,
calculations were by direct proportion with a correction term; adult sizes
could be back-calculated from scales by direct proportion; but back calculation
of parr sizes from scales taken from adults is complicated (L177). Scale
structure and annulus characteristics have been described by several (e.g.,
J88). Scales of fast-growing fingerlings grow faster in comparison to body
growth than do those of slow-growing fingerlings (B224). The b value of log
(FL - 2.4 mm) = 1 + b log. Scale radius was 0.885, 0.928, 0.944, and 1.097 for
progressively faster-growing groups of salmon fingerlings reaching respectively 75, 78, 91, and 99 mm FL in September of age I. The scales were found
to form at FL 24 mm. In Petchora River, Russia, scales appear about July 7
when the fingerlings are about 36 mm (V73). In Maine (W152) scales form at
about TL 33 mm, and scale formation is completed by TL 46-51 mm. Body-
scale regressions gave intercepts of 23 to 28 mm TL in four populations and a
correction factor of 25 mm was recommended by W152. The body-scale relationship changes at the smolt transition stage (L141). Scale resorption during
spawning also may affect back calculation of growth. Scale resorption occurred with spleen enlargement and malfunction (B224). Also when scales are
removed, neighboring scales may show resorption (B224). No scale resorption
was noted when age II parr were starved to death (B224). Checks on scales
similar to annuli were formed, however, when salmon were transferred, measured, or starved and then fed. Sudden change from high to low temperatures
resulted in poorer feeding and narrowing of the circuli (B224). Regenerative
scales were found to reach normal width in 40 days but not normal length until
over 59 days (B269). Scales near an area of regeneration show growth checks
corresponding to the period of most rapid regeneration. Sometimes two
scales, one small, form in the same pocket.
    Growth of fry and fingerlings was more rapid at 10-18C than in colder
water, and probably best at 15.5-18.5C (M396). The weight increase ranged
from 28-42% at 7.8C, from 21-39% at 8.9C, from 44-73% at 11.7C, and 43-62%
at 13.9C (B224). Alevins up to the time of feeding grew more rapidly on
grooved than smooth surfaces and in dark than in light (M498). Less movement was needed on grooved surfaces. Growth of well-fed fingerlings was
more rapid at 12.2C than at 8.9C, but there was little difference in growth at
these two temperatures with poorly fed fingerlings (B224). Higher temperatures in streams result in more rapid growth and in earlier seaward migration (A51). Young parr grew 5 mm per month in January and February in an
Irish pond (M397).

Yearlings from sea-run stock averaged 109 mm (329 fish) when 290 year-lings from lake stock averaged 104 mm under similar conditions (E77). Year-lings from grilse parentage grew at least as rapidly as those from spring salmon parentage (I9). Wild salmon parr grew more rapidly than stocked parr (S328). Impoundment of Ellerslie Brook, Prince Edward Island, did not appear to affect growth of salmon, but delayed their migration and increased fresh-water mortality (S354). Parr growth appeared to be related to local feeding conditions, showing little evidence of active foraging (H320). The age at which the smolt migrates to the sea varies roughly with latitude (P153). Age I smolts predominate in southern France, and age III in Norway. In Labrador and northern Norway some do not transform to smolts until age VII (D188). Lengths at migration to the sea are 127-152 mm in the Gulf of Maine, 127-178 mm in Canada, 127-229 mm in Newfoundland, and 216 mm in Ungava Bay (D188).

Male salmon usually grow faster than females (H241, J68) and reach a larger size (H239). The largest female reported was 22680 g while males of 33566 g were reported. Even in the first year males grow faster than females (D169, O35). In the Moisie region, Quebec, males were larger than females among previously spawned salmon of comparable ages, but not among maiden fish at 2 or 3 years of sea life (B363).

Males usually mature as parr (A51, O35) and the parr may play an impor-tant role in fertilization (D188). Some may be mature at 8 g (E109). While many males and females mature after a little over 1 year at sea, most do not mature until 2 or 3 years and some not until 4+ years. Greater smolt age is associated with slow growth and with later maturity. There is, however, an inverse relation between years in freshwater and years in sea before maturity (A51). Occasional small pre-grilse with only a short time at sea return to rivers at sizes of 221-406 mm (S423). Females are more likely to spawn more than once than males (J88, D188). Some males and females spawned 3 years consecutively, many spawned only in alternate years, and one skipped 3 years (W158). A few fish may spawn 4 seasons (D188). An old female developed secondary male characteristics (D53). The kelt (spawned salmon) are usually listless and may be carried to the ocean by the currents. They often show no attempt to feed even if they are still in the stream in the spring. Most, how-ever, can be fed and fattened (H320). In Maine, 85% of the spawners are maiden (first-spawning) fish (C243). The percentage that survive after spawn-ing to the next spawning ranged from 1 to 34% (D188).

Salmon may come up from the sea at any time of the year, but the peak in many streams is in March and April, and in others may be summer or fall (J88). In Little Codroy River, Newfoundland, most grilse and small salmon migrate in July, but large salmon may continue to enter throughout August (M283). High temperature and absence of freshets delay entrance into the river from estuarine habitats (D188). Spawning usually takes place in October but may be as late as December (N95, C243, J88, E107). Females lost 24.8 $\pm$ 3.1% of their body weight in spawning (W158). Nest building and spawning have been described many times (D188). Spawning is almost always in fresh water, but 2 redds have been recorded in estuaries where the eggs were at least part of the time in brackish water (S485).

		No.	Size of females	Eggs per female
A17		-	-	9000 by stripping
Li		-	-	9363
		-	10292 g	20992
B23	Penobscot R.	-	3629 g	5000-6000
		-	22680 g	15000
B73		-	737-787 mm	6000-8700
		-	889-1016 mm	16000-20000
P24	Britain	+	4536 g	8800-9000
		+	5443 g	10000
		-	9072-9979 g	17000-18000
J68	Czechoslovakia	5	4763-5216 g	13860-16920 mean 15370
C243	Maine	1	4536 g	8000
I9	Ireland, stripped	11 age 1.5	-	3820
		7 older	-	4640
L166	Russia	-	-	8500-10400
B274	Canada	503 2 year sea	-	8848
		31 older	-	13049
R179		534	-	9092
P136	Scotland	11	549 mm FL	2525
		35	550-599 mm	3341
		13	600-649 mm	4192
		19	650-700 mm	4473
		29	701-750 mm	5247
		20	over 750 mm	6865

P136:   log No. of eggs = 2.3345 log FL - 2.9165
        The standard deviation was correlated with the mean, necessitating the
        log transformation. No differences between stripping and ovary counts
        were noted.
Eggs per kg of female were reported as 1819 (R179), 1594-1839 (B274), 922
and 1698 (J88), and 1100-2205 (D188).

Large females usually have larger eggs than small females (D188), but
may have fewer per kg of female.

Eggs develop normally up to 10C (M396). Incubation time was 191 days at
0.5C and 88 days at 5.5C (D188). Hatchery methods have been well developed
(M396). In Sweden, most salmon are reared in hatcheries through the first 2
winters, because dams have destroyed natural habitat (L203, L204). Egg mortality was higher in a fast-growing stock (D169). Nonfertilization resulted in
less than 1% loss and survival to late eyed stage was estimated at 93.2% in
Maine (W175). Sand, silt and clay in the redds had no noticeable effect and su-
perimposition was not serious (W175). Steady flow of water and lack of ice
favored survival in redds (W175). Bird predator control resulted in increased
numbers of smolt (E78, E107, H328, W171, W172). The effects of the birds
were most pronounced in years of low, clear waters (W171). Brook trout appeared to benefit more than salmon by the bird reduction (W172). Water levels
the year prior to seaward migration were correlated with year class abundance
(H327). Competition of young fish in the rivers appeared to be the cause of a
short cycle, with the cycle started by an unusual scarcity or abundance of fish
(H330). Years of low river discharge were associated with small year classes
(L167) and those of high discharge with large year classes. High temperatures
the first summer usually resulted in small year classes (L167). The abundance

of Canadian salmon varied with local conditions and with interactions of short-term (2-4 years) and long-term (8-9 year) cycles (H329).

Sac fry acclimated at 5-6C had 7-day median lethal temperatures of 22 C, while those acclimated at 10 or 20C had median lethal temperatures of 23C (B206). The ultimate limit on temperature was recorded as 27C (S333). In the Mosers River, Nova Scotia, however, fresh-run grilse died at 29.5C, acclimated grilse at 30.5C, large parr at 32.9C and small parr at 33.8C (H314). Greater resistance to high temperatures was noted in warm rather than cool summers.

Survival of eggs to young of year was estimated at 8.9-11.4%, of age 0+ to I+, 41.1-59.4%, and of age I+ to II+, 25.4-38.1% in Cove Brook, Maine (M380). Survival rates of 0.1-7.1% of fall-stocked fingerlings to smolt size were reviewed by S328 and 15% survival from smolt to spawning in Sheepscot River, Maine. Only 33 of 1,000 parr were lost from age 0-fall to II May in a pond where the salmon were well fed and uncrowded (I9). Survival of smolts to mature salmon have ranged from 1 to 10% (D188).

The weight of fry at the end of the season was similar, 0.96 to 1.18 g/m^2 for 4 years in Allt a' Chomair stream and 2.7 to 29.4 g/m^2 for 2 years in Allt doe Mhuicarain (but 4.25 in another year after fertilization). The number of fry surviving was inversely related to the number of salmon and trout parr in the stream (P179). Overwinter survival the first winter is positively correlated with size of fish (L203), and this seems to be particularly true with a group of fish; the smaller fish have a higher mortality rate.

Fry remain in shallow riffles and other shallow areas of the stream during the summer but move into parr habitats, the pools and deep riffles in autumn. Parr usually remain in the same area most of their freshwater life (S486).

As parr change to smolt, the skin becomes tender, the smolts stay up from the bottom and therefore drift to sea (H320). The young salmon must be 100 mm long before migrating to sea (K90). Smolt migration is usually at night in spring or early summer (D188).

Silver wire or plastic telcothene tags were better than hydrostatic or bottom tags (M367).

Insects (ephemerids, chironomids, and trichopterans) were the principal foods of young and yearling parr while smolts fed mostly on fish (*Thymallus*) in Petchora River, Russia (V73). Chironomids, ephemerids, plecopterans and trichopterans were the principal foods of fry, parr and smolts in Scotland (M465), but cladocerans were also common summer foods of fry. Parr over 64 mm ate some fry in April to July. Their food consisted of small items, mostly from riffle areas (T123). Feeding is intermittent at 0-6C and increases to a maximum at 16C (B224). Huntsman (H331) found in one situation salmon parr fed extensively on sucker fry and suggested that an abundance of suckers indicates a reduction of salmon and trout. The sucker fingerlings also were feeding on chironomids, ephemerids, and trichopterans. Diptera, Trichoptera, molluscs, and Ephemeroptera were the principal foods of parr in an Irish stream (P152) and elsewhere (D188). At sea, salmon feed mostly on fish and euphausiids (D188). They eat little in fresh water on spawning migrations (D188).

Several papers in the Salmon and Trout Magazine have not been included. The selections from this periodical were made on the basis of issues available at the time and do not include equally valuable information in other issues.

For more complete review of life histories of Atlantic salmon, see P153, J88, M398, D188, H406. "The Atlantic Salmon Journal" which started publication in 1957 has not been reviewed in this summary. A 62-page bibliography has been prepared (B362).

SEBAGO SALMON, *Salmo salar sebago* (including other "landlocked" forms)
Populations of Atlantic salmon which spend their entire life cycle in fresh-water lakes are found in United States, Canada, and Europe within the range of the ocean-going populations. They have also been introduced into New Zealand (R146) and Argentina (M214).

S415 Finland

FL	No.	Weight		FL	No.	Weight	
470	2	1043	998-1134	701	10	4128	3674-4581
518	3	1950	1814-2223	724	7	4536	3583-4989
556	2	2041	1814-2585	747	4	5262	5080-5715
579	3	2313	2177-2404	767	7	5352	3311-6804
602	7	2994	2404-3992	800	7	6305	5897-6622
625	8	3039	2585-3493	826	3	7802	6985-8119
645	11	3493	2994-3992	851	4	7530	6078-9525
668	9	3810	3084-5080	942	1	9163	-

	TL	No.	Weight	
H280 ME Mt. Desert	470-475	2	1048	1021-1075
	483-498	8	1288	1248-1474
	546	1	1588	-
N89 NY Cayuga L.	785	1	4354	-
G36 ME	914	1	10206	-

H320 NS   Average $K(FL)$ = 0.95 (maximum length 511 mm)
Y13 males   38 untagged $K(TL)$ = 0.91 $\pm$ 0.04   32 tagged 0.86 $\pm$ 0.07
    females   27 untagged   1.03 $\pm$ 0.06   17 tagged 1.03 $\pm$ 0.08
    juveniles   16 untagged   0.80 $\pm$ 0.04   9 tagged 0.86 $\pm$ 0.12

	No.	SL	FL	TL	Range	No.	Mean W	Range
Age 0-June								
M154 ON Duffin C.	+	-	25	-	-	-	-	-
J87 G. Brit.	5	-	36	-	-	-	-	-
Age 0-July								
M154 ON	+	-	43	-	-	-	-	-
J87 G. Brit.	13	-	56	-	-	-	-	-
Age 0-August								
M154 ON	+	-	66	-	-	-	-	-
J87 G. Brit.	9	-	74	-	-	-	-	-
Age 0-September								
M154 ON	+	-	71	-	-	-	-	-
J87 G. Brit.	20	-	86	-	-	-	-	-
Age 0-October								
M154 ON	+	-	81	-	-	-	-	-
J87 G. Brit.	41	-	102	-	-	-	-	-
Age 0-November-December								
J87 G. Brit.	6	-	89	-	-	-	-	-
M154 ON	+	-	94	-	-	-	-	-
Age 0-Fall								
A61 Sweden Kalarne L.	+	-	-	-	51-81	-	-	-
H280 ME Long P.	25	-	-	127	-	23	28	-

	No.	SL	FL	TL	Range	No.	Mean W	Range
Age I								
W163 NY Black P.	250	-	-	76	-	-	-	-
J78 G. Brit. L. Teyrn	107	-	79	-	74-104	-	-	-
A61 Sweden Kalarne L.	+	-	-	-	69-130	-	-	-
J87 G. Brit. L. Dyrnogydd	340	-	112	-	84-147	-	-	-
M154 ON Duffin C.	+	-	-	-	119-142	-	-	-
F113 NY Black P. spring	+	-	-	160	150-163	-	-	-
F131 NY Brandon Park	359	-	-	168	-	-	-	-
H280 ME Long P.	25	-	-	173	-	25	50	-
F113 NY Black P. fall	6	-	-	272	-	6	164	-
Age II								
J87 G. Brit.	528	-	147	-	119-224	-	-	-
A61 Sweden Kalarne L.	+	-	-	-	99-221	-	-	-
M154 ON Duffin C.	556	-	155	-	104-198	-	-	-
G26 NY Schroon L.	-	-	-	178	-	-	-	-
C62, C64, F26 ME	27	238	-	279x	185-439	20	184	57-765
S67 NS L. Jesse	1	-	290	-	-	-	-	-
F113 NY Black P.	22	-	-	325	-	22	293	-
F131 NY Brandon Park	18	-	-	330	312-361	18	298	232-389
H81 NH	7	-	-	356	-	7	663	-
H280 ME Mt. Desert	42	-	-	445	-	33	694	-
G20 NY L. George	17	-	-	518	356-599	17	1565	227-2268
G15, H240, P4, P23, P65, S92 N.Z.	14	-	551	-	457-653	11	1350	1021-2495
Age III								
J87 G. Brit.	349	-	175	-	150-279	-	-	-
P106 N. Can. L. Astray	3	-	239	-	-	-	-	-
C62 ME Rangely	8	257	-	295x	211-391	8	266	94-513
S67 NS L. Jesse	49	-	325	-	-	49	241	-
C64 ME Moosehead L.	14	328	-	386x	325-450	14	589	369-879
F131 NY Brandon Park	38	-	-	388	373-445	38	468	380-677
F113 NY Black P.	25	-	-	391	340-409	24	573	354-646
F26 ME	30	-	-	419	351-538	30	743	181-1814
F26 ME	24	363	-	-	395-476	-	-	-
C62 ME	88	384	-	445x	376-508	4	655	491-808
G26 NY Schroon L.	+	-	-	457	-	+	-	-1588
B58 L. Erie	-	-	-	-	-	+	1247	-
H280 ME Mt. Desert	26	-	-	531	-	24	1389	-
G20 NY L. George	60	-	-	561	462-693	60	1831	907-4536
H240, P4, P65 N.Z.	18	-	582	-	445-668	18	1761	1219-2863
S415 Finland, L. Iso-Saimaa	11	-	589	-	470-749	11	2744	992-5100
Age IV								
J87 G. Brit.	17	-	221	-	175-274	-	-	-
P106 N. Can. L. Aigneau	2	-	269	-	-	-	-	-
R146 G. Brit.	1	-	279	-	-	-	-	-

	No.	SL	FL	TL	Range	No.	Mean W	Range
P106 N. Can. L. Astray	7	-	307	-	-	-	-	-
W163, F113 NY Black P.	52	-	-	432	-	52	615	558-754
C62 ME Rangely L.	25	376	-	429x	295-572	25	898	201-2342
C64 ME Moosehead L.	6	386	-	452x	384-498	6	1021	454-1418
F131 NY Brandon Park	10	-	-	462	452-475	10	765	706-811
F26, C62 ME	174	432	-	500x	384-703	35	1058	340-2041
G26 NY Schroon L.	+	-	-	559	-	-	-	-
T81 QU	1	-	584	-	-	1	1588	-
G20 NY L. George	27	-	-	599	467-696	26	2228	907-4026
G15, H240, P4, P65, R146, S92 N.Z.	28	-	-	615	381-724	22	1724	1134-2948
H280 ME Mt. Desert	2	-	-	640	-	2	2722	-
S415 Finland	35	-	650	-	510-805	31	3629	1786-6492
Age V								
J87 G. Brit.	1	-	305	-	-	-	-	-
P106 N. Can. L. Aigneau	4	-	315	-	-	-	-	-
P106 N. Can. L. Astray	4	-	386	-	-	-	-	-
O13 Sweden ponds	+	-	421	-	-	+	726	-
W163 NY Black P.	1	-	-	437	-	1	369	-
F131 NY Brandon Park	1	-	-	457	-	1	831	-
H280 ME Mt. Desert	1	-	-	470	-	1	1021	-
C62 ME Rangley L.	42	419	-	478x	371-594	42	1173	388-2421
B58 L. Erie	-	-	-	-	-	1	1247	-
C64 ME Moosehead L.	7	455	-	523x	498-554	7	1520	992-2041
C62, F26 ME	39	-	-	574	432-693	-	-	-
F26 ME	4	566	-	-	528-599	-	-	-
G20 NY L. George	27	-	-	605	483-769	27	2310	1134-4082
P4, P65, R146 N.Z.	6	-	-	696	589-775	4	2254	2098-2495
S415 Finland	42	-	706	-	587-851	34	4830	2416-8193
T81 QU	6	-	759	-	737-793	6	3935	3402-4196
N89 NY Cayuga L.	1	-	-	785	-	1	4366	-
M214 Argentina	-	-	-	-	-	1	4479	-
Age VI								
J87 G. Brit.	1	-	279	-	-	-	-	-
P106 N. Can. L. Aigneau	8	-	345	-	-	-	-	-
P106 N. Can. L. Astray	2	-	564	-	-	-	-	-
B58 L. Erie	-	-	-	-	-	1	1247	-
C62, C64, F26 ME	30	518	-	594x	457-635	16	1860	1083-2977
M214 Argentina	-	-	-	-	-	1	6169	-
S415 Finland	12	-	800	-	711-942	12	6294	3799-9526
R146 N.Z.	1	-	864	-	-	1	8165	-
Age VII								
P106 N. Can. L. Aigneau	6	-	376	-	-	-	-	-
P106 N. Can. L. Astray	2	-	505	-	-	-	-	-
C62 ME Rangely L.	10	485	-	544x	518-597	10	1622	1517-2027
G20 NY L. George	1	-	-	762	-	1	4082	-
Age VIII								
P106 N. Can. L. Aigneau	3	-	399	-	-	-	-	-
C62 ME Rangely L.	2	513	-	577x	572-584	2	1980	1820-2163
Age IX								
W153 ME Long L.	1	-	-	711	-	1	3220	-

(This fish was tagged at age VIII when scales were taken, and grew little thereaft

In general, growth in streams is relatively insignificant to growth in lakes and some authors list sizes only in relation to years of lake life, with 1 to 5 years stream life previous.

	No.	FL	Range	No.	Mean W	Range
1 year lake life						
C147 ME	3	401	368-419	5	635	340-1134
C147 N.Z. L. Te Anau	4	538	445-660	4	1531	907-2268
2 years lake life						
C147 ME	2	544	495-592	2	1474	907-2041
C147 N.Z. L. Te Anau	21	635	584-686	21	2586	1928-3175
3 years lake life						
C147 ME	4	572	559-584	2	2098	1701-2495
C147 N.Z. L. Te Anau	8	701	686-762	8	3161	2268-4763
C147 N.Z. L. Te Anau						
4 years lake life	4	752	660-826	4	3770	3175-4309
5 years lake life	3	782	737-813	3	3912	3175-4989
6 years lake life	2	790	780-800	2	3742	3402-4082
7 years lake life	2	749	699-800	2	3175	2268-4082
11 years lake life	1	699	-	1	2268	-
12 years lake life	-	-	-	1	2041	-

		Average calculated FL at each annulus						
	No.	1	2	3	4	5	6	7	8
P106 N. Can. L. Aigneau	23	61	104	155	211	264	310	345	386
P106 N. Can. L. Astray	18	64	114	173	249	328	411	508	
S415 Byglodsfjord (from Dahl)	+	33	71	109	122	213	236	246	
S415 Sweden (from Runstrom)	+	51	122	193	394	556	671		
S415 Ladoga (from Prevdin)	+	51	109	157	386	495	602	696	
S415 Finland L. Iso-Saimaa									
1 year stream life	4	66	249	411	546	686	818		
2 years stream life	23	76	188	437	612	719			
3 years stream life	65	61	142	208	439	599	752		
4 years stream life	8	58	132	206	269	475	620		
P4, P65 N.Z. L. Te Anau									
1 year stream life	15	127	486	607	645	673			
2 years stream life	22	109	175	424	541	635			
3 years stream life	4	107	198	297	493	559			

		Average calculated TL at each annulus							
	No.	1	2	3	4	5	6	7	8	9
C19 NY Raquette R.	+	53	124	315	439	498				
W84 ME	+	150	295	386	460	516	564	597	640	709
H245 ME Long P.										
1952 year class	+	94	203	290	340	373	427			
1954 year class	+	76	137	297	421	495	610			

In most cases lengths at annuli were calculated on a direct proportion basis, but a Fraser modification of 25 mm was used by H245. In some mountain lakes the first annulus is often poorly formed (J87). Annuli formed in March-April, with rapid growth showing on the scales by then (J87). Tagging data indicated wide variation in growth, as did scale data (J87). Good scale pictures are given by J87.

Growth in some lakes approached that in the ocean, but in general growth was much slower and in some cases very slow. Fry from sea-run stock grew about the same rate as landlocked fry when stocked in Black Pond, New York (F113). Increase in smelt is believed to be a cause of the increased growth in Long Pond, Maine, of the 1952 year class in the 6th year and the 1954 year class in the 4th year (H245). Salmon tagged with a mandibular ring grew more slowly than untagged salmon (Y13).

Landlocked-stock fingerlings and age II salmon survived in Black Pond, New York, better than did Atlantic Ocean stock (F113) at the same ages. Some males and females of the landlocked stock matured at age III (F113).

In Maine, most sebago salmon males mature at age III or IV with some at age I, and most females at IV or V with some at III (W84). About 70% of those spawning are maiden fish (W84) and there is evidence of alternate-year spawning (W85). One female was recorded as spawning at age IV, VI, and VIII (W153). Most spawn October 15 to November 30, but may migrate to the streams in September (W84). Eggs hatched in 56 days at 11.7C, in 65 days at 9.3C, and in 104 days at fluctuating temperatures between 0.5 and 3.9C (D189). The yolk sac constricted, removing some of the yolk, in fry raised in warm water.

B73 Average of 4000 eggs per female.

B22 Three females averaged 767 eggs per female, but general average was 4000.

Food of 25 young salmon in October was listed as aquatic insects and alewives (H280) and of 8 adults in July and August, mostly alewives. Cayuga Lake, New York, a 4354 g fish had eaten 15 smelt and 2 shad (N89). The presence of forage fish was considered the key to success of landlocked salmon in Maine (W84).

Hatchery-reared salmon brought a 0 to 2.6% return to anglers in Fish River Lakes, Maine, an insignificant amount compared to natural spawn (W154). In Long Pond, Maine, fingerling stocking brought 0.76 to 6.1% harvest to anglers (H245). Tagged landlocked salmon were caught by anglers at rates of 24 to 36% within 3 years (W85). Annual survival rates of age III to VII salmon were estimated at 13 to 35% (H245).

BROWN TROUT, *Salmo trutta* Linnaeus

Brown trout are native in northern Europe and the British Isles, but have been introduced into North America, Australia, New Zealand, and Africa. The first introductions into North America were in 1883 (D188). Anadromous populations, usually referred to as sea trout, are common in the original range and a few are established in Newfoundland, Nova Scotia, and Maine to New Jersey (D188). Most populations in North America are confined to fresh water

		SL	No. of fish	
N46	NY	20-175	-	TL = 1.161 SL
C37	MN	130-159	10	1.175
		160-169	31	1.146
		170-229	1184	1.134
		230-299	265	1.126
		300-	5	1.106

G112 Poland 27 trout 24-44 mm TL = 1.173 SL = 1.025 FL, FL = 1.144 SL

F140 Poland FL/SL ratio decreased from 1.19 at 200 mm to 1.04 at 1000 mm.

In the following length-weight relationships the lengths indicate the lower limits of each range and not the class center. No obvious differences were noted in the weights at equivalent lengths of sea trout (N55, B323, S484 Great Britain; D49 Norway; A28 Sweden; F140 Poland) and those from fresh waters (T100 Belgium; N80 Sweden; S415 Finland; N46, S20 New York; E105 Maryland; S178 Michigan; P125 Colorado; S179 Utah; B277 Montana; H258 Wyoming; B233, B248 Oregon) except that a selected group of spawning females in a Pennsylvania fish hatchery (B279) and fish from South Africa (H164,166, P39,41,42,92,93, 95,96,99,128) were heavier than most others at the same lengths.

| | | | | Weight | | | |
| | | | | Mean of | Central | | Citation of |
TL	FL	SL	No.	means	50%	Range	smallest and largest
25	25	22	+	0.1	-	-	N46-N46
38	38	33	+	0.5	-	-	N46-N46
51	51	43	9+	2.6	-	2.0-4.3	S179-T100
76	74	65	16+	5.7	-	5.1-6.0	N46-S179
102	99	87	11+	12.2	-	9.9-14.7	T100-N46
127	124	109	33+	28	-	25-40	N46-S20
152	150	130	131+	48	43-51	28-79	F140-P125
178	173	152	79+	74	57-91	51-113	S20-P125
203	198	174	179+	105	85-116	68-170	N92-P125
229	224	195	141+	145	113-170	82-184	N92-B323
254	249	217	71+	184	170-201	113-340	F140-S. Afr.
279	272	239	66+	235	207-267	150-340	N92-S. Afr.
305	297	260	54+	300	258-334	213-510	P125-S. Afr.
330	323	282	66+	406	369-400	255-794	P125-S. Afr.
356	348	304	71+	556	499-519	397-1049	P125-S. Afr.
381	371	326	88+	658	627-760	340-1021	S. Afr.-S. Afr.
406	396	347	19+	683	638-718	567-794	B277-B277
432	421	369	210+	901	828-921	567-1474	B277-S. Afr.
457	447	391	67+	1295	1080-1361	794-2722	S. Afr.-B323
483	470	412	230+	1247	1003-1304	984-1951	A28-B279
508	495	434	21+	1670	1474-1814	1474-2087	H258-S178
533	521	456	397+	1449	1276-1715	680-2540	S. Afr.-B279
559	546	477	25+	2019	1846-1985	1134-2948	S. Afr.-B279
584	569	499	371+	2098	1738-2268	1738-3583	A28-B279
610	594	521	21+	2702	2546-2807	1134-4678	N323-B279
635	620	543	495+	2877	2197-3232	2197-4196	A28-S415
660	645	564	11+	3385	3008-4111	1588-6078	S. Afr.-B279
686	671	586	377	3060	2807-3402	2495-4252	E105-S. Afr.
711	693	608	6	4621	-	3856-5488	S. Afr.-B279
737	719	630	247	4207	-	3549-6407	A28-S. Afr.
762	744	651	6	5046	-	3685-5897	S. Afr.-S415
787	770	673	211	4286	-	3912-5557	F140-S415
813	792	694	3+	6498	-	5897-7076	B279-B279
838	818	716	40	5443	-	5046-7116	A28-S415
864	842	738	119	5594	-	3899-7439	S415-H258
889	869	760	12	6980	-	-	A28-A28
914	892	781	+	9021	-	-	D49-A28
940	917	803	1	6010	-	-	A28-A28
965	942	825	-	-	-	-	-
991	968	846	39	9752	-	7910-17123	F140-S179
1016	991	868	1	12029	-	-	D49-D49
1041	1016	890	-	-	-	-	-
1118	1090	955	6	15025	-	-	F140-F140

	No.	Length-weight formula
S179 UT Logan R.	386	$\log W = -4.6576 + 2.96 \log SL$
B225 Wales	-	$\log W = -4.8 + 2.92 \log FL$
W156 Poland Wdzydze L.	males	$\log W = -2.689 + 2.65 \log FL$
	females	$\log W = -2.908 + 2.89 \log FL$

Slope of the length-weight regression was 3.0 in an Ireland population (S488), and 2.91 in a Wales river (T124) and 2.86 in a Wales lake (G145).

G132 MI Pigeon R.	1003		$\log W = -4.908 + 2.96 \log TL$
S20 NY	wild		$\log W = -5.422 + 3.1891 \log TL$
	tagged		$\log W = -4.970 + 3.003 \log TL$
H258 WY Pathfinder L.	53	343-864 mm	$\log W = -5.976 + 3.159 \log TL$
B210 PA Spruce C.	1139	114-302	$\log W = -3.883 + 2.94 \log TL$
Tionesta C.	191	127-277	$\log W = -3.917 + 2.95 \log TL$
Kettle C.	136	114-277	$\log W = -4.059 + 3.01 \log TL$

	Length	Number	K(SL) mean	Range
N28 North America	-	+	1.16	-
N122 CA Hot C. hatchery	224	50	1.16	-
Convict C. 5 months later	231	25	0.89	-
N34 CA Convict C.	-	+	1.15	-
S21 NY Crystal C.	-	1412	1.44	-
N46 NY	25-178	+	-	1.17-1.61
C23 MN Duschee C.	100-449	53	1.43	-
H1 Germany	-	178	1.08	-
M467 CO Little Beaver C.	-	+	-	0.87-1.17
W134 OR Sand C.	252	1	1.32	-
M340 UT Carter C.	175	2	1.38	-
R151 CA Convict C.	150-800	75	1.587	-
M370 CO wild	231-297	10	1.56	1.45-1.63
hatchery	183-231	10	1.42	1.12-1.60

	Length	Number	K(FL) mean	Range
W1 CA Castle L.	150-350	489	0.92	-
W2 CA Castle L.	-	878	0.95	-
A28 Sweden, sea trout	300-947	2207	0.94	0.87-1.28
T123 Wales R. Teify	-	-	-	0.99-1.18
A27 N.Z.	56-421	2083	1.23	1.06-1.33
F78 England	Age I-VIII	780	1.09	-
T124 Wales R. Teify	Age I-V	+	-	0.93-1.29
N125 Tasmania NW	Age I-VII			
angler-caught		825	-	1.09-1.35
net-caught		133	-	1.00-1.19
N127 Tasmania North				
Esk R. system	Age I-IV	+	-	1.06-1.34
N128 Tasmania S and SE	Age II-VIII	+	-	1.08-1.55
N126 Tasmania South				
Esk R. system	Age I-VII	652	1.12	1.10-1.33
S313 G. Brit.	25-495	333	1.12	0.69-1.65
P150 Ireland sea-run	224-442	351	1.16	1.15-1.24
P152 Ireland	58-272	59	1.35	-

	Length	Number	K(FL) mean	Range
B225 Wales	Age II–V	+	–	.92–1.14
P4 N.Z. Main Drain	–	+	1.14	–
P4 N.Z.	140–698	233	1.16	0.55–1.95
H58 S. Afr.	–	+	1.19	–
P4 Eng. R. Test	–	+	1.16	–
P4 N.Z. Mataura R.	–	+	1.25	–
P4 N.Z. Rakaia R.	–	+	1.25	–
P4 N.Z. Selwyn R.	–	+	1.26	–
P39,41,42,128,99,92,93,95, 96, H166, H164 S. Afr.	254–787	417	1.26	0.72–2.27
G6 G. Brit.	–	+	–	1.12–1.28
P77 G. Brit.	Age VIII	1	1.48	–
V52 G. Brit. reservoir	–	+	1.16	–

			K(TL)	
F138 MT Meadow L.	Age II–V	30	0.95	0.77–1.14
K34 MT Missouri R.	–	127	0.96	–
C132 MI	–	4352	0.97	0.74–1.23
P55 MT	–	102	0.99	0.94–1.11
H258 WY Pathfinder L.	Age II–VI	88	1.10	–
B155 MT	Age I–IV	636	1.05	–
E62 MI Houghton C.	127–381	1800	0.93	–
Enriched area	127–635	1800	0.98	–
B324 CA Little Sardine L.	127–150	+	1.11	–
	279–302	+	–	0.83–0.94
R151 CA Convict C.	150–800	75	1.051	–
R149 CA starved trout	–	+	–	0.85–0.94
N78 CA	127–178	+	–	0.79–1.09
C132 MI	–	4352	1.26	0.97–1.60
B210 PA	114–302	1102	1.24	0.96–1.74
M437 WI at stocking	142–185	400	1.53	1.51–1.54
later	168–371	1925	1.35	1.28–1.44

Males were heavier than females among the brown trout over 597 mm FL, but there were fewer of them, 122 males to 605 females (A28). At size groups from 300–597 mm FL there was no consistent sex difference in weight, but there were 541 males to 453 females (A28). No sex difference in the length-weight relationships of brown trout in Pathfinder Reservoir, Wyoming, was noted (H285). Females had higher condition factors than males in spawning runs (N128).

Condition factors tended to decrease with increase in length of brown trout in the South African data, though the trout over 610 mm FL tended to have relatively high K(FL) values again. A decrease in condition factor with increase in length was also evident in A27, and with increase in age in B225, K34, T124, N125, N126, N128, B155, K(TL) decreased from 1.06 at age I to 1.00 at age IV (B155). No decrease with age was detected by H258. The slopes of the regression lines for H258 and the wild trout in S20 suggest increased condition with increased length while those for W156, B225, and possibly Spruce and Tionesta creeks (B210) indicate decreased condition. In the others there appears to be little change in condition with increased length. Seasonal changes in condition are noted by N78, E62, B210, B225. Highs of condition factors were reported

in April (G145), June (B225, G145), July (B210), and August (T123, T124) with
lows in late winter or early spring. R149 found that the average K(TL) of
brown trout of 0.793 in October dropped 0.628 in 200 days when the trout were
held without food, and those that died by 245 days had average K(TL) of 0.57
while the survivors at 248 days averaged 0.609. Another group, starved for
180 days, dropped from 0.852 to 0.684 and then when fed daily recovered in 56
days to 0.754. Fertilizing increased the average condition of trout in streams
(E62) and in lakes (M379). Tagged trout had lower condition than untagged
trout, as tested by analysis of covariance (S20). Faster-growing trout in
Naddor and in Avon had lower K(FL) values, 1.12 and 1.22, than slow-growing
trout from the same waters, 1.28 and 1.28 (G6). No difference was evident in
condition factors of brown trout lightly and heavily infested with *Bolbophorus*
metacercaria (F138). There was a negative correlation between condition factor and numbers of *Neochinorhynchus rutuli* (T124). Growth rate and amount
of fat were correlated with seasonal differences in condition in a Wales river
(T123, T124). Condition was better in Tasmanian rivers with a steeper gradient than in a stream with a gradient classified by Huet as barbel zone (M26).
Resident trout had higher condition factors than stocked trout (N127) and
stocked trout lost condition after stocking (N122, M437).

    Condition was lower in brown trout at 2650-2700 m elevation than at lower
elevations in Little Beaver Creek, Colorado (M467). Brook trout were at
higher elevations than the brown trout.

	No.	Mean FL	Range	No.	Mean W	Range
Age						
G112 Poland hatching	9	15	13-15	9	.08	-
M252 G. Brit. 5 days	100	17	15-20	-	-	-
G112 Poland 16 days	14	20	18-20	14	.09	-
G112 Poland 36 days	12	25	23-28	12	.14	-
Age 0-Feb., March (First month)						
W4, W5 CA	-	-	-	+	-	0.13-0.2
B266 PA Spruce C.	+	25	-	-	-	-
M466 G. Brit. beck	+	-	20-30	-	-	-
Age 0-April (Second month)						
M252, J3, M466 G. Brit.	409+	23	20-36	+	-	0.08-0.14
W4, W5 CA	-	-	-	+	-	0.16-0.6
W35 MI	-	-	-	+	-	0.23-0.32
A27 N.Z.	75	30	23-33	-	-	-
T27 NY	+	-	30-38	+	1.3	-
Age 0-May (Third month)						
W35 MI	-	-	-	+	-	0.37-0.53
W4, W5 CA	-	-	-	+	-	0.3-1.1
N24 BC	+	25	-	-	-	-
J3, M252 G. Brit.	+	25	20-36	+	-	0.14-0.2
A27 N.Z.	194	51	33-66	57	1.4	.06-2.3
W72 NY	-	-	-	+	1.4	-
Age 0-June (Fourth month)						
W1, W4, W5, N31 CA	+	38	36-41	+	-	0.5-1.6
H37, T28 N.Z.	-	-	38-58	+	0.36	-

	No.	Mean FL	Range	No.	Mean W	Range
B266, B213 PA	+	-	30-71	+	1.5	-
M466, S336 G. Brit.	+	36	20-56	-	-	-
L161 Denmark	+	-	20-71	-	-	-
W35, S37 MI	23	71	-	+	-	0.7-1.2
A27 N.Z.	108	76	56-79	100	6.2	3.7-7.4
Age 0-July (Fifth month)						
W4, W5 CA	-	-	-	+	-	0.7-2.6
W35, S37 MI	90	79	-	+	-	1.6-2.1
S23 NY Crystal C.	-	-	-	408	4.6	-
S336 Ireland	+	38	-	-	-	-
S400 Poland	198	38	25-51	198	0.5	0.01-1.0
M466 G. Brit. beck	+	-	30-61	-	-	-
W125 G. Brit. hatcheries	56	66	46-94	56	2.8	-
A27 N.Z. Horokiwi R.	134	91	66-99	134	10.8	5.1-13.6
Age 0-August (Sixth month)						
H39, W72 NY hatcheries	-	-	-	+	-	0.6-8.0
W1,2,4,5 CA	+	-	43-56	+	-	1.0-3.7
I9, M397, S336 Ireland	+	-	43-48	-	-	-
S400 Poland	192	43	30-71	+	0.8	0.2-1.4
M466 G. Brit. beck	+	-	30-86	-	-	-
S37, W35 MI	48	91	-	+	-	2.7-3.5
S23 NY slow water	-	-	-	74	4.6	-
S23 NY fast water	-	-	-	134	5.8	-
T27 NY	+	-	69-89	+	7.0	-
G34 WV	-	-	-	+	-	7.9-11.1
H55 Germany	+	-	152-160	-	-	-
A27 N.Z.	133	112	84-127	133	18.7	9.9-25.8
B155 MT Prickley Pear C.	98	84	-	-	-	-
Age 0-September (Seventh month)						
S340 Poland	193	51	36-79	193	1.1	0.3-2.7
L161 Denmark	+	-	41-99	-	-	-
I9, M397, S336 Ireland	+	-	56-71	-	-	-
M466 G. Brit. beck	+	-	36-86	-	-	-
S134 Sweden	96	64	46-89	-	-	-
W1 CA Castle L.	+	61	-	-	-	-
P11, H407, H267, T27 NY hatcheries	+	-	81-114	+	-	2.9-20.0
S23 NY Crystal C.	476	79	56-104	+	4.8	-
S23 NY slow water	8	89	-	8	6.2	-
S23 NY fast water	21	97	-	21	6.9	-
B266 PA	+	-	86-127	-	-	-
B155 MT Prickley Pear C.	230	97	-	-	-	-
G34 WV	-	-	-	+	-	7.8-12.8
S37 MI	8	114	-	-	-	-
A27 N.Z.	68	122	107-135	68	23.9	19-31
Age 0-October (Eighth month)						
N24 BC Vancouver	+	-	46-56	-	-	-
A45, M145 Sweden	15339	53	48-71	+	2.0	-

	No.	Mean FL	Range	No.	Mean W	Range
M466 G. Brit. beck	+	-	30-81	-	-	-
S400 Poland	185	56	43-81	185	1.6	0.5-3.7
L161 Denmark	+	-	41-91	-	-	-
I9, M397, S336 Ireland	+	-	66-79	-	-	-
L98 North Eng. streams	+	-	56-71	-	-	-
N26, N31 CA	+	76	-	-	-	-
S33 MI Perry C.	+	89	-	-	-	-
S33 MI Clarey C.	506	102	-	-	-	-
G34 WV	-	-	-	+	-	7.6-15.4
S331 MN	-	-	-	+	13.6	-
T27 NY	+	140	-	+	40.6	19-62
Age 0-November (Ninth month)						
M466 G. Brit. beck	+	-	41-86	-	-	-
S400 Poland	181	38	48-86	181	2.2	-
I9, M397, S336 Ireland	+	-	71-89	-	-	-
M360 PA Young Woman C.	64	81	51-124	-	-	-
M360 PA 5 other streams, means	+	102	86-114	-	-	-
B155 MT Prickley Pear C.	35	107	-	-	-	-
S489 IA Elk C.	+	132	-	-	-	-
S331 MN hatchery	-	-	-	+	21.9	-
A27 N.Z.	558	163	135-185	558	54.0	31-82
Age 0-December						
I9, M397, S336 Ireland	+	-	74-91	-	-	-
W27 NY	-	-	-	+	-	11-28
S331 MN hatchery	-	-	-	+	-	25-31
B266 PA	+	-	99-198	-	-	-
A27 N.Z.	297	175	135-203	297	68.2	28-150
Age 0						
N33 CA	+	-	30-41	+	-	0.2-0.6
C338 G. Brit. lakes	9	43	-	9	0.4	-
L92 Denmark	+	-	36-130	-	-	-
L205 N.Z. Hinds R.	+	58	-	-	-	-
A46 MT	+	-	53-107	-	-	-
G19, R8 NY	96	-	46-112	-	-	-
S488 Ireland	+	-	51-81	334	2.5	-
S5, A61 Sweden	2+	-	51-102	-	-	-
W100, F54 MI, ME	+	-	76-79	-	-	-
S487 CA Lower Sardine L.	+	-	94-124	-	-	-
C286, B210 PA streams	27+	-	64-109	-	-	-
K104 Czecho. Vrica C.	4	89	-	-	-	-
F78 Eng., 3 Dubs L.	-	-	-	52	56.7	-
N125 Tasmania two creeks	49	104	-	49	14.0	-
H166 S. Afr.	1	279	-	1	312.0	-

In the above data on age 0 brown trout, FL and TL have not been separated but are listed together since in fish at most of the sizes considered the differences between the two measurements are less than 2 mm. In the data that follow it is believed that all lengths are FL except those marked TL.

	No.	FL or TL Mean of Means	Central 50%	Range	No.	Weight Mean of Means	Range
Age I							
V53 Kenya	-	89	-	-	-	-	-
B225, F78, J75, L98, C338, N64, P9, P76, R187, M466, T124, S480, S291, W23,58,60, 42, 125 British Isles	584+	102	84-114	41-229	714	24	3-91
N24 BC Vancouver	+	-	-	64-142	-	-	-
W1 CA hatcheries	132	124	-	79-190	132	26	-
E11, H267, B266, P1, S331 hatcheries	37+	-	-	51-241	+	-	20-162
F72, L92, K104, T106 Europe	91+	150	-	36-231	1	26	-
A23, A45, A61, D49, D51, M276, R123, S5 Norway, Sweden	514+	137	69-193	41-251	86	72	8-94
A46, B155, B359, K34, P78, S487, S179 MT, WY, UT, CA rivers	674+	157	132-165	76-216TL	-	-	-
B213, B210, C286, D108, G19, B266, G21, G22, M360, S21, S482, S23, Y13 NY, PA, NH streams	429+	173	147-188	97-241TL	235+	34	8-105
B10, B187, E103, F54, R151, W1, S487, W2 lakes and ponds	613+	170	155-198	102-254TL	+	-	113-227
S37,211,221,489, B278, W100 MI, MN, WI, IA	71+	170	163-183	117-226TL	+	71	17-170
A27, B194, G16, N125,26, 127,128, P4, W205 N.Z., Tasmania	2754	175	107-246	56-381	610+	130	3-491
M29 U.S.	-	-	-	-	+	-	-340
B115 PA selected females	8	368	-	340-401	8	589	408-816
P128, H166, H311 S. Afr.	45+	363	-	165-462	42+	825	340-1332
Age II							
A51 Sweden ponds	6039	132	-	109-165	-	-	-
A23, A45, D49, D51, M145, M276, R123, S5 Sweden, Norway streams	649+	168	152-211	71-241	63+	40	17-60
B225, N63, N64, P9, F115, P76, V52, W23,42,58,60, 125, S484, S488, T124, Y13 British Isles streams	3591+	175	140-206	104-330	292	54	14-181
B225, F78, J75, R140, C338, R187, G145 British Isles lakes	1395+	185	152-196	97-381	500	65	23-142
B84, N23, W1, W2 MT, WY, UT, CA lakes	715+	203	180-224	157-224	-	-	-
F72, H1, L161, K104, L92 Europe	37+	218	173-272	119-315	29	213	159-261
A23, A61, D49, S415 Sweden, Norway, Finland lakes	420+	218	208-246	140-361	64	116	105-176
C286, G19,21,22,25, M360, S482, S21, PA, NY, NH, streams	287+	229	193-246	145-345	184	102	28-389

	No.	FL or TL Mean of means	Central 50%	Range	No.	Weight Mean of means	Range
A46, B155, B359, K34, P78, B277, S178, S487 MT, WY, UT, CA streams	282+	234	190-279	132-356	2	482	-
B278, S37,211,221,489, W100 MN, MI, WI, IA	18+	236	193-259	188-300	+	184	82-272
A27, B194, P2, P4, L205, W90, N125,126,127,128, W205 Australia, N.Z., Tasmania	4451+	272	236-318	234-478	1507	340	11-2750
V53 Kenya	28	287	-	119-457	2.0	417	-
S10 OR streams	47	295	-	-	47	216	-
E103, F54, R151, H258 WY, NY, CA, ME lakes	127+	323	305-366	290-366TL	+	306	272-397
H166, P39,41,93,97,99, 128 S. Afr.	51	368	-	267-495	50	817	255-1474
B195 PA selected females	60	488	-	432-579	60	1588	936-2353
S246 Punjab	1	508	-	-	1	1588	-
K7, H55, N49 rivers with possible sea life	2+	587	-	442-813	1	1616	-

Age III

	No.	FL or TL Mean of means	Central 50%	Range	No.	Weight Mean of means	Range
A51 Sweden ponds	4340	178	-	135-231	-	-	-
A45, A23, D49, D51, R123, S3 Sweden, Finland, Norway streams	531+	190	170-221	119-241	166+	71	14-130
B225, G11, N55, N63, F115, N64, P76, V52, W23, T124, W58, W60, S488, Y13 British streams	1523+	221	178-246	127-419	91+	74	26-210
C338, G145, B225, F78, J75, M213, R140 British lakes	1227	226	208-241	107-406	428	173	37-1134
A51, S415, D49, M276, A23, A61 Sweden, Finland, Norway lakes	513+	264	259-284	160-536	90	255	207-1800
D92, D161, F72, H1, K1, W156 Denmark, Germany, Czechoslovakia, Poland	105+	274	251-310	157-396	104+	326	150-700
C286, G21, G22, M360, S482, S21 NY, PA, NH streams	170+	287	259-292	157-460TL	86+	227	45-882
B278, S489, S37, W100 MI, WI, IA	4+	290	249-297	218-376TL	+	378	193-567
S487, R151, W2 CA lakes	241+	307	-	239-371	-	-	-
B277, A46, B155, B359, K34, P78, S487, S179 UT, MT, CA, WY streams	228+	312	300-348	208-432	21	545	397-737
V53 Kenya	88	338	-	196-508	81	471	-
N24, S10 OR, BC streams	310	340	-	-	10+	411	372-454
A27, B194, G61, P2, P4, L205, W90, N125,126,127, 128 Australia, N.Z., Tasmania	2638+	348	282-358	178-838	144	1326	204-7131
E103, F54, H258, M296 MA, ME, NY, WY lakes	30+	421	-	351-495TL	+	697	624-771
H163, H166, P41,54,95,96, 99,128 S. Afr.	27	475	-	310-724	28	1243	340-4082
B195 PA selected females	36	610	-	531-742TL	36	3220	1633-5171

	No.	FL or TL Mean of means	Central 50%	Range	No.	Weight Mean of means	Range
Age IV							
A51 Sweden ponds	2824	224	-	178-267	-	-	
A23, A45, D1, D49, D51, M145, R189, R123, S5, S415 Finland, Norway, Sweden streams	437+	244	201-264	140-351	151	111	57-173
F115, B225, G11, N55, N64, T124, P76, V52, W23, W58, S488, W60 British streams	259+	259	211-279	150-546	12+	125	57-340
C338, B225, F78, J75, M213, G145, R140 British lakes	446	272	257-300	130-533	270	278	60-1814
A23, A51, A61, D49, M276, S415 Finland, Sweden, Norway lakes	323+	318	310-323	190-597	67	460	346-3050
S487, R151, W2 CA lakes	363+	323	-	251-421	-	-	-
S246 Punjab	1	330	-	-	1	567	-
G19,21,22,25, S482, M360, S21, NY, PA, NH, streams	70+	363	312-368	236-566TL	61	504	213-2495
B277, A46, B155, B359, K34, N24, P78, S487, S179 MT, WY, UT, CA, BC streams	111+	368	348-406	236-445TL	22+	706	284-907
V53 Kenya	45	371	-	229-533	37	669	
F72, H1, L92, W156 Denmark, Czechoslovakia, Germany, Poland	37+	417	-	310-511	37+	1060	389-2400
S489, W100 MI, IA	+	421	-	391-452	-	-	-
B194, P2, P4, L205, N125, N126, N128, W90 Australia, N.Z., Tasmania	860+	421	358-457	239-660	70	1176	454-4082
F54, H258, M296 MA, ME, WY lakes	23+	490	-	399-544TL	+	1452	-
H163, P39,41,95,96,99, 128 S. Afr.	14	513	-	305-737	14	1803	227-5500
B266, B279 PA selected females	5+	696	-	617-813	5	4355	2750-6350
Age V							
A51 Sweden ponds	2262	257	-	206-312	-	-	-
A23, D49, D51, R123, S415 Sweden, Norway, Finland streams	135	274	246-320	170-411	105	119	85-162
B225, G11, N55, F115, P76 British streams	70	307	193-429	168-528	16+	269	76-567
B225, F78, J75, M213, C338, G145, R140 British lakes	91	315	302-320	216-572	89+	428	113-2495
S246 Punjab	8	335	-	267-406	8	609	312-907
S487, W2, R151 CA lakes	115+	348	-	251-493	-	-	-
G19,21,22,25, C286, M360, S482, S21 NH, NY, PA streams	24+	401	358-419	320-574TL	23	913	434-3629
A23, A61, D49, M276, S415 Sweden, Norway, Finland lakes	175+	404	312-404	201-752	30+	1565	250-5897
P78, B155, N24, K34, B359, B277, S179 WY, MT, UT, BC streams	56	429	419-486	320-500TL	7+	947	567-1361

	No.	FL or TL Mean of means	Central 50%	Range	No.	Weight Mean of means	Range
V53 Kenya	16	462	-	310-648	14	1185	-
W100 MI	+	488	-	- TL	-	-	-
P2,23,4, N125,126,128, W90 Australia, N.Z., Tasmania	243+	544	500-635	330-775	40	2643	539-7711
F54, H258, M296 MA, ME, WY lakes	11+	564	-	493-594TL	+	2360	-
H1, F72, K104, W156 Germany, Czechoslova-kia, Poland	15	572	-	234-671	11+	1704	916-2869
P41,95,96,99,128, H163 S. Afr.	12	635	-	478-737	12	3680	794-6407
B266, B279 PA selected females	5	716	-	655-826	5	5216	3203-7530
Age VI							
W2, S487 CA lakes	87+	277	-	264-282	-	-	-
A51 Sweden ponds	1584	284	-	246-340	-	-	-
A23, A61, D49, D51, R123, S5, S415 Sweden, Norway, Finland streams	66+	335	284-371	190-500	56+	485	85-1500
F78, J75, M213, C338, G145, R140 British lakes	26	361	320-455	231-584	25	595	125-2722
N55, G11, F115, P76 British streams	27	394	-	254-615	8+	513	250-907
C286, M360, B23, G21, S482, G22 NH, NY, PA streams	4+	442	-	284-513TL	6+	2942	1191-4706
B155, P78, K34, B359, S179 MT, WY, UT streams	7+	465	-	340-622TL	-	-	-
A23, A61, D49, M276, S415 Sweden, Norway, Finland lakes	12+	490	-	229-849	9+	-	471-7088
V53 Kenya	2	508	-	457-559	1	1928	-
S246 Punjab	1	526	-	-	1	2268	-
W90, P4, N126, N128, P23 Australia, N.Z., Tasmania	24+	554	521-577	345-775	21	2892	1361-8391
F54, H258, W52 ME, WY, NV lakes	9+	645	-	622-665TL	+	1860	1531-2452
H1, W156 Germany, Poland	1	734	-	-	+	2121	700-4244
Age VII							
G79 CA hatchery two-headed trout	1	203	-	-	1	227	-
W2, B324 CA lakes	28+	279	-	277-282	-	-	-
A51 Sweden ponds	1147	297	-	226-351	-	-	-
A23, D49, D51, R123, S5 Sweden, Norway streams	69+	371	-	201-589	17	128	113-150
G11, P76, S300 British streams	9	381	-	259-577	1	567	-
F78, J75, N55, C338, R140 British lakes	4	450	-	249-610	3+	794	181-1814
A23, A61, D49, S415 Sweden, Norway, Finland lakes	38+	483	-	229-820	6+	-	748-5812
N28, G22, B23, W100, P78, B155 CA, NY, MI, WY, MT	3+	549	-	356-800TL	2+	3955	2948-4989

| | | FL or TL | | | | Weight | |
	No.	Mean of means	Central 50%	Range	No.	Mean of means	Range
P98, P99, P128, H163							
S. Afr.	7	620	-	457-693	7	3200	936-4536
P4, P23, N126, N128, W90							
Tasmania, N.Z.	17+	594	-	483-800	16	2611	1134-7031
F54 ME Branch L.	+	681	-	- TL	+	3583	-
H1, W156 Germany, Poland	1	769	-	-	+	3495	1500-5727
Age VIII							
A51 Sweden ponds	761	312	-	264-358	-	-	-
A61, D49, D51, R123, S5							
Sweden, Norway streams	9+	396	-	208-701	13+	857	91-2495
A23, D49, S415 Norway,							
Finland, Sweden lakes	27+	457	-	231-820	+	-	944-5812
W2 CA Castle L.	3	478	-	-	-	-	-
M213, N13, R140, P77,							
C338, S300 British	6	447	-	264-648	12	1922	218-4989
S246 Punjab	1	526	-	-	1	2268	-
P42, P98, P99, H163							
S. Afr.	2	622	-	559-686	2	2863	1871-3856
G22, C286, K34, S179 NY,							
PA, MT, UT streams	4	625	-	404-749 TL	1	3686	-
W156 Poland	-	-	-	-	+	4023	2646-5710
P4, W90, N128							
Tasmania, N.Z.	4+	632	-	584-698	3	5140	4366-6577
F54 ME Branch L.	+	688	-	- TL	+	3629	-
H1 Germany	1	900	-	-	1	7952	-
Age IX							
A51 Sweden ponds	525	328	-	312-348	-	-	-
A23, S5, D49, D51, A61,							
F7 Sweden, Norway	13+	427	-	279-780	5	567	173-833
W90, N128 Tasmania	1+	638	-	630-650	-	-	-
K34, M64, F54, B233							
MT, WI, ME, OR	4	683	-	549-838 TL	2	4550	4111-4990
W156 Poland	-	-	-	-	+	5514	3742-7938
Age X							
S5, N49 Sweden, Norway	+	505	-	361-820	+	1882	-
A51 Sweden ponds	488	343	-	328-371	-	-	-
W156 Poland	-	-	-	-	+	5316	4678-5982
N51 NY Crystal C.	1	483	-	- TL	1	907	-
Age XI							
H1 Germany	1	1029	-	-	1	1500	-
W156 Poland	-	-	-	-	+	5698	-
D49 Norway	+	579	-	391-861	-	-	-
A51 Sweden ponds	87	356	-	-	-	-	-
Age XII							
W156 Poland	-	-	-	-	+	1497	-
D49 Norway	+	599	-	419-912	-	-	-
Age XIII							
D49 Norway	+	-	-	460-929	-	-	-
Age XIV							
D49 Norway	+	-	-	480-960	-	-	-
Age XV							
D49 Norway	+	-	-	-980	-	-	-
Age XVII							
A61, R183 Sweden	2	740	-	580-900	2	-	150-9934
Age XVIII							
N13 G. Brit.	-	-	-	-	1	5670	-

	FL or TL			Weight		
	No.	Mean of means	Range	No.	Mean of means	Range
Sea life + less than one year at sea, "finrock," "whitling"						
N55,60,62,63,64, O19, P150, W23, W58, N129, W60 British Isles	2392	284	196-371	1824	238	99-680
N16 Norway	131	351	-	131	521	255-907
F140 Poland	3	411	371-490	-	-	-
Sea life 1						
J74, N55,60,62,59,64,129, P150, O19, W97,23,58,60 British Isles	1839	409	279-648	675	771	198-3175
A28, N54, N16 Norway, Sweden	82	486	315-516	65	961	454-2041
F140 Poland	68	698	531-940	-	-	-
Sea life 2						
O19, W23, W58, W60, N55, P150, W97, J74, T82, N63,62,64,129,59 British Isles	344	486	330-635	202	1355	369-3402
A28, D54, N16, R189 Norway, Sweden, Finland	86	480	335-516	31+	-	907-4218
F140 Poland	217	861	500-1041	-	-	-
Sea life 3						
N55,59,63,129, O19, W58, W60, W23 British Isles	107+	508	445-660	94	1602	879-3062
A28 Sweden	219	521	-	-	-	-
F140 Poland	77	869	620-1055	-	-	-
Sea life 4						
N55, N59, N63 British Isles	60+	561	516-686	61	2004	1134-3402
A28 Sweden	157	584	-	-	-	-
D54 Norway (an age XII fish)	-	-	-	1	3742	-
F140 Poland	29	914	790-1000	-	-	-
Sea life 5						
D59 Scotland	25	577	-	25	2291	1361-4082
A28 Sweden	183	622	-	-	-	-
F140 Poland	4	945	740-1072	-	-	-
Sea life 6						
D59 Scotland	5	638	-	5	3068	2268-4536
Sea life 7						
D59 Scotland	2	635	-	2	3062	-
A28 Sweden	35	709	-	-	-	-
Sea life 8						
D59 Scotland	2	668	-	2	3402	2722-4082
A28 Sweden	3	777	-	-	-	-
Sea life 9						
A28 Sweden	1	810	-	-	-	-
D54 Norway (a IV 5 fish)	-	-	-	1	7031	-

	No.	FL or TL Mean of means	Range	No.	Weight Mean of means	Range
Sea life with one spawning mark						
Plus less than another year						
N16,60,63,64, W97 British Isles	179	381	292-445	162	646	184-1701
N62, N63, N64, W97 British, plus 1 year sea	92	526	356-592	74	1747	454-4195
N62, W97 British, plus 2 years sea	20	640	-	15	3047	1644-5783
Sea life with 2 SM						
N60,62,63,64, W97 British, plus less than 1 year sea	31	452	333-533	30	1128	454-1747
N62, N64, W97 British, plus 1 year sea	29	622	406-648	25	2962	737-4536
N62 British, plus 2 year sea	3	775	-	3	5574	4366-7938
Sea life with 3 SM						
N62, N63 plus less than 1 year sea	6	521	429-620	6	1720	836-3402
N63, N62 plus 1 year sea	12	729	648-742	12	4686	2268-7258
Sea life with 4 SM						
N62, N63 (plus 2-3.2)	8	752	-	8	4998	3317-7484
Sea life with 5 SM						
N16, N63, N64 (plus 2-5 years)	18	663	-	22	3754	2495-6804
Sea life with 6 SM						
N62 (plus 2-31.1)	6	853	-	6	7031	6124-7711
Sea life with 7 SM						
N62 (plus 3.1)	1	866	-	1	7598	-
Sea life with 8 SM						
N62 (plus 2+)	1	858	-	1	7371	-

As an example of the small effect of river life on size after sea life, the following are adapted from N63:

River years	Sea life + No.	Mean FL	Sea life 1 No.	Mean FL
1	7	239	1	284
2	335	257	41	368
3	200	284	10	378
4	8	287		

	No.	Median of mean calculated FL at each annulus 1	2	3	4	5	6	7	8	9	10
V53 Kenya	194	119	196	325	394	462	508				
P96, P128, H164 S. Afr.	21	188	310	399	508	564	594	627	665		
G60, W90 Tasmania, N.Z., lakes	56+	132	264	388	462	546	577	594	615	635	602
B194, G60, H210, P2, N125, P4 Tasmania, N.Z., Australia streams	362+	122	231	318	384	470	521	582	665		
A44, F78, C277, M213, R140, S85, S313, W124 British lakes	2013+	66	157	216	284	328	358	427	351		

Median of mean calculated FL at each annulus

	No.	1	2	3	4	5	6	7	8	9	10	11	12	13	14	15
G56, G6, R141, R142, T123, P76, N13, T124, G57 British streams	1456+	86	180	246	259	315		391	442	478	414					
F72 Silesia	461	79	127	160	178	196										
B325, F72, K104 Czechoslovakia	115+	109	185	318	406	450		716	824							
W156 Poland lakes	823+	89	183	277	371	457		546	607	683	762	800	850	1030		
S428 Austria, German Alps, lakes	+		152	356	551	732	861	1080								
S415 Finland lakes	+	43	102	183	328	434		544	599	599	739	813	882	927		
S415, R189 Finland streams	33+	43	94	142	178	224		267								
D50, R183, S292 Norway, Sweden	+	51	114	178	241	333	417	475	559	648	696	732	757	851	960	980

N13,60,61,59,62, W23, W58, W60, P150   ---- Sea life ----

	No.	1	2	3	4	5	6	7 (SL 1)	8 (SL 2)	9 (SL 3)	10 (SL 4)
British parr	5510	64	142	193	231	284		312	406	450	516
D17 Poland	+	81	150	208				447	617	703	739

---- Sea life ----   1   2   3   4

	No.	1	2	3	4	5	6	7	8	9	10	11	12	13
D1, D54, D61 Norway	131	53	119	180	211	236		297	323	361	323	411	599	665

Median of mean calculated TL at each annulus

	No.	1	2	3	4	5	6	7	8	9	10
C276 Switz. L. Geneva	+	163	297	411	528	615					
C276 French rivers	+	94	160	206	249	290	323				
C211, B155, E3, M467, K34, P55, M453, P159, R271, S179 U.S. streams	7209+	97	203	282	348	445	495	551	635	549	
F54, F55, H258, P159, M292, P56, F137, R151 U.S. lakes	1326+	107	216	333	394	513	584	605	620	688	701

Minimum calculated FL at each annulus

	1	2	3	4	5	6	7	8	9	10
(From above references)										
G60, S313, S415 lakes	25	61	112	160	183	203	269	279	480	602
G6, S292, S415, R141, R142 rivers	28	71	119	160	201	257	262	384	411	632

Maximum calculated FL at each annulus

	1	2	3	4	5	6	7	8	9	10
G60, W156 lakes	178	401	597	678	802	762	826	872	900	846
P128, P2, R142, R183 streams	188	335	495	589	607	594	627	665	648	696

Minimum mean TL calculated at each annulus

	1	2	3	4	5	6	7	8
C276 French rivers	64	114	155	183	206	231		
C211, E3, K34, P159, M467, S179 U.S. streams	91	142	193	201	394	465	533	564
F54, P156, F137, M292 U.S. lakes	79	145	196	254	439	486	554	620

(From above references)	Maximum mean TL calculated at each annulus							
	1	2	3	4	5	6	7	8
C276 French rivers	109	211	290	371	439	330		
E3, P55, B155, R271,								
S179 U.S. streams	152	224	335	404	572	513	648	703
H258, R151, F54 U.S.								
lakes	140	318	450	516	579	655	635	681

	Calculated weight at each annulus									
	No.	1	2	3	4	5	6	7	8	9	10
F72 Czechoslovakia	11	14	82	545	1291	1670				
P76 British, western	10	23	170	216						
Chalk district	11	17	142	482	992	1503	1956	2013		
Hill Burne	8	4	18	43	99	156	190	204		
F54 ME Branch L.	+	-	-	272	765	1418	2353	2438	3572	3629	4990
H258 WY Pathfinder L.	88	26	369	978	1512	2155	3221			
W156 Poland Wdzydze L.	823	14	91	278	618	1153	1828	2591	3883	5046	6478

	Annual calculated FL and increments						
		1	2	3	4	5	6	7
S291 Scotland	incr.	89	48	25				
	No.	80	68	7				
(after population								
reduction)	incr.	112	69	33				
	No.	87	23	3				
V53 Kenya		117	196	325	381	462	508	
	incr.	117	119	97	71	56	43	
	No.	176	176	151	62	18	2	
T123 Wales R. Teify		76	165	216	254	272		
	incr.	76	89	56	41	15		
	No.	82	82	58	34	10		
C339 Scotland L. Garry		58	140	188	218	241	254	
	incr.	58	82	51	25	23	15	
	No.	2209	2209	772	364	120	32	
(after impoundment)		74	163	216	269	302	318	
	incr.	74	91	71	61	43	30	
	No.	531	524	1887	2113	1697	1225	
F115 G. Brit. Haweswater		51	109	160	193	221		
	incr.	51	58	56	38	33		
	No.	101	101	60	31	5		
(after impoundment)		56	130	206	251	277	287	
	incr.	56	74	81	56	41	43	
	No.	113	113	98	58	22	8	
K104 Czech. Vrica C.		79	127	157	203	218		
	incr.	79	53	30	28	15		
	No.	100	8	2	1	1		

	--- Sea life ---	
	1	2
P150 Ireland	79	193	216	300	366
	incr.	79	64	30	94	76
	No.	351	114	4	163	5

	------ Lake life -------				
	1	2	3	4	5
S415 Finland L. Iso-					
Saimaa	incr.	76	107	99	97	185	150	97	64	56
	No.	35	33	29	3	26	19	16	7	2

Annual calculated TL and increments

		1	2	3	4	5	6	7	8	9
B155, K34 MT		97	196	282	348	419	513	564	564	549
	incr.	97	99	84	58	61	43	36	25	36
	No.	1035	485	303	128	38	10	4	2	1
S179 UT Logan R.		102	175	246	307	396	465	648	703	
	incr.	102	74	74	81	94	61	165	84	
		260	136	44	12	6	4	3	2	
C211 MI Pigeon R.		94	203	279	358	486				
	incr.	94	109	69	66	66				
	No.	1429	660	74	10	2				
H258 WY Pathfinder L.		140	318	450	516	579	655			
	incr.	140	178	119	91	69	69			
	No.	88	88	72	42	19	8			

Additional information on the growth of *Salmo trutta* in the following papers:
British Isles B46, B89, F17, G3,5,6,7,8,9,10,11, N1,2,3,5,6,7,8,9,10,11,12,14,
15,16,17,18,19,20,22, P10, P115, S85, W21,23,24,44.
Europe P22, S27.
New Zealand A24, A25, G35.

Annual increments as percentages of lengths at beginning of the year

TL	S179 UT	C211 MI	B155, K34 MT	V53 Kenya	Range
51-75	-	-	-	170	131-210
76-101	71	122	104	103	48-164
102-126	69	99	95	97	50-144
127-151	55	-	75	118	37-161
152-177	-	-	51	78	36-135
178-202	42	-	41	56	39-108
203-228	29	32	43	51	27-63
229-253	30	-	29	33	18-38
254-278	29	29	19	32	18-90
279-304	46	17	18	27	7-55
305-329	24	-	22	17	13-38
330-355	-	24	17	16	8-28
356-380	11	-	12	23	6-34
381-405	-	-	18	16	8-25
406-431	40	16	17	9	9-40
432-456	16	-	11	9	9-16
457-482	-	-	9	10	9-10
483-507	-	-	8	16	6-17
508-532	28	-	7	10	7-28
533-558	-	-	5	-	5
559-583	-	-	6	-	6
584-609	-	-	4	-	4
610-634	14	-	-	-	14

In Poland (S399) 58 females from one spawning period to the next year increased 0-22% in length and 4.7-81.4% in weight while 23 males increased 1.3-29.0% in length and 17.5-99.0% in weight.

Scales were considered unsatisfactory for accurate aging of trout (A27), but have been used by most of the papers cited above. The failure of scales to show good annuli in Horokiwi Stream, New Zealand (A27), may have been because winter temperatures do not fall below 7.5C (N108). A false annulus the first summer was reported by L206, but tagging indicated that ages determined from scales were otherwise correct.

Most computations of growth from scales of brown trout have been on the basis of direct proportion, dating from the early work of Dahl, 1910 (K102). Some papers have made a Fraser-type correction (B155, F72, M292). Scale formation starts at 30 mm SL and is generally complete at 45 mm SL (G112). Although scales were found to form at about 30 mm (K102), the scales were found to grow faster than body length up to about 100 mm. There appeared to be need for correction, at least during the first year (K102). Body-scale relationship was found to change at 100 mm in Lake Windermere (F101). N108 found that body and scale did not grow exactly proportionally, but thought a direct-proportion assumption valid for general growth studies. S179 and C211 found curvilinear relationships and made corrections accordingly. There was no sex difference in body-scale relationships (K102). Scale growth was found to commence as much as 4 months later than body growth (Y13).

Annuli form from April to June in Poland (W156). R151 reported scale reading difficult because most of the fish were old and in spawning condition with scales deeply in the skin. Scale pictures are given in R151 and B225. In the Au Sable River, Michigan, 26% of the 102-203 mm brown trout had regenerative scales and 94% of those over 500 mm (W130). Brown (B275) found an annual growth cycle despite constant environment, with rapid growth in length alternating with growth in weight. With constant temperature, Swift (S421), however, found no evidence of a growth cycle. In Llyn Tegid, Wales, brown trout grew most rapidly from June to August or September and showed little or no growth from November to March (G145). In rivers in Wales the rapid growth period was from March or April to August or September (T124). In Pennsylvania, half of the year's growth occurred from mid-April to mid-June and none from December through March (B210). However, also in Pennsylvania, brown trout stocked at 150-254 mm grew 0.15-0.20 mm per day from December to April (T110). A formula for specific growth is given by B201: $\dfrac{\log_e Y_{t+n} - \log_e Y_t}{t+n-t}$ where Y = weight in grams, and t = time in days. Specific growth ranged from 2.03 to 5.60, May 9 to July 20, but from 0 to 1.95, July 21 to February 28 and April 25 to May 8, in Spruce Creek, Pennsylvania (B201). In Scotland the greatest growth in length and in weight occurred in late summer after the period of maximum feeding (Y13). In Pennsylvania streams with fairly uniform temperature, growth occurs throughout the year, but is maximum in spring when temperatures are ideal and insect populations at their highest (B266).

In England brown trout under natural temperature and light conditions, but regularly fed, showed maximum growth in spring and autumn (S306). Growth in spring begins before the water temperatures rise, suggesting that day length is a factor. Peak thyroid activity and fat storage was in midsummer (S306). Maturation of gonads began in females in June and in males in July (S306).

In general, the slowest average growth has been reported from Sweden and Norway, and the fastest from New Zealand and South Africa where most of the records are of newly introduced brown trout. The subspecies *lacustris, trutta,* and *fario* showed little difference in growth the first year when reared under similar conditions in hatcheries (S400) and *trutta* and *fario* showed similar growth when introduced into Swedish lakes (A61). Crosses between the sub-

species had growth similar to the parent stock (A51). Growth of rainbow and
brown trout was similar in New South Wales waters (L206).

Growth, in general, is faster in lakes than streams (for example: W156,
B225, A51, R189, A23, S415, S291), and is fastest in the ocean. The advantage
of ocean growth appears to be less than with the rainbow, or the Atlantic
salmon. Trout which remained longer in streams before migrating to lakes or
ocean were smaller at equivalent ages than trout which migrated earlier
(G145).

In many populations males grow faster than females (A51, H258, H267) but
in some, female brown trout grow faster than males after the first years
(W156). Some investigators have found no sex differences in growth (A51,
B225, B277). Males had a slight advantage in growth in weight but not in length
(H267). Growth decreases after maturity is reached (T124). Fry from large
eggs grow faster than those from small eggs (D2, D51), but the size of alevins
at hatching showed no relationship to later growth rate (B89).

Growth is more rapid in southern than northern Wisconsin (B278). Growth
rate increased with temperature from about 2C (S422). At high temperature,
respiratory difficulty probably hindered more rapid growth. Trout reared at a
constant 8.4C were retarded in growth compared to those reared at an average
12.2C temperature (4-18C, range). These retarded fish had softer fat, and had
lower survival rates when stocked, particularly at higher temperatures (E113).
Growth was found to be related to summer temperatures in Sweden and cold
summers resulted in slower annual growth (S292, R242). Growth increased as
water temperatures approached 10C, and continued to increase to 21C, but a
check on the scale indicating slow growth occurred when temperatures exceeded 21C (N108).

Growth was more rapid with less than 12 hours of light per day (B275) but
S422 reported little correlation between growth and photoperiod.

Maximum growth occurred when the condition factor, K(SL), was about
1.10 (B275).

Growth is faster in alkaline than in acid waters (S273, F141, D2). Pentelow (P9) thought that this was due more to population and food supply than
water condition and pointed out that brown trout grow to large size in acid or
alkaline waters. Survival is greater in acid waters due to lack of predation
and to good spawning conditions and this population density is greater and
growth less (C277). Spawning conditions were found to be inversely correlated
with growth, and masked any correlation between growth and alkalinity in lakes
ranging from 1-109 ppm CaCO$_3$ (C277). Slower growth in more acid streams
was associated with a smaller proportion of large, readily accessible foods in
the diet than in faster growing populations (T124). Greater parasitism of trout
by helminths in acid streams was related to the slower growth (S85), but the
reverse was found by T124 where the faster-growing trout in more alkaline
streams were more heavily parasitized.

Growth of brown trout was found to be correlated with total dissolved
solids, but not with food supply (R151). In Pennsylvania streams a high correlation was found between growth of brown trout and water conductivity (M360).
Fertilization of the water increased the growth of brown trout (Y13, L206), and
growth was slower in less fertile than in fertile streams (M430). Trout grow
faster and reach a larger size in the more productive streams (T124).

Population density is believed to be a most important factor controlling
growth rate (C277, T106, C276, B225). Growth increased in a Montana stream
as the population density decreased (B155). An optimum degree of crowding
was suggested (B89, B275) since, when overcrowded, brown trout were found to

have low appetites and low efficiency in food conversion, but when under-crowded, brown trout ate and grew erratically. Growth increments were less as population density increased in various sections of a stream (N127). The most abundant year class showed the slowest growth, but the growth of age I and II brown trout was not affected by the abundance of age 0 (B194). Population reduction was followed by increased growth (S291), but in another case showed little effect on growth (C286). In streams, numerical adjustments in population density may take place instead of growth adjustments to population density (M468). Increasing the population density by stocking trout reduced growth rate (A80).

Brown trout in aquaria established a hierarchy so that the larger individuals had favored position and food and continued to grow more rapidly than the smaller individuals (B89, B275). When sorted by size in a hatchery, the smaller fish and the larger fish grew at the same rate, on the average, the next year (H267). Selection by size had little value in predicting subsequent growth (H267).

Food supply is also an important controlling factor (D51). The addition of smelt in Quabbin Reservoir, Massachusetts, resulted in considerably increased growth of brown trout (M296). Pellets provided better growth than did corn distillers' solids (S331). Good growth in later years of life was believed to be related to feeding on forage fish (R151). Faster growth among trout more heavily parasitized by helminths suggested that the helminths make their intermediate invertebrate hosts more susceptible to predation by trout (T124).

Competition between trout and salmon parr did not seem to affect growth of the trout (T123, F141). Introduction of char into lakes with trout often results in decreased growth of the trout (N130).

Growth of brown trout in Swedish impoundments increased immediately after impoundment due to washed-in food (R242) and dropped somewhat the next year due to drop in bottom fauna. In general, growth was better after impoundment than before (R242, S292), but in some cases was poorer after several years (R242). The presence of char may make the decrease of bottom fauna more severe on the trout growth in some reservoirs (N92). Increased growth with impoundment indicated that the increase in oligochaetes more than made up for the decline in *Gammarus* in one Swedish reservoir (S292). Growth was found to be fastest during floods (Y13). Increasing the water level of Haweswater by 11.6 m was followed by a considerable increase in growth rate of the brown trout (F115). The increased growth was believed to be a result of increased food supply with the flooding and apparently was not maintained after the lake became stabilized.

Reduced mean water levels in Llyn Tegid reduced the bottom fauna by about 5 percent, but decreased the calorific value of the diet by about 10 percent because of changes in species composition, and the growth of brown trout decreased (G145).

After impoundment of Loch Garry (C339), the brown trout over 2 years old showed an immediate increase in growth probably related to availability of the larger terrestrial insects washed in during flooding. Younger trout did not show an increase in growth rate until 2 years after impoundment. Average size of trout in anglers' catch decreased within 4 years after impoundment, but the growth rate remained high and indicated more rapid harvest by the increased angling pressure (C339).

Tags slowed the growth of brown trout in some cases (P78), but not in others (Y13).

Annual mortality in ponds where the brown trout were fed and where there were no predators, ranged from 1.3 to 46.5%, average about 16%, and showed no trend with age (ages II-XI), growth rate, sex, or age at maturity (A51). In streams in New York (S21), California (N32), Great Britain (F78, B258) and New Zealand (A27) annual mortality was reported at 50-95% for ages 0-VIII. In Polish ponds (S400) survival the first five months was 65%. In a new stream (B213), 22.7% survived one year compared to 39.6% rainbows and 0.1% brook trout. Survival from potential eggs to fall fingerlings in Pennsylvania streams ranged from 1.4 to 5.8%, and from fingerlings to fall age I, 22-74%, and from fingerlings to age II, 5-12%, and from fingerlings to age III, 1.5-2.0% (C212). Winter mortality of brown trout in Convict Creek, California, averaged 60%, principally because of drifting ice and snow (N32).

The upper critical temperature for brown trout acclimated at 5C was 22.5C, at 10C was 24.2C, at 20C was 24.8C, and the ultimate limit was 25.3C (S333). Tests with sac fry acclimated at 5-6C indicated that 22C was the lethal limit for 50% in 7 days, and when acclimated at 10 or 20C was 23C (B206). The upper instantaneous lethal temperature differed with age of the brown trout and with season, but was 27-29C in summer, and 25-27C in the autumn (G113, S400).

The minimum oxygen was listed as 4.5 mg/l in winter and 2.5-3.0 mg/l in summer (G113) or as 2 mg/l (S400).

The sex ratio was reported as 50-50 in both soft-water and hard-water streams, with no trend in age 0-VI (M362). Males sometimes mature at age II and the smallest mature male was listed as 155 mm TL, but most do not mature until age IV (A51). Some males may spawn at age I (P76, B278). A few females mature at age III, and the smallest mature female was listed at 206 mm, but most do not mature until age V (A51). In Montana, some females were mature at age II (B277), also in Great Britain (T124, P76), and is Wisconsin (B278). Slow-growing trout tend to mature at an earlier age (II-III) than fast-growing trout (age III-IV) (T124). In some infertile Pennsylvania streams where growth was slow, some male trout matured at 127-150 mm TL and at age I, whereas in fertile streams with good growth the smallest mature males were 178-201 mm (M430). In fertile streams, some females also matured at age I, at 203-226 mm, but in infertile streams not until age II but at 152-175 mm (M430). In a small stream, an age II female, 147 mm TL, had already spawned (C338). In Horokiwi Stream, New Zealand, a high proportion of females mature at the end of their second year (A27). In some populations brown trout do not mature until VI or VII (A51). In Kenya the age at maturity was III-IV compared to II for rainbow trout (V53). After reaching maturity a few males and some (17% in one case) of the females may skip a spawning season (A51, S399). A sea-run brown trout, age IX, had 2 or possibly 3 spawnings (B233).

Brown trout spawn in October in Wisconsin (O40) and Montana (P156), in November in Michigan (G116), and in October to February in Maine (F55). Controlled light periodicity resulted in spawning a month early, but with smaller eggs (D165). In Kenya, Africa, peak spawning occurs in June and December (V53). Spawning runs were stimulated by increase in water flow (M489). Spawning behavior has been described (G116, O40). Larger trout spawned earlier than the smaller ones (O40). When spawning was in the wider and deeper parts of the river survival of eggs and fry was negligible compared to that in small streams (N127). Generally a female will deposit all her eggs in the redd (H408). When sea trout enter fresh water to spawn, the skin of the

male is hard and fibrous which may assist in holding the female during spawning while that of the female is covered with mucoid cells which may protect the
skin during nest building (S491).

Size of females	No. of females	Number of eggs per female Mean	Range	Citations
Not indicated	14	1061	-	N110 CA
Not indicated	+	ca 1400	200-6000	L1, N28, B22, L101
Not indicated	477	1717	220-7850	N124 Tasmania
Not indicated	12	2020	587-3623	N24 BC
Not indicated	4	2553	-	E4 MN
2 year olds	+	-	303-500	B22, B23 NY
3 year olds	6+	ca 900	370-1600	B22, B23 NY, N108 Tasmania
5 year olds	+	-	1500-2000	B23 NY
178-254 mm 71 g	12+	380	144-476	A27, C212, S122, S291
254-330 mm 142-340 g	99+	640	218-1100	A27, C212, S122, S291
330-406 mm 340-624 g	20	1070	512-1300	A27, C212, S291
406-483 mm 624-1247 g	8+	ca 1800	1648-2000	C212, P24, S122
457-532 mm	9	2598	-	C212 MI
533-583 mm	2	3305	-	C212
- 2041 g	+	4000	-	P24 G. Brit.
737 mm 5488 g	1	7750	-	H166 S. Afr.
445-579 mm 1134-2539 g	23	4334	2973-6629	B279 PA, selected 3 year
508-739 mm 1633-5171 g	22	5858	3700-10250	4 year
617-813 mm 2767-5897 g	5	9786	5964-14342	5 year
655-826 mm 3220-7076 g	5	12008	7960-20865	6 year

	Size	No.	Eggs per kg of female	
B195, B270 PA	2 year, 368 mm, 590 g	8	3679	3203-4839
	3 year, 480 mm, 1406 g	24	3386	2513-3992
	3 year, 493 mm, 1678 g	36	2606	1717-3748
	4 year, 610 mm, 3220 g	36	1920	1116-2778

R179 33 eggs per cm of FL, or 2124 eggs per kg of female; egg diameter
$\qquad$ 4.9 mm

N124 Tasmania lakes Number of eggs = 5.33 (FL-133 mm) or 1.09 (W - 410 g)
$\qquad$ rivers log N of eggs = 3.12 (log FL - 1.6)
$\qquad$ N = 1.27 (W - 16 g)

A27   NZ $\qquad$ Number of eggs = 6.124 (FL-156 mm)

The number of eggs per female trout increases with altitude, perhaps because intraspecific competition, which leads to selection for larger and fewer
eggs, is greater at lower altitudes (S523).

In infertile streams, trout were older at first maturity and produced fewer
eggs because the females were smaller (M430). Also trout of the same size
produced fewer eggs in the less fertile streams. This lower basic fecundity in
the less fertile waters may be an important factor in population balance
(M430). Spawning and recruitment were found to be limited in impounded
waters in Sweden (R242).

Fecundity rates were lower in streams with lower fertility, as measured by water conductivity, and were lower in one fertile stream because of slow growth and high population density (M468).

No change in egg diameter or weight was observed during the incubation period (D165). Older trout had larger eggs (D165). Pellet and meat-diet fish had smaller eggs than fish on the "standard diet" (D165). Chemical composition of the eggs was controlled by diet and by age of brood fish (D165). The left ovary was larger and contained more eggs than the right (B277).

Egg development to hatching takes 165 days at 1.5C (C287, E110)

148 days at 1.9C	59 days at 7.9C
120 days at 2.9C	46 days at 9.2C
95 days at 5.0C	38 days at 10.7C
66 days at 7.0C	33 days at 11.2C

Normal development was found to occur up to 10C, compared to 13C for rainbows (M396), and fingerlings grew best at 13C. Embryology was described (C287). The yolk sac is resorbed at 23 mm (G112). Emergence from the nest was reported at 23 mm, from March 28 to early May in Pennsylvania (B210). Wood pulp fiber reduced hatching of eggs from 98% in the control and at 60 ppm to 24% at 250 ppm (K123).

T57 NY hatcheries Healthy hatchery trout 710,000-800,000 red blood cells
per cu. mm

P139 NY hatcheries 789,000 to 1,063,000 red blood cells per cu. mm, the
lower figures from brown trout on poor diet and slow growth

M370 CO 10 wild trout   231-297 mm SL 1,250,000 rbc 1,050,000-1,630,000
         10 hatchery trout 183-231 mm SL 1,100,000 rbc   960,000-1,220,000

M370 CO 10 wild trout   8885 white blood cells per cu. mm 6050-13900
         10 hatchery trout   6250 white blood cells per cu. mm 4450-13050

N122 NV 10 underfed trout at 18 months 1,450,000 rbc 12.8 hemaglobin 41.8
hematocrit
         10 five months after stocking   1,090,000 rbc 10.3 hemaglobin 36.2
hematocrit

Brown trout of lakes and the ocean spawn in streams and the smolts migrate down more in relation to size than age (B258), but usually after 2-3 years in the streams (F55). Many young were found to migrate to the lake and to return to the streams as yearlings (S291). Downstream movement of trout fry was mostly at night (E128). Homing of brood stock was demonstrated (S291). Movement of trout in streams is also described by H292. In the lakes, the brown trout leave the littoral zones for deeper water in April to June when surface temperatures reach about 12C, and the older trout usually move earlier than the young (B258). A few enter tributary streams for the summer. Juvenile trout use covers more in winter than spring, and increased water current resulted in increased nipping and decreased display behavior (H334).

Fingerlings start feeding on minute bottom fauna, not plankton, after the yolk sac is lost (R220, M405). Fingerlings fed on *Simulium*, Ephemeroptera and Plecoptera, in the summer, usually in the current. In the winter they usually remained under the bank overhang and fed on *Gammarus* and *Asellus* (R220). Maximum feeding occurred at 7.7C and the feeding dropped abruptly at 17C (R220).

With mature trout, food intake increased as the water temperature rose from 4C to 5C (March - April in Scotland) and continued to a maximum in July or August (Y13), with a minimum at spawning time (Y13, P125, W157). Others report a drop in August due to low availability of food or to high temperature (T124). Very little plant material is ingested (B278). Brown trout are active and feed more at night than in daylight (B278) and feed throughout the winter. Trout feed at temperatures below 6C (M405). Brown trout in lakes showed more activity in the day than at night, with most at dawn (S432, S490). They were also more active in May and June (S432) or in June and August (S490) than at other times of the year.

In streams, brown trout fed mostly on bottom fauna, insects, and amphipods (G106, O40, F141, M405, T122, C338, L205, T123, M392, C295, I11, V53, P125, P152, G118) with some terrestrial insects (O40, C295, V53, G118, M405, H405, T122, T123, L205). Brown trout fed more on surface foods than brook or rainbow trout (T122). Flooding increased the number of terrestrial organisms eaten (C339, G118). The fact that 60-80% contained inedible items such as stones suggest that brown trout were not very discriminating in their feeding (V53). They were found active and feeding all winter (M466), even in slush ice (M392). In some streams the larger brown trout feed on fish (T123, L205, B324); (C338, when over 74 g); (I11, when over 455 mm); (O40, when over 300 mm, with some fish eaten by 175 mm brown trout), but they are reportedly less predatory than cutthroats (I11). Even age I and II trout fed somewhat on fish, although salmon parr of the same size did not in the same area (T123).

Impoundment resulted in Entomostroca and terrestrial insects becoming more important in the diets of brown trout (N92). In some reservoirs the impoundment resulted in greater predation on char as the bottom fauna decreased (N114). *Chaoborus,* freshwater crabs, molluscs, insects, frogs and small fish were important foods of brown trout in a South African impoundment (P128). Fish, particularly small rainbow trout, were the principal foods of brown trout in a Colorado reservoir (S396). A 381 mm brown trout ate 2.64 rainbow trout 51-102 mm per day at 14.2C (S397).

In lakes chironomids, *Gammarus,* and surface insects are the principal foods for the smaller trout (N13, W167, R151, W1), but fish are the principal foods of larger trout (R151, W1, W157). Brown trout in Castle Lake, California, eat forage fish to a considerable extent after reaching 3 years of age, and to a greater extent than rainbow trout in the same situation (W1). In estuaries, crabs and fish constituted the bulk of the food of brown trout (L208).

When brown trout and char, *Salvelinus alpinus,* are in the same lake and food is in obvious superabundance, their feeding habits are similar and their food is similar when they live in separate populations, but when food is scarce, the trout continue to feed on bottom fauna and surface food, while char turn toward planktonic foods (N92, N130).

The amount of food to be fed brown trout at various sizes and temperatures in hatcheries is reviewed in P177.

Introduction of brown trout into a small lake resulted in elimination of tadpoles and of certain insects, such as *Notonecta,* which are conspicuous and readily caught by trout (M490).

Year class strength was not related to brood stock nor to groundwater levels in the Pigeon River, Michigan (L207).

In 24 populations in Belgium, the average standing crop was 104 kg per hectare with 74 kg per hectare in acid waters and 166 kg per hectare in alkaline (T100). In New Zealand streams, standing crops were from 44-151 kg per hectare (B194) and in German mountain streams 20-250 kg per hectare (T106).

In Pigeon River, Michigan, 59 and 69 brown trout per hectare were found at the end of the 1949 and 1950 fishing seasons (C143). In the Kinnickninic River, Wisconsin, October populations were estimated at 259 brown trout 64-137 mm and 395 larger brown trout per hectare compared to April populations of 84 and 195 small trout and 731 and 672 larger brown trout per hectare in 2 years respectively (B278). In McKenzie Creek, Wisconsin, there were 356 brown trout over 152 mm and 526 smaller brown trout per hectare, or 81 kg per hectare. In Convict Creek, California, brown trout reproduction averaged 1710 fingerlings per kilometer of stream (N111).

Fry stocking was found to be effective in rivers when properly stocked (L172). In Quabbin Reservoir, Massachusetts, 2.4% one year and 22.6% another year of 229-305 mm brown trout stocked were caught in the first year and an additional 10.2% were caught the second year after the 22.6% were caught (M296). In Peshtigo River, Wisconsin, 2% of the fingerlings stocked, 9% of yearling and age II, and 21% of the legal-sized (305 mm and up) stocked were caught by anglers (B208).

In streams of Windermere Lake, 9 fingerlings per square meter at 6 months is the maximum sustained population (F101).

Evaluations of trout stocking are summarized in C300 and N111.

In Tasmania, stocking after successful introduction was unprofitable (N125, N126). Brown trout are more likely to reproduce and maintain themselves without additional stocking than brook or rainbow trout in the larger, warmer streams (S482).

General reviews of life history of brown trout and extensive bibliographies are given in F142, S487, and B278.

ARCTIC CHAR, *Salvelinus alpinus* (Linnaeus)
    The arctic char is circumpolar and widespread in arctic-subarctic regions of North America, Europe, and Asia, both anadromous and freshwater populations. Isolated populations are found in cold lakes as far south as Switzerland, Quebec, and northern New England (B326). The Dolly Varden is often considered a subspecies, *malma*, but listed separately here.

E53 ME *S. a. aquassa*   7 char   225-305 mm SL   FL = 1.093 SL
                                                     TL = 1.168 SL = 1.068 FL
R214   one male char   838 mm FL   FL = 1.09 SL
The caudal fin is usually lacking in a fork and thus FL = TL.

FL	Mean W	Range	No.	Citation
165	41	-	+	N80 Sweden
203-228	77	68-86	2	M401 Sweden, females
229	86	-	+	N80 Sweden
229-253	122	104-150	25	M400, M401 Sweden, females
254	168	-	+	N80 Sweden
262	168	-	1	E53 ME
254-278	159	127-213	53	M400, M401 Sweden, females
251-295	227	-	74	G65 Can.
279-304	204	141-263	50	M400, M401 Sweden, females
318	254	-	1	E53 ME
305	340	-	+	N80 Sweden
318	340	-	1	S165 Hudson Bay
305-329	281	209-358	22	M400, M401 Sweden, females
297-345	272	-	206	G65 Can.
340	281	-	1	E53 ME

FL	Mean W	Range	No.	Citation
330-355	345	218-468	55	M400, M401 Sweden, females
351-394	680	-	+	B326 Spitzbergen
348-396	499	-	162	G65 Can.
356-380	399	254-548	27	M400, M401 Sweden, females
356	426	-	+	N80 Sweden
381-405	468	354-590	17	M400, M401 Sweden, females
371-399	680	454-907	19	A48 Labrador
399-445	771	-	144	G65 Can.
401-445	998	-	+	B326 Spitzbergen
401-429	816	454-998	63	A48 Labrador
432	558	-	1	M401 Sweden, females
432-457	998	680-1860	105	A48 Labrador
450-498	1089	-	242	G65 Can.
424-495	1497	-	+	B326 Spitzbergen
460-489	1225	907-1678	123	A48 Labrador
490-509	1384	1134-1724	83	A48 Labrador
510-539	1724	1225-2177	86	A48 Labrador
500-549	1497	-	226	G65 Can.
500-546	1588	-	+	B326 Spitzbergen
550-599	1996	-	176	G65 Can.
541-559	1950	1134-2359	43	A48 Labrador
551-597	2177	-	+	B326 Spitzbergen
561-589	2313	1678-2994	32	A48 Labrador
590-619	2722	2132-3402	15	A48 Labrador
601-645	2495	-	+	B326 Spitzbergen
620-639	3039	1905-3629	10	A48 Labrador
601-650	2631	-	100	G65 Can.
640-669	3357	2359-4082	13	A48 Labrador
651-699	3266	-	34	G65 Can.
678	3742	-	1	S165 Hudson Bay
700-749	3810	-	14	G65 Can.
700-747	3992	-	-	B326 Spitzbergen
749	5443	-	1	S165 Hudson Bay
751-800	4717	-	8	G65 Can.
762-826	5761	4763-7711	4	S165 Hudson Bay
914	9072	-	1	S165 MB
965	11793	-	1	S165 Hudson Bay

	A48	Labrador	$\log W = 3.125 \log FL - 4.803$
		dressed char	$\log W = 3.268 \log FL - 5.121$

D90 L. Geneva $K(FL) = 0.9979 \pm 0.0094$

     France $K(FL) = 0.9941 \pm 0.0127$ for males and 0.9844 ± 0.0112 for
                 females

N80 Sweden
   L. Blasjön $K(FL) = 0.86 \pm 1.09$

	No.	FL	Range	No.	Mean W	Range
Age group 0						
M176 QU Ungava	5	48	46-56	-	-	-
A45 Sweden	5151	69	58-79	-	-	-
B326 Sweden hatcheries,						
fall	+	97	-	-	-	-

	No	FL	Range	No.	Mean W	Range
Age I						
A45 Sweden	+	-	51-71	-	-	-
L209 Sweden	+	-	71-112	-	-	-
M176 QU Ungava	13	74	41-81	-	-	-
E129 ME	18	117	91-127	-	-	-
L221 Switzerland	+	-	91-196	+	-	3-77
B326 Sweden hatcheries, fall	+	124	-	-	-	-
Age II						
M176 QU Ungava	18	102	76-132	-	-	-
A45 Sweden	+	-	99-122	-	-	-
L221 Switzerland Albinos	+	-	122-198	+	-	17-79
normal	+	-	132-241	+	-	20-85
B326 Sweden hatcheries, fall	+	-	137-216	-	-	-
S165 NT Little Fish L.	3	163	157-173	3	28	-
E129 ME	21	170	135-206	-	-	-
D90, D94 France L. Geneva	2	246	-	-	-	-
Age III						
M176 QU Ungava	3	117	112-119	-	-	-
A45 Sweden	+	-	170-190	-	-	-
S165 NT Little Fish L.	1	165	-	1	43	-
D91 France L. Janet	+	163	-	-	-	-
E129 ME	70	234	168-292	-	-	-
G65 Spitzbergen	2	246	-	-	-	-
D91, D90 France L. Geneva	18	300	-	-	-	-
R184 Sweden Torron	+	-	239-302	+	-	136-252
Age IV						
G65 Baffin Land Sylvia L.	3	89	84-99	-	-	-
M176 QU Ungava	2	152	127-178	-	-	-
S165 NT Little Fish L.	2	188	185-190	1	57	-
G65 NT Herschell's L.	1	201	-	-	-	-
G65 Greenland	+	190	-	-	-	-
D91 France L. Janet	+	183	-	-	-	-
A45 Sweden	+	-	221-259	-	-	-
E129 ME	56	269	190-343	-	-	-
R184 Sweden Torron	+	-	249-348	+	-	164-380
S165 Hudson Bay sea-run	1	343	-	1	340	-
E93 ME Rangely L.	7	343	262-356	4	261	170-340
G65 Spitzbergen	4	340	-	-	-	-
D90, D91 France L. Geneva	35	353	-	-	-	-
Age V						
G65 Baffin Sylvia R.	18	130	109-145	-	-	-
S165 NT Little Fish L.	1	190	-	1	57	-
G65 Greenland	+	213	-	-	-	-
D91 France L. Janet	+	229	-	-	-	-
G65 NT Herschell's L.	3	254	218-274	-	-	-
E129 ME	11	290	262-330	-	-	-
A45 Sweden	+	-	279-300	-	-	-
S165 Hudson Bay sea-run	3	381	361-401	2	680	-
D90, D91 France L. Geneva	34	409	-	-	-	-
G65 Spitzbergen	23	368	-	-	-	-
A48 Labrador	1	445	-	-	-	-

	No.	FL	Range	No.	Mean W	Range
Age VI						
G65 Baffin Sylvia R.	7	140	124-165	-	-	-
S165 NT Little Fish L.	3	231	203-249	3	122	85-142
G65 Greenland	+	282	-	-	-	-
E129 ME	1	272	-	-	-	-
G65 NT Herschell's L.	1	333	-	-	-	-
S165 Hudson Bay sea-run	3	409	386-427	3	907	-
G65 Spitzbergen	19	411	-	-	-	-
A48 Labrador	9	455	399-483	-	-	-
D90, D91 France L. Geneva females	7	564	-	-	-	-
Age VII						
G65 Baffin Sylvia R.	7	173	145-284	-	-	-
M176 QU Ungava	1	231	-	-	-	-
S165 NT Little Fish L.	2	259	246-274	2	128	113-142
G65 Greenland	+	310	-	-	-	-
G65 NT Herschell's L.	2	373	368-378	-	-	-
G65 QU Ungava	6	409	376-434	-	-	-
S165 Hudson Bay sea-run	7	445	424-470	7	1041	907-1134
A48 Labrador	48	467	411-500	-	-	-
G65 Spitzbergen	11	483	-	-	-	-
Age VIII						
S165 NT Little Fish L.	2	274	-	2	198	-
G65 Baffin Sylvia R.	25	302	170-396	-	-	-
M176 QU Ungava	1	282	-	-	-	-
G65 NT Herschell's L.	2	390	388-391	-	-	-
G65 Greenland	+	391	-	-	-	-
S165 Hudson Bay sea-run	14	467	445-483	14	1219	1134-1247
A48 Labrador	79	480	424-556	-	-	-
G65 Spitzbergen	9	533	-	-	-	-
Age IX						
S165 NT Little Fish L.	5	284	267-300	5	227	-
G65 Baffin Sylvia R.	68	338	185-465	-	-	-
G65 NT Herschell's L.	3	386	358-401	-	-	-
M176 QU Ungava	2	419	396-442	-	-	-
G65 QU Ungava	2	434	424-445	-	-	-
A48 Labrador	108	480	442-569	-	-	-
S165 Hudson Bay sea-run	20	493	470-516	20	1534	1361-1701
Age X						
S165 NT Little Fish L.	7	310	292-330	7	255	198-284
G65 Baffin Sylvia R.	68	363	264-521	-	-	-
M176 QU Ungava	2	388	386-391	-	-	-
G65 Greenland	+	475	-	-	-	-
G65 NT Herschell's L.	1	480	-	-	-	-
A48 Labrador	99	495	452-584	-	-	-
S165 Hudson Bay sea-run	17	523	483-605	17	1701	1588-1928
G65 Spitzbergen	4	671	-	-	-	-
Age XI						
S165 NT Little Fish L.	5	307	297-325	5	284	255-340
G65 Baffin Sylvia R.	73	414	295-549	-	-	-
M176 QU Ungava	1	480	-	-	-	-
G65 Baffin, Bay of Two Rivers	2	485	480-490	-	-	-

	No.	FL	Range	No.	Mean W.	Range
A48 Labrador	66	490	450-627	-	-	-
G65 NT Herschell's L.	2	510	500-521	-	-	-
G65 Greenland	+	518	-	-	-	-
S165 Hudson Bay sea-run	16	564	521-605	16	2220	1814-2495
G65 Spitzbergen	1	701	-	-	-	-
Age XII						
S165 NT Little Fish L.	6	351	318-388	6	360	255-567
M176 QU Ungava	1	363	-	-	-	-
G65 Baffin Sylvia R.	85	424	305-569	-	-	-
A48 Labrador	44	505	475-625	-	-	-
G65 Greenland	+	533	-	-	-	-
S165 Hudson Bay sea-run	4	602	584-622	12	2835	2722-3062
Age XIII						
S165 NT Little Fish L.	4	345	343-351	4	406	397-425
M176 QU Ungava	2	472	445-503	-	-	-
G65 Baffin Sylvia R.	96	483	305-650	-	-	-
A48 Labrador	30	513	483-655	-	-	-
G65 Baffin, Bay of Two Rivers	3	541	470-615	-	-	-
G65 Greenland	+	556	-	-	-	-
S165 Hudson Bay sea-run	3	638	589-660	3	3024	2722-3175
Age XIV						
S165 NT Little Fish L.	3	351	343-356	3	445	425-454
M176 QU Ungava	2	493	478-508	-	-	-
G65 Baffin Sylvia R.	71	511	361-630	-	-	-
A48 Labrador	5	531	475-607	-	-	-
G65 Baffin, Bay of Two Rivers	1	589	-	-	-	-
G65 Greenland	+	602	-	-	-	-
S165 Hudson Bay sea-run	2	635	615-655	2	3345	3289-3402
Age XV						
S165 NT Little Fish L.	2	378	376-381	2	567	510-624
G65 Baffin Sylvia R.	48	554	310-676	-	-	-
G65 Greenland	+	541	-	-	-	-
A48 Labrador	5	572	-	-	-	-
Age XVI						
G65 Baffin Sylvia R.	38	556	439-686	-	-	-
G65 Baffin, Bay of Two Rivers	1	660	-	-	-	-
G65 Greenland	+	668	-	-	-	-
S165 Hudson Bay sea-run	1	678	-	1	3742	-
A48 Labrador	2	691	620-767	-	-	-
Age XVII						
G65 Baffin Sylvia R.	31	559	366-701	-	-	-
G65 Baffin, Bay of Two Rivers	2	645	615-676	-	-	-
S165 Hudson Bay sea-run	3	787	762-826	3	5250	4763-5783
Age XVIII						
M176 QU Ungava	1	554	-	-	-	-
G65 Baffin Sylvia R.	19	599	460-706	-	-	-
G65 Baffin, Bay of Two Rivers	1	650	-	-	-	-
S165 Hudson Bay sea-run	1	749	-	1	5443	-
Age XIX						
G65 Baffin Sylvia R.	23	612	465-739	-	-	-
G65 Baffin, Bay of Two Rivers	4	691	676-706	-	-	-

	No.	FL	Range	No.	Mean W	Range
Age XX						
G65 Baffin Sylvia R.	35	648	549-764	-	-	-
Age XXI						
G65 Baffin Sylvia R.	35	645	500-764	-	-	-
Age XXII						
G65 Baffin Sylvia R.	19	658	440-785	-	-	-
G65 Baffin, Bay of Two Rivers	1	645	-	-	-	-
S165 Hudson Bay sea-run	1	820	-	1	7258	-
Age XXIII						
G65 Baffin Sylvia R.	16	678	511-815	-	-	-
G65 Baffin, Bay of Two Rivers	1	744	-	-	-	-
Age XXIV						
G65 Baffin Sylvia R.	9	678	584-785	-	-	-
Age XXV and +						
G65 Baffin Sylvia R.	12	686	511-826	-	-	-

	FL at each annulus											
	1	2	3	4	5	6	7	8	9	10	11	12
R184 Sweden low year	69	135	178	229	249							
best year	119	178	251	302	348							
F116 L. Windermere												
spring spawners	51	104	145	241	290	320	330	340				
fall spawners	51	104	145	226	259	279	290	300				
G65 Baffin Sylvia R.												
(313 fish)	13	25	51	81	117	155	196	241	284	335	381	427

	13	14	15	16	17	18	19	20	21	22	23	24
(continued)												
G65 Baffin	467	503	536	564	587	605	622	638	653	663	676	686

Growth was determined from otoliths by G65 and R255 who thought the scale method unsatisfactory, but scales were used by several others. The scales were reported to be difficult to interpret by N118, but usable as were otoliths, but not gill covers, fin rays or vertebrae. Otoliths were imbedded in methyl methacrylate and then sectioned by grinding (N118). When scales for the same fish showed differing numbers of annuli, it was assumed that the scale with the largest number was correct (N118, H379). Some scales seemed to be missing the first winter ring (N118) and many mature char ceased to form growth zones.

Two races of char with different growth rates were reported in Lake Windermere (F130), in Iceland (L187) and Sweden (N118), and three races in some other lakes (F129, H378, R255, N118).

Growth increased after impoundment in several Swedish lakes where the char were primarily plankton feeders, but declined after a few years in some of the impoundments (R242). At times, however, char may not respond with good growth or good recruitment to impoundment because they do not take advantage of the newly flooded shallows (L209), particularly when trout are also in the same lake (N131).

Optimum temperature for growth of Windermere char was 12 to 16C and differences in oxygen concentration in the water from 50 to 200% saturation had no detectable effect (S492).

The Baffin Island population grew more slowly than other sea-run populations (B326).

In all lakes in Maine where both species were found, the arctic char grew more slowly and lived longer than brook trout (E129).

Males grow faster than females (A51), but not in Lake Geneva (D90). Growth of sea-run char (S165, G65) seems to be most rapid and growth in the southern part of the range more per year than in the far north. Growth is faster where populations are less dense (M400).

In a general review of European literature, Alm (A51) stated that age at maturity was associated with growth and tended to be reached at a certain size. In fast-growing populations maturity is reached in 5-7 years, but in slow-growing populations not until 11-12 years. In Ungava, Quebec, however, arctic char mature at age II and at 102 mm (M176) or males even at 53 mm and females at 64 mm (D45). In Lake Windermere (F116) males may mature at 3 years and usually both sexes at 4.

In Salangen Lake, Norway (N118), the small race spawns at 145 to 221 mm and first matures at age I, the large race spawns at 211 to 381 mm and first matures at age II, and the sea-run trout spawn at 259 to 566 mm and first mature at age III. As many as 12 spawning zones were detected on the scale of one female from the large race.

In a landlocked population in Newfoundland females mature at age I and males at II (Q4). From this lake they spawn in late October or early November. Hudson Bay char spawn in September and October (S165). In Sweden they spawn from September to October with one population reported spawning in February (M400). They may spawn in streams (N13, M400) or in lakes on stony bottoms at 8 to 100 m depth (M400). Higher average egg numbers are found in populations with low density, because of larger females (M400). In Lake Windermere (F116), one population spawns in water 20-30 m deep near shore in February-March and another spawns in 2-3 months, or in streams in November-December. Spawning may occur only every second or third year in anadromous char (G120, S165). Alternate year spawning was also reported for char in Novaya Zemlya, where they spawn in October and November (Y20). Spawning behavior has been described (F117, F118). The female builds a redd in suitable gravel (H409) in lakes or streams (B326). Temperatures of over 8C may cause egg mortality (H409). Incubation period is 60-70 days at 4.5C (B326).

		Number of eggs per female		
Citation	Size of females	No.	Mean	Range
D45 QU Ungava Bay	81-94 mm	3	21	17-24
M176 QU Ungava	age II, 102 mm	-	under 50	-
G65 Baffin Island	465-665 mm	23	3589	2173-7223
Ungava Bay	376-434 mm	6	2726	2352-3039
Y20 Novaya Zemlya	-	-	3500	-
M400, M401 Sweden	201-432 mm	262	988	265-3471
D91 France	mean, 142 g	-	490	-
	mean, 794 g	-	761	-
B326 Greenland	521 mm	-	4620	-
T88 AK	376-546 mm, 595-1843 g	71	1926	1182-4038
T88	log egg no. = 2.843 log FL - 4.226 cm.			

The number of eggs per female increased with the length of the female in several Swedish populations (M400) and in Alaska (T88).

The number of eggs per kilogram of body weight was 1822 for anadromous char and 3838 for Swedish char (S179), 1970 for Alaska (T88), and 1900-2300 for Lake Windermere (F116). The number of eggs per weight of the female decreases with the size of the female (M400). No differences in the number of eggs in the right and left ovaries were noted (T88). Impoundment did not seem to limit spawning of char since they spawned in the shallows and did not need to go upstream as did the brown trout (R242).

In streams in Jasper Park, Alberta, arctic char ate largely bottom and surface food and in lakes were largely bottom feeders (N13). In a Newfoundland lake they were feeding mainly on smelt (Q4). In Lake Windermere they feed mainly on planktonic cladocera, chironomids, and *Chaoborus* (F116). Char in Little Fish Lake, Northwest Territory had sticklebacks and *Sphaerium* as their food. Char depended more on zooplankton after impoundment in Swedish lakes (N114) in contrast to bottom fauna before. Char and brown trout have very similar food habits when not in competition, but when both species are present they tend to segregate, particularly if food is scarce (N131, N130). The char is more adaptable than the trout and shifts toward a pelagic life. In Maine lakes, juvenile char fed only on plankton and older char on plankton and benthic organisms (E129). The char ate fewer terrestrial and aquatic insects than brook trout in the same lakes, and the char were in deeper water. In an Alaskan lake the Arctic char preys on young red salmon (R272). Char feed little if at all in winter (H409). Anadromous char from Baffin Island fed on amphipods and other small planktonic forms, with only a few fish (G65). The sand-lance was the principal food of char in Hudson Bay (S165). Young cod, capelin, and other fishes are dominant foods off Labrador and Greenland in some areas and amphipods and euphausiids at others (G120). The few sea-run char examined in fresh water in winter contained fish or were empty.

Anadromous char may first migrate to the ocean at 1 to 4 years (G120). Most apparently return to fresh water to overwinter.

An extensive bibliography is given by B326.

SUNAPEE TROUT, *Salvelinus alpinus aureolus* (Bean)

This variety from Sunapee Lake and Big Dan Hole Pond, New Hampshire, Averill Lake, Vermont, and Flood Pond, Maine, has not been successfully established elsewhere (N74, E93) and is nearly extinct in Sunapee Lake.

G36 NH	718 mm TL trout weighed 3629 g	
M99 NY	724 mm TL trout weighed 4309 g	

	Age	No.	Mean SL	Range	Mean TL	Range	No.	Mean W	Range
F26, E93 ME									
Flood P.	I	1	130	–	152	–	1	26	–
N74 NH	II	22	–	–	198	–	3	57	–
F26, E93 ME	II	6	211	173-244	249	203-290	6	133	102-176
N74 NH	III	61	–	–	287	–	51	230	–
F26, E93 ME	III	8	277	246-328	325	284-381	8	321	216-474
N74 NH	IV	43	–	–	356	–	40	468	–
F26, E93 ME	IV	7	356	330-391	417	388-455	7	700	539-857
F26, E93 ME	V	1	417	–	495	–	1	933	–
N74 NH	V	13	–	–	508	–	8	1222	–
N74 NH	VI	6	–	–	665	–	6	2442	–
	VII	1	–	–	762	–	1	3686	–
	VIII	1	–	–	800	–	1	3629	–

Presence of smelt as forage favors rapid growth (N74).

Some males mature in the second year but most in the third, some females in the third but most in the fourth (N74, N95). They spawn mid-October to mid-November (N95). They spawn on rubble and boulder shores from a few centimeters to 6 m deep (N74). A female had 1200 eggs (M426). Stripped females produced 938 eggs per female 1951-1955 (N74). Incubation at 7C takes 80 days (N74).

BROOK TROUT, *Salvelinus fontinalis* (Mitchill)

Brook trout inhabit cold-water streams and lakes from northernmost Labrador, Hudson Bay and James Bay, to northern New Jersey and in the Allegheny Mountains to North Carolina and Georgia, and in the Great Lakes Region, and to northeastern Iowa and southeastern Minnesota to eastern Saskatchewan (B326). Brook trout have also been successfully introduced into high-altitude lakes and streams in the Rocky Mountains of United States, Canada, and southern Alaska, and in South America, South Africa, New Zealand and northern Europe. Sea-run populations, often known as salters, have been known from Long Island to Hudson Bay and are still abundant in some areas. In smaller streams and lakes, a 680 g brook trout is a large one, and a 2300 g is large in larger bodies of water, though specimens to 6577 g are known (B326). Sea-run brook trout average larger than the freshwater brook trout but do not exceed the maximum size reported above.

	No.	Length	
N46 NY	-	-	TL = 1.151 SL
B326 PE sea-run	6	209-247 mm	TL = 1.06 SL
C37 MN	46	120-159 mm	TL = 1.136 SL
	431	160-199 mm	TL = 1.124 SL
	23	200-279 mm	TL = 1.102 SL
W75 NS	380	87-428 mm	SL, FL = 4.4 mm + 1.0096 SL
natural position			TL = 10.3 mm + 1.100 SL
I8 ID	34		TL = 1.1536 SL = 1.0326 FL,
			no trend with sex, size, or age
C132, C131 MI			log TL = 0.15037 + 0.964 log SL

In the following data, standard lengths have been converted to total lengths by using the relationship TL = 1.15 SL since this gave values about in the middle of the various conversion factors throughout the range. Fork lengths were converted to total lengths by using the relationship TL = 1.0326 FL.

TL	Mean of means	Weight Central 50%	Range	No. of fish	Citations
25-37	0.12	-	-	+	N46
38-50	0.77	-	0.45-0.85	665+	N46, N73, W139
51-63	1.6	-	1.1-2.0	22+	N46, W139
64-75	2.6	-	1.7-3.7	4429+	N73, A43, N46, S140, W139
76-101	7.7	7.4-8.5	2.8-11.3	582+	N46, W139, A43, N73
102-126	14.2	11-16	10-24	818+	N46, N73, A43, R151, W139
127-151	24.4	23-28	8-43	751+	R202, N46, N73, R151, W139, S380, A43, S161
152-177	45	43-48	16-60	687+	N46, S161, N73, R151, R202, A43, W139, S524

TL	Mean of means	Weight Central 50%	Range	No. of fish	Citations
178-202	68	63-77	45-88	564+	N73, R202, N46, W57, R151, S161, A43, W139, E105, G89, S524
203-228	102	94-111	76-142	864	T24, W57, A43, N73, R151, W139, S161, R202, M135, S524
229-253	139	122-145	113-198	2134	R151, W57, S161, M135, W139, R202, N73, S524
254-278	195	171-210	170-340	772	R151, W57, T24, S161, S524, M135, W139, R202, P54
279-304	273	227-284	201-510	279	R151, T24, M135, S161, W139, W57, R202, P54, P128, S524
305-329	360	312-378	272-500	153	R151, T24, M135, W57, W139, S178, S161, R202, P54, B279
330-355	455	391-510	349-638	104	T24, M135, W139, W57, P128, R202, S178, B279
356-380	573	488-646	454-936	81	R151, M135, T24, W57, W139, S178, B279, P54, P128, H166, R202
381-405	731	567-743	533-1049	41	T24, R151, S178, P128, H166, B279, M135
406-431	884	765-1012	652-1180	29	T24, W139, S178, P54, P128, B279, M135
432-456	1080	1006-1083	851-1679	36	M135, T24, W139, W57, P128, B279
457-482	1355	1077-1497	964-2268	16	M135, T24, W139, B279
483-507	1755	-	1403-2176	7	W139, B279, W198
508-532	2140	-	1814-2676	3	M135, B279, M491
533-558	2211	-	-	1	M135 AT
673	5670	-	-	1	K6 ME
800	6577	-	-	1	E4 ON

W75 NS		Fresh-run sea trout August-September	$\log W = -5.2880 + 3.2117 \log SL$
		sea trout	$\log W = -5.3245 + 3.2128 \log SL$
		Freshwater trout	$\log W = -4.4639 + 2.8526 \log SL$
R151 CA		661 trout	$\log W = -4.6806 + 2.9146 \log SL$
R29 ON Wolf L.			$\log W = -5.433 + 3.16 \log FL$
R29 ON Lac Cassette			$\log W = -5.257 + 3.12 \log FL$
R29 ON Glen Major ponds			$\log W = -5.215 + 3.09 \log FL$
R29 ON Mad R.			$\log W = -4.991 + 3.01 \log FL$
B228 ON Red Rock L.	295 trout May-June		$\log W = -4.674 + 2.862 \log FL$
A43 WY			$\log W = -5.2105 + 3.1106 \log FL$
S524 QU Baldwin P.	2654 trout, Age II		$\log W = -4.523 + 2.781 \log TL$
S194 PE	152-343 mm		$\log W = -5.115 + 3.097 \log FL$
C132 MI	1997 trout		$\log W = -4.874 + 2.94 \log TL$
C281 MI	5212 trout 58-323 mm		$\log W = -4.950 + 2.99 \log TL$
PA hatcheries			$\log W = -5.239 + 3.14 \log TL$
significantly different			$\log W = -5.214 + 3.13 \log TL$
			$\log W = -5.266 + 3.15 \log TL$
			$\log W = -5.287 + 3.15 \log TL$
			$\log W = -5.160 + 3.10 \log TL$

Composite of 3 rapidly growing populations $\log W = -5.257 + 3.154 \log TL$
B170 MA   1303 trout $\log W = -4.708 + 2.89 \log TL$
W206 PA   1088 trout from
                           infertile streams $\log W = -5.095 + 3.04 \log TL$

Location	Size of fish	No.	Mean	Range
				Condition factor K(SL)
Streams				
M467 CO Little Beaver C.	-	83	-	0.79-1.25
G103 CA Sagehen C.	Age II, 142 mm	64	0.94	0.81-1.29
	Age III, 185 mm	34	1.08	0.59-1.33
	Age IV, 203 mm	5	1.13	0.89-1.28
H84 NH	-	-	-	1.12-1.31
C23 MN	150-294 mm	4	1.33	-
H42 NY	-	202	1.37	1.18-1.75
F53 WY	Age I-IV	117	1.41	-
S39 MI Hunt C.	90-249 mm	144	1.47	0.97-2.03
W134 OR Crater L. streams	87-252 mm	149	1.57	1.21-2.18
Lakes				
R151 CA 8 lakes	averages	+	0.96	0.92-0.99
R150 CA	Age 0, 66 mm	100	1.10	-
	I, 119 mm	76	0.87	-
	II, 132 mm	33	0.85	-
	III, 152 mm	27	0.78	-
W2 CA Castle L.	-	389	1.05	-
R202 CO beaver ponds	18 stunted population	761	0.99	0.63-1.39
	10 normal population	271	1.13	0.78-2.19
S65 NK L. Jesse	-	4	1.17	-
R151 CA Mono. Co. lakes	-	661	1.34	-
F53 WY pond	Ages I-V	92	1.45	-
H43 UT Fish L.	-	14	1.45	-
B108 MT Irwin P.	Age II	26	1.81	1.40-2.23
W1 CA Castle L.	140-270 mm	123	1.98	-
I8 ID Henry's L.	-	34	2.03	-
B108 MT Evans P.	Age II	32	2.46	1.81-3.85
Miscellaneous				
D109 QU	Age III, 180-188 mm	90	-	0.90-0.99
G89 CA	183 mm mean	22	1.15	-
S255 NK	-	-	-	1.13-1.26
H42 NY hatchery	-	37	1.56	-
H42 QU	-	15	1.76	-
				Condition factor K(TL)
R273 UT	Spawned-out fish	1	0.45	-
	Lake X25, growth slow	115	0.81	-
	Lake X26	65	0.97	-
	Lake X49, growth rapid	21	1.14	-
R274 UT	Lake X30, growth slow	40	0.89	-
	Fish transplanted to Lake X50	11	1.14	-
Streams and hatcheries				
C132 MI Pigeon R.	-	5075	0.97	0.77-1.29
Oden Hatchery	-	4600	1.09	0.87-1.23

				Condition factor K(TL)
V63 NY wild stock	53-117 mm	+	-	0.80-1.07
hatchery stock	69-135 mm	+	-	1.08-1.50
G45 NY wild stock	Age II	149	0.87	-
hatchery stock	Age II	189	1.12	-
B155 MT Prickley Pear C.	Ages I-III	76	1.04	-
N73, N75 NH at stocking	178-224 mm	1628	1.27	1.22-1.44 (means)
Swift R. stocked fish	178-239 mm	1196	1.07	1.00-1.10 (means)
native trout	-	752	1.00	0.94-1.05 (means)
P55 MT W. Gallatin R.	-	71	1.10	1.00-1.28
R185 ME Sunkhaze Stream	-	1121	1.11	-
M375 WI Lawrence C.	-	+	-	1.07-1.40
L159 VA	178-279 mm	+	-	1.12-1.29
C132 MI Pigeon R.	-	5075	1.27	1.00-1.68
Oden Hatchery	-	4600	1.42	1.14-1.60
C209 MI 3 rivers	-	1924	1.27	0.86-1.75
M437 WI at stocking	127-201 mm	600	1.61	1.52-1.66
later	127-338 mm	1730	1.46	1.35-1.54
H166, P54, P128 S. Afr.	259-445 mm	36	1.47	1.05-2.27
H57 usual range	-	+	-	1.05-1.25
Lakes				
S524 QU Baldwin P.	Age II, 150-301 mm	2654	0.90	0.69-1.06
F134 NY Laramie Rearing P.	Age I	+	-	0.92-1.08
D190 MT 7 mountain lakes	152-455 mm	664	0.98	0.75-1.40
S417 WY Libby L.	Ages 0-V	247	1.06	-
H256 WY Towner L.	Ages 0-VI	226	1.09	1.03-1.47
R216 WY Little Brooklyn L.	Ages 0-V	106	1.11	-
B108 MT Irwin P.	Age II	26	1.11	0.90-1.52
H256 WY Little Telephone L.	Ages I-V	61	1.13	1.00-1.19
Y13 NY 3 ponds	Ages I-II	162	1.16	0.97-1.30
B108 MT Evans P.	Age II	32	1.52	1.14-2.25
Miscellaneous				
K16 VA		354	1.00	-
H57 NY	338 mm	1	1.47	-

The length-weight relationships and condition factors vary much more locally than they do by region or by major type of habitat. Salters are thicker-bodied but less deep than freshwater trout and tend to be somewhat heavier at similar lengths (B326), but these differences did not seem statistically significant (W75). There does not seem to be any consistent increase or decrease in condition with increase in length, though more populations do show weight increasing at a little more than the 3.0 power of the length than less than 3.0. In five Montana lakes condition decreased with increase in length but in two lakes condition increased (D190). The two lakes where condition increased also had the larger trout. In R150 condition decreases with age; in G150 and M375 it increases; and B155, S417, R216 show no trend. Trout from stunted populations showed lower condition factors than normal populations (R202, R273). When trout from a stunted population were transplanted to another lake K(TL) improved from 0.87 to 1.11 in 7 weeks (R274).

Hatchery trout show little seasonal trend in condition (C281) but seasonal trends have been noted in many populations. In Swift River, New Hampshire (N75), K(TL) averaged 1.04 in May and June, but declined to 0.95 by October. Highest average condition occurred in June in Sankhoge Stream, Maine (R185). Extensive data on three Michigan rivers (C209) show low values of K(TL) 1.03-1.11 in winter and spring with high values, 1.27-1.43, from May to August, and slight drops in September and October. October values in Hunt Creek were as high as June and July, however. In Lawrence Creek, Wisconsin, condition was lowest in December and highest July-September with little trend during the latter period (M375). In Wyoming, condition increased gradually from early July, K(TL) = 1.00, to September, 1.08, in Libby Lake (S417) and showed a similar trend in Little Telephone Lake (H256). In Towner Lake, condition built up early in the summer and was better in July than August or September following spawning. Condition was better at elevations below 2590 m than above in Little Beaver Creek, Colorado (M467).

Cortland domestic stock trout showed better condition and growth than two other strains reared for about one year in a pond (F134). Poorer condition was related to increased population density (W193).

Tagging with jaw tags had little consistent effect on condition of trout (Y13, S255). The condition of stocked trout declined after the stocking (S255, M437, N73, N75, V63). After stocking, trout increased in length but lost weight (N75): 825 trout stocked in 1953 gained 6 mm but lost 4.8 g and 371 in 1955 gained 8 mm and lost 4.8 g.

Cooper (C281) suggested that a relative condition factor in which the weight was divided by the expected weight under ideal conditions could be better for most studies. He therefore presented a table of expected weights based upon the average growth of the three best hatchery populations.

Table of expected weights of brook trout at different sizes (C281)
(Weight calculated from the weight-length equation $\log W = -5.257 + 3.154 \log L$)

TL		\multicolumn{10}{c}{Proportionate parts}									
mm	0	3	5	8	10	13	15	18	20	23
inches	0.0	0.1	0.2	0.3	0.4	0.5	0.6	0.7	0.8	0.9
mm inches										
25 1.0	0.149	0.202	0.265	0.341	0.431	0.536	0.657	0.796	0.953	1.13
51 2.0	1.33	1.55	1.79	2.06	2.36	2.69	3.04	3.42	3.84	4.29
76 3.0	4.77	5.29	5.85	6.45	7.08	7.76	8.48	9.25	10.1	10.9
102 4.0	11.8	12.8	13.8	14.9	16.0	17.1	18.4	19.7	21.0	22.4
127 5.0	23.9	25.4	27.1	28.7	30.5	32.3	34.2	36.1	38.2	40.3
152 6.0	42.5	44.8	47.1	49.5	52.1	54.7	57.4	60.2	63.0	66.0
178 7.0	69.1	72.2	75.5	78.9	82.3	85.9	89.5	93.3	97.2	101
203 8.0	105	109	114	118	123	127	132	137	142	147
229 9.0	153	158	164	169	175	181	187	193	200	206
254 10.0	213	220	226	234	241	248	256	263	271	279
279 11.0	287	296	304	312	322	331	340	349	359	368
305 12.0	378	388	398	409	419	430	441	452	464	475
330 13.0	487	499	511	523	536	548	561	574	588	601
356 14.0	615	629	643	657	672	687	702	717	733	748
381 15.0	764	781	797	814	831	848	865	883	901	919

	TL				Weight		
	No.	Mean of means	Central 50%	Range	No.	Mean of means	Range
Age 0-Feb.							
H411 WI Lawrence C.	65	15	-	-	65	0.04	-
W5 CA	-	-	-	-	+	-	0.08-0.1
M5 CT	-	-	-	-	+	-	1.8-2.0
Age 0-Mar.							
H411 WI Lawrence C.	36	23	-	-	36	0.07	-
L25 MI stream	50	23	-	20-25	-	-	-
P146 ON *S. f. aurora*	-	-	-	-	+	0.17	-
Hatcheries, NY (T28, B112) CA (W5)	+	-	-	25-64	+	-	.09-.22
Age 0-April							
H411 WI Lawrence C.	32	28	-	-	32	0.11	-
P146 ON *S. f. aurora*	-	-	-	-	+	-	.08-.45
C209, C128 MI streams	168	-	-	20-38	-	-	-
Hatcheries, CA (W5), MI (W35), NY (P139, T27, B112), CT (M5), PA (C281), VA (L159)	250+	-	-	20-71	+	-	.14-4.0
Age 0-May							
H411 WI Lawrence C.	25	36	-	-	25	0.4	-
P146 ON *S. f. aurora*	-	-	-	-	+	-	0.14-.68
Streams, ON (R29), MI (C209, C128)	149	38	-	20-56	-	-	-
Hatcheries, D4, CA (W5), MI (W35), NY (T27, B112), PA (C281)	250+	-	-	38-84	+	-	0.25-2.24
Age 0-June							
S335 WY streams and lakes	27	20	-	-	-	-	-
W206 PA streams	9	33	-	-	9	0.4	-
P146 ON *S. f. aurora*	-	-	-	-	+	-	.37-15.3
W1 CA Castle L.	+	43	-	-	-	-	-
ON (R29, B228), NS (W57)	21	43	-	36-46	-	-	-
H441, H334 WI Lawrence C.	385	56	-	-	385	2.0	-
MI streams (S37, C209, L169, C128)	422+	56	-	30-94	-	-	-
V63 NY streams	+	58	-	53-58	+	-	1.1-2.8
Hatcheries, L44, W31, D4, W5, T27, W35, M5, V63, B266, C281, P139, F107, B112, F134,135,136,143,144	409+	66	48-94	30-102	+	2.1	0.4-6.2
Age 0-July							
S335, G146 WY streams and lakes	152	-	-	20-33	75	-	0.03-0.39
Lake, NY H279	140	-	-	33-48	-	-	-
Streams, NY G20, NH H84	+	51	-	33-61	-	-	-
Streams, ON R29, PE S194	13+	53	-	28-69	-	-	-
W1 CA Castle L.	+	56	-	-	-	-	-
Stream, PA W206	120	58	-	-	120	2.2	-
Streams, MI C209, C128, L169, S37	240	69	-	36-102	-	-	-
Stream, ID B319	+	-	-	64-127	-	-	-
Hatcheries, T27, M71, L41, D4, W5, E14, D11, H182, C281, B112, F135	281+	86	51-112	41-127	2000+	3.7	0.5-8.5
H411 WI Lawrence C.	289	71	-	-	289	4.5	-
P146 ON *S. f. aurora*	-	-	-	-	+	-	0.5-21.2

		TL				Weight	
	No.	Mean of means	Central 50%	Range	No.	Mean of means	Range
Age 0-Aug.							
Lake, MI C128	85	-	-	33-58	-	-	-
P146 ON *S. f. aurora*	-	-	-	-	-	-	0.8-2.5
Lakes, CA W2, W1, WY R216, G146	261+	51	43-56	33-43	61	0.8	0.4-1.1
Streams, NS W75	45	58	-	56-61	-	-	-
Stream, PA W206	33	61	-	-	33	2.3	-
Streams, NY V63	76+	64	-	56-71	76+	-	1.4-2.5
Streams, MT H215, B155, WY S335	244	71	-	43-104	-	-	-
H411 WI Lawrence C.	317	81	-	-	317	7.5	-
Streams, MI S37, C209, C281, S140	3149	79	-	43-122	2964+	-	3.0-79.0
Stream, ID B319	+	-	-	71-135	-	-	-
Hatcheries, D15, D4, W5, N31, D11, M5, T27, L41, G34, S253, R150, C281, B113	399+	99	-	56-152	1567+	9.5	0.8-22.6
Age 0-Sept.							
Lake, MI C128	83	-	-	33-81	-	-	-
Lakes, CA N35, WY A43, H50, S335	93	58	-	61-84	-	-	-
Stream, PE S194, NS W75	1+	61	-	43-84	-	-	-
Streams, PA W206	343	64	-	-	243	2.6	-
Streams, CA S77, WY B155, S335	2708	71	-	43-97	-	-	-
Streams, MI S37, C209, WI M375, M402, H410, H339, H411	40887+	99	94-102	51-190	472	10.0	-
Hatcheries, D4, D2, P11, D11, M5, M2, G34, S33, L41, T27, S253, H182, B266, C281, P139, B112	1286	132	102-157	61-178	1090+	9.3	1.2-22.3
Age 0-Oct.							
S335 WY lake	68	58	-	48-79	-	-	-
W30, W31 PE streams	811	64	-	38-104	-	-	-
H411 WI Lawrence C.	136	102	-	-	136	13.0	-
S335 WY stream	5	71	-	64-79	-	-	-
MA salters, M270	+	-	-	76-127	-	-	-
PA streams, W206	211	74	-	-	211	4.2	-
MI streams, S33, C209	348+	97	-	61-157	-	-	-
ON ponds, J97	+	127	-	-	+	2.3	-
Hatcheries, A18, M65, S33, T27, V63, B112, C281, F134	455+	142	97-163	64-193	+	25.3	3.0-77.1
NY ponds and lakes, F134, 135,136,143,144, E130	2185+	137	112-173	76-193	539+	56.0	32-77
Age 0-Nov.							
PA stream, W206	60	64	-	-	60	2.4	-
S33 MI Clancy C.	541	127	-	-	-	-	-
ID stream, B319	+	135	-	89-160	-	-	-
Hatcheries, M5, B23, B112, H182, C281	404+	165	-	89-185	+	22.3	2.5-34.0
H411 WI Lawrence C.	168	112	-	-	168	18.0	-
Age 0-Dec.							
PA stream, W206	29	71	-	-	29	3.2	-
Hatcheries, B112, B266, F44, S253	26+	163	-	142-203	-	-	-
H411 WI Lawrence C.	377	114	-	-	377	18.0	-

| | | TL | | | | Weight | |
	No.	Mean of means	Central 50%	Range	No.	Mean of means	Range
Age 8 weeks							
Hatcheries, T29, T30, B50	-	-	-	-	+	-	7.0-12.0
Age 12-14 weeks							
R90, W29, M3, T29, G12	5+	-	-	23-114	+	-	2.6-17.0
Age 16-18 weeks							
M4, T29, T30	-	-	-	-	+	-	6.0-19.0
O39 Japan	+	48	-	-	+	1.1	-
Age 20-24 weeks							
R90, H39, M3, M4, T30, T29	11	53	-	43-61	+	-	1.3-20.0
Age 30-36 weeks							
H39, R90, T29, T30	2+	61	-	-	+	-	2.3-42.0
Age 38 weeks							
V63 NY 3 stocks (means)	+	-	-	94-132	-	-	-
Age 40 weeks							
T29, T30 CT	-	-	-	-	+	-	9.0-63.0
Age 44-52 weeks							
T30 CT	-	-	-	-	+	-	10.0-95.0
Age 0 (dates not indicated)							
MI streams, S39, W100	33+	69	-	56-91	-	-	-
SA lakes, R17	+	69	-	-	-	-	-
Lakes, CO R202, WY S416	62	71	-	23-91	2	7.1	-
A45 Sweden	3424	74	-	53-99	-	-	-
Streams, NK S403, NY K10, G19, NH S324, PA C286, MA B170	1206+	79	66-91	41-99	+	4.5	-
S352 NK Crecy L.	39737	99	-	-	-	-	-
Hatcheries, S252, S254, S257, F113	1108+	-	-	56-193	1103	-	2.8-76.4
Lakes, ME H280, W159, MA M273	173+	97	-	66-236	-	-	-

	No.	Mean TL	Range	No.	Mean Weight	Range
Age I						
L159 VA Big R. overcrowded	+	-	51-76	-	-	-
R271 NM stream	+	102	-	-	-	-
W206 PA 12 streams	1037	102	81-119	1037	11	6-17
M71, K10, K11, G19,22, 24,25, E11 NY	515	109	74-287	100	34	6-293
C286 PA Mud Lick C.	578	109	99-124	-	-	-
R202 CO 23 beaver ponds (normal)	24	137	114-173	-	-	-
20 beaver ponds (stunted)	43	109	84-142	-	-	-
W159 ME	399	112	86-152	-	-	-
H43, A50, G146, H256, R216, S335, S416, WY lakes	506	91-132	56-140	404	11-26	-
B170 MA 4 rivers	318	112-147	-	-	-	-
North Mashpee R.	17	173-188	-	-	-	-
C209, C281 MI Hunt C.	549	114	58-190	+	23	-
Pigeon R.	712	124	71-218	+	34	-
Au Sable R.	823	137	64-244	+	68	-
N73, S324 NH 11 rivers	3927	107-130	76-188	-	-	-

	No.	Mean TL	Range	No.	Mean Weight	Range
W75 NS	50	117	104-127	-	-	-
R150 CA Bunny L. (stunted)	76	119	91-150	76	16	-
R29 ON Mad R.	14	130	-	-	-	-
K104 Czechoslovakia	7	140	-	-	-	-
S257 PE Ellerslie R.	303	163	-	-	-	-
smolts	160	150	-	-	-	-
H212, B155 MT streams	140	155	119-231	-	-	-
W1, W2, W82 CA						
Castle L.	6883	160	102-178	163	43	-
R185 ME Sunkhaze R.	489	165	152-201	-	-	-
S107 VA St. Mary's R.	107	175	-	-	-	-
S250, S253, S254 NK						
Crecy L.	3990	175	160-302	3566	60	45-142
J97 ON ponds	+	178	-	+	91	14-170
F53 WY mill pond	3	185	-	3	65	-
S482 NH Ammonoosuc	4	190	145-234	3	91	-
H411, M375 WI						
Lawrence C.	18153	190	140-264	2065	-	14-65
B319 ID stream	+	-	160-251	-	-	-
S33, A62 MI hatcheries	567	157-226	-	300	113	-
P146 ON hatchery, *S. f.*						
aurora	63	201	-	25	105	-
S524 QU Baldwin P.	77	205	165-244	77	79	36-170
H285 NY 4 lakes	2625	206-224	135-292	2739	99-139	3-284
E130, F131,134,135,136,						
143,144,145, W197,199,						
200,207, NY ponds	3687+	224	124-307	3946+	170	31-329
E124, C62 ME lakes	110	183-226	130-267	62	142	68-258
Z2 NY reclaimed ponds	292	229	152-325	292	187	43-578
B235 OR Skookum L. (in-						
troduced)	134	315	-	134	454	-
B23,195,266,279, H41,						
C281, F113 NY, PA						
hatcheries, selected stock	560+	282-340	267-378	829	233-545	233-726
R17, R67 AT Maligne L.	+	163	-	+	34	-
When first introduced	+	-	366-472	+	-	680-907
P41, P52 S. Afr.	+	-	229-381	+	604	-
Age II						
R273 UT lake, stunted	3	122	-	-	-	-
R150 CA Bunny L., stunted	33	132	109-160	33	20	-
W206 PA 12 streams	465	135	119-142	475	26	14-31
C286 PA Mud Lick C.	61	142	117-160	-	-	-
W159 ME	172	142	122-241	-	-	-
G19,22,24,25, K11, K10,						
M71, NY streams	362	152	66-287	94	60	17-250
A51 Sweden ponds	251	137-165		-	-	-
C209 MI Hunt C.,						
Jan.-Mar.	103	147	117-190	-	-	-
Pigeon R., Jan.-Mar.	105	157	114-211	-	-	-
Au Sable R., Jan.-Mar.	120	201	155-249	-	-	-
L159, VA Big R., over-						
crowded	+	-	102-201	-	-	-
When less crowded	+	-	152-277	-	-	-

	No.	Mean TL	Range	No.	Mean Weight	Range
G103 CA Sagehen C.	64	168	-	-	-	-
W1, W2, W82, CA Castle L.	6487	168	157-208	-	-	-
R271 NM stream	+	170	-	-	-	-
W75 NS	178	173	130-231	-	-	-
S257 PE Ellerslie R.,	4539	175	-	-	-	-
smolts	159	160	-	-	-	-
R202 CO 23 beaver ponds,						
normal	43	188	142-259	-	-	-
20 beaver ponds, stunted	62	165	124-208	-	-	-
K104 Czechoslovakia	15	180	-	-	-	-
B170 MA 4 rivers	478	147-188	-	-	-	-
North Mashpee R.	38	201-224	-	-	-	-
N73, S324 NH 11 rivers	449	152-196	127-272	-	-	-
M491 CA Horseshoe R.	+	185	-	-	-	-
A43, A50, G146, H256,						
R216, S335, S416,						
WY 5 lakes	584	155-188	112-234	539	45-71	-
R185 ME Sunkhaze R.	945	190	152-251	-	-	-
F53 WY mill pond	23	201	-	23	85	-
S482 NH Ammonoosuc	35	206	130-272	35	94	-
R29 ON Mad R.	146	206	-	-	-	-
S107 VA St. Mary's R.	10	206	-	-	-	-
B155, H215 MT streams	35	221	201-251	-	-	-
J97 ON ponds	+	229	-	+	142	48-369
S256 NK Charlotte Co.						
lakes	361	229	140-300	269	130	85-176
C64 ME Mooseheart P.	52	236	206-297	52	142	96-295
H411, M375 WI						
Lawrence C.	3578	241	178-290	1214	-	48-113
S524 QU Baldwin P.	2990	245	155-310	2990	133	28-255
E129, C62 ME lakes	180	244-249	175-325	98	164	91-343
H285 NY 4 lakes	1566	229-287	218-328	1566	108-252	85-318
F134,135,136,131,143						
144,145, E130, W193,						
197,198,199,200,207						
NY ponds	2668+	251	155-343	1685+	238	119-408
S33 MI Grayling hatchery	400	262	-	-	-	-
S250, S253, S254 NK						
Crecy L.	3352	264	221-343	3160	210	113-510
A62 MI East Fish L., in-						
troduced	222	267	249-358	222	210	173-595
Z2 NY reclaimed ponds	47	297	226-442	47	414	128-1474
B195, B266, B279 PA se-						
lected hatchery stock	28	439	401-488	28	1173	816-1679
Age III						
R150 CA Bunny L.	46	150	135-180	-	-	-
W206 PA 10 streams	87	163	150-211	87	45	31-62
C286 PA Mud Lick C.	3	170	147-185	-	-	-
K11, G19,20,21,25 NY						
streams	62	175	102-381	14	111	60-321
R274 UT lake, stunted	30	185	-	30	57	-
Transplanted 7 weeks	5	241	-	5	235	-

	No.	Mean TL	Range	No.	Mean Weight	Range
W139 NS Mosers R. sea						
trout	170	221	-	-	-	-
Freshwater	51	178	-	-	-	-
Small streams	27	147	-	-	-	-
S107 VA St. Mary's R.	8	183	-	-	-	-
D109 QU 3 lakes	90	185-193	-	90	57-65	-
W2, W82 CA Castle L.	3065	196	175-295	-	-	-
R271 NM stream	+	196	-	-	-	-
R273 UT lake	8	203	-	-	-	-
S257 PE Ellerslie R.	751	224	-	-	-	-
smolts	19	178	-	-	-	-
G103 CA Sagehen C.	34	213	-	-	-	-
A43, A50, F53, H256, R216,						
S355, S416 WY 6 lakes	192	190-241	155-290	176	105-136	-
L159 VA Big R. over-						
crowded	+	-	127-251	-	-	-
When less crowded	+	-	203-328	-	-	-
B170 MA 4 rivers	95	188-216	-	+	105	-
North Mashpee R.	17	251-259	-	-		
S482, N73, S324 NH rivers	68	198-246	165-335	39	184	-
W75, S162 NS	250	234	163-284	2	213	198-227
R185 ME Sunkhaze R.	325	241	190-328	-	-	-
K104 Czechoslovakia	1	241	-	-	-	-
R202 CO beaver ponds,						
stunted	19	196	152-234	-	-	-
Beaver ponds, normal	61	272	180-373	-	-	-
R29 ON Wolf L.	9	262	-	-	-	-
M491 CA Horseshoe R.	+	264	-	-	-	-
Z2 NY reclaimed ponds	24	267	-	24	252	-
B155 MT Prickley Pear C.	5	277	267-307	-	-	-
E129, C62 ME lakes	194	277-305	183-419	133	306	102-893
H285 NY 4 lakes	183	262-310	259-318	238	162-300	159-323
R29 ON Mad R.	21	282	-	-	-	-
B228 ON Red Rock L.	131	290	229-376	-	-	-
E130, F131,134,136,143,144,						
145, W193,197,198,199,						
207 NY ponds	781+	300	203-373	605+	321	170-612
S256 NK Charlotte Co.						
lakes	648	300	226-353	569	264	179-425
R17 AT Maligne L.	+	302	-	+	327	-2722
H411, M375 WI						
Lawrence C.	289	305	267-318	196	-	74-181
C64 ME Moosehead L.	28	307	249-356	28	326	173-533
S250, S253, S254 NK						
Crecy L.	208	318	297-384	174	326	235-646
D35 MB rivers	45	399	-	45	581	-
S149 CO Battle L.	1	432	-	1	1588	-
B266, B279 PA hatcheries,						
selected females	9	475	432-521	9	1905	1361-2676
Age IV						
R150 CA Bunny L.	27	152	137-168	27	26	-
R274 UT lake, stunted	20	198	-	20	71	-
Transplanted 1 year	6	348	-	6	354	-

	No.	Mean TL	Range	No.	Mean Weight	Range
R273 UT 3 lakes	52	188-297	-	-	-	-
W139 NS Mosers R. sea						
trout	163	279	-	-	-	-
Freshwater	147	208	-	-	-	-
Streams	19	160	-	-	-	-
W2, W82 CA Castle L.	982	234	183-274	-	-	-
K11, G20, G25 NY streams	11	236	132-391	-	-	-
F53, H256, S335, S416 WY						
4 lakes	70	267-284	236-290	70	187-250	-
S257 PE Ellerslie R.	29	277	-	-	-	-
W75 NS	77	279	193-315	-	-	-
B170 MA 4 rivers	17	262-312	-	+	227	-
R185 ME Sunkhaze R.	50	300	267-404	-	-	-
S482 NH Ammonoosuc	2	312	295-330	-	-	-
F131,134,136,144,145,						
W193,197,198,199 NY						
ponds	369	320	218-396	354	378	233-765
R202 CO beaver ponds	18	323	267-381	-	-	-
S156 NB Charlotte Co.						
lakes	152	333	307-394	146	360	298-652
M491 CA Horseshoe R.	+	335	-	-	-	-
R17 AT Maligne L.	+	340	-	+	460	-
R29 ON Wolf L.	15	353	-	-	-	-
C62 ME	51	356	196-516	44	658	116-1611
M375 WI Lawrence C.	14	356	-	-	-	-
S253 NK Crecy L.	31	361	315-421	31	533	312-971
C64 ME Moosehead L.	25	363	315-442	25	556	312-851
B228 ON Red Rock L.	91	373	325-432	-	-	-
D35 MB rivers	29	455	-	29	848	-
D29 WA introduced	+	660	-	+	-	2722-3629
Age V						
R150 CA Bunny L.	37	163	142-185	-	-	-
S140 MI Hunt C.	2	186	180-193	2	57	
R273 UT 3 lakes	20	213-302	-	-	-	-
W82 CA Castle L.	489	251	236-302	-	-	-
W139 NS Mosers R. sea						
trout	37	318	-	-	-	-
Freshwater	37	267	-	-	-	-
Streams	1	193	-	-	-	-
H256, F53, R216, S416						
WY 4 lakes	18	277-328	257-328	18	204-363	-
R185 ME Sunkhaze R.	8	325	297-361	-	-	-
F131,144,145, W193,197,						
198 NY ponds	81	353	328-404	77	457	369-652
M375 WI Lawrence C.	2	366	-	-	-	-
W75 NS	9	368	315-384	-	-	-
R17 AT Maligne L.	+	381	-	+	652	-
A51 Sweden ponds	108	351-419	-	-	-	-
G20 NY stream	6	396	348-470	1	474	-
M491 CA Horseshoe R.	+	414	-	-	-	-
C62 ME	14	427	249-531	12	1233	740-3516
D190 MT Bald Knob L.	5	-	394-450	5	-	907-1276

	No.	Mean TL	Range	No.	Mean Weight	Range
B228 ON Red Rock L.	10	429	419-470	-	-	-
C64 ME Moosehead L.	8	495	419-561	8	1381	509-1985
D35 MB rivers	9	500	-	9	1236	-
Age VI						
W82 CA Castle L.	144	259	244-274	-	-	-
A61 Sweden Kalarne L.	30	-	310-361	-	-	-
W139 NS Mosers R. sea-						
run	6	386	-	-	-	-
Freshwater	2	335	-	-	-	-
M375 WI Lawrence C.	2	373	-	-	-	-
R17 AT Maligne L.	+	406	-	+	779	-
W193 NY Honnega L.	11	419	-	11	907	-
C62, C64, R185 ME	8	516	411-597	7	1647	1192-2410
B233 OR L. of Woods	2	516	513-521	2	1588	1585-1591
D35, W75 Canada rivers	4	526	455-549	-	-	-
R29 ON L. Nipigon	2	579	577-582	-	-	-
Age VII						
A61 Sweden Kalarne L.	4	-	340-486	-	-	-
W139 NS Mosers R. sea-						
run	1	455	-	-	-	-
W193 NY Honnega L.	1	460	-	1	1276	-
Age VIII						
E19 MI Lost L., introduced	+	559	-	-	-	-
W81 CA Castle L.	2	307	297-315	-	-	-
Age IX						
W81 CA Castle L.	11	284	193-297	-	-	-

Summaries of Length at Capture

Age Group	Median TL	Central 50%	Range	Weight Range
I	155	127-190	51-472	1.4-907
II	196	168-241	66-488	17-1679
III	241	201-282	102-521	57-2722
IV	323	274-356	132-660	26-3629
V	368	267-396	142-561	57-3516

Sizes at various ages are given in the following papers (those starred include calculated lengths):

CA	G89, G103, N31, N35, R150, S77, W1, W2, W82, M491*
CO	L39, R151*, R202*, S149
CT	M5, T29, T30
ID	B156, I8*, B319
ME	C62, C64, F26, F69, H187*, H280, R185, W159, E129
MA	B170, M273, M270
MI	A62, C128, C209, C211*, C281, E15, E19, L25, L169, S33, S37, S39*, S140, S221, S223*, W35 W100
MN	M299, E2*, E3*
MT	B108, B155*, H215, P55*, S429
NH	H48*, H84, N73, S324, S482
NM	R271
NY	A66, B23, B112, E11, E103, E104, E112*, F107, F113, F134, F135, F136, F143, F144, F145, G19, G20, G22, G24, G25, H41, H42*, H182, H279, H285, K10, K11, M71, P139, T28, T30, V63, W91, Y13, Z2, W193, W197, W199, W207, W208, W200, W198

NC	L160
OR	B84, B233, B235, C149, C150
PA	B195, B245*, B266, B279, C281, C286, D108, W206
UT	H43*, S237*, R274, R273*
VA	L159, S107
WA	D29
WI	M65, M161, M375, M402, H410, H339
WY	A43, A50, C283, F53, H256*, R216*, S335*, S416*, G146

Hatcheries not designated   B23, B73, L18, P1, D4, L41, T27, W29, G34, F44,
                               G12, M3

AT SA	E99, R17, R67, R98
MB	D35
NK	C43, R99, S65, S250, S252, S253, S254, S255, S256, S352, S403, W31
NS	S40, S162, S373, W75, S57, S139
ON	B228, H15, M315*, M330*, M334*, M371*, P146, R29, R90, J97
PE	S194, S257
QU	D109
Argentina	T34
Japan	O39
S. Afr.	P41, P47, P54, P128
Sweden	A45, A51, A61
Czech.	K104*

		Calculated TL and increment at each annulus			
		1	2	3	4	5
K104 Czechoslovakia		84	137	198		
	Increments	84	51	56		
	No.	23	16	1		
C211 MI Pigeon R.		91	150	201	226	
	Increments	91	74	58	36	
	No.	4361	1924	121	1	
S416 WY Libby L.		71	145	203	251	282
	Increments	71	81	56	41	28
	Increments in g	4	28	54	77	65
	No.	291	247	102	26	5
H256 WY Telephone L.		58	140	203	259	300
	Increments	58	81	61	48	33
	No.	61	60	54	30	9
S335 WY Towner L.		81	147	208	264	
	Increments	81	66	53	43	
	No.	79	52	30	5	
R216 WY Lower Brooklyn L.		97	170	218	259	297
	Increments	97	61	61	58	38
	No.	83	73	12	2	2
B155 MT Prickley Pear C.		104	178	244		
	Increments	104	71	64		
	No.	101	25	5		
I8 ID Henry's L.		112	216	328	419	
	Increments	112	104	114	71	
	No.	34	28	18	9	

			Calculated TL at each annulus					
	No.	1	2	3	4	5	6	7
H48 NH smallest	+	58	81	114				
average	580	79	109	137				
largest	+	104	124	160				
R273 UT Lake X25,								
stunted	31	71	102	135	163	183		
Lake X26	40	79	117	157	190	203		
Lake X49	12	81	147	198	244	274		
S39 MI Hunt C.	90	76	127	173				
R202 CO beaver ponds,								
stunted	37	86	132	173				
normal population	71	99	160	221	259			
M467 CO Little Beaver C.	+	124	152	188				
F137 CO Granby Res.	+	117	173	196	254			
H42 NY	202	109	155	190				
C211 MI Pigeon R.,								
electric	2610	86	145	196	226			
angling	1829	91	157	206				
S223 MI Au Sable R.,								
electric	942	97	170	221	244			
angling	1290	102	188	251	279			
D190 MT Beartooth Mt.								
lakes	655	64	135	193	239	302	361	
Other MT lakes	418	97	175	236	361	391		
MT streams	1250	76	135	203	272	406		
M453, P159 MT lakes	2095	84	157	211	259	381	384	442
M453, P159 MT streams	2566	84	145	211	272	406		
E2, E3 MN	22	89	165					
M334 ON Glen Major P.								
1959	+	109	155	211				
1953	+	130	183	218				
M371 ON pond 1960	+	107	168	206	229			
1953	+	107	180	226	249			
S237 UT Kay's L.	21	107	178	221				
R151 CA Dorothy's R.	93	81	142	203				
CA Bright Dot L.	18	127	231	318				
8 lakes (including								
2 above)	594	89	157	234				
S483 OR Waldo L.	100	112	203	335				
Otter R.	1	160	333	411				
M330 ON	301	127	196	239				
H43 UT Fish L.	15	117	211	269	328			
P55 MT Bridge Spring	86	104	168	277				
W. Gallatin R.	71	119	216	300				
E112 NY farm ponds	+	206	282	343	361			
H187 ME lakes	+	-	259	333	404	475	518	
M315 ON Marmot L.	+	104	305	401	447	472		

Weight at each annulus

M334 ON Glen Major P.

1959	+	17	37	99	
1953	+	28	62	108	
E112 NY farm ponds	+	113	284	482	624

Annual increment as a percentage of length at beginning of year

TL	S416 WY Libby L.	B155 MT P.P. C.	C211 MI Pigeon R.	Range
51-75	119		85	85-126
76-101		67	89	57-95
102-126		66		60-74
127-151	42	63	41	27-65
152-177		37		37
178-202		40	19	19-40
203-228	20	26		20-26
229-253				-
254-278	11			11

Good pictures of scales showing characteristics of annuli are given in R151 and S335. The best areas for taking scales have been reported as above the lateral line even with the adipose fin (S429, S335) and from the caudal peduncle (R202). Scale platelets were first noticeable at 46 mm TL (S335). Annuli form in April to June in Michigan (C128), from May 17-June 25 and thus later than rainbow trout in Montana (B175), and from June 24 to July 7 in Wyoming (S335). Younger brook trout formed annuli earlier than older trout even of the same size (M161). In 4 New York lakes (H279) 65-90% of the trout showed false annuli, not detectable except by position and knowledge that the fish could not be as old as indicated. A46 and R273 reported brook trout scales as difficult to interpret beyond the 3rd year, and R202 listed many papers reporting difficulty. In Bunny Lake, California (R150) growth was so slow that there was no evidence of annuli after number II, even though the fish could be followed for 5 years. Some brook trout failed to form annuli the first winter in Montana mountain lakes (D190).

Direct proportion was used in calculating growth from scales in many studies, but some used a Fraser-type correction factor (B155, H42, H48, S416, R216, S335) usually from 28 to 30 mm. In two New York lakes the body-scale relationship appeared to be rectilinear but in two other lakes it was curved (H285). Empirical body-scale relationships were used by C131, C211 and S223.

Growth may occur throughout winter, particularly of fish at the end of their first year (B112, C209, C281, A62, T110). In Lawrence Creek, Wisconsin (M375) growth in weight was detectable by mid-January, but growth in length not until early March. Age 0 brook trout continued to grow through September, but growth of age I and II trout in this creek was completed by September. In two high mountain lakes in Wyoming, brook trout started growing in mid-June and completed most of the season's growth before the end of August (G146). A longer growing season in 1962 was related to warmer temperatures in August. Trout stocks in December had grown 0.10 to 0.15 mm per day by April in Pennsylvania streams (T110). Brook trout stocked October 14 in East Fish Lake, Michigan, at age I (A62) at 226 mm and 113 g grew as follows:

38 in January averaged	251 mm	172 g
54 in March averaged	259 mm	186 g
and 42 in April averaged	267 mm	204 g

In Hunt Creek, Michigan (C209):

Age 0 Oct.-Nov.	92 trout	85 mm	61-114 g
I Jan.-Mar.	178 trout	89 mm	58-135 g
I April	77 trout	99 mm	66-122 g

In Michigan most growth of age I trout occurred from March to June with little from July to September, but some growth occurred in the winter (C209). In California, where temperatures in spring and early summer are low, growth of age I was most rapid from June to August and of age II in October (M491). Growth declines in the fall, with onset of maturity (C281, G146). "Salter" brook trout in Massachusetts grew at a fairly constant rate throughout the year (M270). In New York hatcheries there was a marked decline in growth of sub-yearling brook trout from about October 1 to late January (P180). In Lawrence Creek, Wisconsin, instantaneous growth rates were lowest in November or December and highest in May (H411). Similar data on a daily increment basis are given in H247.

Average growth in New York hatcheries increased as the water temperatures increased from 7.8 to 10.5C (H183). For brook trout 38-152 mm the expected growth in mm per month was:

Water temperature	4.5	5.6	6.7	7.8	8.9	10	11.1	12.2
Expected growth	2.2	5.5	8.6	11.9	15.2	18.8	21.8	24.6

Growth tends to be faster in warm than in cold water (G103, L159), and therefore growth is faster in Virginia than in New Hampshire streams (L159). Slower growth of brook trout in the Beartooth Mountains at elevations of 2400-3000 m was believed related to the elevation and shorter growing season (D190). Growth of S. f. aurora fry was much slower at 6C than at 10C (P146). Most favorable temperatures for brook trout growth have been reported as 7-18C (L18, R216). High midsummer temperature resulted in slow growth in one stream when the trout were crowded into water of suitable temperature (M270).

No general geographic effects on growth are evident from the growth tabulations. In general, maximum size of brook trout seems correlated with the size of the body of water in which they live (R29). In small streams, trout are often overabundant and slow growing (B326). Growth in lakes was in general better than in streams (from the tabulations above). Slowest growth was usually related to overcrowding (R202, R150, R274, L159, G103).

In Mud Lick Creek, Pennsylvania (C286) growth was faster downstream where the population was 33.6 kg per hectare. Population reduction resulted in faster growth, but was not considered practical. Slower growth was also noted in upper parts of streams in New Hampshire (S324), but population densities were not given. In Maine (W159), growth was faster after DDT spraying despite decreases in food availability, but probably because of population thinning. In Bunny Lake, California (R150) slow growth was believed to be due to overstocking and to food shortage. Even as the population was reduced, little growth occurred. When some trout from Bunny Lake were transferred to the laboratory and fed, they grew 15 mm and 13 g in 3 months (N109). In New York ponds, growth was slow if more than 250 fall fingerlings were stocked per hectare (Z4). Up to age I and II, growth was slower where populations were over 250 trout per hectare (E104). In 7 lakes in the Beartooth Mountains, average weight of brook trout was negatively correlated with population density as measured by gill net catches (D190). Good growth in one lake was related to low population density because of lack of spawning areas (D190). Growth was more rapid when population density was lower (W193). An inverse relationship was noted between growth and number of trout per hectare in Ontario ponds, but the numbers were not necessarily related to stocking rates (J97). In an Ontario pond, slow growth was believed to result from high sucker population density (M371). Competition of rough fish also reduced growth in New York ponds (Z4), and in other Ontario ponds (J97).

Fish introduced at low population density showed rapid growth (A62, R67, B235, Z2, D29). Trout from a stunted population grew well when transplanted to a new habitat (G73, R274).

Growth was somewhat faster in sea-run brook trout than in those remaining in the Mosers River, New Brunswick (W139), but not as much as with various sea-run *Salmo* spp. Sea-run brook trout, however, remain at sea relatively short times. Those of the Mosers River average 42-84 days (W165). Growth was less in the tributaries and small streams than in the river (W139). Smolts were in general smaller than trout of the same age remaining in the streams (S257). In the spring of age I, 43 trout from tributary streams averaged 99 mm while 27 from East Lake, New York, averaged 175 mm (H285). Growth was much faster in tidal areas than in strictly freshwater areas in Massachusetts, perhaps due to the abundance of small elvers, alewives, etc. (M270).

In some populations no sex difference in growth was noted (A43, B228, D190, H285), but in several, males grew faster than females (M375, C286, H182, W206, N73, Y13, B228), a difference, though slight, usually evident by the end of the first year. Females lived longer than males on the average (M375, M491, M270).

Selective breeding resulted in faster growth (B195, B266, B279, D15), and fish bred from a fast-growing population grew faster than those bred from a slow-growing population when both were kept under identical conditions (F107). At 44 weeks brook trout of the Berlin, New Hampshire, strain averaged 35-53 g compared to 19.7-20.3 g for a Pennsylvania strain (P86). Males seemed to have more genetic effect than females in breeding for growth (D15). Inbred trout showed slower growth than randomly bred trout (C281). Cortland stock and wild stock showed the same growth and survival rates in New York ponds (E130).

Tagging slowed the growth of brook trout in S255, Y13, G103, but not in H285. Fin regeneration was greater in brook than in rainbow or brown trout (W3).

No change in growth of brook trout was noted following stream improvement of Hayes Brook, New Brunswick (S403).

Growth of brook trout with various diets in hatcheries has shown big differences (B245, P154, and others).

Food poisoning slowed the growth of *S. f. aurora* fry for 20 days in an Ontario hatchery (S403). In nature growth is more rapid where fish or crayfish comprise a greater part of the diet (D110, R29). Where fish-food organisms are abundant, growth may be rapid even in alpine lakes with short growing seasons (R274). Bottom fauna is extremely rare in Bunny Lake, which has a stunted brook trout population (N122).

Brook trout caught by anglers showed more rapid growth than those caught by electric shocking in Michigan streams (C211, S223).

Installation of a 229 mm size limit in a Wisconsin stream reduced the catch, the catch per hour, and fishing pressure; it increased normal summer and winter mortality, and increased egg production, but probably not effective reproduction; and it resulted in slower growth of the trout (H322).

Hatchery-stocked trout grew faster than wild fish in New York, but were subject to higher fishing and natural mortality (H285). Domesticated hatchery stock grew faster than fry from wild stock reared under the same conditions, but the wild stock fingerlings showed better survival after stocking, less emigration, and less overwinter weight loss (F120).

	0-June		0-Oct.		0-Oct. (in ponds)		I-May	
	No.	TL	No.	TL	No.	TL	No.	TL
Domestic stock	1425	66	150	124	678	130	270	168
Wild stock	1725	51	298	71	1212	109	547	150

Grading of brook trout in the hatchery did not increase the growth of smaller brook trout (P137, P181) as it did for brown trout (B89). (Removal of the larger brown trout destroyed the hierarchy which had a limiting effect on the smaller individuals.) Nor was there evidence that grading separated brook trout capable of faster growth (P181).

Growth of brook trout in New York ponds did not differ from that of other trout nor with the water supply, size, or age of the pond (E104). Chemical quality of the water did have an effect on growth of brook trout in reclaimed lakes (Z4). Total dissolved solids and growth of brook trout were found to be correlated (R151), but no correlations could be detected with food supply.

Growth was better in unstratified than in stratified ponds in New York (H285), but annual production was about the same, because growth slowed down earlier in life in the unstratified ponds (perhaps because the carrying capacity was reached).

Growth was somewhat slower when brook trout were stocked in lakes within a year of treatment with rotenone than if stocked a bit later, probably because of time needed to build up a bottom fauna (Z4).

Growth was believed to be related to size rather than age (C281). In nature temperature limits growth in winter, but food limits growth in late summer (C281).

In general, brook trout are short-lived, few exceeding 4 years of age (M491). The maximum age, at least 15 years, was found in Bunny Lake, California, where growth was very slow; also there is other evidence that slow-growing populations show a greater longevity (M491). Faster-growing fingerlings of hatchery stock have a shorter life span (W193).

Survival from potential eggs to fall fingerlings in Michigan streams was 2.5, 3.3, 4.1, 4.3, 4.5, and 5.1% (C209), and in Lawrence Creek, 0.5-2.2% (H282) and from egg to fall fingerlings in Hunt Creek, 2.7-6.7% and in confinement 4.5-31.7% (S382), and in Convict Creek 14% (S77). Of naturally spawned eggs, 29% were viable at eyed stage in a stream, but only 0.9-1.5% were viable in sand and silt compared to 70% in washed gravel (H285). Floods washed out 99-100% of the fry in some Japanese streams (O39). Wild trout were better able to survive flash floods in Virginia than stocked fingerlings even though the wild fingerlings were smaller (L159). Under identical conditions wild fingerlings were more active, nervous, and resistant to metabolites and high temperature and showed greater swimming stamina and stayed nearer the bottom than hatchery fingerlings (V63). Stocked fingerlings from wild stock showed survival rates of 65 to 76% compared to domestic stock at 43 to 53%; and 81-95% compared to 59-67%; and 24-94% compared to 32-78% (F120). Wild trout showed more stamina in swimming tests than hatchery trout at 11 months (G123).

Survival from 0 September to I September in Michigan streams was 14,

23, 32, 39, 47, and 62% (C209); in Hayes Brook, New Brunswick, 77% (S403); in
Hunt Creek, Michigan, 66% (S39); in a "new" stream 0.1% compared to 39.6%
for rainbow trout (B213).

   Survival from 0 September to II September in Michigan streams was 1.3,
2.6, 3.3, and 3.5%, and to September III, 0.5% (C209).

   Survival from I to II in New Hampshire streams was 6.1, 7.1, 10.0, 6.8,
2.1, 24.0, 0.8, 1.5, 21.4, 15.2, 3.2, and 5.8%, and from I to III 1.0 and 0.8%, and
from II to III 1.4 and 4.4% (S324).

   In Hunt Creek, Michigan, survival was 63% from I to II and 14% from II to
III (S39). In Crecy Lake, New Brunswick, survival from fingerling stocking to
creel was 35 and 42% when fishing started April 1 but 23% when fishing started
June 1 (S352). Survival of hatchery trout to anglers was 1.4, 2.9, 3.3, 6.4, 7.6,
and 19.1% (S446). Trout stocked at 150 to 200 mm in September brought a
14.6% return to anglers compared to 22.5% for those stocked in March in Massachusetts (M270). About 0.6% were caught over one year after stocking.
Brook trout stocked at 178-229 mm in fall returned less than 1% to anglers
compared to a maximum of 19.6% in spring, in Michigan (S60).

   Yearling brook trout stocked in estuaries brought a return to anglers of
27 and 29% compared to 6 and 19% of yearlings stocked in the stream and to 4%
of younger fish stocked in the stream (S466). Survival of brook trout stocked
in Honnedaga Lake, New York, was lower in periods when the water was acid
and high in heavy metals than at other periods (S493). However, trout from a
hatchery with hard water and high zinc content survived better during these
periods than did trout from a soft-water hatchery and another hard-water
hatchery.

   Survival rates in New York ponds were the same for brook trout, rainbow
trout, and brook-brown trout hybrids. The survival rates were not related to
number, biomass, or size at introduction, but duration of bottom temperatures
above 23C seemed the most important factor. Mortality of spring-stocked fingerlings was variable but fall stocking was more dependable. Mean natural
mortality rates 1, 2, and 3 years after stocking as fall fingerlings were 0.60,
0.80, and 0.80 respectively (E130).

   Instantaneous mortality rates per day were 0.024 for March-June, 0.007
for June 26-July 24, and 0.005 for July 24-Sept. 19 in Pigeon River, Michigan
(L169) and 0.003-0.004 in East Fish Lake, Michigan, except during the spawning season, when it was 0.015 and in April-May when it was 0.04 to 0.06 mostly
due to fishing (A62).

   Instantaneous mortality rates during the summer in New York lakes averaged 1.34, range 0.23-3.02, and during the winter 0.56, range 0.09-3.02 (H285).
In Lawrence Creek, Wisconsin, mortality rates ranged from 0.75-0.96 (M375)
and in Ford and Hemlock Lakes, Michigan, 0.88-2.50 (L170).

   Male brook trout may mature as fingerlings at 94 mm and females at age I
at 104 mm (M375). Males of hatchery strains all reached maturity the first
fall, but not all wild stock in a New York pond (F134). Males also matured at
132 mm in the first December of their lives in a Vermont hatchery (F44). In
Ungava Bay, a female matured at 119 mm TL (D45). Both males and females
matured at age I at 102-124 mm in stunted populations in Colorado while they
did not mature until age II at 152-175 mm in normal populations (R202). In
Virginia, the stunted trout matured at 102-127 mm (L159). The stunted population in Bunny Lake, California, apparently never spawned successfully (N109).
Histological studies indicated that the gonads were underdeveloped or atro-
phied, but that other tissues were normal (H403). Some gonads showed anoma-
lies, however (H122). In Idaho (B156), brook trout mature at 127-203 mm in

the 3rd or 4th year. In Maine they may spawn at 76-102 mm in cold streams
or first at 305-356 mm in productive lakes, maturing at age II (H187). In
Sweden, the smallest mature males were age II and 140 mm and females were
age III or IV and 300 mm (A51). In an Ontario pond, males and females mature
at 127-152 mm at the end of the second year (M371), but in Red Rock Lake
most females do not mature until age III (B228). Where spawning grounds are
not suitable, eggs may be retained over winter and resorbed (R202). A few
males and females mature at age I in Wyoming (A43). In infertile Pennsylvania
streams, one male was mature at age 0, 86 mm long, but most did not mature
until age II and one age III male of 170 mm was still immature (W206); two females matured at age I, 91 mm, but most not until age II.

Brook trout spawn from September to December in Maine (H187), California (M491), and New Hampshire (N95), in late October and November in Wisconsin (O40) and Michigan (G116), from September to early November in
Pennsylvania (W206), in October in Ontario (P146). In "salter" streams in
Massachusetts (M270), brook trout mature at age I, spawning in late October
or early November and often going to salt water after this, but not before.
Brook trout may spawn from August to November (K118). Reproductive cycle
is controlled by the photoperiod (H335). Spawning behavior has been described
in N96, H90, G116, O40, B170. Large trout spawn earlier than smaller individuals (O40, W206). There is some evidence of homing to spawn, and spawning
on newly prepared gravel areas was secured only by "seeding" so that the offspring spawn in these areas (W160).

Eggs (E110) hatch in 142 days at 1.6C
                      105 days at 3.9C
                      (90 days at 4.5C, H187)
                      65 days at 7.8C
                      45 days at 10C
                      32 days at 13.4C
          and 28 days at 15C

Eggs develop normally up to 13C and fingerlings grow best at about that
temperature (M396). Greater egg loss was reported in heated water, 10C,
than at 6C although the latter took 20 days longer to hatch (P146).

	No.	TL	
W206 PA Tomtit Run	63	91-196	no. eggs = -12.47 + 3.25 W
			log no. eggs = -5.074 + 3.23 log TL
Bobb's C.	17	-	log no. eggs = -3.922 + 2.68 log TL
Lingle Stream	54	-	log no. eggs = -4.435 + 2.96 log TL
Cherry C.	17	-	log no. eggs = -4.609 + 3.14 log TL
S382 MI streams	20	160-234	no. eggs = -246 + 3.0 TL
lakes	36	193-417	no. eggs = -1925 + 10.4 TL
H285 NY East L.	33	178-274	no. eggs = -111 + 5.0 W
M375 WI Lawrence C.	81	102-251	no. eggs = -425 + 4.56 TL
			(r = 0.847)
A43 WY	14	124-198 FL	no. eggs = -286 + 3.34 FL

	Size of females	No. of females	No. of eggs per female	
			Mean	Range
W206 PA infertile stream	91-196 TL	63	-	18-277
T24 CT	203-228 TL	2	349	130-560
	229-278	3	485	308-656
	269-329	9	748	410-1312
	330-380	5	1252	923-1845
	381-431	6	1846	923-2563
	432-482	5	2583	2358-2665
M375 WI Lawrence C.	102-126 TL	11	-	50-160
	127-151	32	-	120-370
	152-177	15	-	205-395
	178-202	10	-	240-615
	203-228	3	-	350-580
	229-253	2	672	605-740
S71 PE	140-559	54	-	24-5630
B279 PA selected females				
2 yrs.	307-366	28	1828	860-2920
3 yrs.	401-488	28	4380	1972-6894
4 yrs.	432-521	9	5699	3850-6811
F144 NY pond Age I,	277	31	1066	-
W197 NY Panther L.	203-343	52	1225	332-1962
V61 QU	140-162 FL	6	106	88-117
Mature eggs only	163-192 FL	25	163	108-211
	193-229 FL	54	266	112-424
	230-280 FL	57	386	236-588
	281-328	12	884	581-1349
	340-439	8	1710	975-2444
	541	1	4765	-
C209 MI	102-126 TL	38	104	-
	127-151	91	169	-
	152-177	59	268	-
	178-202	24	395	-
	203-228	15	525	-
	229-253	8	643	-
	254-278	4	753	-
M491 CA Mt. Whitney stock	356-432	-	-	1493-4140
H336	211-228 FL	-	-	667-761
	229-251 FL	-	-	869-1261
	274-290 FL	-	-	1340-2410
K118 Review	-	-	-	80-17000

The number of eggs per kg of female was 3329 for 2-year-olds and 3887 for 3-year-olds in a Pennsylvania hatchery (B195, B270), and 5476 (R179). No consistent differences in numbers of eggs in left *vs* right ovary was detected (W208). The lower fecundity in Quebec than Wisconsin was attributed to shorter growing season and less fertile water (M375). Low fecundity in Tomtit Run and Lingle Stream, Pennsylvania (W206) was related to high population density and low productivity of the streams. The number of ova decreases as the eggs ripen and, therefore, some fecundity data are misleading (V61), but this decrease is reported by H336 not to be as much as calculated in V61 since part of the decrease is due to increase in size of female as the female reaches the maturation period. Additional data on fecundity is given in A43, B23, B73, D109, D12, E31, F16, F39, H83, H256, H285, K6, L1, L16, N25, N28, P47, P146, R29, R179, S40, S146, V28, V29, W34, W134.

Ripe eggs varied from 3.35 to 5.0 mm and there was little correlation between age of females and size of eggs, but the size of the eggs was correlated with TL of female (W206): Mean diameter of ripe ova in mm = 2.71 + 0.009 TL.

Average egg diameters of wild stock were 4.8 mm, of hatchery stock 4.39 mm, and of first generation hybrids 4.19 mm (V63); of 2-year-olds 4.17 mm and of 3-year-olds 4.39 mm (B195); and generally range from 4.0-4.4 mm (R179).

P59,60,56,58,139, T57 NY Cortland
  healthy hatchery trout 850,000-1,520,000 r.b.c. per cu mm
  anemic trout   down to 451,000 r.b.c.
P57 r.b.c. of 1,090,000 increased to 1,302,000 after trout put in water low in
     dissolved oxygen

Upper lethal temperatures (B201, B202, B203, F93, H253, R216, S333) ranged from 21C (B203) to 26.6C (B201). A temperature of 26C can be tolerated 24 hours and 28.3C for one hour (B170). The maximum lethal temperature could be raised 0.5-2.0C by acclimating the fish at temperatures near the maximum (B202, H253, S333). Optimum temperature was listed at 12.8C (S333) at 7-18C (L18), and preferred temperatures at 14-19C (G121). Maximum movement to a stimulus occurred at 10C and decreased in both directions at least to 3C and to 25C (E108). Stocked trout move more if stocked at below 10C than at higher temperatures (N106). Food consumption doubled with a 3C rise up to 12.8C when it was 50% of the body weight per week. At 17C food consumption decreased and at 21C it was only 6% of body weight per week (B280). Spontaneous activity of brook trout increased with temperature and then decreased, but increased again at some point above the preferred temperatures (S430). Near the upper temperature limits brook trout may lose weight although fed all they will eat.

Brook trout have a greater cold tolerance than other trout (B281). A minimum threshold temperature in hatcheries was considered to be 3.6C (H183). Brook trout are sluggish at low temperatures. Brook trout were in pools more than fingerling salmon, were more apt to move to cooler water areas, were usually under cover, and were more active in the evening than in midday (G150).

Movements and behavior of sea-run brook trout are reported by 257, W31, W165, W166, B326.

Brook trout begin to feed 23-35 days after hatching (B78). Up to 25-38 mm they feed mainly on Entomostraca (C295, R234). Food of brook trout is mostly of insects and aquatic invertebrates (S417, S402, W1, L168, M491, N104, C295, L164, R151, G106, G146, H287, H285, I8, E99, B228, R273, W134, A50, H256, R216, W31). Plankton may also be quite important (W1, H43,287,285,256, I8). Even adults may do well on plankton if it is sufficiently abundant (N95). Maturing female brook trout fed more on mollusks *(Physa)* than did other trout (G146).

In Crecy Lake, New Brunswick, 794 age I brook trout ate mostly benthic insects, though flying ants, fish, and snails were sometimes important (S402). Very few *Hyalella* and sphaeriids, which were increased by fertilization, were utilized.

DDT spray reduced the food of brook trout in streams in Maine (W159), particularly the caddisworms, stoneflies, and black flies, but there was an increase in utilization of snails.

Terrestrial insects taken from the surface may be important in lakes (W1, R151) or in streams (H405, O40, L168, T122, W134). A shrew, *Blarina brevicauda*, was taken by one brook trout (B228).

In general brook trout are not particularly piscivorous (B170). Fish and frogs were the most important foods of brook trout over 150 mm in Brule River, Wisconsin (O40), although not taken by 96 smaller trout. In New York streams a few fish were even taken by trout of 50-100 mm (C295). There was no evidence of predation on trout fry even where fry were abundant (L164). In Maine lakes, brook trout fed mainly on alewives (H280). Snails and fish *(Umbra* and *Lepomis)* were the important foods of trout 200-300 mm in Ford Lake, Michigan (K81). In Red Rock Lake, Ontario (B228), perch fingerlings were almost the exclusive food in July and August when the brook trout grew very rapidly. There was little feeding in late August and September. Sea-run trout travel along the shore, feeding on elvers, other small fishes and crustacea (W166, B326, W31), but may also feed primarily on amphipods and isopods (B170). "Salters" rarely eat in fresh water (B326). During spawning season, 60% of the brook trout had eaten brook trout eggs (I8). Brook trout showed a greater diversity in feeding than rainbow trout (R15).

The average stomach contents of brook trout in stunted populations was less than in normal populations in beaver ponds in Colorado (R202). The available bottom fauna was also less. Stunted fish had about twice as much terrestrial food and plant material as normal fish. Of stunted fish, 79% contained food compared to 97% of the normal population.

The size of beetles eaten increased with the size of the brook trout (A50).

Diets for trout in hatcheries are reviewed in many papers (P177, P182).

Stocking has been discussed and evaluated in many papers (C300). In stocking alpine lakes, it should be recognized that they are low in productivity (R274). Size limits were found to be the "best and most reliable single regulation for managing wild brook trout in Wisconsin" (H410). Management plans for brook trout in California lakes are given in M491.

Evaluations of production in trout populations are given in J97, H411, E130. Good life history summaries: B168, B170, B326, M491.

DOLLY VARDEN, *Salvelinus malma* (Walbaum)

The Dolly Varden is sometimes considered a subspecies of the arctic char (B326). Dolly Varden occur in Pacific drainages from northern California to the Alaskan and Siberian shores of the Arctic Ocean and southward to Japan and Korea (M492). Anadromous and freshwater populations are known and they range inland to northern Nevada, Idaho, Montana, and Alberta (M492).

Length	No.	Weight	Range	Citation
168 TL	1	45	-	L155 AK
231 FL	1	119	-	M165 BC
305 FL	1	318	-	S363 OR
455 TL	1	737	-	B113 MT
475 TL	1	1150	-	L155 AK
490 TL	1	1021	-	B113 MT
533-558 TL	2	1404	1389-1418	B113 MT
559-583 TL	4	1729	1361-1928	B113 MT
584-609 TL	4	1877	1644-2155	B113 MT

Length	No.	Weight	Range	Citation
610-634TL	3	2211	2013-2325	B113 MT
635TL	1	1814	-	R275 AK
635-659TL	4	2693	2466-3033	B113 MT
660-685TL	7	3008	2750-3402	B113 MT
660FL	1	3402	-	M491 CA
686TL	1	3288	-	B113 MT
711TL	1	3969	-	B113 MT
876TL	1	9072	-	R275 MT
1029TL	1	14515	-	G36 ID

	No.	Length	Mean	Range
W134 OR Sun Creek	16	117-200 mm K(SL)	1.39	1.16-1.80
H412 AK Eva C.				
to-sea migrant	1311	173-373 mm K(FL)	-	0.79-0.86
from-sea migrant		211-373 mm K(FL)	-	1.02-1.18

The condition factor showed a tendency to decrease with increase in age (H412) and to increase during the feeding period in the ocean and decrease during the winter stay in Eva Lake (H412).

	No.	FL	TL	Range	No.	Mean W	Range
Age 0							
H412 AK Eva C. August	86	53	-	20-43	-	-	-
B10, N23 AT Jasper Park	+	X	-	61-76	-	-	-
D61 AK July	109	41	-	-	-	-	-
August	13	58	-	-	-	-	-
September	220	41	-	-	-	-	-
Age I							
H412 AK Eva C.	46	61	-	51-74	-	-	-
D61 AK	+	X	-	43-104	-	-	-
M113 AT Elbow R.	+	165	-	-	+	57	-
B10, N23 AT Jasper Park	+	X	-	201-330	-	-	-
M165 BC Upper Campbell	1	231	-	-	1	119	-
Age II							
H412 AK Eva C.	33	79	-	61-99	-	-	-
B156 ID Priest L. streams	7	-	117	-	-	-	-
H412 AK to-sea migrants	3	112	-	-	3	11	-
from-sea migrants	1	132	-	-	1	26	-
M113 AT Elbow R.	+	211	-	-	+	99	-
M165 BC Upper Campbell	6	239	-	-	6	142	-
B10, N23 AT Jasper Park	+	X	-	244-300	-	-	-
B10 AT Athabaska R.	1	371	-	-	-	-	-
Age III							
H412 AK Eva C.	9	99	-	86-107	-	-	-
B156 ID Priest L. streams	18	-	157	-	-	-	-
H412 AK to-sea migrants	101	147	-	-	101	28	-
B156 ID Upper Priest L.	5	-	180	-	-	-	-
H412 AK from-sea migrants	74	208	-	-	74	108	-
M113 AT Elbow R.	+	246	-	-	+	159	-
B156 ID Priest L.	9	-	257	-	-	-	-
M165 BC Upper Campbell	3	290	-	-	3	269	-
B10, N23 AT Jasper Park	+	X	-	300-401	-	-	-

	No.	FL	TL	Range	No.	Mean W	Range
Age IV							
H412 AK to-sea migrants	486	206	-	-	486	74	-
M113 AT Elbow R.	+	269	-	-	+	216	-
H412 AK from-sea migrants	354	269	-	-	354	216	-
B156 ID Upper Priest L.	14	-	318	-	-	-	-
B156 ID Priest L.	15	-	363	-	-	-	-
B10, N23 AT Jasper Park	+	X	-	411-460	-	-	-
Age V							
H412 AK to-sea migrants	388	249	-	-	388	128	-
M113 AT Elbow R.	+	320	-	-	+	340	-
H412 AK from-sea migrants	378	297	-	-	378	278	-
B156 ID Upper Priest L.	7	-	396	-	-	-	-
B156 ID Priest L.	24	-	460	-	-	-	-
B10, N23 AT Jasper Park	+	X	-	500-541	-	-	-
Age VI							
H412 AK to-sea migrants	245	315	-	-	245	258	-
from-sea migrants	164	333	-	-	164	378	-
M113 AT Elbow R.	+	335	-	-	+	369	-
B156 ID Upper Priest L.	8	-	566	-	-	-	-
B156 ID Priest L.	9	-	574	-	-	-	-
B10, N23 AT Jasper Park	+	X	-	541-620	-	-	-
Age VII							
H412 AK to-sea migrants	91	371	-	-	91	408	-
from-sea migrants	77	371	-	-	77	524	-
R275 AK	1	-	635	-	1	1814	-
B156 ID Upper Priest L.	5	-	668	-	-	-	-
B156 ID Priest L.	4	-	698	-	-	-	-
Age VIII							
H412 AK to-sea migrants	28	394	-	-	28	482	-
from-sea migrants	22	421	-	-	22	760	-
B156 ID Upper Priest L.	2	-	732	-	-	-	-
R275 MT Flathead L.	1	-	876	-	1	9072	-
Age IX							
H412 AK migrants	18	447	-	-	18	765	-
M113 AT Elbow R.	+	404	-	-	+	680	-
Age X							
H412 AK migrants	9	439	-	-	9	737	-
Age XI							
H412 AK migrants	2	442	-	-	2	635	476-746
Age XX							
S30 CA	-	-	-	-	+	6123	-

Lengths have been calculated from scale measurements using a straight-line regression (R275) and with an intercept of 10 mm (L155) or using a second-degree parabola (B156, B238). Otoliths gave better results than scales in the Eva Creek, Alaska, study (H412). In less than 2% of 5300 pairs of otoliths examined did the number of annuli disagree.

		TL at each annulus							
	No.	1	2	3	4	5	6	7	8
P159 MT lakes	650	71	142	218	312	406	536	625	632
P159 MT streams	360	81	145	224	333	401	366	424	

DOLLY VARDEN

		TL and annual increment at each annulus							
		1	2	3	4	5	6	7	8
B156 ID Priest L. streams		66	104	142					
	incr.	66	38	41					
	no.	25	25	18					
B156,238 Upper Priest L.		66	102	155	239	358	462	546	612
	incr.	66	36	53	84	104	109	130	102
	no.	41	41	41	36	22	15	7	2
B156,238 Priest L.		71	114	183	310	424	516	605	
	incr.	71	43	69	130	127	112	71	
	no.	61	61	61	52	37	13	4	
R275 MT Flathead L.		71	140	208	323	452	594	724	876
	incr.	71	69	69	99	99	117	109	99
	no.	289	289	289	245	203	80	14	1
L155 AK Port Walter		94	157	229	310	401			
	incr.	94	66	74	71				
	no.	100	80	40	20				

	Median annual increase as a percentage of TL at start				
TL	B156 ID streams	B156 Upper Priest L.	B156 Priest L.	R275 Flathead L.	Range
51-75	55	54	65	100	52-103
76-101	-	50	-	88	41-94
102-126	41	54	60	58	18-66
127-151	-	57	-	49	43-68
152-177	-	-	74	43	40-77
178-202	-	53	70	62	53-89
203-228	-	44	-	42	39-55
229-253	-	43	-	27	27-43
279-304	-	42	-	-	42
305-329	-	32	42	37	27-42
356-380	-	-	-	49	49
381-405	-	30	26	-	26-30
406-431	-	32	-	-	24-35
432-456	-	-	-	27	26-29
457-482	-	-	30	-	30
508-532	-	20	-	26	20-26
533-558	-	-	14	-	14
559-583	-	-	-	20	20
660-685	-	-	-	16	16
762-786	-	-	-	13	13

Dolly Varden grow more slowly during their first 2 years in stream residency than after entering the lake (R275, H412). They grew little during the winter in Eva Lake but grew in the sea from May to September (H412). No difference was found between the growth of male and female Dolly Varden (H412).

Most Dolly Varden spend 2 to 3 years in fresh water before migrating to sea, but fry less than 28 mm TL have been seen migrating into salt water in early May in Alaska (R276).

Some Dolly Varden mature in their 4th year but usually in the 5th or 6th in Idaho (B156) and in Alaska (H412) and spawn in September in Idaho (B156) and in October and early November at 5.5 to 6.6C in Alaska (N132). November is reported as the spawning time by K118 and minimum size at maturity as 457 mm.

	Size of female	No.	Number of eggs per female	
			Mean	Range
W134 OR Sun C.	163 mm FL	1	74	-
	190 mm FL	2	337	-
M113 AT	Age III	3	529	486-395
B113 MT	737 g	1	1337	-
	1021 g	1	2672	-
	1361-1813 g	6	3136	2880-3400
	1814-2267 g	5	4148	3816-4556
	2268-2721 g	4	4982	4288-5539
	2722-3174 g	8	6355	4844-8322
	3175-3629 g	2	7396	7382-7399
	4422 g	1	8845	-
N132 AK Baronof I.	439 mm, 907 g	68	3873	1230-5968

Dolly Varden from Sun Creek, Oregon, had eaten insects, mostly aquatic, in July-August (W134). Insects were also the principal food of Dolly Varden in northern Saskatchewan lakes (R193) but 9% had eaten sockeye salmon and 9%, crustaceans. Of 5050 examined 52.8% were empty, with the percentage lower in small than large fish. Predation on sockeye salmon was greatest in the rapids below the outlet and was greatest from intermediate-sized (150-380 mm) Dolly Varden (R276). Although Dolly Varden ate little during their freshwater stay, they did eat salmon eggs during the spawning season, but these were only drifting eggs not lodged in redds (R276). Kokanee were the principal food of Dolly Varden in Idaho lakes (J53). In Port Walter estuary, fish were 93% of the diet, and 28% of 143 Dolly Varden stomachs were empty (L155). At 13.3C Dolly Varden digested 43 mm sockeye salmon in 20 to 24 hours (A81).

Of 95 Dolly Varden tagged and recovered in Idaho, 68 had traveled over 1.6 kilometers and the average was 35.7 kilometers (B284). The general trend was upstream in the spring and downstream in the fall.

In 19 mm (stretch) gill nets, 8 Dolly Varden 94-190 mm FL were caught (H313).

General reviews and bibliographies are in M492, A82.

LAKE TROUT, *Salvelinus namaycush* (Walbaum)

The lake trout is found throughout Canada except Newfoundland, portions of the prairie provinces and the coastal region of British Columbia. In the United States, it is limited to the Great Lakes drainage and parts of northern New England, New York, Wisconsin, Minnesota, and Montana. It is the only freshwater species that ranges into northern Canada and Alaska but does not extend into Siberia (L210). Lake trout have been introduced in high lakes of Nevada, California, and Colorado and in Europe. They are confined to cool deep lakes in the southern part of the range, but also occur in shallower lakes and in rivers in the far north. They may also enter brackish water in the far north. Predatory lampreys may be a factor limiting the distribution of lake trout (L210). Sea lampreys practically eliminated lake trout from the Great Lakes (E132).

Data are also included here for the siscowet, *S. n. siscowet* (Walbaum), a variety found only in Lake Superior, usually at depths over 90 m (E115). These data are indicated in the tables. It is a deeper-bodied fish, too oily for most cooking methods, and thus usually smoked for sale. Fat content of siscowets ranged from 32.5 to 88.8% compared to 6.6 to 52.3% in lake trout (E131). Cross-breeding indicated that the fat content was genetically determined.

	Length (SL)	No.	TL
N26 NY young fish	-	-	1.177 SL
W154, W93 NY Cayuga L.	-	-	1.082 FL + 1.1 mm
V18 Great Lakes	-	-	1.192 SL
V51 L. Michigan	140–240 mm	539	1.200 SL
	240–340	9644	1.186
	340–440	1149	1.181
	450–490	125	1.175
	490–600	86	1.168
	600+	28	1.160

K98 (and K93) questions an 1836 record in Lake Huron of 54.4 kg, but accepts 39.9 kg from Lake Michigan in 1864, 30.16 kg from Great Slave Lake, and an angling record 28.62 kg, 1207 mm (D175). T98 records a 46267 g, 1270 mm FL, 1118 mm girth from Lake Athabaska, Saskatchewan.

N46	TL-weight relationship, New York hatcheries									
TL	25	38	51	64	76	89	102	114	127	152	178
No. per pound	4371	1303	519	255	143	92	58	40	28	16	10
No. per kilo	9636	2873	1144	562	315	203	128	88	62	35	22

TL	Citation	No.	Mean W	Range
152–177	F46, H144, V51	5+	32	27–41
	E131 L. Superior	212	32	27–36
178–202	C145, F46, H144, T93, V51	10+	45	27–59
	E131, R277 L. Superior	182	45	-
203–228	E131,132 L. Superior	268	59	45–77
	K57, R211, F46, T93	+	68	45–91
	H144, C236, H294	7	73	59–86
	C145, V51 L. Michigan	13	77	54–86
	R277 L. Superior "humper"	4	91	-
229–253	K57, R211, T93, F46, W208	+	91	68–113
	E131 L. Superior	92	100	-
	H144, C236, W144, F26	7	113	77–127
	C145, V51 L. Michigan	18	118	104–204
	M315 L. Louisa, ON	18	141	-
254–278	E131,132 L. Superior	300	136	-
	H144, C236, F30, G21	9	141	109–172
	K57, R211, T93, V13, F46, W208	+	145	113–177
	C145, V51 L. Michigan	240	154	136–168
	R277 L. Superior "humper"	6	172	-
	M315 L. Louisa, ON	9	200	-
279–305	K57, R211, V13, F46, W208	+	181	159–204
	E131 L. Superior	121	181	-
	R277 L. Superior "humper"	9	181	-
	H144, C236, F30, H294	24	186	141–254
	C145, V51 L. Michigan	1287	195	177–213

TL	Citation	No.	Mean W	Range
	E131 L. Superior siscowet	1	272	-
	M315 L. Louisa, ON	30	313	-
306-329	K57, R211, T93, V13, F46, W208	+	231	209-254
	E131,132 L. Superior	291	231	-
	C145, V51 L. Michigan	2818	236	231-272
	C236 Watertown L., SA	2	245	227-268
	R277 L. Superior "humper"	13	249	-
	F30 ON	24	249	-
	H144 Green L., WI	10	259	200-313
	E131 L. Superior siscowet	1	318	-
	M315 L. Louisa, ON	54	367	-
330-355	H144 Green L., WI	11	272	227-340
	K57, R211, T93, V13, F46, W208	+	290	227-381
	C145, V51 L. Michigan	3405	295	281-340
	F30 ON	48	299	-
	E131 L. Superior	149	304	-
	R277 L. Superior "humper"	20	318	-
	C236, H144, C20, S178, H294	12	318	240-381
	M315 L. Louisa, ON	106	426	-
	E131 L. Superior siscowet	3	486	-
356-380	H144 Green L., WI	14	345	286-399
	E131,132 L. Superior	265	345	-
	C145, V51 L. Michigan	2235	363	354-426
	K57, R211, T93, V13, S62, F46, W208	+	367	340-426
	C236, W144, S178, H294	8	417	354-468
	F30 ON	77	417	-
	R277 L. Superior "humper"	26	454	-
	E131 L. Superior siscowet	6	522	454-538
	M315 L. Louisa, ON	146	541	-
381-405	K57, R211, T93, V13, F46, W208	+	449	408-477
	C145, V51 L. Michigan	1239	459	449-547
	E131 L. Superior	119	463	-
	H144 Green L., WI	12	477	367-567
	M176, C236, W144, C20, F26	17	486	399-635
	F30 ON	142	535	-
	R277 L. Superior "humper"	34	544	-
	E131 L. Superior siscowet	13	599	587-613
	M315 L. Louisa, ON	56	622	-
406-431	C20 MN	7	535	-
	E131,132 L. Superior	220	541	-
	K57, R211, T93, V13, S62, M136, F46, W208	+	544	499-635
	C145, V51 L. Michigan	566	553	502-653
	H144 Green L., WI	11	596	486-680
	C236 Watertown L., SA	10	622	340-768
	F30 ON	243	677	-
	R277 L. Superior "humper"	43	703	-
	M315 L. Louisa, ON	20	768	-
	E131 L. Superior siscowet	21	810	707-880
432-456	K57, R211, T93, V13, F46, W208	+	644	590-694
	C145, V51 L. Michigan	280	671	622-834

TL	Citation	No.	Mean W	Range
	E131 L. Superior	142	712	-
	C236 Watertown L., SA	13	762	653-907
	H144, M176, M315, S178	17	762	595-907
	F30 ON	358	780	-
	C20 MN	15	807	-
	R277 L. Superior "humper"	40	862	-
	E131 L. Superior siscowet	44	930	902-953
457-482	C145, V51 L. Michigan	139	807	785-1043
	E131,132 L. Superior	268	816	-
	S178, K57, R211, T93, V13, S62, M136, F46, W208	+	825	731-953
	H144, W144	10	830	680-1134
	F30 ON	572	902	-
	R277 L. Superior "humper"	22	1089	-
	E131 L. Superior siscowet	71	1121	1038-1197
483-507	R211, V13, F46	+	921	857-953
	C145, V51 L. Michigan	67	1003	907-1356
	E131 L. Superior	160	1012	-
	H144, M176, W144, T93, C20, S178	21	1025	794-1565
	C236 Watertown L., SA	34	1043	508-1406
	R277 L. Superior "humper"	21	1225	-
	E131 L. Superior siscowet	78	1352	1265-1397
508-532	F30 ON	797	1038	-
	K57, T93, S62, M136, V13, F46	+	1094	1007-1179
	C145, V51 L. Michigan	34	1129	1021-1161
	E131,132 L. Superior	381	1139	-
	H144, M176, W144, C20, S178, C64	22	1148	907-1333
	R277 L. Superior "humper"	20	1451	-
	E131 L. Superior siscowet	51	1574	1515-1619
533-558	F30 ON	787	1161	-
	K57, R211, T93, V13	+	1279	1139-1451
	H144, M176, W144, S178	11	1320	907-1787
	E131 L. Superior	186	1352	-
	C145, V51 L. Michigan	34	1394	1171-1570
	C236 Watertown L., SA	12	1394	1270-1588
	E131 L. Superior siscowet	25	1700	1628-1751
	R277 L. Superior "humper"	4	1814	-
559-583	F30 ON	540	1366	-
	H144, M176, W144, C20, S178	19	1388	1347-1928
	K57, R211, T93, S62, M136, V13	+	1478	1225-1724
	E131,132 L. Superior	241	1487	-
	C145, V51 L. Michigan	20	1502	1279-1565
	C236 Watertown L., SA	29	1542	1361-1928
	R277 L. Superior "humper"	2	2041	-
	E131 L. Superior siscowet	17	2168	2146-2200
584-609	F30 ON	343	1547	-
	V51, C236, W144, C20	22	1760	1451-2064
	E131 L. Superior	66	1783	-
	K57, R211, T93, V13	+	1805	1602-1996
	E131 L. Superior siscowet	15	2395	2345-2449
610-634	F30 ON	168	1800	-
	K57, R211, T93, S62, M136, V13	+	1987	1633-2223

TL	Citation	No.	Mean W	Range
	E131,132 L. Superior	103	2036	1950-2168
	V51, M176, W144, C20	24	2064	1905-2522
	C236 Watertown L., SA	11	2263	1928-2767
	E131 L. Superior siscowet	12	2876	2753-3116
635-659	F30 ON	76	2241	-
	K57, R211, T93, V13	+	2359	2168-2540
	E131 L. Superior	43	2386	-
	C236 Watertown L., SA	19	2509	2019-3175
	H144, V51, W144, C20, V13, C64	22	2558	1945-3538
	E131 L. Superior siscowet	11	2717	2853-3556
660-685	K57, R211, T93, S62, M136, V13	+	2540	1996-2858
	E131,132 L. Superior	94	2649	2540-2849
	F30 ON	46	2785	-
	C236 Watertown L., SA	18	2881	2359-3357
	H144, V51, M176, W144	19	2948	2435-4536
	E131 L. Superior siscowet	9	3719	3569-3796
686-710	V51, M176, W144, C20, R122	10	2785	2531-3574
	K57, R211, T93, V13	+	2912	2762-3266
	E131 L. Superior	63	3175	3075-3252
	F30 ON	24	3402	-
	C236 Watertown L., SA	10	3474	3084-4536
	H144 Green L., WI	10	3810	3175-4581
	E131 L. Superior siscowet	2	4604	-
711-736	V51, V13, V39 Great Lakes	11	3111	2948-4536
	K57, R211, T93, S62, M136, V13	+	3289	2948-3719
	E131,132 L. Superior	175	3375	3311-3520
	F30 ON	22	3770	-
	H144 Green L., WI	22	4082	2722-7620
	C236 Watertown L., SA	12	4128	3220-4581
	E131 L. Superior siscowet	5	5271	5048-5602
737-761	E131 L. Superior	72	3842	-
	K57, R211, T93, V13	+	3964	3674-4196
	V51, C236, W144, C20, V13, B230	26	3992	3429-5216
	F30 ON	23	4142	-
	H144 Green L., WI	39	4717	3175-7620
	E131 L. Superior siscowet	1	5670	-
762-786	E131,132 L. Superior	153	4128	3992-4372
	V51, C236, W144, V13	20	4304	3901-5648
	K57, R211, T93, S62, M136, V13	+	4391	3901-5125
	F30 ON	25	4368	-
	H144 Green L., WI	41	5125	3856-8074
	E131 L. Superior siscowet	2	5529	-
787-812	E131 L. Superior	50	4604	-
	K57, R211, T93, V13	+	4989	4445-5670
	C145, C236, V13, C135	13	5194	4536-6396
	F30 ON	15	5302	-
	H144 Green L., WI	62	5670	3992-8391
	E131 L. Superior siscowet	2	7076	6985-7167
813-837	S62, M136, V13	+	4853	4627-5125
	E131,132 L. Superior	78	5035	4853-5466
	C236, F30, V13, G21	17	5574	5171-7212
	H144 Green L., WI	58	6078	4627-7847
	E131 L. Superior siscowet	2	8210	7257-9163

TL	Citation	No.	Mean W	Range
838-863	M350, C20, V13, C126	12	5529	5243-5784
	K57, R211, V13	+	5625	4899-5987
	E131 L. Superior	33	5665	5548-5906
	H144 Green L., WI	90	6985	4899-10024
864-888	K57, R211, S62, M136, V13	+	6069	5352-6758
	E131,132 L. Superior	45	6169	6033-6486
	V51, C236, F30, V13, S165, B230	31	6577	6260-7711
	H144 Green L., WI	52	7121	5625-10433
	E131 L. Superior siscowet	1	9208	-
889-913	M136, C236, F30, V13, M63	22	6967	5897-8165
	E131 L. Superior	16	7040	6758-7394
	K57, R211, T93	+	7303	5897-8391
	H144 Green L., WI	29	7620	5715-9888
914-939	E131,132 L. Superior	19	7530	7439-7666
	K57, R211, T93, V13	+	7711	6441-8890
	C236, C20, F30, M63	7	8522	7257-9525
	H144 Green L., WI	22	9208	7031-11793
940-964	K57 Great Slave L.	+	6985	
	E131 L. Superior	6	8377	8074-8936
	C236, R211, W144, F30, V13	15	8709	7666-13835
	H144 Green L., WI	24	9843	7620-12701
965-990	K57 Great Slave L.	+	7530	-
	E131,132 L. Superior	10	8278	-
	C236, F30, V13, M63	18	9258	5897-11567
	H144 Green L., WI	11	9571	7847-11975
991-1015	K57 Great Slave L.	+	8165	-
	E131 L. Superior	2	9639	9344-9934
	H144, C236, R211, R98, M63	11	11263	8845-16420
1016-1040	K57 Great Slave L.	+	8845	-
	E132 L. Superior	16	10478	-
	H144, R211, V13, R104	19	10705	9661-13835
1041-1066	K57 Great Slave L.	+	9616	-
	E131 L. Superior	1	11521	-
	H144, R211, V13, M63, R139	27	13676	11249-17236
1067-1091	K57 Great Slave L.	+	10251	-
	E132 L. Superior	14	12882	-
	H144, R211, V13	13	13381	11476-16420
1092-1117	H144, R211	9	15149	12383-17463
1118-1142	H144, R211, V13, R139	10	14333	12927-16783
1143-1167	R211, V13	8	15740	14968-18597
1194-1218	G36 MB	1	28576	-
1219-1244	T93	1	20412	-
1270-1294	R211	1	18824	-
1295-1320	R211	1	16783	-
1321-1345	R211	1	19504	-
1372-1396	T98	1	46267	-

K57 Great Slave L.	immature trout	$\log W = -5.406 + 3.16 \log FL$
	mature trout	$\log W = -4.970 + 3.02 \log FL$
C145 L. Michigan	1197 trout	$\log W = -5.391 + 3.1125 \log TL$
V51 L. Michigan		$\log W = -5.465 + 3.1377 \log TL$
E131 L. Superior	3284 trout	$\log W = -5.592 + 3.191 \log TL$
	393 siscowets	$\log W = -5.992 + 3.387 \log TL$
R277 L. Superior	268 humpers	$\log W = -5.770 + 3.282 \log TL$

		No.	Length	Average K(SL)	Range
C23 MN Greenwood L.		60	(150-750 mm SL)	1.18	-
E3 MN		80	(279-812 mm SL)	1.43	-
C23 MN		8	(250-450 mm SL)	1.48	-
H144 WI Green L.		118	(152-533 mm TL)	1.26	1.05-1.47
		484	(533-1118 mm TL)	1.90	1.56-2.28
V51 L. Michigan	1930	783	-	1.34	-
	1931	4640	-	1.28	-
Upper L. Michigan	1932	3138	-	1.20	-

				K(FL)	
S168 N.Z.		-	584 mm FL	1.23	-
C236 SA Watertown L.		248	(193-927 mm)	1.19	0.87-1.55

				K(TL)	
E89		1370	203-305 mm	0.84	-
E85 L. Superior trout		221	610-912 mm	0.91	0.78-1.10
siscowets		5	427-582 mm	1.15	1.04-1.21
siscowets		7	643-869 mm	1.45	1.17-1.69

Average K(SL) increased from 0.95 to 1.30 as length increased from 25 to 178 mm in New York (N46).

No sex differences in length-weight or condition factor were noted by K57 and V51, but in Lake Superior (E85) condition factors of males appeared to be lower than those for females (no test of significance was run, however). No increase in condition factor was noted as the summer progressed (V51, E85) even though the female gonads increased significantly during the time (E85). Fat content was not necessarily related to condition factor (E131).

Condition factors increased with increase in length of trout (V51, C236, H144), and in the length-weight regressions weight appears to increase at more than the cube of the length in all except the mature trout of Great Slave Lake (K57). In the Watertown Lakes, deepwater lake trout, perhaps a separate race, were in poor condition (C236). Average condition factors of fish in the same size-range increased with the mesh size in gill nets used in the capture of the fish (V51).

At hatching, lake trout fry are 15 mm SL or 23 mm TL (F5, L. Erie).

	No.	Mean TL	Range	No.	Mean W	Range
Age 0-April						
E86 L. Superior	2	23	-	-	-	-
Age 0-June, 3 months						
E86 L. Superior	29	30	28-33	29	0.1	-
B73 hatcheries	+	-	25-51	-	-	-
Age 0-July						
E86 L. Superior	18	33	30-36	18	0.2	-
Age 0-August, 5 months						
A22 ON	+	53	-	+	1.1	-
E86 L. Superior	178	53	36-69	123	0.9	-
V13, S112 MI, MN	+	-	76-102	-	-	-
Age 0-September, 6 months						
E86 L. Superior	73	66	56-81	73	1.7	-
S142 MI hatchery	+	76	-	+	2.8	-
H131 NY	-	-	-	+	-	2.3-4.3

	No.	Mean TL	Range	No.	Mean W	Range
Age 0-October						
S178 MI	3000	61	38-76	-	-	-
E86 L. Superior	65	74	53-94	65	2.8	-
F41 Switzerland	+	-	76-127	-	-	-
H294 WI Green L.	+	140	-	+	11.3	-
Age 0-November						
A16 MI Birch L.	46	81	71-99	-	-	-
P1, B73, S112	+	-	102-254	+	-	8.5-62.4
Age 0						
M63, N23 AT, SA	+	-	30-51	-	-	-
A18 ME	-	-	-	+	-	6.8-9.9
B156 ID Priest L.	+	-	102	-	-	-
Age I						
M63 SA Great Bear L.	29	71	-	20	2	-
S178, S142 MI						
spring-stocked	12237	-	51-119	5808	6	-
E86 L. Superior	163	127	89-152	163	14	-
A22, M142, R138 Can.	3+	142	109-221	+	8	-
S142, S178 MI						
fall plantings	5379	165	94-185	5045	40	-
A16 MI Birch L.	13	168	137-221	-	-	-
W128, W93, W143 NY						
L. Cayuga	50	175	163-190	-	-	-
B156, B230, H132, H294,						
H413, S112, V13, V39,						
G143	1+	203	64-356	+	-	8-340
B250 hatcheries						
18 months	5449	213	175-282	-	-	-
B211 ON hatchery	89	229	-	-	-	-
W2 CA Castle L.	78	290	-	-	-	-
R211 SA L. La Ronge	14	290	213-363	-	-	-
Age II						
M176 QU	1	86	-	-	-	-
M63, M59 SA Great						
Bear L.	10	107	-	10	9	-
R277 L. Superior						
"humper"	1	124	-	-	-	-
E86 L. Superior	335	178	142-234	335	41	-
S142 MI spring stocking	4667	190	-	4667	50	-
E99, R138, M142, R98,						
R139 Can. lakes	13+	198	140-343	+	-	45-91
H294, H413 WI Green L.	98	208	145-361	95	73	18-502
E89 MI hatcheries	600	216	-	600	82	-
W199 NY Little						
Moose L.	8	239	-	8	100	-
T93 ON Georgian Bay	14	241	216-251	-	-	-
S142 MI fall stocking	3460	246	-	3460	113	-
G148 Switzerland	10	254	198-287	-	-	-
C64 ME Moosehead L.	13	254	201-371	13	145	100-400
B250 hatcheries						
25-30 months	6757	262	188-391	-	-	-
V51, C145 L. Michigan	45	262	183-320	42	136	32-240

	No.	Mean TL	Range	No.	Mean W	Range
W93, W128, W143 NY						
Cayuga L.	559	264	249-284	310	136	-
B156, S112, V13, V39,						
M296 U. S. lakes	+	274	178-457	+	141	-
F80, F81 ON South Bay	21	277	-	21	191	-
V39 MI	29	284	-	24	186	-
G143,147 OR	2+	305	221-386	-	-	-
B230, G21, F26, V39						
U. S. lakes	14	302	234-373	13	231	86-499
S76 L. Michigan	127	325	-	127	281	-
B211 ON South Bay	15	335	-	-	-	-
R211 SA L. La Ronge	21	391	267-472	-	-	-
Age III						
M176 QU Ungava	1	122	-	-	-	-
R277 L. Superior						
"humper"	13	175	152-201	-	-	-
C149 OR	10	193	-	-	-	-
M142, R89, F29, H15,						
A15, C233 Can. lakes	+	249	206-386	+	113	91-136
W199 NY Little						
Moose L.	33	269	-	33	132	-
S142 MI hatcheries						
spring	1954	277	-	1954	168	-
C64 ME Moosehead L.	13	279	221-376	13	191	104-449
M63, R138, C236, E99						
Can. lakes	34	292	147-427	17	122	27-272
H413, H294 WI Green L.	504+	302	198-421	504+	259	41-880
F29, F30 ON L. Opeongo	46	318	249-358	46	281	-
V39, V51, C145						
L. Michigan	463	318	246-566	463	240	100-907
T93 ON Georgian Bay	148	318	239-333	-	-	-
G148 Switzerland	19	320	241-498	-	-	-
B250, E89 hatcheries,						
MI, MN, WI	1927	320	208-445	729	177	127-268
S142 MI hatcheries fall	868	338	-	868	322	-
F80, F81 ON South Bay	142	351	343-414	32	331	-
A16, E3, F26 U. S. lakes	11	358	295-536	11	395	286-567
W128, W93, W143 NY						
Cayuga L.	932	363	330-386	575	454	-
V39 L. Michigan	131	366	-	108	404	-
B156, B197, M296, S112,						
V39 U. S. lakes	+	368	254-559	+	-	222-544
G143,147 OR	3+	371	335-455	-	-	-
B230 UT Fish L.	10	376	356-432	10	417	454-726
F41 Switz. introduced	70	-	356-457	70	-	1248-1361
S76 L. Michigan	42	414	-	42	680	-
B211 ON South Bay	15	421	-	-	-	-
R211 SA L. La Ronge						
& Hunter Bay	16	432	351-516	-	-	-
Age IV						
R277 L. Superior						
"humper"	36	221	178-267	-	-	-

	No.	Mean TL	Range	No.	Mean W	Range
R139, C233, M142, R98, M59, F29, H15 Can. lakes	+	284	196-358	+	358	181-816
G148 Switz. Ergstlensee	14	292	264-330	-	-	-
W199 NY Little Moose L.	56	333	-	56	263	-
G143,147 OR	20+	340	277-505	-	-	-
T93 ON Georgian Bay	243	356	348-417	-	-	-
V39 WI	660	356	-	-	-	-
H294, H143 WI Green L.	443+	363	226-737	443	468	59-2835
S76, V39, V51, C145 L. Michigan	1089	363	267-533	1005	345	132-1358
F29, F30 ON L. Opeongo	143	363	277-414	143	445	-
S142 MI hatcheries spring	472	366	-	472	413	-
M63, E99, K57, R138, C236 Can. lakes	18	368	198-467	11	486	363-907
C64 ME Moosehead L.	32	371	525-472	32	431	286-880
G148 Switz. Arnensee	5	381	520-467	-	-	-
A16 MI Birch L.	11	381	211-450	11	481	-
D65 ME	95	391	345-450	95	526	363-862
V39 MI	44	394	-	37	494	-
E3 MN	44	406	353-566	44	626	-
S142 MI hatcheries fall	439	421	-	439	876	-
B230 UT Fish L.	44	427	376-445	44	653	544-907
F80, F81, B211 ON South Bay	558	480	404-508	342	971	-
B156, B197, M296, V39, S112 U. S. lakes	+	475	396-660	+	-	454-1814
A16, F26, G51 U. S. lakes	14	483	262-605	14	1193	109-2014
W128, W93, W143 NY Cayuga L.	380	483	460-508	232	998	-
R211 SA L. La Ronge	24	511	363-612	-	-	-
S168 N.Z. introduced	1	632	-	-	-	-
Age V						
R277 L. Superior "humper"	59	287	229-343	-	-	-
G148 Switz. Engstlensee	6	295	264-315	-	-	-
C233, M142, R98, F29, H15 Can. lakes	+	345	264-439	+	1089	272-1270
H294, H413 WI Green L.	512+	373	279-653	512+	513	141-2586
W199 NY Little Moose L.	102	376	-	102	381	-
F29, F30 ON L. Opeongo	268	404	305-467	268	621	-
V39, V51, C145 L. Michigan	410	401	254-559	337	581	145-1371
B162, G143,147 OR	114+	429	277-607	-	-	-
C64 ME Moosehead L.	43	409	343-551	43	600	313-164
A16, F26, G21, V39 U. S. lakes	26	432	264-551	21	907	286-201
D65 ME	330	437	366-480	330	703	408-953

	No.	Mean TL	Range	No.	Mean W	Range
E99, K57, C236,						
R138 Can. lakes	32	439	335-592	17	771	318-1814
S142, E89 MI hatcheries	899	450	-523	899	798	-1134
H132 NY Raquette	25	450	-	25	785	-
E3 MN	41	452	391-597	41	1021	-
T93 ON Georgian Bay	242	457	445-500	-	-	-
B230 UT Fish L.	66	490	450-533	66	848	726-1134
F80, F81, B211 ON						
South Bay	1775	518	495-625	832	1300	-
G145 Switz. Arnensee	6	551	376-742	6	1497	227-3765
S76 MI	26	564	-	26	1769	-
W128, W93, W143 NY						
Cayuga L.	289	589	541-643	129	1678	-
R211 SA L. La Ronge	72	577	378-721	-	-	-
B156, B197, R122						
U. S. lakes	+	622	533-762	+	-	1814-6350
Age VI						
R277 L. Superior	98	348	279-378	-	-	-
W199 NY						
Little Moose L.	82	417	-	82	508	-
V51, C145 L. Michigan	38	432	310-513	30	902	322-1034
R139, C233, L119, F81,						
H15, M142, R98, F29						
Can. lakes	+	439	305-632	+	862	408-1769
F29, F30 ON L. Opeongo	294	450	386-579	294	830	-
H294, H413 WI Green L.	475+	452	297-721	475+	998	195-3516
C236 AT Watertown L.	27	460	371-599	27	862	408-1633
D65, D173, C64 ME	1179	467	366-566	526	925	400-1788
G147 OR Odell L.	34	470	358-688	-	-	-
H132 NY Raquette	84	480	452-490	84	958	834-1084
B230 UT Fish L.	99	516	457-572	99	1206	907-1678
K57 Great Slave L.	-	-	-	49	1306	907-2722
A16, F26 U. S. lakes	11	531	361-640	11	1914	1814-2495
R138, E99 Can. lakes	21	-	467-625	-	-	-
T93 ON Georgian Bay	361	533	518-564	-	-	-
E89 MI hatchery	198	541	-	198	1361	-
E3 MN	27	549	450-653	27	1729	-
F80 ON South Bay	968	566	-	249	2204	-
A16 MI Crystal L.	48	579	533-660	48	2182	680-2522
V39, S76 L. Michigan	43	592	-	43	2041	-
R211 SA L. La Ronge	113	622	472-734	-	-	-
W93, W128, W143 NY						
Cayuga L.	206	645	617-665	75	2313	-
M350 OR introduced	1	787	-	1	5784	-
B156 ID	+	-	762-889	+	-	6350-9979
Age VII						
R277 L. Superior						
"humper"	129	401	343-445	-	-	-
C233, L119, F29, H15,						
M142, R98 Can. lakes	+	478	330-711	+	1611	680-2495
W199 NY						
Little Moose L.	38	480	-	38	834	-

	No.	Mean TL	Range	No.	Mean W	Range
F29, F30 ON L. Opeongo	1042	493	305-769	1042	1029	-
C64, F26, D65, D173 ME	604	505	460-773	293	1239	816-5075
H294, H411 WI Green L.	268+	541	300-739	268+	1656	227-4368
H132 NY Raquette L.	13	544	-	13	1583	-
C236 AT Watertown L.	22	549	445-701	22	1678	771-3583
R138, E99 Can. lakes	13	-	503-645	-	-	-
B230 UT Fish L.	33	572	533-640	33	1700	1089-2359
E89 MI hatcheries	162	574	-	162	1678	-
T93 ON Georgian Bay	228	582	533-627	-	-	-
E3 MN	12	594	536-681	12	1474	-
V39, V51, S76 L. Michigan	38	594	505-625	38	1945	1161-2228
K57 Great Slave L.	-	-	-	289	1950	454-4082
F80 ON South Bay	38	625	-	5	3062	-
R211 SA L. La Ronge	148	655	460-789	-	-	-
G147 OR Odell L.	72	671	386-818	-	-	-
W93, W128, W143 NY Cayuga L.	117	673	655-688	31	2722	-
V13 NY	+	686	-	+	9196	-
A16, G21 U. S. lakes	5	688	627-785	4	2580	2223-3175
B156 ID Priest L.	+	-	838-991	+	-	8165-1179

Age VIII

	No.	Mean TL	Range	No.	Mean W	Range
R277 L. Superior "humper"	110	452	394-521	-	-	-
F29, H14, S165, M142, R98, R137, C233 Can. lakes	+	480	333-612	+	2182	254-3719
F29, F30 ON L. Opeongo	1135	523	358-907	1135	1206	-
W199 NY Little Moose L.	26	546	-	26	1352	-
C64, D65, D173 ME	135	561	478-714	124	1515	967-4536
E99, F80, M63, R138 Can. lakes	15	589	358-851	4	399	-
E89 MI hatcheries	132	599	-	132	1905	-
C236 AT Watertown L.	12	599	483-757	12	2268	953-453
K57 Great Slave L.	-	-	-	931	2327	907-635
T93 ON Georgian Bay	48	625	589-808	3	-	2177-535
S76, V39, V51 L. Michigan	10	632	437-739	10	2572	703-412
H294, H411 WI Green L.	131+	640	424-769	131+	2504	468-456
B230 UT Fish L.	11	676	635-711	11	2812	2041-340
E3 MN	12	681	508-764	12	2495	-
R211 SA L. La Ronge	135	691	488-818	-	-	-
W93, W128, W143 NY Cayuga L.	44	701	691-716	7	2994	-
G147 OR Odell L.	56	714	414-947	-	-	-
L119 ON rivers	+	749	660-787	-	-	-
B156, V13 WY, ID	+	-	914-1067	+	-	5330-16

Age IX

	No.	Mean TL	Range	No.	Mean W	Range
M176 QU	1	297	-	-	-	-
M63 SA Great Bear L.	10	384	-	10	513	-
R277 L. Superior "humper"	93	503	445-556	-	-	-
F29, F30 ON L. Opeongo	631	554	467-798	631	1401	

	No.	Mean TL	Range	No.	Mean W	Range
F29, H15, M142, R98,						
C233 Can. lakes	+	566	439-676	+	4568	3261-5488
W199 NY						
Little Moose L.	3	579	-	3	2087	1361-3357
C64, D173, D65 ME	53	582	546-709	28	2069	1361-3357
E89 MI hatcheries	128	615	-	128	2041	-
K57 Great Slave L.	-	-	-	1448	2585	907-6350
V13, V39, S76						
L. Michigan	5	673	610-734	5	3084	1860-4536
H294, H411 WI Green L.	85	676	536-767	85	2917	1220-4789
T93 ON Georgian Bay	19	693	-929	5	-	5125-8890
B230 UT Fish L.	5	701	610-762	5	3932	2948-5080
W93, W143 NY						
Cayuga L.	4	714	701-744	+	3084	-
G147 OR Odell L.	71	714	442-991	-	-	-
R211 SA L. La Ronge	115	709	556-844	-	-	-
E99, F80, C236, R138						
Can. lakes	12	732	607-902	5	5398	4196-7257
L119 ON rivers	+	747	671-824	-	-	-
E3, V13 MN	7	789	597-914	6	5262	4763-7756
B156 ID Priest L.	+	-	1067-1118	+	-	16329-19051
Age X						
M59, M63 SA Great						
Bear L.	7	406	-	7	567	-
R277 L. Superior						
"humper"	26	561	533-582	-	-	-
F29, F30 ON Opeongo L.	244	594	439-769	244	1832	-
C64, D173 ME	10	620	528-681	3	3089	1333-3992
K57 Great Slave L.	-	-	-	1483	3034	907-11793
H15, M142, R98, R139,						
C233 Can. lakes	+	645	564-742	+	5425	2631-7439
T93, V51 Great Lakes	10	693	668-810	3	3728	2563-5987
H413 WI Green L.	+	-	686-729	+	-	3017-3538
R211 SA L. La Ronge	78	739	638-871	-	-	-
L119 ON rivers	+	782	724-841	-	-	-
B230, E3, G21	8	747	622-813	5	5570	-
G147 OR Odell L.	27	851	770-963	-	-	-
C135, C236, E99						
Can. lakes	8	884	798-947	7	7167	5443-9616
Age XI						
M63 SA Great Bear L.	14	445	-	14	740	-
M176 QU	2	486	455-518	-	-	-
W199 NY Little Moose L.	1	528	-	-	3719	-
R277 L. Superior						
"humper"	8	589	572-632	-	-	-
F29, F30 ON Opeongo L.	86	640	523-826	86	2467	-
K57 Great Slave L. west	-	-	-	808	3039	907-7257
Great Slave L. east	-	-	-	479	4236	1814-10886
H413 WI Green L.	+	706	-	+	3306	-
H15, M142, R98, C233						
Can. lakes	+	706	549-813	+	7756	5330-9979
B230, C64, E3, V13						
U. S. lakes	7	762	691-886	5	5148	3516-6577

	No.	Mean TL	Range	No.	Mean W	Range
R211 SA L. La Ronge	47	770	681-914	-	-	-
L119 ON rivers	+	795	742-848	-	-	-
G147 OR Odell L.	15	881	826-935	-	-	-
V13, T93 Great Lakes	2	881	838-925	1	5216	-
E99, C236 AT lakes	6	909	820-942	4	9435	7756-11340
Age XII						
M176 QU	5	424	388-447	4	608	471-766
M63 SA Great Bear L.	11	445	-	11	766	-
R20, M142, S165,						
R139 Can. lakes	+	638	495-813	+	2132	1089-3946
K57 Great Slave L. west	-	-	-	709	3348	1361-10886
Great Slave L. east	-	-	-	362	5176	907-13608
L119 ON rivers	+	810	770-851	-	-	-
R211 SA L. La Ronge	53	810	721-940	-	-	-
B230, C64, V13						
U. S. lakes	4	831	638-912	4	6100	2772-7711
C236, E99 AT lakes	3	897	846-927	2	8936	8346-9525
G147 OR Odell L.	7	970	884-1100	-	-	-
Age XIII						
M63 SA Great Bear L.	13	457	-	13	821	-
M176 QU	4	483	455-526	3	1134	794-1565
K57 Great Slave L. west	-	-	-	489	3629	1814-1224
Great Slave L. east	-	-	-	308	6364	2268-1451
B230 UT Fish L.	2	828	-	2	6486	-
R211 SA L. La Ronge	20	856	775-914	-	-	-
G147 OR Odell L.	2	980	975-988	-	-	-
C236, E99 AT lakes	2	983	975-991	2	10206	8845-1156
V13 Great Lakes	2	1130	1118-1143	3	16195	13154-1950
Age XIV						
M176 QU	2	465	457-472	1	794	-
M63 SA Great Bear L.	17	488	-	17	993	-
R19, R20, R139 SA lakes	+	643	536-798	+	2576	1248-489
K57 Great Slave L. west	-	-	-	298	3997	1814-140
Great Slave L. east	-	-	-	228	7298	2268-167
B230 UT Fish L.	1	762	-	-	-	-
F30 ON Opeongo L.	27	833	632-963	27	5162	-
L119 ON rivers	+	-	846-848	-	-	-
R211 SA L. La Ronge	10	879	831-914	-	-	-
Age XV						
M176 QU	6	513	480-536	3	1170	1161-118
M63 SA Great Bear L.	27	528	-	27	1361	-
K57 Great Slave L. west	-	-	-	187	5262	2268-145
Great Slave L. east	-	-	-	161	7938	3175-158
B230 UT Fish L.	2	889	-	2	6436	-
R211 SA L. La Ronge	16	914	818-968	-	-	-
C236 AT Watertown L.	1	963	-	1	10886	-
G147 OR Odell L.	1	1087	-	-	-	-
Age XVI						
M176 QU	1	513	-	1	1107	-
M63 SA Great Bear L.	32	559	-	32	1615	-
R139, C236, R19, R20						
SA, AT	+	777	592-1003	+	5466	1814-99

	No.	Mean TL	Range	No.	Mean W	Range
K57 Great Slave L. west	-	-	-	102	6350	2268-13608
Great Slave L. east	-	-	-	87	8890	3175-16783
B230 UT Fish L.	3	927	864-1016	3	7203	6895-10070
R211 SA L. La Ronge	10	955	886-1024	-	-	-
Age XVII						
M176 QU	1	594	-	1	1642	-
M63 SA Great Bear L.	33	602	-	33	1928	-
B230 UT Fish L.	1	876	-	1	7031	-
K57 Great Slave L. west	-	-	-	61	7802	1814-15422
Great Slave L. east	-	-	-	59	9117	4082-16783
R211 SA L. La Ronge	3	968	953-996	-	-	-
Age XVIII						
M176 QU	1	617	-	-	-	-
M63 SA Great Bear L.	23	622	-	23	2404	-
R19, R20, R139 SA lakes	+	762	653-930	+	5239	2495-9253
K57 Great Slave L. west	-	-	-	24	9253	2722-17236
Great Slave L. east	-	-	-	25	9934	5443-14515
R211 SA Hunter Bay	6	980	968-996	-	-	-
T93 ON Georgian Bay	1	1240	-	1	20412	-
Age XIX						
M63 SA Great Bear L.	30	658	-	30	2636	-
M176 QU	1	714	-	-	-	-
K57 Great Slave L. west	-	-	-	20	7439	2722-12701
Great Slave L. east	-	-	-	20	10659	5443-16329
R211 SA Hunter Bay	5	1019	996-1052	-	-	-
Age XX						
M63 SA Great Bear L.	22	678	-	22	2863	-
M58, R19, R20, R139 SA	+	772	660-991	+	6622	3175-12065
K57 Great Slave L. west	-	-	-	10	7394	3629-12701
Great Slave L. east	-	-	-	8	10342	4082-15422
R211 SA Hunter Bay	5	1049	1024-1064	-	-	-
Age XXI						
M63 SA Great Bear L.	10	711	-	10	2921	-
M176 QU	2	892	886-897	1	5897	-
K57 Great Slave L. west	-	-	-	3	6804	4989-9979
Great Slave L. east	-	-	-	5	14152	9525-24947
R211 SA Hunter Bay	1	1113	-	-	-	-
Age XXII						
M63 SA Great Bear L.	9	759	-	9	3601	-
T112 CT	1	813	-	1	4082	-
R19, R20, R146 SA lakes	+	879	754-1057	+	8029	3856-14787
K57 Great Slave L.	-	-	-	6	12111	6804-20412
Age XXIII						
M63 SA Great Bear L.	10	754	-	10	3801	-
R139 SA	1	1115	-	1	15649	-
Age XXIV						
M63 SA Great Bear L.	6	701	-	6	3969	-
R19 SA Great Slave L.	+	792	-	-	-	-
R139 SA	+	1128	-	+	16874	-
Age XXV						
M63 SA Great Bear L.	7	831	-	7	4904	-
R57 Great Slave L.	-	-	-	2	8391	8165-8618

	No.	Mean TL	Range	No.	Mean W	Range
Age XXVI						
M63 SA Great Bear L.	8	838	-	8	5302	-
Age XXVII						
M63 SA Great Bear L.	2	843	-	2	4763	-
Age XXIX						
M63 SA Great Bear L.	3	864	-	3	6265	-
Age XXX						
M63 SA Great Bear L.	3	874	-	3	6265	-
Age XXXI						
M63 SA Great Bear L.	1	907	-	1	5897	-
Age XXXII						
M63 SA Great Bear L.	1	1006	-	1	10319	-
Age XXXIII						
M63 SA Great Bear L.	1	983	-	1	5897	-
Age XXXV						
M63 SA Great Bear L.	1	914	-	1	7257	-
Age XXXVII						
M63 SA Great Bear L.	1	1046	-	1	15422	-
Age XLI						
S165 NT	1	879	-	1	6804	-

(See growth data, page 287.)

Average calculated weight at each annulus

	No.	1	2	3	4	5	6	7	8	9	10	11	12	13	14
W117 BC Atlin L.	20	-	-	-	320	360	730	950	1860	2000					
R138 SA Cree L.	120	-	113	231	431	680	998	1361	1905	2404	2948	3493	3992	4763	5488
Wollaston L. 1956	78	68	254	426	621	1134	1615	2096	2889	3910	5103				
R277 L. Superior "humpers"	573	5	14	41	77	150	318	549	853	1256	1792	2458			

Average annual TL increments

		1	2	3	4	5	6	7	8	9	10	11	12	13	14	15	16	17
B230 UT Fish L.		190	102	79	64	51	46	46	44	33	28	25	20	20	18	18	15	18
	No.	295	294	285	275	231	165	66	33	22	17	13	12	10	8	6	4	1
from scales		140	76	84	81	74	61	66	66	56	53	33	46	33				
W128 NY Cayuga L.		122	74	89	107	94	76	58	38									
	No.	238	232	182	95	64	34	10	4									
R277 L. Superior "humpers"		89	41	46	41	48	66	61	56	56	58	56						
	No.	573	573	572	559	523	464	366	237	127	34	8						

Age and growth have usually been determined by scales, but branchiostegal rays were believed to be more satisfactory by some (M296, B223, B230). The ray closest to the operculum was viewed without sectioning or staining. Annuli appeared as narrow transparent lines on the posterior end of the ray or as a slight ripple with a change in the direction of growth (B223). The ages checked with known ages and showed general agreement with ages from scales except that some of the larger fish showed more annuli on rays than on scales. Growth was computed from measurements of the branchiostegals on the basis of a straight line with an intercept at 15 mm TL (B230). A good discussion of the scale method with pictures of scales is given in V51.

Ages as determined from scales agreed quite well with known ages up to 5 years of age (C145, V51, F81), and with growth of tagged lake trout (W93). However, the ages of some fish were misassigned from scale readings. One lake trout caught 21 years after fin-clipping showed the proper number of annuli (T112). Age determinations from scales were considered unreliable after

Average calculated TL at each annulus

	No.	1	2	3	4	5	6	7	8	9	10	11	12	13	14	15	16	17
E2, E3, S62, M258, M136 MN	297	137	229	315	396	467	518	587	648	709	719	737						
V13 L. Michigan	+	127	178	229	318	368	457	546	615	655								
V39 L. Michigan	97	124	216	302	391	467	536	587	622	711								
V51 L. Michigan 1930-32	811	84	180	257	307	356	427	472	508	635	686							
C145 L. Michigan	1319	150	221	284	345	394	386											
E87 L. Michigan 1947-51	+	142	218	277	338	401	452											
M152 UT Bear L.	44	221	343	434	500	554	592	630	673	693	726	747	765	803				
B230 UT Fish L. (from scales)	295	137	213	295	376	445	508	587	691	770	823	813	838	871				
(from branchiostegals)	295	190	290	368	429	486	538	620	703	754	798	831	856	879	886	917	945	914
W128 NY Cayuga L.	238	122	193	282	386	478	554	610	650									
M292 MA Quabbin L. 1946-52	299	124	218	318	396													
1953-7 after smelt introduced	207	—	269	368	465													
W117 BC Atlin L.	20	69	124	185	244	330	371	432	503	528	549							
S. Tagish L.	+	66	109	170	226	292	358	421	467									
Central Tagish L.	+	84	124	193	295	371	427	472	518									
R138 SA Cree L.	120	165	246	307	371	429	490	541	594	632	665	698	737	775	818			
Wollaston L. 1949	+	—	—	373	414	467	526	577	632	676	721	754	813	866	922	978	1057	
1956	78	208	312	378	417	511	574	635	696	752	813							
D64 ME Moosehead L.	+	—	—	—	279	371	409	467	536	569								
Cold Stream L.	+	—	—	—	421	445	488	523	577	615								
Brand L.	+	—	—	—	391	460	513	577	658									
M453, P159 MT lakes	35	74	147	249	358	508	635	747	808	874								
R277 L. Superior "humpers"	573	89	130	175	216	272	335	394	447	500	554	589						
F137 CO Granby L.	+	99	170	236	282													
M493 CA L. Tahoe	+	130	201	272	348	414	465	514	574	630	683	729	770	792	853	889	912	
M494 ON 29 lakes	12239	102	173	236	269	335	386	424	467	513	561	610	700	765	843	876	909	698
lowest mean		—	124	178	208	279	335	366	399	427	513	564	688	765	792	823		
highest mean		—	208	330	343	439	437	486	538	622	668	696	777	879	894	907		

The mean calculated lengths given in M494 are apparently calculated lengths only at the last annulus since no data are given for early annuli on some lakes. An unweighted mean of all lakes, except Lake Opeongo, the data for which were quoted from F30, is given in the above table at each annulus.

6 years of age in Cayuga Lake, New York (W93) with a strong tendency to underestimate the age. For less than 50% of those over age VI scale readings agreed with known ages. Second readings agreed with the first about 75% of the time in the Algonquin Park studies (M494).

Annulus formation extended from May through August in young lake trout in Lake Michigan (V51).

In most studies growth was computed from scale measurements by the direct proportion method. M292 used an arbitrary 32 mm as a correction factor. Curved body-scale relationships were demonstrated and used for growth computations by W128 and two straight lines were used by R277. Calculated lengths at annuli 5 through 7 showed Lee's phenomenon in the "humper" data (R277). The "humper" race was slower growing than other Great Lakes lake trout populations studied.

Though known-age males were larger than females at age V, VII, VIII, and IX, the differences were small (E89), and in Raquette River, New York, males were smaller than females at age VI (H132). No sex differences in growth were reported by R211, T93, W93. Growth slows with onset of sexual maturity at about age V-VI in Cayuga Lake and such is true of other lakes where lake trout depend primarily on alewives for food, but is not true in lakes where larger fish such as ciscoes or suckers serve as the principal food of large trout (W93). A greater number of age VIII-XI "humpers" were females rather than males (R277).

In general, growth is slower in the northern part of the range than in the south, but the trend is not very consistent. Except for a 28-year-old hatchery-stock trout in Minnesota (E114), the oldest lake trout reported from the United States was age XVII (B230), and no lake trout older than age XXI were reported except from northern Saskatchewan and the Northwest Territories. Growth was reported to be slower in colder Hunter Bay than in Lake La Ronge (R211) and to be slower in deeper, colder Upper Watertown Lake than in the Lower (C236). The growing season was listed as June to September in Great Slave Lake (K57), March to August in Lake Michigan (E114). Some growth may occur even in January (C145, E86).

Lake trout introduced into new habitats usually showed good growth (S168, G148, H294), but not always (B162, F137). Faster growth in Cree Lake in 1956 than in 1949 was believed to be the result of removal of some of the fish (R138).

Average growth rate and life span increased as the percentages of plankton in the food of 254-505 mm lake trout decreased in 30 Algonquin Park lakes (M494). When plankton-feeding lake trout were transferred from Lake Louisa to Lake Opeongo they changed to fish eating and grew more rapidly and lived longer (M494). Introduction of ciscoes as forage fish in Lake Opeongo resulted in more rapid growing, better conditioned lake trout (M494). Introduction of ciscoes may reduce spawning escapement, however.

Reduction in calorie content of diet in hatchery lake trout reduced growth rate (P183). Replacement of calories with corn or fish oil increased growth again but also increased fat content of flesh. Tests with introduction of water at the trough bottom rather than at the end did not result in changes of growth rate (P183).

Growth improved in Quabbin Reservoir, Massachusetts, after smelt were introduced as food (M296) and was more rapid in Green Lake, Wisconsin, in years when *Mysis* were abundant (H294).

Tags retarded growth (E89, D173). In Green Lake, Wisconsin, one tagged fish lost 3629 g in 3 years but stayed the same length. This fish was then age XX and the other fish of this age in Green Lake had large heads and slender bodies indicating weight lost (H413). In Maine and Massachusetts ponds, growth was usually only fair when 10-24% of the water volume was suitable for lake trout during the summer, good when 25-39% was suitable and excellent when over 40% was suitable.

A literature review (A51) indicated that age at first spawning is associated with growth. Where growth is slow, lake trout may not mature until the 13-17th year, whereas when growth is rapid, males may mature at 5 years and females at 6 to 8 years. However, the slower growing plankton-feeding lake trout matured at a younger age and smaller size than fish-eating lake trout in a series of Algonquin Park lakes (M494). Mature lake trout at age IV were taken only in plankton-feeding populations. Male lake trout mature a year earlier than females in some lakes (M494).

The slow-growing "humper" lake trout in Lake Superior did not mature until age VI and the smallest mature male was 323 mm and female 373 mm (R277). Rapidly growing lake trout of Seneca Lake do not mature until 660-760 mm long, but in nearby Keuka Lake trout matured at 457-610 mm (R236). Minimum size at maturity in Ontario lakes varied from 356-457 mm (F21). A small number of eggs, 962 per female, were stripped from female lake trout raised in a hatchery, 4 years, 6 months old and 457-660 mm long, but only 10 of 2000 fish spawned at this age (S112). Both males and females were reported as mature at the Watertown Lakes at age V (C236). In Maine, males were mature at age IV and 330-381 mm TL while females matured at age VI-VII and 445-483 mm TL (D64). In Minnesota some mature when about 279 mm TL, to spawn the next season (S385) and in New Hampshire some lake trout mature in the third and fourth years (N95). In Utah, gonads showed signs of maturing at age IV, but the lake trout probably do not spawn until age V (B223). In Georgian Bay, Ontario, a few males mature at age V and a few females at VI (T93). The smallest mature male in Green Lake, Wisconsin, was 549 mm TL and age IX (H294). Another male was mature at age VI and 612 mm TL. The smallest mature female was 625 mm and age IX with another mature at age VII and 643 mm TL. In Lake Superior some males matured at age VI and females at age VII, but most a year later (L119). The smallest mature male siscowet was 400 mm and mature female, 419 mm (E115).

Lake trout apparently do not always spawn each year after reaching maturity. In Great Bear Lake they spawn about once in 3 years, in Great Slave Lake about once in 2 years, and in Lake La Ronge about 8% of the mature females do not spawn in a given year (R211). Scale markings suggest spawning only every third year in Great Slave Lake (R19). In Watertown Lakes they probably spawn alternate years (C236). In Maine and New York, lake trout were believed to spawn each year after reaching maturity (D64, R236). In Lake Opeongo, no lake trout of over 560 mm FL were believed to be capable of spawning (F23). Only those from 432-560 mm developed eggs.

The average number of eggs for female lake trout was reported as 984 per mm of length or 1469 per kg of weight (R179).

	Size of females	No. of females	No. of eggs per female Mean	Range
N45 NY Seneca L.	-	-	-	3732-6578
V13 Great Lakes	-	-	6000	-
N25, L18, B23, L1, J13, M65 NY	10886 g	1	14943	-
D32 ON	2722-4536 g	12	6354	2542-11931

	Size of females	No. of females	No. of eggs per female Mean	No. of eggs per female Range
D32 ON	4536-5443 g	10	7694	4646-10578
D32 ON	6800-11340 g	3	14793	11043-18051
N39 Great Lakes	2722 g	1	8000	-
D12	-	-	10000	-
M65 Great Lakes siscowets	2268 g	3	3191	2796-3756
L1, B73	-	-	-	5000-6000
E4 MN	508 mm	16	-	1500-2300
E4 MN siscowet	-	5	5000	3294-
V21 MN	381-483 mm	3	1779	1518-2234
T93 ON Georgian Bay	518-643 mm FL	7	2684	2068-3124
E88 MI L. Superior	Age V, 1043 g	-	1007	-
	Age VI, 1361 g	-	1890	-
	Age VII, 1547 g	-	2093	-
	Age VIII, 1633 g	-	2621	-
	Age IX, 1905 g	-	2964	-
E85 L. Superior	635-685 mm TL, 2812 g	9	3383	2476-5645
	686-786 mm TL, 3810 g	28	4597	3017-7170
	787-888 mm TL, 5580 g	25	8735	5414-14260
	889-990 mm TL, 7800 g	10	12086	9062-17119
E85 L. Michigan	622-914 mm TL, 1814-7030 g	16	-	2970-14648
E85 Siscowet,	427-457 mm TL, 862-1134 g	3	1184	1093-1211
L. Superior	511-582 mm TL, 1588-2041 g	2	1444	1220-1628
	643-648 mm TL, 3765 g	2	3360	2859-3861
	714-762 mm TL, 5080-7350 g	4	6263	3283-8243
	816-870 mm TL, 9163-9208 g	2	8955	7435-10476
H294 WI Green Lake	Age IX and X	206	-	600-2000
	Over Age XIX	169	-	67-6120
R277 L. Superior "humpers"	381-431 mm TL, 590-726 g	6	697	411-1033
	432-532 mm TL, 953-1497 g	13	1320	757-1565
	533-583 mm TL, 1633-1996 g	6	2072	1615-2640

Average hatchery production per stripped female was reported at 1590 eggs per kg of female for Lake Ontario (D32) and 1424 eggs per kg for Lake Superior and 1710 eggs per kg for Lake Michigan (E114). Average production per female siscowet was 1025 eggs per kg of female (E115), and of female "humper" was 1138 with a range of 628 to 1515 (R277).

Sexually mature male lake trout do not normally develop the hooked jaw common in other trout species and male and female lake trout cannot be readily distinguished, except that in the water at spawning time males display a prominent black stripe along the sides (D175). One male lake trout with a well-developed kype has been reported (R236), although lake trout jaws and snouts do not usually modify as do those of salmon.

Spawning period of lake trout in the Great Lakes is September to November (E114) but various races of siscowets in Lake Superior may spawn from June through November (E115). In Watertown Lakes, lake trout spawn from October 20 to November 10 (C236); in New York, spawning may extend into December (E117); in New Hampshire, October and November (N95); in Maine, mid-October to mid-November (D64); in Lake La Ronge at 55°N., first week in October; in Great Slave Lake at 62°N., in mid-September; and in Great Bear Lake at 66°N., in mid-August (R211). In Green Lake, Wisconsin, spawning peaks about October 25 at 13C for trout under age X but about November 28 at 7C for older trout (H294). In Simcoe Lake, Ontario, lake trout spawn at 11-14C with strong onshore winds being a necessary stimulus (M155). The spawning lasted 9-16 days.

Spawning time at Raquette Lake, New York, was found to be earlier with lower temperatures and greater number of cloudy days, suggesting that temperature drop and light both affect spawning time (R236). Royce also found a good correlation between spawning time and maximum depth of the lake in New York. In lakes less than 45 m deep spawning usually peaked by October 25 but might peak as late as December in deeper lakes. In deep Seneca Lake, however, the peak is early October, suggesting a racial difference. Length of spawning season seemed to increase with size of lake.

In southern Lake Michigan, lake trout spawn at 55 m or more on clay bottoms (V13); in Green Lake, Wisconsin, at 18-30 m on silt, hardpan clay, marl or gravel (H294); in Cayuga Lake, New York, there is no evidence of spawning in shallow water (G114); in Simcoe Lake, Ontario, lake trout move into the shoals to spawn, but show no evidence of homing (M155); in Lake Superior, spawning is usually at less than 36 m over stony bottoms and may occur in tributary streams (E85); in Lake La Ronge, spawning occurs over stony reefs in 1-3 m of water, and there is evidence of homing to the same reefs (R211). Homing was also noted in Lake Superior (E85). Siscowets usually spawn in water over 90 m but at least one population in Lake Superior spawns on a rocky reef at 15-27 m (E115). Males are reported to clear rubble areas for spawning but make no nests (D64, E132). Spawning is at night (D175).

In Green Lake, where spawning takes place over unprotected mud bottom, egg predation by *Necturus* and by fish may be fairly high (D175). In one sample of eggs collected from a shallow spawning-grounds in Otsego Lake, New York, 94% were fertilized and 79% alive after one month (R236).

Lake trout are usually solitary except at spawning season and do not form schools (E114, D175).

Lake trout usually remain in waters of 4-18C (E114) or of 7-13C in the summer (G114). In spring and fall, when surface temperatures are near 10C lake trout may be found in shallow water (G114, F21). Siscowets spend most of their time in water under 4.5C (E117) and were sometimes caught near the surface at night in June and July (T107). Lethal temperatures were found to be 23.5C for lake trout acclimated at 15 or 20C (S333).

Hatching time for lake trout eggs was 50 days at 10C, 67-85 days at 7.5C, 108-117 days at 5C, and 141-156 days at 2.5C (G84). Hatching time was shorter at 10.4 ppm of dissolved oxygen than at 2.6 ppm with intermediate times at 3.7 and 4.5 ppm. Similar hatching times were reported by E110: 49 days at 10C, 59 days at 8.5C, 80 days at 6.7C, 92 days at 5.7C, 86 days at 5.1C, 106 days at 4.5C, and 162 days at 1.8C.

Red blood cell counts for healthy lake trout in New York hatcheries were 610,000-670,000 r.b.c. per cu. mm (T57); in Minnesota hatcheries, 800,000-890,000; and in age III-V lake trout from Lake Superior, 820,000 (P144). Hemoglobin readings were 5.9-6.3 g/100 ml for hatchery trout and 6.4 for Lake Superior trout (P144). Hematocrit readings for hatchery trout were 26.5-33.0 (P144).

Zooplankton and dipterous larvae and pupae are the principal foods of young lake trout (H287, N13, D175). Occasionally even 38 mm lake trout start eating small fish (D175).

Mysis were the most important food of lake trout through age II in Lake Superior (E86) and of some trout up to 660-686 mm in Green Lake, Wisconsin (H294). *Mysis* and *Pontoporeia* were important at all sizes in Lake La Ronge (R211), in young trout in Cayuga Lake (W143), of lake trout 152-178 mm long in Keuka Lake (R236), and of 127-254 mm in Cayuga Lake (W93). *Mysis* continued to be the major food of some of the large lake trout in Green Lake, Wisconsin, even when forage fish were abundant (H113).

In several lakes, lake trout continue to feed primarily on zooplankton throughout life (M494).

Larger lake trout feed primarily upon fish in most lakes (G54, H287, C236, R138, V1, L165, S234, T93, H271, K77) including ciscoes (V18, D32, W144, R211, H294, H271, R19). *Cottus* (C303, R236, W93, V18, W144, R211, H294, W143, M325, M493), *Prosopium* (B163, G53, G147, M493), *Perca* (W144, H294), *Notropis hudsonius* (H294), *Richardsonius* (M493), *Siphateles* (S234, M325, K77, M493), *Catostomus* (M325, M493), alewives (D32, W143, C303, R236, W93), *Hesperoleucas* sp. (B163, G53), *Pungitius* (R211, R236), brook trout (H43 — some up to 250 mm long), *Thymallus* (N13), *Percopsis* (W93), *Osmerus* (H271, C303, W93), and kokanee (M493, B156).

In Green Lake, Wisconsin, it was noted that ciscoes were not eaten by trout under 432 mm TL but the abundance of *Mysis* may have been a factor (H294). In Great Bear Lake, fish comprised only 44% of the food of trout over 1360 g — a low percentage. Terrestrial insects comprised 33%, bottom fauna 25%, and plankton 22% (M60). Terrestrial insects were frequently taken by even the largest trout in Georgian Bay (T93). There appeared to be a group feeding at the surface and one at the bottom, with little overlap. The bottom feeders, feeding largely on *Mysis,* were usually pink fleshed and fishermen claimed they could taste the difference (T93). Surface insects were also reported from Lake Superior (H271) and Donner Lake (K77). Even siscowets contained flies and bees in their food, some individuals having surface insects and deep bottom fish (T107). In Lake Minnewonka, Banff, water level changes separated lake trout from the forage fish and resulted in small, slow-growing lake trout (P121). Even salamanders (G147) and shrews and yellow warblers have been found in lake trout stomachs (D175).

The feeding rate of adults decreases greatly immediately preceding and during spawning (E85, T93).

Siscowet fillets may be 67% oil, the highest known of any fish, while typical lake trout fillets do not exceed 20% oil. Siscowets do not blow up when taken from 180 m whereas lake trout do if brought up from 72 m. Siscowets have small visceral cavities (T107).

In a test of tags, 7.5% of 40 jaw tags were returned, including 2 after 5.5 years, 10.9% of 175 dart tags, all less than 2 years, and 15.6% of 847 Petersen tags, some up to 8 years (H294). In another experiment a 12.4% return of Petersen tags was reported compared to 3.9 to 4.8% for jaw and streamer tags. In Lake Superior, 10.7% of lake trout marked with aluminum tags on the lower jaw, 14.0% of those tagged with monel tags in the upper jaw, 19.8% of those with streamer tags and 45.4% of those with Petersen discs were recaptured the first year (E116). The Petersen disc apparently increased vulnerability of the fish to gill nets. Most lake trout were caught within 80 km of the release point, but some travelled over 320 km. One tagged in 1941 was recaptured over 480 km away in 1949. Of about 3700 lake trout tagged in South Bay, Lake Huron, in 1949-51, only 3 were recaptured outside the bay (F80). Angling returns accounted for 5.4% of the marked fish compared to 8.6% returns by netting in Lake La Ronge over a 3-year period (R211). In Lake Michigan, 15.4% of 1416 tagged lake trout were recaptured within 3 years (S76). One travelled 860 km.

No. 3 strap tags on the dorsal fin gave satisfactory recovery rates on lake trout over 508 mm long but not on smaller (W93). Button tags on the preopercle gave low returns. Rings and wire dangler tags on the mandible gave good returns, but the wire dangler gave enough irritation to be abandoned (W93).

Ventral fins were more subject to regeneration after clipping than pectoral fins, but it was believed that most lake trout clipped as fingerlings or yearlings could still be recognized for several years (W93).

Spring planting of fingerling lake trout resulted in better returns than fall stocking in Maine (A64). Little difference was noted in returns from stocking 18-, 25-, 30-, and 37-month lake trout in Lake Superior (B250). While yearling plantings brought about 4 times the survival as fingerling plantings in Cayuga Lake, costs per adult fish still favored fingerling plantings (W93). No advantage could be demonstrated for scatter planting over spot planting of fingerlings.

Annual mortality in Lake Huron was estimated as 70% from age IV to V and 90% from V to VI with lamprey predation a factor (B197). In Great Slave Lake, annual mortality increased gradually from 37% at age XIII-XIV to 65% at age XXV (K57). In South Bay, Ontario, it was estimated as 70% for age IV-V with natural mortality estimated at 25% (F80). In Cayuga Lake, annual mortality after age V averaged 54% (W93).

Nylon gill net was found to be 2.5 times as efficient as cotton in catching lake trout and also entangled smaller fish (P157).

		Average TL	Range	Central 80%
25 mm nylon net	778	480	152-734	401-554
25 mm cotton	209	488	203-683	404-556

In Alaska, 19 mm stretch gill net took 3 lake trout 124-140 mm TL (H313).

R97 NT						
Mesh size	38	51	76	102	127	140
Number of lake trout	72	164	127	96	115	103
Mean FL	432	490	500	587	640	676
Range	119-851	140-889	211-902	290-859	401-965	401-940
Mean weight	2087	2449	2177	3266	4173	4944
W93 NY Cayuga L.						
Mesh size	32	38	51	64	76	89
Predominant FL	152	152	203-229	254-305	305-381	356-406
Number	37	48	227	378	374	415
V51 L. Michigan						
Mesh size	60	64	67	70	76	
Mean TL	310	335	345	361	386	
Number	1806	3650	3684	3633	1641	
Central 80%	267-343	292-368	318-394	318-394	343-419	
Mean weight	235	295	331	380	504	

In Lake Ontario, whitefish, burbot, and lake trout have varied in abundance relatively synchronously from 1880 to 1960 (C303). In Lake Opeongo (F30) a positive correlation was found between spawning stock and year class strength, but not in some other lakes (M494). Stocking of ciscoes may reduce year class success of lake trout (M494). Year class strength seems more uneven in plankton-feeding populations (M494).

Summaries and bibliographies are in L210, D175, M493.

HYBRID TROUT, *Salmo clarki* X *Salmo aquabonita*

At 246-323 days hybrids of this cross were 84-160 mm TL in Montana (G149).

HYBRID TROUT, *Salmo clarki* X *Salmo gairdneri*

I8 ID 18 trout TL = 1.154 SL = 1.031 FL. No sex, age, or size difference in relationship noted.

H258 WY Pathfinder L. log W = -4.240 + 2.911 log TL

K72 MT Grebe L. log W = -4.539 + 2.542 log TL

K72 TL	Weight	TL	Weight		H258 TL	No.	Weight
102-126	26	254-278	204		361	2	454
127-151	43	279-304	204		451	3	1097
152-177	57	305-329	295		483	1	907
178-202	74	330-355	369		513	4	1446
203-228	113	356-380	386		538	2	1418
229-253	142	381-405	559		559	1	1588

I8 ID   18 trout   K(SL) = 1.83

H258 WY Pathfinder L.   13 trout   K(TL) = 1.01   range 0.81-1.15

P55 MT W. Gallatin R. 117 trout   K(TL) = 1.07   range 0.80-1.16

	No.	TL	Range	No.	Mean Weight	Range
Age I						
M113 AT Willow C.	+	180	-	+	57	-
M113 AT Burmis L.	16	241	-	16	156	-
Age II						
G95 AT Willow C.	6	168	155-175	-	-	-
K72 MT Grebe L.	32	183	-	-	-	-
M113 AT Willow C.	+	236	190-284	-	156	142-261
G95 AT Bow R.	20	292	226-330	-	-	-
Age III						
G95 AT Willow C.	9	211	185-226	-	-	-
K72 MT Grebe L. males	45	249	-	-	-	-
M113 AT Willow C.	+	279	-	+	250	-
K72 MT Grebe L. females	30	282	-	-	-	-
G95 AT Kananaskis L.	1	290	-	-	-	-
G95 AT Bow R.	18	335	234-396	-	-	-
Age IV						
G95 AT Willow C.	7	241	211-284	-	-	-
K72 MT Grebe L. males	42	310	-	-	-	-
K72 MT Grebe L. females	53	343	-	-	-	-
G95 AT Bow R.	21	381	323-432	-	-	-
M113 AT Willow C.	+	439	-	+	907	-

		No.	TL	Range	No.	Mean Weight	Range
Age V							
G95	AT Kananaskis L.	3	366	333-409	-	-	-
K72	MT Grebe L. females	18	373	-	-	-	-
K72	MT Grebe L. male	1	388	-	-	-	-
G95	AT Bow R.	36	417	335-490	-	-	-
M113	AT Willow C.	1	508	-	1	1106	-
Age VI							
G95	AT Kananaskis L.	1	437	-	-	-	-
K72	MT Grebe L. females	4	455	-	-	-	-
G95	AT Bow R.	14	450	381-513	-	-	-
Age VII							
G95	AT Bow R.	12	486	417-538	-	-	-
G95	AT Kananaskis L.	2	513	493-531	-	-	-
Age VIII							
G95	AT Bow R.	6	500	465-559	-	-	-
G95	AT Kananaskis L.	1	528	-	-	-	-
Age IX							
G95	AT Bow R.	5	526	505-605	-	-	-
G95	AT Kananaskis L.	3	541	488-635	-	-	-
Age X							
G95	AT Kananaskis L.	1	688	-	-	-	-

		Average calculated TL at each annulus								
	No.	1	2	3	4	5	6	7	8	9
P55 MT Gallatin R.										
elevation 6000 ft.	21	89	175	279						
5000 ft.	29	99	185	269	368					
4000 ft.	71	89	188	292	384	424				
K72 MT Grebe L.	329	94	183	267	325	378	439	445		
P159 MT lakes	38	84	201	241	269	272				
P159 MT streams	468	79	157	241	320	348	460			
G95 AT Bow R.	132	84	168	234	325	386	432	467	498	536
I8 ID Henry's L.	20	140	320	460	528	594	658	716		

	Average annual TL increments								
	1	2	3	4	5	6	7	8	9
G95 AT Bow R.	84	84	89	79	71	53	38	33	25
Number	132	132	112	94	73	37	23	11	5
I8 ID Henry's L.	140	180	145	71	56	46	30		
Number	20	20	18	9	6	2	1		
K72 MT Grebe L.	97	89	81	48	36	30	41		
Number	329	317	262	143	34	6	2		

Annual increment as percentage of length, K72 Grebe L.													
TL	76	102	127	152	178	203	254	279	305	330	356	382	406
Median	93	103	68	58	47	46	18	15	15	13	10	9	7
Range	75-113	-	-	-	38-49	25-62	17-24	-	11-26	12-15	-	7-10	4-9
No.	313	2	2	6	250	6	91	48	26	8	2	6	4

Growth was computed from scale measurements by direct proportion, except in K72 where a correction for late scale formation was used. Difficulties in aging the fish from scales were discussed in K72. Females were larger than males in some age classes in K72 but no other sex differences in growth were noted.

Males were found to mature at age II and females at age III in Grebe Lake (K72). In Willow Creek, Alberta (M113), males were also found to be mature at age II, but even age IV females examined were immature and possibly sterile. The average female had 780 eggs in Grebe Lake (K72). Spawning was in mid-May to late June. M455 reported no evidence that rainbow-cutthroat hybrids were fertile.

Food was found to be mostly insects (K72, M113), but snails comprised 50% of the food of 10 trout 432-711 mm TL in Henry's Lake, Idaho (I8).

TROUT HYBRIDS, *Salmo gairdneri* X *Salmo trutta*
Rainbow X brown crosses failed to survive to fry stage at Cornell and Benner Springs Hatcheries (E130).

B186 PA	Age IV males	587 mm	2990 g
	Age IV females	627 mm	3175 g

Salmo gairdneri X *Salvelinus fontinalis*
Brook X rainbow hybrids failed to survive to fry stage at Cornell, but one batch showed a 0.6% survival at Benner Springs (E130).

S218 QU	2 months	25-38 mm	0.2 g
B186 PA	Age II	males and females both averaged 381 mm	

Salmo gairdneri X *splake (Salvelinus fontinalis* X *S. namaycush)*

B186 PA	Age II males	381 mm
	Age II females	368 mm

Salmo gairdneri X *Salvelinus namaycush*

S218 QU	9 months average	152 mm	26-46 g

SALMON-TROUT HYBRID, *Salmo salar* X *Salmo trutta*
The following compares FL growth of hybrids with each species in Ireland (I9).

	Hybrid	Salmon	Trout
0-Sept.	69	51	56
0-Oct.	89	56	69
0-Nov.	104	61	79
0-Dec.	109	61	84

	No.	TL	Range	No.	Weight
0-May					
P184 Ireland	+	25	-	-	-
Age 0-Fall					
A45 Sweden	5899	48	46-71	-	-
P184 Ireland backcross with *S. salar*	+	71	-97	-	-

	No.	TL	Range	No.	Weight
R125 Sweden F$_2$ generation	-	-	71-74	-	-
S484 Ireland	+	89	-	-	-
J12 G. Brit.	8	145	-	8	42
J12 backcross with *S. salar*	10	94	-	-	-
P184 Ireland Ballinlough	+	130	-	-	-
Age I					
A45 Sweden *S. salmo* X *S. t. fario*	700	66	-	-	-
A45 Sweden *S. salmo* X *S. t. trutta*	1741	69	46-97	-	-
P184 Ireland backcross with *S. trutta* (April)	+	84	79-89	-	-
J12 G. Brit. backcross with *S. salar* (March)	12	89	-	-	-
S484 Ireland ponds	+	183	-	-	-
M397 Ireland Sept.	100	190	-	100	93
P184 Ireland	+	-	79-257	-	-
M397 Ireland Nov.	110	241	-	+	to 284
J12 G. Brit. Dec.	7	218	-	7	113
S484 Ireland	+	257	-305	-	-
Age II					
A45 Sweden X *S. t. fario*	659	99	97-132	-	-
A45 Sweden X *S. t. trutta*	1348	117	99-135	-	-
P184 Ireland females	+	353	-404	+	550
males	+	376	-429	+	680
Age III					
A45 Sweden X *S. t. fario*	221	127	124-180	-	-
A45 Sweden X *S. t. trutta*	1074	157	130-246	-	-
A51 Sweden, from *S. trutta* females	436	124	-	-	-
A51 Sweden, from *S. salmo* females	440	147	-	-	-
A51 Sweden, from *S. trutta* females	429	170	-	-	-
P184 Ireland females	+	388	-432	+	709
males	+	427	-465	+	990
Age IV					
A45 Sweden X *S. t. fario*	158	178	-	-	-
A45 Sweden X *S. t. trutta*	800	206	175-282	-	-
A51 Sweden from *S. salmo* females	293	249	-	-	-
J12 G. Brit.	2	330	-	2	411
S484 Ireland	+	439	-	-	-
P184 Ireland females	+	414	-467	+	900
males	+	470	-549	+	1500
Age V					
A45 Sweden X *S. t. fario*	150	193	-	-	-
A51 Sweden, from *S. trutta* females	388	241	-	-	-
A45 Sweden X *S. t. trutta*	644	292	274-343	-	-
A51 Sweden, from *S. salar* females	216	305	-	-	-
Age VI					
A45 Sweden X *S. t. fario*	54	211	-	-	-
A51 Sweden, from *S. trutta* females	50	325	-	-	-
A51 Sweden, from *S. salar* females	55	345	-	-	-
A45 Sweden X *S. t. trutta*	112	356	290-401	-	-
Age VII					
A45 Sweden X *S. t. fario*	26	241	-	-	-
A45 Sweden X *S. t. trutta*	10	371	333-526	-	-

	No.	TL	Range	No.	Weight
Age VIII					
A45 Sweden X *S. t. fario*	11	267	-	-	-
A45 Sweden X *S. t. trutta*	6	368	-	-	-
A45 Sweden X *S. t. fario*					
Age IX	10	282	-	-	-
Age X	4	290	-	-	-
Age XI	2	307	-	-	-

Some males mature at age I (M397, P184), or first at age II (A45) and the smallest mature males from *S. salar* females were age III and 260 mm and from *S. trutta* females were age IV and 210 mm (A51). In Ireland some females were reported to mature at age II (M397, P184), in Sweden at age III (A45) and the smallest mature females from *S. salar* females were age V and 409 mm and from *S. trutta* females were age V and 328 mm (A51). Females in stripping averaged 1134 to 1604 eggs (P184) and fertilization rate was high.

The salmon X brown trout hybrids grew more rapidly in acid water than either parent and proved fertile and not unduly susceptible to disease (P184). Growth was much more rapid in Ballinlough Lake than in the hatchery (P184).

HYBRID TROUT, *Salmo trutta fario* X *Salmo trutta lacustris*

		No.	TL	Grams
A61 Sweden Age group	I	11	160-190	60
	II	16	220-290	187
	III	21	260-360	500
	IV	3	330-380	600
	IX	2	300	435
	X	1	300	435

HYBRID TROUT, *Salmo trutta* X *Salvelinus alpinus*

R124 Sweden F$_2$ generation   0-Fall   70-80 mm FL

"TIGER TROUT," *Salmo trutta* X *Salvelinus fontinalis*

Brown X brook and brook X brown hybrids both showed about 16% survival to fry stage at Cornell Hatchery and the brook X brown hybrid a 4 to 5% survival at Benner Springs, where the brown X brook hybrid showed only 0.5% survival (E130). This cross appears to be sterile (W198).

E130 NY	74 brook X brown hybrids			TL = 1.062 FL - 0.3 mm		
E130 NY	TL 125-149	150-177	178-202	203-228	229-253	254-278	279-304
Mean weight	34	65	85	142	173	270	340
H311 S. Afr.	55 hybrid trout 305-406 mm TL K(TL) = 1.33 range 1.02-1.55					

Age group	No.	TL	Range	No.	Weight	Range
0-May						
E130 NY	+	48	41-64	-	-	-
0-Oct.						
H311 S. Afr.	+	-	76-127	-	-	-
A45 Sweden	554	71	58-94	-	-	-
S218 QU	+	-	69-117	+	-	7.7-12.2
E130 NY	+	109	-	-	-	-

Age group	No.	TL	Range	No.	Weight	Range
Age I						
A45 Sweden	225	91	66-117	-	-	-
W198 NY	45	315	206-381	-	-	-
E130 NY	+	340	-	-	-	-
H311 S. Afr.	28	363	305-414	27	644	454-652
Age II						
A45 Sweden	106	178	142-185	-	-	-
B186 PA females	+	267	-	-	-	-
males	+	279	-	-	-	-
Age III						
A45 Sweden	84	254	178-262	-	-	-
Age IV						
A45 Sweden	39	330	208-358	-	-	-
Age V						
A45 Sweden	33	348	274-388	-	-	-
B186 PA males	+	546	-	+	2310	-
females	+	671	-	+	4627	-
Age VI						
A45 Sweden	32	388	343-414	-	-	-
Age VII						
A45 Sweden	18	411	384-434	-	-	-

HYBRID TROUT, *Salmo trutta* X *Salvelinus salvelinus*

		No.	TL	Weight
A61 Sweden	Age I	19	170-210	88
	Age II	5	260-290	222
	Age III	11	250-330	255
	Age IV	1	370	500

HYBRID TROUT, *Salmo trutta* X *Salvelinus namaycush*

S218 QU   at 9 months   50-140 mm   5 g

HYBRID TROUT, *Salmo trutta* X *splake (Salvelinus fontinalis* X *S. namaycush)*

B186 PA   Age II   males 380 mm;   females 356 mm

HYBRID TROUT, *Salmo fario* X *Salvelinus alpinus*

		No.	TL Mean	Range
A45 Sweden	Age 0-Fall	3129	64	61-71
	Age I	1357	74	69-94
	Age II	618	135	122-168
	Age III	460	196	150-218
	Age IV	113	251	201-292
	Age V	47	315	292-325
	Age VI	12	363	343-409
	Age VII	18	409	394-455

HYBRID TROUT, *Salvelinus alpinus* X *Salvelinus fontinalis*

	Age group	No.	TL mean	Range of means	No.	TL mean	Range of means
						F_2 generation	
A45 Sweden	0-Fall	2170	64	56-84	3230	76	61-97
	I	990	79	71-91	1910	91	69-117
	II	636	130	127-132	1102	145	94-178
	III	523	193	183-211	299	174	173-175
	IV	311	274	251-295	228	259	244-290
	V	-	-	-	150	315	300-335
	VI	133	386	-	83	363	348-386
	VII	119	411	-	13	391	-
	VIII	52	462	-	-	-	-
	X	19	490	-	-	-	-
S218 QU	Age II	168 g	-	-	-	-	-

SPLAKE, *Salvelinus fontinalis* X *Salvelinus namaycush*

I include here both crosses although Sowards (S274) used the term "brookinaw" when the brook trout was the female parent and splake when the lake trout was the female. Survival rates of either cross were 75-79% to free-swimming fry stage at Cornell (E130).

S274 WY	23 trout	TL = 1.054 FL	TL = 1.181 SL
E130 NY	38 brookinaw		TL = -2 mm + 1.073 FL
	36 splake		TL = -2.8 mm + 1.051 FL

M315 ON Jack's Lake

FL	203	229	254	279	305	330	356	381	406	432	457	483	508
g	140	140	200	310	370	450	540	710	850	1020	1105	1360	1815
No.	2	3	8	4	8	25	41	30	15	13	5	4	3

V63 NY Laramie P.   mean (FL) 0.83, 56 mm;   1.04, 109 mm

	No.	TL	Range	No.	Weight	Range
Age 0-Jan.						
S242 ON hatchery	40	23	-	40	0.11	-
0-Feb.						
S242 ON hatchery	30	28	-	30	0.14	-
S280 Banff hatchery	-	33	-	-	-	-
0-Mar.						
S218 QU	-	-	-	-	0.14	-
S280 Banff 2nd generation	-	25	-	-	-	-
S242 ON hatchery	30	33	-	30	0.2	-
0-Apr.						
S242 ON hatchery	20	36	-	20	0.28	-
0-May						
W199 NY backcross with brook trout	+	18	-	-	-	-
S280 Banff 2nd generation	-	30	-	-	-	-
S280 Banff 1st generation	-	41	-	-	-	-
S242 ON hatchery	20	41	-	20	0.48	-
E130 NY	+	41	38-43	-	-	-
0-June						
S242 ON hatchery	20	43	-	20	0.62	-
F120, V63 NY	625+	56	-	-	1.4	-

	No.	TL	Range	No.	Weight	Range
0-July						
S242 ON hatchery	10	51	-	10	0.96	-
0-Aug.						
S242 ON hatchery	70	58	-	70	1.6	-
B254 CO Parvin L.	100	71	-	-	-	-
0-Sept.						
S242 ON hatchery	58	66	-	58	2.7	-
B254 CO Parvin L.	11	94	-	-	-	-
0-Oct.						
S242 ON hatchery	25	64	-	25	2.5	-
F120, V63 NY Laramie P.	388	109	-	388	13.6	-
0-Nov. - I Dec.						
S242 ON hatchery	88	79	-	88	3.9	-
Age I						
S218 QU 2nd generation	-	-	114-170	-	19.0	-
B212 ON South Bay	-	102	-	-	-	-
F120 NY Laramie P.	75	140	-	-	-	-
S274 WY "brookinaw"	154	-	152-254	-	50	-
B254 CO Parvin L. "brookinaw"	434	206	170-229	-	-	-
W199,208 NY backcross with brook trout	21	234	226-246	21	145	113-184
B254 CO Parvin L. splake	1031	213	178-234	-	-	-
E130 NY 3 ponds	+	-	297-305	-	-	-
S132 AT Banff	+	-	305-356	-	-	-
Age II						
C252 2nd generation	+	137	86-254	-	-	-
S218 QU	-	-	-	-	74	-
W208, W199 NY splake June	+	229	-	+	-	82-85
B254 CO Parvin L.	1048	262	251-295	-	-	-
S274 WY splake	1	279	-	-	-	-
W200,208 NY splake	8	312	259-376	8	272	-
B364 Great Lakes	+	362	302-445	-	-	-
W208 NY backcross with brook trout	1	373	-	1	550	-
B212 ON South Bay	+	373	-	-	-	-
Age III						
S274 WY "brookinaw"	23	345	305-384	11	357	-
A66, W193,207 NY splake	19	447	427-475	19	865	-1106
B186 PA backcross with *S. fontinalis*, males	+	470	-	-	-	-
females	+	498	-	-	-	-
B212 ON South Bay	+	523	-	-	-	-
Age IV						
W200,208 NY splake	+	429	-	+	-	836-1758
S218 QU	-	-	-	-	1389	-
B212 ON South Bay	-	582	-	-	-	-
A66 NY Green L.	1	559	-	1	1545	-
C274 WI Little Bass L.	1	635	-	1	3232	-
Age V						
W199,209 NY splake	5	521	-	7	1625	992-2622
B186 PA males	-	643	-	-	2948	-
females	-	665	-	-	2722	-
Age VII						
W207 NY splake	1	559	-	1	1542	-

| | Calculated FL at each annulus | | | | |
	1	2	3	4	5
M315 ON Redrock R.	102	295	399		
Jack's L.	102	295	417	500	523
Sproul L.	102	318	447		

Growth of splake was faster than that of either parent in Algonquin Park lakes (M315) but slower than that of domestic or wild-stock brook trout in New York ponds (F120). Rapid growth in B212, C274, and indeed in most lakes may be the result of recent introduction. Slow growth in C252 was believed to be related to low water temperatures. Lethal temperatures were 23.5-24.0C (S333). In Algonquin Park lakes splake were found between 8-20C usually near the thermocline, similar to brook trout in lakes (M315). In laboratory experiments, however, splake showed temperature preferences nearer the lake trout, 12C, than the brook trout, 16-19C. Splake show strong schooling behavior (M315). These hybrids were not well-suited to New York farm ponds because of low temperature tolerances (E130).

Spawning takes place in October-November (M315) on gravel at 8-12 foot depths like lake trout or in streams like brook trout. They stay on the spawning beds all night like lake trout but also all day like brook trout. Ten age III females produced an average of 1246 eggs when stripped (S274). First generation eggs gave hatches of 4.8 to 94.1% (S274).

	Age	TL	g	No.	Eggs per female	Range
M315 ON Jack's L.	II	343-406	425-822	15	1233	855-1939
	III	400-495	992-1389	8	2830	1508-6075

In Jack's Lake, Ontario, 65% of 700 splake stocked in 1954 were caught by anglers in 1956-57 (M315). Elsewhere less than 2% were caught but a higher percentage than of brook trout stocked at the same time.

Winter food of 86 splake, 180-380 mm TL, included the following in declining order of abundance: isopods, decapods, amphipods, midge larvae, algae, odonates, and fish (B254). A variety of foods were found in spring, particularly mayfly nymphs and crayfish, but also including blue-spotted salamanders (M315). Feeding seemed to drop off in summer (M315). Larger splake feed on fish more than do rainbow trout of the same size (W207).

HYBRID TROUT, *Salvelinus fontinalis* X *Salvelinus salvelinus*

	No.	TL	Range	No.	Weight
Age I					
A61 Sweden	99	-	140-240	99	62
Age II					
A61 Sweden	74	-	170-340	74	94
A51 Sweden backcross with					
S. fontinalis	614	-	97-137	-	-
A51 Sweden	867	-	130-163	-	-
A51 Sweden one lake	39	-	193-213	-	-
A51 Sweden another lake	10	-	280-350	-	-
Age III					
A61 Sweden	42	-	200-360	42	136
A51 Sweden backcross with					
S. fontinalis	285	-	165-190	-	-
A51 Sweden	643	-	188-274	-	-

	No.	TL	Range	No.	Weight
Age IV					
A51 Sweden backcross with					
S. fontinalis	222	-	250-295	-	-
A51 Sweden	614	-	224-340	-	-
A61 Sweden	15	-	230-420	15	221
Age V					
A51 Sweden backcross with					
S. fontinalis	146	-	300-340	-	-
A51 Sweden	476	-	272-363	-	-
Age VI					
A51 Sweden backcross with					
S. fontinalis	83	-	335-386	-	-
A51 Sweden	240	-	348-384	-	-
Age VII					
A51 Sweden	170	-	378-421	-	-

The smallest mature males were 165 mm and age II and the smallest mature females 160 mm and age II. In backcrosses with brook trout, the smallest mature males were 193 mm and age III and females 206 mm and age III (A51).

HYBRID TROUT, *Salvelinus malma* X *Salvelinus namaycush*

S218 QU   Age II   57 g

INCONNU, *Stenodus leucichthys* (Güldenstadt)

The inconnu inhabits the Caspian Sea and the Arctic Ocean and the rivers flowing into these (N133). Most are anadromous but some lakes have populations which remain in fresh water. Inconnu are found in most of the large rivers of Arctic Canada and those of Alaska entering the Arctic Sea (S471). They reach a maximum of 40 kg in U.S.S.R. and 28.6 kg in North America (K98) and of 25 kg in Great Slave Lake (K98).

FL	Weight	Citation
451	1134	C126 BC Teslin L.
826	5556	C126 BC Teslin L.
1270	24811	K98 Great Slave L.
1505	28576	D34 Canada

F95 Great Slave L.   278 inconnu   mean K(SL) = 29.9

F83, F95 Great Slave L.

		SL in mm		Standard	Weight in grams	
Age group	No.	Mean	Range	Deviation	Mean	Std. Deviation
0	2	-	150-165	-	-	-
I	17	146	100-228	31	54	-
II	7	247	210-305	30	200	100
III	5	315	225-405	61	395	222
IV	16	403	318-470	31	735	186
V	15	473	395-530	36	1361	553
VI	17	529	460-640	52	1478	576
VII	48	572	440-690	54	2549	685

F83, F95 Great Slave L. (cont.)

Age group	No.	SL in mm Mean	SL in mm Range	Standard Deviation	Weight in grams Mean	Weight in grams Std. Deviation
VIII	60	610	490-750	44	3688	771
IX	29	657	545-760	43	3769	794
X	21	688	640-770	45	4423	880
XI	10	727	670-770	39	5012	934
XII	17	-	680-805	-	-	-
XIII	7	-	670-890	-	-	-
XVI	6	-	730-790	-	-	-
XV	8	-	750-880	-	-	-
XVI	5	-	770-820	-	-	-
XVII	4	-	830-995	-	-	-
XVIII	1	920	-	-	-	-
XIX	1	1085	-	-	-	-

Age group	FL	Weight	Citation
I	30	-	W78 Mackenzie R.
IV	300	-	D34 (from B45) Russia
	419	907	N133 Yenisei R. Russia
	856	5670	N133 Caspian Sea
V	470	1406	N133 Yenisei R.
	889	6350	N133 Caspian Sea
VI	437	-	D34 Russia
	554	1678	N133 Yenisei R.
	927	7575	N133 Caspian Sea
VII	676	2087	N133 Yenisei R.
	970	9072	N133 Caspian Sea
VIII	572	-	D34 Russia
	721	3720	N133 Yenisei R.
	1011	10478	N133 Caspian Sea
VIII-X	660-737	-	D34 Canada
X	676	-	D34 (from B45) Russia
XII	777	-	D34 Russia
XIV	846	-	D34 Russia
XV	914	7258	S435 Great Slave L.
XVI	929	-	D34 Russia

Standard lengths at various annuli were computed on a direct-proportion basis with an intercept at 20 mm in the following:

F95 Great Slave L.

Year	1	2	3	4	5	6	7	8	9	10	11	12	13	14	15	16
Mean SL	124	208	288	360	426	488	547	595	634	679	707	732	762	772	795	801
Increment	124	85	80	72	68	65	62	51	47	42	32	28	27	10	14	13
Number	278	262	257	254	238	220	210	185	120	78	55	48	27	20	12	4

In older age groups, females tended to be heavier than males but not longer (F83). Females tended to be longer in some Russian populations at ages IV-VIII (N133).

In Great Slave Lake inconnu are reported to spawn in May-June (R19) or in late September (F95) and possibly into October (S435) normally in streams but sometimes in the lake (F95). Young usually stay in the streams 2-3 years but may stay 1 to 5 (F95). Males mature at about 550 mm SL, age VI, and females at 650 mm, age VIII (F83, F95). In the Mackenzie River, inconnu mature at 508 mm FL, age VII (W78). Individuals spawn at intervals of 3 to 4 years (N133). Fecundity of the Yenisei River inconnu ranges from 125,000 to 325,000 eggs (N133).

Food of small inconnu was mostly mayfly nymphs (R19) but fish are the principal foods of the larger inconnu (C126, R19, F95, D34). They may eat large numbers of their own young (F83).

ARCTIC GRAYLING, *Thymallus arcticus* (Pallas)

The arctic grayling is found in northern Siberia and North America (N133, S471). It is also found southward in the Rocky Mountains to Montana. A similar race or species was once in Michigan but is extinct. Grayling have since been stocked in these Michigan waters. The maximum size recorded for North America is 2270 g (K93,98).

K72 MT under 18 mm TL = FL; 18-22 mm TL = 1.06 FL; 22-400 mm TL = 1.097]
B76 MI   92 grayling   140-289 mm TL = 1.178 SL, no consistent trend
W133 Athabaska   9 grayling   131-199 mm FL = 1.11 SL
                17 grayling   200-248 mm FL = 1.10 SL
                38 grayling   257-420 mm FL = 1.09 SL

In the tabulations which follow standard and fork lengths have been converted to total lengths using the above conversion factors.

TL	No.	Weight	Range	Citation
76-101	+	8	-	K72 MT Grebe L.
102-126	+	17	-	K72
127-151	+	28	-	K72
152-177	8+	51	45-54	B76, K72 MT Grebe L.
178-202	4	68	62-71	B76 MT lakes
	+	74	-	K72
203-228	+	85	-	K72
	9	105	105-108	B76 MT lakes
229-253	20	128	105-147	B76
	+	133	-	K72
254-278	9	150	119-190	B76
	+	162	-	K72
279-304	5	201	156-278	W133 Athabaska females
	+	204	-	K72
	10	227	170-340	B76
305-329	+	255	-	K72
	2	275	255-284	W133
	15	278	170-454	B76
330-355	+	255	-	K72
	30	363	198-414	B76
356-380	+	386	-	K72
	13	430	340-451	B76
	2	454	425-482	W133
381-405	9	609	510-680	B76
406-431	2	836	-	B76
445	1	950	-	S146 WY

K72 MT Grebe L.			log W = -4.5248 + 2.768 log TL			
N70 MT Red Rock Creek			353 grayling average K(TL) = 0.89			

	No.	TL	Range	No.	Weight	Range
Age 0-June						
N70 MT	34	13	10-15	-	-	-
T62 MI	+	36	-	-	-	-
Age 0-July						
K72 MT Grebe L.	20	20	13-28	-	-	-
B74 MI	+	25	-	-	-	-
B70 MT	79	36	20-33	-	-	-
M58 AT Great Bear L.	44	30	23-46	-	-	-
Age 0-Aug.						
K72 MT Grebe L.	10	48	43-53	-	-	-
L36 VT	+	41	-	-	0.6	-
N70 MT	21	69	56-79	-	-	-
Age 0-Sept.						
F28 MT	-	-	-	-	-	0.8-1.1
L36 VT	-	-	-	+	1.1	-
K72 MT Grebe L.	1	64	-	-	-	-
N70 MT	6	104	97-107	-	-	-
Age 0-Oct.						
B74 MI	+	64	-	-	-	-
M358 MN Twin L.	50	152	140-165	14	40	-
Age 0-Dec.						
B74 MI	+	178	-	-	-	-
Age 0						
T62 MI	+	122	112-137	-	-	-
N23 AT Jasper Park	2	48	38-58	-	-	-
C74 MI	1	112	-	-	-	-
Age I						
K72 MT Grebe L.	25	91	66-122	-	-	-
M58 AT	5	152	-	5	60	-
B76 MI Ford L.	2	150	102-193	1	68	-
N47 MI	-	152	-	-	-	-
W133 Athabaska	13	163	178-224	13	54	-
C74 MI Otter R.	5	201	178-224	-	-	-
B76 MT	40	218	178-338	40	119	60-460
N70 MT	73	229	-	-	-	-
C74 MT	1	302	-	-	-	-
Age II						
M113 AT Athabaska R.	14	218	206-218	14	74	71-77
W133 Athabaska drainage	15	221	-	15	96	-
B76 MI Ford L.	5	249	226-290	3	88	-
N47 MI	+	-	254-305	-	-	-
M58 AT	22	277	254-315	22	190	170-227
C74 MI	6	272	264-305	-	-	-
B76 MT	28	302	274-333	28	250	173-360
M358 MN Twin L.	251	315	254-368	237	281	-
M70 MT	78	320	-	-	-	-
C74 MT	4	353	295-371	-	-	-
Age III						
M113 Athabaska R.	24	246	211-259	24	122	71-170

	No.	TL	Range	No.	Weight	Range
W133 Athabaska drainage	54	264	-	54	159	-
C74 MI	1	287	-	-	-	-
B76 MI Ford L.	4	315	295-330	3	252	-
M58 AT	16	320	-	16	278	-
N47 MI	+	-	330-381	-	-	-
B76 MT	31	333	297-356	31	357	179-448
N70 MT	160	361	-	-	-	-
D6 MT	-	-	-	+	-	907-1361
Age IV						
M113 AT Athabaska	8	300	282-330	8	218	170-284
W133 Athabaska drainage	75	310	-	75	244	-
B76 MI Ford L.	1	343	-	1	357	-
M58 NT Great Bear L.	18	348	-	18	384	-
B76 MT	22	358	338-409	22	439	247-757
N70 MT	34	391	-	-	-	-
C74 MT	2	394	391-396	-	-	-
N47 MI	+	-	406-432	-	-	-
Age V						
M113 AT Athabaska R.	1	305	-	1	227	-
W133 Athabaska drainage	32	338	-	32	446	-
B76 MT	5	363	335-432	5	405	187-765
M58 AT	25	399	-	25	575	-
N70 MT	6	414	-	-	-	-
R89 SA Reindeer R.	+	472	-	-	-	-
R89 SA Rocky Falls	+	500	-	-	-	-
Age VI						
W133 Athabaska drainage	16	368	-	16	454	-
B76 MT	1	363	-	1	460	-
M58 AT	21	414	-	21	663	-
N70 MT	2	411	-	-	-	-
Age VII						
W133 Athabaska R.	2	368	-	2	460	-
M58 AT	17	434	-	17	703	-
Age VIII						
S165 NT	1	340	-	1	425	-
M58 NT Great Bear L.	6	455	-	6	794	-
Age IX						
S165 NT	1	394	-	1	454	-
M58 NT Great Bear L.	5	465	-	5	936	-
Age X						
S165 NT	1	404	-	1	680	-
M58 NT Great Bear L.	2	480	-	2	1063	-
Age XI						
M58,59 NT Great Bear L.	1	508	-	1	848	-

Average calculated TL and annual increments at each annulus

	1	2	3	4	5	6	7	8	9	10	11
W133 Athabaska drainage	122	185	246	292	323	345	356				
Incr.	122	66	61	43	30	25	20				
No.	207	194	179	125	50	18	2				
M58 Great Bear L.	104	168	254	312	356	386	414	439	450	475	498
Incr.	104	64	86	61	46	36	28	25	18	25	15
No.	100	99	96	82	64	47	30	14	8	7	2

Average calculated TL and annual increments at each annulus (cont.)

		1	2	3	4	5	6	7	8	9	10	11
M58 Great Slave L.		117	203	279	345	394	424	465				
	Incr.	117	89	76	66	48	33	30				
	No.	23	19	6	6	6	2	1				
M58 Athabaska L.		122	208	279	333	381	411					
	Incr.	122	86	79	66	48	13					
	No.	15	15	9	7	7	3					
B76 MI Ford L.		89	218	305	335							
	Incr.	89	127	81	33							
	No.	169	84	6	3							
B76 MT lakes		147	274	330	356	358	363					
	Incr.	147	119	56	30	15	33					
	No.	129	87	59	28	6	1					

Average calculated TL at each annulus

	No.	1	2	3	4	5	6	7
W19 AK Yukon R.	677	99	145	206	254	300	318	335
W117 BC Tagish L.	-	84	188	251	343	378		
W117 BC Atlin R.	44	84	165	257	307	368		
C74 MI rivers	13	94	208	229				
K72 MT Grebe L.	366	112	224	284				
P159 MT lakes	108	97	221	284	302	323	353	
C74 MT Georgetown L.	5	107	282					
N70 MT Red Rock Cr.	353	155	282	343	373	396	406	
C74 MT Rogers L.	2	188	345	368	391			

Annual increments expressed as percentage of size at beginning of year

TL	B76 MI Ford L.	B76 MT	M58 NT	Range
76-101	142	144	63	31-158
102-126	-	103	72	60-124
127-151	-	-	44	35-65
152-177	-	82	46	42-110
178-202	-	62	-	51-64
203-228	38	23	27	17-40
229-253	-	29	26	25-33
254-278	-	-	26	23-35
279-304	11	16	18	7-27
305-329	-	7	16	3-17
330-355	-	-	12	9-14
356-380	-	5	9	5-10
381-405	-	7	8	6-9
406-431	-	-	5	4-7
432-456	-	-	5	4-7
457-482	-	-	3	3

Data on growth of grayling *(Thymallus thymallus)* in the British Isles are given in H103.

Characteristics of annuli are discussed in B76 and K72, with some difficulties in recognizing annuli in some fish reported by both authors. Grayling raised in uniform temperature and food conditions showed no annuli on their scales (B76). A warm August may have resulted in false annuli on grayling in

Ford Lake, Michigan (B76). In Michigan, annuli usually form April 15 to May 15, and in Montana lakes from April 20 to July 1 (B76). In Grebe Lake, Wyoming, yearlings formed annuli in mid-June, age II graylings in early June, and age III in July and August (K72). After age III there was little growth and no annuli formed (K72).

In several studies growth was computed from scale measurements by direct proportion. B76 reported that scales formed at 29 mm SL or 36 mm TL but that the best straight line had a 51.5 mm SL intercept, which was used as the correction factor for growth computations. W133 used 50 mm FL as a correction factor and K72 used 27 mm TL for the 1953 collections and 36 mm for the 1954. B76 reported that S436 found the body-scale relationship of grayling in Europe and Asia to be parabolic.

In general growth appears to be faster in Montana than in more northern waters. Growth is also greater where the grayling are introduced into new waters (D6). In the Athabaska drainage (W133), growth is better in lakes than streams, growth seems to be better where bottom fauna is more abundant, and there is a slight tendency for males to grow faster than females. In Grebe Lake (K72) males were larger than females but because of difficulty in aging older fish it was not determined whether there was a difference in growth rates.

Grayling usually first mature at age III (K72, M58, W133) but a few males mature at age II even in the Athabaska (W133) and both males and females sometimes spawn at age II in Michigan (B74). In the Yukon grayling first mature at age IV (W180). Spawning runs may be from March 15 to June 20 in different areas of Michigan and Montana with temperature probably being the most important factor determining the time (B74). In Grebe Lake, Montana, grayling spawn from mid-May to late June at temperatures of 4.5-10C (K72), in the Athabaska drainage in May at temperatures of 4.5-11C (W133). In the Yukon grayling spawned from May 13 to June 16 in 1952 (W180). The males usually move to spawning areas in streams earlier than the females (K72) and establish territories but do not build nests. Spawning has been described (B74, K72, F123). The brood stock return to Great Bear Lake in mid-June (M60).

	Size of females	No. of females	No. of eggs per female Mean	Range
B74 MT Grebe L.	142 g	2	832	416-1248
	340 g	1	5563	-
	average	-	1650	-
B74 MT Rogers L.	325-368 mm	11	6029	-9059
W28 MI	255 g	1	3555	-
	567 g	1	5200	-
H51 MI	305 mm	1	3000	-
R89 SA	-	-	-	4000-10000
B73 L1	-	-	-	3000-4000
L17	reports 6.6 eggs per g of female			
B74 Georgetown L.	907 g	3	-	8135-12946
K72 MT Grebe L.	under 275 mm	11	1889	1348-2166
	279-302 mm	13	2344	1307-2928
	over 305 mm	13	2781	1836-4166
W133 Athabaska	254-277 mm	5	2278	574-3433
	287-292 mm	2	4230	3831-4630
	325-343 mm	2	6501	5963-7039

The eggs hatch in 20 days at 6-7C (W133), in 16 days at 10C (B74), in 16-20 days at 8-16C with a mean of 10.5C (W146), and in 11-22 days at temperatures not indicated (B74). Embryology is described and fry are 5.2-8.8 mm body length at hatching (W146). Losses at various fry stages were estimated at Grebe Lake (K72). Yolk sac is absorbed in 16-18 days at 7-11C (R89) or in 7-10 days (H340).

Food of young is primarily *Daphnia* and Diptera (K72) but insects in general are the principal foods of older grayling (R138, K72, W133, N13, B74, L179, L180, M60, M58). Terrestrial insects were more important than aquatic insects in the far north (M58, M60, R138, W133). A few grayling were found to eat fish (R138, M60, M58) including a 50 mm grayling (W133) and one had eaten a lemming (M58, M60).

Standing crops of grayling in Grebe Lake were estimated as 78.6 kg/ha in 1953 and 79.4 kg/ha in 1954 by marking and recovery methods (K72). Grayling proved to be less susceptible to trapping than trout (K72).

SURF SMELT, *Hypomesus pretiosus* (Girard)

The surf smelt spawns on sandy beaches from northern California to southeastern Alaska (C254) from June to September.

Scales are formed so late that the first annulus is not represented. Most spawning fish have a single annulus (age II) but some may spawn at age I or age III (H416). A female may produce 2500 to 37000 eggs.

AMERICAN SMELT, *Osmerus mordax* (Mitchill)

Smelt are anadromous along the Atlantic Coast from Labrador to New Jersey with the center of abundance in the southern Gulf of the St. Lawrence (H414). A closely related species (or same species, M495) is found in the Atlantic of northern Europe. Smelt rarely go far from the coast (M196, M496). Freshwater populations are found in many lakes in New England and the Maritime Provinces and smelt have become established in the Great Lakes since 1920 (H414).

B27 MI	241 fish 80-230 mm SL	TL = 1.165 SL
W8 NH		TL = 1.12 SL to 1.17 SL
M495 St. Lawrence	160 smelt	TL = 1.053 + 5.2 mm
M162 Length to end of scales on caudal peduncle = 1.02 SL = 0.94 FL		

M162 NK Miramichi   Relation of length to end of scales in mm to weight in grams. There appeared to be no consistent sex difference but maturing fish averaged higher than immature fish at the same size.

Length	No.	Weight in grams
80-89	19	6.9
90-99	16	7.9
110-109	73	11.4
110-119 imm.	467	14.6
mat.	42	15.7
120-129 imm.	339	17.3
mat.	377	19.7
130-139 imm.	47	21.1
mat.	1146	24.7
140-149 imm.	8	27.8
mat.	1951	30.7

Length		No.	Weight in grams
150-159	imm.	5	34.7
	mat.	1813	37.3
160-169	imm.	5	41.1
	mat.	1343	45.3
170-179	imm.	3	49.1
	mat.	784	54.8
180-189	mat.	368	66.6
190-199		187	78.9
200-209		66	92.1
210-219		29	99.4
220-229		13	127.1
230-239		8	131.1
240-249		1	132.0

B27 MI $\log W = -4.59918 + 2.8095 \log SL$
C71 MI Total length-weight relationship

TL	197	210	241	279
Weight	42	57	85	142

B327 L. Superior $\log W = -5.276 + 2.952 \log TL$

TL	81	94	124	135	147	163	175	188	203	216	231	247	262	274	290
Weight	3	6	9	14	17	23	28	40	54	65	79	91	108	119	133
No.	4	3	3	32	80	70	22	73	74	67	64	71	40	9	2

B95 Green Bay

males $W = -53.319 + 3.892\, SL + 8.012\, \text{depth} + 3.036\, \text{thickness} \pm 2.07\,g$

$W = -53.887 + 4.108\, SL + 8.907\, \text{depth} \pm 2.11\,g$

$W = -55.072 + 5.765\, SL \pm 2.51\,g$

$W = 0.0118\, SL^{2.9025} \pm 2.81\,g$

$W = 0.009133\, SL^{3} \pm 2.82\,g$

females $W = -69.945 + 6.814 \pm 3.00\,g$

$W = 0.0068\, SL\, 3.1142 \pm 2.98\,g$

	No.	TL	Range	No.	Weight	Range
Age 0-July						
G105 L. Michigan	50	-	28-48	-	-	-
K86 L. Erie	6	36	30-41	-	-	-
M162 NK Miramichi	+	-	56-71	-	-	-
Age 0-Aug.						
G105 L. Michigan	48	-	30-58	-	-	-
R278 ME Shagg	+	58	-	-	-	-
Age 0-Sept.						
G105 L. Michigan	30	-	43-58	-	-	-
K86 L. Erie	5	48	41-53	-	-	-
Age 0-Oct.						
G105 L. Michigan	31	-	43-64	-	-	-
C275 L. Erie	-	61	-	-	-	-
Age 0-Nov., Dec.						
S156 NK Gibson L.	9	71	66-79	-	-	-
K86 L. Erie	28	74	56-86	-	-	-

	No.	TL	Range	No.	Weight	Range
Age 0						
G29 NY Finger lakes	53	109	-	-	-	-
G29 NY L. Champlain	17	157	-	-	-	-
G29 NY Saranac L.	1	196	-	-	-	-
Age I						
H271 L. Superior	23	58	46-76	-	-	-
B327 L. Superior	181	66	38-97	-	-	-
C73 MI Mt. L.	32	84	-	-	-	-
W8 NH	10	86	-	10	2	-
C71 Great Lakes	+	-	76-102	-	-	-
R278 ME lakes	+	91	71-137	-	-	-
B27 MI Crystal L.	3	127	-	3	14	8-23
V16 Green Bay 1941	1	135	-	1	14	-
G29 NY Finger lakes	119	137	-	-	-	-
V16 Green Bay 1944-45	6	147	-	6	23	-
G29 NY Champlain dwarfs	19	147	-	-	-	-
V7 L. Erie	2	150	142-157	-	-	-
M162 NK	18569	160	-	7929	28	-
Z3 NY	29	173	-	-	-	-
S7 WI	+	178	165-216	-	-	-
G29 NY Saranac L.	98	203	-	-	-	-
G29 NY L. Champlain	118	208	-	-	-	-
Age II						
R192 ME Mooselookmeguntic L.	67	127	102-168	-	-	-
R278 ME lakes	+	135	71-213	-	-	-
G29 NY Finger lakes	4	137	-	-	-	-
W8 NH	189	145	-	189	17	-
B91 L. Huron	54	150	-	-	-	-
G29 NY	83	145	-	-	-	-
B327 L. Superior	199	152	177-208	199	20	-
H271 L. Superior	324	155	117-201	-	-	-
M496 NK Miramichi males	4172	155	-	-	-	-
females	1506	157	-	-	-	-
V16 Green Bay 1941	71	173	-	71	31	-
B27 MI Crystal L.	6	178	175-180	6	34	31-37
C71 Great Lakes	+	-	165-190	-	-	-
M162 NK Miramichi	9128	188	-	4196	51	-
V16 Green Bay 1944-45	87	186	-	87	57	-
G29 NY Saranac L.	3	216	-	-	-	-
Z3 NY L. Champlain	183	224	-	-	-	-
G29 NY L. Champlain	83	239	-	-	-	-
V7 L. Erie	1	241	-	-	-	-
S7 WI	+	279	229-292	-	-	-
Age III						
G29 NY Champlain dwarfs	13	157	-	-	-	-
B91 L. Huron	38	168	-	-	-	-
W8 NH	85	170	-	85	31	-
R278 ME lakes	+	173	130-239	-	-	-
M496 NK Miramichi males	2723	175	-	-	-	-
R192 ME	69	178	114-244	-	-	-
H271 L. Superior	74	180	145-221	-	-	-
M496 NK Miramichi females	868	183	-	-	-	-

	No.	TL	Range	No.	Weight	Range
B327 L. Superior males	45	185	152-221	45	40	-
V16 Green Bay 1941	23	196	-	23	45	-
B327 L. Superior females	64	196	157-226	64	48	-
B27 MI Crystal L.	6	198	180-226	6	51	40-71
M162 NK Miramichi	1451	213	-	560	74	-
V16 Green Bay 1944-45	21	218	-	21	79	-
C71 Great Lakes	-	229	-	-	-	-
Z3 NY L. Champlain	267	246	-	-	-	-
G29 NY L. Champlain	31	269	-	-	-	-
S7 WI	+	305	279-330	-	-	-
Age IV						
V16 Green Bay 1941	1	193	-	1	40	-
M496 NK Miramichi males	380	193	-	-	-	-
B91 L. Huron	8	198	-	-	-	-
B327 L. Superior males	8	201	183-213	8	54	-
H271 L. Superior	99	206	183-213	-	-	-
B27 MI Crystal L.	3	206	203-211	3	54	51-60
M496 NK Miramichi females	150	208	-	-	-	-
R278 ME lakes	+	213	203-224	-	-	-
B327 L. Superior females	84	221	180-272	84	71	-
W8 NH	3	221	-	3	54	-
R192 ME Mooselookmeguntic L.	22	229	183-202	-	-	-
M162 NK Miramichi	170	236	-	45	91	-
C7 Great Lakes	-	254	-	-	-	-
V16 Green Bay 1944-45	2	267	-	2	139	-
Z3 NY L. Champlain	194	290	-	-	-	-
G29 NY L. Champlain	14	290	-	-	-	-
S7 WI	+	356	-	-	-	-
Age V						
M496 NK Miramichi males	46	208	-	-	-	-
B27 MI Crystal L.	2	211	205-217	2	60	54-62
B327 L. Superior males	3	218	208-234	3	65	-
H271 L. Superior	19	234	221-249	-	-	-
M496 NK Miramichi females	10	236	-	-	-	-
B327 L. Superior females	35	239	206-284	35	91	-
M162 NK Miramichi	11	251	-	-	-	-
R192 ME Mooselookmeguntic L.	5	251	216-269	-	-	-
C71 Great Lakes	-	-	274-305	-	-	-
G29 NY	1	295	-	-	-	-
Z3 NY L. Champlain	124	310	-	-	-	-
Age VI						
B327 L. Superior females	9	249	208-272	9	108	-
H271 L. Erie	1	259	-	-	-	-

	Average TL and increments at each annulus						
	1	2	3	4	5	6	7
B327 L. Superior males	66	150	183	203	218		
Increment	66	84	41	20	10		
Number	306	306	124	14	3		
females	66	150	168	221	239	257	310
Increment	66	84	46	25	18	13	13
Number	428	428	303	157	47	11	1

	Average calculated weight at each annulus						
	1	2	3	4	5	6	7
B327 L. Superior males	17	20	41	56	64		
females	17	20	45	64	80	94	108

Scale structure and the characteristics of annuli are discussed by M162, B327, and R278. Scales first form when smelt are 20-25 mm SL. A shiny line, associated with a break in the circuli pattern, seems to be the best criterion of an annulus. In smelt with slow growth the first year, the first annulus may occur before any ridges form on the scale. Growth the second year probably starts in April, but in the third year growth is delayed by the spawning season and does not begin until the fish are back in brackish water in late May or June. In Lake Superior, annulus formation was from mid-June to mid-August and was later for older fish than for age I fish (B327). Lengths at previous years were calculated from scale measurements only in one study (B327) and a straight line with an intercept of 23 mm was used. In Lake Erie, growth of young smelt ends early in October at water temperatures of 18.4C (C275). Size at the end of the second year seemed to be related to the earliness of spawning of the year class and thus the length of the first growing season (M162) and earliness of spawning was related to spring temperatures. Female smelt grow somewhat faster than males (Z3, B27, M162, H271, B327). No males older than age III were found in Lake Superior although females reached age VI (H271), and females were strongly predominant in ages IV-VII in another study (B327).

No relationship was noted between abundance of a year class and its growth rate (M162).

In Lake Champlain two races of smelt were detected, the slower growing race comprising 30% of the smelt taken in 1929 but only 7% and 4% in 1948 and 1950 (Z3). Great variation in average size of smelt was found in Maine lakes (R192, R186). In some lakes the maximum length was 125 mm TL and in others, 355 mm TL. Transplantation from some of these lakes indicated that the growth differences were not primarily genetic, nor dependent upon population density (R278). In Lake Erie, yearling smelt taken in July offshore averaged larger (107 mm) than those near shore (81 mm TL) where they were more abundant (F147). Frequency of older smelt increased in South Bay, Lake Huron, when an intensive dip net fishery was discontinued (F146). Abundant young have been taken only in even-numbered years in Lake Erie in recent years (F147). Abundance of year classes was not related to the number of larvae produced in the Miramichi River (M496).

Although smelt are usually considered coldwater fish, they are also found in warm-water unstratified ponds in Maine (R186). Lethal high temperatures were found to range from 21.5-28.5C (H254). In Cayuga Lake, New York, smelt are usually found on the flats at 30-37 m deep, at below 13C (G114). In Lake Erie, young were common in shallow water and in the epilimnion at over 21C in the summer, but adults were restricted to water cooler than 15.5C and were more abundant at temperatures below 7C (F147). Some adults may migrate upward or inshore into warmer water at night, following *Daphnia* and *Gammarus* (F147). Light appears to be the factor controlling the migration rather than the direct effect of feeding, however.

Smelt spawn in streams, or in shallows in lakes (R192, R186) mostly at night. In Maine, spawning runs start from February 23 to May 16 or from 60 days before to 19 days after the ice goes out, averaging 5 days after the ice goes out. The spawning run in Cold Stream Pond was usually from February 15 to March 10, about 2 months earlier than ice breakup and before most of the

Maine spawning runs (R278). When smelt from this population were transplanted to a lake where the previous population spawned in mid- or late May,
after ice was gone, the introduced smelt retained their early spawning season.
A population which normally spawned in streams spawned successfully on the
shore area when transplanted to a lake without tributary streams (R278). Stream
temperatures are usually 4.5-7C but the major stimulus is probably not temperature since the runs may occur before the ice goes out (R192, M496).
Spawning may occur under the ice (R186). The average run lasts 10 days but
may extend 1-28 nights (R192). In New Brunswick smelt spawn from late
April to late July with hatching from the third week of May to late July (M162,
M496). Average size of the spawning individuals decreases as the spawning
season progresses (M496, B327). In Lake Superior smelt spawn April 14-30
at temperatures of 2.2-14.5C (H271); in Lake Erie, May 1-15 at 10C (C275);
and in New Hampshire, April and May (N95). The males have tubercles on the
scales at spawning time (H414, M496).

	Size of females	No. of females	No. of eggs per female Mean	Range
V31 MI	185-196 mm	-	25000	-
L14 L. Erie	241 mm	1	57910	-
N25, B73, K6 NY	57 g	-	-	46000-50000
H82 NH	104 mm SL	-	-	1700-7000
K6 ME	117 mm	1	5893	-
K9 ME	-	-	17300	-
L10 L. Erie	185-195 mm SL	14	25102	-
	232 mm SL	1	43125	-
K118 review	-	-	-	1700-57910
H414 Canada	-	-	30000	over 60000
M496 NK Miramichi	145-239 mm	9	-	8500-69600
B327 L. Superior	185-224 mm	10	31338	21534-40894
L. Huron	140-224 mm	5	20570	-
S91 ON Lake Huron	140 mm TL	1	9650	-
	160 mm TL	1	16200	-
	178-224 mm TL	3	25667	22600-27600
M496 NK Miramichi	log N of eggs = 4.4 log SL + k			

Both male and female smelt mature at age II or III (H271, M162, M496,
N95). Maximum survival of eggs was found in areas of moderate egg deposition
about 11745 eggs per square foot (R208). Mean eggs survival was estimated at
24% with survival to the prolarval stage 0.5% (R208). Incubation time was
listed as 19-20 days at 5-8C (H271), 10 days at 15C (H82), 29 days at 6-7C, 19
days at 9-10C and 11 days at 12C (M496). Crowding of eggs reduced survival
(M496) with the maximum survival at about 12000 eggs per square foot. Percentage survival increased as density decreased at least to about 5400 per
square m (M22). Spawning in flooded shores also results in loss of eggs as
waters recede, so that smelt larval production in the Miramichi River is
negatively correlated with rainfall during the spawning period (M396).

Young smelt fed primarily on copepods and cladocerans (G105, K86). In
Lake Erie adult smelt fed primarily on *Daphnia* and *Gammarus* and some fingernail clams and smelt young in August 1963-64 (F147) and primarily on
emerald shiners in May (K86). Shiners were also the principal food of smelt
in February in Michigan, but 122 of 210 smelt were empty (B27). Shiners,
young smelt, *Hexagenia*, midge larvae, *Daphnia*, and *Pontoporeia* were the food
of adult smelt in Lake Michigan (G105). *Mysis, Pontoporeia* and young smelt
were the foods of smelt in Lake Superior (H271).

Additional data on smelt is given in M497 which was not seen.

LONGFIN SMELT, *Spirinchus dilatus* Schultz and Chapman
The longfin smelt is limited to Puget Sound and the southern part of the Strait of Georgia. They spawn in fresh water in October to December (C254). The young fish do not form scales before the first winter. Fish in the next season of growth were 61-72 mm SL in the Fraser River (H416). They spawn at the end of the second year (age II) (C254).

SACRAMENTO SMELT, *Spirinchus thaleichthys* (Ayres)
Sacramento smelt is anadromous on the Pacific Coast from central California to Prince William Sound, Alaska, and some freshwater populations are found in the same region (D191). In all populations studied, the smelt mature at age II and live after spawning early in the spring (D191, A92).

	FL
Age 0-April	
D191 L. Washington	under 10
Age 0-Fall	
G137 CA San Pablo Bay	61-81
Age I-Fall	
G137 CA San Pablo Bay	102-130
Age II-Spring	
D192 WA Harrison L.	slightly over 76
D191 anadromous populations	127-152

Growth rate in Lake Washington equals or exceeds that of anadromous populations (D192). Population density is high in Harrison Lake where growth is stunted (D192).

Young smelt feed on copepods and cladocerans and adults feed on mysids and chironomids in Lake Washington (D192).

EULACHON, *Thaleichthys pacificus* (Richardson)
Anadromous on the Pacific Coast from northern California to Bering Sea (C254).
B335   133 eulachon   FL = 1.134 SL; TL = 1.225 SL = 1.082 FL

		FL	
	No.	Mean	Range
B335 BC			
0 April	24	23	18-32
0 Nov.-Dec.	67	53	46-65
I Jan.-April	47	62	50-71
I Nov.-Dec.	25	82	71-91
II Jan.-April	35	83	72-95
II July-Aug.	27	102	91-111
III Jan.-April	5	111	104-119
III July-Aug.	14	131	121-141
IV	15	143	129-155
V	2	152	140-165
S498 Columbia R.			
III males	+	161	130-200
IV males	+	170	135-200
II in ocean	32	115	-

Ages have been determined from length frequencies (B335), scales (H416) and otoliths (S498).

Most eulachon first mature at age II and others at age III in the Fraser River (H416), but B335 found none spawning before age III in British Columbia and they spawn at age III and IV in the Columbia River (S498). Although most probably die after spawning (S498), some may survive to spawn a second time (B335). The spawning migration is from mid-March to mid-May in British Columbia (H416, C254). Spawning is over gravel at night at water temperatures of 4.5-8C. The eggs attach to rocks singly or in clusters by a short peduncle. Eggs hatch in 30-40 days and the young fish are soon swept out to sea (S498, B335).

	Size	No. of females	No. of eggs per female	
			Mean	Range
H416 BC	145 mm	2	17450	17300-17600
	158-185	8	32250	20600-39600
S498 Columbia R.	140-195	18	-	20000-60000

The number of eggs approximately doubled with a 40 mm increase in length over the range covered in the Columbia River sample (S498).

Not much is known of their food at sea but they probably feed on euphausiids (H416, C254). No food is found in stomachs of eulachon taken in fresh water

GOLDEYE, *Hiodon alosoides* (Rafinesque)

The goldeye is found in Canada in the Hudson Bay and Arctic drainages from Quebec to the southern parts of Northwest Territories (S471) and in the Ohio, Mississippi, and Missouri River drainage south to Tennessee and Oklahoma (E119, T113), but not in the Great Lakes drainage. It is quite tolerant of clay turbidity but not of industrial pollution (T113).

G87 MN Red L.   361 goldeye   SL = 0.815 TL   FL = 0.907 TL
M175 OK L. Texoma   TL = 1.226 SL
G22 MN Lake of Woods   5 goldeye   FL = 1.10 SL

Data originally listed as FL (R97, M113, B207, R230) and as SL (E3, B12) are converted to TL according to the formula above in G87.

M431 QU	395 mm FL, 624 g;	400 mm, 737 g;	450 mm FL, 1021 g

	No.	64	89	114	140	165	190	216	241	267	292	318	343	368	394	419	445
						Mean weight by TL groups centering at											
M175 OK L. Texoma	996	-	-	-	-	40	45	94	136	184	232	278	349	485	510	-	-
SD Ft. Randall S226, S225, S228, S230, S393	644	-	-	-	14	34	57	88	130	181	244	312	386	493	-	-	-
SD Gavins L. S276, S227, S229, N94, N115	225	-	-	-	-	-	-	-	-	186	238	309	395	493	615	681	779
SD Oahe F100, F105, F124, F125	916	14	23	34	45	45	65	99	139	198	258	349	468	562	700	776	-

C30 IA   2 goldeye   381-404 mm TL, 575-652 g;   457 mm TL goldeye 962 g
S146 WY   406 mm TL goldeye 1219 g
K93, K98 lists maximum weights as 1180 g in Great Slave L. and 1406 g in Ohio.

M175 OK L. Texoma	996 goldeye	log W = -4.544 + 2.891 log TL
G87 MN Red L.	1156 goldeye	log W = -4.638 + 2.844 log TL
S225 SD Ft. Randall	40 goldeye	log W = -5.430 + 3.178 log TL
V20 MN Red L.	Average K(SL) of 10 fish (252-342 mm SL)	1.73
C22 MN L. of Woods	Average K(SL) of 3 fish	2.24

		K(TL)	
	No.	Mean	Range of means
SD Ft. Randall S225,226,228,230,393	652	0.94	0.88-0.98
SD Gavins Point S227, S229, S276,			
N94, N115	225	1.04	0.92-1.14
SD Oahe F100,105,124,125	1271	1.15	1.02-1.29

Condition factors increased with length of the fish (S227, S228, S393) but no such trend was evident in S225, S229, S276. No sex difference was evident in condition factors in S227, S228, F124; males had higher condition factors than females in F100, F105, and S226 and females higher than males in F125 and S393 but the differences were probably not significant. Goldeye from Ft. Randall Reservoir had higher condition factors than those from the tailwaters (S225, S226).

	No.	TL	Range	No.	Weight	Range
Age 0-June						
B207 AT Claire L.	229	18	10-36	229	.04	.01-.1
S227 SD Ft. Randall	9	33	20-51	-	-	-
Age 0-July						
B207 AT Manitoba	142	38	25-76	142	.4	.1-2.3
S227 SD Ft. Randall	22	71	38-94	-	-	-
R230 SA rivers	+	-	38-46	-	-	-
Age 0-Aug.						
B207 MB	488	71	51-109	488	2.8	2.3-5.7
R230 SA rivers	+	-	66-94	-	-	-
Age 0-Sept.						
B207 MB	125	114	89-130	125	11	-
R230 SA rivers	+	114	-	-	-	-
S393 SD Ft. Randall	9	102	-	-	-	-
Age 0-Fall						
T113 OH	+	-	127-165	-	-	-
H342 IA	+	-	76-127	-	-	-
Age I						
B207 AT	75	79	-	67	3	-
B37 MT Ft. Peck	+	-	140-163	-	-	-
R230 SA rivers	+	168	-	+	28	-
M175 OK L. Texoma	4	175	165-196	4	34	-
I10 IA Snyder Bend	1	244	-	1	167	-
Age II						
S437 AT Claire L.	3	175	-	3	28	-
S437 MB L. Winnipegosis	8	185	-	8	60	-
R230 SA rivers	+	211	-	+	71	-

	No.	TL	Range	No.	Weight	Range
P37 MT Ft. Peck	+	-	216-239	-	-	-
M175 OK L. Texoma	148	231	206-325	148	150	-
J31 OK Grand L.	2	244	-	-	-	-
Age III						
R97 NT Great Slave L.	2	145	135-155	2	43	-
S437 AT L. Claire	89	198	-	89	62	-
S437 MB L. Winnipegosis	130	221	-	130	130	-
P37 MT Ft. Peck	+	282	-	-	-	-
R230 SA rivers	+	297	-	+	227	-
M175 OK L. Texoma	247	305	254-356	247	258	-
Age IV						
R97 NT Great Slave L.	1	165	-	1	57	-
S437 AT L. Claire	4	259	-	4	150	-
S437 MB L. Winnipegosis	194	300	-	194	252	-
P37 MT Ft. Peck	+	312	-	-	-	-
R230 SA rivers	+	323	-	+	278	-
M175 OK L. Texoma	423	333	254-376	423	309	-
Age V						
R97 NT Great Slave L.	2	216	201-231	2	99	85-113
S437 ON Sandy L.	1	267	-	1	142	-
S437 AT L. Claire	1	290	-	1	227	-
S437 MB L. Winnipegosis	27	328	-	27	335	-
R230 SA rivers	+	330	-	+	312	-
P37 MT Ft. Peck	+	333	-	-	-	-
M175 OK L. Texoma	67	343	305-376	67	351	-
Age VI						
R97 NT Great Slave L.	2	218	206-234	2	108	85-127
S437 ON Sandy L.	21	290	-	21	210	-
S437 AT L. Claire	50	302	-	50	258	-
S437 MB L. Winnipegosis	420	335	-	420	351	-
R230 SA rivers	+	366	-	+	448	-
M113 AT Bassano L.	+	376	-	+	567	-
M175 OK L. Texoma	2	363	356-371	2	471	-
M215 KS Big Blue R.	1	394	-	-	-	-
Age VII						
S437 ON Sandy L.	11	323	-	11	290	-
S437 AT L. Claire	219	338	-	219	346	-
S437 MB L. Winnipegosis	266	348	-	266	405	-
R230 SA rivers	+	378	-	+	496	-
M113 AT Bassano L.	+	409	-	+	567	-
Age VIII						
S437 ON Sandy L.	12	312	-	12	278	-
S437 AT L. Claire	268	363	-	268	436	-
S437 MB	16	361	-	16	471	-
R230 SA rivers	+	386	-	+	519	-
M113 AT Bassano L.	+	432	-	+	822	-
Age IX						
S437 ON Sandy L.	24	325	-	24	307	-
S437 AT L. Claire	106	396	-	106	550	-
R230 SA rivers	+	409	-	+	630	-
M113 AT Bassano L.	+	437	-	+	851	-

	No.	TL	Range	No.	Weight	Range
Age X						
S437 ON Sandy L.	37	330	-	37	318	-
S437 AT L. Claire	71	409	-	71	581	-
R230 SA rivers	+	427	-	+	694	-
M113 AT Bassano L.	+	457	-	+	1106	-
Age XI						
S437 ON Sandy L.	26	333	-	26	326	-
S437 AT L. Claire	53	414	-	53	618	-
Age XII						
S437 ON Sandy L.	7	335	-	7	366	-
S437 AT L. Claire	23	432	-	23	691	-
Age XIII						
S437 ON Sandy L.	3	340	-	3	378	-
S437 AT L. Claire	2	470	-	2	964	-
Age XIV						
S437 ON Sandy L.	2	353	-	2	439	-
S437 AT L. Claire	1	486	-	1	964	-
Age XV						
S437 ON Sandy L.	1	406	-	1	709	-
Ages VII-XIII						
B207 MB	15	-	335-434	15	-	369-510

		Calculated TL at each annulus						
	No.	1	2	3	4	5	6	7
E3 MN	625	74	168	249	290	328	356	
B12 MB	-	122	183	251	282	356	404	424
V20 MN Red L.	10	79	201	279	315	343	366	
I10 IA Snyder Bend	1	91						
P159 MT streams	17	84	175	246	277	292	310	
P159 MT lakes	324	117	224	284	312	338	358	
M175 OK L. Texoma	891	190	226	284	320	328	361	
J45 OK Verdigris R.	4	117	173	221	272	310		

	Mean TL and annual increments at each annulus								
	1	2	3	4	5	6	7	8	9
G87 MN Red L.	102	211	287	328	358	381	401	421	437
Increment	104	107	76	41	30	18	10	20	13
Number	1165	1165	1165	1165	1043	853	423	165	70
S225,226,228,230,393									
SD Ft. Randall	122	241	297	323					
Increment	122	119	56	36					
Number	436	340	116	14					
S227,229,276, N94,115									
SD Gavins Pt.	114	216	274	320	356	376	406		
Increment	114	102	69	51	43	33	20		
Number	204	203	188	98	38	13	2		
F100,105,124,125									
SD Oahe	109	213	277	302	325	333	340		
Increment	109	109	66	46	38	36	36		
Number	641	581	478	196	71	22	4		

G87 MN Red L. Annual increment as percentage of length at beginning of year

TL	Median	Range
76-101	118(81)	115(48)-122(9)
102-126	101(1084)	55(122)-107(22)
178-202	39(122)	-
203-228	37(1043)	32(9)-39(258)
254-278	16(131)	15(9)-16(122)
279-304	15(1034)	9(190)-17(62)
305-329	7(909)	5(258)-11(22)
330-355	7(831)	4(430)-9(134)
356-380	6(423)	1(258)-7(119)
381-405	5(187)	4(31)-5(156)
406-431	4(204)	3(72)-6(62)

Cutting-over was described as the best criterion of an annulus (G87). In South Dakota the annulus forms in mid-May (S226, S228). Most growth studies have used lengths computed on the basis of direct proportion between scale and body length, but G87 used a curvilinear relationship.

In general, growth is slower in the far north than in the southern part of the range. Growth was faster in Saskatchewan rivers than lakes (R230), but in Montana (P159) growth was faster in lakes than streams. In Red Lake, Minnesota (G87) growth was more rapid in summers with higher temperatures. Growth was much more rapid in Ft. Randall immediately after impoundment than before (S225). The average increments for the 1st, 2nd and 3rd years of life were 155 mm (31 fish), 137 mm (6), and 66 mm (3) compared to 74 mm (9) and 97 mm (3) for the first 2 years of life before impoundment.

Females were found to grow faster than male goldeye by M175, G87 and S227 but no sex difference in growth was noted by S226 and S228. The more rapid growth of females did not appear until the third year in Lake Texoma (M175) and until the fourth year in Red Lake (G87).

Some males were found to mature at age I in South Dakota (S228, S226) but not until age III or IV in Red Lake, Minnesota (G87) and southern Manitoba. (Dymond, quoted in G87) and age VI in Lake Claire, Alberta (B207). Females mature at age II in South Dakota (S393, S228, S226, S229) but not until age IV in Red Lake and Manitoba (G87) and age VII in Alberta (B207). Adult males can be distinguished by a distinct elongated lobe on the anterior portion of the anal fin, which appears as the male approaches its first spawning season (S437). In South Dakota, spawning occurs from May 25 to June 26 (S228, S229) though ripe goldeye were found in late July (S393). In Saskatchewan they also spawn in June and in Lake Claire, Alberta, they spawn at 10C and the eggs float on the water surface (R230). Eggs spawned June 7 in Lake Mamawi, Alberta, hatched June 19 and the fry were 16-17.1 mm long (B207).

B207 MB   15 females   370-510 g   14150 eggs per female 5800-25200

Small fingerlings feed on plankton and small aquatic insects (S437). Adult goldeye feed mainly on aquatic insects (R20, G87, R97), mainly mayfly nymphs (R230, H243). A few fish are eaten (R230, R20, E4). In August 1939, I found goldeye feeding exclusively on grasshoppers in Upper Red Lake, Minnesota. Goldeye were found to forage mostly at night (G87, S437) and mostly near the surface (B12, S437).

MOONEYE, *Hiodon tergisus* LeSueur

The mooneye is found in the St. Lawrence and Lake Champlain drainages, the lower Great Lakes, Ohio, Mississippi and Missouri drainages south to Tennessee and Arkansas, and in the northern drainages from northwest Quebec to Saskatchewan (S471, T113). The mooneye is less tolerant of silt and turbidity than is the goldeye (T113).

V67 L. Huron   TL = 1.206 SL

TL	No.	Weight	Range	Citation
76-101	23	0.9	0.2-6.8	S472 AL
102-126	107	7.7	2.8-11.3	S472 AL
127-151	123	11	-	S472 AL
	2	21	-	V67 L. Erie
152-177	33	28	17-37	S472 AL
	7	35	-	V67 L. Erie
178-202	32	51	45-79	S472 AL
	9	71	-	V67 L. Erie
203-228	8	85	-	W132 TN
	17	85	57-94	S472 AL
	43	88	-	V67 L. Erie
229-253	5	91	-	W132 TN
	29	113	-	V67 L. Erie
	67	119	99-181	S472 AL
254-278	13	156	-	V67 L. Erie
	104	162	91-227	S472 AL
	9	167	-	W132 TN
279-304	1	181	-	W132 TN
	35	233	170-298	S472 AL
	13	235	-	V67 L. Erie
	2	247	227-269	C30 IA
305-329	11	301	-	V67 L. Erie
	8	310	227-366	S472 AL
	4	326	290-363	C30 IA
330-355	7	351	-	V67 L. Erie
	1	391	-	C30 IA
356	1	403	-	V67 L. Huron
368	1	418	-	V67 L. Erie
305 (FL)	1	425	-	M431 QU

S472 AL	550 mooneye, 76-152 mm TL	log W = -5.63 + 3.27 log TL
E28 TN		
Chickamauga L.	17 mooneye Age II Average K(SL) 1.5	
S472 AL	286 mooneye, 76-152 mm Mean K(TL) 0.68, range 0.57-.85	
	263 mooneye, 178-305 mm Mean K(TL) 0.98, range 0.91-1.0	
F5 NY L. Erie	Age 0-June 15 mm TL	
E28 TN		
Chickamauga L.	17 fish Age II 199 mm SL, 244 mm TL	

		TL	g
V67 L. Erie			
Age I	31	213	82
Age II	5	244	120

Age III	9	295	229
Age IV	5	310	292
Age V	7	320	317
Age VII	1	333	361
V67 L. Huron			
Age VIII	1	356	402 only mooneye known from Lake Huron

		Average TL at each annulus						
	No.	1	2	3	4	5	6	7
E28 TN Chickamauga L.	17	112	213					
V67 L. Erie	58	124	206	257	287	307	315	325
Increment		124	84	48	38	23	13	10
Number		58	27	22	13	8	1	1

Lengths were calculated from measurements of anterior radius and lateral diameters of scales (V67); that of the diameters is given above. The lengths computed from anterior radii at the first 2 annuli were 104 and 196 respectively but at other annuli the differences were not over 3 mm.

In Lake Erie, mooneyes spawn in April and May (V67). Females produce 10000-20000 eggs per season (L80). The food of adults was aquatic insects, mostly *Hexagenia*, in a study on the Mississippi River (H243).

ALASKA BLACKFISH, *Dallia pectoralis* Bean
    The blackfish is found in fresh waters of northern Alaska and northeast Siberia.

TL	25-50	51-75	76-101	102-106	127-151	152-177
B260 AK Eldorado C.						
Weight	-	1-4	4-12	7-21	14-32	28-37
Number	-	30	23	44	21	3
O43 AK Point Barrow						
Weight	0.2-1.5	1.4-6.0	4.3-9.5	9.1-14.5		
Number	16	over 100	over 50	10		

B260 AK Eldorado C.	Variance from regression
30 females (gonads not developed) log W = -4.984 + 2.971 log TL	0.004
33 males = -6.519 + 3.737 log TL	0.002

Age	Number	TL	Range
I	28	64	51-91
II	58	107	79-140
III	22	127	94-165

Females mature at 79 mm (TL), 5 g. Two sizes of eggs were intermingled in the ovary. Some blackfish collected in June were partially spawned. Spawning had been completed by August (B260). Ostracods, cladocerans, and diptera larvae are the principal foods (O43).

	Lengths in gill nets				
H313 AK					
Mesh size, mm	10	13	19	22	25
Mean FL	41	56	66	84	89
Range	38-43	46-58	61-71	79-86	86-91
Number	6	20	2	2	3

CENTRAL MUDMINNOW, *Umbra limi* (Kirtland)

The central mudminnow is found from Manitoba south to eastern Kansas, northern Arkansas, Reelfoot Lake, Ohio, to the Lake Champlain basin and Quebec (H382). They require soft bottoms into which they burrow tailfirst (T113). Dredging, draining, and increased turbidity have eliminated suitable habitat in much of their range (T113). The largest specimen was 132 mm (T113).

	No.	SL	TL =
A14 MI	122	under 50	1.256 SL
	219	50–69	1.249 SL
	47	70–79	1.242 SL
	34	over 79	1.232 SL

W144   The anal fins of males are longer than those of females.

E105 MD   109 mm TL   14 g

	No.	TL	Range
Age 0-May			
P128 IN Judy C.	37	17.8	10–23
Age 0-June			
P128 IN Judy C.	20	20.8	16–28
W144 NY Red C.	132	-	9–32
Age 0-July			
W144 NY Red C.	71	-	17–43
W144 NY Round P.	88	-	19–37
P128 IN Judy C.	275	37.0	24–48
Age 0-Aug.			
P128 IN Judy C.	135	42.8	27–54
Age 0-Sept.			
W144 NY Red C.	27	-	23–53
P128 IN Judy C.	187	46.3	32–55
A14 MI Kimes L.	1413	37.0	25–45 (mode)
Age 0-Nov.-March			
P128 IN Judy C.	98	55.8	32–70
Age I			
W144 NY Red C.	25	51.0	-
A14 MI Kimes L.	1268	52.0	-
P128 IN Judy C.	137	69.4	46–94
Age II			
W144 NY Red C.	11	64.0	-
P128 IN Judy C.	28	79.8	71–91
Age III			
W144 NY Red C.	6	76.0	-
P128 IN Judy C.	9	86.5	73–117
Age IV			
W144 NY Red C.	3	85.0	-

Scales do not give valid evidence of age (A14, P128, W144) though P128 reports that Pavlov, 1953, found scales usable in *Umbra krameri*. Otoliths were used in aging mudminnows by P128 and W144. A14 used only the length frequency method. Opercular bones were also examined by W144 but were reportedly unsatisfactory. Females grew faster than males in New York (W144), and A14 reports more females among the larger size groups, but P128 found no consistent growth differences except that females may have been larger at age III. The females did show greater variability in growth (P128).

Male mudminnows and some females mature at age I (W144, A14). They spawned April 18-25 at 13C in flood vegetation in New York (W144) and April 9-29 in Indiana (P128, P185). Females guarded the nest (W144). Ova were 1.6 mm diameter and were attached singly on aquatic plants (P128, P185). Young hatched on the 6th day and were 5 mm at hatching (P128, P185).

	TL	No.	Eggs per female Mean	Range
E32 IN	-	-	-	425-450
P128 IN	52-57	6	339	220-571
	60-69	13	449	202-758
	71-79	7	737	304-1175
	81-91	10	957	379-1489
P128 MI	77-92	4	773	487-903
W144 NY	46-49	4	268	199-335
	50-62	9	329	189-456
	63-75	11	560	431-812
	76-87	3	898	623-1262
	88-100	9	1483	1201-1983
	101-108	5	1895	1672-2286

Mudminnows were found to be active under the ice (P128) although reported by Abbott in 1870 to hibernate. Stomachs were more apt to be empty during the breeding season than in the winter (P128). Ostracods were the principal foods of young mudminnows (P128, P185) but insects, molluscs, amphipods, and isopods were the principal foods after about 20 mm (P128, P185, W144, K125, N104).

EASTERN MUDMINNOW, *Umbra pygmaea* (DeKay)

The eastern mudminnow is found in swamps and sluggish streams of the Atlantic coastal plain from Long Island to Florida (E119).

Five specimens were found to have eaten copepods and trichopteran larvae (F148).

REDFIN PICKEREL, *Esox americanus americanus* Gmelin

The redfin pickerel is found in streams and lakes along the Atlantic seaboard from southern Maine to Florida and in the Lake Champlain drainage and in the St. Lawrence River near where the Lake Champlain drainage joins the St. Lawrence (C340). It intergrades with the grass pickerel in Florida, Georgia, Alabama, and Mississippi. The reported maximum length of 483 mm TL is suspected (C340) of referring to a hybrid between the redfin and chain pickerels.

R210   TL = 1.136 SL

L157 NC

FL	Weight	No.
25-50	.03-.11	14
51-75	.08-.11	10
87-80	.34	1
101-120	.8-2.0	16
121-140	1.1-2.8	13
141-170	1.7-48	37
171-190	34-74	32
191-220	43-99	38
221-231	79-116	7
239	142	1
269	176-210	2
287	187	1

T113 OH　　　381 mm TL　　　397 g

C290 NC

	No.	FL	Range
Age 0-March	+	-	8-10
Age I	5	112	91-130
II	49	140	117-163
III	71	178	140-206
IV	30	211	193-241
V	4	246	211-272
VI	1	284	-

Little difference was noted in growth of male and female redfin pickerel the first four years of their life, but, at age V, 2 males averaged 218 mm and 2 females 272 mm, and the only age VI pickerel was a 284 mm female (C290). Fred Dickinson was quoted as saying redfin pickerel reach 454 g within the first year in south Georgia (C290). In North Carolina redfins are usually in sluggish streams and backwaters and less commonly in lakes and ponds where chain pickerel are more common.

Redfins spawn at 10C in March and February in North Carolina (C290). Submandibular pigmentation may be useful in separating sexes. The smallest spent female and the smallest mature male were each 130 mm FL. Some males were still immature at 165 mm FL. Redfin pickerel have 2 sizes of eggs (unlike northern pike).

		Eggs per female	
		Mean	Range
C290 NC	10 females, Age III-IV, 213-231 mm	3716	722-4364
	counting only large eggs for current spawning	269	186-542

Eggs hatch in 12-14 days, and the fry feed on plankton within a week (C290). Cladocera, amphipods, and immature insects constitute the food until 2-3 inches long (C290) and after that, fish, crayfish, and dragonfly nymphs are the principal foods (C290, P132, C298).

A literature review is given by B328, C340.

GRASS PICKEREL, *Esox americanus vermiculatus* LeSueur
Grass pickerel are found from the St. Lawrence River at Montreal, along

Lake Ontario, southern Michigan, Wisconsin, Nebraska, south to the Gulf of Mexico in Texas. It intergrades with the redfin in Mississippi to Florida (C340). The grass pickerel requires nonturbid, densely vegetated waters, and since such habitat has been destroyed by siltation and draining, pickerel are scarce in many areas where they were once abundant (T113).

C299 ON   129-352 mm SL   SL = .8826 FL   TL = 1.0582 FL
           Live FL = 1.029 formalin FL;   Live weight = 0.94 formalin weight
           For pickerel up to 70 mm   FL = TL

| | | Weight | | | | | | |
| | | C299 ON | V44, B328 LA | | | K124 WI | | |
FL	TL	Range	No.	Mean	Range	No.	Mean	Range
25-50	-	.02-1.1	-	-	-	-	-	-
51-75	-	.06-3.7	-	-	-	-	-	-
76-101	-	.3-9.4	-	-	-	-	-	-
102-126	107-134	8.5-17	-	-	-	13	8.5	2.8-14
127-151	135-159	11-37	-	-	-	17	14	11-23
152-177	160-187	25-65	-	-	-	32	31	20-43
178-202	188-215	48-91	-	-	-	62	48	34-65
203-228	216-240	65-128	2	82	-	62	65	34-94
229-253	241-268	99-164	-	-	-	29	99	74-181
254-267	269-281	150-196	-	-	-	7	119	105-136
	282-304	-	-	-	-	13	139	112-170
	305-329	-	1	233	-	10	204	162-193
	330-355	-	1	235	-	1	227	-
	356-380	-	3	312	284-326	-	-	-

Weights given in R52 seemed out of line and are not included here.

K124 WI Pleasant L.   123 pickerel $\log W = -4.696 + 2.752 \log TL$
     WI Eagle Spring L.   143 pickerel $\log W = -5.765 + 3.206 \log TL$

	No.	TL	Range
Age 0-April			
K124 WI	+	10	8-15
Age 0-May			
L3 MI	28	53	18-43
R27 IL	+	-	15-64
K124 WI	+	-	10-38
Age 0-June			
C299 ON Jones C.	+	25	-
L3 MI	14	36	25-43
K124 WI	+	-	28-86
Age 0-July			
C299 ON Jones C.	+	71	-
L3 MI	3	53	-
Age 0-Aug.			
L3 MI	1	61	-
C299 ON Jones C.	+	79	-
Age 0-Sept.			
C299 ON Jones C.	+	94	-
C13 MI Deep R.	300	104	64-130

	No.	TL	Range	No.	Grams	Range
Age 0-Oct.						
K124 WI males	62	137	81-196			
females	56	152	114-198			
V44 LA Springhill L.	4	312	229-376	4	232	82-326
Age 0-Nov.						
L3 MI Fall spawned	12	83	23-43			
L3 MI	3	137	109-163			
Age I						
C299 ON Jones C. males	13	109	89-132			
K124 WI May Fall spawned	6	61	30-86			
Oct. males	43	206	165-272			
females	44	211	170-284			
Age II						
C299 ON Jones C. males	6	132	107-140			
ON Jones C. females	5	147	132-170			
K124 WI males	23	236	193-284			
females	37	259	203-323			
Age III						
C299 ON Jones C. males	26	170	150-206			
ON Jones C. females	30	180	132-216			
K124 WI females	14	287	257-323			
P31 MO	1	305	-	1	170	-
Age IV						
C299 ON Jones C. males	50	185	165-213			
ON Jones C. females	55	196	132-229			
C215 MI Whitmore L.	3	307	-			
K124 WI female	1	356	-			
Age V						
C299 ON Jones C. males	20	206	180-249			
ON Jones C. females	28	218	183-246			
Age VI						
C299 ON Jones C. males	7	234	208-254			
ON Jones C. females	10	251	226-284			
Age VII						
C299 ON Jones C. females	3	274	267-284			

	Calculated TL at each annulus				
	No.	1	2	3	4
J64 OK Little R.	213	97	190	257	300
R52 OH	-	140	216	279	

Growth during the first month was 0.96 mm per day in Portage Lake, Michigan (H86).

Growth in Louisiana (V44) is much more rapid than in the other studies, probably because of the longer growing season. Growth in Jones Creek, Ontario (C299), from which most of the data are available, is probably relatively slow. Females show a greater size than males at age II (C299) and at ages 0-II (K124). Females also had a greater longevity (K124). The body-scale relationship was described by two straight lines crossing at about 140 mm FL (C299).

The grass pickerel is a fish of slow-moving heavily vegetated streams and sometimes lakes. The preferred temperature is 25.5C and cruising speed in-

creases with acclimation temperatures (C299). They usually spawn in April in
Ontario (C299) and Wisconsin (K124) but may occasionally spawn again in late
August or September (C299, K124, L3). Ripe males were found in Pennsylvania
in October (M387). They spawn in shallow water over vegetation, nearer shore
than northern pike (C299). The smallest mature male at Jones Creek, Ontario,
was 140 mm FL, and the smallest mature females, 157 mm (C299). In Nebraska,
however, mature pickerel at 102 mm TL have been seen in spawning runs
(M293). Three sizes of eggs were found in pickerel ovaries:

C299 ON   12 females   756 primary + 2380 secondary + ? tertiary ova per
female.

C9 MI   a 157 mm TL female,   15732 eggs (sizes not indicated)
K124 WI   log No. of eggs = -2.819 + 2.544 log TL (counting only the mature
           eggs).

TL	No. of females	No. of mature eggs	
		Mean	Range
160-218	3	759	625-843
241-295	5	2515	2050-3176
325	1	4584	-

Young remained in the general area of spawning throughout the first summer's growth and showed little dispersion (K124). Mature eggs of grass pickerel ranged from 1.5 to 2.4 mm compared to 2.4 to 3.2 mm for northern pike.
At hatching grass pickerel were 5.0 to 6.0 mm and northern pike were rarely
less than 6.5 mm. Pickerel have well-developed mouths at 9 mm and begin
feeding at 10 mm, northern pike at 11 mm and 12 mm respectively (K124).
Fingerling pickerel were less easily observed and captured in ponds than fingerling pike at similar population densities (K124).

Under 50 mm pickerel eat Cladocera, ostracods, amphipods, and some insect larvae; from 51-100 mm mostly insect larvae, a few crayfish and small
fish; and at larger sizes mostly fish, crayfish, and occasional dragonfly nymphs
(C299). Only 1 tadpole was taken although they were abundant (C299). Fish,
mostly *Lepomis*, constituted most of the food of grass pickerel, even at 51-100
mm in Oklahoma (M435). In Wisconsin ponds, pickerel from 10-15 mm ate
cladocerans and copepods; up to 38 mm they continued to eat these but also included chironomids and small fish; and thereafter insects and fish (K124).
Feeding reached its peak in late afternoon and minimum at night (K124). In
October all stomachs were empty (C299). In Ontario, the grass pickerel
seems to always be associated with *Umbra limi* and to feed mostly on them
even though other species are present (C298).

Fall populations in Pleasant Lake, Wisconsin, were estimated at 159 pickerel or 3.9 kg/ha (K124).

CHAIN PICKEREL, *Esox niger* LeSueur
Chain pickerel are found along the eastern seaboard from the St. Lawrence
River through Florida and in the Gulf of Mexico drainage north to southern
Missouri and west to eastern Oklahoma (T113). They have been introduced into
a few scattered areas of the United States.

S289 MA   1340 pickerel   60-579 mm SL   TL = 0.0462 SL
M383 MD Severn R.   TL = 14 mm + 1.165 SL (r = 0.99)

Weight in grams

TL	No.	Mean of means	Central 50%	Range	Citations
76-101	3	6	-	1-9	S472, W108
102-126	50	7	-	5-9	S472, W108
127-151	185	14	-	10-17	S472, W108
152-177	426	23	-	17-28	S472, U6, W108
178-202	367	37	28-40	17-40	S472, F9, U6, W108
203-228	324	51	40-60	31-60	S472, F9, U6, W108
229-253	344	70	62-77	54-77	S472, F9, H202, U6, W108
254-278	285+	102	99-108	68-119	S472, F9, H61, W108, U6
279-304	254+	130	113-144	99-156	S472, H61, F9, W108, H202
305-329	263+	187	164-232	128-312	S472, H61, F9, W108, U6, H202
330-355	227	221	201-244	170-269	S472, F9, H202, W108, U6
356-380	175	298	278-312	218-397	S472, F9, H202, U6, W108, S361
381-405	146	363	351-386	269-545	F9, H61, S472, U6, W108, S361
406-431	101+	414	380-471	287-499	H61, F9, S472, H202, S361, W108
432-456	129	533	485-590	428-604	F9, H202, U6, S361, W108, S472
457-482	149	627	584-672	480-737	H202, U6, F9, S472, W108, S361
483-507	106+	748	691-768	553-936	F9, S361, U6, S472, H202, W108
508-532	71	869	805-896	653-981	F9, S361, S472, H202, U6, M242, W108
533-558	50	987	927-1032	822-1361	S472, F9, S361, U6, H202, W108
559-583	28	1168	1077-1329	794-1435	H202, S361, S472, F9, W108
584-609	17	1304	1258-1573	1247-1673	S472, S361, U6, W108, M242
610-634	17	1661	-	1264-1769	S472, S361, W108
635-659	5	1655	-	1531-1814	S472, S361
660-685	1	1588	-	-	S361 MD
686-710	1	1928	-	-	E105 MD
711-736	1	2948	-	-	V6 NY
762-786	1	4082	-	-	G36 NJ

S186 RI   257 pickerel   log W = -5.510 + 3.130 log FL
H202 FL   87 pickerel;   229-566 mm;   log W = -5.491 + 3.098 log TL
(The 3.098 was not significantly different from 3.0 at 95% confidence level.)
239 fish average K(TL) = 0.31

At 18, 36, 54, and 72 days after hatching, chain pickerel averaged 13, 20, 61, and 94 mm respectively in an Ohio hatchery (A89).

	No.	TL	Range	No.	Weight	Range
Age 0-Hatching						
U6 NY	14	10	-	-	-	-
Age 0-April						
H61 MD	90	-	33-104	-	-	-
Age 0-May						
U6 NY	9	24	-	-	-	-
Age 0-June						
U6 NY	50	28	-	-	-	-
R4 NY	54	41	30-66	-	-	-
M383 MD Severn R.	33	89	58-124	-	-	-
Age 0-July						
U6 NY	79	56	-	-	-	-
A1 NY Oneida L.	7	58	43-76	-	-	-
R4 NY	36	66	58-76	-	-	-
M383 MD Severn R.	56	132	71-196	-	-	-
Age 0-Aug.						
U6 NY	37	66	-	-	-	-
R4 NY	22	94	74-114	-	-	-
M383 MD Severn R.	123	163	79-262	-	-	-
Age 0-Sept.						
R4 NY	8	91	79-119	-	-	-
Age 0-Oct.-Dec.						
M89, L20, K4	+	114	-	+	14	-
E9 NY	+	-	102-127	-	-	-
U6 NY	26	168	-	-	-	-
R129 MA	7	160	119-188	-	-	-
Age 0						
S259 NJ 2 lakes	+	-	89-114	-	-	-
Age I						
E9 NY	+	-	152-178	-	-	-
M1, M87, S289 MA	179	173	145-201	77	34	-
M147, W108 MA Paradise P.	4	175	-	-	-	-
M89, L20, K4 MA	+	178	-	+	43	-
U6 NY	54	178	168-196	-	-	-
M147, W108 MA Paradise P. reclaimed	3	203	-	-	-	-
M146 MA	9	206	-	9	48	-
S259 NJ 3 lakes	+	224	203-241	-	-	-
G24 NY Long Island	5	244	193-295	2	57	40-77
S361 MD Severn R. males	7	396	381-427	7	414	369-482
S361 MD Severn R. females	2	434	432-437	2	539	-
Age II						
F9 CT	29	196	-	-	-	-
M147, W108 MA Paradise P.	20	213	-	-	-	-
S289 MA ponds males	130	236	-	-	-	-
S289 MA ponds females	129	241	-	-	-	-
M89, L20, K4 MA	+	254	-	+	113	-
M1 MA	63	254	193-267	63	85	-
M146 MA	76	272	-	76	125	-

Age II (cont.)	No.	TL	Range	No.	Weight	Range
E9, G24, U6 NY	267+	282	239-394	6	147	85-207
M147, W108 MA Paradise P. reclaimed	8	333	-	-	-	-
S258 NJ L. Hopatcong females	6	394	-	-	-	-
S361 MD Severn R. males	7	460	432-490	7	578	482-680
S361 MD Severn R. females	26	508	457-546	7	822	652-964
M89 MA	-	-	-	-	-	1814-2268
Age III						
M147, W108 MA Paradise P.	8	257	-	-	-	-
S289 MA ponds males	107	284	-	-	-	-
S289 MA ponds females	171	292	-	-	-	-
E93 ME	+	295	-	-	-	-
M1 MA	17	297	287-312	17	159	-
M147 ME	13	312	-	-	-	-
F9 CT	75	312	-	-	-	-
M87 MA	75	323	-	75	213	-
M147, W108 MA Paradise P. reclaimed	2	330	-	-	-	-
M146 MA	47	330	-	47	233	-
S259 NJ 3 lakes	+	335	269-381	-	-	-
U6 NY	65	356	302-486	-	-	-
M89, L20, K4 MA	+	-	343-368	-	-	-
S259 NJ L. Hopatcong females	145	427	-	-	-	-
S361 MD Severn R. males	10	508	470-546	10	723	595-936
S361 MD Severn R. females	3	579	546-547	10	1332	992-1673
Age IV						
M147, W108 MA Paradise P.	6	295	-	-	-	-
T96 NJ infertile pond	+	305	-	-	-	-
S258 NJ acid lakes (pH 4.2)	-	343	-	-	-	-
S289 MA ponds males	35	345	-	-	-	-
S289 MA ponds females	109	361	-	-	-	-
M147, W108 MA Paradise P. reclaimed	10	368	-	-	-	-
E93 ME	+	386	-	-	-	-
M1, M82, M146 MA	142	394	325-486	142	413	-
U6 NY	66	414	338-526	-	-	-
M89, L20, K4 MA	+	445	-	+	680	-
S259 NJ 2 lakes	+	498	470-526	-	-	-
S259 NJ L. Hopatcong females	116	488	-	-	-	-
S361 MD Severn R. males	11	561	538-584	11	1032	652-1247
S361 MD Severn R. females	12	627	597-655	12	729	1531-1814
T96 NJ fertile waters	+	673	-	-	-	-
Age V						
S289 MA ponds males	16	373	-	-	-	-
S289 MA ponds females	72	414	-	-	-	-
G24 NY Long Island	3	417	335-521	3	530	224-1030
M146, M1, M87 MA	55	452	442-559	55	627	567-680
M147 ME	76	455	-	-	-	-
F9 CT	13	445	-	-	-	-
U6 NY	31	460	378-556	-	-	-
E93 ME	+	462	-	-	-	-
M89, L20, K4 MA	+	508	-	+	1134	-

Age V (cont.)	No.	TL	Range	No.	Weight	Range
S259 NJ L. Hopatcong males	+	526	-	-	-	-
S259 NJ L. Hopatcong females	43	577	-	-	-	-
S361 MD Severn R.	1	660	-	1	1588	-
Age VI						
S289 MA ponds males	14	424	-	-	-	-
S289 MA ponds females	36	470	-	-	-	-
U6 NY	8	475	460-521	-	-	-
F9, K106 CT	5	498	-	1	1134	-
M1, M87, M146 MA	24	508	-	24	1006	-
M147 ME	80	508	-	-	-	-
E93 ME	+	521	-	-	-	-
S259 NJ L. Hopatcong males	+	605	-	-	-	-
S259 NJ L. Hopatcong females	21	653	-	-	-	-
Age VII						
S289 MA ponds males	3	470	-	-	-	-
S289 MA ponds females	14	500	-	-	-	-
F9 CT	5	486	-	-	-	-
G24, U6 NY	2	516	500-533	1	779	-
M147 ME	24	561	-	-	-	-
M1, M146 MA	5	589	-	5	1503	-
E93 ME	+	597	-	-	-	-
S259 NJ L. Hopatcong males	+	653	-	-	-	-
S259 NJ L. Hopatcong females	7	691	-	-	-	-
Age VIII						
S289 MA ponds males	2	470	-	-	-	-
S289 MA ponds females	5	564	-	-	-	-
M147 ME	9	572	-	-	-	-
M1 MA	1	594	-	1	1520	-
S259 NJ L. Hopatcong females	2	719	-	-	-	-
Age IX						
S289 MA ponds males	2	475	-	-	-	-
S289 MA ponds females	1	630	-	-	-	-
C173 MA	1	749	-	1	4224	-

		Calculated FL at each annulus							
	No.	1	2	3	4	5	6	7	8
S186 RI	257	168	267	315	358	406	437	478	516

		Calculated TL at each annulus							
	No.	1	2	3	4	5	6	7	8
S186 RI	257	183	290	345	394	445	472	513	556
V15 typical	+	203	279	356	406	457	533		
K126 typical	+	180	259	328	391	450	498	533	561
T12, W13 MA	118	180	282	356	417	467	514		
N50 NJ	329	140	254	356	419	533	610	673	729
S259 NJ Cranberry R. males	+	244	310	373	414	437			
S259 NJ Cranberry R. females	+	244	338	409	450	467			
H158 DE	+	292	340	394	434				
G83 MA 3 reclaimed ponds	145	206	338	424					
S259 NJ L. Hopatcong males	+	193	323	414	483	526	544	554	

Calculated TL at each annulus (cont.)

	No.	1	2	3	4	5	6	7	8
S259 NJ L. Hopatcong									
females	+	193	323	445	533	599	643	676	
M147, W108 MA 120 ponds	1056	196	234	290	353	394	447		
slow growth	38	109	175	224	274				
rapid growth, 6 ponds	676	218	345	417	478	503	589		
Billington Sea	158	292	429	521	569				
M292 MA Quabbin R.									
1946-52	123	152	290	401	486	551	599	630	653
1953-57 after smelt	440	173	343	439	508	538			

Calculated average weight at each annulus

	1	2	3	4	5	6	7	8
V15 typical	45	136	272	454	680	1134		
K126 typical	45	91	227	408	680	907	1179	1497

Annuli are often difficult to distinguish perhaps because feeding and some growth occurs year around (W108).

Growth was usually calculated from scale measurements by direct proportion, but a correction factor was introduced in S186, M292, G83, and S188.

No regional trends in growth are evident in the tabulated data. Growth in streams tends to be slow (W108). Females grow faster than males which is evident at age I (S361) or not until age II or III (S259, S289). Growth was slower in water of 6 ppm total alkalinity (age IV pickerel averaged 305 mm) than in water of 35 ppm (age IV pickerel, 673 mm) (T96). Growth was also slow in an acid lake, pH 4.2, in New Jersey (S258). Although the average sizes at the 1st through 3rd annuli were somewhat higher, M292 did not believe that there was evidence that the growth of chain pickerel was better in Quabbin Reservoir, Massachusetts, after the introduction and establishment of smelt in 1953 than before. Population reduction in Paradise Pond, Massachusetts, where a slow-growing population was present, resulted in increased growth (M147, W108).

	Calculated TL at each annulus			
	No.	1	2	3	4
M147, W108 MA Paradise P. 1958	38	109	175	224	274
Increments		109	66	56	51
Number		38	34	14	6
1957 after reclamation	23	127	246	251	368
1957 increment		203	203	99	132
Number		3	8	2	10

Although gonads may mature in age I chain pickerel toward the end of the year, spawning probably does not occur before age II (W108) and may not occur until age III or IV (U6). A slow-growing population will mature at a smaller size than will fast-growing pickerel (W108). In Lake Erie, no mature females under age III were taken and the smallest mature male was age II and was 356 mm long (M147). In an Alabama study, the smallest mature female was 43 g and the smallest mature male, 71 g (D60). Most populations appear to have a preponderance of females (S188, W108). Pickerel usually enter swampy or flooded areas with abundant vegetation soon after the ice leaves and begin to spawn when water temperatures are about 8C (W108). In Alabama, they are reported to spawn at 16C (D60). In Maryland, they spawn March 5-11 at 6C (S361); in New York in April (U6); in Connecticut in April and May (W13) and in Massachusetts in March to May (S289). In Pennsylvania chain pickerel were stripped in September and successfully hatched. Fry 25-38 mm long were found in late November in New Jersey (M147). The eggs are laid in glutinous strings, sometimes 3 m long (W108) which tend to be adhesive and stick to vegetation.

A 900 g female may contain about 30000 eggs (N25).

S188 RI	307 mm female	6102 eggs
	330 mm female	7146 eggs
	361 mm female	8140 eggs

In the fall stripping, 735 eggs were taken from a 522 g female and 1653 from a 1188 g female (M387). These gave a 72% and 64% hatch. Both eggs and young may be stranded as water levels drop in the shallow breeding areas (S114) or they may be washed ashore during a storm (A1). Eggs hatch in 6-12 days depending upon water temperature. The sac fry usually remain inactive, attached to submerged debris or vegetation for 6-8 days (U6).

The chain pickerel is a solitary fish. Larger fish move into shallow water at night and back to deeper water during the day (W108). Young are seldom in water over 0.5 m deep (R4). Chain pickerel may live in brackish water up to about 1.5% salt (S188, M147), in acid water to pH 3.8 (M147) and in temperatures up to 36.7C (H158).

Invertebrates (*Gammarus*, chironomids, and others) were the principal foods of pickerel under 100 mm TL, and fish, of those larger (M383, F148). TL 150 mm is listed as the dividing line by W108 and H383. Some golden shiners were eaten by 50 mm pickerel, and sunfish by 100 mm pickerel (R4). *Daphnia* were the principal foods in June in New York (R4). Larger pickerel feed mostly on fish (R4, M383, S361, G47) but will take frogs (R4), crayfish (R4), snakes and almost anything moving of the proper size (W108). H383 reported 37 species of fish preyed upon. Sizes of various species of fish suitable as forage for pickerel are given by L149 which states that pickerel can swallow a forage fish whose depth is equal to its own depth, without distention. Since white perch fingerlings were smaller than yellow perch they were eaten earlier by young pickerel (M383). Pickerel seemed to avoid bullheads smaller than 200 mm SL (R4). Large pickerel are rather lazy feeders showing a preference for crayfish and slow-moving fishes. No active feeding was observed at water temperatures above 20C, although pickerel survived at 30C (S361). Active feeding occurred at 7-15C with most feeding in early morning in the summer, but in afternoon in the winter (S361). Another study indicated most active feeding in midday in June to August (M383). Individuals tended to feed upon one type of organism at a time (R4). A 495 mm pickerel had eaten a 318 mm pickerel (S188).

Standing crops of chain pickerel in Massachusetts ponds at time of reclamation ranged from 1.2 to 11.5 kg/ha, average 5.6, and the chain pickerel constituted 2.1 to 12.7% of the total weight of fish (W108). Annual harvest was estimated at 0.03 to 0.16 kg/ha from 1954-1958 in Quabbin Reservoir, Massachusetts (M296).

		Sizes in gill net meshes						
Mesh size (stretch)	38	51	57	64	70	76	102	
W66 CT Mean TL	226	-	340	371	378	-	-	-
L216 LA No.	-	18	-	-	-	17	2	1
Mean TL	-	404	-	-	-	478	552	472
Range	-	330-521	-	-	-	432-559	546-559	-

NORTHERN PIKE, *Esox lucius* Linnaeus

The northern pike is found in northern Europe, Asia, and North America in cool to moderately warm, generally weedy, lakes, ponds, and rivers. In North America it is found as far south as eastern New York, the northern part of the Ohio Valley, Missouri, eastern Nebraska, and northern Montana, but not west of the great plains region. It has been introduced into some areas outside its normal range.

	Number	Size	FL/SL	TL/SL
C35 MN L. Vermilion	-	under 400 SL	1.089	1.148
	-	401-500 SL	1.083	1.146
	-	over 500 SL	1.076	1.136
C37 MN	1405	200-800 SL	1.078	-
C30 IA	16	-	1.09	-
C37 MN	894	200-499 SL	-	1.141
	571	over 499 SL	-	1.133
V3 WI	41	-	-	1.13
B30 MI	1034	under 507 TL	-	1.162
	457	508-787 TL	-	1.153
	22	over 788 TL	-	1.140
C30 IA	8	under 430 SL	-	1.18
	35	over 430 SL	-	1.15
R153 IA Clear Lake	390	203-455 TL	1.10	1.16
B332 Czechoslovakia	-	-	1.07	1.14
C17 MN L. of Woods	1015	FL = 1.064 + 8 mm; TL = 1.108 SL + 12 mm		
F77 G. Brit.	157	50-1000 FL TL = 1.06 FL		

S438 MN   Length-weight of small pike

TL in mm	No.	Wt. in grams	TL in mm	No.	Wt. in grams
9.1	277	0.009	75.0	59	2.576
12.0	489	0.014	84.6	59	3.650
17.2	148	0.032	94.3	39	5.037
24.0	91	0.095	104.1	34	6.774
34.9	70	0.276	113.8	20	9.423
44.9	90	0.583	122.8	6	12.115
54.3	62	1.035	134.2	10	15.989
64.7	73	1.679	145.2	5	20.940

Summary of length-weight data on pike

In compiling this table it is assumed that standard length equals 0.86 TL, through TL of 508 mm; 0.868 TL, for total lengths of 508-762 mm; and 0.877 TL for larger fish. For the same size ranges, fork length equals 0.937 TL, 0.939 TL, and 0.943 TL. The lengths given in the first three columns are not class centers, but the lower length in the range covered on that line.

SL	FL	TL	Median	Weight Central 50%	Range	No. of fish	References
66	71	76	10	-	5-36	181	B30, 332; S438; F105
109	119	127	25	-	18-50	27	S438; B30, 332; F105
153	168	178	64	45-90	27-254	43	C30; B30, 332; F105, 125; R153; S209; C20
197	213	229	141	100-159	50-340	174+	J15; B30, 256, 332; C20; F109, 124, 125; G42; H1, 62; J15; M20, 499; O30; R136, 153; S62, 208
262	287	305	254	231-272	122-408	379+	R153; B30, 256, 332; C10, 20, 30, 35; F100, 105, 124, 125; G42; H1, 62; J15; M20, 499; O30; R136; S62, 208
328	358	381	463	435-486	249-1021	964+	R153; B30, 256, 331, 332; C9, 17, 20, 30, 35; F100, 105, 124, 125; G42; M20, 113, 499; O30; R23, 136; S62, 83, 208; U8; J15
393	429	457	749	712-794	481-1361	1694+	R153; B30, 256, 331, 332; C9, 17, 20, 30, 35; F100, 105, 124, 125; G42; M20, 113, 499; O30; R23, 136; S62, 83, 208; U8; J15
459	500	533	1134	1098-1157	567-2722	1628+	S62; B30, 256, 331, 332; C9, 17, 20, 30, 35; F100, 105, 124, 125; G42; M20, 113, 499; O30; R23, 136, 153; S62, 83, 208; U8; J15
529	572	610	1642	1556-1660	794-2803	1163+	J15; B30, 256, 331, 332; C9, 20, 30, 35; F100, 105, 124, 125; G42; M20, 499; R58, 136, 153; S62, 83, 208; U8; C17
595	645	686	2254	2105-2400	907-4309	387+	J15; B30, 256, 331, 332; C9, 20, 30, 35; F100, 124, 125; G42; M20; O30; R136, 153; S62, 83, 208; U8; C17
661	716	762	3135	2971-3266	1361-5216	216+	J15; B30, 331, 332, 256; C17, 20, 30; F124, 125; G42; M20, 499; O30; R136, 153; S62, 83, 208; U8; J15
735	789	838	4309	3992-4536	1588-8391	159+	J15; B30, 256, 331, 332; C9, 17, 20, 30; F124, 125; G42; M20, 499; R58, 136; S62, 83, 208; U8; W13; J15
802	862	914	5126	4763-5444	3629-8164	58	J15; B30, 331, 332; C20; G42; M20; O30; R136, 152; S62, 83; V3; J15
868	934	991	6958	6441-7485	4536-9072	39+	J15; B30, 332; C17, 20; E105; H1, 47; N37; O30; R58; S62, 208, 398; J15
936	1005	1066	7298	6985-8436	5053-19390	24+	J15, C20, N52, S62, B273
1003	1076	1144	9370	9130-13600	6849-16780	17+	J15, C20, E102, M20, S62, B273
1069	1150	1220	11500	9070-16012	6804-20750	13+	J15, C20, R58, E102, F101, R139, B273
1136	1222	1296	19050	-	11340-24040	4	J15, G36, B273

The first and last reference listed for each size group are the references with the lightest and the heaviest weights recorded.

C17	MN L. of Woods	$\log W = -6.2076 + 3.0657 \log SL$
C35	MN L. Vermilion	$\log W = -5.622 + 3.223 \log SL$
M431	QU	$\log W = -4.951 + 2.926 \log FL$
J99	England L. Windermere males	$\log W = -5.615 + 3.176 \log FL$
	females	$\log W = -5.866 + 3.271 \log FL$
B331	OH L. Erie males	$\log W = -4.826 + 2.902 \log TL$
	females	$\log W = -4.579 + 2.779 \log TL$
B30	MI	$\log W = -5.236 + 3.066 \log SL$
M281	Scotland L. Choin	$\log W = -5.406 + 3.077 \log TL$
R153	IA Clear L. 390 pike, 203-955 mm	$\log W = -5.552 + 3.122 \log TL$
	Ventura Marsh, young of year	$\log W = -4.533 + 2.744 \log TL$
F114	MN L. George prolarvae 1955	$\log W = -3.699 + 1.744 \log TL$
	1956	$\log W = -3.924 + 1.965 \log TL$
	1957	$\log W = -3.524 + 1.542 \log TL$
	Alevin-juvenile stage 1955	$\log W = -4.651 + 2.693 \log TL$
	1956	$\log W = -5.119 + 2.940 \log TL$
	1957	$\log W = -5.179 + 2.996 \log TL$

Condition factors

	Number	SL	Mean K(SL)	Range
B30 MI	-	-	-	0.78-0.94
V3 WI	480	-	0.64	-
C34 MN Leech L.	128	-	0.85	-
V20 MN Red L.	1	468	0.86	-
C23 MN L. of Woods 1941-43	1111	-	0.89	-
C17 MN L. of Woods 1939-40	901	-	0.90	-
C35 MN L. Vermilion	65	-	0.90	-
E3 MN	1869	178-1066	0.94	-
C9 MI females	30	338-775	0.95	0.81-1.28
C36 IA	30	173-781	0.97	0.55-1.19
O30 Czechoslovakia	-	200-1000	-	0.72-1.65

	Number	SL	Mean K(FL)	Range
H1 Germany	35	-	0.83	-

	Number	TL	Mean K(TL)	Range
W130 MI	-	305-450	0.47	-
	-	495-538	0.56	-
B30 MI	-	-	-	0.50-0.63
M499 IL	38	178-610	0.60	0.50-0.78
C23 MN	-	-	0.61	-
S228, 393 SD Ft. Randall	22	-	0.63	-
F100,105,124,125 SD Oahe	1251	203-889	0.65	-
R153 IA Clear L.	440	-	0.63	-
R58 OH	-	-	0.69	-

The largest pike reported, except for the fabled "Emperor's pike" and some other unfounded stories, weighed 25.9 kg in Scandinavia in 1892 (B273) and the largest in North America weighed 22.2 kg, St. John's River, Quebec, in 1890 (K98, 93). In most populations there is little evidence of increase in condition with increase in length, or with deviation from the "cube law." Condition showed an increase with length in O30 and J99. The pronounced decrease in condition with increase in length in Ventura Marsh (R153) may have been caused by crowding toward the end of the growing season. Females were significantly heavier than males at equivalent lengths in Ohio spawning runs (B331) and the length-weight relationships were different for males and females in Lake Windermere (J99), but no sex differences in condition were noted in Illinois (M499) or South Dakota (F124). Jaw-tagged pike had lower condition factors (W130).

	No.	TL	Range	No.	Wt.	Range
Age 2-7 days after hatch				-	-	-
S438 MN L. George	238	9	8-10	-	-	-
M149 NB	+	10	-	-	-	-
E102 NY	+	-	9-14	-	-	-
Age 1-2.9 weeks						
S438 MN L. George	658	-	9-19	-	-	-
M149 NB	+	30	-38	-	-	-
E102 NY	+	-	15-23	+	-	.03-.09
A84 OH	+	46	-	-	-	-

	No.	TL	Range	No.	Wt.	Range
Age 3-4.9 weeks						
S438 MN L. George	252	-	15-23	-	-	-
M149 NB	+	43	-84	-	-	-
E102 NY	+	-	17-42	+	-	.26-.57
A84 OH	+	127	-	-	-	-
Age 5-6.9 weeks						
S438 MN L. George	143	-	43-85	-	-	-
M149 NB	+	61	38-86	-	-	-
H225 France	+	-	69-180	-	-	-
E102 NY	+	-	23-58	+	-	-1.0
Age 54 days						
A84 OH	+	148	-	-	-	-
Age 72 days						
A84 OH	+	216	-	-	-	-
Age 0-May						
S438 MN	910	-	9-58	-	-	-
C7, W130 MI	521+	23	8-76	-	-	-
S122 Germany	+	-	23-36	-	-	-
A58 Poland	19	36	28-46	-	-	-
E102 NY	+	-	9-38	-	-	-
C185 OH	7	38	-	-	-	-
M149 NB	+	-	30-84	-	-	-
Age 0-June						
M149 NB	+	58	-84	-	-	-
E102 NY	+	-	25-58	-	-	-
T53 WI	+	-	48-79	-	-	-
S83 SA	2	66	-	-	-	-
S122 Germany	+	61	-	-	-	-
S438 MN L. George	239	-	43-92	-	-	-
L132 MN Linwood L.	2	84	61-109	-	-	-
A58 Poland	7	84	66-104	-	-	-
C7, W130 MI	679	79	28-216	+	-	-14
H225 France	410	112	104-160	-	-	-
H43 MN	+	-	76-178	-	-	-
Age 0-July						
A1 NY Oneida R.	5	94	43-114	-	-	-
A58 Poland	7	89	81-102	-	-	-
S83 SA	24	107	76-165	-	-	-
S438 MN L. George	91	-	85-124	-	-	-
C10 MI Ortonville	21	107	61-221	21	14	0.9-82
L132 MN Linwood L.	5	122	97-137	-	-	-
T53 WI	+	-	117-132	-	-	-
C7, W130 MI	717	157	48-229	-	-	-
C185 OH	66	196	-381	-	-	-
C102 IA Clear L.	15	208	140-239	-	-	-
Age 0-Aug.						
A58 Poland	6	117	102-122	-	-	-
C10 MI Ortonville	12	170	79-264	12	45	2-104
H137 MN Demming L.	39	168	-	39	32	-
S83 SA	3	183	-	-	-	-
C7, W130 MI	7	196	180-226	-	-	-
L132 MN Linwood L.	3	244	218-254	-	-	-
C102 IA Clear L.	20	251	203-315	-	-	-

	No.	TL	Range	No.	Wt.	Range
Age 0-Sept.						
A58 Poland	19	147	104-183	-	-	-
K124 WI ponds	+	168	147-188	-	-	-
L147 Denmark peat pit	+	-	112-264	-	-	-
C35 MN L. Vermilion	1	203	-	1	59	-
W130 MI	78	277	-	-	-	-
L132 MN Linwood L.	2	279	-	-	-	-
F105 SD Oahe	1	381	-	-	-	-
Age 0-Oct.						
H252 Belgium	237	97	71-180	-	-	-
M281 Scotland	+	-	89-160	-	-	-
E102 NY	+	180	-	+	27	-
C10 MI Ortonville	362	208	84-447	362	77	2-459
L132 MN Linwood L.	4	284	-	-	-	-
C185 OH	150	305	-	-	-	-
W130 MI	30+	312	259-421	-	-	-
H398 IA Coralville L.	+	-	330-432	-	-	-
B256 PA introduced	1	447	-	-	-	-
Age 0-Dec.						
C10 MI Belle Isle	+	246	208-305	+	64	-
H250 WI	+	343	229-457	-	-	-
W210 MN	+	376	305-432	+	363	-
Age 0						
L71 MI Seney L.	215	99	-	-	-	-
B332 Czech.	10	99	91-157	-	-	-
H31 Sweden	+	-	91-188	-	-	-
E9 NY	+	-	135-163	-	-	-
H30 England	+	168	-	-	-	-
C17 MN L. of Woods	5	173	147-206	-	-	-
S122 Germany	+	-	117-348	-	-	-
J15, V3 WI	18	272	203-285	-	-	-
C11 MI	+	-	203-305	-	-	-
C14 BC	+	-	203-305	-	-	-
H304 IA Winnebago R.	+	279	-	-	-	-
R153 IA Clear L.	62	290	203-404	-	-	-
M150 NB	+	-	287-335	-	-	-
Age I						
R15, S83 SA 2 lakes	7	137	109-203	-	-	-
C10 MI "small fish" May	+	180	160-198	+	32	-
H30, H31 England	6+	203	107-300	-	-	-
M281 Scotland	+	-	160-251	-	-	-
E102, E9 NY	+	216	203-297	+	91	-
H15 ON	+	-	188-269	+	113	-
H31, R124, K4 Sweden	2+	236	183-391	2	50	-
S208 SA Mud C.	3	239	188-269	3	91	45-136
H30 England L. Windermere	+	267	-	-	-	-
C17 MN L. of Woods	1	267	-	-	-	-
L20 U.S. hatchery	+	-	254-305	-	-	-
B23, G25 NY	2+	287	249-323	1	191	-
H137 MN Demming L.	80	297	-	80	163	-
L71 MI Seney L.	74	302	-	-	-	-

	No.	TL	Range	No.	Wt.	Range
B332 Czech.	18	307	292-356	-	-	-
C10 MI "small fish" Nov.	+	307	254-338	+	145	-
L147 Denmark peat pit	+	-	264-363	-	-	-
C10 MI Belle Isle	+	330	272-386	+	204	-
J15 WI	8	333	272-480	-	-	-
C10 MI "large fish" May	+	340	297-386	+	227	-
C11, W130, C215 MI	267+	348	244-478	-	-	-
H1 Germany	2	353	320-366	2	209	-
C157 IA Little Wall L.	7	373	-	-	-	-
C35, E3 MN	32	386	267-531	6+	308	132-566
C10 MI "large fish" Nov.	+	394	348-536	+	322	-
R153 IA Clear L.	140	404	323-546	-	-	-
S464 MN Grove L., Maple L.	+	-	401-447	-	-	-
N37 WI L. Geneva	13	417	-	-	-	-
D89, S267 WI Murphy Flowage	+	-	368-508	-	-	-
M150 NB	14	-	432-490	-	-	-
S226 SD Ft. Randall	+	467	-	-	-	-
L129 IL	+	483	-	-	-	-
H177, H304, I10 IA rivers	4+	498	406-599	1	1588	-
V3 WI	51	505	-	-	-	-
W210 MN Grace L.	+	-	381-686	-	-	-
U11 SD Oahe L.	454	627	490-627	454	1724	-
Age II						
R15, R138, S83 SA 3 lakes	7	239	152-315	-	-	-
H30, H31 England	13+	267	170-429	-	-	-
H15 ON	+	353	-	+	200	-
S208, R139 SA 2 lakes	1+	371	-	1+	363	-
H31, K4 Sweden	+	-	287-460	-	-	-
L20 U.S. hatcheries	+	-	356-406	-	-	-
H1 Germany	5	386	373-406	5	299	249-349
C10 MI	+	388	259-653	+	286	-
G25, E9 NY	6+	424	351-475	6	431	191-590
E3, C17, C35, H137 MN	361	427	295-645	37+	495	286-989
L71, C215, W130 MI	304	462	376-561	-	-	-
R124 Sweden	2	483	427-564	2	608	249-967
B332 Czech.	28	495	437-511	-	-	-
R153, M236 IA Clear L.	156	505	394-653	28	953	499-1451
V3, M20, J15, D89 WI	148	531	421-653	17	477	-
C11, W130 MI	+	-	508-660	-	-	-
N37 WI L. Geneva	32	592	-	-	-	-
S226 SD Ft. Randall	1	594	-	-	-	-
L129 IL	+	597	-	-	-	-
W210 MN Grace L.	+	630	483-711	+	1724	-
U11 SD Oahe L.	155	688	638-764	155	2223	1769-3447
Age III						
R138, R203 SA	2	236	213-259	-	-	-
H30, H31 England	44+	318	236-467	-	-	-
H62 IN	1	353	-	1	227	-
M113 AT 3 lakes	6+	366	338-429	6	277	-
R15 SA Waskesiu R.	6	373	353-417	-	-	-

	No.	TL	Range	No.	Wt.	Range
M20 WI south	72	396	-	72	390	-
S83 SA males	541	399	-	-	-	-
S83 SA females	516	401	-	-	-	-
H1 Germany	22	447	394-561	19	408	304-499
W130 MI lakes	70	478	409-574	-	-	-
R124 Sweden	4	480	475-523	4	571	245-731
R208, R230 SA	1+	488	472-503	1+	785	771-794
H137 MN Demming L.	14	488	-	14	767	-
G25 NY	9	508	450-536	7	716	539-871
E3, C35, C17 MN	1285	511	325-703	65+	735	426-1161
B332 Czech.	59	531	500-594	-	-	-
H15 ON	+	541	-	-	-	-
M150 NB	+	-	521-559	-	-	-
R153 IA Clear L.	66	572	452-708	-	-	-
J15, V3, D89 WI	182+	574	356-802	-	-	-
L20 U.S.	+	-	559-610	-	-	-
K4 Sweden	+	-	587-638	-	-	-
C215 MI	45	615	-	-	-	-
N37 WI L. Geneva	2	640	-	-	-	-
L129 IL	+	640	-	-	-	-
U11 SD Oahe L.	20	912	714-940	20	5580	2631-6124
Age IV						
R138 SA 2 lakes	7	-	229-300	-	-	-
H30, H31 England	21+	371	277-467	-	-	-
R203 SA Otter R.	2	394	386-399	-	-	-
R15 SA Waskesiu L.	6	439	432-503	-	-	-
M20 WI south	104	445	-	104	590	-
M113 AT 4 lakes	+	467	447-518	+	658	567-853
S83 SA males	142	488	-	-	-	-
S83 SA females	106	508	-	-	-	-
H1 Germany	20	513	447-574	20	721	449-953
H31 Sweden	+	-	478-566	-	-	-
H62 IN	6	549	-	4	1016	-
G25 NY	7	556	439-815	7	1206	459-4133
H15 ON	+	566	-	+	680	-
R208, R139, R230 SA	4+	569	488-577	4+	1025	726-1134
C17, C35, E3 MN	1046	574	414-785	891	1025	626-1700
B332 Czech.	55	577	546-665	-	-	-
O2 Finland	0	-	-	+	1814	-
L71, C215 MI	18	640	630-671	-	-	-
J15, D89, V3 WI	158+	643	432-773	-	-	-
R153 IA Clear L.	18	660	630-671	-	-	-
L129 IL	+	706	-	-	-	-
U11 SD Oahe L.	23	967	931-1028	23	6305	5579-7030
Age V						
R138 SA 2 lakes	9	-	259-378	-	-	-
H30, H31 England	5+	432	363-490	-	-	-
R15, R203 SA	11	467	411-574	-	-	-
M113 AT 3 lakes	+	495	472-541	+	862	707-1021
M20 WI south	102	516	-	102	694	-
S83 SA males	135	579	-	-	-	-
S83 SA females	41	607	-	-	-	-

	No.	TL	Range	No.	Wt.	Range
H31 Sweden	+	-	531-650	-	-	-
S208 SA	13	594	549-655	13	1225	998-1406
H15 ON	+	622	-	+	1134	-
H1 Germany	10	630	561-798	10	1393	953-2400
C215, L71, W130 MI	10+	638	559-785	-	-	-
C17, C35, E3 MN	546	645	445-785	457	1560	853-2100
G25 NY Ontario R.	2	650	587-716	2	1692	1112-2276
J15, V3, D89 WI	111+	671	399-950	-	-	-
R230 SA rivers	+	688	-	+	2327	-
R153 IA Clear L.	5	691	544-820	-	-	-
H62 IN	6	701	-	4	1580	-
L129 IL	+	721	-	-	-	-
B332 Czech.	8	747	602-934	-	-	-
W13 CT	1	864	-	1	4082	-
C11 MI	1	991	-	-	-	-
U11 SD Oahe L.	25	1052	1024-1095	25	8391	7800-9070
Age VI						
R138 SA 2 lakes	6	-	325-378	-	-	-
H30, 31 England	1+	493	429-554	-	-	-
R203, R15 SA	15	511	439-574	11	707	-
M113 AT 4 lakes	+	566	505-643	+	1275	1048-1505
M20 WI south	83	579	-	83	1003	-
S83 SA males	84	610	-	-	-	-
S208 SA	31	640	566-703	31	1678	1179-2585
S83 SA females	29	640	-	-	-	-
H15 ON	+	648	-	+	1361	-
B332 Czech.	6	678	554-912	-	2105	1474-2812
C215, L71 MI	3	698	678-739	-	-	-
H1 Germany	10	703	617-795	10	2132	1352-4000
R139 SA L. La Ronge	+	716	-	+	2132	-
C17, C35, E3 MN	176	726	521-896	20+	2105	1474-2812
J15, V3, N37, D89 WI	101+	744	432-1069	-	-	-
H62 IN	2	787	630-942	2	2050	1375-2722
L129 IL	+	789	-	-	-	-
O5 Finland	-	-	-	+	2500	-
V3 IL	1	947	-	1	4309	-
R153 IA Clear L.	1	955	-	-	-	-
L20 U.S.	+	991	-	-	-	-
K4 Sweden	+	1062	-	-	-	-
Age VII						
R138 SA 2 lakes	6	-	358-480	-	-	-
H31 England E. Anglia	3	-	384-620	-	-	-
M113 AT 2 lakes	+	546	513-582	2	1157	948-1361
R15, R203 SA	16	584	523-632	-	-	-
M20 WI south	65	610	-	65	1361	-
S83 SA males	33	622	-	-	-	-
C215, L71 MI	2	638	630-645	-	-	-
R208 SA	8	706	663-795	8	2177	1724-3042
S83 SA females	25	709	-	-	-	-
H15 ON	+	732	-	+	1814	-
J15, V3 WI	89	782	411-1141	-	-	-
C17, C35, E3 MN	92	792	673-813	3+	2449	2014-3200

	No.	TL	Range	No.	Wt.	Range
H1 Germany	7	815	737-1062	7	3833	2300-6000
L129 IL	+	826	-	-	-	-
B332 Czech.	4	913	-	-	-	-
N34 WI L. Geneva	1	1014	-	1	6350	-
Age VIII						
R138 SA Cree L.	5	-	437-493	-	-	-
R15, R203 SA	13	653	605-714	-	-	-
M20 WI south	37	703	-	37	2092	-
S83 SA males	14	686	-	-	-	-
S83 SA females	15	777	-	-	-	-
G25 NY L. Ontario	1	775	-	1	2812	-
H15 ON	+	782	-	+	2268	-
O2 Finland males	-	-	-	+	2400	-
O2 Finland females	-	-	-	+	3606	-
R139, S208 SA	8+	795	663-837	8+	2948	1996-3266
C17, E3 MN	32	853	703-1041	+	3520	-
J15, V3 WI	44	871	526-1247	-	-	-
H1 Germany	7	890	764-1062	3	4540	2750-6010
B332 Czech.	1	980	-	-	-	-
Age IX						
R138 SA 2 lakes	3	528	460-582	-	-	-
M113 AT Skeleton L.	+	569	-	+	1278	-
R15, R203 SA	10	693	560-762	-	-	-
M20 WI south	21	773	-	21	2881	-
S83 SA females	6	795	-	-	-	-
B332 Czech.	1	828	-	-	-	-
S208 SA	4	837	-	4	3765	3538-3992
S83 SA males	1	889	-	-	-	-
H15 ON	+	889	-	-	-	-
O2 Finland	-	-	-	+	4900	-
C17, C35, E3 MN	20	945	757-1130	18	5593	-
J15, V3 WI	25	967	610-1130	-	-	-
H1 Germany	1	1100	-	1	7500	-
Age X						
R138 SA 2 lakes	9	-	503-681	-	-	-
R15, R203 SA	9	764	714-813	-	-	-
M20 WI south	8	786	-	8	3100	-
S83 SA females	4	846	-	-	-	-
R139 SA L. La Ronge	+	929	-	+	5035	-
E3 MN	15	955	813-1072	15	7462	-
H15 ON	+	971	-	+	4082	-
M113 AT Elkwater L.	+	991	-	+	7257	-
J15, V3, D89 WI	18+	1000	767-1125	-	-	-
Age XI						
R138 SA Cree L.	6	-	541-635	-	-	-
R15, R138, R203 SA	13	813	709-909	-	-	-
M20 WI south	2	856	-	2	4028	-
S83 SA females	2	980	-	-	-	-
H15 ON	+	1024	-	+	5443	-
C35, E3 MN	7	1034	927-1190	+	8618	-
B332 Czech.	1	1036	-	-	-	-
J15, V2 WI	9	1085	711-1366	-	-	-
C11 MI	1	1346	-	-	-	-

	No.	TL	Range	No.	Wt.	Range
Age XII						
R138 SA 2 lakes	10	-	622-742	-	-	-
R15, R203 SA	9	864	795-932	5	4536	-
S83, S208, R139 SA	3+	983	958-1024	1+	-	5670-6395
J15, M20 WI	10	1000	615-1092	5	4880	-
H15 ON	+	1130	-	+	6350	-
E3 MN	1	1158	-	1	9070	-
L20 U.S.	+	1344	-	-	-	-
K4 Sweden	+	1350	-	-	-	-
Age XIII						
R138 SA 2 lakes	12	-	648-837	-	-	-
F149 Windermere males	2	724	711-737	2	3856	3175-4536
R15, R203 SA	9	907	858-971	-	-	-
M20 WI south	2	914	-	2	4536	-
S83 SA female	1	1050	-	-	-	-
E3 MN	1	1156	-	-	-	-
H15 ON	+	1186	-	+	7257	-
J15, V2 WI	3	1196	1156-1245	-	-	-
E102 England	1	1240	-	1	15876	-
Age XIV						
R138 SA 2 lakes	10	-	-	-	-	-
F149 Windermere male	1	-	-	1	4763	-
E3 MN	1	-	-	-	-	-
S83, R203, R139 SA	4+	-	-	+	8164	-
F149 Windermere females	2	-	-	1	12700	-
F101 England L. Windermere	1	-	-	1	15876	-
H15 ON	+	-	-	-	-	-
Age XV						
F149 Windermere male	1	-	-	1	6123	-
R15, R138 SA 3 lakes	6	-	-	1	-	-
Age XVI						
F149 Windermere males	2	-	-	2	4763	3400-6123
R138, R139 SA 3 lakes	2+	-	-	+	9980	-
R149 Windermere females	5	-	-	6	10886	9618-14515
Age XVII						
R138 SA Cree L.	7	-	836-942	-	-	-
F149 Windermere females	2	953	940-965	2	9526	8164-10886
Age XVIII						
R138 SA Cree L.	8	-	849-958	-	-	-
Age XIX						
R138 SA Cree L.	8	-	902-1064	-	-	-
Age XX						
R138 SA 2 lakes	7	-	916-1036	-	-	-
Age XXII						
R138 SA Cree L.	2	-	1077-1184	-	-	-
Age XXIII						
R138 SA Cree L.	1	1265	-	-	-	-
Age XXIV						
R138 SA Cree L.	1	1292	-	-	-	-

Average calculated TL at each annulus

	No.	1	2	3	4	5	6	7	8	9	10	11	12	13	14	15	16	17	18	19	20	21	22	23	24
A58 Poland 20 lakes	613	272	401	508	626	721	884																		
smallest mean	-	163	256	349	554	680	858																		
largest mean	-	465	599	714	737	770	907																		
A58, 030 European review, 24 studies	-	192	324	436	536	615	661	737	734	809															
smallest mean	-	94	184	267	323	377	427	576	605	756															
largest mean	-	329	468	605	734	866	804	887	844	872															
B325, 332 Czech.	187	270	395	493	547	680	776	862	863	971	1002														
minimum	-	177	337	319	371	393	517	765	809																
maximum	-	409	597	667	763	906	1004	1049	917																
R138 SA Cree L.	88	81	150	206	257	312	361	406	452	505	551	605	655	698	747	795	846	866	938	985	1036	1080	1125	1175	1245
R138 SA Wollaston L.	47	-	152	206	264	323	381	434	495	551	610	668	726	777	828	878	929	971	1015	1040	1077				
M62 NT Great Slave L.	73	114	170	234	302	361	419	472	531	574	617	660	714	762	820	902	945	978							
M62 NT Athabaska L.	65	107	165	236	310	373	439	508	566	622	673	726	777	824	864	894	932	973	1016	1039	1107	1120	1146	1200	1230
R136 SA Churchill L.	-	132	216	269	320	394	434	498	546	594	643	693	757												
M62 NT Great Bear L.	70	102	163	249	335	414	480	541	599	650	698	747	782	811	858	891	950	989	1011						
M62 NT Little Slave L.	64	117	226	315	391	450	526	572	622	686															
B12 MB	+	213	363	429	503	574	594	671	691	724	747														
C35 MN L. Vermilion	127	193	325	427	505	572	650	737	831	878	938	988													
E2, 3, M258, 562 MN	8198+	180	318	442	531	622	711	777	851	909	922	964	1016												
R58 OH	+	152	330	483	635	762	864	965	1041	1092	1143	1168	1194	1220	1245										
K126 "typical"	+	251	396	533	630	703	777	838	896	951	996	1046	1090	1133	1171										
P159 MT	41	282	411	488	610	686																			
C270 MN Mississippi R.	60	226	399	528																					
B256 L. Erie	130	300	486	607	703	789																			
B256 PA 3 lakes	352	277	488	584	648	719	862																		

Average calculated TL and increments at each annulus

	1	2	3	4	5	6	7	8	9	10	11	12	13	14	15
F77 Eng. L.															
Windermere males	244	424	551	625	671	693	698	721	724	747	762	757	754	767	777
Increment	244	185	127	74	51	30	25	25	20	20	18	18	15	13	10
Number	74	74	73	72	54	33	13	8	4	4	4	3	1	1	1
females	241	421	587	683	739	802	844	894	929	958	983	1022	1077	1085	
Increment	241	180	165	102	69	58	46	38	36	28	25	18	15	8	
Number	116	116	115	106	62	32	20	17	11	11	8	6	1	1	
C186 OH															
East Harbor males	305	437	503	554	589	612	643	698							
Increment	305	152	79	43	25	20	15	18							
Number	289	237	183	109	77	45	22	4							
females	287	457	554	620	678	696	757	773	793	833					
Increment	287	163	114	61	41	28	23	20	15	18					
Number	399	394	330	215	151	109	64	31	12	4					
V3 IL	251	445	533	599	902	939									
Increment	251	193	102	84	46	36									
Number	73	53	21	4	1	1									
C17 MN L. of Woods	206	318	414	498	564	643	693	764	837						
Increment	206	112	102	91	86	89	86	86	84						
Number	350	349	323	229	98	39	13	7	1						
V3 WI	257	465	584	683	764	833	904	967	1016	1107	1133	1168	1176		
Increment	257	211	124	84	61	43	38	33	25	30	30	10	10		
Number	515	464	341	232	145	97	56	31	19	6	3	1	1		
M281 Scotland															
L. Choin males	71	183	292	363	417	465	498	533							
Increment	71	109	109	81	58	46	38	30							
Number	172	114	59	41	26	21	19	9							
females	81	193	312	414	475	554	574	635	701	869	900	929			
Increment	81	109	119	91	71	61	43	41	30	41	30	30			
Number	132	104	42	36	27	21	15	7	3	1	1	1			
R153 IA Clear L.	307	421	518	617	693	922									
Increment	307	122	107	102	97	69									
Number	190	142	90	24	6	1									
R153 Ventura Marsh	307	467													
Increment	307	157													
Number	168	76													
S226, S228, S393 SD															
Ft. Randall	300	470	632	701	795										
Increment	300	170	163	97	56										
Number	24	22	17	5	2										
F100,105,124,125															
SD Oahe	323	475	605	693	782	844									
Increment	323	155	137	81	104	61									
Number	557	441	152	23	1	1									

Median annual increments as percentages of TL at beginning of year

TL	M62 NT	V3 WI	V3 IL	Range	TL	M62 NT	V3 WI	V3 IL	Range
76–101	57	–	–	38–67	457–507	12	19	–	4–31
102–126	50	–	–	29–75	508–558	11	16	15	8–32
127–151	52	–	–	28–67	559–609	9	15	–	6–28
152–177	43	144	–	23–150	610–659	8	13	17	6–20
178–202	34	104	–	23–164	660–710	8	10	–	4–23
203–228	34	93	–	26–132	711–761	6	10	14	4–15
229–253	30	87	78	11–112	762–812	5	5	–	3–11
254–278	25	80	72	22–93	813–863	5	7	9	4–11
279–304	23	84	–	10–115	864–913	5	4	4	3–10
305–329	24	75	102	13–102	914–964	3	4	–	2–6
330–355	21	62	–	14–62	965–1015	4	2	–	1–6
356–380	18	30	–	14–31	1016–1117	4	3	–	2–6
381–405	18	35	–	11–72	1118–1218	3	2	–	1–5
406–456	16	33	24	9–47					

Mean calculated weight at each annulus

	No.	1	2	3	4	5	6	7	8	9	10	11	12	13	14	15	16	17	18	19	20	21	22	23	24
M62 NT Great Slave L.	73	28	57	85	170	255	397	567	794	1049	1304	1503	2041	2466	3175	4253	4734	5245							
R138 SA Cree R.	88	-	28	57	85	170	284	454	595	765	1049	1361	1871	2268	2807	3685	4536	5386	5783	7087	8000	8650	9440	10260	11113
R138 SA Wollaston L.	47	-	28	85	170	255	425	624	851	1134	1446	1956	2552	3260	4026	4820	5360	6040	6975	8165	9525				
M62 NT Athabaska L.	65	28	57	85	198	284	510	624	1077	1474	1843	2381	2892	3460	3970	4394	5100	5810	7090	7800	9925	10490	10915	13040	14175
M62 NT Great Bear L.	70	57	142	198	284	454	652	907	1276	1700	2070	2552	3120	3345	4110	4540	5528	6380							
K126 "typical"	-	85	369	595	936	1361	1900	2440	3090	3910	4700	5670	6720	7880	9157										
B325, 332 Czech.	187	142	425	822	1220	2240	3490	4590	4620	5780	6720	7170													
S494 Poland	-	170	400	940	1300	2210	3910	7655																	
F149 Eng. L. Winder-mere males	75	85	454	1135	1814	2268	2722	3060	3400	3855	4080	4310	4536												
females	117	85	567	1700	2722	3742	4763	5785	6575	7370	7825	8730	9300												
A58 European review, 4 waters	-	113	369	794	1247	2353	3205																		
A58 Poland	-	170	454	992	1814																				
V15 "typical"	-	113	567	1304	1700	2610	3970	4650	5330	5900	7710	8390	9070	9750											
C146 Spain	-	710	2500	4540	6800																				

C12 MI Growth of tagged pike

	out	1 year, an increase of	5-157 mm TL
		2	18-175
		3	61-254
		8	366

F114 MN L. George first 20 days young pike grew 0.5-0.7 mm per day
                     next 30 days   1.9-2.3 mm per day
                     50th-90th days   0.9-1.2 mm per day
F100 SD Oahe young pike grew 25 mm per 16 days
F105 SD Oahe   25 mm per 8 days

Carbine (C7) found so much variation between ages as determined from scales and between known ages of pike that he did not consider the method valid. Most of the papers listed above used scales for age and growth determinations. More detailed study of scale characteristics were given by W130 and F77. The first annulus was reported (W130) as usually unique with a chainlike pattern of irregular or incomplete circuli. Occasionally the first annulus was represented only by a change from slow to more rapid growth. The "unique" pattern of the first annulus was not found in Lake Windermere pike (F77), but the first annulus was missing on some fish (F106). The second annulus was usually distinguished by several close discontinuous circuli (W130) and each later annulus appeared as a white line, with cutting-over of circuli on the lateral regions and also usually with closely spaced irregular circuli. Where 2 or more marks coalesced on the scales they were considered a single annulus (M281). The annuli can often be more clearly recognized as white lines on the posteriolateral angle of the scale than on the anterior field (W130, F77). Difficulty in defining the exact location of annuli was reported by F77. On the posterior field these difficulties were such that calculations of growth were not attempted, although age determinations were believed to be more accurate than scale readings on the anterior field alone. On the anterior field the body-scale relationship was (F77)

$$\log L = 0.836 + 0.749 \log S$$

and this correction for allometry permitted fairly satisfactory growth computations. Most scale studies on pike have assumed isometry and computed lengths from scale measurements by direct proportion. Some, however, used a correction value for late scale formation and then used direct proportion (R153). A parabolic body-scale relationship was used by C17:

$$SL = 25 mm + 6.88S - 0.0448S^2 + 0.0004S^3$$

Scale measurements have been made of the anterior field except for O30 and M62 which used the lateral field. M62 used a log L-log S relationship with a zero S intercept at FL 50 mm. The first scales form along the lateral line when the young pike are 32-35 mm and complete scalation may occur at 66 mm (F89).

A false annulus frequently appears within the first year as a result of the change from insects to fish as food (W130, B332, F77). In other years false annuli may form in midsummer, but these can usually be detected by the greater number of anteriolateral than posteriolateral circuli (W130, F77). Late summer growth often results in incomplete circuli and if the number of anterior circuli is over twice that of the posterior circuli between a midsummer annulus and the next annulus, it can usually be assumed that the inner annulus was a false one (M30, F77). Brofeldt (B288) was reported by W130 to be able to age pike only to age VI because of the profusion of false checks. R15 reported a check on scales of female but not male pike at the beginning of the 4th year, which he attributed to first spawning. Jaw-tagged pike which were growing slowly failed to form annuli (W130). The best scales for reading were found to be those just above or below the lateral line on the central part of the trunk (W130).

Opercular bones were found to be more satisfactory for determining age and growth of pike than were scales, if it were recognized that the first one or two annuli may become obscured and if the size at these annuli is then determined by applying a Walford line (F77, F106).

Annuli form on scales from mid-February to mid-May in Lake Windermere, England (F77); early March to June 1st in southern Michigan or late March to late June in northern Michigan (W130); in May to June in Finland (G126) and in Czechoslovakia (B332); about May 1st in South Dakota (F105).

Age I pike may form annuli earlier than older pike which do not form annuli until after spawning (W130). Some age I pike had annuli in March in Ohio but no older fish (C309). Scales first form when pike are about 36 mm TL (F77), at 38 mm TL (B332), or at 33 mm, with 6 to 10 rows at 41-43 mm and complete scalation at 66 mm (F89).

In Michigan the following percentages of scales were regenerative (W130):
    791 scales from 51 pike under 610 mm   6.6% regenerative
    579 scales from 22 male pike over 635 mm   16.2% regenerative
    482 scales from 28 female pike over 760 mm   12.8% regenerative
Lee's phenomenon was evident in gill net-caught pike, presumably because of size selectivity in the sampling (F77).

Annual growth rate of pike is generally slower in the northern part of the range than in the south but the life span is longer in the north (M62, V3). The fastest growth reported was from Spain (C146). Variation in growth is quite striking in many populations. Pike known to be the same age in a pond in Michigan varied from 84-447 mm in October of the first year (C10). In Poland, growth is reported to be better in larger lakes (about 1100 hectares or larger) than in smaller ones, and in lakes with more varied environments (A58). Growth in a Nebraska Sandhills lake was believed to be slowed due to the high alkalinity of up to 1000 ppm (M388). Slower growth of age I pike in one year was associated with highly turbid waters, perhaps interfering with sight feeding (C186). Growth rates were more rapid in lakes with higher total dissolved solids and carbonates (E3, E94) and in lakes where growth of other species was rapid.

Rapid growth in Grace Lake, Minnesota, was associated with an abundance of small perch (W210). Intensive gillnetting reduced the numbers of large and old pike in Lake Windermere, so that few pike were over 7 years old whereas older pike constituted about one third of the adult population prior to the netting (F149). Mass mortality in a Czechoslovakian reservoir was followed by increased growth rate (B332). Growth was more rapid in streams than lakes and seemed to be related to the abundance of other fish (D193).

Pike were found to show growth compensation in some populations but not in others (E3).

Male pike usually grow more slowly than the females (M236, S83, O2, M149, O30) particularly after the 2nd year (F77, F149, C186). However, in the Northwest Territory lakes, where growth was slow, differences in growth rates of the sexes were not evident (M62) though females lived longer than males: oldest male XVIII, oldest female XXIII. In a slow-growing Czechoslovakian population there was also little sex difference in growth except that females lived longer (B332).

In Michigan, over 50% of the annual growth was completed by July 1 (W130); in Finland most growth took place from June through August (C126).

In Lake Windermere, growth began in March, was most rapid in June, and declined through late summer and fall and ceased at midwinter (J98). Nonmature fish had a longer growing season than mature fish (D193).

Pike were eliminated from Devil's Lake, North Dakota, when salinity increased from 0.8 to 1.5%, but are present in Quill Lake, Saskatchewan, with 1.6% salinity. In Nebraska they are found up to 1000 ppm total alkalinity and to a pH of 9.5, although they are somewhat bleached in color in waters of high alkalinity (M388). No spawning occurred when alkalinity increased 40% in 10 months in Big Alkali Lake, Nebraska (M388). Fry are apparently more sensitive to alkalinity than adult pike. A small number of "pugheaded" (shortened head) pike were found in Manitoba and have been reported from Europe (L211).

A review of European literature (A51) indicates that male pike usually mature at age II or III and females at age III but gonads were well developed at the end of the first year in one pond in Germany and age I pike, 320-330 mm long, were mature in Austria. In Spain, both males and females were mature at age I where growth was rapid (C146). In South Dakota, male pike (M150) mature at age I and females at age II (S393, F124); in Ohio, age II females produced eggs (C310); in Wisconsin, both males and females sometimes mature at age II (W181); in Nebraska, females at age IV (M150); in Great Bear Lake, Northwest Territories, males mature at age VI and females at age VI (M62). The smallest mature male reported from Cayuga Lake, New York, was 406 mm and weighed 311 g (E102), though spawning males have been reported at 135 mm in Sweden and 349 mm in Pennsylvania (B289), and 305 mm in Wisconsin (T125). The smallest mature female in Pennsylvania was reported as 325 mm (B289).

	Size of females	No. of females	No. of eggs per female Mean	Range
N25 NY	-	-	100,000	-
L13 OH	597 mm	1	48950	-
V15 -	-	-	35000	-100,000
W40 MI	508 mm	-	-	8000-12000
W40 MI	1016 mm	1	88000	-
H44 MI	-	-	32200	-
V21 MI	330-380 mm	8	7500	2000-11000
	381-430	10	11000	2000-17000
	431-480	6	22000	9000-35000
	481-530	14	27000	17000-48000
	531-580	4	31000	23000-35000
	581-635	4	50000	15000-62000
K4	-	-	80000	-100,000
	2720 g	1	135,000	-
	12700 g	1	292,320	-
	14500 g	1	595,200	-
E4 MN	1814 g	26	60000	-
H3 Europe	-	-	-	80388-272,160
C9 MI	338-775 mm SL	30	32200	7691-97273

	Size of females	No. of females	No. of eggs per female Mean	Range
P24 G. Brit.	2950 g	-	136,000	-
	4540 g	-	149,000	-
	10430 g	-	224,000	-
	14500 g	595,000		-
E102 NY Cayuga L.	495 mm	1	22000	-
M225, 229, 236 IA	860-1450 g stripped	-	11250	-22745
V21	19800 eggs per kg of female			
F116 G. Brit.	22000-44000 eggs per kg of female			
F114 MN	6 females No. of eggs = 179 TL - 66245			

Spawning begins at 7C (B43), 8C (A1), 9C (C311), 9-14C (M410), 13C (R15), and 5-11C (E102).

Adults begin moving to streams or flooded marshes at temperatures of 1-4.5C (F114) or under the ice before breakup (C310). Most of the movement is at night and there is no evidence of homing (F114).

In the British Isles, pike spawn from March to April (F116); in Poland, spawning near shore occurred March 26-May 9 but on shoals in mid-lake where temperatures rose more slowly, April 16-May 31 (B285); in New York, March 19-23 (E102) or March and April (A1); in Ohio, March 28 to April 8 (C310); in Houghton Lake, Michigan, April 2-25 (C7); in Minnesota, April 10-early May (B43); in Lake Waskesiu, May 1-15 (R15); in Wisconsin, peak in first week of April (W181); in Alaska, July (B289).

No nest is made but eggs are strewn in shallow areas over vegetation (S439, F114, C311, C7). Artificial propagation methods are reviewed by H252, H225, B43, H298, S497, B289. Temperatures of 9-11C are reported as best for incubation (H298), and eggs hatch in 12-14 days (T125). Eggs hatched in 23-29 days at 6C; 16-19 at 8C; 8-12 at 12C; and 4-5 days at 18-20C (S496). Embryology and early development are described by F114, S438. Eggs are 2.5-3 mm in diameter and the fry at hatching are 6.5-8 mm (F114, F94) and average 7.3 mm (E102). In nature 52-99% of the eggs collected are fertile and in hatcheries 75-85% (T125). Hatching success of natural spawning were reported as 64, 67, and 90% in different years in Lake George, Minnesota, 25-76% in Sweden, and 94-97% in Poland (F114).

Pike at hatching averaged 9.3-9.5 mm and started feeding at 13.3-15.1 mm with the pike at 11.5C feeding at a smaller size than those raised at 7C (B333). Early feeding behavior was described by B333. The fry may start feeding at 10.3 mm, but most not until 12 mm and at 35-40 mm (about 25-30 days) they change from alevin to juvenile, assuming the general form of adults (F114). The tail and jaw grow more rapidly than the rest of the body from 2.5-6.5 mm (F94).

From egg to fingerling stage, mortality was listed as 99.56-99.93% (C9). The most critical stages seem to be from fertilization to the closing of the blastopore and from hatching to the end of the alevin stage (F114). Toxic iron compounds and rapid temperature change were believed to be most critical, with little evidence that oxygen or carbon dioxide concentrations were involved (F114). The brood stock and year class strength did not seem to be related but the abundance of a year class was established prior to leaving the nursery slough (F114, S438). High water levels in the spring maintained at least a month after spawning are favorable for good year classes (S438).

Annual mortality rate for ages III-VIII in Ball Club Lake, Minnesota, was estimated at 0.6 (J57); at 0.769 and 0.708 for 2 central Minnesota lakes (G110) and for ages I-VI in a southern Minnesota lake, at 0.497 (S213). Angler harvest of 41% of pike stocked as fingerlings in a Nebraska lake was reported by M150. In Oahe Reservoir, South Dakota, 14.7% of 292 pike with opercular tags were reported caught by anglers within a year (F100). Jaw-tagged pike grew only 45% as much as untagged pike (W130). Number 1, 3, and 8 monel metal strap tags on opercles showed evidence of shedding in a short time but 18.9% were recaptured; Number 3 and 18 monel strap tags on the lower jaw gave a 26.6% return but many of the fish were in poor condition; plastic streamer tags through the back gave a 24.3% return but also indicated shedding (W130).

In a Denmark peat pit, the pike population was estimated as 1 per 90 sq. M. compared to 1 per 110 and 526 square meters in Swedish lakes and 1 per 2700 and 3680 square meters in Minnesota lakes (L147). Intensive removal of 1.3-5.3 kg of pike per hectare from Heming Lake, Ontario, each year from 1945 to 1960 did not cause a drop in the catch of pike, but the pike were mostly young at the end of the period (L146). The number of pike in Grove Lake, Minnesota, was estimated at 7.6 over 250 mm TL per hectare; and in Maple Lake at 4.8 over 355 mm TL (S464); and in Grace Lake, 0.6 per hectare (W120).

The first food of young pike are microcrustaceans (M313, F114, H298, H343). Fish larvae may be eaten before reaching 13 mm, and, by 34 mm, the food may be almost exclusively fish (M313). Insects may be important as an intermediate stage, 20-40 mm (F114). In Nebraska fingerling pike feed on anostracan and notostracan shrimp (M388). Beyond 50-65 mm, pike are almost exclusively predatory with fish as the primary food (H298, D183, A83, F114, G151, B334, K125, L127, F101, M313, N117, M465, J63, R153, R138, D176, S495, H343) but they may also feed on salamanders (M388), crayfish (P162) and mayflies (G150, M465). The relationship of pike as predators of waterfowl has been discussed (S83, R241, L71). Little cannibalism or little feeding on tadpoles was reported in some rearing areas (F114) but cannibalism may be extreme.

In general, the size of the prey increases with the size of the pike (S495, F150). Most food items were not over 2/3 the width of the pike mouth but some pike ate food 1.33 the width of the mouth (measured while closed) (J63). Size of the prey rather than species seems to be important (C298). Size of prey also seems to be a function of abundance, with pike often taking smaller prey if it is more abundant (L127). Strongest pressure was on immature fish in Polish lakes (A83), but on age II trout in a trout stream (H415). The smaller trout were in shallower water, and less available to the pike. Rainbow trout were more susceptible to pike predation than were brown trout (H415). Trout and char were eaten in Lake Windermere in the winter and predation on them was higher than that on perch (F150). Feeding is almost exclusively in the daytime (M313, J63). Of 1000 pike caught by angling 55% were empty and 92% were less than one-third full (J63). There is some evidence that they do not feed during the spawning season (F116). Feeding was heaviest in spring and fall and about 20 hours was required to digest 50% of the stomach content (S464).

Pike require 1.385 g of minnow to maintain each gram of body weight a year (J98). In June the requirement is 45 mg/g/week and from October to April, the requirement is 25 mg/g/week. There seemed to be no trend in the maintenance requirements over the range of 21 to 110 g in body weight. After maintenance requirements are met, 1 g food produces 0.485 g of growth. Compared to other species, pike have relatively low maintenance requirements and relatively efficient conversion of food to growth (J98). Ten years of intensive fishing on pike in Lake Windermere reduced the biomass of pike by 47% and the consumption of fish by the pike by 25% (J99).

In much of the early literature, pike was considered an undesirable predator, but over most of the range it is now recognized as an important game species and as an important factor in preventing overpopulation and crowding of stunted fish (B289).

A mutant, silver pike, first reported near Nevis, Minnesota, about 1930 is known from as far north as Beaverlodge Lake, east of Great Bear Lake, from 3 lakes in Manitoba, one near Ottawa, Canada, and from Sweden (B289). The silver pike appears to be hardier than other pike (L181). A silver pike survived handling, transportation, dissection, and lack of food for 30 days.

	Sizes in gill net mesh										
mesh size (stretch)	38	51	64	73	76	83	86	95	102	127	140
F97 NT no.	2	29	-	-	31	-	-	-	44	9	1
mean FL	584	486	-	-	526	-	-	-	617	706	650
mean weight	1633	1134	-	-	1315	-	-	-	1950	3040	2360
C17 MN L. of Woods no.	10	20	54	26	103	53	32	15	333	-	-
mean FL	447	401	480	486	493	503	513	569	566	-	-
central 50%	381-533	305-457	406-508	432-533	457-508	457-533	457-559	508-610	533-635	-	-

The pike was considered a poor bioassay animal (G156) because it remains a suitable size only a short part of the year, requires live food and is excitable under test conditions.

A review of literature was prepared by B289.

MUSKELLUNGE, *Esox masquinongy* Mitchill

The range of the muskellunge includes 4 areas: 1) the upper Ohio River drainage from southwestern New York to northern Georgia, 2) the Great Lakes drainage, 3) the upper Mississippi drainage in Minnesota and Wisconsin, and 4) the southern part of the Hudson Bay drainage (K126, B336).

	FL	No.		
H138, H219 QU, ON	301-500	2	FL = 1.11	SL = 0.919 TL
	501-700	30	1.084	0.9315
	701-900	186	1.083	0.9408
	901-1100	100	1.076	0.9472
	1101-1300	32	1.072	0.9538

Approximate value TL = 1.08 FL
TL was measured with the caudal fin in normal position

M392 ON   1   711 mm TL   TL = 1.09 FL

TL in inches	No.	Weight Mean	Range	References
51-75	8	5	-	F47 ON
75-194	3	23	20-27	G119 WI
221	1	45	-	G119 WI
254-278	2	86	-	K29, M69 NY
305-380	9	240	200-367	K29, J15, H138
381-456	24	417	263-730	K29, J15, M69, H417
457-532	58	771	454-1330	K29, J15, H138, H417
533-609	30	1206	1134-1805	K29, J15, H138, B257, H417

TL in inches	No.	Mean	Weight Range	References
610-685	155	1542	816-2994	K29, J15, H138, B257, H417
686-761	228	2943	1730-5100	K29, J15, H138, B257 (a 749 mm, 13380 g B5, Iowa not included) H417
762-837	184	3588	2404-5897	K29, J15, W39, W86, B257, H417
838-913	320	4636	3400-14515	K29, J15, W39, B257, W86, H138, H417
914-990	239	5797	1814-13608	K29, J15, W39, H138, W86, B257, W37, H417
991-1066	174	7466	3316-13894	K29, J15, W39, H138, W86, B257, M69, B329
1067-1142	174	9467	4990-17240	K29, J15, W39, H138, B257, W86, M69, G23
1143-1210	107	11135	6491-16785	K29, J15, W39, H138, B257, G23, H417
1211-1269	107	13950	4536-19050	K29, J15, W39, G23, B257, H138, W37
1270-1346	55	15590	9980-23590	K29, J15, W39, H138, A30, B257
1347-1421	8	16162	11340-19730	J15, W39
1422-1523	14	25420	18710-31865	J15, W39, G36, S3, J59, B255, W37
1524-1600	4	25220	20410-30620	W37, B255
1615	1	31610	-	B255, M99
1638	1	31725	-	B255

W37 MI The reported 2235 mm muskie that weighed 49.9 kg proved to be a hoax (John Williams, conversation, Dec. 1950).

J61 WI young of year		$\log W = -5.5362 + 3.093 \log TL$
	adult females	$\log W = -6.3056 + 3.399 \log TL$
	adult males	$\log W = -6.4914 + 3.459 \log TL$
M329 ON Nogies C. 40 muskies		$\log W = -5.778 + 3.285 \log FL$
	Nogies Hatchery muskies	$\log W = -5.365 + 3.074 \log FL$
	central Canada muskies	$\log W = -5.728 + 3.204 \log FL$
H138 east and central Canada		$\log W = -5.462 + 3.259 \log FL$

No significant difference was noted in the length-weight relationship of muskellunge from the east and central regions of Canada, but those from the western region (near Lake of the Woods, Ontario) were heavier at all lengths (H138).

C22 MN L. of Woods	2 fish	737-965 mm	Average K(SL) 1.14
R45 OH	-	-	Average K(TL) 0.66-0.89

At 18, 36, 54, 72 days after hatching, muskellunge averaged 23, 38, 81, and 114 mm respectively in Ohio hatcheries (A84).

	No.	TL	Range	No.	Weight	Range
Age 0-May						
O5, O26 WI hatchery	+	-	13-18	-	-	-
Age 0-June						
E7, M18 ON	+	-	15-64	-	-	-
F47 ON	8	61	-	8	5	-
O5, O26, T53 WI	+	-	15-97	-	-	-
P69 TN streams	8	79	61-94	-	-	-
Age 0-July						
E7, M18 ON	+	-	61-109	-	-	-
O5, O26, T53, J59 WI	+	-	61-178	+	9	-
P69 TN streams	12	94	91-127	-	-	-

	No.	TL	Range	No.	Weight	Range
O20 WI ponds	1503	130	99-203	1503	7	-
N43 NY	11	152	-	-	-	-
Age 0-Aug.						
E7 ON	+	-	109-188	-	-	-
O5 WI	+	-	43-254	-	-	-
P69 TN streams	74	150	102-216	-	-	-
O20 WI ponds	2530	178	140-267	2530	30	-
N43 NY	100	216	-	-	-	-
Age 0-Sept.						
E7 ON	+	-	109-234	-	-	-
E56, E57 MN ponds	+	-	208-356	-	-	-
P69 TN streams	14	155	135-203	-	-	-
Age 0-Oct.						
E7 ON	+	-	124-262	-	-	-
W12, J59, O26 WI	+	-	127-368	+	-	136-227
F47 ON	8	170	-	8	17	-
H417 WI Little Green L.	173	272	206-333	-	-	-
Age 0-Nov.						
E7, H15 ON	+	-	218-333	+	200	-
G114 WI	24	170	-	24	17	-
Age 0-Fall						
J78 WI	+	203	152-305	+	91	-
G119 WI (I Jan.)	24	241	-	2.4	59	-
E106 OH	1	356	-	-	-	-
M329 ON Nogies C.	193	396	-	-	-	-
Nogies Hatchery	6	421	-	-	-	-
Age I						
O20 WI ponds April	599	218	201-262	599	45	-
May	170	249	241-305	170	59	-
J78 WI	+	254	229-305	+	136	-
K69, H257 WI High L., May	488	-	190-330	-	-	-
G152 WI George L., Corrine L.	151	300	178-371	34	113	-
O20 WI ponds June	730	310	254-368	730	150	-
H417 WI April	7	315	267-353	-	-	-
M69, G23 NY Chautauqua	4	318	274-368	3	240	86-367
E7, H15 ON	3+	361	333-386	+	907	-
S2 WI	21	394	-	-	-	-
H417 WI Oct.	47	470	325-516	-	-	-
M329, C227 ON Nogies C.	646	483	-	-	-	-
E106 OH Deer Creek L.	25	533	178-711	-	-	-
Age II						
M58 NY Chautauqua	1	381	-	1	263	-
G152, S499 WI Corrine L.	185	406	325-584	185	177	-
G152, S499 WI George L.	122	432	318-597	122	295	-
J15 WI	113	437	330-559	-	-	-
J78 WI	+	457	381-559	+	454	-
C227 ON Nogies C.	+	478	-	-	-	-
S2 WI	24	486	-	-	-	-
H417 WI Little Green L.	18	505	445-528	-	-	-
H138 ON, QU	4	508	-	4	431	227-635
E7, H15 ON	+	-	483-599	+	1814	-

	No.	TL	Range	No.	Weight	Range
M329 ON Nogies C.	1009	549	445-528	-	-	-
B272 PA	+	660	-	-	-	-
E106 OH Deer Creek L.	23	757	648-838	-	-	-
Age III						
G152, S499 WI Corrine L.	91	470	373-630	91	367	-
G152, S499 WI George L.	96	483	356-559	96	413	-
J15 WI	11	538	406-775	-	-	-
S2 WI	9	602	-	-	-	-
C227 ON Nogies C.	+	605	-	-	-	-
H138 ON	6	617	607-630	6	1315	1179-1451
M329 ON Nogies C.	923	630	-	-	-	-
J78 WI	+	635	508-711	+	1588	-
H417 WI Little Green L.	154	663	546-724	-	-	-
G23 NY Chautauqua L.	25	671	594-762	5	1959	1048-2722
H15 ON	+	703	-	+	2268	-
M329 Elephant L.	1	800	-	-	-	-
E106 OH Deer Creek L.	11	907	838-965	-	-	-
Age IV						
G152, S499 WI 2 lakes	78	584	421-864	78	862	-
C227, M329 ON Nogies C.	633	693	-	-	-	-
G23, M69 NY Chautauqua L.	18	706	561-813	17	2263	1107-3856
J78 WI	+	711	559-813	+	2268	-
H15, H138 ON, QU	26+	749	734-759	25	2631	-
H417 WI Little Green L.	80	749	635-844	-	-	-
J15 WI	15	795	584-951	-	-	-
H257, K69, S367 WI	16	800	-	-	-	-
S2 WI	12	807	-	-	-	-
I13 IA Clear L.	2	813	800-826	2	4173	3856-4536
B272 PA	2	864	785-940	1	5670	-
E106 OH Deer Creek L.	25	909	813-1002	-	-	-
Age V						
G152 WI George L.	19	559	523-610	-	-	-
C227, M329 ON Nogies C.	286	767	-	-	-	-
M29, G23 NY Chautauqua L.	33	777	665-851	32	3053	1928-4082
J15, J78 WI	25+	811	470-1041	+	3175	-
H15, H138 ON, QU	32+	813	773-849	29	3493	2812-3946
W37 MI	+	813	-	+	-	3175-3629
S2 WI	20	833	-	-	-	-
B272 PA	1	953	-	1	8029	-
E106 OH Deer Creek L.	24	960	864-1068	-	-	-
Age VI						
M329 ON Nogies C.	92	800	691-	-	-	-
H138 ON	10	811	716-820	8	3680	3447-4763
C227 ON Nogies C.	+	823	-	-	-	-
G23, M69 NY Chautauqua L.	48	833	739-967	48	3780	2608-5176
H15 ON	+	864	-	+	5443	-
H138 QU	46	874	-	47	4491	-
J15, J78 WI	43+	891	660-1069	+	4309	-
S2 WI	36	922	-	-	-	-
Age VII						
S499 WI Corrine L.	38	620	511-726	38	1315	726-2087
H138 ON	13	820	-	14	3810	3630-4040

Age VII (cont.)	No.	TL	Range	No.	Weight	Range
C227, M329 ON Nogies C.	19	866	-	-	-	-
G23, M69 NY Chautauqua L.	31	869	759-1040	31	4170	3175-6900
J15, J78 WI	85+	912	559-1219	+	5900	-
H15 ON	+	914	-	+	5900	-
H138 QU	51	927	-	54	5350	-
S2 WI	54	953	-	-	-	-
B272 PA	59	996	-	59	8210	-
Age VIII						
S499 WI George L.	14	635	594-686	6	1134	880-1450
S499 WI Corrine L.	23	660	505-929	12	4037	-
G23, M69 NY Chautauqua L.	22	876	782-1105	22	4537	2835-9415
C227, M329 ON Nogies L.	4	963	-	-	-	-
H15 ON	+	965	-	+	6800	-
H138 west ON	11	973	-	12	4037	-
S2, J15, J78 WI	115+	980	693-1257	+	6800	-
H138 east ON	4	991	-	4	6350	-
H138 QU	41	1011	-	43	7170	-
B272 PA	-	-	-	+	-	9980-11340
Age IX						
S499 WI George L.	3	709	640-744	2	1814	1451-2177
H138 west ON	21	889	-	19	5400	-
G23 NY Chautauqua L.	15	973	831-1130	13	6214	4140-9180
C227 ON Nogies C.	+	1002	-	-	-	-
H138 east ON	5	1011	-	5	7620	-
H115 ON	+	1020	-	+	8620	-
J15, J78 WI	78+	1034	767-1360	+	9070	-
S2 WI	39	1085	-	-	-	-
H138 QU	21	1085	-	19	9660	-
Age X						
H138 west ON	7	864	-	7	4580	-
G23 NY Chautauqua L.	3	1064	866-1220	3	8620	4765-12020
W37 MI	-	-	-	+	-	9070-11340
J15, J78 WI	42+	1182	818-1325	+	12250	-
H15 ON	+	1095	-	+	9525	-
H138 east ON	4	1123	-	4	9707	-
S2 WI	39	1140	-	-	-	-
H138 QU	12	1195	-	17	11475	-
Age XI						
H138 west ON	13	940	-	11	6350	-
S2 WI	26	1077	-	-	-	-
J15 WI	41	1113	762-1468	-	-	-
H15 ON	+	1146	-	+	10886	-
G23, M69 NY Chautauqua L.	4	1176	1118-1219	4	11594	9526-13240
H138 QU	7	1180	-	9	12700	-
Age XII						
H138 west ON	5	947	-	4	6760	-
H138 QU	7	1146	-	6	11295	-
H15 ON	+	1171	-	+	12700	-
G23, M69 NY Chautauqua L.	2	1200	1118-1283	2	12590	8845-16330
J15 WI	31	1206	991-1367	-	-	-
Age XIII						
H138 west ON	7	1020	-	7	7983	-

	No.	TL	Range	No.	Weight	Range
G23 NY Chautauqua L.	1	1194	-	1	12700	-
J15 WI	225	1220	1067-1372	-	-	-
S2 WI	6	1224	-	-	-	-
H15 ON	+	1224	-	+	13608	-
Age XIV						
H138 west ON	1	945	-	1	6260	-
J15 WI	10	1273	851-1500	-	-	-
H15 ON	-	1278	-	-	14515	-
S2 WI	2	1295	-	-	-	-
Age XV						
J15 WI	3	1224	1156-1448	-	-	-
G23 NY Chautauqua L.	2	1245	1194-1295	2	14060	12980-15200
H15 ON	-	1328	-	-	-	-
W37 MI	-	-	-	-	22680	-
S3 WI	1	1448	-	1	23590	-
S2 WI	2	1494	-	-	-	-
Age XVI						
J15 WI	1	1156	-	-	-	-
S2 WI	1	1217	-	-	-	-
H15 ON	-	1354	-	-	-	-
Age XVII						
J15 WI	1	1372	-	-	-	-
H15 ON	-	1380	-	-	-	-
S2 WI	2	1488	-	-	-	-
Age XVIII						
J15 WI	1	1322	-	-	-	-
H15 ON	-	1430	-	-	-	-
S2 WI	1	1463	-	-	-	-
Age XIX						
J15 WI	1	1328	-	-	-	-
S2 WI	1	1510	-	-	-	-
Age XXX +						
O26 WI	-	-	-	1	31640	-

(See page 360.)

F6 reports a specimen kept in captivity 10 years.
F86 TN Dale Hollow A tagged muskie grew from 1134 to 3290 g in 11 months.

		Increments on tagged fish TL					
M421 ON	age at recapture	IV	V	VI	VII	VIII	IX
	1 year increment	38	43	28	25	23	25
	number of fish	11	18	52	27	17	7
	2 year increment			107	102	58	51
	number of fish			5	9	3	2

The second year shows greater growth than the first year after tagging.

Average calculated TL at each annulus

	No.	1	2	3	4	5	6	7	8	9	10	11	12	13	14	15	16	17	18	19	20
E2, 3, M258, S62 MN	48+	175	318	434	546	655	737	849	993	1062	1105	1156									
R45 OH	+	254	432	584	762	940	991	1016	1067	1143	1245	1245									
B257 PA	152	198	437	622	754	862	958	1036	1105	1133	1163	1200	1234	1280	1336	1372					
K126 typical	+	267	432	569	671	769	849	922	991	1041	1087	1123	1163	1201	1237	1267	1295	1318	1344	1367	1382

Calculated TL and increments at each annulus

	No.	1	2	3	4	5	6	7	8	9	10	11	12	13	14	15	16	17	18	19	20
S2 WI		198	424	579	726	818	904	978	1054	1100	1148	1191	1247	1285	1321	1372	1412	1435	1440	1483	
Increment		198	211	170	132	107	81	64	53	43	43	36	28	28	25	20	18	20	23	18	
Number		351	330	306	297	285	265	229	175	132	93										
E106 OH Deer Creek L.		330	645	815	907	951															
Increment		330	267	180	91	38															
Number		84	78	57	45	22															
P69 TN streams		170	348	488	640	754	826														
Increment		170	193	178	160	109	91														
Number		33	25	16	11	3	1														

Calculated weight at each annulus

	No.	1	2	3	4	5	6	7	8	9	10	11	12	13	14	15	16	17	18	19	20
K126 typical		227	680	1724	2500	3220	4580	5900	7030	8350	9660	10890	12060	13560	14880						

Annual increments in percentage of length at beginning of year
S2 WI   (Numbers for each mean in parentheses)

SL	Median	Range	SL	Median	Range
127-151	164(1)	-	635-685	14(154)	10(80)-20(12)
152-177	106(245)	96(20)-120(24)	686-736	10(162)	9(81)-22(2)
178-202	101(82)	87(2)-116(2)	737-786	9(186)	7(109)-12(2)
279-304	98(1)	-	787-837	7(71)	3(1)-9(3)
305-329	44(9)	-	838-888	6(200)	3(26)-9(2)
330-355	41(213)	39(6)-46(13)	889-939	5(132)	e(27)-9(2)
356-380	41(80)	30(1)-44(2)	940-990	4(73)	2(1)-7(1)
381-405	33(3)	30(2)-37(1)	991-1040	4(36)	1(1)-8(1)
457-482	25(33)	22(6)-32(1)	1041-1091	4(36)	2(15)-5(4)
483-507	24(158)	20(2)-24(117)	1092-1142	3(8)	2(2)-4(3)
508-532	26(92)	21(36)-33(2)	1143-1193	2(7)	1(1)-4(3)
533-583	13(8)	9(1)-19(6)	1194-1244	2(10)	1(2)-3(1)
584-634	14(223)	7(1)-17(41)	1245-1294	1(11)	1(6)-2(5)

Rings on vertebrae were used, as well as scales, for determining age by J78, but the other growth studies used scale examination. The scale growth has been assumed to be directly proportional to growth in length. In Tennessee, 4% of the scales were regenerative (P69).

Growth of muskellunge was slower in western Ontario around Lake of the Woods than in eastern Ontario or Quebec (H138). The Quebec muskies tended to be slightly faster growing in weight and perhaps in length than the eastern Ontario muskellunge. Growth is often faster in recently introduced muskellunge (E106, M329) than in established populations.

Hatchery fish, fin-clipped, grew more slowly in later years of life than native muskies, perhaps being somewhat less competitive (M329). Growth in Nogies Creek, Ontario (M329) was believed to be fairly rapid in the early years of life because of an abundance of minnows, but to be slow in later years because of the scarcity of suckers and perch. Greatest growth occurs in early summer (June) and early fall (September) when available forage is at a maximum and when water temperatures are more favorable, about 20C (O26).

Growth of muskellunge was slow in two Wisconsin lakes (G152, S499) when the predation had almost eliminated the age II and older yellow perch. Lack of sufficiently large prey also resulted in slowed growth of older muskies in Ontario (M329).

Female muskie usually grow more rapidly than males (O26, J61, H138) particularly after 5 or 6 years of age. Females also usually live longer than males (O26).

In Tennessee, male muskie first mature at age III and 560 mm and females at age III-IV and 635 mm (P69). The smallest mature male in Wisconsin was 457 mm, 340 g, and age II. The smallest mature female was 648 mm, 1814 g, age IV (J91) but muskie usually do not mature until age III to V in Wisconsin (O26).

	Size of females	No. of females	No. of eggs per female Mean	Range
N40 WI	18145 g	1	225,000	-
N45 NY Chautauqua L.	-	-	-	36908-77231
C47 NY	14515 g	1	265,000	-
V15	-	-	-	-250,000
B24 OH	15875 g	1	255,000	-
H36 WI	900-1170 mm	11	-	22092-164,112
W42 WI	-	30	-	10000-15000
W42 WI	1320 mm	1	260,000	-

| | Size of females | No. of females | No. of eggs per female | |
			Mean	Range
W38 WI	900–1375 mm	13	107,224	-
N25, B73, L1, K4	15875 g	1	265,000	-
B23 NY Chautauqua L.	-	1	60000	-
K4	-	190	6315	-

Water-hardened eggs range from 0.13 to 0.10 inch diameter. Large muskie tend to produce large eggs but not always. In some lakes muskie eggs are typically smaller than in others (J62). Eggs from females over 1260 mm often result in lower hatching rates (J62).

Muskellunge culture is reviewed in E7, M410, B336, J62, S497. Spawning is in April in Tennessee (P69) and from mid-April through late May in Wisconsin (O26). Spawning may occur at 9.5-15.5C but 13C is about optimum (O26). Eggs are deposited indiscriminately over several hundred yards of shoreline in water 150 to 760 mm deep. Adult spawners return to the same spawning ground in consecutive years (O26). Hatching occurs in 8 to 14 days at temperatures of 12-17C (O26) or 13 to 21 days at 10-21C (J62). At 20C, eggs hatch in 6 days (J62). Among eggs found in lakes, only 34% were fertile (J91). In hatcheries 85-95% of the eggs hatch (J62).

Factors limiting reproduction were listed as (O26):

1. Cold water temperatures and fluctuating water levels at spawning time. Muskie eggs and fry seem particularly sensitive to water temperature fluctuation.
2. Predation by other fish and invertebrates on eggs and fry, including *Dytiscus* and *Notonecta*.
3. Quantity and size of live zooplankton and forage fish available to muskie fry at critical stages.
4. Hybridization with northern pike.

Fry usually require 10 to 14 days to absorb the yolk sac and start feeding (J62). Muskie fry feed on live zooplankton the first 4 days of feeding and may continue on it for over 30 days if forage fish are not present. Live forage fish are preferred after 4 days and usually required after muskies are over 40 mm (O26). Fingerling feeding drops off as water temperature exceeds 30C. Turbid waters interfere with feeding. Probably there is little feeding after dark. It took 2.73 g of live minnows per gram of increase (range 2.22-3.22) and muskies consumed 6.4% of their body weight per day (G119). Adults eat mostly fish (O26, H138) but may also take muskrats (H138), waterfowl, shrews, and other live animals (O26, H138). One 635 mm muskie had eight 127 mm salamanders and a 710 mm muskie had eaten a 406 mm muskie (P69).

Introduction of muskellunge in two Wisconsin lakes (G152, S499) eliminated most of the yellow perch and largemouth bass over 100 mm long but enough of these remained, apparently in secure habitats, to reproduce successfully each year. Smallmouth bass were apparently relatively secure from muskellunge predation and increased in abundance as largemouth bass decreased.

Of 488 muskellunge tagged as yearlings and stocked in High Lake, Wisconsin, 17% were caught by anglers at ages IV through X (H257). In Fishtrap Lake, only 1.7% return was reported.

In Ontario (M421) the percentages recaptured were as follows:

Age transplanted	II	III	IV	V	VI	VII	VIII	IX
Number transplanted	167	240	369	387	219	97	38	6
% recaptured	5	7	11	20	22	26	26	33

Fry loss and differential mortality of tagged fish had to be corrected for making estimates of population and exploitation in Ontario (M500).

HYBRID, *Esox americanus vermiculatus* **X** *E. lucius*
S404 PA   age I   318-333 mm TL
M293 NB Watts L.   4 females age II, 462-508 mm TL, mean 483

HYBRID, *Esox americanus vermiculatus* **X** *E. masquinongy*
T101 OH   age 6 months   10 fish   292 mm TL   156 g
                        age I, Oct.   6 fish   305-406 mm   170-510 g
These fish were feeding on dragonfly nymphs and other invertebrates.

HYBRID, *Esox lucius* **X** *E. masquinongy*
N52 WI   1002 mm TL hybrid   weighed 7120 g

	No.	TL	Range	No.	Weight
Age 0-June					
E56 MN	-	74	-	-	-
Age 0-Sept.					
E56, 57 MN	-	-	295-356	-	-
J100 IA	60	295	267-307	-	-
B57 WI	136	305	-356	-	-
Age I					
C117 WI	12	363	-	-	-
Age II					
C117 WI	28	526	-	-	-
Age III					
C117 WI	29	538	-	-	-
Age IV					
C117 WI	10	632	-	-	-
Age IX					
H138 west ON	1	967	-	1	6800
Age X					
H138 west ON	1	1026	-	1	9070
Age XI					
H138 west ON	1	1046	-	1	9070
Age XII					
H138 west ON	2	1019	-	2	8664

Hybrid females are fertile but hybrid males are probably sterile (J91). The Iowa hybrids (J100) were between a muskellunge male and a female silver pike. F_1 hybrids were more resistant to thermal stress than the muskellunge and pike parents (S500).

HYBRID, *Esox lucius* **X** *E. niger*
E10 NY   6 months   127 mm   range 91-152 mm
U2 NY   I-Sept.   407 mm SL
A84 OH

		TL		
Days after hatching	Northern pike	Hybrids	Chain pickerel	Muskellunge
18	71	30	13	23
36	127	51	20	58
54	147	99	61	81
72	216	132	94	114

MEXICAN TETRA, *Astyanax mexicanus* (Filippi)
The northern range of this species includes the Rio Grande in Texas and the lower Colorado River drainage in Arizona and New Mexico (E119). It has been introduced somewhat farther north including Lake Texoma where nine young in June were 18-33 mm TL (D107).

CHISELMOUTH, *Acrocheilus alutaceus* Agassiz and Pickering
Columbia River drainage of Washington, Oregon, and Nevada (E119).
S363 OR   2   178 mm   39 g

STONEROLLER, *Campostoma anomalum* (Rafinesque)
The stoneroller, in its various subspecies, is found from southern North Dakota to Texas east to the Appalachians and north to western New York (T113).

G82 IL	88 stonerollers	38-143 mm SL	TL = 1.19 SL

S472 AL	TL	51	76	102	127	152
	Weight	3	5	11	18	45
	Number	33	29	21	5	1
89 fish 50-155 mm				log W = -3.72 + 2.39 TL		

Average K(TL) = 1.04 for 56 stonerollers 76-155 mm. The average C(TL) reported for 51 mm fish, 2.62, is obviously too high and suggests that these weights were not valid, nor is the length-weight regression.

	No.	SL	TL
R144 PA at hatching	50	5.6	-
Age 0-Aug.			
L108 IL Big C.	+	32	-
G82 Roaring C.	7	40	-
L110 IL Clear C.	+	46	-
L133 NC Great Smokies	10	-	58-99
Age 0-Oct.			
T113 OH	+	-	28-66
Age I			
L108 IL Big. C.	+	52	-
T113 OH	+	-	33-81
G82 IL Roaring C.	11	59	-
L110 IL Hutchins C., Clear C.	+	61	-
L133 NC Great Smokies	32	-	79-130
Age II			
L110 IL Clear C., Hutchins C.	+	78	-
G82 IL Roaring C.	59	74	-
G130 IA Clear L.	1	-	122
L133 NC Great Smokies	87	-	102-165
Age III			
G82 IL Roaring C.	13	88	-
L133 NC Great Smokies	96	-	122-206
Age IV			
L133 NC Great Smokies	42	-	152-239
Age V			
L133 NC Great Smokies	7	-	173-226

Calculated TL at each annulus

		No.	1	2	3
R56 OH	males	+	69	104	119
	females	+	69	102	135
G82 IL Roaring Springs C.		83	51	79	99

Lengths at each annulus were computed with a Fraser-type correction factor in G82.

Reducing the population density increased averaged length and weights (L133). Males grow faster than females (R56, P120, M386). The largest female was 188 mm compared to 287 mm for a male (L133).

Some males and females mature at age III, with a majority of the females maturing at this age but only a few males (L133, P120). The smallest mature male was 102 mm and female, 91 mm (L133). Eggs at spawning are 2.0 mm in diameter and 2.4 mm when water-hardened (R144). After fertilization the eggs turn bright yellow. They hatched in 69-70 hours at 21C (R144). Spawning and territorial behavior were described (M386, S449). Males build the nests, starting in mid-April in New York at 13-16C and spawn until early June at 24C (M432). In Illinois nest building started in mid-April at 12C and spawning continued to June 2 at 24 to 27C (S449). The stonerollers compete with rainbow trout for spawning sites and may destroy trout redds (L133, P120). High mortality from *Columnaris* was reported (P120). During winter stonerollers may congregate under stones (M432).

In streams in the Great Smokies (L133), stoneroller populations ranged from 701-2834 per hectare, or 19-80 kg per hectare. They comprised 21-62 percent of the weight of fish in these six streams. The average gradient of the streams was 4.4%.

Stonerollers fed primarily on diatoms, bluegreen algae, and a few small tendipedids in Kentucky streams (M404), in the Smokies (L104) and in Ohio (K107).

Stonerollers hybridize with the common shiner (T113) and southern redbelly dace (H381). Additional life history data are in a thesis (S444).

GOLDFISH, *Carassius auratus* (Linnaeus)

The goldfish, a native of Asia, has been introduced into many American waters.

C33 IA		197 fantailed goldfish			FL = 1.221 SL
					TL = 1.499 SL

S501 IL	65 females	62-74 mm SL	6.9-12.6 g	avg. 67.6 mm	9.10 g
	75 males	61-73 mm SL	6.9-10.7 g	avg. 69.5 mm	8.02 g

			Weight	Depth mm	K(TL)	
TL	No.	Mean	Range	(from L97)	(from S472)	Citations
15-50	9866+	0.5	0.4-3.7	8	3.35	L97, S472
51-75	89297+	3.4	0.4-9	13	2.80	L97, P109, S472
76-101	123,897+	8	4.5-11	25	1.58	L97, P109, S472
102-126	21451+	16	9-28	28	1.41	L97, P109, S472
	7	22	-	-	-	F31 IA
127-151	2957	29	17-68	33	1.52	L97, S472
	2	24	-	-	-	F31

TL	No.	Mean	Weight Range	Depth mm (from L97)	K(TL) (from S472)	Citations
152-177	537	59	31-91	-	1.66	S472
178-202	4	82	-	-	-	F31
	1336+	111	54-128	-	2.11	P109, S472
203-228	31	111	-	-	-	F31
	1319+	170	99-218	-	2.19	P109, S472
229-253	46	139	-	-	-	F31
	286+	216	176-258	-	1.83	P109, S472
254-278	31	184	-	-	-	F31
	126+	315	204-400	-	1.99	P109, S472
279-304	11	255	-	-	-	F31
	90+	408	340-485	-	2.02	P109, S472
305-329	36	505	227-697	-	1.77	S472
330	30	658	284-771	-	1.83	S472
356	5	833	771-	-	1.86	S472
376-381	2	1942	1644-2240	-	-	E105 MD
381	2	862	635-1088	-	1.55	S472
406	1	1452	-	-	2.16	S472

S472 AL   251,236 fish   15-406 mm $\log W = -4.53 + 2.90 \log TL$
but the slope is undoubtedly depressed by the overly high weights of the 25-51 mm fish.

F31 IA   132 goldfish $\log W = -4.342 + 2.732 \log TL$

K1 Weight in grams equals 0.0309 times the length in cm. cubed, 569 2-month fish.

K1 Weight in grams equals 0.0303 times the length in cm. cubed, 708 older fish.

C33 IA   197 fish   79-228 mm SL   average K(SL) 3.54

In captivity goldfish lived to 25 years (F7) and 30 years (E122).

The length-weight data given for goldfish in Iowa (C33, F31) refers to the fantailed variety.

	No.	SL	FL	TL	Range
Age 0-Hatching					
K1 Japan	+	5	-	5	-
Age 0-10 days					
K1 Japan	+	6	-	8	-
Age 0-May					
B22 DC	+	-	-	25	-
Age 0-30 days					
K1 Japan ironfish	+	34	-	-	-
Age 0-50 days					
H73 China	+	22	-	-	-
Age 0-70 days					
K1 Japan ironfish	+	40	-	-	-
Age 0-100 days					
H73 China	+	37	-	-	-
Age 0-150 days	+	46	-	-	-
H73 China	+	46	-	-	-

	No.	SL	FL	TL	Range
Age 0					
E9 NY	+	-	64	-	-
Age I					
K1 Japan ironfish	+	61	-	-	-
E9 NY	+	-	-	89	-
I2 India	+	-	-	-	203-254
Age II					
K1 Japan ironfish	+	81	-	-	-
E9 NY	+	-	-	X	127-152
Age III					
K1 Japan ironfish	+	X	-	-	86-94
I2 India	+	-	-	X	-457 (to 1360 g)
Age IV					
K1 Japan ironfish	+	114	-	-	-

R28 additional data on the growth of goldfish in Germany.

Scale structure was described by Y18 who reported that any decrease in growth may result in an annulus-like mark. Dylox medication did not seem to hinger growth or reproduction of goldfish (C342).

Goldfish may mature at 100-185 mm TL (I2). Gonad development is described (S451). In Alabama they will spawn from February through November if not repressed by excretion from a high population density of goldfish (S390). In Poland goldfish begin spawning June 5 to July 7, peaking about June 22 to July 13 (B285). In Japan there are three spawning peaks, May to early August, with surface temperatures of 18-29.8C, mostly 18.4-24.9C (M316). The first hatch occurs when water temperatures are 15.5-18.4C, 150 mm from the surface (S334). In May (M316) log no. mature eggs = $-2.3925 + 3.179$ log SL.

In India an average of 14000 eggs per female was reported (I2).

	Lethal temperatures		
If acclimated at	Maximum	Minimum	Citation
5C	29C	-	B202, S333
10C	30.8, 31C	0C	S333, B202
15C	32.8C	-	S333
17.5C	32.5C	-	F92
20C	34, 34.8C	2C	S333, F92, B202
25C	36.5, 36.6C	6C	F92, S333
30C	38, 38.6C	9C	F92, S333, B202
35C	40.5C	-	F92
ultimate	41.0C	-	S333

Goldfish can withstand a rapid change from 0 to 1.5% salt (B204).

The food of goldfish is mostly phytoplankton (S356). The digestive system and enzymes of the goldfish were studied (S450). No decline in feeding or condition factor was noted with seasons (S356).

S151 IL   19 goldfish   1,240,000 ± 466,120 red blood cells per cu. mm
hemoglobin 30.4 ± 13.7

Additional blood characteristics were reported by S501.

A bibliography on goldfish was prepared (C341).

NORTHERN REDBELLY DACE, *Chrosomus eos* Cope
   Northern British Columbia to Hudson Bay drainage to New Brunswick,
south to Maryland, Michigan, Minnesota, sand hills of Nebraska, and Colorado.
Typical of bog ponds and creeks (H382).

C273 MI	348 dace				
TL	15-24	25-37	38-50	51-63	64-75
Weight	.03-.1	.11-.38	.35-1.5	0.9-2.6	1.7-3.0

	Age	No.	TL	Range
C273 MI	0-July	19	20	-
	0-Aug.	219	28	20-33
	0-Sept.	9	28	23-30
C56 MI	0-Oct.	485	33	10-48
C273 MI	I-June	216	48	-
	I-July	300	48	38-53
	I-Aug.	810	48	41-56
	I-Sept.	557	48	41-56
	I-Oct.	137	53	-

C273 MI young of year	Pond A	grew 0.28 mm per day
	B	0.27 mm per day
	Reservoir	0.23 mm per day

D41 Mature at 1 year.   Maximum length 64 mm TL.
B201 When acclimated at 21C the lethal low temperature was 2.7C
          25-26C lethal high temperature   33.1C

SOUTHERN REDBELLY DACE, *Chrosomus erythrogaster* (Rafinesque)
   Southern Minnesota to Pennsylvania and south to northern Alabama and
northeastern Oklahoma (E119, T113).

T113 OH	0-Oct.	18-38 mm
	I	25-46 mm
	maximum	76 mm

   When frightened, redbelly dace form compact schools in mid-water and
are extremely vulnerable to seining (T113). This is a good bioassay animal
and aquarium species if handled with reasonable care (W214). Spawning habits
are described (S462).
   Hybrids with *Campostomum anomalum* are reported (H381).

MOUNTAIN REDBELLY DACE, *Chrosomus oreas* Cope
   In the upper James, Roanoke and Kanawha Rivers (E119).
   These dace are herbivorous, feeding on microscopic plants (F148).

REDSIDE DACE, *Clinostomus elongatus* (Kirtland)
   Redside dace are typically in clear gravelly creeks, with a discontinuous
range from northeastern Iowa and Minnesota to New York and Kentucky (H382).

	No.	SL	Males Standard deviation	Weight	No.	SL	Females Standard deviation	Weight
Age 0								
S212 PA Linesville C.	25	39	3.6	1.0	93	40	5.5	1.2
Age I								
K18 NY	8	38	2.5	-	7	37	2.8	-
S212 PA Linesville C.	12	57	4.5	3.1	66	55	5.2	2.9
Age II								
K18 NY	6	56	2.9	-	10	57	2.7	-
S212 PA Linesville C.	12	66	2.9	4.8	19	69	2.2	5.8
Age III								
K18 NY	5	62	3.4	-	8	69	3.2	-
S212 PA Linesville C.	2	75	3.9	5.7	4	75	4.4	7.3
Age IV								
K18 NY	1	65	-	-	2	77	-	-

Annuli were distinguished on the anterior field by a clear zone between 2 circuli, with no cutting-over (S212).

Females had 409-1526 eggs per female (K18).

Food was mostly insects with terrestrial insects constituting 77%. There was some increase in size of insect eaten with increase in size of dace (S212).

ROSYSIDE DACE, *Clinostomus funduloides* Girard

Headwater streams from Chesapeake Bay to North Carolina (E119).

B67 DC Oxon Run   Age I - Jan.   20-40 mm SL
                       Age II - Jan.   44-64 mm SL

Food is mostly insects (F148).

GRASS CARP, *Ctenopharyngodon idellus* (Vall.)

The grass carp, a native of Asia, has been experimentally reared in Arkansas (S502) as a possible control on aquatic plants. The carp are omnivorous but may be largely herbivorous when other food is scarce. In Arkansas growth was as follows:

Age in months	6	9	12	15	18
Mean TL	81	119	279	450	500
Mean weight	4	21	372	1270	1814

First maturity is reported to be at

10-14 months, at 2040-5000 g, in Malaysia
4 years, 6800 g, in south China
5-8 years, 9070-11340 g, in Israel
8-10 years, 2700-3630 g, in Russia

The grass carp survived 5 months of ice cover in Arkansas.

CARP, *Cyprinus carpio* Linnaeus

Originally Asiatic in distribution, carp have been introduced into many parts of the world, including some waters in most parts of the United States and southern Canada. No attempt has been made to cover the extensive literature on this species outside the United States and Canada. An excellent bioassay animal (B349).

		Length	No.	FL	No.	TL
C89	IA Ruth's P.	-	5	1.089 SL	5	1.215 SL
C30	IA	-	58	1.10 SL	155	1.23 SL
S15	TN Reelfoot L.	-	-	-	-	1.218 SL
M141	UT	-	19	1.086 SL	-	1.2512 SL
M302	IA	53-406	93	-	-	1.25 SL
		407-737	59	-	-	1.23 SL
R148	IA	178-278	31	0.876 TL	-	1.27 SL
		279-380	40	0.885	-	1.26 SL
		381-583	53	0.892	-	1.25 SL
		584-765	19	0.899	-	1.24 SL

F98 WI At SL	50	100	150	200	250	300	350	400,
TL/SL =	1.300	1.260	1.247	1.240	1.237	1.233	1.231	1.230.

In papers which give data as SL or FL the lengths have been converted to TL, unless otherwise stated, by a combination of the above correction factors. Equivalent lengths are given in the summary of length-weight data which follows. The lengths given in the first three columns are not class centers, but the lower length in the range covered on that line.

SL (in mm)	FL	TL	Median	Weight Central 50%	Range	No. of fish	References for lightest and heaviest
20	23	25	1	-	-	3	S472
40	46	51	6	-	2-31	1+	S472
60	66	76	15	9-23	5-45	23+	C100, S472
80	89	102	32	18-36	10-73	98+	S472
101	112	127	41	32-45	19-104	214	S472
121	135	152	64	54-73	27-136	641	S472
142	155	178	95	86-104	45-318	775+	W132, S49
163	178	203	141	127-145	59-195	447	S472, Y4
184	201	229	195	177-209	113-363	472	C20, M302
205	224	254	254	227-281	181-422	835	S472, M302
225	246	279	340	304-367	227-567	896	Y19, S49
246	269	305	422	395-454	290-862	1025	M302,
267	292	330	531	495-554	308-1361	1470+	C208, M302
288	315	356	645	613-690	463-971	1031+	C208, W132
309	340	381	781	726-885	449-2722	906	C208, S49
330	363	406	934	903-1042	680-1724	704	C20, M302
351	386	432	1125	1034-1225	708-3629	600	M302, S49
371	409	457	1361	1256-1452	907-2268	424	S472, S49
392	429	483	1574	1438-1679	943-2994	281+	S472, K127
413	452	508	1850	1692-1973	1197-3424	234	H1, S472
434	475	533	2104	1932-2304	1420-3733	256	M302, S472
454	498	559	2504	2313-2858	1873-4536	314	S472, S49
475	523	584	2800	2495-3098	1787-5443	296+	P112, S49
496	546	610	3175	3026-3383	2172-6830	530+	N115, M302
537	592	660	4381	3751-5171	3084-8165	314+	G39, S49
578	635	711	5371	4790-6028	3865-9072	119+	G39, S49
620	681	762	6827	5693-7716	3570-11567	41+	G39, M302
661	726	813	7121	6718-8732	4432-9752	22	G39, S49
702	769	864	9980		4133-16556	4	G39, P128

SL (in mm)	FL	TL	Median	Weight Central 50%	Range	No. of fish	References for lightest and heaviest
743	815	914	12111	-	10569-14515	3	P66, B337
785	858	965	-	-	-	-	-
826	902	1016	13470	-	9800-17327	3	C20, S49
867	945	1067	19052	-	-	1	G36
-	-	1219	26990	-	-	1	I5

The above data came from the following references: Israel Y4; Egypt K127; South Africa P128; Germany H1; Alabama S472; California W71, B337; Colorado C208, L126; Illinois C55, M499; Indiana G39, U1; Iowa B150, C30,89,100, E47, I5, I10, M302, R148, Y19; Louisiana V44; Minnesota C20, S62; Missouri P112; New York S49; Ohio H156, O7; Oklahoma J31; South Dakota F100,105, 124,125, N94,115, S225-30,276,330,393; Tennessee K71, W132; Utah M340; Virginia G36; Wisconsin F98, N52; other F7.

		No.	Range mm	$\log W = a + b \log L$	
S236 UT Bear L.	June-July	109	169-539	-4.714 + 3.041	SL
	Oct.-April	322	123-830	-4.004 + 2.791	SL
Ogden Bay		305	259-663	-4.020 + 2.811	SL
Cache Valley P.		191	68-408	-4.270 + 2.876	SL
Clear L.		-	125-730	-4.440 + 2.914	SL
G39 IN		171	293-765	-4.801 + 2.624*	FL
W71 CA		53	190-445	-4.475 + 2.888	FL
S472 AL Israeli carp		3272	100-430	-4.81 + 3.00	TL
Israeli carp		490	431-740	-3.64 + 2.62	TL
scaled carp		31344	25-355	-3.95 + 2.63	TL
scaled carp		352	356-770	-5.07 + 3.07	TL
K127 Egypt mirror carp		680	90-535	-3.510 + 2.640	TL
V75 Indonesia		-	-	-4.5 + 3.0	TL
S225 SD Ft. Randall L.		-	127-457	-4.239 + 2.736	TL
H156 OK Canton L.		274	100-440	-5.102 + 2.768	TL
J34 OK Fairfax L.		-	-	-4.964 + 2.709	TL
Ardmore L.		144	100-240	-5.694 + 2.792	TL
J31 OK Grand L.		351	125-720	-5.652 + 2.927	TL
P112 MO Salt R.		-	-	-4.579 + 2.894	TL
E47 IA Clear L. males		133	196-762	-4.224 + 2.746	TL
females		165	241-828	-4.182 + 2.746	TL
R184 IA Des Moines R. 1958		517	180-770	-4.937 + 3.025	TL
Y19 IA Des Moines R. 1961 males		99	272-764	-5.603 + 3.253*	TL
females		74	290-813	-6.226 + 3.477*	TL
Izaak Walton L. 1961 males		53	376-538	-3.982 + 2.664*	TL
females		78	366-544	-3.982 + 2.664*	TL
1962 males		6	452-523	-5.823 + 3.340	TL
females		17	406-587	-6.083 + 3.467	TL
S356 Japan		51	270-350	-3.419 + 2.432*	TL
		23	300-435	-3.445 + 2.453*	TL
		30	273-425	-3.850 + 2.621*	TL
		44	272-413	-3.893 + 2.642*	TL
		69	240-404	-4.043 + 2.695	TL
		47	316-425	-4.125 + 2.734	TL

	No.	Range mm	log W = a + b log L
S356 Japan (cont.)	30	283-357	-4.249 + 2.758 TL
	31	130-414	-4.317 + 2.759* TL
	9	330-495	-4.212 + 2.770 TL
	53	303-460	-4.265 + 2.774 TL
	50	321-440	-4.493 + 2.857 TL
	44	300-520	-4.502 + 2.875 TL
	24	300-520	-4.532 + 2.886 TL
	48	248-420	-4.573 + 2.890 TL
	98	250-438	-4.722 + 2.938 TL
	50	230-414	-4.731 + 2.951 TL
	43	305-501	-4.736 + 2.978 TL
	58	235-459	-4.872 + 2.995 TL
	67	165-365	-4.776 + 3.015 TL
	34	228-356	-5.099 + 3.099 TL
	68	201-451	-5.127 + 3.115 TL
	88	315-570	-5.430 + 3.210 TL
	35	340-404	-5.693 + 3.319 TL
	32	260-375	-6.126 + 3.524* TL

In all cases of the data from S356 and Y19 except those marked * the 95% confidence interval for the regression coefficient includes 3.0. The regression values for one pond (S356) in different years were 2.642, 2.695, 2.978, 2.734, and for another pond 2.938, 3.099, 2.995, 2.886, and for a third 3.210, 3.319, 3.524, and a fourth 2.770, 2.453, 2.432. There was, therefore, a tendency for the regressions to be somewhat typical of the ponds.

	Size	No.	K(SL) Mean	K(SL) Range
E3 MN	127-736 mm SL	77	2.37	-
E28 TN Chickamauga	-	239	2.8	-
C36 IA	140-826 mm SL	191	2.71	1.69-3.77
R115 IA Lost Island L.	-	131	2.62	-
W105 IL Crab Orchard	340-1588 g	8	2.75	2.28-3.06
O28 Czech.	50-500 mm SL	5	2.6	2.3-2.8
M302 IA lakes	51-813 mm TL	875	2.72	1.38-3.85
F98 WI males	-	865	2.56	-
females	-	869	2.63	-
M340 UT Green R.	334-443 mm SL	2	2.52	2.30-2.75
D179 TX	103-532 mm SL	431	2.58	1.13-5.95
after shad removal	105-486 mm SL	297	2.80	1.11-5.18
W178 Poland ponds	-	-	-	2.9-3.8
S236 UT Bear L. June-July	169-539 mm SL	109	2.41	2.20-2.71
October-April	123-830 mm SL	322	2.91	1.72-3.51
Ogden Bay	259-663 mm SL	305	2.92	2.76-3.35
Clear L.	125-730 mm SL	-	-	1.90-2.47
Cache Valley P.	68-408 mm SL	191	-	2.46-3.30
			K(TL)	
H1 Germany	-	10	1.41	-
S122 Germany	-	+	1.8	-
S356 Japan	130-570 mm TL	1126	1.48	1.31-1.85
Y10 Japan	-	1290	1.71	0.9-2.6

	Size	No.	K(SL) Mean	K(SL) Range
C89 IA Ruth's P.	457-762 mm TL	14	1.30	1.11-1.58
C100 IA William's P.	89-564 mm TL	166	1.23	0.58-1.63
E47 IA Clear L.	178-813 mm TL	326	1.33	-
B94 IL	371-394 mm TL	350	1.45	0.79-1.97
	421-445 mm TL	336	1.41	0.73-1.83
R148 IA Des Moines R.	178-787 mm TL	517	1.37	-
S472 AL scaled carp	76-762 mm TL	26744	1.39	1.11-2.02
mirror carp	102-737 mm TL	3762	1.83	1.16-2.60
C96 OK Rod & Gun Club L.	-	+	1.66	-
B150 IA Little Wall L.	201-353 mm TL	165	1.47	-
Y19 IA Izaak Walton L.	356-584 mm TL	159	1.39	-
Des Moines R.	254-559 mm TL	170	1.29	-
H156 OK Canton L.	102-432 mm TL	274	1.26	-
S225,226,228,230,330,393			Annual means	
SD Ft. Randall	127-533 mm TL	4307	1.28	1.17-1.38
S226, S228 SD tailwaters	-	32	1.26	1.25-1.26
S227,229,276, N94,115				
SD Gavins Pt.	102-610 mm TL	2858	1.32	1.22-1.41
F100,105,125,125 SD Oahe	76-635 mm TL	3906	1.45	1.32-1.60

The maximum weight reported for carp was 37.4 kg in South Africa and 27 kg in North America (S463).

No general regional differences in length-weight relationships were noted. Scaled carp were generally lower in weight at equivalent lengths than mirror carp, in Alabama (S472). In most of the population where the regressions were computed the regression slopes were less than 3.0, though few deviated significantly from this value. A decrease in ponderal index values with increase in length was noted in F98, S228, S236, and H156 but not in the other populations. Within the same age and sex groups, slow-growing individuals had higher K factors (F98). In Israel (Y4) carp from ponds were generally heavier for their lengths than those from lakes. In Ft. Randall, Gavins Point, and Oahe reservoirs, South Dakota (references above) there was a general decline in condition factors with years after impoundment.

Males averaged slightly heavier than females in one Utah population (S236) but females were heavier than males at the same lengths from 406 to 508 mm in Wisconsin (F98) and no males were over 508 mm. Regressions of male and female carp differed in some Iowa populations (I19) but not in an Indiana population (G39). Females showed higher condition factors when they were mature but otherwise the values were the same for males and females in Wisconsin lakes (F98). No sexual differences in ponderal indexes were noted in the South Dakota reservoirs (references above) or in R148. The condition factors decreased as the summer progressed in 1957 in Ft. Randall (S230) and in 1958 in Gavins Point (S276) but not in the other summers for which data are available. In the Des Moines River (R148) it was noted that condition was highest in July and decreased into the fall and spring. In Japan feeding was found to decline during the season with a resulting decline in K(TL) values (S356). No seasonal trends were noted in data from Israel (Y10). A definite seasonal cycle in K associated with gonad development was noted (F98). Ponderal indexes were found to be an inadequate index to changes in fat content of carp fry (W178).

In "hunger-fish" the head equals 40% of the total weight compared to 20% in "fattened fish" (W47, W190). "Hunger-fish" also have larger eyes and longer fins. Whereas the body length was 2.7 to 2.85 times the body depth in marketable carp, it was 3 to 3.08 times the body depth in the more polluted portions of the Illinois River (M501). In some parts of the river in 1963 almost 100% of the carp were "knotheads." The lack of fingernail clams in the polluted areas may be part of the reason for the thinness of the carp.

	No.	Mean	TL Range	No.	Mean	Weight Range
Age 0-May						
S101 NY	-	-	3-8	-	-	-
R26 Illinois R.	-	-	5-13	-	-	-
B22 DC	-	-	25-38	-	-	-
Age 0-June						
F5 L. Erie	11	-	10-18	-	-	-
K51 Japan	522	36	-	522	5	-
E28 TN Chickamauga	36	46	20-69	-	-	-
A20 France	-	-	38-51	-	-	-
B183, C271 OK	-	-	13-61	-	-	-
H150 OK Tenkiller L.	11	99	89-107	-	-	-
P141 Israel uncrowded	-	-	-	+	18	-
J34 OK Ardmore L.	-	-	81-244	-	-	-
Age 0-July						
S49, S50 NY	-	-	13-51	-	-	-
R26 Illinois R.	-	-	18-51	-	-	-
K51 Japan crowded pond	348	48	-	348	18	-
uncrowded pond	174	53	-	174	18	-
C284 OH	-	51	-	-	-	-
O28 Bohemia	31	-	43-64	-	-	-
S224, S225 SD Ft. Randall	9+	64	23-	-	-	-
B2, B3, C91, C109, E47, M228 IA lakes	59	66	25-94	43	5	5-14
E28 TN Chickamauga	113	66	46-91	-	-	-
C134 OK Stillwater	2	85	81-89	-	-	-
V44 LA Spring Hill L.	9	124	102-155	9	27	14-45
J41 OK Gibson L.	17	135	-	-	-	-
H150 OK Tenkiller L.	344	198	104-282	-	-	-
W112 Israel select stock	-	-	-	-	45	-
P141 Israel uncrowded	-	-	-	-	-	54-109
Age 0-Aug.						
A1, S49, S50 NY	-	-	48-99	-	-	-
B183, C271 OK	-	56	20-89	-	-	-
S224, S225 SD Ft. Randall	-	58	56-74	-	-	-
G125 Poland best pond	-	-	-	-	23	-
K51 Japan crowded pond	348	58	-	348	27	-
uncrowded pond	174	66	-	174	41	-
C284 OH	-	69	-	-	-	-
E28 TN Chickamauga	391	86	58-100	-	-	-
J34 OK Ardmore L.	-	89	-	-	-	-
B2, B3, C91, C109, E47, M228 IA lakes	263	89	56-183	23	54	9-154
J31 OK Grand L.	7	150	132-160	7	45	-

	TL			Weight		
	No.	Mean	Range	No.	Mean	Range
Age 0-Sept.						
K51 Japan crowded pond	348	69	-	348	45	-
uncrowded	174	79	-	174	64	-
S49, S50 NY	-	-	91-140	-	-	-
G75 OR	6	91	89-99	-	-	-
F109 SD Oahe May-spawned	-	99	-	-	-	-
June-spawned	-	152	-	-	-	-
E28 TN Chickamauga	474	102	69-150	-	-	-
M265, M266 WI	+	-	69-226	-	-	-
S224, S225 SD Ft. Randall	2+	109	81-	-	-	-
E47, M228 IA	30	130	76-160	3	77	27-86
Age 0-Oct.						
W137 UT	1	64	-	-	-	-
B183, C271 OK	-	69	48-94	-	-	-
J1, W47 Germany starved	-	71	64-76	-	9	-
S122, W47 Germany normal	-	-	114-152	-	-	32-54
G125 Poland	-	-	-	-	-	27-32
S49, S50, E9 NY	-	-	64-140	-	-	-
A20 France	-	-	64-127	-	-	-
K57 Japan crowded pond	1044	79	-	1044	64	-
uncrowded	522	94	-	522	100	-
P63 IA pond	77	132	64-175	-	-	-
S390 AL	-	-	-	40	132	-
T113 OH	-	-	18-203	-	-	-
S57 NC	-	-	102-178	-	-	-
L193 NV	+	-	76-229	-	-	-
S101 L. Erie	2	152	-	-	-	-
H4 G. Brit.	-	-	-	-	113	-
S364 MN	202	198	140-279	-	-	-
S340 KS	-	180	-	-	-	-
F98 WI 1936 year class	3796	-	71-328	-	-	-
M340 NV L. Mead	-	216	-	-	-	-
M228 IA lakes	-	-	-	42	281	41-481
J34 OK Ardmore	-	246	-	-	-	-
V44 LA Springhill L.	2	246	201-290	1	313	-
H77 AR	-	-	-	-	-	168-680
L16, B66 U.S.	-	-	-	-	-	340-907
H150 OK Tenkiller L.	17	368	320-424	-	-	-
W112 Israel select stock	-	-	-	-	993	-
Age 0						
M503 IA Coralville L.	3	79	71-94	3	32	14-41
R148, P63 IA streams	22	94	64-140	-	-	-
L126 CO	1	104	-	-	-	-
H398 IA Coralville L.	+	-	76-152	-	-	-
F98 WI	30	127	41-251	-	-	-
K127 Egypt end of 1st year	802	163	91-262	802	86	-
H156, C271 OK	37	198	152-216	-	-	-
Age I						
M215 KS Big Blue R.	4	84	-	-	-	-
O28 Bohemia	32	-	56-91	-	-	-
H150 OK Illinois R.	3	135	-	-	-	-

		TL			Weight	
Age I (cont.)	No.	Mean	Range	No.	Mean	Range
J8 TX	-	-	102-204	-	-	-
G75 OR	5	160	142-180	-	-	-
S207 KS Fall River L.	1	163	-	1	59	-
C208, L126 CO	12	165	145-236	-	-	-
O8, L13 OH	23	165	-	-	-	-
E9, S49 NY	-	-	152-254	-	-	-
M503 IA Coralville R.	12	175	147-218	12	77	36-127
S236 UT Bear R.	13	178	-	-	-	-
J34, B183 OK	15+	185	135-234	+	82	-
P63 IA Squaw C.	134	190	127-272	-	-	-
W137 UT pond	5	190	137-338	-	-	-
L129 IL	+	198	-	-	-	-
M418 IA Chariton R.	72	198	-	72	100	-
C270 MN Mississippi R. backwaters	3	201	183-216	-	-	-
M228 IA April	-	-	-	16	109	23-186
M112 WI	3	-	206-269	3	-	136-318
S15 TN Reelfoot L.	2	224	221-226	2	145	141-154
H156 OK Canton L.	34	226	-	-	-	-
L16 U.S.	-	229	-	-	227	-
S390 AL March	-	-	-	-	227	-
W47 Germany starved	-	249	-	-	127	-
normal	-	259	-	-	290	-
T10 Illinois R. knothead	-	-	-	2	154	86-227
normal	-	-	-	3	540	-794
B97 IL Gale L.	-	-	-	1331	200	-
S122 Germany starved	-	-	-	3	340	290-422
normal	-	-	-	3	426	399-445
K99 Israel March	-	-	-	-	-	113-240
May	-	-	-	753	486	-
R148 IA Des Moines R.	11	246	213-292	-	-	-
E3 MN	85	251	97-437	-	-	-
E28 TN Chickamauga	151	257	-	-	-	-
Y4 Israel lake	29	262	216-305	-	-	-
E47 IA Clear L.	14	269	-	-	-	-
U11 SD Oahe L.	25	277	-	25	136	-
M141, S236 UT Ogden Bay	33	287	-	-	-	-
Y19 IA	13	292	251-335	13	318	181-499
B150 IA Little Wall L.	165	297	201-348	165	381	118-581
Y9 Japan	-	-	137-305	-	-	50-567
L115 China	-	-	-	-	399	-
H4 G. Brit.	-	-	-	-	-	454-680
C270 MN Mississippi R.	12	305	267-366	-	-	-
I10 IA Missouri R.	1	305	-	1	313	-
F98 WI 1936 year class	2005	-	170-452	-	-	-
other year classes	440	320	221-470	-	-	-
N86, S356 Japan 24 ponds	-	-	-	-	689	331-971
J25, H150, J41, W7, O37 OK	3+	345	284-419	3	758	680-907
V44 LA	3	358	351-363	3	667	580-794
B66 U.S.	-	-	-	-	-	313-2268
H56 Germany	-	-	-	-	-	1250-1475

	TL			Weight		
	No.	Mean	Range	No.	Mean	Range
I2 India	-	-	406-508	-	-	907-1361
S504 AR	+	472	-	+	2041	-
L65 Haiti	-	-	-	52	2041	1814-2268
W112 Israel June select	-	-	-	-	1982	1306-2835
Nov. select	-	-	-	-	3600	2096-2676
C55 IL	1	559	-	1	4196	-
Age II						
M340 UT Glen Canyon	8	102	81-122	-	-	-
M215 KS Big Blue R.	10	119	-	-	-	-
C53 BC Okanagan L.	1	160	-	-	-	-
G128 Germany	-	173	-	-	64	-
C208 CO	27	173	-	-	-	-
T44 OK Grand L.	27	221	-	27	141	-
O8, L13 OH	485	221	-	-	-	-
M503 IA Coralville L.	21	231	226-297	21	195	150-308
E28 TN Chickamauga	246	262	-	-	-	-
S15 TN Reelfoot L.	1	269	-	1	263	-
P63, R148 IA rivers	342	277	208-366	-	-	-
S102 NY Cayuga L.	-	-	-	40	340	113-907
U11 SD Oahe	68	297	274-363	68	318	227-590
O28 Bohemia	11	284	178-399	5	585	100-1152
M448 IA Chariton R.	37	300	-	37	322	-
B66, L16 U.S. introduced	-	305	-	-	-	426-508
S207 KS	41	310	267-381	41	358	-
B183, H150, H156, C271, O37 OK	122+	315	-	2	1361	-
L126 CO	6	323	305-353	-	-	-
L129 IL	-	328	-	-	-	-
L115 China	-	-	-	-	650	-
W47 Germany	-	-	328-406	-	-	417-1624
Y9 Japan	-	-	229-421	-	-	177-1284
T10 Illinois R. knothead	-	-	-	6	513	340-740
normal	-	-	-	6	853	626-1193
E9 NY	-	-	343-427	-	-	-
I10 IA Missouri R.	8	351	307-396	8	550	313-771
S236 UT Bear R.	11	356	-	-	-	-
M112 WI	3	-	318-419	3	-	499-1089
Y4 Israel	17	358	279-396	-	-	-
W142 IN Little Grassy P.	35	373	335-399	35	707	495-1320
Y19 IA	62	376	264-457	62	771	272-1179
M141, S236 UT Ogden Bay	37	414	-	-	-	-
E3 MN	188	417	-	-	-	-
H4 G. Brit.	-	-	-	-	-	907-1134
H56, S122 Ger.	-	-	-	-	-	1252-1502
C270 MN Mississippi R.	7	432	396-462	-	-	-
E47 IA Clear L.	186	437	-	-	-	-
F12, F98 WI 1936 year class	931	-	302-498	-	-	-
F98 WI lakes other year classes	66	483	318-538	-	-	-
I2 India	-	610	-	-	2268	-

		TL			Weight	
Age II (cont.)	No.	Mean	Range	No.	Mean	Range
L65 Japan	-	-	-	-	2495	-
V44 LA Springhill L.	3	620	528-737	3	3470	1814-5775
L65 India	-	-	-	1	3856	-
Age III						
M340 UT Glen Canyon	5	114	94-132	-	-	-
M215 KS Big Blue R.	10	178	-	-	-	-
C208 CO	114	229	-	-	-	-
S236 UT Bear L.	21	231	-	-	-	-
O8, L14 OH	287	244	-	-	-	-
J25 OK Tenkiller L.	-	251	-	-	-	-
G128 Ger.	-	259	-	-	231	-
M503 IA Coralville L.	86	323	272-363	86	299	191-658
H1 Ger.	3	330	-	3	308	-
B23, S102 NY	-	-	-	41	703	286-1361
O28 Bohemia	10	330	216-442	3	962	181-1502
Y9 Japan	-	-	287-399	-	-	404-1170
L129, T10 IL	-	358	-	29	1134	539-3006
M448 IA Chariton R.	49	358	-	49	567	-
L126 CO	4	366	318-406	-	-	-
H156 OK Canton L.	14	371	-	-	-	-
S15 TN Reelfoot L.	14	373	345-414	14	821	680-1021
T76, F12 WI	44	394	-	1	544	-
L16 U.S.	-	381	-	-	1021	-
R148, I10 IA rivers	32	401	284-533	2	717	481-953
Y19 IA	66	447	312-594	66	1134	408-2041
U11 SD Oahe L.	55	450	432-460	55	1179	998-1406
S236 UT Bear R.	14	450	-	-	-	-
S122 Ger.	-	-	-	6	1483	1239-1660
E3, C270 MN	132	536	406-625	-	-	-
H56 Europe	-	-	-	-	-	1361-1588
L115 China	-	-	-	-	1497	-
F98 WI 1936 year class	425	-	351-549	-	-	-
other year classes	37	495	409-587	-	-	-
C215 MI	5	503	-	-	-	-
S102 NY Cayuga L.	-	-	-	19	2177	1361-3175
S122 Ger.	-	-	-	-	2513	-
V44 LA Springhill L.	6	584	511-643	3	2694	1760-3290
S236, M141 UT Ogden Bay	69	607	-	-	-	-
E47 IA Clear L.	43	630	-	-	-	-
Age IV						
M340 UT Glen Canyon	3	185	130-229	-	-	-
M215 KS	3	229	-	-	-	-
C208 CO	10	277	-	-	-	-
S236 UT Bear Lake	13	284	-	-	-	-
G128 Ger.	-	330	-	-	399	-
O8 OH	44	353	-	-	-	-
L129 IL	-	381	-	-	-	-
T76 WI	29	381	356-399	29	698	499-998
M503 IA Coralville L.	24	386	356-414	24	640	499-894
O28 Bohemia	5	394	249-486	3	1302	780-1805
T44 OK Grand L.	20	386	-	20	871	-

	TL			Weight		
	No.	Mean	Range	No.	Mean	Range
L126 CO	15	394	318-470	-	-	-
S15, S203 TN Reelfoot L.	34	394	356-445	34	1061	934-
T10 IL rivers knotheads	-	-	-	31	767	313-1306
normal	-	-	-	14	1193	626-1928
S503 ON females	5	419	391-445	5	1116	794-1361
H156, J31 OK	9	445	427-470	6	1025	-
M448 IA Chariton R.	30	437	-	30	1248	-
Y19 IA	52	472	406-574	52	1270	907-2177
R148 IA Des Moines R.	17	503	432-582	-	-	-
S102 NY Cayuga L.	-	-	-	11	1361	454-2268
U1 IN	1	478	-	1	1814	-
D9 France	-	-	-	-	1996	-
U11 SD Oahe L.	16	505	478-531	16	1814	1497-2223
F98 WI	834	538	475-559	-	-	-
S236 UT Bear R.	27	561	-	-	-	-
C270 MN Mississippi R.	9	566	531-632	-	-	-
V44 LA Springhill L.	4	579	566-592	4	2636	2268-3290
E3 MN	86	643	437-749	-	-	-
L16 U.S.	-	508	-	-	2495	-
S102 NY Canoga L.	-	-	-	55	2540	1361-3629
C215 MI	1	671	-	-	-	-
M141 UT Ogden Bay	64	706	-	-	-	-
E47 IA Clear L.	17	737	-	-	-	-
Age V						
M101 CA deformed carp	1	183	-	-	-	-
L118 NY	22	203	142-251	-	-	-
M215 KS Big Blue R.	4	287	-	-	-	-
O8 OH	44	353	-	-	-	-
S236 UT Bear R.	43	358	-	-	-	-
G128 Ger.	+	368	-	+	526	-
T76 WI	14	391	325-541	14	807	590-1678
C206, L126 CO	4	404	371-495	-	-	-
L129 IL	+	406	-	-	-	-
H420 TN Watts Bar L.	7	409	-	-	-	-
O28 Bohemia	1	424	-	1	1420	-
H156, T44 OK	49	429	-	48	1016	-
L13 OH	4	439	-	-	-	-
S15, S203 TN Reelfoot L.	40	429	381-470	40	1134	962-1331
S503 ON females	13	475	457-495	13	1561	1248-1928
T10 Illinois R. knotheads	-	-	-	17	1080	426-1928
normal	-	-	-	4	2041	852-3770
M503 IA Coralville L.	7	480	463-497	7	1239	1148-1265
S102 NY Cayuga L.	-	-	-	16	1814	1134-2268
M448 IA Chariton R.	2	488	-	2	1588	-
J31 OK Grand L.	6	541	526-554	6	1565	-
R148 IA Des Moines R.	7	521	437-627	-	-	-
S49 NY Oneida L.	62	-	406-660	62	-	1814-4990
S102 NY Canoga L.	-	-	-	39	3689	2268-4536
Y19 IA	17	561	419-610	17	2585	953-3946
U11 SD Oahe L.	4	572	526-589	4	2631	2041-2812
F98 WI	132	582	486-627	-	-	-

Age V (cont.)	No.	TL Mean	Range	No.	Weight Mean	Range
S236 UT Bear R.	15	660	-	-	-	-
C270, E3 MN	53	726	625-874	-	-	-
M141 UT Ogden Bay	81	744	-	-	-	-
E47 IA Clear L.	2	789	-	-	-	-
Age VI						
T76 WI	2	368	315-421	2	680	454-907
S236 UT Bear L.	14	399	-	-	-	-
O8 OH	4	399	-	-	-	-
H420 TN Watts Bar L.	7	432	-	-	-	-
L129 IL	+	439	-	-	-	-
S15, S203 TN Reelfoot L.	56	462	394-528	56	1388	1220-1928
M215 KS Big Blue R.	1	472	-	-	-	-
T10 Illinois R. knotheads	-	-	-	5	1275	880-2268
normal	-	-	-	1	3062	-
H1 Ger.	1	486	-	1	948	-
S503 ON females	11	511	493-526	11	1928	1700-2436
C270 MN river backwaters	1	528	-	-	-	-
S102 NY Cayuga L.	-	-	-	15	2268	1361-3175
U1 IN	2	526	-	2	2440	-
L13 OH	5	523	-	-	-	-
J31, T44 OK Grand L.	7	546	533-594	5	2141	-
U11 SD Oahe L.	2	638	615-663	2	3810	3538-3992
R148, Y19 IA Des Moines R.	5	653	589-734	+	5806	-
F98 WI	13	655	615-686	-	-	-
S236 UT Bear R.	12	709	-	-	-	-
M141 UT Ogden Bay	33	754	-	-	-	-
E3 MN	13	798	625-939	-	-	-
E47 IA Clear L.	1	833	-	-	-	-
S102 NY Canoga L.	-	-	-	33	4109	3175-5443
Age VII						
T76 WI	1	363	-	1	595	-
S236 UT Bear L.	6	434	-	-	-	-
H420 TN Watts Bar L.	13	442	-	-	-	-
C53 BC Okanagan L.	+	455	-	-	-	-
M215 KS Big Blue R.	3	480	-	-	-	-
S15, S203 TN Reelfoot L.	59	503	419-577	59	1778	1588-2495
H1 Ger.	3	538	-	3	1166	-
T10 Illinois R. knotheads	-	-	-	6	1700	513-2268
L129 IL	+	541	-	-	-	-
U1 IN	3	549	-	3	2722	-
S503 ON females	3	576	574-577	3	2917	2830-3402
J31 OK Grand L.	1	615	-	1	2268	-
O8, L13 OH	5	620	-	-	-	-
R148, Y19 IA Des Moines R.	5	686	582-785	2	5466	5171-5761
F98 WI	15	650	627-691	-	-	-
S236 UT Bear R.	1	747	-	-	-	-
M141 UT Ogden Bay	11	754	-	-	-	-
E3 MN	4	813	749-967	-	-	-
S102 NY Canoga L.	-	-	-	25	4935	4082-5897
Y19 IA Spirit L.	2	727	-	2	5761	5670-5851

		TL			Weight	
	No.	Mean	Range	No.	Mean	Range
Age VIII						
T10 Illinois R. knothead	-	-	-	1	740	-
H420 TN Watts Bar L.	7	410	-	-	-	-
S236 UT Bear L.	5	480	-	-	-	-
C53 BC Okanagan L.	+	513	-	-	-	-
H1 Ger.	1	516	-	1	1197	-
S15, S203 TN Reelfoot L.	27	561	521-615	27	2572	-2863
S503 ON females	5	612	599-627	5	3810	3411-4717
L129 IL	+	660	-	-	-	-
U11 SD Oahe L.	1	683	-	1	4082	-
U1 IN	1	706	-	1	4990	-
F98 WI	7	703	665-742	-	-	-
O28 Bohemia	1	709	-	1	4091	-
M141 UT Ogden Bay	2	749	-	-	-	-
E3 MN	1	874	-	-	-	-
P44 England	-	-	-	1	6804	-
Y19 IA Des Moines R.	1	769	-	1	5443	-
Age IX						
S236 UT Bear L.	3	483	-	-	-	-
H420 TN Watts Bar L.	1	498	-	-	-	-
M213 KS Big Blue R.	3	523	-	-	-	-
T10 Illinois R. knotheads	-	-	-	2	1433	1361-1502
S15, S203 TN Reelfoot L.	21	587	-655	21	3107	2722-3629
I10 IA Missouri R.	1	615	-	1	2994	-
S503 ON females	6	643	635-648	6	4672	3629-6532
F98 WI	4	676	-	-	-	-
H1 Ger.	2	798	-	2	4128	-
L129 IL	+	818	-	-	-	-
Age X						
S236 UT Bear L.	2	455	-	-	-	-
M215 KS Big Blue R.	2	486	-	-	-	-
H420 TN Watts Bar L.	3	511	-	-	-	-
S15, S203 TN Reelfoot L.	18	645	615-688	18	3800	-4702
S503 ON females	5	668	655-673	5	4990	4309-5625
F98 WI	6	703	688-782	-	-	-
L129 IL	+	896	-	-	-	-
Age XI						
H420 TN Watts Bar L.	1	523	-	-	-	-
M215 KS Big Blue R.	1	536	-	-	-	-
S15, S203 TN Reelfoot L.	8	681	660-732	8	4463	-4877
S503 ON females	6	701	686-724	6	5670	5171-6760
F98 WI	1	711	-	-	-	-
T10 Illinois R.	-	-	-	1	9980	-
Age XII						
S236 UT Bear L.	1	460	-	-	-	-
M215 KS Big Blue R.	1	559	-	-	-	-
F98 WI	4	691	-	-	-	-
S15, S203 TN Reelfoot L.	10	732	698-775	10	5593	5262-5806
Age XIII						
M215 KS Big Blue R.	2	612	-	-	-	-
S236 UT Bear L.	1	671	-	-	-	-
S15 TN Reelfoot L.	3	798	-	3	6718	6074-6804

	TL			Weight		
	No.	Mean	Range	No.	Mean	Range
Age XIV						
I4 IA	1	1219	-	1	26990	-
Age XV						
S503 ON female	1	787	-	1	11068	-
H56 Ger.	-	-	-	2	22000	19050-24948
Age XVI						
F7	-	-	-	1	8500	-
S503 ON female	1	851	-	1	10115	-
Age XVII						
H164 S. Afr.	-	-	-	1	14060	-
Age XXXVIII						
F7	1	831	-	1	9400	-

F7 reports a carp that lived for 47 years in captivity.
S505 Egypt rice ponds

107 mm, 18 g to 190 mm, 113 g in 62 days
145 mm, 54 g to 262 mm, 240 g in 62 days
122 mm, 32 g to 201 mm, 118 g in 46 days

	No.	1	2	3	4	5	6	7	8	9	10	11	12	13
						Average calculated TL at each annulus								
W192 SD Lewis & Clark L.	+	84	142	170	211	234	267	307	371	401				
L195 OK Cimarron R.	22	89	142	188	216	224	224	282	307					
Arkansas R.	100	122	196	264	310	356	388	429	470	551	599			
B134 CO Loveland L.	5	127	213	315										
M152 UT Bear L.	109	61	142	211	279	338	381	419	452	467	483	531	554	678
M141, M151 UT Ogden Bay	330	173	353	518	635	691	716	724	721					
S391 OK Rock Creek L.	17	132	201	246	302	351	371	399	421	457	472			
J41 OK Ft. Gibson	205	127	234	297	373	470								
E61 OK Salt C.	7	147												
J27 OK Lower Spavinaw	22	170	335	427	516	655	696	742	785	813				
J31 OK Grand L.	279	203	351	437	503	556	605	643	719					
J101 OK Spavinaw L.	119	236	353	404	455	500	541	582	671					
J101 OK L. Eucha	123	290	429	505	549	574	648	696						
J37 OK Rod & Gun L.	20	399	612	678										
J34 OK Ardmore L.	30	445	622	703	782	831								
P113 MO statewide	668	165	279	361	424	457	493							
Poorest average		124	246	333	378	406	432	465	490	493	465	486	513	
Best average		190	307	396	478	523	546	582	653	518	579			
P73 MO L. Wappapello	128	168	241	343	388	488								
P112 MO Salt R. middle	102	124	269	363	434	460	513	551						
lower	119	147	295	388	445	488	531	582	653					
K71 TN White Oak L.	78	99	160	224	239	213								
slowest	30	91	127	170										
fastest		173	239	351	414									
E28 TN Chickamauga	397	185	246											
L129 IL	145	198	328	358	381	406	439	541	660	818	896			
M197 IA L. Keomah	135	259	343	406	470	561	630							
M195 IA L. McBride	22	213	356	488	643									
E47 IA Clear L.	263	218	472	610	696	754	798							
E2, E3, S62, M258 MN	1082+	178	368	465	564	676	754	769	820					
O28 Bohemia ponds	83	79	183	241	297	310								
22 other waters, poorest		61	117	170	201	272	264							
best		147	279	386	455	505	561	493						
B290 Caspian Sea males	+	183	320	465	589	676	759	965						
females	+	188	330	478	615	836	955	1026	1102	1161	1219			
P163 L. Aral	+	155	297	409	493	559	632							

Calculated TL and increments at each annulus

	1	2	3	4	5	6	7	8	9	10	11	12	13	14	15
B183, C271 OK Canton L.	201	279	381	434	445										
Increment	201	84	97	84	86										
Number	172														
H156 OK Canton L.	178	251	325	376	424										
Increment	178	69	61	43	33										
Number	127	86	18	4	1										
T44 OK Grand L.	173	254	315	391	432	528									
Increment	173	81	61	46	36	33									
Number	100	100	100	73	53	5									
H303 OK Lawtonka L.	262	353	445	538	569	620									
Increment	262	91	56	23	30	41									
Number	21	10	6	3	3	1									
C270 MN Mississippi R. backwaters	196	376	486	561	655										
Increment	196	175	109	76	43										
Number	54	42	35	13	4										
S236 UT Bear L.	61	142	211	279	338	381	419	452	467	483	531	554	678		
Increment	61	81	69	64	58	43	36	33	28	33	20	23	36		
Number	109	109	109	88	75	32	18	12	7	4	2	2	1		
S236 UT Bear R.	147	295	419	521	605	655	696								
Increment	147	147	130	102	86	51	25								
Number	93	80	69	55	28	13	1								
S236 UT Ogden Bay	173	353	518	635	691	716	724	721							
Increment	173	180	178	119	66	48	33	30							
Number	330	297	260	191	127	46	13	2							
H420 TN Watts Bar L.	150	213	269	300	356	386	411	437	460	483	503				
Increment	150	64	56	46	38	33	28	28	25	25	25				
I10 IA Missouri R.	155	287	351	450	503	523	546	566	582						
Increment	155	135	71	43	53	20	23	20	15						
Number	12	11	3	1	1	1	1	1	1						
R148 IA Des Moines R.	127	236	361	457	518	599	610								
Increment	127	109	112	94	74	51	38								
Number	413														
S227,229,276, N94, N115 SD Gavins Point	89	175	226	254	287	323	361	409	439	480	518	549	577	599	643
Increment	89	86	53	36	30	25	25	23	20	18	15	15	13	13	10
Number	1364	1094	899	767	612	417	292	177	112	71	39	26	16	9	1
S224,225,228,393 SD Ft. Randall	165	264	348	437	523	599	648								
Increment	165	130	91	97	84	74	15								
Number	881	695	376	70	13	5	1								
S224,225 SD Ft. Randall tailwaters	145	231	318	399	483										
Increment	145	104	107	74	89										
Number	41	19	8	6	2										
F100,105,124,125 SD Oahe	84	168	249	305	340	368	401	406	442	488	526	569	587	610	
Increment	84	89	84	79	64	56	48	38	43	46	46	41	28	23	
Number	1047	816	669	369	186	98	49	23	14	10	7	3	1	1	
F98 WI L. Monona	117	348	450	495	564	635	658	724	686	696	701				
Increment	117	241	107	48	41	41	28	25	8	10	5				
Number	525	437	417	411	32	13	11	4	1	1	1				
WI L. Waubesa	132	338	452	505	577	627	653	655							
Increment	132	226	122	56	41	41	25	13							
Number	641	324	311	244	13	2	2	1							
WI L. Kegonsa	122	338	462	500	544	632	660	696	742	767					
Increment	122	216	127	43	43	38	33	13	13	28					
Number	315	304	273	261	89	11	4	2	1	1					
WI L. Wingra	152	409	513	549	572	610	630	643	655	663	665	678			
Increment	152	251	97	36	25	23	18	15	18	15	18	15			
Number	101	77	57	55	53	24	20	15	13	9	1	1			
Y19 IA	224	310	396	462	561	688	724	729							
Increment	224	86	74	61	51	38	33	58							
Number	193	180	141	75	23	6	5	1							

	No.	1	2	3	4	5	6	7	8	9	10	11	12
					Calculated weight at each annulus								
R148 IA Des Moines R.	413	27	172	630	1284	1878	2921	3071					
M197 IA L. Keomah	135	227	454	680	1048	2041	3856						
F98 WI Waubesa L.													
males	622	82	495	780	1152								
females	577	82	495	939	1451								
Kegonsa L.													
males	258	113	531	843	1084								
females	291	113	531	1025	1288								
Monona L.													
males	158	109	431	721	990								
females	176	109	431	839	1130								
Madison lakes	240	1052	1338	1700	2160	4468	3661	5438					
B209 Caspian Sea	54	417	1570	2413	3130	4436	7420	8210	9072	10340	12338	13926	

OK lakes J31, T44, H156 Annual increments as percentages of length at beginning (number of specimens in parentheses)

TL	Median	Range
76-100	74(27)	-
152-177	39(28)	37(27)-42(1)
178-202	45(142)	37(68)-60(6)
203-228	29(25)	24(3)-57(2)
229-253	56(7)	51(6)-62(1)
254-278	26(17)	21(14)-32(3)
279-304	24(80)	19(20)-27(6)
330-355	15(70)	11(20)-26(2)
356-380	13(22)	9(4)-19(7)
381-404	8(48)	-
405-431	15(11)	13(6)-18(5)
432-456	16(3)	14(1)-18(2)
457-482	10(6)	-
483-507	7(5)	-
508-532	6(3)	6(2)-7(1)
533-558	5(3)	4(2)-7(1)
559-583	5(1)	-

			Weight increment (g per day)
Y3 Israel	Nursery stage 10-60 g	Israel race	.66-.354
		Punten race	.306-.415
		crossbreed	.341-.782
	Fattening stage 300-390 g	Israel race	2.2-3.0
		Punten race	2.5-3.1
		crossbreed	2.6-3.2
Y2 Israel	40-50 g carp, May		1.9-9.2
	When kept from bottom, feeding only on plankton		0.35-3.82
	70-90 g carp, April, May		3.1-5.5
Y5 Israel	Winter, no food or feeding,		
	carp alone		0.47-0.53
	carp with *Tilapia*		0.43-0.65
	With feed or fertilizer,		
	carp alone		0.43-1.24
	carp with *Tilapia*		2.50-2.58

				Weight increment (g per day)
Y7 Israel	One 366 g carp, September 4-30			18.7
W112 Israel	July, 1951-Nov., 1952			7.6
S240 Israel				0.1-18.1

				mm per day
S228, S227	SD Ft. Randall	Age	I	1.3-1.5
			II	0.9
			III	0.5-0.6
S229	SD Gavins Point		I	0.7
			II	0.6
J34	OK Ardmore L., April-June	Age	0	2.5
H150	OK Tenkiller, May-Oct.	Age	0	2.5

Scales were used for most of the age and growth studies, and it was with carp that the scale method was first clearly demonstrated by Hoffbauer (H75, H345). The annuli may usually be distinguished by the cutting-over or anastomosis of the circuli in the lateral fields (F98, C271). On the anterior field the circuli are usually close together before the annulus forms and then show a distinct change of pattern (C271). Radii on the anterior field often originate at annuli (F98). In Wisconsin the first annulus formed in May, but the 2nd and 3rd not until mid-June through July (F98). Females formed annuli earlier than males and smaller fish formed annuli earlier than larger fish of the same age. Some carp did not grow in some years and failed to form annuli (F12). In Oklahoma annuli form in late May or June (C271), in South Dakota in late May at about 17-18.5C (S393,229,225,227,226) or by mid-June (S228), and in Lake Aral, U.S.S.R. in April or May (P163).

Most growth computations were made on the assumption of direct proportion between scale and body growth, but C270, C271, P73, P112, and O28 used a Fraser-type correction factor of about 25 mm. The true relationship between the anterior radius of the scale and body length was found to be sigmoid (F98) and the diameter was not usable because of opacity of the scale and the lateral radii were unusable because of frequent resorption.

Accessory marks or false annuli tend to be frequent and often appear as a dark band (C271).

Opercular bones were believed to be superior to scales for age and growth determinations by M141, S236, E47, R148, and Y19. The opercle-body relationship was found to be curvilinear but more consistent than the body-scale relationship (M141).

Carp have been raised extensively in ponds in many parts of the world. It has not been possible to review here the extensive information on growth. This summary attempts to include all the growth data published in the United States, plus a sampling of that in other countries.

Within the United States there is little evidence of regional differences in growth rates. Very slow growth rates were reported from Glen Canyon, Utah (M340), Big Blue River, Kansas (M215), Okanagan Lake, British Columbia (C53), and Wisconsin (T76), but rapid growth was reported from neighboring waters in each of these areas (except British Columbia, where other studies on carp have not been published). Very fast growth was reported from Illinois (C55), Clear Lake, Iowa (E47), and Spring Hill Lake, Louisiana (V44). Growth over several years of life seems to be excellent in the Caspian Sea (B290). Rapid growth in Egypt was believed to be the result of long periods of high temperature (K127) and the same is true of Israel and other semitropical and tropical areas.

No sex differences in growth were noted by N115, S229, S393, S276, R148, S226, Y19, but females were found to grow faster than males in B290, F98, C271, K99, S228, S227, W189, and Y9, particularly after the second year of life.

Selective breeding has developed some fast-growing races of carp in other parts of the world. In Israel (W112) carp from selected stock weighed over 2 pounds in October of the first year. Growth of scaly and mirror races did not differ in some tests (W151), but survival of the scaly race was better than that of mirror carp in crowded ponds. A lethal gene (N) found in some carp retards growth in the heterozygous condition and is lethal in the homozygous (W211).

Most of the season's growth in Oahe Lake, South Dakota, was in late summer (F105). In Israel, where water temperatures are 11-19C, carp may grow even in the winter (Y2, Y5, W112). In Japan (K51) there appeared to be no growth beyond November. Growth per day was positively correlated with water temperature at temperatures of 14.5-18.5C for early fry stages, but the effect was less evident after feeding started (T126).

Individual differences in growth are quite pronounced. G88 noted that when growth is rapid the weights of carp seem to be normally distributed but that when growth is slow the distribution spreads, suggesting individual differences. A deformed carp grew very slowly though it lived 5 years before being caught (M101) at which time it was only 183 mm. Knothead carp, a condition due to vitamin D deficiency, grew more slowly than normal carp (T10).

Well-fed carp grow more rapidly than those on starvation diets (W47), and carp which could feed on the bottom as well as on plankton grew better than those dependent upon plankton (Y2). Quality and quantity of food of the right size were important for growth and survival of carp fry when they started to feed the 4th day after hatching (C343).

Population density seems to be the most significant factor controlling growth of carp (K51, V75, N86, W182). Even with excess food, population density controls growth (K51). With carp fingerlings of 1 g stocked June 4, the average weight on July 4 was (V75):

Fingerlings stocked per hectare	Mean weight, July 4
5500	18
8325	12.5
11000	8.5
19425	6.5
36075	3.5
55500	2

In Poland (W178) the average weights in grams were as follows at the time of harvest:

Fry stocked per hectare	2500	7500	15000	22500	30000
Natural foods only	85	45	24	18	11 g
Fed barley	101	98	60	46	29 g

The population density perhaps controls growth through the buildup of excretory wastes (S390), nitrogen (K51), or antithiaminase (Y6). Carp in separate boxes grew better than carp in ponds with the same space per fish, so that space *per se* is not the only factor (Y2). During the winter in Israel no difference in growth was noted over the range of 680 to 2600 fish per hectare (Y5). In Poland the maximum increase, when fed or on natural food, was at

15000 fry per hectare (W178) with densities of 2500 to 37000 tried. The biomass of fry in ponds where the fish were fed was 3 times that in ponds depending upon natural foods, and the dry weight was 4 times that in the unfed
population (W178).

Weight increase in pens in the Bear River, Utah, was negatively correlated
with the population density (R222).

$$
\begin{array}{lll}
 & & r_{xy} \\
1959 \quad \% \text{ weight increase} = 19.6 - 0.023X & -.87 \\
1960 \qquad\qquad\qquad\qquad = 7.5 - 0.0036X & -.74 \\
1961 \qquad\qquad\qquad\qquad = 10.0 - 0.009X & -.85 \\
\end{array}
$$

where X = kg/ha

The effect was more pronounced in 1959 when the carp averaged 2500 g
than in 1960 or 1961 when they averaged 3900 and 3500 g.

The growth of the very abundant 1936-year class in several Wisconsin
lakes was slower than that of the less abundant year classes (F98).

Growth of young carp was much slower (they averaged 69 mm in October)
in cutoff pools than in Canton Reservoir, Oklahoma, where they averaged 203
mm (C271). Growth improved the first year of impoundment (C271, F105,
S225), but decreased with impoundment age at Lewis and Clark Lake (W192).

In Fort Randall Reservoir, South Dakota (S225), the average annual increments were much higher the first year after impoundment than before.

Year of life	1	2	3	4	5	6
Increment (mm) before	89	89	89	71	84	-
Number	67	43	17	2	1	-
Increment after	165	170	155	163	107	132
Number	136	24	26	15	1	1

Ammonia in the water reduced the growth of carp and increased their
oxygen consumption at higher temperatures (K91). Carp raised in water draining from irrigated lands in Israel showed less growth and poorer food conversion than those in fresher water, although the chloride content did not reach
the 2000 mg of C1/1 usually regarded as inhibiting growth of carp (M502).
Sulfates, magnesium or other ions may be limiting in this situation. Growth of
young carp was slowed when dissolved oxygen was below 3 ppm (C344).

A carp tagged and released at 269 mm, 240 g in March, 1954, in Missouri
was recaptured 1085 kilometers up the Missouri River in South Dakota, 28
months later, at 584 mm, 2068 g (S236).

The following temperature preferences for carp were given by P103:

acclimated at °C	10	15	20	25	30	35
preferred °C	17	25	27	31	31	32

Carp were found swimming freely at 34-35.5C (T102) and at 33.4C (R262).

The upper lethal temperatures were found to be 31-34C when acclimated
at 20C or 35.7C when acclimated at 26C (B200). The lower lethal temperature
was found to be a body temperature of 0.7C (B202). Aquarium carp were inactive at 1.7C and showed some wariness but did not feed at 2.2C (S236). Carp at
1.7-4.5C did not feed. Aquarium carp were frantic at 26.7C when alarmed in
contrast to much less activity at 15C (S236).

Carp can stand a rapid change to 1.0% salt solution but 1.5-3.0% is lethal
in 1.5 days (B204).

Blood counts for 2 carp (S151) gave 1,310,000 ± 1950 red blood cells per cu mm and a hemoglobin of 42.0.

A literature review (A51) indicates fast-growing males may mature at age I but mostly at ages II to IV. Male carp may mature at age II in Wisconsin (M112, F98), at age III in Ontario (S503), at age I in Oklahoma (C271), at age I-IV in South Dakota (N115, S393, S228, S231, S224, S276, S226, S227, S229), and at age I if they weigh 225 g in March in Alabama (S390). The literature review (A51) indicated that female carp mature at ages III-V. Female carp may mature at age III (with a few at age II, F98) in Wisconsin (M112, F98), at age IV in Ontario (S503), at age II in Oklahoma (C271), and at age II-V in South Dakota (N115, S393, S228, S231, S224, S276, S226, S227, S229).

In Israel carp may spawn during their first year of life. Those hatched in August or September will not spawn the next growing season, however, and can be used in ponds which are to be drained only once a year (L212). The usual spawning period in Israel is March to May but spawning may be promoted from February to September.

Males produce semen by the time they reach 50 g and this is used in separating sexes for monosex culture in Israel (K99). The smallest mature males reported elsewhere are 226 mm (B183, C271) and 213 mm (B94). The smallest mature females were 239 mm (B183, C271). Overfed carp mature later than well-fed carp, but good growth is usually accompanied by early maturity (A51).

Spawning starts at 14.5-17C but is most active at 18.5-20C and may spread over a considerable time (S236). Eggs are scattered at random over plant beds, debris, and rubble in shallow water (S503).

In Alabama carp may spawn in March (S390); in Oklahoma, in May to late August (C271); in Illinois, April to June (R26); May to July in Nevada (L193); in South Dakota most spawning is early June at about 18.5C (S226, S393, W192); in Wisconsin, mid-May to mid-August (B294, F98); Michigan, early June (L142). Spawning occurred both before and after closure of the dam at Gavins Point on July 31, 1955 (S227). Two spawnings, in May and June, were evident from two size classes in September in Lake Oahe, South Dakota (F105), and another year spawning occurred in May and late July (F100). Females may spawn a second time in the season, releasing about 80% of the eggs in the first period (S503). Females which completed spawning on June 25, spawned again on July 20 when kept in an isolated pond. Spawning occurred shortly after water levels rose inundating vegetated areas, June 22-28 in 1958 and May 27-31 in 1957 (S330). The first hatch occurred when the water 150 mm from the surface was 23.4C (S334).

	Size of females	No. of females	No. of eggs per female Mean	No. of eggs per female Range
B23, H56, G13, S122	1800-2300 g	-	-	400,000-500,000
M66 IA Okoboji L.	11340 g	1	1,700,000	-
L13 OH	622 mm	1	660,000	-
L16 U.S.	-	-	-	300,000-700,000
H3 Europe	2990 g	1	342,180	-
G13 U.S.	7485 g	1	2,000,000	-
C55 IL	3630 g	1	600,000	-
	7600 g	1	101,200	-
	11570 g	1	203,109	-
	9750 g	1	1,310,750	-
	7485 g	1	2,059,750	-
H4 U.S.	5-year olds	-	500,000	-

	Size of females	No. of females	No. of eggs per female Mean	No. of eggs per female Range
U1 IN	4 years, 1800 g	1	115,500	-
	6 years, 2300 g	2	103,950	79200-128,700
	7 years, 2700 g	3	180,400	145,200-211,200
	8 years, 5000 g	1	366,300	-
C53 BC	528 mm	1	300,000	-
S122 Ger.	11-year old	1	860,000	-
P24 G. Brit.	1814 g	2	300,000	200,000-400,000
	6580 g	1	633,000	-
L13, L80 L. Erie	Age V, VI, 4309 g.	2	1,696,000	1,663,000-1,730,000
R148 IA	Age III, 590-1406 g.	3	101,000	39000-137,000
	Age IV, 1089-2268 g.	5	181,000	138,000-215,000
	Age V, 4425 g.	1	335,000	-
Y9 Japan	247-369 g	4	56950	52600-59400
	967 g	1	125,000	-
F98 WI	Age IV-V	4	-	506,000-1,085,000
S503 ON	Age IV, 771-1361 g	5	56463	36000-72000
	Age V, 1225-1905 g	13	233,150	72000-347,000
	Age VI, 1678-2450 g	11	326,385	226,000-470,000
	Age VII, 2900-3400 g	3	664,350	579,000-750,000
	Age VIII-X, 3400-6530 g	16	917,140	576,000-1,389,000
	Age XI, 5170-6760 g	6	1,127,500	816,000-1,389,000
	Age XV, 11070 g	1	1,851,000	-
	Age XVI, 10115 g	1	2,208,000	-

Spawning may be inhibited by excretions or hormones in the water when the population density is high (S390). Big year classes were produced the first year of impoundment in Canton Reservoir, Oklahoma (C271), and in Lewis and Clark Lake, South Dakota (B276). The big year class of 1934 in Lake Chautauqua, New York, was believed to be due to lack of carp removal in 1932 and 1933 (M414). Reproduction was negligible in Lake Wingra from 1953-55 when the population declined (N68). Spawning success was poorer in ponds with bass, bluegills, and crappies (M266) than when carp were alone (M265).

Artificial fertilization and hatching of eggs in jars is described in W174 which also describes some of the embryology. Eggs hatch in 4 to 8 days at 16.7-18.4C (S101), in 50-76 hours at 11-32C (S503), and in less than 4 days at 22C (E46). They are about 3 mm at hatching (S101).

In Lake Wingra, Wisconsin, the carp population was estimated as 15.0-48.8 kg/ha (H250) and 54-470 kg/ha (N68). In fertilized ponds in Alabama 303 to 504 kg of carp were raised per hectare and with feeding the weight ranged from 1120-1457 kg/ha (S325). In natural ponds in Israel the maximum is 4480 kg/ha with an average of 2800 (Y14). Raising of nutria, to control aquatic vegetation, increased carp population in Polish ponds (E133). Aspects of carp production make up a large literature (C317, L16, A69, H418, Y2, Y3, Y4, Y5, Y14, P166, S78, T118). Extensive summaries are given in S122, W190, H363, H364.

The basic food of carp is bottom fauna, primarily chironomids, zooplankton, and phytoplankton, and plant remains (G125, S357, R148, R115, M157, S356, N115, E121, M228, S236). Young carp fed on zooplankton but avoided algae, and feeding intensity increased with zooplankton abundance (A68, R26). Large fry may be cannibalistic even when zooplankton is abundant (B337). The young fish were rarely satiated. With small carp, 40-400 mg, it took 0.13 of the body weight per day of daphnids or 0.03 of tubifex worms to maintain body weight and for increase in body weight took 10-16 weights of daphnids or 3-6 of tubifex (C343). S356 reports that few adults feed on zooplankton although zooplankton are an important food for the young. Even in carp over 400 mm long, however, entomostracans constituted 13% of the food volume (S357). Fingernail clams were found to be a common food item and their lack in polluted areas was believed to be a cause of poor carp growth and knothead conditions (M501). Sand as one fifth of the intestine contents was taken as evidence of poor feeding conditions and a possible cause of slow growth in Bear Lake, Utah (M157, S236). Some feeding occurred at water temperatures of 3F (F98). Feeding fish characteristically root around the vegetation in shallow water (S102).

The effect of carp populations on rooted vegetation is sometimes very significant (W89, C318, C319, H365, J93, B293, K128, T115, T116, T117, M415, R253, S236, S455).

Literature on carp control is extensive, and the following list is not comprehensive (M414, S50, S49, S101, S102, S454, S456, T49, W189, W191, B295, E84, C320, D85, B294, B296, G131, L185, L184, L186, S231, S455).

Carp are not found in chara-type lakes (T115) nor in Minnesota north of the June 18.4C isotherm (S455); the 15.6C mean July isotherm was considered a limit in Manitoba and Saskatchewan (A67). Carp were not found in Minnesota waters with less than 0.05 ppm total phosphorus (S455).

Carp were designated as "mouth tasters" by E123 because the vagal and facial lobes of the brain are large.

Hybrids have been reported with *Carassius carassius* (K114, T113).

ROUNDNOSE MINNOW, *Dionda episcopa* Girard
Oklahoma, Texas and Mexico (E119).
On April 15, 1951, these fish were observed spawning in about 25 mm of water in two seep springs at 17-18C. The eggs were heavy but nonadhesive and lodged in the gravel of the spring. The spawning appeared to be a mass action (H366).

OZARK MINNOW, *Dionda nubila* (Forbes)
Wyoming to Illinois and south to the Ozarks (E119).
This species was classed as a fair bioassay animal (B349), but it does not eat laboratory foods readily, nor are specimens readily available.

DESERT DACE, *Eremichthys acros* Hubbs and Miller
This species appears to be limited to Soldier Meadows, Nevada, in warm springs and creeks, 29-38C (H370) or 20-38C (L193). The largest males were 41 mm SL and females, 60 mm. A nuptial male, 38 mm, was collected July 5 (H370). They eat small fish and insects (L193).

SILVERJAW MINNOW, *Ericymba buccata* Cope
Western Pennsylvania to southeastern Missouri and south to the Gulf of Mexico (T113).

S472 AL	TL	No.	Weight Mean	Range	K(TL)
	25	82	.36	.23-	2.30
	51	371	.9	.7-1.5	.72
	76	42	2.7	1.8-5.0	.61
	102	1	6.4	-	.61

Because the weights of the 25 mm fish are obviously too high, the length-weight regression in S472 is not valid.

T113 OH   0-Oct.   25-46 mm TL;   I, 38-64 mm;   adults 43-97 mm

H8 IL   Age 0-Aug.   18 mm TL
                        0-Dec.   31 mm TL

Silverjaw minnow is a sand-inhabiting species, abundant in brooks and small streams of moderate gradient. It appears tolerant of turbidity and pollution if these do not cover the sand (T113).

Silverjaw minnows were observed spawning on sandy areas in late April in Illinois and ripe fish of both sexes were taken on June 7 (H8). Young fish reared in aquaria were 19 mm in August and 31 mm in December (H8).

They feed on insects (S461).

CUTLIPS MINNOW, *Exoglossum maxillingua* (LeSueur)
St. Lawrence and Lake Ontario south into Virginia (E119).

Maximum about 150 mm. Males are larger and more robust than females (V76).

In Maryland they spawn in May and build stone nests, like *Semotilus*, in running streams (S461). In New York they spawn in late May to early July (M432, V76). Young remain in the nest about 6 days, with males guarding throughout the period (V76). The more aggressive common shiners often use the nests built by cutlips minnows (V76).

They feed on molluscs and bottom insects (S461). Cutlips minnows stay near the bottom and are sluggish (V76). They are usually near flat rocks.

UTAH CHUB, *Gila atraria* (Girard)
Native to Bonneville drainage, Nevada, but introduced into other western streams (E119).

C259 UT   440 chubs

SL	30	40	50	60	70	80	90	100	110	120	130	140	150
Weight	1	3	5	7	9	12	17	23	32	43	50	65	82

SL	160	170	180	190	200	210	220	230	240	250	260
W	100	120	141	177	190	204	246	283	313	342	387

SL	290	310	330
W	542	673	756

The average K decreased with increase in age in Fish Lake, Utah (O41).

C259 UT   SL in each age group

Age group	I	II	III	IV	V	VI	VII	VIII	IX	X	XI
Plateau L. SL	-	64	94	127	142						
No.	-	1	50	27	11						
Bear River L.	43	75	99	167	210	184					
No.	3	9	8	2	1	1					
Utah L.	55	88	107	134	157	217	220	245	310	252	269
No.	13	34	23	24	17	7	3	5	4	3	1

SL in each age group (cont.)

Age group	I	II	III	IV	V	VI	VII	VIII	IX	X	XI
Fish L.	-	-	132	143	158	173					
No.	-	-	7	12	6	1					
Strawberry L.	-	126	166	177	217	-	-	233	227	226	252
No.	-	43	42	6	2	-	-	3	2	2	3
N71 UT Panguitch L.	52	126	127	154	172	182	194	199	207	233	230
No.	7	5	2	25	48	30	21	27	6	2	1
N71 UT Navajo L.	-	101	130	155	169	184	190	218			
No.	-	36	21	59	18	11	7	3			

G108 MT Hebgen L.

	No.	TL Mean	Range			No.	TL Mean	Range
Age 0-July	450	10	6-18	Age V males		56	236	-
0-Aug.	620	23	10-53	females		44	246	-
0-Sept.	422	30	15-51	VI males		88	267	-
0-Oct.	62	33	-	females		101	279	-
I	38	71	-	VII males		7	300	-
II	2	114	-	females		27	318	-
III	53	157	-	VIII females		4	345	-
IV	53	193	-					

Average calculated SL at each annulus

	No.	1	2	3	4	5	6	7	8	9	10	11
N71 UT Panguitch L.	174	34	59	92	124	146	161	173	185	201	218	223
Navajo L.	155	40	67	107	137	154	169	181	204			
C259 UT lakes	575	53	102	122	136	162	201	206	224	181	251	252

C259 UT — Average calculated SL and increments at each annulus

		1	2	3	4	5	6	7	8	9	10
Plateau L.		35	89	109	131						
	Incr.	35	54	20	21						
	No.	89	88	39	11						
Panguitch L.		31	70	111	148	168	189	208			
	Incr.	31	29	40	37	20	20	19			
	No.	29	27	22	13	10	5	1			
Otter Creek L.		29	80	123	170	192					
	Incr.	29	50	43	47	21					
	No.	6	6	3	2	1					
Utah L.		52	97	127	155	182	199	223	248	234	
	Incr.	52	44	30	29	27	17	24	25	15	
	No.	119	83	60	36	19	13	10	5	2	
Fish L.		44	93	131	154	173	194	207			
	Incr.	44	49	38	24	19	18	14			
	No.	108	92	61	25	6	3	2			
Bear River L.		51	92	147	175	175					
	Incr.	51	40	55	28	1					
	No.	21	12	4	2	1					
Strawberry L.		51	109	150	177	195	216	229	239	226	242
	Incr.	51	58	41	26	18	21	13	15	13	16
	No.	113	68	22	15	12	11	11	8	5	3

Calculated TL at each annulus

	No.	1	2	3	4	5	6	7	8	9
H43 UT Fish L.	26	33	89	140	168	188				
M152 UT Bear L.	206	56	99	147	188	224	254	277	302	356
P159, G108 MT Hegben L.	475	41	89	155	208	246	277	315	345	
Incr.		41	48	66	53	36	25	23	20	
No.		475	437	435	382	329	227	38	4	

A maximum size of 560 mm or 1360 g is reported from Bear Lake, Utah, but specimens over 406 mm are not known elsewhere (S463).

In most studies the growth was computed assuming direct proportion between body and scale growth, but N71 applied a Fraser-type correction factor. The annulus forms October to March (C259). Scales first form, along the lateral line, at 20-21.5 mm SL (C259). The more rapid growth in Bear Lake than in Utah Lake was attributed to the lower population density (M152).

Male chubs mature at age III and females at age IV in Hegben Lake, Montana (G108), and both sexes mature at ages IV to VI in Two Ocean Lake, Wyoming (J55).

			Eggs per female	
		No.	Mean	Range
G108 MT	224-351 mm TL	7	40750	15900-62400
O42 UT	108-809 mm SL	22	25282	10470-38123

In Montana they spawn in June to mid-July with a few in May and early August (G108), and in Utah spawning may occur from April to August with most in June (S463, C259). Spawning may start at 12C but peaks at 16C, and spawning takes place in 300-1200 mm of water near shore (S463, G108) in vegetated areas (J55). In Two Ocean Lake the peak occurred about July 9 but continued into early August (J55).

About 1% of the chubs at age III-V in Two Ocean Lake had spinal curvatures, and these fish had a lower survival rate than those without (J55, J92). The ages given in J92 should be one year older than reported (J55).

Utah chubs feed on zooplankton until about 180 mm FL (J55) but from then on tend to be omnivorous, feeding heavily on branches of *Ceratophyllum* and *Myriophyllum*, midge larvae and crustaceans (M152, S234, G108, C259). D. S. Jordan reported that they fed heavily on young trout (C259). Little change in food was noted with increase in size from 135 to 368 mm (G108).

TUI CHUB, *Gila bicolor* (Girard)

The 1960 American Fisheries Society list of names (A13) lists this species as *Siphateles bicolor* (Girard) but the genus *Siphateles* has been placed in the genus *Gila* (B287). The blue chub, *Gila bicolor* (Girard), in the A13 list now becomes *Gila coerulea* (Girard), found in the Klamath Lake drainage of California and Oregon (B287). The tui chub is found in the Columbia, Klamath, and Sacramento river systems and a number of interior basins in California, Nevada, Washington, and Oregon (E119, B287, B338).

K58 CA Eagle R.

SL	50-100	100-150	150-200	200-250	250-300	300-350
No.	8	11	5	16	52	12
Wt.	10	20-50	60-120	160-320	290-535	450-715

K63 CA Big Sage L.   log W = -4.553 + 2.903 log FL

B233 OR L. of Woods

FL	50	90	175	215	245	265	285
No.	1	1	2	5	18	22	28
Wt.	1.4	8	82	119	190	244	292

K58 CA Eagle L.   139 chub 124-349 mm;   average K(SL) = 1.92, with a range
of 1.04-2.94 and a tendency to increase with length.

Age			No.	Mean	Range	Length type	Weight
0-June	B233 OR	L. of Woods	4	20	-	FL	-
0-July			74	28	-	FL	-
0-Aug.			1	48	-	FL	1.4
0-Sept.	K58 CA	Eagle L.	-	-	22-42	SL	-
I	K62 CA	Susan R.	32	-	38-66	FL	-
	K58 CA	Eagle L.	63	-	37-98	SL	-
	B233 OR	L. of Woods	1	89	-	FL	8
II	K58 CA	Eagle L.	110	-	76-157	SL	-
III	K58 CA	Eagle L.	162	-	110-208	SL	-
	B233 OR	L. of Woods	2	193	-	FL	102
IV	K58 CA	Eagle L.	218	-	153-248	SL	-
	B233 OR	L. of Woods	8	229	-	FL	162
V	K58 CA	Eagle L.	248	-	220-304	SL	-
	B233 OR	L. of Woods	26	259	-	FL	230
VI	K58 CA	Eagle L.	248	-	240-309	SL	-
	B233 OR	L. of Woods	19	277	-	FL	270
VII	K58 CA	Eagle L.	328	-	305-350	SL	-
	B233 OR	L. of Woods	6	274	-	FL	267

Calculated SL at each annulus

	1	2	3	4	5	6	7
K58 CA Eagle L.	66	112	167	216	249	281	325

Calculated FL at each annulus

	No.	1	2	3	4	5	6
K63 CA Big Sage L.	58	58	102	140	157	180	206

The maximum size reported was 409 mm SL (K58).

Scales form at 20-25 mm SL (K58). The number of regenerative scales
was high in Eagle Lake, probably because of wounds inflicted by bird predators
(K58). Scales were easily read through age V but were difficult beyond that because of erosion and spawning checks.

Both sexes of tui chub typically first spawn at age II (K58, B233) in May
and June (K58, L193, M325) or late April (B233). They spawn in vegetated
areas near shore at about 15.5C (K58). Newly extruded eggs are 1.5-1.9 mm in
diameter, increasing slightly in size after fertilization. They are pale orange
to light straw-yellow and are adhesive (K58). Embryology is described by
H351.

	No.	TL	Eggs per female
K58 CA	1	279	11200
B233 OR L. of Woods	2	152-165	4140-4400
	2	272-279	22380-25200

Newly hatched chubs feed on rotifers, diatoms, and other algae (K58), soon taking Entomostraca and small midge larvae (B233). Animal plankton seem to be the major food in lakes (M325, K58, B233), and some surface insects are also taken (L193). In streams bottom fauna is the major food (M325, K62). They may even be cannibalistic (B338).

LEATHERSIDE CHUB, *Gila copei* (Jordan and Gilbert)
Bonneville and Snake River drainages of Nevada, Utah, and Wyoming (E119).
This species was listed as *Snyderichthys copei* in A13, but has been transferred to *Gila* (S463). The maximum length is 152 mm. These chubs inhabit cool to cold creeks and rivers, and spawn in June to August (S463).

MOHAVE CHUB, *Gila mohavensis* (Snyder)
Mohave River, California (E119).
Mohave chub hybridized with *Gila orcutti* (H350).

RIO GRANDE CHUB, *Gila nigrescens* (Girard)
Rio Grande River drainage of Colorado, New Mexico, and Texas, and introduced into the lower Colorado River (E119). Hybrids have been reported with *Rhinichthys cataractae* (C324).

ARROYO CHUB, *Gila orcutti* (Eigenmann and Eigenmann)
Coastal streams of southern California from Santa Ynez to San Luis Rey (E119). Hybrids with *Gila mohavensis* have been reported (H350).

BONYTAIL, *Gila robusta* Baird and Girard
In the Colorado River drainage (E119).

M340 UT Green R. The value of K decreases with increase in length because the caudal peduncle narrows and elongates.

SL	145	167	186	206	250	270	290	308	343	360
W	47	67	78	120	222	254	350	351	488	555
K	1.55	1.43	1.20	1.37	1.41	1.30	1.44	1.20	1.21	1.19
Number	11	12	4	2	2	4	4	1	3	1

M340 UT Green R.	421 mm TL	544 g
B240 WY	488 mm TL	907 g

A 305 mm female from Lake Mohave produced 10000 eggs, spawning in May (M340). Eggs are broadcast over rocks up to 10 m deep (S463). Ripe females were found over a gravel shoal in March in Nevada (L193). They were 305-356 mm long.
Bonytails feed mostly on insect larvae, plants and small fish (L193, M340).

CALIFORNIA ROACH, *Hesperoleucus symmetricus* (Baird and Girard)
Central and northern California (E119).

			No.	Mean SL	Range
F19 CA Coyote C.	Age	0-July	50	19	10-32
		0-Aug.	100	30	21-41
		0-Oct.	100	31	21-43
		0-Dec.	280	36	21-48
		I-May	50	40	29-55
		I-Oct.	20	61	43-72
		II-Apr.	25	65	40-85
		II-Oct.	40	76	70-96
F19 CA Guadalupe C.	II		-	71	48-96
F19 CA San Anselmo C.	II		-	53	40-68
F19 CA Coyote C.	III-Apr.		14	87	74-105

	FL at capture in various age groups				
	II	III	IV	V	VI
C197 OR Gerber R.	152	178	262	267	292
C198 OR Gerber L.	-	-	-	254	254
O34 OR L. of Woods	173	211	239	300	-

Roaches spawn from March through June and the adhesive eggs are deposited between rocks in running water (F19).

F19 CA	55 mm female	250 eggs	75 mm female	600 eggs
	85 mm female	800 eggs	95 mm female	900 eggs

Incubation takes about 10 days and the larvae are about 5 mm. Embryology is described (F19).

Scales form at 17 mm, but the body-scale correction of 6 mm is better because the scales soon make up for later formation (F19). Scale characteristics are described and photographed (F19). The dorsal ventral diameter was found to be the best scale measurement. Growth was more rapid where there was an abundance of sunlight. Growth slowed when drying of the stream caused crowding in the fall and is rather slow in winter and spring (F19). Rapid-growing individuals seem to be shorter-lived. Females average about 6% larger than males.

They first mature at age II with the smallest mature male being 46 mm and female 50 mm.

Roach under 40 mm fed mostly on crustacea and aquatic insects; larger roach ate quantities of *Spirogyra*.

BRASSY MINNOW, *Hybognathus hankinsoni* Hubbs

From Montana to Lake Champlain and southward to Missouri and Colorado (B298, E119) the brassy minnow is typically a minnow of creeks and small streams (whereas the silvery is in larger rivers, backwaters and bayous). They spawn in June in Iowa (S446). The brassy minnow feeds mostly on bottom ooze (S445).

D41 MN   mature at age II at 64-76 mm TL

A hybrid has been reported with *Notropis heterolepis* (H380). These minnows do not transport or hold in tanks readily and were classed as poor bioassay animals (G156).

CYPRESS MINNOW, *Hybognathus hayi* Jordan
   The cypress minnow seems to be limited to quiet water over soft bottoms (K116), from southern Indiana to Tennessee and Mississippi (E119).

SILVERY MINNOW, *Hybognathus nuchalis* Agassiz
   From Montana to Lake Champlain and south to the Gulf (E119). Data on the plains minnow, *H. placita,* in C134, are placed here, since the plains minnow is now considered a form of the silvery minnow (E119).

S472 AL

TL	No.	Weight Mean	Range	K(TL)
25	3	0.15	-	.91
51	95	1.4	0.9-2.1	1.08
76	42	2.7	-	.66
102	5	4.5	-	.44

The decline in K(TL) values is such that the weights reported for the 25-51 mm fish appear too high, and the regression reported in S472 is probably not valid.

	No.	SL	TL	Range
Age 0-May				
R3 NY	+	5	5	-
C134 OK Stillwater C.	+	13	-	-
Age 0-June				
R5 NY	+	-	30	-
Age 0-July				
R3 NY	+	X	-	51-76
C134 OK Stillwater C.	+	X	-	46-102
R5, F66 NY	+	-	X	46-61
Age 0-Aug.				
R3, F66 NY	+	40	69	-
Age 0-Sept.				
R5 NY	6650	-	61	41-71
F66 NY	+	-	X	51-74
Age 0-Oct.				
F66 NY	+	-	X	56-81
Age I				
R3 NY	+	X	-	50-55
Age I-Sept.				
R5 NY	+	-	84	64-89
Age II-Sept.				
R5 NY males	+	-	81	76-89
R5 NY females	+	-	89	79-97

R3 NY	60 mm female, 2000 eggs;	68-70 mm females, 3000 eggs;
	90 mm female, 6600 eggs	

Fast-growing silvery minnows mature at age I in May in New York (F67). The spawning season is prolonged and probably intermittent (C315). Sexually developed minnows have been taken from April to mid-August in Kansas (C315) and in June in Illinois (F10). Spawning has been reported as early as May (C134) and June to July for Oklahoma (C315). Eggs are scattered in calm water near shore (R3). Hybrids have been reported with *Notropis girardi* (C134). Propagation is described in R5. In Kansas silvery minnows are mostly in the sandy larger streams (M215). Food is bottom ooze and algae (K109, F148).

SPECKLED CHUB, *Hybopsis aestivalis* (Girard)

Mississippi River drainage from southwestern Pennsylvania to South Dakota south to the Gulf of Mexico and in Texas and neighboring areas of Mexico (T113). Speckled chub inhabit the clean sand and fine gravel bottoms of the swifter portions of large rivers, usually in water over 4 feet deep in the daytime but coming into shallows at night (T113, D194).

S472 AL			Weight		
Length	Number	Mean	Range		K(TL)
25	2	0.5	-		2.77
51	23	1.1	0.9-1.4		.83
76	7	1.7	-		.39

The condition factors suggest that the weights of the smaller fish are in error and therefore the length-weight equation reported in S472 for these fish is also believed to be erroneous. However, S472 also gives an equation for 65 speckled chub 76-127 mm long, without giving the actual lengths and weights:

$$\log W = -4.61 + 2.77 \log TL$$

S446 IA Des Moines R.	Age from length frequencies	
	0-Oct.	18-31 SL
	I-May	18-53
	I-July	24-49
	I-Oct.	34-51 SL
T113 OH	0-Oct.	28-46 TL
	Breeding adults	46-76 TL

Very few speckled chub attain an age over 1.5 years (S446).

Mostly in larger and sandier streams in Kansas (M215) and mostly in the channel, even in the winter (S460). They spawn in late July and August in Iowa (S446). They feed mostly on Diptera larvae throughout the year (S445).

Speckled chub are adapted for bottom dwelling. The species shows more morphological plasticity than any other *Hybopsis* (D194). Clear-water forms have larger eyes and shorter barbels than those from turbid water, which often have more than one pair of barbels. Methods of securing food probably vary in these forms.

BIGEYE CHUB, *Hybopsis amblops* (Rafinesque)

Lake Ontario and Lake Erie and Mississippi River drainage from southwestern New York to southeastern Kansas and south to the Gulf in Louisiana to Alabama (T113, D194).

Bigeye chub are found in moderate- or small-sized streams with clean sand and fine gravel. Streams of high gradient had low populations, if sand did not accumulate (T113). Bigeye chub transport easily and feed well in captivity (B349).

S472 AL

	Length	Number	Weight Mean	Weight Range	K(TL)
	25	65	.37	.09-.54	2.27
	51	341	.91	.68-2.3	.69
	76	108	4.1	2.3-5.0	.91
	102	37	7.3	4.5-7.7	.72
	127	1	15.0	-	.72

The weights for the 25 mm fish are apparently in error and therefore the length-weight equation reported in S472 is also erroneous.

Barbels are probably little used in feeding and vision is depended upon in feeding (D194).

T113 OH

Age 0-Oct.	25-58 TL	
I	38-64	
Breeding adults	58-89	
Largest	99	

ALABAMA CHUB, *Hybopsis bellica* (Girard)
From the Savannah River in Georgia to the Gulf Coast drainage in Louisiana (E119). Brain development and paucity of taste buds on the barbels indicate that this species feeds largely by sight (D194).

HORNYHEAD CHUB, *Hybopsis biguttata* (Kirtland)
Western New York to North Dakota, Wyoming, Oklahoma and Texas (T113, D194), in small- to moderate-sized clear streams with gravel bottoms.

		No.	Mean SL	Range
L52, L53, L70 NY creeks				
Age 0-July		43	33	27-37
0-Sept.	females	10	49	36-59
	males	29	56	37-67
I	females	157	71	30-95
	males	121	83	40-110
II	females	34	96	76-112
	males	6	117	108-124
III	female	1	118	-

		No.	Mean TL	Range	Weight No.	Weight Mean
J25 OK Illinois R.						
Age I		1	89	-	-	-
II		20	135	132-147	20	23
III		10	160	152-165	10	31
IV	females	6	175	173-180	6	48
	males	8	203	198-208	8	85
T113 OH						
0-Oct.		-	-	25-76	-	-

Calculated TL at each annulus

	No.	1	2	3	4
J25 OK Illinois R.	45	51	104	145	180
J42 OK Ft. Gibson L.	22	64	119	160	-

Maximum 236 mm, 170 g (T113). The largest are males (T113).

Females mature at age II in New York and 4 females 80-89 mm SL had 460-725 eggs (L70).

Hornyheads spawn in streams in riffle areas 300-600 mm deep (H91, H360). In Michigan, they spawn late May or early June (H91). The male makes a pile of gravel and protects it from other fish. The male uses a finer gravel than *H. micropogon* (H91). A male common shiner is sometimes allowed to share a nest, both males being present and taking part in the guarding but only the hornyhead in the construction of the nest (H359). Other fish which may use the pile of gravel made by a hornyhead include blacknose dace, rosyface shiner, southern redbelly dace, stoneroller, and blackside darter (H91).

These chub are absent from muddy, silt-bottomed and stagnant waters H91).

Ostracods and cladocerans constituted much of the food of young hornyhead chub to 37 mm SL but benthic organisms, mostly *Simulium, Chironomus, Helicopsyche,* mayflies and crayfish were the major foods of older chub (L53). Filamentous algae and vascular plants were taken in some quantities, probably in feeding on the animals. Feeding is mostly by sight and the barbels have few taste buds (D194).

SLENDER CHUB, *Hybopsis cahni* Hubbs and Crowe

Only taken from the Clinch and Powell rivers of eastern Tennessee (D194). Brain morphology and taste bud arrangement suggest that the slender chub locates its food primarily by sight, enhanced by the cutaneous taste buds. The body form indicates that it is a bottom dweller in fast-moving streams (D194).

OREGON CHUB, *Hybopsis crameri* Snyder

This chub is the only west coast *Hybopsis* and it is apparently restricted to the Willamette and Umpqua rivers of Oregon. The paucity of cutaneous taste buds and the development of the optic lobes suggest that this species is an obligatory sight feeder (D194).

STREAMLINE CHUB, *Hybopsis dissimilis* (Kirtland)

Tennessee and Ohio river systems from New York to North Carolina and Indiana and also in the Ozarks of southern Missouri and northern Arkansas (D194). The streamline chub inhabits riffles and bars of moderate-sized streams with clean gravel and sand and 300-1200 mm of water (T113).

T113 OH	Age 0-Oct.	33-61 mm	
	I	51-71 mm	
	Breeding adults	64-102 mm	Maximum 107

Brain morphology and cutaneous taste bud distribution suggest that streamline chubs feed largely by sight with some aid from the taste buds (D194).

Hybrids have been reported with *H. x-punctata* (T113).

STURGEON CHUB, *Hybopsis gelida* (Girard)
Missouri River drainage from Montana to the Mississippi River and in the Mississippi River drainage in southern Illinois, normally over gravel in the channels of larger rivers (D194). It is tolerant of high turbidity. On the basis of small optic lobes, large barbels, and density of cutaneous taste buds, these chub find their food by taste.

FLATHEAD CHUB, *Hybopsis gracilis* (Richardson)
From the Mackenzie and Saskatchewan river systems and Lake Winnipeg south to the Rio Grande in New Mexico and east to the Mississippi River in Missouri and Arkansas (D194). Tolerant of high turbidity.

F100, F124, F125 SD Oahe L.		(10 fish)		average K(TL) = .71
TL	203	229	254	
Weight	91	113	145	

With these few fish the females were heavier for their lengths than the males.
Flathead chub mostly inhabit alkaline streams, often turbid, with shifting sand bottoms and fluctuating water levels (O38). They may withstand temperatures of 33.3C (O38). They spawn in July to September as the water levels recede. A warm water subspecies, *H. g. gulonella,* was reported to mature at 65 mm SL compared to 85 mm for the *H. g. gracilis* (O38).
In South Dakota, Gavins Lake, 6 flatheads were feeding mostly on copepods (N94); in Jasper Park, Alberta, flatheads fed on cladocerans, copepods, and *Chironomus* (N13) but terrestrial insects were the principal foods of 54 chub taken from New Mexico to Saskatchewan (O38). Apparently flathead chub feed both by sight and taste (D194).

REDEYE CHUB, *Hybopsis harperi* (Fowler)
In northern Florida and adjacent Alabama areas (D194), redeye chub often live in springs and underwater caves where the temperature is fairly constant at 18-25C. Ripe fish have been taken throughout the year in these constant-temperature environments (M417).
Redeye chub feed on insects, crustacea and small fish (M417). Brain morphology and cutaneous taste buds indicate that these chub feed mostly by sight with some help from the tastebuds (D194).

H201 FL	12 females	25-31 mm	65-126 eggs per female

HIGHBACK CHUB, *Hybopsis hypsinotus* (Cope)
Santee River basin in North and South Carolina (D194). Lack of taste buds on the barbels indicates that these chubs feed largely by sight.

BLOTCHED CHUB, *Hybopsis insignis* Hubbs and Crowe
Cumberland River, Kentucky, to Tennessee River system in northern Alabama (D194). It is normally in clear-water riffles over coarse gravel in moderate to large rivers. Taste bud distribution indicates that the lips are used for food detection.

THICKLIP CHUB, *Hybopsis labrosa* (Cope)
Santee River system in North and South Carolina (D194). Taste buds on barbels and on pectoral fins indicate that these chubs rely to a considerable extent on taste to locate food.

BLUEHEAD CHUB, *Hybopsis leptocephala* (Girard)
   Piedmont and mountain areas of Virginia, North and South Carolina (E119, D194). Clear waters of moderate to rapid flow over clean gravel.

Age				No.	Mean	SL Range	Standard deviation
0-June	L70	VA James R.		58	16	10-20	2.6
I	L70, L52	VA James R.	females	382	40	28-61	6.0
			males	82	47	32-68	6.2
	L70	NC	females	68	35	29-45	3.7
			males	101	37	30-51	3.6
II	L70, L52	VA James R.	females	87	62	50-80	7.3
			males	48	77	59-104	11.6
	L70	NC	females	6	55	51-57	-
			male	1	63	-	-
III	L70, L52	VA James R.	females	12	86	75-102	-
			male	1	116	-	-
	L70	NC	females	2	71	-	-
			male	1	95	-	-

   In Virginia bluehead chub mature at age III and 4 females, 82-92 mm SL, had 710-800 eggs per female (L70).
   Although omnivorous, bluehead chub showed a predominance of plant food (F148). The intestine is longer than in most *Hybopsis* species (D194). The lips are well supplied with taste buds.
   Hybrids have been reported with *Compostoma anomalum* (R246).

SICKLEFIN CHUB, *Hybopsis meeki* Jordan and Evermann
   Missouri River system from Montana to Missouri, Mississippi River in southeastern Missouri, and lower parts of the Kansas rivers (D194). In strong currents of large silty rivers over fine sand or silt bottoms. Abundance of taste buds in the pharyngeal region indicate that this species probably ingests, but sorts, detritus materials. The eyes are small, indicating that vision is of little importance in feeding (D194).

RIVER CHUB, *Hybopsis micropogon* (Cope)
   Northern Alabama, Indiana, Michigan, Lake Ontario Basin, southern New York, Maryland, Virginia, to northern Georgia (D194, E113).

Age				No.	SL Mean	Range
0-July	L70, L52	NY Sandy C.		9	34	30-38
0-Aug.	L70, L52	NY Sandy C.		1	45	-
0-Sept.	L52, L70	NY Sandy C.		118	60	43-74
0-Oct.	L53, L70	NY	females	51	57	43-68
			males	67	63	46-75
I	L70	VA	females	14	60	49-70
			males	22	71	55-102
	L52, L70	NY	females	85	78	54-107
			males	111	89	57-133
II	L70	VA	females	31	82	65-110
			males	8	107	89-125

			No.	SL Mean	Range
	L52, L70 NY	females	20	102	70-125
		males	22	126	86-150
III	L70 VA	females	2	116	115-116
	L52, L70 NY	females	5	115	106-127
		males	5	149	130-167
IV	L70 NY	male	1	155	-

T113 OH	0-Oct.	25-89 mm TL	Largest 282 mm, 284 g

River chub mature at age III and 4 females, 92-100 mm SL, had 400-625 eggs per female (L70). River chub spawn in New York in late May or June (M432), in Illinois from late May to mid-August (H8) at temperatures of 19-28C, usually in water less than 300 mm deep, or at 300-430 mm (M432). The male carries stones in his mouth to build a pile for nesting (H359), taking 20 to 30 hours to prepare the nest (M432). Food habits of the river chub were very similar to those of the hornyhead (L53) and the brain morphology of the two species is similar (D194).

Hybrids have been reported with *Rhinichthys cataractae* (R25) and *Notropis cornutus* (T113). There is little aggressive action at the nests and nests are often used by other fish (M432).

LAKE CHUB, *Hybopsis plumbea* (Agassiz)

From the Delaware and Hudson River systems to the Fraser system in British Columbia, in cold lakes, streams, and irrigation canals (D194).

S156 NK Gibson L.	509 fish	53-109 FL	FL = 1.16 SL
			TL = 1.25 SL

Age		No.	FL Mean	Range
0-Aug.	S156 NK Gibson L.	6	38	-
0	E99, R98 SA Pyramid L.	4	28	20-38
I	E99, R98 SA Pyramid L.	4	43	36-48
	S156 NK Gibson L., summer	49	66	-
	Nov.	509	81	-
II	E99, R98 SA Pyramid L.	4	61	58-69
IV	E99, R98 SA Pyramid L.	+	-	102-124

Food of young lake chub consisted of cladocerans and copepods, and that of older chubs, insects (E99, S146). Paucity of external taste buds indicates that the lake chub is almost an obligatory sight feeder (D194).

Hybrids have been reported with *Rhinichthys cataractae* (H371).

Hybridization with *Semotilus margarita* reported by S442 was indicated as not proven (H377).

ROSYFACE CHUB, *Hybopsis rubrifrons* (Jordan)

Altamaha River system of Georgia and South Carolina (D194). The sparsity of taste buds on the barbels suggests that feeding is mostly by sight (D194).

SILVER CHUB, *Hybopsis storeriana* (Kirtland)

Ontario to Wyoming south to the Red River in Texas and east to the Alabama River system (D194). Usually in large silty rivers but it seeks refuge in smaller tributaries during floods.

K86 L. Erie   TL = 1.8 mm + 1.227 SL

A39 AL	TL	25	51	76	102	127	152
	Weight	.6	1.6	5.0	-	120	-
	Number	6	26	9	-	2	-
S472 AL	Weight	.36	1.0	4.0	7.7	18	28
	Number	1402	3583	1114	782	30	8
	K(TL)	2.22	.75	.89	.75	.89	.78

Since the weights of the 25 mm fish are too high, the length-weight regression given in S472 appears to be erroneous.

K86 L. Erie   log W = -4.876 + 3.062 log SL

				No.	SL Mean	Range
F5	L. Erie	Age	0-June-July	6	-	5-8
K86	L. Erie		0-July	4	29	25-30
			0-Aug.	2	40	35-45
			0-Sept.	7	49	40-60
			0-Nov.-Dec.	36	43	30-60
			I-Mar.-May	37	45	30-65
			I-July	10	67	55-80
			I-Aug.	14	96	80-115
			I-Sept.-Dec.	206	107	80-145
			II-Jan.-June	746	116	85-150
			II-July-Dec.	417	136	110-155
			III-Jan.-June	357	144	125-170
			III-July-Dec.	60	153	145-170
			IV-Jan.-June	24	164	155-180

T113 OH   0-Oct.   25-76 mm TL;   Age I 51-114,   largest 231 mm, 170 g

Annuli are formed the last week of May to mid-June in Lake Erie (K86). Silver chub spawn in June in Iowa (S446).

Spawning may start at 18C but most occurs above 21C (K86). The number of eggs was determined as 365 + 746 times the ovary weight in grams. Ovaries of 60 g females ranged from 3-15 grams (K86). Log of ovary weight = -4.876 + 3.062 log SL. They probably spawn in open water (K86). Dead chub often appear along the shore of Lake Erie after spawning (T113).

Cladocerans, copepods, and chironomids were the food of young chub in Lake Erie (K86), and *Hexagenia* mayflies constituted much of the food of the adults until the big die-off of these mayflies (B299). Thereafter, the chub made greater use of the chironomids and of *Gammarus*. Chub were estimated to eat 10% of their weight per day or an average of 8.6 *Hexagenia*. This would mean that 1.3 to 13% of the *Hexagenia* population was eaten by the chub per year (K86).

Large eyes and paucity of taste buds on the barbels indicate that silver chub feed largely by sight (D194). In aquaria silver chub seem easily excited.

GRAVEL CHUB, *Hybopsis x-punctata* Hubbs and Crowe
   Ontario, Pennsylvania, and Kentucky to Minnesota, Kansas, and Oklahoma. Gravel chub usually inhabit deeper, slower moving waters than streamline chub (T113). As silting increases, gravel chub move to shallow swift waters (D194).

T113 OH	Age 0-Oct.	28-61 mm	
	I	43-71	
	Breeding adults	64-97	maximum 99 mm

   Taste buds on the barbels are unusually large and feeding is probably accomplished by probing under rocks and in crevices (D194).
   Hybrids have been reported with *Hybopsis dissimilis* (T113).

GOLDEN IDE, *Leuciscus idus* (Linnaeus)
   A European species established at times a few places in Texas and on the Pacific coast.

B73	76 mm TL at 6 months	152 mm at 1 year
F6	Specimen lived 29 years in captivity	

LEAST CHUB, *Iotichthys phlegethontis* (Cope)
   Tributaries of Great Salt Lake and Lake Sevier, Utah (E119).
   These fish seldom exceed 64 mm from Cottonwood Creek near Salt Lake, Utah. In the laboratory they fed on mosquito larvae, ostracods and to a lesser extent, *Hyalella* (P167).

HITCH, *Lavinia exilicauda* Baird and Girard
   Streams and lakes of central California, but introduced as a bait fish throughout California (E119, B339). They avoid swift water except when spawning.

M86 CA Clear L.

FL-Weight relationship

FL	No.	Weight Mean	Weight Range	FL	No.	Weight Mean	Weight Range
152	1	28	-	254	13	227	113-340
178	5	57	57-85	279	24	312	198-397
203	2	85	57-85	305	11	369	312-454
229	7	142	85-227	330	4	482	454-539

M86 CA Clear L.

		No.	FL Mean	FL Range
Age 0-May		+	23	-
Age 0-June		+	38	-
Age 0-July		+	43	-
Age I-April	males	8	99	79-114
Age II-April	males	10	145	114-180
	female	1	175	-
Age III-April	males	12	185	145-226
	females	24	259	201-325
Age IV-April	males	5	234	216-246
	females	37	284	246-330
Age V-April	male	1	259	-
	females	2	302	279-325
Age VI-April	female	1	340	-

Males first spawn at age I and females at age III (M86). They move from Clear Lake into tributary streams in late March and spawn in early April (M86, C327) and have been observed to spawn alongshore in the lake in May (K111). No care is given the eggs, which lodge in spaces in the gravel. The eggs are nonadhesive (M86) and are 0.8 to 1.4 mm in diameter on extrusion. After water hardening, the eggs are about 2 mm in diameter (K111). Larvae become free-swimming in 20 days.

Adult hitch feed largely on plankton (M86) but up to 76 mm long they feed upon eggs, larvae, and adult *Chaoborus* (L183).

K76 proposed that although hitch may spawn on beaches, the predators which build up there more than in streams may prevent survival.

WHITE RIVER SPINEDACE, *Lepidomeda albivallis* Miller and Hubbs
This species occurs in cool springs (18-22C) and in the White River in eastern Nevada. The largest specimen is 103 mm SL or nearly 127 mm TL. Postnuptial males with remnants of tubercles were found August 27 (M413).

PAHRANAGAT SPINEDACE, *Lepidomeda altivelis* Miller and Hubbs
In Upper Pahranagat Lake and nearby Ash Spring, Nevada, as late as 1938 but probably now extinct. Maximum SL 66 mm, TL 95 mm (M413).

MIDDLE COLORADO SPINEDACE, *Lepidomeda mollispinis* Miller and Hubbs
In Virgin River system, Utah, Arizona, and Nevada. Maximum SL 88 mm, TL 108 mm. Nuptial tubercles on males in June and July (M413).

LITTLE COLORADO SPINEDACE, *Lepidomeda vittata* Cope
Little Colorado River system, Arizona. Maximum 84 mm SL, about 102 mm TL (M413).

SPIKEDACE, *Meda fulgida* Girard
Gila River system, New Mexico and Arizona, in moving water 600-1200 mm deep. Young occur in backwater areas over silt and sand. Maximum 60 mm SL, 76 mm TL. They spawn in late spring and early summer. Some 4 mm fry were collected May 6 (M413).

MOAPA DACE, *Moapa coriacea* Hubbs and Miller
Moapa dace were collected only from warm springs and creeks 30.5-32C in Nevada. Males were mature at 32 mm SL and females at 33 mm SL. The largest male was 58 mm and female, 74 mm. A ripe female was collected July 12-13 (H370). They mostly eat insects (L193).

PEAMOUTH, *Mylocheilus caurinus* (Richardson)
Lower Columbia River drainage (E119).

S359 MT

Standard length-weight relationship

SL	Males			Females		
	No.	Weight	Range	No.	Weight	Range
134-140	1	43	-	1	44	-
160-169	3	75	71-81	1	65	-
170-179	2	91	-	1	88	-

Standard length-weight relationship

	Males			Females		
SL	No.	Weight	Range	No.	Weight	Range
180-189	6	98	91-109	4	101	90-108
190-199	8	120	112-134	4	135	124-154
200-209	3	135	129-140	3	138	118-170
210-219	4	150	140-173	1	187	-
220-229	3	157	135-176	4	197	170-217
230-239	-	-	-	4	221	205-244
240-249	-	-	-	4	249	235-264
250-259	-	-	-	1	305	-

S359 MT   19 males avg. $K(SL) = 1.56$ range 1.35-1.74

There was no evident trend in the value of K to increase with increase in length, but the K values were higher in October-December than in June-July.

					SL			Mean	
				No.	Mean	Range	No.	Weight	Range
W184 MT Flathead L.		Age I	6	51	44-55	-	-	-	
S359 MT		I	7	51	44-55	-	-	-	
		III	10	185	169-195	10	102	72-120	
		IV	7	204	181-214	7	126	97-139	
		V	2	223	222-224	2	172	169-175	

						No.		Range
N82 BC Nicola L.	Age	0-July	FL	1	25	-		
H337 MT L. Placid		0-Aug.	TL	22	33	28-43		
		0-Sept.		30	56	51-61		
R275 MT Flathead L.		0-Sept.	TL	+	-	36-72		
C14 BC		V	FL	1	247	-		
R275 MT Flathead L.		XIII	TL	1	356	-		

Average calculated FL at each annulus

	1	2	3	4	5
C53 BC Okanagan	51	107	142	170	198

Average TL at each annulus

	No.	1	2	3	4	5	6	7	8	9	10	11	12	13
P159 MT streams	36	61	119	155	213									
H337, P159 MT L. Placid	300	66	117	170	211	244	269	292	320	343				
Increment		66	51	53	46	33	28	23	23	20				
Number		300	300	294	200	176	125	71	12	1				
R275 MT Flathead L.	192	48	84	114	147	178	208	234	259	284	305	318	333	343
Increment		48	36	30	33	33	30	30	25	25	23	18	13	13
Number		170	170	166	156	123	95	72	52	25	15	8	3	1

Females grow slightly faster than male peamouth and tend to be longer-lived (S359, H337). At age V in Lake Placid, Montana (H337), females averaged 5 mm longer than males, and at age VII 18 mm. No males older than VII were found.

Male peamouth mature at age III (H337, S359) at a minimum length of 165 mm TL and females at age IV (H337) at a minimum length of 213 mm.

H337 MT Seeley L.	7 females	300-325 mm	15600 eggs per female
			11800-18900

Peamouth in Montana spawn from late May to early July at temperatures of 11-22C (H337, S359). In a cold year spawning did not occur until July 5-19 (S359). In May and June at 12-19C in Washington, they spawn in shallow water along the lake shore over gravel (S441). Breeding males averaged 198 mm SL (184-219) and weighed 124 g (100-168) compared to females at 225 mm (194-270) and 204 g (108-377) (S441).

Lethal upper temperatures for peamouth acclimated at 10C and at 14C were determined as 27 to 27.1C (B200).

Snails, and to a lesser extent aquatic insects, constitute the major foods of peamouth (S359, H337, J53). Ants were abundant in stomachs of one collection (H337). Peamouth apparently feed as normally, during the breeding season (S359).

Hybrid: *Mylocheilus caurinus* X *Ptychocheilus oregonensis*

W184 MT Flathead L.	Age I-July	41 fish	43 mm SL	37-65 mm

HARDHEAD, *Mylopharodon conocephalus* (Baird and Girard)
Sacramento-San Joaquin drainage and Russian River, California (B340), in clear foothill streams and reservoirs.

		FL at each annulus				
	1	2	3	4	5	6
R279 CA American R.	79	147	224	282	356	421
Weight, g	-	27	113	240	500	900

Hardheads migrate upstream to spawn in the spring. Most mature at 2 years. A 394 mm female contained 21800 eggs (B340).

Juveniles feed largely on cladocerans, aquatic insects and snails; adults, primarily on green algae and higher plants (B340).

GOLDEN SHINER, *Notemigonus crysoleucas* (Mitchill)
Manitoba to Quebec and south to Florida and Mexico (E119). Introduced west of the Rockies (M504).

	No.	Range SL	FL	TL
C33 IA ponds	6	79-107	1.133 SL	1.233 SL
L29 IA East L.	69	40-169	1.138 SL	-
	97	10-169	-	1.255 SL
B86 IL Onized L.	+	-	-	1.25 SL
C30 IA	58	-	1.14 SL	1.26 SL
S65 NS	+	-	1.144 SL	-

	mm	SL
M312 MI Lower Loch Alpine	126-163	1.272
	164-209	1.280

In the following tabulations, data originally published as SL or FL are converted to TL using the factors TL = 1.26 SL = 1.11 FL.

SL	TL	No.	Weight (grams) Mean of means	Range	Citations
20	25	+	0.1	-	D81
30	38	+	0.4	-	D81
40	51	+	1.4	1.05-1.8	D81, P109
50	64	+	2.3	2.1-2.45	D81, P109
56	71	1	3.0	-	L29
60-80	76-101	21+	5.4	2.7-8.0	F31, C30, C33, L29, D81, P109, L149, S472
81-100	102-126	31+	12.8	4.5-25	F31, C30, C33, L29, D81, P109, L149, S472
101-120	127-151	72+	23.7	11-38	F31, C30, C33, H61, L29, D81, P109, L149, S472
121-140	152-177	611+	39	23-62	F31, C30, H61, L29, L149, S472
141-160	178-202	823+	66	32-100	N52, H61, L29, C30, L149, S472
161-180	203-228	190	89	64-135	C30, L29, H61, S472
-	229	2	114	91-136	S472
212	267	1	277	-	E105
234	295	1	289	-	S398

More than 1000 fish were included in the 76, 102, and 127 mm groups from S472.

L29 IA East L.	log W = -5.306 + 3.294 log SL
S65 NS L. Jesse	log W = -6.521 + 3.31 log FL

P122 WI Flora L.	1952	8 shiners	log W = -1.757 + 1.568 log TL
	1956	8 shiners	log W = -5.975 + 3.344 log TL

These (P122) regressions seem very unusual and may be in error, or just the result of small and unusual samples. Because weights for 25 and 51 mm golden shiners were obviously high, the regression given in S472 is erroneous. The weight increased more rapidly than the cube of the length for golden shiners, 76 to 229 mm (S472).

	No.	Length		
C23 MN	2	121-140 SL	average K 1.49	
C36 IA	44	56-158 SL	average K 1.74	range 1.03-2.37
T5 AL	3283	25-76 TL	average K 1.7	
	187	76-127 TL	average K 2.0	
C33 IA ponds	6	79-107 SL	average K 1.96	
L29 IA East L.	65	110-169 SL	average K 2.12	
S65 NS	108	-	average K (FL) 1.27	
S472 AL	106,608	76-230 TL	mean K (TL) from .65 to .95, increasing with length of fish	

D81 MN ponds   Growth during early life

Age in days	10	20	30	40	50	60	70
TL	8	15	23	30	38	46	53

	No.	Mean TL	Range	No.	Mean Weight
Age 0-June					
C60 MI	1121	18	-	1121	0.05
L29 IA East L.	9	23	15-25	-	-
H195 MI ponds	+	-	13-23	-	-
C134 OK Stillwater L.	+	-	20-30	-	-
D41 MN ponds	-	25	-	-	-
W105 IL Crab Orchard L.	14	-	13-48	-	-
B86 IL Onized L.	8	56	25-89	-	-
Age 0-July					
L29 IA East L.	17	30	23-38	-	-
C60 MI	603	43	-	603	0.9
W105 IL Crab Orchard L.	6	-	33-66	-	-
H195 MI ponds	+	-	33-48	-	-
D41 MN ponds	+	56	-	-	-
Age 0-Aug.					
L29 IA East L.	1	25	-	-	-
T5 AL	3283	-	25-84	3283	1.5
H195 MI ponds	+	-	46-58	-	-
F66 NY ponds	+	-	46-51	-	-
M243 NY L. George	10	36	-	-	-
D41 MN ponds	+	89	-	-	-
Age 0-Sept.					
M234 NY L. George	10	51	-	-	-
F66, F67, R239 NY ponds	+	-	25-89	-	-
Age 0-Oct.					
D41 MI ponds	+	41	-	-	-
R239 NY bass ponds	+	-	38-58	-	-
L13 OH	+	-	43-64	-	-
F66 NY ponds	+	-	53-74	-	-
M314 MI Lower Loch Alpine	6	76	-	-	-
C60 MI	682	74	-	682	3.3
M504 CA	+	76	-	-	-
D41 WI ponds	+	76	-	-	-
T113 OH	+	-	18-102	-	-
Age 0					
C57, E9 NY	24+	38	-	-	-
C57, C56, C58 MI	246+	46	23-76	-	-
G130 IA Clear L.	6	-	36-66	-	-
M25 IA	+	102	-	-	-
B218 MD Redington P.	+	-	41-86	-	-
A10 OK	-	-	-	+	15.0
H141 ON	5	56	48-61	-	-
H141 OH	23	58	41-71	-	-
H141 FL	16	86	38-117	-	-
Age I					
E9, F66, C57 NY	+	-	53-91	-	-
H141 ON	+	66	56-74	-	-
T5 AL	173	-	56-97	173	4.9
B97 IL Gale L.	+	76	-	-	-
L29 IA East L.	6	71	-	-	-
G130 IA Clear L.	4	-	71-94	-	-

	No.	Mean TL	Range	No.	Mean Weight
C57, C58, D41 MI	532+	81	58-114	-	-
H98 NC March	+	-	33-102	-	-
H141 OH	23	91	76-117	-	-
M312 MI Lower Loch Alpine	28	122	-	28	16.5
B218 MD Redington P.	+	-	86-97	-	-
R239 NY ponds	+	-	81-127	-	-
B86 IL Onized L.	38	135	102-165	-	-
J41 OK Ft. Gibson L.	9	112	-	-	-
H141 FL	15	119	97-152	-	-
V44 LA Springhill L.	1	155	-	1	31.1
Age II					
D41 MI ponds	-	84	-	-	-
E9 NY	-	-	69-84	-	-
H141 ON	16	89	79-99	-	-
B218 MD Redington P.	-	102	-	-	-
T5 AL	9	-	94-127	9	17.2
C57 MI	254	107	97-147	-	-
C57 NY	2	109	-	-	-
G130 IA Clear L.	1	122	-	-	-
L13 OH	36	127	-	-	-
J41 OK Ft. Gibson L.	1	137	-	-	-
M312 MI Lower Loch Alpine	94	150	-	-	-
L29 IA East L.	1	152	-	-	-
B86 IL Onized L.	38	183	165-216	-	-
H141 FL	18	185	127-224	-	-
Age III					
H141 ON	6	99	84-109	-	-
T5 AL	1	117	-	1	23.5
C57 MI	146	117	107-152	-	-
C57 NY	20	119	-	-	-
H141 OH	3	132	-	-	-
M312 MI Lower Loch Alpine	7	168	-	7	48.0
L29 IA East L.	33	178	-	-	-
B86 IL Onized L.	6	198	165-229	-	-
H141 FL	9	203	173-264	-	-
Age IV					
T5 AL	4	-	127-140	4	34.2
H141 ON	2	117	-	-	-
H141 OH	2	142	-	-	-
C57 MI	48	127	102-145	-	-
M312 MI Lower Loch Alpine	10	180	-	10	62
L29 IA East L.	15	193	-	-	-
H141 FL	3	249	-	-	-
Age V					
H141 ON	1	122	-	-	-
C58 MI	-	140	-	-	-
C57 MI	21	152	-	-	-
H141 OH	1	173	-	-	-
M312 MI Lower Loch Alpine	39	190	-	39	77
L29 IA East L.	5	196	-	-	-
C57 NY	24	211	-	-	-

	No.	Mean TL	Range	No.	Mean Weight
Age VI					
C57 MI	7	190	-	-	-
M312 MI Lower Loch Alpine	21	196	-	21	86
C57 NY	3	208	-	-	-
W13 CT	1	249	-	1	198
Age VII					
H141 ON	1	140	-	-	-
C57 MI	2	168	-	-	-
M312 MI Lower Loch Alpine	2	201	195-203	2	91
Age IX					
C57 MI	1	259	-	-	-

		Average calculated TL at each annulus			
	No.	1	2	3	4	5
L29 IA East L.	60	46	109	170	188	196
F62 OK Sub Prison L.	22	99	145	168	180	
After restocking	92	94	163			
B86 IL Onized L.	87	94	173	190		
P122 WI Flora L.	22	89	127	160		
After harvest	176	91	119	147		

		Calculated TL and increment at each annulus						
	1	2	3	4	5	6	7	8
M312 MI Lower Loch Alpine	71	122	146	168	180	188	193	198
Increments	71	51	30	18	10	8	8	8
Number	210	204	176	82	75	65	26	5

T113 OH, maximum TL 267 mm (340 g)

In most studies lengths were calculated on the basis of direct proportion between scales and lengths, but P122 and M312 used Fraser-type correction factors of 64 and 38 mm, respectively. In general, growth is faster in the southern than in the northern part of the range with fast growth in Louisiana (V44) and Florida (H141) but the variations in each area are great. Growth is faster in southern than northern Michigan in the cooler waters (H91). Females were slightly larger than males in most age groups in Lower Loch Alpine, Michigan (M312), but the differences were not entirely consistent. Females outnumbered males in all age groups above III. Growth was faster during the second year of life in ponds with largemouth bass than in ponds with only golden shiners, presumably because of the reduced population density due to bass predation (R239).

In Alabama, golden shiners may spawn at 7 months (P107), in the south at age I and 64 mm TL (D41) and in New York both males and females will spawn in June at age I if over 64 mm TL or later in the summer or when at age II if they grew more slowly (F66). They spawn from May through August in New York with most of the spawning at 20-21C (F67) or May to July in Illinois (F10). The first hatch came when water temperatures first reached 20C, 150 mm from the surface (S334). The eggs are adhesive and usually scattered over vegetation (H91). Golden shiners spawning in largemouth bass nests had greater reproductive success than when spawning in unprotected areas (K75).

Young shiners feed principally on Entomostraca and phytoplankton while adults feed largely on algae (R247, F148). On the other hand, zooplankton was found to be more important in food of golden shiners than phytoplankton (S447, H91, M412, E120, S448, P165). Zooplankton and insect larvae were the food of shiners, 23-40 mm TL, in Ohio (E131). Insects and Cladocera were eaten by shiners in Minnesota (N117) and in Ontario (K125) where filamentous algae was important food in summer only. Where molluscs are abundant, they may be the main food (F10).

Golden shiners were classed as sight feeders by E123 because the brain did not show development for tasting. Golden shiners ate insect larvae, hydrachnids, cladocerans, and algae in Clear Lake, Iowa (G130). Because golden shiners were primarily plankton feeders, M412 recommended them for reservoirs.

Lethal temperatures °C

Acclimated at	Lethal low	Lethal high	Citation
15	1.5	30.5	S333
17.1-17.5	-	31-31.6	B201
20	4.0	31.9	S333
21	3.4	-	B201
22.8	-	32.7	B201
25	7.0	33.2	S333
25-26	-	33.2	B201
30	11.2	34.7	S333

Golden shiners were found freely swimming in water at 30C, and live specimens with body temperatures of 33-35C were found (T102). When frightened in waters of 27-35C, some shiners died but others regrouped (T102).

Golden shiners are easily damaged by handling and are very excitable and thus not recommended as bioassay animals (W214). They will jump out of the water at a sharp noise from a boat or shore.

Methods of propagation have been outlined in D41, D81, P107, R247, H185, H195, P109, and F67. In Flora Lake, Wisconsin, the population of golden shiners was estimated at different times as 25 to 287 fish or 0.9 to 11.2 kg/ha (P122). In New York ponds the production was 213 kg/ha in one year or 470 kg/ha in 2 years (F67).

		Sizes in gill net meshes				
Mesh size	(stretch)	38	51	57	64	70
W66 CT	Mean TL	147	-	198	203	213
L216 LA	No.		1			
	TL		190			

WHITEMOUTH SHINER, *Notropis alborus* Hubbs and Raney
From streams up to 18 m wide, in piedmont areas of Virginia and North and South Carolina (E119). Maximum SL 50 mm (H375).

HIGHFIN SHINER, *Notropis altipinnis* (Cope)
In piedmont streams of North and South Carolina. Maximum SL 45 mm (H362). One subspecies is found in the Neuse River, North Carolina coastal plain, where its range overlaps with the similar *Notropis cummingsae* (H372, H374).

PALLID SHINER, *Notropis amnis* Hubbs and Greene

In the northern part of the range, Minnesota to Missouri, the pallid shiner is principally in large lowland rivers; in the south to Texas it is mostly in small to medium streams. Turbidity may be a factor in eliminating large rivers in the south and temperature in eliminating smaller streams to the north (H373). Maximum SL 55 mm.

SATINFIN SHINER, *Notropis analostanus* (Girard)

Coastal drainage from the St. Lawrence to North Carolina (E119).

Satinfin shiners live in larger streams and rivers but upstream from the emerald shiners. Usually they are in shallows, with little vegetation and with moderate to rapid current (S370). They may frequently be found in 35C water (T102).

At hatching satinfin shiners are 5 mm TL, and scales first form at 15-16 mm on the caudal peduncle (S370).

S370 NY		TL			TL	
Age	No.	Mean	Range	No.	Mean	Range
0-July, Cayuga L.	145	30.7	22-40			
0-Aug., Cayuga L.	473	31.8	12-53			
		Males			Females	
0-Sept., pond	106	51.0	34-67	101	41.0	25-53
0-Oct., Fall C.	163	38.1	26-50	137	37.7	25-52
I	362	67.1	50-78	274	56.5	45-65
II	99	72.4	54-71	197	62.6	54-71
III	8	84.0	76-91	60	70.8	65-74
IV	-	-	-	10	74.7	71-77

The annulus forms in early May (S370). Males tend to grow more rapidly than females.

Both males and females may mature at age I though some do not mature until age II (S370). The smallest mature male was 53 mm and the smallest mature female was 47 mm. The largest were 93 mm and 79 mm, respectively.

Satinfins spawn from late May to mid-August, in the daytime, at 18-27C (S370). They probably spawn twice a season. Eggs are attached to branches, stumps, and logs. No care of eggs or young but the male guards the territory of the nest during the spawning season (G20, S370).

S370 NY		No. of	Number of eggs per female	
	TL	females	Mean	Range
	47	1	55	-
	51-52	3	142	98-202
	60-69	5	468	167-851
	70-79	5	510	157-864

Eggs hatch in 8 to 11 days at 17-25C (S370).

Insects, particularly Diptera (S370) or Ephemeroptera (F148), are the principal foods of satinfin shiners.

PUGNOSE SHINER, *Notropis anogenus* Forbes

The pugnose shiner, while rather rare, is found from New York to North

Dakota in clear, vegetated lakes (B302). Maximum SL 48 mm (B302) or TL 56 mm (T113).

B329 WI   2 fish   TL = 1.25 SL   52-53 mm TL   1.27-1.5 g

Females full of eggs were taken in May and June in Illinois (F10).
Pugnose shiners seem to be extremely intolerant of turbidity (T113).

ROSEFIN SHINER, *Notropis ardens* (Cope)
Roanoke River in Virginia and Ohio River system in Ohio, Kentucky, Tennessee, and northern Alabama (E119, T113).

T113 OH   0-Oct.   18-51 mm TL;   I, 33-58 TL;   adults 46-89 TL

Males (maximum 76 mm) tend to be slightly larger than females (maximum 64 mm) (R246). Males tended to maintain territory over spawning gravel.
Rosefin shiners spawn in fast riffles over gravel, often over nests of *Hybopsis* (R246). They are intolerant of silting (T113).
Insects, principally terrestrial forms, were found to be the principal foods of age I and II rosefin shiners (M205).

BURRHEAD SHINER, *Notropis asperifrons* Suttkus and Raney
Found in clear streams, small to moderate size, in Mobile Bay drainage.
Maximum SL, 60 mm. Females reach a greater length than males. The pelvic fins reach beyond the anus in mature males but not in females (S457).

EMERALD SHINER, *Notropis atherinoides* Rafinesque
Lake Athabasca to Lake St. John in Quebec, Lake Champlain, south to the Potomac, Tennessee, Louisiana, and Texas (H382, T113).

A39, S472 AL   Weight

TL	Number	Mean	Range	K(TL)
25	2133+	.36	.02-1.8	2.27 (obviously too high)
51	10025	.9	.6-1.8	.69
76	3468	2.7	1.3-4.4	.61
102	133	6.4	2.0-11.3	.64

51-102 mm   13623 shiners   log W = -4.71 + 2.73 log TL

	No.	SL Range	Mean TL	Range
Age 0-July				
G100 L. Erie	+	8-22	-	-
F5 L. Erie	979	-	-	5-23
Age 0-Aug.				
F5 L. Erie	994	-	-	5-33
Age 0-Sept.				
F5 L. Erie	3	-	18	-
C275 L. Erie	+	-	61	-
Age 0-Oct.				
T113 OH	+	-	-	25-58
B330 WI	+	-	-	41-53
Age 0				
H87 MI	+	26-60	-	-

	No.	SL Range	Mean TL	Range
Age I				
D41, D81 IL	+	-	44	-
G100 L. Erie	47+	25-52	-	-
H87 MI	+	70-91	-	-
T113 OH	+	-	-	33-71
Age II				
D41, D81 IL	+	-	76	-
G100 L. Erie males	54+	46-65	-	-
females	58+	50-75	-	-
T113 maximum 99 mm TL				

Growth of young emerald shiners in Lake Erie ends in the last half of September when water temperatures are still 21C (C275). The growth during the first season averaged 5.6 mm per week.

In Lake Erie they spawn from June 25 to July 28 or even August 15, near the surface in open water (G100).

Emerald shiners were considered early spawners in the Des Moines River, Iowa, by Starrett (S446) as females collected in mid-July were spent. Females ready to spawn were taken in Illinois in mid-May to early June (F10). Spawning occurs at 24C and the eggs may hatch in less than 24 hours (C275).

Acclimated	Lethal low	Lethal high	Citation
5C	-	23-23.2	H253, S333
10	-	26.7-27	H253, S333
15	1.6	28.9	S333
20	5-5.2	30.7-31	H253, S333
25	-	30.7	S333
Ultimate	-	30.7	S333

In a Kentucky stream (M404) emerald shiners fed mostly on terrestrial insects in the summer and on amphipods, mayfly naiads, and caddis worms in the winter. In Lake Erie, winter foods consisted mostly of microcrustacea with most of the feeding, which appeared to be rather light, in late afternoon (G100). One of two emerald shiners taken from an alewife spawning area had fed on eggs (E118). Insects, aquatic and terrestrial, seem to be the major foods (F10).

ROUGH SHINER, *Notropis baileyi* Suttkus and Raney
The rough shiner is from small wooded streams, with sand and gravel bottoms in Alabama and Mississippi. Maximum SL 63 mm (S458).

RED RIVER SHINER, *Notropis bairdi* Hubbs and Ortenburger
Oklahoma and Texas (E119).
They probably spawn in the current and let the eggs be carried in the current as in *N. girardi* (M72).

BRIDLE SHINER, *Notropis bifrenatus* (Cope)
Atlantic drainage, Maine to Virginia (E119).

	No.	Mean SL	Range	TL
Age 0-June				
H16 NH	-	-	4-6	5-8
Age 0-July				
H16 NH	-	-	8-14	10-20
H17 NH	667	12	5-21	-
Age 0-Aug.				
H17 NH	117	16	11-22	-
Age I				
H17 NY March	1751	-	14-33	-
H17 NH June	677	-	23-38	-
Age II				
H17 NY March	326	-	34-45	-
H17 NH June	43	-	37-43	-

No sexual differences in growth were noted except that the largest fish were mostly females, which apparently live longer. Apparently none live to be age III, however (H352). They spawn at age I in late May into July in New Hampshire (H352), at 14-27C, mostly 7 a.m. to 2 p.m. Spawning is in still water, at about 600 mm depth, over *Myriophyllum* or *Chara*. The spawned eggs fall into the dense weed beds (H353). A female may spawn about 8 days during the season (H354). Counts of eggs with yolks gave 370 for a 25 mm female, 1062 for a 34 mm female, and 2110 eggs in a 44 mm female (H17). However, only eggs over 0.6 mm in diameter had a micropyle and were mature. This gave 260 mature eggs for the 44 mm female (H354). Aquarium studies indicated 278 eggs spawned per female in one day (H354). Embryology and early growth are outlined in H16. Spawning can be secured early in the laboratory by lengthening the days in the winter (H356, H357).

Bridle shiners feed mostly on insects and crustacea taken from plant surfaces and on plankton (H355).

RIVER SHINER, *Notropis blennius* (Girard)
Northeast Texas to southern Alberta, southern Manitoba, Wisconsin, Ohio, and northern Louisiana (E119).

T113 OH   Age 0-Oct. 20-56 mm TL;   I 25-64 mm;   adults 52-132 mm

	No.	Mean SL	Range
Age 0-July			
H88 MI	101	14	10-17
Age 0-Aug.			
H88 MI	-	-	11-39
Age 0			
C22 MN L. of Woods	8	31	28-36
Age I			
C22 MN L. of Woods	2	49	48-50
Age II			
C22 MN L. of Woods	4	70	59-79

River shiners spawn in late July and August in Iowa (S446) and throughout the summer to late August in Ohio (T113) over sand and gravel.

BIGEYE SHINER, *Notropis boops* Gilbert
Most of Ohio River drainage and southwest to southern Kansas and eastern Oklahoma (E119, T113).

T113 OH   0–Oct. 25–38 mm TL;   I 33–61 mm;   adults 46–91 mm

Bigeye shiners inhabit only clear streams. They feed by sight in mid-water, at the surface, or jumping for insects (T113). They transport well and are easily kept in tanks (W214).

GHOST SHINER, *Notropis buchanani* Meek
Ohio River drainage and southeastern Minnesota to Kansas, Texas, and Louisiana (E119, T113).

T113 OH   0–Oct. 20–38 mm TL;   I 28–58 mm;   adults 33–64 mm

Ghost shiners were found in breeding condition on August 14 in a Kansas stream (M215). They are found mostly in larger streams. In Ohio they are in clear, quiet waters with clean sand, gravel bottoms and some aquatic vegetation (T113).

BLUESTRIPE SHINER, *Notropis callitaenia* Bailey and Gibbs
The bluestripe shiner is found in large rivers with sandy bottom in Alabama and Georgia. Maximum SL 68 mm.
Hybrids have been reported with *N. venustus* (B300).

BLUNTFACE SHINER, *Notropis camurus* (Jordan and Meek)
Arkansas River drainage (E119).
These shiners transport easily and feed well in the laboratory but are very excitable during bioassay tests (B349).

CRESCENT SHINER, *Notropis cerasinus* (Cope)
Upper Roanoke and Kanawha River drainages, Virginia and West Virginia (E119).
In the Roanoke River, Virginia, the life span of crescent shiners is 4 years, with males growing faster than females (S426). The average SL of males is:

Age	0	I	II	III
	35	55	65	80

Crescent shiners were observed spawning in June in Virginia (R246). Males were up to 102 mm long and few females were over 76 mm. Spawning was over gravel in riffles, often over *Hybopsis leptocephala* nests. Males attempted to hold territories over the spawning area.
Chironomids, simuliids, trichopterans, and coleopterans were the principal foods (S426).

IRONCOLOR SHINER, *Notropis chalybaeus* (Cope)
Coastal lowlands, New Jersey to Texas and north in the Mississippi drainage to Iowa.

M27 FL

Age in days	No.	SL	TL	Age in days	No.	SL	TL
0	2	2.2	2.3	26	1	5.7	6.2
3	1	3.7	3.9	33	1	6.3	7.0
9	1	4.3	4.6	47	1	7.4	9.2
19	1	5.4	5.7	69	1	11.6	14.8

Scales did not give satisfactory age criteria (M27). In Maryland they spawn in late spring (S461). The breeding season in Florida is from early April through September. The spawning males average smaller than the females and there is no nesting or care of eggs, which sink to the bottom and adhere to sand and other particles. Eggs hatched in 54 hours at 17C. These minnows tend to school throughout the year. They feed, by sight, on Entomostraca and insects. The plant material in the intestines probably represents the food of their prey (M27).

WARPAINT SHINER, *Notropis coccogenis* (Cope)
Upper Tennessee River system in Virginia, North Carolina, Georgia, Alabama, and Tennessee and headwaters of New and Catawba rivers in North Carolina (O44).

O44 NC

Age	Number	Mean SL	Range
0-July	33	27.5	23-32
0-Aug.	40	40.9	31-53
0-Sept.	90	43.5	30-59
0-Oct. 18	36	51.1	45-60
I	260	48.9	21-86
II	72	67.1	50-89
III	16	84.3	73-98
IV	5	100.0	92-106

Males were usually slightly longer than females at the same age, but there was a tendency for more females to reach ages II and III (O44). Annuli were formed in late April.

Males and females first mature at age II and females age II-IV had 300 to 1636 mature-sized ova per female. Males maintained territories over nests of *Hybopsis micropogon* in June (O44) at water temperatures of 20-24C.

Food was largely of insects, primarily ephemeropterans (O44), but terrestrial insects were important in late summer and fall.

COMMON SHINER, *Notropis cornutus* (Mitchill)
Southern Alberta to Nova Scotia south to Virginia, Mississippi, Louisiana, Oklahoma, and Colorado (T113).
Gilbert (G129) considers the eastern form (Michigan, south and east) as a distinct species, *Notropis chrysocephalus*, with both species in Illinois, Indiana, Michigan, and Missouri.

L108 IL	517 shiners 17-132 mm	TL = 1.27 SL
R210 NY	593 shiners	TL = 1.302 SL

B130 OH

SL	30	35	45	55	65	75	85	95	105	115	125	135	145	150
Weight	.25	.35	.85	1.84	3.71	4.27	7.44	10.0	14.1	21.6	22.6	29.9	50.9	64.1

B150 IA A 140 mm TL shiner weighed 25 g.

S472 AL	Length	Number	Weight Mean	Range	K(TL)
	76	38	3	3-7	.72
	102	7	13	8-19	1.22
	127	7	22	19-23	1.05
	152	1	45	-	1.30

53 shiners   75-150 mm $\log W = -5.56 + 3.29 \log TL$

	SL (except as marked TL)		
	No.	Mean	Range
Age 0-June			
F5 L. Erie	1	6	-
B167 OH	-	-	14-19
Age 0-July			
F5 L. Erie	9	6	-
B167 OH	-	-	9-24
Age 0-Aug.			
B167 OH	-	-	20-35
L108 IL Big C.	-	27	-
Age 0-Oct.			
B70 DC	+	30	-
D81, H91 MI	+	51 TL	-
T113 OH	+	-	TL 25-64
Age 0			
M26 OH	63	41	25-51
C57 MI	+	51 TL	-
G130 IA Clear L.	11	-	TL 36-79
Age I			
R210 NY females	11	30	23-34
males	10	30	23-34
B70 DC	-	-	26-42
H141 ON	11	46	39-60
H141 TN	16	54	46-67
L108 IL Big C.	3	54	-
M26 OH	42	59	39-87
G130 IA Clear L.	3	-	TL 71-96
T113 OH *N. c. frontalis*	+	-	TL 33-81
N. c. chrysocephalus	+	-	TL 38-89
Age II			
R210 NY females	35	43	30-45
males	28	46	35-56
B70 DC	+	54	-
H141 ON	10	69	61-74
L108 IL Big C.	30	78	-
M26 OH	25	83	53-105
G130 IA Clear L.	1	104 TL	-
Age III			
R210 NY females	56	55	40-62
males	44	63	47-79
H141 ON	11	83	70-90
L108 IL Big C.	7	98	-
H141 TN	8	100	82-135
M26 OH	16	102	86-122

	SL (except as marked TL)		
	No.	Mean	Range
Age IV			
R210 NY females	12	73	60-90
males	18	85	45-99
L108 IL Big C.	1	103	-
H141 ON	1	104	-
M26 OH	2	106	104-107
H141 TN	1	145	-
Age V			
R210 NY females	2	93	92-94
males	9	102	94-115
L108 IL Big C.	1	132	-

T113 OH maximum TL *N. c. frontalis* 208 mm; *N. c. chrysocephalus* 236 mm

	Average calculated TL at each annulus					
	1	2	3	4	5	6
R56 OH females	51	99	119	137	152	
males	51	99	130	168	196	211
L108 IL Big C.	36	74	102	122	165	
Increments	36	38	25	33	28	
Number	42	39	9	2	1	

Males grow faster than female common shiners (R210, R56). Scales do not form until September of the first summer (B167). In New York the annulus was formed in June (R210) and in Ohio in May (M26). The annulus is characterized by anastomosis, branching, and irregular breaking of circuli on the lateral fields (M26). The anterior field is much narrower than the posterior field.

Growth during the first month of life was reported as 0.45 mm per day in Portage Lake, Michigan (H86).

Common shiners mature at age II with the smallest mature females at 37 mm SL and males at 40 mm (R210). The smallest mature females were 75 mm SL and males were 77 mm in an Ohio study (B130). In Iowa they spawn mid-April to late May (P63); in Michigan late May to June (H91); in Minnesota in June (N116); and in New York in May to July (M432).

			Eggs per female	
	Size of female	No.	Mean	Range
B130 OH	83-93 mm SL, 7-9 g	2	1025	900-1150
	95-100 mm			

Common shiners spawn only on gravel bottoms in stream riffles (H91). Spawning is similar to the creek chub (H91, H360, R252). Common shiners may use the gravel nests of creek chub and hornyhead chub (H395). When necessary, males may fan away sand but usually depend upon the current or other fish to clear the gravel (H91). Males defend territories (M432). They did not spawn until temperatures were 19-21C (N116) or 14.5-25.5C (M432).

Acclimation temperature	Lethal low	Lethal high	Citation
5C	-	26.7-27	H253, S333
10	-	28.6-29	H253, S333
15	0C	30.3	S333
20	3.7-4.0	31-32.3	H253, S333
25	7.8	33.5	S333
25-26	-	32	B201

Common shiners feed on algae, rotifers, Entomostraca, insects and are generally omnivorous (N104, B130, B70, F148, S447, G130, R249, N117, S445). No cessation of feeding occurs in the breeding season or the winter although the quantities eaten in the winter are low (B130).

Common shiners transport and hold well in tanks and readily take dried food, but they are rather excitable for bioassay animals (B349).

Hybrids have been reported with *Notropis rubellus* (B250), *Chrosomus erythrogaster* (C324, T113), *Hybopsis plumbea* (T113), *Hybopsis micropogon* (T113), *Compostoma anomalus* (T113), *Notropis photogenis* (T113), *Hybopsis biguttata* (T113), *Semotilus atromaculatus* (C324).

DUSKY SHINER, *Notropis cummingsae* Myers
The dusky shiner is usually found in the austroriparian and coastal plain of Alabama to North Carolina and is similar to *N. altipinnis* of the piedmont area. Maximum SL 60 mm (H372). Females average larger and stouter than males. In Georgia 16 nuptial males collected April 27 were 30-38 mm and averaged 34 mm, and 20 gravid females were 32-42 mm and averaged 37 mm.

BIGMOUTH SHINER, *Notropis dorsalis* (Agassiz)
Montana and Kansas east to Minnesota and Illinois and in scattered localities of the Great Lakes drainage to New York (E119, H382, T113).

	No.	Mean TL	Range
Age 0-July			
K131 IA Des Moines R.	6	30	25-33
Age 0-Aug.			
P63 IA Squaw C.	+	20	-
K131 IA Des Moines R.	15	33	25-36
Age 0-July-Aug.			
G130 IA Clear L.	27	-	28-58
Age 0-Oct.			
P63 IA Squaw C.	+	-	25-43
T113 OH	+	-	28-51
Age I			
K131 IA Des Moines R.	296	36	25-48
P63 IA Squaw C.	120	46	36-56
T113 OH	+	-	33-64
G130 IA Clear L.	8	-	53-66
Age II			
K131 IA Des Moines R.	658	43	33-61
P63 IA Squaw C.	52	58	51-74
Age III			
K131 IA Des Moines R.	333	53	46-69

Scales were used in distinguishing the age I and II fish (P63), but length frequencies were used by K131. In Michigan, bigmouth shiners reach age IV (W212).

Growth in the Des Moines River was slow in late summer, perhaps due to crowding as water levels declined (K131).

In Iowa, bigmouth shiners spawn in June (K131) to mid-July (P63) and July-August (S446). High water levels in June may delay spawning (K131). They probably spawn in midstream and let the current carry the eggs, as in *N. girardi* (M72). They fed on insect larvae, hydrachnids, cladocerans, and algae (G130, S445).

No sex differences were noted in several meristic characteristics (U10). These shiners are usually found in smaller streams, and many populations in Minnesota showed significant differences in meristic characteristics, suggesting some isolation (U10). Although primarily a pioneer fish in small streams, the bigmouth shiner was also abundant in the Des Moines River, Iowa (S460). In Ohio bigmouth and silverjaw minnows are similar in habitat and in methods of schooling, feeding, resting, and retreating when frightened (T113). The bigmouth remained the dominant as long as it was abundant, but when it became rare, silverjaws became abundant.

Bigmouth shiners were hard to hold in tanks and too excitable to be recommended as bioassay animals (B349).

RIBBON SHINER, *Notropis fumeus* Evermann
Texas (E119).
Hybrids have been reported with *Notropis venustus* (H384).

WHITETAIL SHINER, *Notropis galacturus* (Cope)
Cumberland River in Tennessee and Kentucky, upper Tennessee River system, and White River system in Arkansas and Missouri (O45).

O45 NC, VA, TN		SL	
	No.	Mean	Range
Age 0-Aug. 31-Oct. 18	479	33.6	19-54
I	2139	39.4	22-76
II	398	64.4	37-101
III	30	78.4	60-100
IV	1	80.0	-

Males usually were slightly larger than females of the same age.

Males and females first spawned at age II and the females had 404 to 1815 mature eggs (O45). Spawning was observed from May 25 to June 28 at temperatures of 24-28C. Eggs adhere to the underside of rocks and other surfaces. At hatching larvae were 5.2 mm TL.

Insects were the principal food items but some small fish were also eaten (O45).

ARKANSAS RIVER SHINER, *Notropis girardi* Hubbs and Ortenburger
Arkansas River drainage in Arkansas and Oklahoma (E119).

	Days old	No.	TL		Days old	No.	TL	
M72 OK	4	+	5.4	M72 OK	59	+	28.0	
	7	+	7.7		6.5 months	+	34.5	
	22	+	16.0		7.5 months	13	35.4	28-40
	40	+	24.5					

These shiners transport well and eat dry food readily and are recommended as excellent bioassay animals (G156).

The maximum SL reported is a 54 mm female (M72). They spawn in July in Oklahoma and rainfall seems to stimulate the spawning. Spawning apparently takes place in the main channels, with the eggs being fertilized and then developing as they are swept downstream in the current. The eggs are transparent and slightly over 1 mm in diameter.

BLACKCHIN SHINER, *Notropis heterodon* (Cope)
North Dakota to Quebec and south to New York and Iowa (E119).

| T113 OH | 0-Oct. | 18-36 mm; | I 25-51 mm; | adults 41-71 mm |

Blackchin shiners spawn in May and June in central Illinois (F10).

Their food includes many Entomostraca and small insects (F10, K125,129). Some filamentous algae were taken in midsummer, but this minnow is a specialized feeder (K129).

They were found in Lake Erie and other lakes where waters were clear and where there were dense beds of submerged vegetation. They disappeared quickly when waters became turbid, the bottom silty, or when the vegetation vanished (T113).

BLACKNOSE SHINER, *Notropis heterolepis* Eigenmann and Eigenmann
Southeast Alberta to Hudson Bay, Quebec, Ohio, Illinois, and Nebraska (T113).

T113 OH	0-Oct.	28-46 mm TL;	I 38-64 mm;	adults 43-69 mm
M243 NY L. George	0-Aug.	25 fish 27 mm SL 19-38		
	0-Sept.	3 fish 38 mm SL		

A maximum length of 81 mm SL is reported (H371).

A hybrid is reported with *Hybognathus hankinsoni* (H380).

Blacknose shiners feed on algae, sponges, and insects (N104, D183).

SPOTTAIL SHINER, *Notropis hudsonius* (Clinton)
Alberta, Hudson Bay, Quebec south along coast to northern Georgia and in the Mississippi Valley to Missouri and Kansas (T113).

| M290 IA | 85 shiners | 28-117 mm TL | SL = 0.805 TL + 0.08 mm | r = 0.999 |

Standard lengths in the original references have been converted to TL by the above formula in some of the following tabulations.

E105 MD	A 137 mm TL shiner weighed 23 g.			
A40 AL	TL	51	76	102	127
	W	0.5	3.6	8.2	13.6
	Number	286	24	40	14
S433 MN Red L.	log W = -2.044 + 2.989 log TL			

Female shiners tended to have higher condition factors than males but the difference was not significant (S433). Trawl-caught shiners in Red Lake had higher condition factors than shiners taken in shore seines (S433).

	No.	Mean TL	Range
Age 0-June			
H88 MI	-	-	5-23
L132 MN Linwood L.	82	15	-
M290 IA Clear L.	50	19	13-28
K86 L. Erie	+	-	10-20
S433 MN Red L.—means	+	-	10-17
Age 0-July			
F5 L. Erie	1	5	-
H88 MI	+	-	10-30
L132 MN Linwood L.	67	30	-
M290, G130 IA Clear L.	699	41	15-58
K86 L. Erie	13	37	-
S433 MN Red L.—means	-	-	13-29
Age 0-Aug.			
P186 SA Nemeiben L.	+	23	-
C22 MN L. of Woods	2	48	46-51
L132 MN Linwood L.	3	53	-
M290, G130 IA Clear L.	1241	58	28-79
S433 MN Red L.—means	+	-	25-44
Age 0-Sept.			
L132 MN Linwood L.	5	64	-
M290, G130 IA Clear L.	345	69	25-81
K86 L. Erie	19	54	-
Age 0-Oct.			
L132 MN Linwood L.	12	64	-
M290 IA Clear L.	114	77	69-84
C275 L. Erie	+	76	-
T113 OH	+	-	51-76
Age 0			
C19 MN	29	36	23-51
Age I			
P186 SA Nemeiben L. 55° N lat.	+	38	-
P186 SA Crooked L. 50° N lat.	+	46	-
C19 MN	44	53	41-66
C22 MN L. of Woods	3	66	64-69
M290, G130 IA Clear L.	979	-	76-109
E118 L. Michigan	199	76	64-91
T113 OH	+	-	64-84
Age II			
P186 SA Nemeiben L. 55° N lat.	+	61	-
P186 SA Crooked L. 50° N lat.	+	69	-
C19 MN	12	74	64-84
C22 MN L. of Woods	1	89	-
M290, G130 IA Clear L.	136	-	94-112
E118 L. Michigan	12	102	84-112
Age III			
P186 SA Nemeiben L. 55° N lat.	+	84	-
P186 SA Crooked L. 50° N lat.	+	91	-
C19 MN	6	91	84-94
M290, G130 IA Clear L.	26	-	104-117

		Calculated TL at each annulus			
	No.	1	2	3	4
C19 MN	116	36	58	76	
S433 MN Red L. males	-	56	86	99	104
females	-	58	91	107	114
M290 IA Clear L.	1137	76	99	109	
Increments	-	76	23	10	
Number	1137	152	152	22	

T113 OH   maximum 147 mm TL

Scale platelets were found on spottail shiners at 17-19 mm TL in Clear Lake (M290) but the smallest spottails with scales in Lake of the Woods were 22 mm SL and some did not have scales at 23 mm (C19). The body-scale relationship in Clear Lake showed an intercept of 14 mm which was used in computing lengths at various annuli (M290). This relationship gave a correlation coefficient $r = 0.975$. In Red Lake the body-scale relationship was found not to differ in different parts of the lake or in different years but was found to be different for the two sexes, with an intercept of 11 mm for males and 17 mm for females (S433).

The first annulus forms in late June in Red Lake but later annuli form in the second week of July after spawning (S433). Evidence for the validity of the annuli is given in S433, M290. Growth of young shiners ends early October in Lake Erie when water temperatures are at 18C (C275). During the first season they grow 4.6 mm per week in Lake Erie (C275), and during the first month they grew 3.6 mm per week in Portage Lake, Michigan (H86).

In general growth is slower to the north (P186).

In Clear Lake, Iowa (M290, G130), the average increment of young shiners from July 24 to August 28 was 0.63, 0.68, 0.65 and 0.55 mm TL per day in 1956, 1957, 1958 and 1961, respectively. The more rapid growth was associated with higher temperatures. Yearling shiners grew 0.18 mm per day from June through September, and .022 mm per day from October through June (M290). In Red Lake, Minnesota (S433), the correlation coefficients for temperature and growth of spottail shiners were 0.902 for females and 0.967 for males, both showing significantly faster growth in years with higher temperatures.

Spottail shiners first spawn at age III in Nemeiben Lake, Saskatchewan (P186), at age II in Red Lake (S433), but may mature at age I in Clear Lake, Iowa (M290). Apparently spottail shiners do not mature until they are over 66 mm at the spawning season.

	Size of females	No.	Eggs per female
M290 IA Clear L.	70-90 mm SL, Age I	10	100-1400
	Age II-III	20	1300-2600

In Red Lake, Minnesota, spottails spawn in late June and early July (S433); in Lake Erie they spawn June 1-15 at about 20C (C275) or in late June and early July (F5); in Clear Lake, Iowa, they spawn in early May through mid-June (M290, G130) and a second spawning in early August, 1961, was reported (G130). Strength of the year classes seemed to be associated with the abundance of the brood stock in Red Lake (S433). They spawn in closely packed groups, with no evidence of nesting (G28).

Some spottails were found to be active in water at 35C (T102).

Rotifers and algae were the principal foods of shiners under 10 mm and Entomostraca for shiners up to 70 mm, while larger shiners ate *Hexagenia,*

Mollusca, and small shiners in Red Lake (S433). In Clear Lake, insect larvae, Cladocera, algae, and Hydracarina were the major foods, with no evidence of change of food with growth (M290, G130). In Lake Erie, spottails appeared to feed upon fingernail clams to a greater extent than many other species of fish (P134). In Lake Michigan a number of spottails collected from an alewife spawning area were found to feed upon the eggs (E118). The smaller shiners in the group seemed to feed more on eggs than the large ones. Many of these shiners also had eaten shiner scales apparently picked up while eating eggs.

HIGHSCALE SHINER, *Notropis hypsilepis* Suttkus and Raney
The highscale shiner is apparently endemic to the Apalachicola River system, Georgia and Alabama, in piedmont streams with sand bottoms. Maximum SL is 51 mm. They apparently spawn in June (S459).

SILVERBAND SHINER, *Notropis illecebrosus* (Girard)
Illinois, south into Texas (E119).

A39 AL	TL	25	51	76	102
	Weight	0.4	0.5	4.5	6.6
	Number	5	25	79	8

They spawn in late May in Illinois (F10).

PLATEAU SHINER, *Notropis lepidus* (Girard)
Hybrids are reported with *Notropis venustus* (H348), *Notropis proserpinus* (H368).

LONGNOSE SHINER, *Notropis longirostris* (Hay)
Coastal streams from eastern Louisiana to southeastern Georgia (E119).
Longnose shiners live in sandy areas of streams in the Gulf states, in the shallow water, and with *Ericymba buccata* in deeper waters of the same type (H315).
The breeding behavior of longnose shiners on April 11 in Alabama is described (H315). No sexual size-dimorphism was noted but there were color differences (H315). They spawn on sandy bottoms and do not defend territories nor build redds. The eggs may be carried in the current as in *N. girardi* (M72).

RED SHINER, *Notropis lutrensis* (Baird and Girard)
Kentucky to South Dakota and Wyoming south to Mississippi and Mexico (E119, K130). Introduced into Arizona (K130).

C134 OK Stillwater C.	Age 0-July	12-19 mm SL

	Age	SL
C315 KS	0	25-30
	I	30-40
	II	40-50
	III	50-60

Red shiners mature as yearlings (C315) and spawn from June to August in Oklahoma (C271) and in Kansas (M215) or May through October in Kansas

(C315) and May to June in Illinois (F10). Courtship display is described (M215) and males in breeding colors guarded the nests among newly flooded weeds (C271). The dorsal fin which is actively used in the display phase of breeding behavior was higher in males than in females (K130). Red shiners are almost ubiquitous in the Big Blue River system in Kansas, particularly in the muddy rivers (M215). They spawn both in lakes and streams (C315). They were taken in a warm spring in New Mexico, 39.5C (B297), the highest temperature reported for cyprinids (H370).

Algae, insects, and crustaceans are eaten (K109).

Red shiners are nervous but can with reasonable care be handled and used as bioassay animals (W214).

Hybrids have been reported with *Notropis venustus* (H348, H366, H385), *Notropis spilopterus* (H342), and *Notropis lepidus* (H349).

PLAINS SHINER, *Notropis percobromus* (Cope)
In Kansas, plains shiners are found mostly in the larger, sandy streams (M215). Their food consists mostly of cladocerans, copepods, and a few chironomids (N93, N94).

SILVER SHINER, *Notropis photogenis* (Cope)
Most of the Ohio River drainage (E119).

T113 OH	0-Oct.	38-61 mm TL	
	I	51-76	
	Adults	69-109	maximum 130 mm

Silver shiners are in clear water, moderate to large streams with clean gravel and boulder bottoms, usually in riffles. They normally school and often jump for flying insects (T113).

Hybrids are reported with *Notropis cornutus* (T113).

SWALLOWTAIL SHINER, *Notropis procne* (Cope)
Delaware River to South Carolina (E119).

R6 NY	Age I	20 fish	37 mm SL avg.	range 33-39 mm

An occasional large individual may approach 76 mm TL (R6). No sex differentiation in growth was noted. They spawn in June and July in shallow riffle areas over fine gravel and sand (R6). Males guard territories during spawning but do not build nests. Females spawn only a few eggs at a time. The embrace at spawning is like that of *Rhinichthys atratulus*.

CHERRYFIN SHINER, *Notropis roseipinnis* (Hay)
Coastal streams, Louisiana to Florida (E119).

S472 AL	Length	No.	Mean weight	K(TL)
	25	2421	0.05	.34
	51	3115	0.7	.57
	76	12	2.3	.51
5548 shiners	25-76 mm		$\log W = -6.02 + 3.42 \log TL$	

ROSYFACE SHINER, *Notropis rubellus* (Agassiz)
   North Dakota to Quebec south to Virginia, Tennessee and Arkansas (E119, T113). In moderate-sized streams with relatively clear water (T113).

T113 OH   0-Oct.   25-51 mm TL;   I 33-64 mm;   adults 46-89 mm

			No.	Mean	Range
				SL	
Age I					
P85 NY Five Mile C.	females		134	47	39-58
	males		145	52	40-61
R143 PA streams	males		1146	50	38-60
	females		978	52	38-60
G130 IA Clear L.			1	58 TL	
Age II					
R143 PA streams	males		393	59	50-68
	females		375	60	52-68
P85 NY Five Mile C.	males		155	63	58-69
	females		127	64	56-70
Age III					
R143 PA streams	males		50	66	58-76
	females		74	68	60-78
P85 NY Five Mile C.	males		22	65	61-71
	females		47	67	61-75

   Scales first form when the shiners are 20-24 mm (R144). Females grow a little faster than males (R143, R250, P85).

P85 NY	Age	No.	SL	Eggs per female	
				Mean	Range
	I	10	50-57	600	450-754
	II	10	52-67	1090	675-1460
	III	8	64-75	1175	783-1482

   The eggs at spawning are 1.2 mm in diameter and when water hardened are 1.5 mm (P85, R144). Fertilized eggs turn bright yellow. They hatch in 57-59 hours at 21C and the fry average 5.1 mm (R144). Spawning is over cleared gravel, but the rosyface shiners do not clear any nest areas (P85, R250, M432, H360). They often spawn over nesting *Notropis cornutus*, *Hybopsis bigutta*, and *H. micropogon*. They spawn in June in Michigan (H359), May to July in Illinois and New York (P85, M432, R257), and July in Iowa (S446). Spawning occurs only at temperatures over 21C (M432).
   In Kansas rosyface shiners are found mostly over gravel bottoms (M215). They tend to be omnivorous as young or adults (R144), but mostly eat insects (F148).
   Hermophoditic specimens have been found (R256).
   Hybrids have been reported with *Notropis cornutus* (R250, T113). The hybrids are probably fertile (R250). Hybrids have also been reported with *N. zonatus* (C321) and *N. volucellus* (B303).
   Rosyface shiners feed on aquatic and terrestrial insects, arachnids, vegetation, and fish eggs, feeding both on the bottom and at the surface (P85). Breeding fish rarely ate (P85).
   Rosyface shiners are not readily kept in tanks and are not recommended as bioassay animals (B349).

SAFFRON SHINER, *Notropis rubricroceus* (Cope)
Headwaters of the Tennessee River (E119).

O45 NC, VA, TN

		SL	
	No.	Mean	Range
Age 0-July	7	20.9	19-23
0-Aug. 30-Sept. 27	293	29.5	20-44
0-Oct.-Nov.	220	28.2	20-40
I	1007	38.0	22-56
II	537	50.3	36-61
III	136	55.8	49-65
IV	12	57.6	52-67
V	2	62.0	60-64

There was no evident sex difference in growth except the females live longer. The age IV and V fish were all females.

Males and females first spawned at age II and females had 445 to 1174 mature eggs (O45). Males maintained territories over *Hybopsis micropogon* nests. Spawning was observed from May 25 to July 21 at water temperatures of 19-30C.

Insects constituted much of the food (O45).

SABINE SHINER, *Notropis sabinae* Jordan and Gilbert
Southeastern Missouri into eastern Texas (E119).
They probably spawn in midstream and let the current carry the eggs as in *N. girardi* (M72).

SANDBAR SHINER, *Notropis scepticus* (Jordan and Gilbert)
In streams of North and South Carolina. Maximum 73 mm SL (H362).

FLAGFIN SHINER, *Notropis signipinnis* Bailey and Suttkus
Flagfin shiner is an inhabitant of tiny to moderate-sized creeks, up to 6 m wide, with considerable gradient in coastal areas of Florida to Louisiana. They school in mid-water in clear, deep pools with substantial current. Maximum 56 mm SL (B301).

SPOTFIN SHINER, *Notropis spilopterus* (Cope)
Eastern North Dakota to Missouri to Lake Champlain and the Potomac River (E119).

S472 AL

Length	Number	Mean Weight	K(TL)
25	2665	0.6	.49
51	19	5.0	1.08

The regression in S472 is not believed to be valid.

T113 OH   0-Oct. 23-61 mm TL;   I 41-71 mm;   adults 51-97 mm

			Males			Females	
		No.	Mean TL	Range	No.	Mean TL	Range
S370 NY	0-Sept. pond	71	51.6	27-70	163	43.8	23-57
	0-Oct. Fall C.	32	53.1	31-70	34	44.0	27-61
	I two rivers	23	56.9	43-82	18	52.7	48-63
	L. Chautauqua	29	60.9	40-91	14	47.9	30-65
	L. Saratoga	148	67.8	38-87	152	53.1	34-74
	Fall C.	81	75.3	65-84	63	62.3	55-67
	II two rivers	13	81.0	59-91	10	69.8	60-78
	Fall C.	60	82.2	69-90	55	66.6	61-72
	L. Saratoga	7	82.7	73-93	4	77.3	71-81
	L. Chautauqua	36	86.2	72-98	50	70.1	59-79
	III NY	12	93.7	81-102	44	76.4	69-89
	IV NY	2	105.0	104-106	18	82.8	73-93

S446 IA Des Moines R.

Age from length frequencies	0-Sept.-Nov.	14-27 SL
	I-May	14-31 SL
	I-Aug.	20-37 SL

K131 IA Des Moines R.

		SL (1946-7)			TL (1960-61)		
		No.	Mean	Range	No.	Mean	Range
Age 0-Aug.		96	18	8-25	23	25	20-30
Age 0-Sept.		460	22	12-33	-	-	-
Age 0-Oct.		273	27	18-40	130	28	20-36
Age 0-Nov.		864	24	15-43	-	-	-
Age I Mar.-July 10		4405	27	12-53	674	38	23-51
July 11-Nov.		4523	36	20-65	543	43	25-58
Age II		234	51	37-68	215	58	43-69

It was estimated that some were age V and 70-75 mm (S446).

Scales first form at 15-16 mm, on the caudal peduncle (S370). Annuli form in early May. High water reduced growth rate during the early part of the summer and in mid-August (K131). There was also some evidence of increased growth with increased water temperatures. The males were larger than the females in each age group in each population (S370), and there was a tendency for more females to reach ages of III and IV.

Both sexes mature at age I though some individuals may not spawn until age II. The smallest mature female was 47 mm SL and male, 53 mm. They spawn in late July and August in Iowa (S446) and in June in Maryland (S461). They spawn from early June to late August in New York (S370), probably twice in the season. Eggs are attached to branches and logs, usually in clusters (T113), and thus are more tolerant of siltation than some other species.

	SL	No. of females	Eggs per female Mean	Range
E32 IN	-	2	715	686-745
S370 NY	54	1	225	-
	61-65	3	457	316-604
	71-78	7	595	207-871
	80-82	3	925	469-1155
	91	1	1580	-

The food consists mostly of insects (S370, S445, M404). In a Kentucky stream and in Iowa, terrestrial insects and seeds were quite abundant as food items of spotfin shiners (M404, S445).

Spotfin shiners were not readily kept in tanks nor did they eat dry food readily. They were also too excitable as bioassay animals (B349).

SILVERSTRIPE SHINER, *Notropis stilbius* (Jordan)
Alabama River system (E119).

S472 AL	Length	Number	Weight Mean	Weight Range	K(TL)
	25	1363	.3	.2-.4	1.94
	51	1961	.7	.6-1.8	0.53
	76	558	3.0	1.9-4.5	0.66
	102	627	5.9	3.0-8.6	0.58

Since the weights of the 25 mm fish appear to be too high, the regression given in S472 is probably not valid.

SAND SHINER, *Notropis stramineus* (Cope)
North Dakota to the St. Lawrence and Ohio River drainages and south to Mexico (E119).

	No.	SL	TL Mean	TL Range
Age 0-July				
F5 L. Erie	8	-	-	5-13
K131 IA Des Moines R.	5	-	-	20-25
Age 0-Aug.				
F5 L. Erie	131	-	-	5-15
K131 IA Des Moines R. 1947	26	20	-	-
K131 IA Des Moines R.	33	-	28	20-38
Age 0-Sept.				
K131 IA Des Moines R. 1947	64	24	-	-
Age 0-Oct.				
S90, S446 IA Des Moines R.	1642	16-33	-	-
K131 IA Des Moines R.	81	-	25	20-33
T113 OH	+	-	-	33-41
Age I				
S90 IA Des Moines R., March	212	18-37	-	-
S90, S446 IA Des Moines R., July-Aug.	1044	22-49	-	-
Oct.-Nov.	443	34-47	-	-
T113 OH	+	-	-	36-58
K131 IA Des Moines R.	2657	-	36	23-48
Age II				
S90, S446 IA Des Moines R., March	23	38-47	-	-
July	57	42-53	-	-
Oct.-Nov.	143	47-57	-	-
K131 IA Des Moines R.	1106	-	51	33-64
Age III				
S90, S446 IA Des Moines R., March	+	48-55	-	-

Growth of age I and II fish continued into November in Iowa (K131) but age 0 fish quit growing in October. High-water levels may delay growth in spring.

In Kansas, *N. s. deliciosus* is found in clear gravel streams while *N. s. missuriensis* is in sandy areas (M215). They spawn in late June (K131) or in late July and August in Iowa (S446) with eggs scattered over clean gravel and sand (H91). High water during these periods reduced reproduction in the Des Moines River in 1955 and high population density reduced reproduction in 1947 and 1948 (S446).

Age I females had about 250 eggs; age II, 1100 eggs; and age III, 1800 eggs (S446).

Sand shiners feed mostly on aquatic insects, terrestrial insects, bottom ooze and are considered omnivorous (S445). Young sand shiners feed mostly on bottom ooze diatoms (S445). They seem tolerant of some inorganic pollutants such as mine wastes as long as the sand and gravel are not covered (T113).

Sand shiners are excellent bioassay animals (G156). They transport and hold well, eat dry food readily and withstand low dissolved oxygen conditions. In more concentrated toxic solutions they have a tendency to jump out and covers may be needed on the containers.

WEED SHINER, *Notropis texanus* (Girard)

A39 AL	25 TL	3	0.3 g
	51 TL	10	0.5 g

S472 AL	TL	Number	Mean Weight	K(TL)
	25	67	.44	2.69
	51	112	1.95	1.50
	76	153	3.76	.86
	102	3	8.2	.78

Since the weights for the 25 and possibly 51 mm fish are obviously high, the regression given in S472 is misleading.

TOPEKA SHINER, *Notropis topeka* Gilbert
South Dakota and southern Minnesota to Kansas and Missouri (E119).

M418 KS	Age 0-Oct.	50 fish	28 TL	20-43
	I	13	48	43-56
	II	3	58	51-64

The Topeka shiner is primarily a minnow of smaller streams and usually in pool areas with moderate to little flow. During wet cycles it seems to be limited to headwater tributaries in Kansas but during drought is more widespread (M418).

In breeding condition July 14 in Kansas (M215).

REDFIN SHINER, *Notropis umbratilis* (Girard)
Southern Minnesota and eastern Kansas to West Virginia and Lake Ontario (E119).

T113 OH   0-Oct.   18-51 mm TL;   I   33-56 mm;   adults   46-81 mm

Females ready to spawn have been taken about June first in Illinois and tuberculate males from mid-May to August 1 (F10). They spawn over sand and gravel in sluggish riffles. When not spawning, they are rather tolerant of turbidity and silt bottom but prefer submerged vegetation (T113).

BLACKTAIL SHINER, *Notropis venustus* (Girard)
Southern Missouri to Texas and Florida (E119).

	TL	25	51	76
A39 AL	Weight	.5	.9	3.2

			Weight		
	TL	No.	Mean	Range	K(TL)
S472 AL	25	92	0.7	0.1-2.3	4.18
	51	669	1.2	.3-5.0	.94
	76	981	4.1	1.4-10.9	.94
	102	892	7.7	2.0-20	.75
	127	283	15.4	5-23	.75
	152	4	28.6	18-32	.80

2829 shiners 51-152 mm   log W = -4.57 + 2.72 log TL

Blacktail shiners are excitable and require constant feeding in tanks and thus are only fair bioassay animals (B349).

Hybrids have been reported with *Notropis lutrensis* (H385, H366, H349), *N. lepsidus* (H348), *N. callitaenia* (B300), *N. whipplei* (C324), *N. fumeus* (H384).

MIMIC SHINER, *Notropis volucellus* (Cope)
Ontario and Quebec south to North Carolina, Alabama and Texas (E119, T113).

T113 OH   0-Oct.   20-40 mm TL;   I   30-64 mm;   adults   38-76 mm

B56 IN	515 fish	TL = 1.255 SL
B56 IN Shriner's L.	Age I 40-50 SL	
	Age II 55-57	
B56 IN	367 eggs per female	

Females average slightly larger than males (B56). They spawn in late June and early July in Indiana and eggs are broadcast over weed beds (B56). Larger entomostracans, *Chaoborus*, and other insect larvae, and algae seem to be the main foods (B56).

When mimic and sand shiners are in the same streams, mimic inhabit the slow-moving riffles and pools while sand shiners are in the swifter currents (T113).

Hybrids have been reported with *Notropis rubellus* (B303).

STEELCOLOR SHINER, *Notropis whipplei* (Girard)

T113 OH   0-Oct.   25-64 mm TL;   I   41-76 mm;   adults   58-135 mm

Age		No.	Mean SL	Range
I L134 IL Crab Orchard L.		655	38	29-65
L134 IN		2	65	63-68
II L134 IL Crab Orchard L.		74	53	40-68
L134 IN		15	74	67-82

			SL and increments at each annulus	
			1	2
L134 IL Crab Orchard L.		1958	19	37
		Increments	19	18
		Number	654	70
	Crab Orchard L.	1956	31	53
		Increments	31	21
		Number	63	4
L134 IN			22	53
		Increments	22	31
		Number	17	15

The spawning season may be from May to August (H91).

In Illinois age I fish spawn in late August and early September, and age II fish spawn in July (L134). Ten females, 42-51 mm SL, averaged 1049 eggs with a range of 610 to 1308. The eggs adhere in clusters to vegetation, logs, etc. (H361, T113). They spawn under almost identical conditions as *N. spilopterus* but do not seem to hybridize with them (T113). Hybrids have been reported with *N. venustus* (C324).

Insects, mostly terrestrial, were the principal food found in 30 steelcolor shiners (L134).

Steelcolor shiners tend to be excitable for the first 2 weeks in tanks and will not take dry food but they are otherwise fair bioassay animals (G156).

BLEEDING SHINER, *Notropis zonatus* (Agassiz)
Ozark streams of Missouri and Arkansas (E119).

Hybrids have been reported with *Notropis rubellus* (C321). These shiners spawn over gravel, frequently using nests of other species (M72).

PUGNOSE MINNOW, *Opsopoeodus emiliae* Hay
Southern Minnesota to Michigan south to Florida and Texas (E119).

A39 AL	TL	25	51	76	102
	Weight	.3	1.4	2.7	4.5
	Number	66	8	1	1
S472 AL	Weight	.4	.9	2.2	-
	Number	147	62	5	-
	K(TL)	2.30	.72	.50	-

The sharp decline in K(TL) suggests that the weights of the 25 mm fish are too high and that the length-weight regression in S472 is not valid.

T113 OH   0-Oct.   25-43 mm TL;   I   33-51 mm;   adults   38-64 mm

Spawning in Illinois is in June (F10).

SACRAMENTO BLACKFISH, *Orthodon microlepidotus* (Ayres)
Sacramento River system, Clear Lake, and streams tributary to Monterey Bay, California. Introduced into southern California and Nevada (B341).

M139 CA Clear L.   FL-weight relationship

FL	No.	Weight	Range	FL	No.	Weight	Range
91-101	3	14	-	356-380	16	740	600-850
102-122	5	28	14-42	381-405	28	900	700-1250
127-151	9	42	28-56	406-431	33	1130	800-1400
152-175	2	60	54-65	432-456	10	1420	1050-1560
345-355	2	600	539-652	457-480	4	1560	1530-1620

	Age Group	No.	Mean FL	Range
M139 CA Clear L.	0-May	50	20	13-28
	0-June	415	36	15-66
	0-July	50	56	41-91
	0-Aug.	50	64	43-94
	0-Sept.	45	97	79-124
	I-July	19	145	130-170

			Calculated FL at each annulus				
		No.	1	2	3	4	5
M139 CA Clear L.	males	51	107	254	345	391	
	females	54	107	259	363	414	437

M139 CA Clear L.	430 mm FL female	350,000 eggs

Blackfish mature at age III, and some at age II (B341), and spawn April through June in weed beds at a depth of about 1 m (M139). The eggs are adhesive and hatched in 14 days at 14.5C.

Plankton and bottom ooze appear to be the major source of food (M139). The detritus was believed to be suspended rather than bottom ooze by C327.

They have hybridized with tui chub, *Gila bicolor* (M341).

TONGUETIED MINNOW, *Parexoglossum laurae* Hubbs

Kanawha River system, Virginia, West Virginia, and North Carolina; Upper Allegheny River and Genesee River, New York and Pennsylvania; Great Miami River system, Ohio (T113).

T113 OH		
	0-Oct.	33-58 mm
	I	46-64
	Breeding adults	69-147
	Largest	155 mm, 43 g

Tonguetied minnows build large nests, usually rectangular, of small rounded stones, with the long axis at right angles to the current (T113). River and hornyhead chub nests are usually parallel with the current. The nests may be in quiet water, often near a log or some other shelter (R258). Males guard the nest but often rather futilely (R258). Nesting is in June in Pennsylvania.

RIFFLE MINNOW, *Phenacobius catostomus* Jordan

S472 AL

Length	No.	Weight	K(TL)
51	22	2.5	1.88
76	35	5.9	1.30
102	24	10.4	1.00

SUCKERMOUTH MINNOW, *Phenacobius mirabilis* (Girard)
Colorado and South Dakota to western Ohio, Louisiana, and Texas (E119).

C134 OK Stillwater C.	Age 0-June	18-31 mm SL
	0-July	up to 41 mm SL
C284 OH St. Mary's L.	0-Aug.	19 mm TL
T113 OH	0-Oct.	38-71 mm TL
	I	51-74 mm TL
	adults	64-122 mm TL

Suckermouth minnows mature as yearlings and spawn late May to mid-July in Iowa (S446). In Illinois they spawn late May to early June (F10), and suckermouths in breeding condition were found in March in Oklahoma (C134). In Iowa they spawned in late June and early July (S446). There are no published observations on spawning site or reproductive behavior (C315).

Suckermouth minnows are reported as almost ubiquitous in the Blue River system, Kansas, most frequent over riffles and young most commonly in backwaters (M215). They feed on insect larvae, principally Diptera (S445, F10).

BLUNTNOSE MINNOW, *Pimephales notatus* (Rafinesque)
Southern Manitoba to Quebec, south to North Carolina, Alabama, Louisiana, and Oklahoma (T113).

S472 AL

		Weight		
Length	Number	Mean	Range	K(TL)
51	170	1.5	1.2-2.9	1.11
76	73	3.5	2.3-6.8	.78
102	10	9.0	-	.86

The regression given in S472 is not believed to be valid because the weights of the 51 mm fish appear high.

	No.	SL	TL	Range
Age 0-June				
F5 L. Erie	+	-	X	5-8
W26 NY	+	12	-	-
L132 MN Linwood L.	4	-	15	-
R82 OH L. Alma	12	-	18	-
H86 MI Portage L. 0.48 mm growth per day for the month				
Age 0-July				
K131 IA Des Moines R.	270	-	23	13-33
L132 MN Linwood L.	82	-	23	-
R82 OH L. Alma	7	-	28	-

	No.	SL	TL	Range
Age 0-Aug.				
K131 IA Des Moines R.	67	-	28	13-38
L132 MN Linwood L.	116	-	36	-
R82 OH L. Alma	1	-	36	-
L108 IL Big C.	+	27	-	-
Age 0-Sept.				
L132 MN Linwood L.	121	-	43	-
R82 OH L. Alma	3	-	48	-
D41 WI	+	-	51	-
Age 0-Oct.				
K131 IA Des Moines R.	212	-	28	18-38
T113 OH	+	-	X	13-71
C56 MI	1700	-	36	20-38
L132 MN Linwood L.	12	-	41	-
F66 NY ponds	+	-	X	38-53
Age 0-Nov.				
S446 IA Des Moines R. 1947	851	25	-	18-33
Age 0-Dec.				
W26 NY	+	X	-	15-55
Age I				
V1 IL April	+	X	-	13-45
K131 IA Des Moines R.	1776	-	41	20-58
L108 IL Big. C.	-	52	-	-
W26 NY female	4	50	-	-
male	7	57	-	-
D41 MN	+	-	57	-
T113 OH	+	-	X	25-84
Age II				
K131 IA Des Moines R.	380	-	56	41-66
V1 IL April female	+	50	-	45-56
male	+	62	-	50-70
D41 MN	+	-	76	-
Age III				
V1 IL April male	+	X	-	70-79

	Calculated TL at each annulus		
	1	2	3
R60 OH female	51	74	79
male	51	79	97

Maximum TL reported 112 mm (T113).

Stunting may occur in crowded ponds even if there is adequate food (H358). When propagation was in a pond where food is scarce, growth of young was slow (H91). Growth was more rapid in 1961 in the Des Moines River than in 1960 perhaps due to higher water levels in 1960 (K131).

Males apparently grow somewhat faster than females. The maximum age is reported as IV (D41).

In New York females mature at age II and males at age III (W26). The eggs are placed on the lower surface of stones or other objects and attended by the male (H8, H359). In Illinois eggs were found May 1 to August 2 (H8) at temperatures of 21-26C. As many as 5000 eggs were found in a single nest (H91).

Bluntnose minnows were classed as intermittent spawners (S446).

In Michigan bluntnose minnows spawn from late May to late August at water temperatures above 21C (H91, H359) and in Ohio from early April to early September (T113). The average of 39 nests contained 2447 eggs and covered 13550 sq. mm. of surface. The eggs are usually in one layer but may be doubled if space is limited. Eggs of various stages of development are included, indicating spawning over a period of time. Eggs hatch in 7-14 days and the new fry are 5 mm long (H91). The nests that males guard may be close together.

Early hatched fish under favorable conditions may mature and spawn early the next year (H91).

Propagation is reviewed in C56, W188.

Males were caused to form tubercles by injection of pituitary extracts (R254).

Hybrids are reported with *Pimephales promelas* and *P. vigilax* (T113).

C56 MI	200-500 eggs per spawning, 2 or more times a year.
W26 NY	One female produced 2298 eggs in 12 spawnings in 2 months. 10 females had an average of 2005 eggs in various stages of development. Range 1743-2223.

The fry average 5 mm at hatching (W26).

When acclimated at	Lethal temperatures Low	High	Citation
5C	-	26C	S333, H253
10	-	28-28.3	H253, S333
15	1.0C	30.6	S333
20	4-4.2	31.7-32	H253, S333
25	7.5	33.3	S333

Bluntnose minnows feed mostly on bottom ooze diatoms (S445).

Bluntnose minnows were found to be an important destroyer of eggs of other fish in Walnut Lake, Michigan (H347). The fact that they fed primarily on plankton was the reason M412 recommended bluntnose minnows for reservoirs. Most studies indicate that bluntnose minnows feed on small organisms and debris about equally from the bottom and from plankton (H91, K129, K125, K112, P165, S445, E121).

FATHEAD MINNOW, *Pimephales promelas* Rafinesque

Great Slave Lake to southern Chihuahua, Mexico, from the Rockies to the Appalachians (V85). Also in the Pacific drainage, Rio Yaqui, in Mexico (V85). Introduced elsewhere west of Rockies (E119).

I15 MN Horseshoe L.

SL	16-18	18-20	20-22	22-24	24-26	26-28	28-30
W	.081	.117	.166	.233	.302	.373	.459
No.	21	24	32	45	57	22	17
K(SL)	1.65	1.71	1.79	1.92	1.93	1.90	1.88

C36 IA Clear L.   6 fish 30-61 mm SL,   average K 1.90,   range 1.54-2.26

Mean weight (numbers in parentheses)

TL	C345 IA Skunk R.	Des Moines R.	MN D81	P109	AL A39,40	S472
25	-	-	0.17	-	0.6 (19)	0.36 (88, O49)
29-31	.36 (2)	-	-	-	-	-
32-34	.46 (5)	-	-	-	-	-
35-37	.61 (26)	-	-	-	-	-
38-40	.75 (56)	.67 (2)	.6	.9	-	-
41-43	.96 (59)	.80 (9)	-	-	-	-
44-46	1.19 (70)	1.07 (50)	-	-	-	-
47-49	1.48 (76)	1.27 (130)	-	-	-	-
50-52	1.73 (60)	1.51 (116)	1.5	1.6	1.36 (47)	1.0 (71,7121)
53-55	2.13 (34)	1.80 (77)	-	-	-	-
56-58	2.52 (24)	2.11 (26)	-	-	-	-
59-61	2.93 (19)	2.40 (16)	-	-	-	-
62-67	3.88 (11)	2.92 (8)	3.0	3.9	-	-
71-76	5.41 (3)	4.24 (1)	5.4	6.8	3.6 (8)	3.1 (25)
89	-	-	8.6	9.7	-	-
102	-	-	13.2	-	9.0 (1)	-

C345 IA Des Moines R. July 16-30   males $\log W = -2.926 + 3.006 \log TL$
                                    females $\log W = -4.176 + 3.141 \log TL$
                                    combined $\log W = -4.334 + 3.238 \log TL$
          Skunk R.   July 16-30   combined $\log W = -4.272 + 3.219 \log TL$

S472 AL   K(TL) for 25, 51, and 76 mm fish, 2.19, 0.80, 0.69 respectively.

Because the K values for the 25 mm fish were so high, suggesting errors in weighing these small fish, the regression given in S472 is not valid.

Average weights at equivalent lengths were usually higher during the last half of July than earlier in the season in the Des Moines River in 1965 but were highest in the last half of June, 1966, in the Skunk River (C345). Differences in spawning seasons may have been the reason. While the length-weight relationships differed by sex, neither sex was consistently heavier at equivalent lengths, but most of the larger individuals were males. In Horseshoe Lake, Minnesota (I15), weights were highest at equivalent lengths in January, particularly for females.

D81 MN   Average TL at the end of 10-day periods, first summer

Days	10	20	30	40	50	60	70	80	90	100	110	120
TL	5	10	15	20	25	29	34	39	45	50	55	59

	No.	SL	TL	Range
Age 0-May				
M23 IA	+	4.7	-	-
W9 OH	+	-	6	-
D41 MN	+	-	25	-
Age 0-June				
I15 MN Horseshoe L.	200	11	-	7.5-18.5
W9 OH	+	-	13	-
D41 MN	+	-	36	-
H195 MI	+	-	X	13-28

		No.	SL	TL	Range
Age 0-July					
C134 OK Stillwater C.		+	X	-	9-32
W9 OH		+	-	22	-
I15 MN Horseshoe L.		1063	22	-	7.5-33.5
D41 MN		+	-	56	-
H195 MI		+	-	X	30-46
C157 IA Little Wall L.		45	-	30	23-41
Age 0-Aug.					
L35 IA		+	X	-	8-40
W9 OH		+	-	32	-
D41 MI		+	-	36	-
H195 MI		+	-	X	36-46
F66 NY ponds		+	-	X	36-53
Age 0-Sept.					
W9 OH		+	-	38	-
H195 MI		+	-	28	15-58
F66 NY ponds		+	-	X	41-58
Age 0-Oct.					
W9 OH		+	-	51	-
F66 NY ponds		+	-	X	46-56
T113 OH		+	-	X	13-64
Age 0-Nov.					
M23 IA		+	X	-	20-55
P63 IA pond		+	-	X	25-64
Age 0-Dec.					
A9 OK		+	-	61	-
Age I					
T113 OH		+	-	X	25-76
C134 OK Stillwater C.	July	1	39	-	-
C345 IA Skunk R.		298	-	43	28-61
C345 IA Des Moines R.		95	-	48	36-61
P63 IA pond		+	-	X	64-74
F66 NY ponds		+	-	X	41-58
C157 IA Little Wall L.	Jan.	83	-	61	23-56
Feb. -	female	77	-	66	56-76
	male	93	-	71	46-84
June -	female	8	-	71	66-71
	male	6	-	79	74-81
July -	female	23	-	53	43-71
	male	23	-	58	53-66
Age II					
C345 IA Des Moines R.		336	-	51	41-71
Skunk R.		150	-	53	36-74
Age III					
C345 IA Des Moines R.		1	-	64	-

Growth may be slow in crowded ponds even if there is adequate food (H358). Young fatheads grew slightly in winter in Iowa, but adults lost weight (M23). In Alabama small fatheads grew well in the winter at temperatures of 2-7C if well fed and not crowded (P108). No growth occurred after November in Minnesota (I15). Annulus formation in two Iowa rivers was not until late July or August (C345).

The maximum TL reported is 89 mm (D41, T113) with males being the largest (I15, C345). The smaller size at age I in July than June in Little Wall Lake, Iowa (C157), suggests that the life span is about one year, with the age I July fish being from the later broods of the previous year. There is heavy mortality of males after spawning (H358, P108).

A similar decline in average length of age I fatheads was noted in Skunk River (C345). In most populations fatheads mature and die when about one year old but some fatheads in some populations do not mature until age II (H91, I15, C345). In the Des Moines River most of the spawners in 1965 were age II (C345). One age III male is reported (C345) and a few males were taken entering the 3rd winter (I15). Early hatched fish may spawn later the same year, even as far north as Iowa and Minnesota (I15, L35).

Spawning begins when water temperatures reach 15.6C and continues until water temperatures drop to 15.6-18.4C in the fall (P108) from May through August (L35, H91, M23, H195). The first hatch occurs when water temperatures are 15.6-18.4C, 150 mm from the surface (S334). Spawning was interrupted in Minnesota when water temperatures dropped from 27 to 21C, June 10-13 (I15). Late July was the period of peak spawning in two Iowa rivers, where water levels may have more effect than water temperatures (C345). One female produced 4144 offspring in 11 weeks in 12 spawnings (D41, M23, D81). Since counts of ovaries ranged from 800-1000 eggs per female, H358 questions this 4144.

However, egg production in the season may be considerably greater than the number of eggs detectable in the ovaries at any time (I15). In Horseshoe Lake, Minnesota, egg production per female was estimated at 255 in 1959, but at 2400 eggs in 1960 when the population density was lower. The number of eggs in 10 females, 41-51 mm SL, averaged 950 with a range of 636-1338 (I15). Of these 355 were ripe on the average with a range of 186-494. Eggs counts on 4 females 47-55 mm TL from Skunk River were 802-2622, average 1888, of which about one-third were ripe (C345).

The eggs are attached to the lower side of lily pads, rocks, shingles (L6) or other surfaces and the male attends the eggs, occasionally stroking them with a thickened pod on the back of the neck (W186). Single nests may contain eggs in various stages of development, indicating repeated spawning in the same nest (H91). The eggs are 1.3 mm in diameter and newly hatched larvae are 4.8 mm (W186). Eggs hatched in 4.5 to 6 days (H358) and 5 days at 25C (H195). Seven nests had 4500 to 12124 eggs (H195). In Wisconsin, 80% of 320 nests were at a depth of 2.5 feet. Stone objects were selected for egg attachment twice as frequently as wood (H358). One nest contained 2740 eggs but the average of 44 nests was 480 eggs (H358).

Propagation of fathead minnows is reviewed in R247, L6, H358, W187, M23, P108, W9, W188, H195.

Hybrids have been reported with *Pimephales notatus* (T113).

Fathead minnows are tolerant to both clear and turbid water and to a wide range of pH. They seldom occur where bluntnose minnows are abundant. Fathead minnows are too excitable for good bioassay animals, but transport and eat well (G156).

Acclimated temperature	Lethal low	Lethal high	Citation
10C	-	28-28.2	H253, S333
20	1.5-2	31.7-32	H253, S333
21	21	-	B201
25-6	-	32.3	B201
30	10.5-11	33-33.2	H253, S333

Fatheads do not succeed under severe competition with other fish (H91, C345, S460). They are often found in lakes where most other species winter-kill (C157) and may survive in muskrat burrows at such times.

In Nebraska, fathead minnows were found at pH of 9.8 and total alkalinity of 1400 ppm, and at 29000 ppm in Saskatchewan (M388).

Diatoms and bottom algae are the major foods of fatheads (M404, R247, C316, S445). Plankton-feeding of the fathead was considered an important factor in recommending it for reservoirs (M412) and in rearing ponds (L35). Females will eat their own eggs if the eggs are not guarded by the male (I15).

BULLHEAD MINNOW, *Pimephales vigilax* (Baird and Girard)

South Dakota to southwest Pennsylvania south to northern Alabama, Mississippi and northern Mexico (E119, T113).

	Length	Number	Mean	Weight Range	K(TL)
S472 AL	25	2641	0.5	.03–1.5	2.80
	51	1519	1.4	.6–3.6	1.02

The 25 mm fish seem to be too heavy and thus the regression in S472 is probably not valid.

T113 OH	0-Oct., 20-51 mm TL;	I, 30-64 mm;	adults, 38-94 mm

S446 IA Des Moines R.	0- Aug.	12-17 SL
	0- Nov.	16-30
	I- Apr.	16-35
	I- Aug.	26-41
	II and older	34-69

Few exceed 3 years of age though some may live as long as 5 years (S446).

This species is most abundant in backwaters and pools of large and medium-sized rivers (H91, S460). They spawn in late July and August in Iowa (S446). Gravid females have been found in May and June in Illinois (F10) and in September (G156). They feed on aquatic and terrestrial insects, bottom ooze, and a variety of foods (S445, G156). Young feed primarily on bottom ooze diatoms.

Bullhead minnows transport and hold well in tanks, eat dry food readily, but are too nervous to be good bioassay animals (G156).

Hybrids are reported with *Pimephales notatus* (T113). Additional data on life history are given in P187 which was not seen.

WOUNDFIN, *Plagopterus argentissimus* Cope

Virgin River system in Arizona, Utah and Nevada. In swift shallows over stone, sand, or mud bottom. Apparently able to tolerate high turbidity. Maximum SL 71 mm, about 90 mm TL (M413).

SACRAMENTO SQUAWFISH, *Ptychocheilus grandis* (Ayres)

Streams and lakes of Sacramento River system, Russian River and streams flowing into Monterey Bay, California (B342). One of the largest was 1145 mm and 14500 g (B342).

T39 CA   25-430 mm FL   FL = 1.122 SL + 3 mm

B342 CA American R.

FL	104	132	150	173	211	257	315	373	401	452	577
Wt.	14	18	36	45	100	154	281	560	622	962	2250

T39 CA Coyote C.

Age Group	Mean SL	Range
0-June	30	16-35
0-July	36	30-40
0-Oct.	39	35-47
0-Nov.	44	39-51
I-Feb.-April	58	44-74

Average calculated SL at each annulus

	1	2	3	4	5	6
T39 CA Coyote C.	52	112	190	250	290	
T39 CA Sacramento R.	60	142	206	260	320	380

In the Sacramento River they may live to age IX (T39). Maturity is probably reached at age II or III (K110). A 500 mm female had 17730 eggs (B342).

They spawn in April to June in California, scattering the eggs over gravel (T114, K110) at temperatures of about 14C. The eggs are adhesive and cling to rocks (T39).

Insect larvae were the principal foods of squawfish 33-180 mm (T39). Larger squawfish are predatory (T114) and are important predators on young sockeye salmon (R245, F126). A 250 mm sucker was found in the stomach of a 740 mm squawfish (B342).

COLORADO SQUAWFISH, *Ptychocheilus lucius* Girard
     Colorado River drainage (M420).

		No.	Weight	K(SL)
M340 UT Green R.	340 SL	+	535	1.36
	465	+	1250	1.24
	544 TL	+	1225	-
B240 WY	610 TL	+	1588	-
S146 WY	572 TL	+	2040	-
S463 UT	343 TL	1	426	-
	400	1	426	-
	445	1	580	-
	464	1	794	-
	490	3	816	(735-867)
	607	4	1575	(1490-1673)

Colorado squawfish have been reported to reach 1.8 m and 45 kg but few weighing as much as 18 kg have been taken since 1930 (S463).

The squawfish is largely predatory (S463).

A 32 mm specimen was taken in May in southern Arizona and a 16 mm specimen in late August in Utah, suggesting that spawning may take place throughout the summer (S463).

NORTHERN SQUAWFISH, *Ptychocheilus oregonensis* (Richardson)
Columbia River drainage and in coastal streams of Oregon and Washington (E119).

J53 ID	50-120 TL	TL = 1.114 FL = 1.15 SL
	120-300	TL = 1.087 = 1.094 SL

J53 ID	Weight	450	910	1360	1810	2270	2720	3175	3630
	Average TL	366	452	503	554	602	640	673	700

	No.	Mean	Range	Length type
Age 0-July				
N82 BC Nicola L.	18	23	18-25	FL
Age 0-Aug.				
H337 MT Clearwater R.	23	25	20-38	TL
L. Placid	17	38	33-43	TL
Age 0-Sept.				
R275 MT Flathead R.	+	-	30-46	TL
H337 MT L. Placid	15	53	41-61	TL
Age I				
W184 MT Flathead L.	20	42	32-56	SL
G75 OR borrow pit	1	168	-	FL
Age II				
L193 NV	1	236	-	TL
Age V				
Z6 Columbia R.	261	315	193-396	TL
Age VI				
C14 BC	+	-	254-305	FL
Age VII				
L193 NV	1	427	-	TL
Age XI				
C14 BC	+	457	-	FL
Age XIII				
J52 ID Hayden L.	1	673	-	TL, 4082 g
Age XV				
H337 MT	2	452	442-465	TL

Average calculated FL at each annulus

	1	2	3	4	5	6	7	8	9	10	11
C53 BC Okanagan L.	46	89	130	160	201	239					
C53 BC L. of Woods	48	99	147	198	226	274	302	345	366	386	411

Average calculated TL at each annulus

	No.	1	2	3	4	5	6	7	8	9	10	11	12	13	14	15	16	17
J53 ID	123	81	152	211	267	305	348	378	406	434	470							
P159 MT lakes	284	53	84	117	150	183	213	244	267	292	312	340	366					
streams	112	46	89	132	183	229	269	310	325	373	409	442	467					
H337 MT Seely L.	72	56	89	127	165	201	234	262	290	320	353	376	455					
Alva L.	267	53	84	117	150	183	213	241	267	292	312	34C	366					
Increment		53	30	33	33	33	30	28	25	25	23	25	25					
Number		267	267	267	267	263	233	180	113	73	38	16	9					
R275 MT																		
Flathead L.	200	30	53	81	112	145	175	203	234	262	290	318	345	373	404	434	457	486
Increment		30	23	28	36	36	33	33	33	30	28	25	25	23	23	23	20	20
Number		200	200	198	197	187	181	168	147	127	115	98	82	65	41	27	11	6

Female squawfish grow slightly faster and live longer than males (H337, J53). The oldest male was age XI and female, age XV (H337). Sex was not given for older fish in R275. At age IX females were 18 mm TL longer than males in Alva Lake, Montana (H337). A squawfish in the Columbia River tagged in July was 323 mm TL and 368 mm when recaptured in November (Z6).

A few males mature at age III (J52) in Hayden Lake, Idaho, but not until age VI and 250 mm TL in Seeley Lake, Montana (H337). In the latter lake females did not mature until age VII and 254 mm TL. They spawn in June at water temperatures of 12-22C, on rubble under 50-300 mm of water (H337). The eggs are broadcast, but adhere to rocks and there is no care of eggs or young (J53). Eggs hatch in 7 to 8 days and the young school (J53).

		Females		Eggs per female	
		Size	No.	Mean	Range
H337 MT Seeley L.		287-351 mm TL	7	17400	6700-27500
J52 ID Hayden L.		406-521	4	-	37000-83000
J53 ID		406 mm, 680 g	1	12000	-

At the first spawning female squawfish produce 3600 to 4560 eggs per kg of female and at later spawnings, 5400 to 7260 (J53). They spawn at 18C and the eggs hatch in about 4 days (L193).

Squawfish are usually in vegetated areas of the lakes (H337). When acclimated at 19-22C, the upper lethal temperature was 29.3C (B200).

Hybrids have been reported with *Acrocheilus alutaceus* (P168), *Mylocheilus caurinus* and *Richardsonius balteatus* (W184).

Tagging suggested little movement in the Columbia River (Z2).

Larger squawfish, over 300 mm, feed mostly on fish, and squawfish may start predation on other fishes when only 100 mm long (J53, T70). Insects are the most important foods for squawfish up to 100 mm (J53, H337) or to 200 mm (T70, L192) and crayfish for intermediate-sized squawfish (J53). Surface feeding is not unusual. Their relationships as predators on young salmon have been studied by C313, R244, R245, T70.

BLACKNOSE DACE, *Rhinichthys atratulus* (Hermann)
North Dakota to Maritime provinces south to North Carolina, Tennessee and Nebraska (E119, T113).

T113 OH 0-Oct. 20-51 mm TL; age I, 25-61 mm; breeding adults, 51-97 mm

	Age Group	No.	Mean SL	Range	Mean Weight	Range
T32 NY	0-July	39	17.5	14-24	0.09	0.03-0.23
	0-Aug.	55	20.5	13-30	0.17	0.06-0.5
	0-Sept.	8	25.2	22-28	0.3	0.2-0.4
	0-Oct.	3	29.5	27-33	0.33	0.2-0.55
	0-Nov.	27	29.3	21-33	0.30	0.1-0.5
	I-Jan.	2	25.3	24-26	0.12	0.1-0.15
	I-March	29	32.6	25-39	0.52	0.2-0.8
	I-April	3	35.0	31-38	0.70	0.4-1.0
	I-May	11	35.2	31-40	0.70	0.4-1.0
	I-Nov.	-	-	37-44	-	0.7-1.4

Year classes overlap such that length frequencies are not very helpful in aging dace (T32). Males are generally smaller than females of the same age

(C314, S305). They probably mature at age II (H91). The smallest mature
male was 44 mm (T32). The pectoral fin of males occupies one half of the
pectoral-pelvic space while that of females occupies only one third. In males
the pelvic fin is square while that of the female is pointed and longer (S305).
Spawning of *R. a. obtusus* was observed at 15.5-18C (S305), of *R. a. meleagris*
at 21C (R243), and of *R. a. atratulus* at 22C (T32). *R. a. meleagris* is reported
to nest; the others apparently do not. *R. a. obtusus* have been observed to
spawn only in pools whereas the others use both pools and riffles. In New York
and Pennsylvania they spawn in late May and early June (T32, R243) but were
reported as in spawning condition in early September (C314). Spawning was observed in late June in West Virginia (S305) and reached its peak at midday. An
average of 746 eggs per female was found in 6 specimens (T32).

Blacknose dace show a pronounced selection of small streams with steep
gradient (S460). They may be forced into larger streams as water flow fails
(T113).

Acclimated temperature	Lethal low	Lethal high	Citation
5C	-	25-26.5	H253, S333
10	-	27-28.8	H253, S333
15	-	29.6	S333
20	1-2.2	29.3-30	H253, S333
30	5	-	S333

In a Kentucky stream blacknose dace fed on amphipods, isopods, mayfly
and stonefly nymphs, oligochaetes and chironomids (M404). Diatoms may contribute about 25% of the food but insects are the major foods (T32, B70, P164,
S447, M71).

Plants were about 64% of the diet in one study (F148).

Additional life history data are in a thesis, not seen (S506).

LONGNOSE DACE, *Rhinichthys cataractae* (Valenciennes)
British Columbia to Newfoundland, south to Pennsylvania and Oregon ex-
cept for extensions in the Appalachians to South Carolina and in the Rockies to
New Mexico and west Texas (T113).

S463 ID		0-July		11 mm SL	
			No.	TL	Range
Age 0-June-July					
F5 L. Erie			-	-	13-20
Age 0-Nov.					
T113 OH			-	-	25-58
Age I					
R145 PA Little Sandy C.	male		117	58	48-66
	female		113	58	51-66
K44 MN Thompson C.	male		13	61	-
	female		9	58	-
T113 OH			-	-	38-64
Age II					
R145 PA Little Sandy C.	male		103	76	64-84
	female		113	76	66-84
K44 MN Thompson C.	male		19	74	-
	female		41	76	-

		No.	TL	Range
Age III				
R145 PA Little Sandy C.	male	132	81	74-89
	female	137	84	76-91
K44 MN Thompson C.	male	37	81	-
	female	54	84	-
Age IV				
T145 PA Little Sandy C.	male	76	86	81-97
	female	160	91	84-99
K44 MN Thompson C.	male	10	91	-
	female	13	94	-
Age V				
R145 PA Little Sandy C.	female	82	102	94-119
R44 MN Thompson C.	female	8	107	-

			Calculated TL at each annulus				
		No.	1	2	3	4	5
K44 MN Whitewater R.	middle	61	48	61	74	81	99
K44 MN Whitewater R.	north	53	51	64	76	89	99
K44 MN Thompson C.	males	79	48	61	74	84	
	Increments		48	15	10	10	
	Number		79	66	47	10	
	females	125	46	61	76	89	99
	Increments		46	15	13	10	10
	Number		125	116	75	21	8

The average growth for southern Minnesota trout streams given in D81 is a summary of the data from K44. Female longnose dace appear to grow slightly faster than the males after the first 2 years and to live longer (R145, K44).

The spawning is from late June to August in southeast Minnesota and females have 160 to 680 eggs (K108). They were considered early spawners by H91 and S446. They are reported to spawn in May in Maryland (S461).

Longnose dace are usually found in water currents of over 0.5 m per second and at depths of less than 0.3 m (G124). Aquatic insects are the principal foods, particularly Diptera (G124, K44, R145) or Ephemeroptera (G153). In riffle areas in Pennsylvania, the population was estimated as 1 per 1.11 square meter (R145). They were classed as skin tasters by E123 because the facial lobe of the brain is enlarged.

Hybrids have been reported with *Hybopsis micropogon* (R251), *Hybopsis plumbea* (H321), *Gila nigrescens* (C324).

LEOPARD DACE, *Rhinichthys falcatus* (Eigenmann and Eigenmann)

	Age Group	FL (mm)	Grams
G124 BC (Petersen method)	0-Aug.	9-18	-
	I-June	18-36	0.1
	II-June	44-61	1.6
	III-June	60-80	4.5
	IV and over	80-120	8.4 males, 10.4 females

No difference in growth of males and females was evident except that the larger fish were females (G124).

Leopard dace have a larger swim bladder than longnose dace and are found in slower currents (usually less than 0.5 m per second) and deeper water (over 0.3 m) (G124). They spawn in early July. When acclimated at 14C, the upper lethal temperature was 28.3C (B200).

Aquatic insects were the principal foods although adult leopard dace also ate large numbers of *Lumbricus* (G124).

SPECKLED DACE, *Rhinichthys osculus* (Girard)

West of Rockies in Columbia River drainage and other streams from Washington to southern California, Lahontan Basin, and Colorado River drainage (E119).

			Weight	
	FL	No.	Mean	Range
J82 CA Trinity R.	23-30	4	.15	.10-.19
	32-40	5	.33	.18-.42
	41-47	5	.56	.44-.70
	51-57	3	1.23	1.12-1.30
	60-62	3	1.40	1.65-1.88
	66-71	3	2.51	2.14-3.16
	80	1	4.1	-
J82	$\log W = -4.29 + 3.09 \log FL$			$r = .99$

Annuli were characterized by cutting over, with one set of circuli tangent to those of the previous year. In some cases annuli showed 1 or 2 narrow weakly marked circuli (J82). Although scale pictures were given, J82 did not believe the scales to be reliable indicators of age and the following data are from length-frequencies:

J82 CA Trinity R.

Age			No.	Peak FL	Range
0-July			837	8.9	6.5-18.5
0-Aug.			311	23.7	8.5-34.5
0-Sept.			177	28.5	14.5-40.5
I			-	-	32.5-40.5
II	males		-	-	44.5-48.5
	females		-	-	50.5-52.5
III	males		-	-	54.5-58.5
	females		-	-	62.5-66.5

In the Trinity River, California, speckled dace breed from June through August with a peak in late June (J82). In the Green River, Utah, they were observed spawning in a riffle area on July 11 and gravid females were taken July 27 (M340). Females with eggs were taken in June 3-5 in Jasper Park, Alberta (B10), and in June in Nevada (L193).

In Lake Tahoe feeding was primarily at night and benthos constituted about 80% of the food with the rest mostly plankton (M325). A few fish eggs were taken. Summer foods were primarily winged insects in Trinity River (J82) and in Heenan Lake (C265). Fall foods were mostly plant materials and filamentous algae; winter and spring, aquatic insects (J82). In Alberta, chironomids, cladocerans, copepods, *Gammarus* and insects constituted the food (N23).

Hybrids are reported with *Richardsonius egregius* (C322).

REDSIDE SHINER, *Richardsonius balteatus* (Richardson)
Columbia River drainage and Salt Lake basin (E119).

Age	Citation	No.	Mean	Range	Length type
Hatching	W56 MT	+	-	3.8-5.6	TL
0-June	L176 BC Sixteen Mile L.	+	-	5-10	FL
1.5 months	W56 MT	+	-	12.7-13.5	TL
0-July	N82 BC Nicola L.	1	25	-	FL
3 months	W56 MT	+	-	16.8-21.6	TL
0-Aug.	C228 BC Paul L.	+	19	-	FL
0-Sept.	C229 BC Paul L.	+	-	25-38	FL
	L182 BC	+	-	17-27	SL
0	C229 BC Pinanton L.	+	28	-	SL
I	C229 BC Pinanton L.	+	56	-	SL
	L176 BC Sixteen Mile L.	+	-	25-56	FL
	C229 BC Paul L.	+	-	38-76	FL
	L182 BC	+	-	34-55	SL
	W56 MT	+	-	30-58	TL
II	W56 MT	+	-	61-86	TL
	L176 BC Sixteen Mile L.	+	-	56-71	FL
	L182 BC	+	-	55-75	SL
	C229 BC Pinanton L.	+	74	-	SL
	C229 BC Paul L.	+	-	76-102	FL
V or VI	L182 BC	1	141	-	SL

The maximum size in Paul Lake, British Columbia, is about 170 mm FL (L90) and maximum age about VI (L176). Scales tend to be difficult to read (L182). There tend to be more large females than large males (L176), few over age II being males (L182).

W56 MT	6 females	84-104 mm	1852 eggs per female range 829-3602

Most redside shiners first spawn at age II (W185) or at age III (L176) in British Columbia in late May or June with eggs broadcast over vegetation (L90), some even spawning in mid-July (C229, L176). The eggs are adhesive and spawning occurs at night (C229).

In Montana, redside shiners were found spawning in a warm spring (17-18C) in early April (W56). Elsewhere in Montana they spawn in late May and June (W56, S442) and in late June and early July in Wyoming (S146). At Sixteen Mile Lake the shiners enter streams to spawn when water temperatures first reach 10C, with males migrating earlier than females (L176).

No territorial or courting behavior was evident in observation of spawning schools (W56). Pectoral fins in mature males extend beyond the origin of the pelvics but not in females (L182). The crimson and gold of males is more vivid than that of females (W56, L182).

Freshly fertilized eggs measure 1.9-2.2 mm in diameter and have a pale yellow yolk (W56). Eggs hatched in 7 days at 21C, 8 days at 18C, 11 days at 15C, and 15 days at 12C (L182) or at 3-7 days at 21-23C (W56). Embryology is given by W56. At hatching the fry are 5.3 mm TL and 8.7 mm at 10 days when they start feeding in aquaria (W56). In nature, hatching may be at a slightly larger size (L176). Fry appear about 10 days after the spawning at 12-15C (L176).

Acclimated temperature	Lethal high	Citation
9-11C	25C	B200
14	27	B200

Food consists of plankton, bottom fauna, surface insects, and some trout fry (L62, L90, L182, J84, I8, W56). In Pinanton Lake, British Columbia, at night most of the food was *Daphnia* and in the daytime, algae (J84). The shiners moved offshore at night. They feed on their own eggs (W56), eggs of suckers (S412) and of grayling (S146).

Effect of the introduction of redside shiners on rainbow trout in Paul Lake, British Columbia, has been studied intensively (L90, J84, C228, C229, C230).

Hybrids with the following species have been reported: *Mylocheilus caurinum, Ptychocheilus oregonensis, Rhinichthys osculus*, and *Rhinichthys cataractae* (S443, W183, W184).

LAHONTAN REDSIDE, *Richardsonius egregius* (Girard)

Lahontan Basin and related waters of Nevada and California (E119).

They spawn in May and June in Nevada (L193) and in Lake Tahoe, in June and July, over gravel along the shore in 200-450 mm of water at a temperature of about 17C (M325). Redsides seem to feed mostly at night on the surface but they also eat some plankton, bottom fauna and fish eggs (L193, M325).

Hybrids are reported with *Rhinichthys osculus* (C322).

CREEK CHUB, *Semotilus atromaculatus* (Mitchill)

Southeastern Alberta to Newfoundland south to northern Florida and the Gulf of Mexico (E119, T113).

G81, L110 IL	145 chub	38-196 mm SL	TL = 1.19 SL

This conversion factor was used when data were reported in SL and changed to TL for these summaries.

		Weight		
TL	No.	Mean of means	Range	Citation
20-9	15+	.27	.14-.3	D81 MN, D195 IA
30-9	34+	.47	.28-.65	D81,195
40-9	14+	.77	.11-1.2	G67 NY, D41,195
50-9	91+	1.33	0.8-1.9	G67, D41,81,195, S472 AL
60-9	25+	2.1	1.0-3.0	G67, D41,81,195
70-9	43+	4.1	2-6.6	G67, D41,85,195, S472
80-9	11+	6.5	4.8-8	D81,195
90-0	13+	8.2	5.7-10.6	G67, D41,195
100-9	14+	13.0	10-15	D41,81,195, S472
110-9	9+	14.9	9.1-19	G67, D81,195
120-9	21+	21.6	10-25.8	G67, D41,81,195, S472
130-9	6	21.6	14-33	G67, D195
140-9	10+	28.6	17-37	G67, D81,195
150-9	6+	36.3	23-45	G67, D81,195
160-9	2	58.0	57-60	D195
170-9	3	68.0	42-82	G67, S472
196-8	2	72.0	57-85	G67, E105 MD
216	2	79.0	74-85	G67
262	1	153.0	-	G67
302	1	340.0	-	T113 OH

D195 IA Des Moines R.		log W = -3.972 + 2.98 log TL		
L110 IL Hutchins C.		log W = -4.49 + 2.86 log SL r = 0.995		
43 chub	45–196 mm	mean K(SL) = 1.75		
G67 NY	93 chub	mean K(SL) = 1.02		
D195 IA Des Moines R.	151 chub	mean K(TL) = 1.02		

	TL	25	51	76	102	127	179
S472 AL	No.	71	74	27	6	6	1
	K(TL)	2.94	1.22	1.25	1.39	1.08	1.44

Since the weights at 25 mm are obviously too high, the regression given in S472 is not included.

| E83 NY | 7.9 mm at 7 days | | and | | 35 mm at 67 days | |

	No.	Mean TL	Range	No.	Mean Weight	Range
Age 0						
D195 IA Des Moines R.	62	33	23–51	62	0.05	–
G67 NC and SC	3	53	43–61	–	–	–
G67 AL	17	58	48–84	–	–	–
C58, H91, W10 MI	+	–	51–124	–	–	–
L24 QU	2	–	81–99	2	11	8–14
Age 0–June						
K49 OH Lytle C., Polluted	27	28	–	–	–	–
Clean zone	14	30	–	–	–	–
C49 OH	+	30	–	–	–	–
D41, H195 MI ponds	+	25	23–36	–	–	–
Age 0–July						
D195 IA Des Moines R.	+	–	23–43	–	–	–
K49 OH Lytle C., Polluted	61	48	–	–	–	–
Clean zone	20	41	–	–	–	–
D41, H195 MI ponds	+	38	36–51	–	–	–
Age 0–Aug.						
G67 NY	6	30	25–33	–	–	–
D195 IA Des Moines R.	+	–	36–51	–	–	–
C49 OH	+	46	–	–	–	–
K49 OH Lytle C., Clean zone	251	51	–	–	–	–
Polluted	304	61	–	–	–	–
D41, H195 MI ponds	+	64	53–76	–	–	–
G81, L110 IL streams	16	69	–	6	0.1	–
Age 0–Sept.						
K49, OH Lytle C., Clean zone	192	53	–	–	–	–
Polluted	117	71	–	–	–	–
B222 NY Wolf L.	3	56	53–58	3	2.3	–
B16 MI Drayton Hatchery	+	64	–	+	–	2.0–2.3
C49 OH	+	69	–	–	–	–
D41, H195 MI ponds	+	81	46–97	–	–	–
M24 IL	+	89	–	–	–	–
D81 MN	+	89	–	–	–	–
B16 MI Hastings	+	–	79–124	+	–	5–25

	No.	Mean TL	Range	No.	Mean Weight	Range
Age 0-Oct.						
K49 OH Lytle C., Clean zone	105	53	-	-	-	-
Polluted zone	36	71	-	-	-	-
T113 OH	+	-	33-71	-	-	-
Age 0-Oct.-Dec.						
G67 NY	11	46	33-51	-	-	-
C49 OH	+	81	-	-	-	-
Age I						
T113 OH	+	-	46-89	-	-	-
G67 NY	173	51	28-69	-	-	-
K49 OH Lytle C., Clean zone	366	56	-	-	-	-
Polluted zone	233	71	-	-	-	-
H141 ON	13	64	53-71	-	-	-
E83 NY	+	64	-	-	-	-
H141 TN	8	69	56-86	-	-	-
B222 NY Wolf L.	37	79	53-97	37	6	-
D195, IA Des Moines R.	56	76	56-104	56	5	-
G81, L110 IL streams	97	89	-	16	14	-
G67 AL	6	91	76-102	-	-	-
G67 GA	1	102	-	-	-	-
D41 MI	+	127	-	-	-	-
Age II						
G67 NY	126	66	46-102	-	-	-
H141 ON	7	81	74-89	-	-	-
H141 TN	12	91	71-114	-	-	-
G67 AL	5	94	76-104	-	-	-
B222 NY Wolf L.	36	94	64-129	36	8	-
G67 GA, NC	5	104	99-112	-	-	-
G81 IL Roaring Springs C.	54	112	-	-	-	-
D195 IA Des Moines R.	24	117	91-140	24	18	-
L24 QU	32	137	99-173	32	31	14-57
L110 IL Hutchins C.	8	165	-	8	51	-
D41 MI	+	178	-	-	-	-
Age III						
G67 AL	2	97	89-102	-	-	-
H141 ON	4	112	109-114	-	-	-
G67 NY	51	114	76-196	-	-	-
B222 NY Wolf L.	66	114	91-137	66	17	-
H91 MI	+	-	102-178	-	-	-
H141 TN	4	130	89-168	-	-	-
G81 IL Roaring Springs C.	8	145	-	-	-	-
D195 IA Des Moines R.	9	150	137-163	9	47	-
G67 NC	1	150	-	-	-	-
L24 QU	40	173	140-196	40	60	28-96
G67 KS, SD	7	173	160-185	-	-	-
G67 MI, WI	13	188	137-234	-	-	-
L110 IL Hutchins C.	5	198	-	5	77	-
Age IV						
H141 ON	1	127	-	-	-	-
B222 NY Wolf L.	7	137	130-145	7	20	-
G67 NY	58	152	102-272	-	-	-

	No.	Mean TL	Range	No.	Mean Weight	Range
G67 KS	4	160	130-178	-	-	-
G81 IL Roaring Springs C.	1	170	-	-	-	-
G67 WY	1	178	-	-	-	-
H141 TN	2	185	-	-	-	-
L24 QU	1	203	-	1	105	-
G67 SD	2	224	-	-	-	-
G67 MI, WI	9	229	178-272	-	-	-
G67 IA	2	272	262-284	-	-	-
Age V						
G67 GA	1	107	-	-	-	-
G67 NC	2	132	-	-	-	-
G67 NY	28	178	132-249	-	-	-
G67 MI	10	229	155-307	-	-	-
G67 IA	1	292	-	-	-	-
Age VI						
G67 NC	1	132	-	-	-	-
G67 NY	10	193	147-231	-	-	-
G67 MI	3	249	203-320	-	-	-
Age VII						
G67 NY	1	254	-	-	-	-

			Average calculated TL at each annulus					
		No.	1	2	3	4	5	6	7
R60 OH	females	-	51	102	119	127	130	157	
	males	-	51	102	140	173	196	213	221
J41 OK Ft. Gibson	females	5	48	94	135	165			
	males	8	66	122	173				
G81 IL Roaring Springs C.		144	58	94	132	160			
L108 IL Big C.		17	97	114					
L110 IL Hutchins C.		35	64	132	170				
Increment			64	58	41				
Number			29	13	5				
Increments in grams			3	16	23				
D195 IA Des Moines R.		151	58	94	127				
Increment			58	33	36				
Number			89	33	9				

In most papers lengths were calculated on a direct-proportion basis, but D195 used a 8 mm intercept. There is little evidence of regional difference in growth of the creek chub, except for slower growth in some populations in the southeast and New York (G67). Some of these were listed in G67 as *S. a. thoreauianus*, a subspecies. Most rapid growth is in the Midwest. Males grow more rapidly than females (j41, R60, G67). All chub over 240 mm in Ohio were males (T113). Better growth of young creek chub was noted in a polluted but recovering zone in Lytle Creek, Ohio, than in the area above the pollution (H91).

Males mature at age IV and females at III in Michigan (H91).

In Illinois creek chub spawn from April to as late as mid-May (H8).

Spawning runs into tributary streams occur in late March in Iowa (P63) but in 1956 spawning was delayed until June because of low water levels. In

Michigan spawning may occur from late April to July (H81, R248). Males carry stones upstream to make nests for spawning (H8, H91, M432, R248) and they guard the territory.

Propagation is described by M24, W10. Injection of pituitary extract permitted the stripping of creek chub whereas this is usually difficult (B292). Eggs were also secured 6 weeks early but the prolonged growing season did not provide much benefit in rearing bait minnows.

W10 MI	28 females	76-203 mm TL	4250 eggs per female
C49 OH	2 females	114-121 mm TL	4193-4671 eggs per female
G67 NY	Age III, female	119 mm, 126 gm	2820 eggs per female
	Age V	168 mm, 35.2 gm	4042 eggs per female

Acclimated temperature	Lethal low	Lethal high	Citation
5C	-	24.7-25C	H253, S333
10	-	27-27.3	H253, S333
15	-	29.3	S333
17.1-17.5	-	30.1-30.5	H253, S333
20	0.7-1.0	30-30.5	B201
21-21.9	1.7	31.8	B201
22.8	-	32.1	B201
25	4.5	31.5	S333
25-26	-	32.6	B201

Food habits of creek chub are quite varied (S447, D195, M71). Larger chub may feed on small fish (S447, D195, M404) and smaller chub mostly on insects. Chub include quite a lot of vegetable matter in their food (M404). Creek chub were classified as sight feeders by E123 because the taste lobes of the brain were not enlarged.

Hybridization with *Hybopsis plumbea* suggested by S442 was indicated as not proved (H377). Hybrids have been reported with *Clinostomus elongatus* (T113, C324), *Compostoma anomalum plumbeum* (C324), *Notropis cornutus frontalis* (C324).

FALLFISH, *Semotilus corporalis* (Mitchill)
Eastern Canada into James Bay drainage and south on the east side of the Appalachians to Virginia (E119).

E105 MD	a	368 mm TL fallfish weighed 386 g.

M431 QU	log W = -4.063 + 2.698 log FL

B70 DC	Age 0-Sept.	26 mm SL	
B70 DC	Age I-Jan.-June	30-38 mm SL	
W13 CT	Age VI 1 fish	360 mm SL	445 mm TL 907 g

Fallfish are rarely found in water over 28C (T102).

The stone-pile nests may be 1.6 m in diameter and 600-900 mm high (S461). The first nests in a New York study were observed in early May (M432). Insects and fishes constituted most of the food (F148).

PEARL DACE, *Semotilus margarita* (Cope)
Most of Canada south of the tundra and east of the Rockies, south to the

Sand Hills of Nebraska, northern two-thirds of Michigan, Pennsylvania, and east of the Alleghenies to Virginia (E119, H382).

D81	MN		Age 0- Fall	-	89 TL	-
D41	WI		I	-	89 TL	-
B10	AT	Jasper Park lakes	I	4 fish	79 FL	71-81
			II	4 fish	100 FL	99-102
L7	OH	114 mm female	1686 eggs			

Males build spawning areas in gravel (L7) and maintain territories.
Pearl dace were feeding on copepods, cladocerans, and chironomids (N13).

TENCH, *Tinca tinca* (Linnaeus)

A European species introduced into southeastern and western United States (E119).

S122	Germany		Age I fall	50-100 g
			II fall	150-400 g
S122	Germany	K(TL) average 1.3		

Tench may reach 460 mm and 900 to 1400 g in Maryland (S461). They deposit their eggs on vegetation. They are scavengers, feeding on decaying matter, snails, etc. (S461).

Additional growth and life history data on tench in Europe: F127, F128, K113, Z5.

RIVER CARPSUCKER, *Carpiodes carpio* (Rafinesque)

Montana to Pennsylvania and south to Tennessee and Texas (E119).

	Number	Length
B169 MO L. of Ozarks	600	34-45 SL TL = 1.336 SL
B180 IA Des Moines R.	53	TL = 1.326 SL = 1.123 FL
		FL = 1.179 SL
M433 NB Missouri R.	148	61-429 TL TL = 1.29 SL

TL	No.	Mean of means	Weight Central 50%	Range	Citation for lightest and heaviest
38-50	10	2	-	-	B180
51-75	20+	6	-	2-9	B180, F105
76-101	17+	10	-	5-14	B180, F105
102-126	20+	20	18-23	5-27	B180
127-151	20+	35	32-36	14-45	B180
152-177	22+	61	59-68	50-73	B180, Gavins
178-202	19+	90	86-91	59-113	B180, H156
203-228	81+	130	127-132	113-159	H156, Oahe
229-253	69+	177	168-177	145-218	H156, Oahe
254-278	213+	236	227-240	186-281	B180
279-304	170+	308	295-318	227-376	B180, Oahe
305-329	200+	404	381-408	304-495	H156, B180
330-355	229+	513	490-540	367-785	B180
356-380	92+	626	599-667	508-733	H156, Oahe
381-405	41+	753	708-808	513-961	B180

TL	No.	Mean of means	Weight Central 50%	Range	Citation for lightest and heaviest
406-431	30+	934	910-950	672-1134	H156, S472
432-456	10+	1111	1025-1220	1007-1315	S472, Oahe
457-482	10+	1320	1270-1451	816-1656	B216, Oahe
483-507	5+	1642	1606-1673	1411-2092	M505
508-532	2+	1955	-	1883-2001	Oahe, M505
533-558	+	2318	-	-	H419
559-583	+	2671	-	-	H419
584-609	+	3062	-	-	H419

The above weights come from Alabama S472; Iowa B180, C30, C89, I10; Oklahoma B216, H156, H419; Tennessee M505; Texas D179 and the following reservoirs in South Dakota: Ft. Randall S225,226,228,230,330,393; Gavins N94, 115, S227,229,276; and Oahe F100,105,124,125. The numbers at each length were not given for the South Dakota reservoirs but the totals were 950 in Ft. Randall, 2247 in Gavins Point, and 614 in Oahe Reservoir.

B180 IA Des Moines R.		$\log W = -4.256 + 2.809 \log TL$
S225 SD Ft. Randall L.		$\log W = -4.863 + 2.984 \log TL$
H156 OK Canton L.		$\log W = -4.139 + 2.892 \log TL$
B169 MO L. of Ozarks		$\log W = -4.776 + 2.953 \log TL$
D162 MO Perche C.	561 fish	$\log W = -5.69 + 2.75 \log TL$
M433 NB Missouri R.		$\log W = -5.115 + 3.099 \log TL$
H419 OK		$\log W = -3.592 + 3.126 \log TL$
S472 AL	54 fish	$\log W = -3.77 + 2.58 \log TL$
B344 IA Des Moines R.		$\log W = -3.463 + 2.866 \log TL$
H420 TN Watts Bar L.	males 390-590 mm	$\log W = -2.109 + 1.971 \log TL$; $r = .65$
	females 450-600 mm	$\log W = -4.897 + 3.010 \log TL$; $r = .82$

The length-weight regression given in B216 did not appear to be a good fit.

D179 TX 2 carpsuckers 209-245 mm SL K(SL) = 3.55 and 2.82 respectively
B169 MO L. of Ozarks Average K(SL) varied with the season from 3.1 to 3.5.

	K(TL)		
	No.	Mean	Range
N94,115, S227,229,276 SD Gavins Point L.	2247	1.20	1.15-1.26
S225,226,228,230,330,393 SD Ft. Randall L.	1035	1.30	1.21-1.41
S225,226 SD Ft. Randall tailwaters	30	1.11	-
F100,105,124,125 SD Oahe L.	810	1.45	1.31-1.49
B216, B307 OK L. Texoma	604	1.25	1.09-1.45
H156 OK Canton L.	375	1.21	.91-1.66
M433 NB Missouri R.	+	-	1.00-1.50
B180 IA Des Moines R.	209	1.40	1.14-2.33
S472 AL	53	1.39	1.33-1.58
C89 IA Ruth's Pond	11	1.33	1.08-1.52
B344 IA Des Moines R.	+	-	1.43-1.99

It was reported that no sex difference was evident in condition factors (F100, S227,229,225,226,228,230). No sexual differences in length-weight relationships were noted by B180, but B343 found that the females were rather consistently heavier than the males at least during the summer months. The slopes of regressions were parallel for males and females but that for the females was above (B343).

Condition decreased with increase in length in H156, and a downward trend is evident from the regression slopes in B180, B344, H156, D162, and S472. The regression slope was less than 3.0 in most of the regressions computed to show seasonal variation in condition in the Des Moines River, 1960-65 (only 7 of 81 slopes were over 3.0) (B344). Six of the 7 slopes above 3.0 were taken at times when some of the larger fish were ripe, which may have tilted the slopes upward. In most cases the slope was lower for carpsuckers over 279 mm than for those 229-279 mm, suggesting that in the Des Moines River the larger fish decreased in condition with increase in length more than the fish in the 229-279 mm range even though there was a decrease in condition with increase in length in this range also (B344).

Condition increased with increase in length (S229) and similar trends are suggested by the regression slope in M433 and H419. No trends in condition with increase in length were noted in B216, S225,226,227,228,229,230, and M433.

Some decrease in condition as the summer progressed was reported by S229 but was not noted in S226, S227, and S230. In the Des Moines River (B344) the change of condition from May to October differed each year 1960-1965, but in general there was an increase in average condition from May to July (B344). Attempts to relate changes in condition to average water levels or to water temperatures failed. The average lengths of the carpsuckers varied in the years of study, probably due to differences in year class abundance, and, in the years when the average length was high, the condition of these fish was also higher than in the other years, comparing the condition at the same lengths (B344).

	No.	Mean TL	Range	No.	Mean Weight	Range
Age 0-July						
S225 SD Ft. Randall L.	15	18	-	-	-	-
M215 KS Big Blue R.	+	20	-	-	-	-
O37 OK Heyburn L.	1	51	-	-	-	-
B169 MO L. of Ozarks	81	53	30-89	-	-	-
Age 0-Aug.						
J31 OK Grand L.	69	81	-	-	-	-
B169 MO L. of Ozarks	144	86	71-132	-	-	-
Age 0-Sept.						
F105 SD Oahe L.	+	66	-	-	-	-
B183 OK Canton L.	41	-	81-135	-	-	-
Age 0-Oct.						
B183 OK Canton L.	6	56	46-69	-	-	-
B169 MO L. of Ozarks	12	109	69-140	-	-	-
Age 0-Nov.						
B169 MO L. of Ozarks	26	97	69-130	-	-	-
Age 0						
T44 OK Salt Plains	+	130	-	-	-	-
T44 OK Grand L.	11	140	-	11	28	-
Age I						
M215 KS Big Blue R.	10	48	-	-	-	-
D162 MO Perche C.	24	91	64-119	-	-	-
A85, K131 IA Des Moines R.						
1960-63	110	91	51-150	83	11	-
M433 NB Missouri R.	70	99	61-213	-	-	-

	No.	Mean TL	Range	No.	Mean Weight	Range
B180 IA Des Moines R.	109	127	-	-	-	-
B183 OK Canton L.	19	-	99-175	-	-	-
B216 OK L. Texoma	2	157	130-185	-	-	-
J31 OK Grand L.	+	168	-	-	-	-
B169 MO L. of Ozarks	42	170	-	-	-	-
D166 KS	+	178	-	-	-	-
T44 OK Grand L.	18	190	-	18	91	-
H156 OK Canton L.	5	206	-	-	-	-
U11 SD Oahe L.	19	206	-	19	91	-
E28 TN Chickamauga L.	77	224	-	-	-	-
J41 OK Ft. Gibson L.	5	226	-	-	-	-
I10 IA Missouri R.	2	231	196-267	2	255	198-312
T44 OK Salt Plains	+	287	-	-	-	-
W7 OK Duncan L.	1	406	-	1	99	-
Age II						
M215 KS Big Blue R.	5	99	-	-	-	-
J25 OK Illinois R.	37	150	107-231	37	43	-
A85, K131 IA Des Moines R.						
1960-63	509	155	91-284	259	40	-
D162 MO Perche C.	53	165	122-208	-	-	-
B169 MO L. of Ozarks	11	201	-	-	-	-
M433 NB Missouri R.	46	208	132-259	-	-	-
B180 IA Des Moines R.	26	221	-	-	-	-
B216 OK L. Texoma	133	224	168-284	-	-	-
T44 OK Grand L.	80	234	-	80	159	-
O37 OK Heyburn L.	+	-	254-274	-	-	-
U11 SD Oahe L.	22	269	216-292	22	232	-
H156 OK Canton L.	69	302	-	-	-	-
T44 OK Salt Plains	+	371	-	-	-	-
Age III						
M215 KS Big Blue R.	8	135	-	-	-	-
A85, K131 IA Des Moines R.				-	-	
1960-63	362	196	124-338	129	91	-
J25 OK Illinois R.	21	229	196-269	21	170	-
D162 MO Perche C.	39	231	180-295	-	-	-
B180 IA Des Moines R.	13	236	-	-	-	-
M433 NB Missouri R.	55	244	196-305	-	-	-
B169 MO L of Ozarks	1	272	-	-	-	-
B216, B307 OK L. Texoma	327	282	175-376	-	-	-
I10 IA Missouri R.	10	300	282-320	9	357	260-482
O37 OK Heyburn L.	+	-	292-315	-	-	-
T44 OK Grand L.	107	305	-	107	306	-
H156 OK Canton L.	44	335	-	-	-	-
U11 SD Oahe L.	25	381	363-396	25	779	-
T44 OK Salt Plains	+	429	-	-	-	-
Age IV						
M215 KS Big Blue R.	5	196	-	-	-	-
A85, K131 IA Des Moines R.						
1960-63	621	229	157-333	412	176	-
M433 NB Missouri R.	55	274	226-330	-	-	-
D162 MO Perche C.	53	284	241-338	-	-	-

	No.	Mean TL	Range	No.	Mean Weight	Range
Age IV (cont.)						
B180 IA Des Moines R.	14	295	-	-	-	-
J25 OK Illinois R.	70	320	229-358	70	394	-
T44 OK Grand L.	65	325	-	65	408	-
B216 OK L. Texoma	121	338	254-401	-	-	-
B169 MO L. of Ozarks	11	351	-	-	-	-
S330 SD Ft. Randall L.	8	368	-	-	-	-
H156 OK Canton L.	8	378	-	-	-	-
U11 SD Oahe L.	99	421	401-437	99	1134	936-1219
T44 OK Salt Plains	+	470	-	-	-	-
H420 TN Watts Bar L.	4	445	-	-	-	-
Age V						
A85, K131 IA Des Moines R.						
1960-63	118	269	201-368	79	193	-
M215 KS Big Blue R.	2	302	-	-	-	-
M433 NB Missouri R.	17	307	274-366	-	-	-
D162 MO Perche C.	66	312	246-323	-	-	-
B180 IA Des Moines R.	24	325	-	-	-	-
J25 OK Illinois R.	33	366	345-401	33	621	-
T44 OK Grand L.	3	376	-	3	652	-
B169 MO L. of Ozarks	72	381	-	-	-	-
B216 OK L. Texoma	19	381	315-462	-	-	-
S330 SD Ft. Randall L.	10	404	-	-	-	-
U11 SD Oahe L.	117	439	427-462	117	1247	1191-1503
T44 OK Salt Plains	+	495	-	-	-	-
H420 TN Watts Bar L.	11	490	-	-	-	-
Age VI						
M215 KS Big Blue R.	7	295	-	-	-	-
A85, K131 IA Des Moines R.						
1960-63	36	325	269-373	19	363	-
M433 NB Missouri R.	3	333	305-356	-	-	-
B180 IA Des Moines R.	39	338	-	-	-	-
D162 MO Perche C.	34	343	287-394	-	-	-
T44 OK Grand L.	7	386	-	7	757	-
B169 MO L. of Ozarks	85	391	-	-	-	-
J25 OK Illinois R.	3	394	386-399	3	830	-
B216 OK L. Texoma	1	417	-	-	-	-
S330 SD Ft. Randall L.	2	421	-	-	-	-
U11 SD Oahe L.	86	457	442-528	86	1474	1304-2155
T44 OK Salt Plains	+	508	-	-	-	-
H420 TN Watts Bar L.	26	508	-	-	-	-
Age VII						
M215 KS Big Blue R.	6	325	-	-	-	-
A85, K131 IA Des Moines R.						
1960-63	8	343	312-376	4	400	-
B180 IA Des Moines R.	24	348	-	-	-	-
D162 MO Perche C.	15	368	340-399	-	-	-
B169 MO L. of Ozarks	41	404	-	-	-	-
M433 NB Missouri R.	1	424	-	-	-	-
U11 SD Oahe L.	35	475	470-500	35	1616	1588-1786
H420 TN Watts Bar L.	22	516	-	-	-	-
Age VIII						
M215 KS Big Blue R.	1	333	-	-	-	-

	No.	Mean TL	Range	No.	Mean Weight	Range
B180 IA Des Moines R.	11	356	-	-	-	-
A85 IA Des Moines R. 1963	1	381	-	-	-	-
D162 MO Perche C.	3	381	361-394	-	-	-
B169 MO L. of Ozarks	10	417	-	-	-	-
U11 SD Oahe L.	5	465	455-508	5	1559	-
H420 TN Watts Bar L.	1	561	-	-	-	-
W7 OK Duncan L.	1	643	-	1	3442	-
Age IX						
K131 IA Des Moines R. 1961	1	361	-	1	510	-
M215 KS Big Blue R.	2	378	-	-	-	-
D162 MO Perche C.	3	411	-	-	-	-
B169 MO L. of Ozarks	2	432	-	-	-	-
B180 IA Des Moines R.	1	445	-	-	-	-
U11 SD Oahe L.	2	508	490-523	2	2041	1786-2296
H420 TN Watts Bar L.	1	541	-	-	-	-
Age X						
M215 KS Big Blue R.	8	401	-	-	-	-
B169 MO L. of Ozarks	1	455	-	-	-	-
U11 SD Oahe L.	1	531	-	1	1786	-
H420 TN Watts Bar L.	1	569	-	-	-	-
Age XI						
M215 KS Big Blue R.	1	447	-	-	-	-
U11 SD Oahe L.	1	513	-	1	1899	-
H420 TN Watts Bar L.	1	594	-	-	-	-

		Average TL at each annulus							
	No.	1	2	3	4	5	6	7	8	9
E28 TN Chickamauga	77	119								
J25 OK Illinois R.	164	79	157	241	305	348	376			
S391 OK Rock C.	35	76	152	221	269	323				
J45 OK Verdigris R.	47	124	218	297	351	394	450	526	566	
tributaries	36	107	175	231	262					
J41 OK Ft. Gibson L.	17	97	221	318	356					
H150 OK Illinois R.	195	-	122	208	305	351	373			
E61 OK Salt C.	3	112	183							
P112 MO Salt R. upper	233	86	160	190	241	267				
middle	190	81	163	234	279	305	378			
lower	325	79	155	229	279	312	338	394	421	
P113 MO combined	1100	81	165	229	279	312	348			
headwaters	522	81	155	206	257	287	343			
middle	198	84	178	254	292	320	345			
lower	380	179	163	239	292	330	353			
poorest mean	-	76	150	190	241	267	338	368	371	
best mean	-	86	190	274	305	345	366	394	391	411
B183 OK Canton L.	169	135	254	343	391					
L195 OK Arkansas R.	87	86	155	216	264	305	307	328	307	
L195 OK Cimarron R.	77	99	183	254	297	287	292	328	333	
L195 OK state average	+	135	254	323	373	450	503			
W192 SD Lewis & Clark L.	+	56	117	178	216	259	292	345		
H419 OK slowest	+	51	107	160	231	318	366	488	577	594
state average	+	135	257	328	378	432	462	554	592	607
fastest	+	340	432	498	541	589	582	594	605	640
J101 OK L. Eucha	10	107	300	401						
J101 OK Spavinaw L.	3	84	318	414	455					

Average TL and increments at each year

		1	2	3	4	5	6	7	8	9	10	11
T44, J31 OK Grand L.		112	213	272	307	325	371	396				
	Incr.	112	107	64	46	23	20	18				
	No.	693	538	405	186	75	10	7				
H156 OK Canton L.		127	236	297	356							
	Incr.	127	112	74	58							
	No.	126	121	52	8							
B169 MO L. of Ozarks		79	150	213	259	292	328	351	381	409	429	
	Incr.	79	71	66	48	43	46	30	41	33	28	
	No.	276	234	223	222	211	139	54	13	3	1	
M433 NB Missouri R.		76	160	221	269	312	356	432				
	Incr.	76	84	71	56	48	20	20				
K131 IA Des Moines R. 1960-61		71	127	178	218	264	312	340	340	361		
	Incr.	71	56	53	43	36	23	25	20	20		
	No.	1006	916	655	526	111	31	7	1	1		
A85 IA Des Moines R. 1963		69	132	185	239	287	323	320	361			
	Incr.	69	64	53	51	43	38	25	36			
	No.	760	740	492	259	53	15	3	1			
H420 TN Watts Bar L.		157	272	351	394	442	467	486	513	516	554	577
	Incr.	157	114	79	43	46	30	30	36	13	56	23
	No.	67	67	67	67	63	52	26	4	3	2	1
B180 IA Des Moines R. 1956		74	152	203	254	284	312	333	345	417		
	Incr.	74	81	56	53	38	33	25	20	33		
	No.	261	152	126	113	99	75	36	12	1		
I10 IA Missouri R.		79	155	244								
	Incr.	79	84	89								
	No.	12	10	10								
B216 OK L. Texoma		91	173	239	302	366	381					
	Incr.	91	81	69	61	53	43					
	No.	604	602	469	142	20	1					
D162 MO Perche C.		76	150	218	272	305	343	368	391	411		
	Incr.	76	76	74	53	38	38	23	23	13		
	No.	290	265	212	173	120	54	20	6	3		
H303 OK Lawtonka L.		112	282	363	417	460	493	538				
	Incr.	112	170	81	53	38	25	30				
	No.	49	44	38	25	16	6	3				
S226,228,230,393 SD Randall L.		53	168	259	328	371	409	406	434	429	457	
	Incr.	53	114	97	71	48	43	36	46	41	28	
	No.	530	518	397	150	35	15	5	3	1	1	
N94,115, S227,229,276 SD Gavins L.		64	137	190	231	259	307	373	396			
	Incr.	64	79	56	46	36	30	30	25			
	No.	877	786	635	500	396	106	13	2			
S226 SD Ft. Randall headwaters		53	147	244	282	353						
	Incr.	53	94	102	48	51						
	No.	12	12	11	8	2						
F100,105,124,125 SD Oahe L.		76	165	241	302	345	371	399	419	452		
	Incr.	76	86	76	71	53	48	36	30	38		
	No.	480	402	239	155	98	55	32	14	3		

Average calculated weight at each annulus

		1	2	3	4	5	6	7	8	9
B180 IA Des Moines R.		5	50	109	204	281	367	435	486	816
	Incr.	5	45	64	100	95	100	91	77	168
B216 OK L. Texoma		5	50	154	363	537	853			
H419 OK	slowest	1	14	50	159	426	662	1628	2750	3002
	state average	29	218	468	740	1112	1380	2422	2980	3230
	fastest	531	1112	1740	2254	2943	2826	3003	3643	3810

Annual increments as percentage of TL at beginning of year

TL	B180, IA Des Moines R.	J31, T44 OK Grand L.	H156 OK Canton L.	Range
51-75	113	-	-	86-129
76-101	115	108	-	82-133
102-126	37	91	87	37-120
127-151	37	59	82	31-90
152-177	42	35	-	35-42
178-202	21	30	-	20-31
203-228	23	35	32	15-48
229-253	16	34	-	15-35
254-278	13	17	-	12-18
279-304	12	11	20	8-20
305-329	7	10	-	6-24
330-355	9	6	-	4-9
356-380	-	5	-	-
381-405	9	-	-	-

Photos and description of scales and annuli are given by A85 and D162. Difficulty in identifying annuli was reported by A85, M433, and K131. The great variation in growth reported even from the same waters (note A85, B180, and K131) may be partly the result of faulty aging. A false annulus was found in mid-August on the scales of many carpsuckers in the Des Moines River, presenting problems in proper assignment of year classes (A85). The annulus forms in late May or early June in Missouri (D162) and South Dakota (S229, 225,226,393) and in May in Iowa (B180). Growth was believed to start at about 15.5C (S228). Of 16 river carpsuckers, none showed aberrant scalation in populations where some other suckers showed fairly common aberrations (M407). At least 44% of the fish from which scales were collected in the Des Moines River had some regenerative scales in the sample (A85).

Females grew slightly faster than males (B169, B216) but no sex difference in growth could be detected by B180, S393, S229.

In most studies growth was calculated on a direct-proportion basis but some used Fraser-type corrections (B180, H156, B169, P112, D162). The body-scale relationship of Lake Texoma carpsuckers had an intercept of only -0.15 mm (B216) which indicated no need for deviating from direct proportion. Lee's phenomenon was evident in D162, B180.

Growth increments the first year after impoundment were much higher than prior to impoundment at Ft. Randall (S226):

Year of life	1	2	3	4	5	6
Before impoundment	51	76	89	64	43	
Number	107	15	12	2	2	
First year	132	127	99	97	-	43
Number	3	92	3	10	-	2

Trends in later years after impoundment do not seem to be striking nor consistent in the various reservoirs where carpsuckers have been studied. Growth decreased with increasing age of impoundment in Lewis and Clark Lake (W192), a lake with poor growth even at first. Growth was more rapid in cutoff lakes and sloughs than in the mainstream Missouri River (M433). There was some evidence that high water levels caused reduced growth of carpsuckers in the Des Moines River (K131) and some correlation between higher water temperatures and more rapid growth.

In most populations the numbers drop off abruptly after age V or VI, even though specimens up to age X are reported.

Females first mature at age III (S228, N115, S94) or IV (M433, B343, S230, S229, S393) and males may mature at age III (M433, S393). One 244 mm male and one 254 mm female were mature at age II in the Missouri River (M433) and a 241 mm female at age II in the Des Moines River (B180). Two females were ripening at 208 and 224 mm in the Des Moines River in 1965 but most were at least 240 mm before maturing (B344).

B180 IA Des Moines R. 1957. The average number of eggs per 19 females was 102,766 ranging from 4828 in a 244 mm age II female to 149,744 eggs in a 358 mm age V female.

$$\log N = -11.636 + 6.549 \log TL$$

gave a better estimate of number of eggs (N) than

$$N = 141,539 + 0.714 \ TL.$$

B344 IA Des Moines R. 1964-5. The number of eggs per female ranged from 4431 in a 184 g female to 154,038 in a 765 g female.

1964 egg number in thousands = -4.18 + 0.149 weight
1965 egg number in thousands = -12.7 + 0.181 weight

The regression coefficients (r) were 0.84 and 0.73 for 1964 and 1965 respectively.

In addition to the eggs reported above most ovaries included many atretic eggs which would not be spawned at the next spawning (B343, 344). Some carpsuckers probably spawn more than once in the year (B344).

Carpsuckers spawn in May and early June in Missouri (D162), in Kansas (D166) and in South Dakota (S227, S228) but not until mid-June in one year (F105) or July 1 (N94). There were two spawnings in 1955 at Gavins Point, one before and one after the dam closed (N94). In the Des Moines River, Iowa, spawning is from late April into July and appears to be somewhat intermittent (B180, B343, B344). Reproduction was sharply reduced in Heyburn Reservoir, Oklahoma, when turbidity was high (O37).

Young carpsuckers fed mostly on diatoms and a few copepods, ostracods, rotifers, chironomids, and filamentous algae (N93, N94, N115), which is very similar to the food of adults (B180).

The sizes taken in gill nets in L. Texoma, Oklahoma (B216):

Bar measure (mm)	19	25	32	38	51	64
TL mean	152	211	234	269	333	419
Range	130-175	124-315	185-295	168-417	185-462	140-434
Number	2	61	22	314	196	6

A 467 mm river carpsucker was taken in 38 mm bar measure gill net (L216). Samples taken with different types of gear (gill nets, hoop nets, rotenone, and shocker) differed in mean lengths of carpsuckers (L195).

QUILLBACK, *Carpiodes cyprinus* (LeSueur)
Southeastern Manitoba to Kansas and western Florida to Virginia and Quebec (E119, T113).

C22 MN L. of Woods	14 fish	330-399 mm SL	FL = 1.100 SL
	21 fish	400-483 mm SL	FL = 1.098 SL
C22 MN L. of Woods	6 fish	391-435 mm SL	TL = 1.22 SL
C30 IA	5 fish		TL = 1.32 SL = 1.112 FL

TL	No.	Weight Mean	Weight Range	Citation
51	6	1.7	-	S472 AL
76-101	6	7	6.8-7.3	S472, V70 IA
102-126	8	17	9-20	S472, V70
127-151	9	38	30-45	V70, S472
152-177	26	45	32-91	S472, V70
178-202	53	77	50-91	V70, S472
203-228	162	113	95-118	S472, V70
229-253	169	154	136-181	V70, S472
254-278	98	218	204-318	V70, S472
279-304	202	308	254-340	S472, V70, C89 IA
305-329	169	408	363-431	S472, V70, C89
330-355	59	508	440-567	S472, V70, C89
356-380	17	649	635-680	S472, V70, C89
381-405	19	766	513-907	S472, V70, C30, C89
406-431	21	952	816-1080	S472, V70
432-456	9	1067	934-1250	S472, V70, C89, C30
457-482	7	1179	567-1500	S472, V70
483-507	3	1628	1306-1955	S472, C30
518-523	2	2041	1814-2268	C30, E105 MD

| V70 IA Des Moines R. | | | log W = -5.238 + 3.134 log TL |
| S472 AL | 958 fish | 25-483 mm | log W = -5.44 + 2.83 log TL |

The latter regression appears to be affected by inclusion of 25 mm fish at too high a weight.

E28 TN Chickamauga L.	421 fish	-	average K(SL) 3.0	-
C22 MN L. of Woods	26	330-483 mm	3.12	-
C36 IA	4	293-390 mm	3.19	
			range 2.83-3.62	
C89 IA Ruth's P.	13	300-442 mm average K(TL) 1.22		
			range .75-1.50	
V70 IA Des Moines R.	652	76-457 mm	1.16	-
S472 AL	956	50-483 mm	1.39	
			range 1.16-2.22	

No trend with size but some increase in condition June to August (V70).

		TL			Weight	
	No.	Mean	Range	No.	Mean	Range
Age 0-July						
C283 OH St. Mary's L.	+	57	-	-	-	-
Age 0- Aug.						
C284 OH St. Mary's L.	+	38	-	-	-	-
Age I						
M503 IA Coralville L.	3	244	241-246	3	181	150-386

		TL			Weight	
	No.	Mean	Range	No.	Mean	Range
Age II						
E28 TN Chickamauga L.	423	221	-	-	-	-
M503 IA Coralville L.	4	340	277-391	4	261	150-386
Age III						
M503 IA Coralville L.	2	386	378-419	2	439	369-510

Average calculated TL and increments at each annulus

	1	2	3	4	5	6	7	8	9	10
E28 TN Chickamauga L.	122	196								
Incr.	122	74								
No.	423	423								
V70 IA Des Moines R.	124	165	201	234	269	340	361	399	386	419
Incr.	124	38	36	38	36	46	28	38	28	33
No.	149	122	116	73	20	11	7	3	1	1
g	20	48	96	156	235	499	604	763	740	960

The Iowa data were computed with a Fraser-type correction of 69 mm (V70).

		Sizes in gill nets		
Mesh size stretch measure	73	83	95	102
C22 MN L. of Woods				
FL	152-226	229-277	229-277	343
Number	3	8	6	1

HIGHFIN CARPSUCKER, *Carpiodes velifer* (Rafinesque)

Minnesota to Pennsylvania south to Oklahoma and the Gulf of Mexico (E119, T113).

	V70 IA		W132 TN			S472 AL Weight	
TL	No.	Weight	No.	Weight	No.	Mean	Range
89	5	8	-	-	-	-	-
119	5	21	-	-	-	-	-
135	5	33	-	-	-	-	-
160	5	57	-	-	-	-	-
178-202	5	73	-	-	39	77	-
203-228	5	119	4	147	3	113	85-127
229-253	5	167	3	196	5	190	176-201
254-278	5	235	2	250	3	272	-
279-304	5	289	2	363	8	318	287-340
305-329	3	386	4	408	10	386	340-454
330-355	1	422	12	507	2	454	-
356-380	-	-	14	590	11	618	516-680
381-405	-	-	7	731	7	729	606-794
406-431	-	-	5	907	5	970	862-998
432-456	-	-	5	1150	5	1285	1233-1332
457-482	-	-	3	1194	3	1435	1361-1477

V70 IA Des Moines R.

399 fish   76-330 mm

$\log W = -4.920 + 2.997 \log TL$

mean K(TL) = 1.19

S472 AL   101 fish   178-460 mm   log W = -4.89 + 3.02 log TL
                                            mean K(TL) = 1.43   range 1.25-1.67

There was no trend in K(TL) with increase in length, but the K values increased from June to August (V70).

J42 OK Ft. Gibson   Age I   11 carpsuckers   190 mm TL

J25 OK Illinois R.

Age Group	No.	TL	Range	Weight	TL at last annulus
I	1	213	-	142	71
II	9	287	259-302	315	216
III	2	312	305-318	425	274
IV	5	328	318-343	454	312
V	5	361	348-388	680	343

Calculated TL at each annulus

	No.	1	2	3	4	5	6	7	8
J25 OK Illinois R.	22	102	203	262	315	343			
H150 OK Illinois R.	135	71	218	284	323	356			
V70 IA Des Moines R.	97	89	132	165	196	221	259	279	305
Incr.		89	38	30	30	25	25	10	28
No.		97	93	68	45	18	7	3	1
g		0.6	28	54	94	128	210	255	326

The Iowa data were calculated with a Fraser-type correction of 64 mm (V70).

UTAH SUCKER, *Catostomus ardens* Jordan and Gilbert
Found in the Bonneville Basin, Utah (E119).

Average TL at each annulus

	No.	1	2	3	4	5	6	7	8	9	10
M152 UT Bear L.	189	38	130	213	277	333	371	419	447	505	559

Utah suckers spawn in late May to early June on rocky shoals and in tributary streams (L193, M152). In Bear Lake these suckers were highly parasitized by *Ligula*. Both young and adults fed on bottom fauna (M152).

LONGNOSE SUCKER, *Catostomus catostomus* (Forster)
In colder lakes and streams from Alaska to Labrador south to Washington, Montana, northern Minnesota, Great Lakes drainage, and northern New England (E119, H382).

H269 Great Stone Lake   84 suckers   SL = 0.88 FL

In the following compilation of length-weight data, it is assumed that the relationships between SL, FL, and TL are as in the white sucker. The lengths given in the first three columns are not class centers but the lower length in the range. The first and last reference in each group includes the smallest and largest fish, respectively.

SL	FL	TL	Weight Median	Weight Range	Number of fish	References
127	142	150	34	28-37	3	H307, H269
129	145	152	43	34-57	9+	H268, H269, R97
151	168	178	71	51-88	16	H268, H269
173	193	203	113	79-170	31+	H268, B226, R97
194	216	229	144	111-261	99	H268, B226
215	239	254	196	142-284	89+	H268, B226, H269, R97
237	262	279	264	184-326	60	H268, H269, B226
258	287	305	329	244-454	27+	H268, B226, H269, R97
279	310	330	446	436-457	10	H269
301	333	356	615	548-737	14+	H269, R97
323	356	381	748	729-776	22	H269
344	378	406	930	845-960	40	H269
366	401	432	1106	822-1140	45+	R98, R97, H269
387	424	457	1287	1230-1315	104	H269
409	447	483	1457	1412-1506	82+	H269, R97
430	472	508	1687	1627-1735	110	H269
452	495	533	1852	1814-1956	83+	H269, R97
473	518	559	2016	1980-2064	69	H269
495	544	584	2336	2257-2510	38	H269
517	566	610	-	-	-	-
-	643	-	3312	-	1	K93

The data from H269, Great Slave Lake, were in SL; from R97 and K93, Great Slave Lake, and R98, Alberta, were in FL; and from B226 and H268, Colorado, Shadow Mountain Lake, were in TL.

H269 Great Slave L.		
	1944-46 350 suckers	$\log W = -4.457 + 2.88 \log SL$
	1950-51	$\log W = -4.401 + 2.88 \log SL$
M431 QU		$\log W = -4.453 + 2.842 \log FL$
H268 CO Shadow Mt. L.		$\log W = -5.0685 + 3.0225 \log TL$
H269 Great Slave L.	529 suckers	mean $K(SL) = 1.90$, range 1.73-2.04
C22 MN L. of Woods	6 suckers	$K(SL) = 1.96$

Weights of *C. catostomus griseus* were higher in winter than in summer at equivalent lengths in Shadow Mountain Reservoir, Colorado (B226).

	No.	FL	TL	Range	No.	Weight	Range
Age 0-July							
B174 WY Yellowstone	+	-	13	10-15	-	-	-
R230 N. Saskatchewan R.	+	X	-	30-51	-	-	-
Age 0-Aug.							
H269, H307 Great Slave L.	380	X	-	33-81	-	-	-
W98 WY Guernsey L.	+	-	X	-135	-	-	-
Age 0-Sept.							
B174 WY Yellowstone	+	-	23	18-30	-	-	-
Age 0-Oct.							
B174 WY Yellowstone	+	-	30	25-38	-	-	-
Age 0							
R97 NT Great Slave L.	5	48	-	36-58	-	-	-
N23 AT Jasper Park	+	X	-	-61	-	-	-

	No.	FL	TL	Range	No.	Weight	Range
Age I							
R98 SA Pyramid L.	+	51	-	-	-	-	-
N23 AT Jasper Park	+	X	-	69-76	-	-	-
B174 WY Yellowstone	44	-	119	-	-	-	-
R97 NT Great Slave L.	7	142	-	97-178	-	-	-
H258 WY Pathfinder L.	4	-	183	-	4	57	-
W98 WY Guernsey L.	+	-	X	155-234	-	-	-
Age II							
R98 SA Pyramid L.	+	84	-	-	-	-	-
N23 AT Jasper Park	+	91	-	-	-	-	-
H143 CO *C. c. griseus*	2	-	135	130-140	2	40	31-51
H269 NT Great Slave L.	3	157	-	-	-	-	-
H307, H269 Great Slave L., N.	4	152	-	-	-	-	-
S.	3	178	-	-	-	-	-
R97 NT Great Slave L.	6	175	-	152-211	-	-	-
B174 WY Yellowstone	6	-	183	-	-	-	-
H258 WY Pathfinder L.	5	-	269	-	5	221	-
W98 WY Guernsey L.	+	-	X	236-335	-	-	-
Age III							
N23 AT Jasper Park	+	99	-	-	-	-	-
R98 SA Pyramid L.	+	107	-	-	+	22	-
E99 SA Patricia L.	3	X	-	114-147	-	-	-
H143 CO *C. c. griseus*	1	-	180	-	1	85	-
H307, H269 Great Slave L., N.	2	170	-	-	-	-	-
S.	12	203	-	-	-	-	-
H269 NT Great Slave L.	10	178	-	-	-	-	-
R97 NT Great Slave L.	7	218	-	198-244	-	-	-
B174 WY Yellowstone	1	-	269	-	-	-	-
C53 BC Okanagan	1	257	-	-	-	-	-
K34 MT Missouri R.	3	-	328	-	-	-	-
H258 WY Pathfinder L.	6	-	328	-	6	422	-
R203 S. Saskatchewan R.	2	295	-	-	2	284	-
N. Saskatchewan R.	1	323	-	-	1	397	-
W98 WY Guernsey L.	+	-	X	338-368	-	-	-
Age IV							
R98 SA Pyramid L.	+	132	-	112-145	+	37	-
E99 SA Pyramid L.	10	X	-	102-137	10	23	-
Patricia L.	4	X	-	132-168	-	-	-
H307, H269 Great Slave L., N.	1	254	-	-	-	-	-
S.	7	239	-	-	-	-	-
H269 Great Slave L. 1944-46	6	224	-	-	-	-	-
R97 Great Slave L. 1944-46	9	251	-	229-279	-	-	-
B174 WY Yellowstone males	50	-	318	-	-	-	-
female	1	-	386	-	-	-	-
K34 MT Missouri R.	10	-	340	-	-	-	-
H258 WY Pathfinder L.	2	-	399	-	2	672	-
R230 S. Saskatchewan R.	7	356	-	-	7	510	-
N. Saskatchewan R.	6	384	-	-	6	652	-
C53 BC Okanagan L.	1	386	-	-	-	-	-
Age V							
R98 SA Pyramid L.	+	152	-	142-165	+	57	-
E99 SA Pyramid L.	19	X	-	119-168	19	40	-
Patricia L.	6	X	-	150-198	-	-	-

	No.	FL	TL	Range	No.	Weight	Range
Age V (cont.)							
H269, H307 Great Slave L.							
1944-46	6	249	-	-	-	-	-
H269, H307 Great Slave L., N.	4	290	-	-	-	-	-
S.	4	305	-	-	-	-	-
R97 Great Slave L. 1944-46	6	292	-	262-338	-	-	-
B174 WY Yellowstone males	130	-	340	-	-	-	-
females	33	-	396	-	-	-	-
K34 MT Missouri R.	19	-	373	-	-	-	-
R230 S. Saskatchewan R.	20	368	-	-	20	567	-
N. Saskatchewan R.	13	427	-	-	13	992	-
Age VI							
R98 SA Pyramid L.	+	178	-	-	+	91	-
E99 SA Pyramid L.	17	X	-	132-183	17	71	-
Patricia L.	6	X	-	198-226	-	-	-
H269, H307 Great Slave L.							
1944-46	9	287	-	-	-	-	-
H269, H307 Great Slave L., N.	3	315	-	-	-	-	-
S.	5	307	-	-	-	-	-
R97 Great Slave L. 1944-46	5	343	-	300-391	-	-	-
K34 MT Missouri R.	28	-	388	-	-	-	-
B174 WY Yellowstone males	64	-	388	-	-	-	-
females	100	-	424	-	-	-	-
R203 S. Saskatchewan R.	9	401	-	-	9	709	-
N. Saskatchewan R.	7	439	-	-	7	1106	-
Age VII							
R98 SA Pyramid L.	+	203	-	-	+	113	-
E99 SA Pyramid L.	8	X	-	170-185	8	113	-
Patricia L.	4	X	-	234-249	-	-	-
H269, H309 Great Slave L.							
1944-46	9	320	-	-	-	-	-
H269, H309 Great Slave L., N.	10	330	-	-	-	-	-
S.	11	363	-	-	-	-	-
R97 Great Slave L. 1944-46	10	366	-	325-417	-	-	-
K34 MT Missouri R.	19	-	399	-	-	-	-
B174 WY Yellowstone males	8	-	417	-	-	-	-
females	22	-	450	-	-	-	-
Age VIII							
R98 SA Pyramid L.	+	224	-	-	+	159	-
E99 SA Pyramid L.	13	X	-	180-229	13	128	-
Patricia L.	6	X	-	262-295	-	-	-
H269, H309 Great Slave L.							
1944-46	9	386	-	-	-	-	-
R97 Great Slave L. 1944-46	7	401	-	363-470	-	-	-
H269, H309 Great Slave L., N.	13	371	-	-	-	-	-
S.	19	411	-	-	-	-	-
K34 MT Missouri R.	8	-	414	-	-	-	-
B174 WY Yellowstone females	6	-	462	-	-	-	-
Age IX							
R98 SA Pyramid L.	+	244	-	-	+	198	-
E99 SA Pyramid L.	3	X	-	201-244	-	-	-
Patricia L.	5	X	-	305-330	-	-	-

	No.	FL	TL	Range	No.	Weight	Range
H269, H307 Great Slave L.							
1944-46	5	414	-	-	-	-	-
H269, H307 Great Slave L., N.	31	404	-	-	-	-	-
S.	64	478	-	-	-	-	-
R97 Great Slave L. 1944-46	5	457	-	445-467	-	-	-
Age X							
R98 SA Pyramid L.	+	264	-	-	+	241	-
H269, H307 Great Slave L.							
1944-46	6	457	-	-	-	-	-
H269, H307 Great Slave L., N.	41	434	-	-	-	-	-
S.	101	495	-	-	-	-	-
R97 Great Slave L. 1944-46	9	513	-	472-556	-	-	-
Age XI							
R98 SA Pyramid L.	+	290	-	-	+	306	-
E99 SA Pyramid L.	3	X	-	259-279	-	-	-
H269, H307 Great Slave L.							
1944-46	12	475	-	-	-	-	-
H269, H307 Great Slave L., N.	31	465	-	-	-	-	-
S.	128	495	-	-	-	-	-
R97 Great Slave L. 1944-46	5	523	-	478-569	-	-	-
Age XII							
R98 SA Pyramid L.	+	315	-	-	+	380	-
E99 SA Pyramid L.	2	X	-	292-295	-	-	-
H269, H307 Great Slave L., N.	14	498	-	-	-	-	-
S.	66	518	-	-	-	-	-
R97 Great Slave L. 1944-46	6	544	-	493-589	-	-	-
H269, H307 Great Slave L.							
1944-46	2	551	-	-	-	-	-
Age XIII							
R97 Great Slave L. 1944-46	2	584	-	577-589	-	-	-
H269, H307 Great Slave L.							
1944-46	7	538	-	-	-	-	-
Great Slave L., N.	13	511	-	-	-	-	-
S.	50	531	-	-	-	-	-

	Age XIV	XV	XVI	XVII	XVIII	XIX
H269, H307 Great Slave L., N.	14 518	4 546	4 528	2 579	2 607	1 643
S.	32 541	18 556	7 574	5 587		

			Calculated SL at each annulus			
	No.	1	2	3	4	5
E3 MN	20	99	174	242	270	307

		Calculated FL at each annulus			
	1	2	3	4	5
R136 SA Churchill L.	190	318	409	465	505

				Calculated TL at each annulus						
	No.	1	2	3	4	5	6	7	8	9	10
K34 MT Missouri R.	87	76	140	206	264	312	348	373	394		
Incr.		76	64	66	58	56	46	41	36		
No.		87	87	87	84	74	55	27	8		

Calculated TL at each annulus

	No.	1	2	3	4	5	6	7	8	9	10
B174 WY Yellowstone											
males	253	48	119	213	292	340	381	417			
Incr.		48	71	94	79	53	36	33			
No.		253	253	253	252	202	72	8			
females	222	51	124	221	302	366	409	445	460		
Incr.		51	74	97	81	64	46	33	30		
No.		222	222	222	222	221	188	88	6		
sexes combined	575	51	122	216	297	345	401	442	460		
H258 WY Pathfinder L.	17	84	190	267	325						
Incr.		84	91	84	74						
No.		17	13	8	2						
H268 CO Shadow Mt. L.											
C. c. griseus											
males	42	76	130	185	231	264					
females	79	84	152	211	251	264					
M453, P159 MT lakes	136	43	109	145	244	297	356	401	439		
streams	345	43	99	170	241	297	345	368	394	424	434

Calculated weight at each annulus

	1	2	3	4	5
R136 SA Churchill L.	85	454	1106	1588	1928

Annual increment as percentage of TL at beginning of year

	B174 WY Yellowstone L.			
TL	Males	Females	K34 MT Missouri R.	Range
25	144	150	-	137-164
51	168	152	81	57-168
76	127	-	91	72-127
102	83	85	48	43-89
127	64	72	45	45-76
152	-	-	45	33-69
178	43	44	24	24-44
203	38	37	28	25-41
229	26	-	29	26-30
254	23	36	19	18-36
279	19	20	18	18-21
305	-	20	14	14-21
330	11	-	12	9-14
356	-	12	10	10-13
381	9	10	-	9-10
406	-	8	-	7-8

In most cases growth was computed by direct proportion, but H268 used a Fraser-type correction of 25 mm. The smallest found with scales was 38 mm TL (B174) or 33 mm FL (H307). Some may go through the first winter without scales (B174). Scales from marked spawning fish indicate that scale reading may underestimate the age of older fish. Three fish were recovered 8 years after being marked as spawning fish, but the scales indicated only age X (G154).

In general, growth was slower and life span longer in the northern part of the range than in the southern. However, growth was rapid in the North Saskatchewan River (R230) and faster in Great Slave Lake than in the smaller, more southern Saskatchewan and Alberta lakes. Growth was faster in southern Great Slave Lake than in the northern part (H269, H307). Annual growth of adult suckers in a British Columbia lake was 15-20 mm FL per year according to tag returns and length frequencies (G154).

Females grew more rapidly than males in Colorado (H268) and in Yellowstone and also had longer life spans (B174), but no sex differences in growth rates were noted by H269, H307, E99 except that the oldest fish, and usually the largest, were females. Males averaged smaller than females in spawning runs in British Columbia (G154). Females continued to return in subsequent years more than males, suggesting a greater longevity. Spawning mortality was estimated at 11-28% (G154). Annual mortality after age XI was estimated at 55% in Great Slave Lake (H307, H269).

Males spawn at age II and at 100-125 mm TL in Colorado (H268), but usually not until age IV in Yellowstone (B174) and Saskatchewan (E99), age V in British Columbia (G154) and age X in Great Slave Lake (H269, B307). Females mature at age III, 203 mm, in Colorado (H268), at age IV in Saskatchewan (E99), at age V in Yellowstone (B174) and at age IX in Great Slave Lake (H269, B307).

H268, H191 CO *C. c. griseus*	2 females,	307-320 mm TL, 10270-12380 eggs/female
B307, H269 Great Slave L.	1 female,	450 mm FL, 1089 g, 17525 eggs
	5 females,	465-523 mm FL, 1547-1633 g, 31955 eggs, 20817-38746
	2 females,	526-569 mm FL, 1950-2041 g, 42430-60307 eggs

They spawn in streams on riffles scoured clean by current in May in British Columbia (G154); in late May and June in Colorado (H268); in late June and early July in Yellowstone (B174); in mid-June to early July in Pyramid Lake, Saskatchewan (E99); and from breakup in mid-May to mid-June in Great Slave Lake (H269). Spawning occurred from 0600 to 2130 hr (G154). Females remained in the tributary stream for spawning a shorter time than the males (G154). Many suckers spawned several years in a row; others appeared to skip a season or two (G154). Population thinning may have resulted in greater survival of young (E99). Fry were 10-12 mm at first downstream movement (G154).

Young fish, 50-150 mm long, remain in weed-bed areas (H268) but older fish show a definite offshore movement during daylight hours. Tagging indicated a tendency to return to the same areas (H268).

Suckers acclimated at 11.5C had an upper lethal temperature of 27C (B200).

H143 CO 3 suckers, 130-180 mm TL, 1,262,000 red blood cells per cu. mm, range 1,027,000-1,570,000; hemoglobin 11.0-11.9, avg. 11.4.

The population in Pyramid Lake, Saskatchewan, was estimated at 158 fish or 10.2 kg per hectare in 1943 and 43.7 kg per hectare in 1945 (E99).

	Sizes in gill nets						
Mesh size	38	51	76	102	114	127	140
R97 Great Slave L.							
Number	36	90	119	49	-	30	25
Mean FL	152	239	315	396	-	434	450
Mean weight	68	268	500	1043	-	1315	1450
H307, H269 Great Slave L.							
Number	13	19	33	47	5	10	497
Mean FL	163	241	338	409	442	457	503
Mean weight	60	200	544	862	1043	1360	1815

The food of longnose suckers is mostly of algae, chironomid larvae, amphipods, and other bottom organisms (B174, N13, R230, B226, H268, E99, H256). From 11-18 mm they feed only on plankton and from 20-90 mm graze upon weeds and solid surfaces taking no mud (H268). Chironomids were the principal food of longnose suckers in the Saskatchewan River in contrast to redhorses which fed mostly on mayflies (R230).

WHITE SUCKER, *Catostomus commersoni* (Lacépède)

East of the Rockies from northern Alberta to southern Labrador and south to northern Georgia, Oklahoma and Colorado (E119, T113).

	No. of fish	SL	$\frac{FL}{SL}$	No. of fish	SL	$\frac{TL}{SL}$
C35 MN L. Vermilion	-	under 300	1.108	-	under 300	1.187
	-	over 300	1.098	-	301-350	1.184
C22 MN L. of Woods	-	under 100	1.160	-	-	-
	-	101-200	1.119	-	-	1.174
	-	201-300	1.113	-	-	-
	-	301-400	1.103	-	-	-
	-	over 400	1.096	-	-	-
C37 MN	22	100-199	1.117	22	100-199	1.193
	121	200-299	1.107	571	200-399	1.177
	450	300-399	1.099	-	-	-
	104	over 400	1.095	104	over 400	1.169
C30 IA	2	-	1.12	63	-	1.19
S156 NK	27	262-380 mm FL	1.16	-	-	-
L110 IL	-	-	-	62	42-263	1.19

In the data which follow, SL and FL are converted, where necessary, to TL according to the system described for the length-weight summary.

Summary of length-weight data on white sucker

In compiling this table it is assumed that SL equals 0.847 TL, and that FL equals 1.14 SL when SL is under 100 mm, 1.12 SL when SL is 100-200 mm, 1.11 SL when SL is 200-300 mm, 1.10 SL when SL is 300-400 mm, and 1.096 SL when SL is over 400 mm. The lengths given in the first three columns are not class centers, but the lower length in the range. For example, the first line includes suckers with standard lengths from 21 to 42 mm, fork lengths from 25-47 mm, and total lengths from 25-50 mm.

SL	FL	TL	Median	Weight Central 50%	Range	Number of fish	References of lightest and heaviest
21	25	25	.37	.28-.4	.09-.85	336+	D40, S89
43	48	51	2	-	1.7-3.0	54+	D40, D82
65	74	76	8	5-9	5-11	49+	D40, D82
86	99	102	15	-	-	+	D81
108	122	127	33	29-42	21-42	67+	H268, C258
129	145	152	48	45-48	34-71	189+	C30, R97
151	168	178	74	68-74	54-113	164+	H268, C20
173	193	203	113	105-122	68-142	342+	C258, R97
194	216	229	145	139-150	85-213	523	C258, C20
215	239	254	198	190-201	113-298	913+	C258, R97
237	262	279	261	244-289	190-375	775	H268, M499
258	287	305	351	323-378	195-539	513+	H268, R97
279	310	370	425	406-460	255-709	466	R24, C22
301	333	356	581	518-627	378-879	230+	H268, C22
323	356	381	720	652-785	419-978	305+	H268, C30
344	378	406	865	768-947	624-1219	352+	H268, C258
366	401	432	1063	964-1157	765-1474	123+	R14, C22
387	424	457	1230	1140-1315	907-1786	172+	C258, C22
409	447	483	1534	1361-1600	907-2353	141+	C258, C30
430	472	508	1743	1588-1789	1304-2722	57+	C258, C30
452	495	533	2070	1701-2183	1474-2325	9+	C258, C35
473	518	559	2000	1925-2254	1860-2608	15+	S62, E93
495	544	584	1922	-	1276-2877	3	C22, C30
517	566	610	-	-	-	-	-
538	589	635	3047	-	2948-3147	2	E93, W13

The above length-weight data were given originally in SL in C258, Saskatchewan; C20, C35 Minnesota; R24 Michigan; S89 Wisconsin; B222 New York; C30 Iowa; in FL in C22 Minnesota; R97 Great Slave Lake; and in TL in E93 Maine; W13 Connecticut; R14 New York; R51 Ohio; D40, D81, D82, S62 Minnesota; B226, H268 Colorado.

C35 MN L. Vermilion	$\log W = -4.987 + 3.113 \log SL$
L110 IL	$\log W = -4.94 + 3.08 \log SL$
R14 NY ripe males	$\log W = -3.4193 + 2.4224 \log TL$
spent males	$\log W = -3.885 + 2.5914 \log TL$
ripe females	$\log W = -2.822 + 2.2303 \log TL$
spent females	$\log W = -3.671 + 2.5070 \log TL$
M431 QU	$\log W = -3.849 + 2.629 \log FL$
P122 WI Flora L. 1952 73 suckers	$\log W = -4.785 + 2.914 \log TL$
1956	$\log W = -5.051 + 3.040 \log TL$
H268 CO Shadow Mt. L.	$\log W = -5.395 + 3.223 \log TL$

	No.	Length	Type	K	Range
C23 MN standards: poor			SL	under 1.7	-
excellent			SL	over 2.2	-
L110 IL	62	42-263	SL	1.72	-
S89 WI	2369	-	SL	1.70	-
B222 NY	112	178-356	SL	1.8	1.7-1.9
C34 MN Leech L.	67	200-499	SL	1.85	-

	No.	Length	Type	K	Range
C23 MN 1941-43	486	100-499	SL	1.94	-
C35 MN L. Vermilion	83	-	SL	2.02	-
E3 MN 1936-41	474	102-482	SL	2.07	-
V20 MN Red L.	11	291-400	SL	2.12	-
C22 MN L. of Woods	393	150-520	SL	2.15	-
C36 IA	64	77-505	SL	2.26	1.54-3.15
S373 NS	-	335-439	FL	1.15	-
R50 OH	-	-	TL	1.11	-
S393 SD Ft. Randall L.	-	-	TL	1.13	-
M499 IL	117	140-457	TL	1.14	1.02-1.30
C23 MN standards: poor			TL	under 1.02	
excellent			TL	over 1.27	

There was no evidence of sexual difference in length-weight relationship or condition factor (S89, B226). No trend in condition factor with change in age or length was evident in S89, M499, B222. Condition factor improved from May to July in Muskellunge Lake, Wisconsin (S89), and older suckers showed better condition in winter than in summer in Shadow Mountain Reservoir, Colorado (B226).

	No.	TL	Range	No.	Weight	Range
Age 0-May						
C258 SA	66	10	-	-	-	-
D82 MN	14	15	-	-	-	-
S91, C70, D41 NY	+	-	10-25	-	-	-
D81 MN	+	-	18-28	-	-	-
D41, C8 MI	+	-	23-25	-	-	-
Age 0-June						
C258 SA	459	-	10-18	-	-	-
F5 L. Erie	33	-	13-20	-	-	-
B49 ON L. Nipigon	60	23	-	-	-	-
H92 MI Douglas L.	655	20	15-36	-	-	-
D41, H195 MI ponds	+	-	25-38	-	-	-
D40, D81, D82 MN ponds	76+	-	20-61	28	0.3	-
Age 0-July						
H331 NS	+	-	13-15	-	-	-
D10 NY _C. c. utawana_	+	-	18-25	-	-	-
F5 L. Erie	53	-	20-25	-	-	-
C258 SA	292	-	20-33	-	-	-
B49 ON L. Nipigon	171	30	23-36	-	-	-
A1 NY Oneida L.	+	-	25-38	-	-	-
C35 MN L. Vermilion	3	41	25-51	-	-	-
H92 MI Douglas L.	102	-	28-64	-	-	-
H195, D41 MI ponds	+	-	43-71	-	-	-
D41 WI ponds	+	51	-	-	-	-
D40, D41, D82 MN ponds	28+	53	46-64	28	1.7	-
H76 WV Back C.	6	58	-	-	-	-
Age 0-Aug.						
D10 NY _C. c. utawana_	+	38	23-46	-	-	-
B49 ON L. Nipigon	82	38	30-43	-	-	-
C258 SA	277	-	36-56	-	-	-
H92 MI Douglas L.	94	-	46-74	-	-	-

	No.	TL	Range	No.	Weight	Range
L110 IL creek	8	64	-	8	3	-
D41, F66 NY ponds	+	-	46-117	-	-	-
D41, H195 MI ponds	+	-	53-97	-	-	-
D40, D41, D82, S113 MN ponds	22+	79	64-91	22	5	-
W98 WY lake	+	-	-152	-	-	-
Age 0-Sept.						
C258 SA	11	56	-	-	-	-
A1 NY Oneida L.	+	-	51-76	-	-	-
D40, D82, S113 MN ponds	8+	94	-	8	11	-
D41, F66 NY ponds	+	-	56-142	-	-	-
D41, H195 MI ponds	+	-	46-127	-	-	-
B16 MI Wolf L.	+	-	84-208	+	-	6-90
0-Oct.-Nov.						
C8 MI	+	86	53-163	-	-	-
A1 NY Oneida L.	+	-	76-102	-	-	-
F66 NY ponds	+	-	66-142	-	-	-
P112 MO Salt R.	15	104	-	-	-	-
Age 0						
E9 NY	+	56	-	-	-	-
Age I						
H314 NS Mosers R.	2	48	43-51	-	-	-
B20 MN Superior N. F.	+	51	25-89	-	-	-
C258 SA	96	58	36-79	-	-	-
B49 ON L. Nipigon	3	61	-	-	-	-
D10 NY *C. c. utawana*	+	-	38-102	-	-	-
M215 KS Big Blue R.	4	71	-	-	-	-
R14, S91 NY lakes	+	-	71-76	-	-	-
E9, F66 NY ponds	+	89	71-117	-	-	-
D41, S113 MN ponds	+	-	102-114	-	-	-
B192 OH Massie C.	+	130	-	-	-	-
L110 IL creek	20	137	-	20	14	-
K34 MT Missouri R.						
C. c. suckleyi	9	142	-	-	-	-
S89 WI Muskellunge L.	23	150	-	-	-	-
W98, H258 WY lakes	+	-	145-216	1	77	-
L129 IL	+	170	-	-	-	-
C35, E3 MN	6	201	150-300	2	95	91-99
W91 NY pond	+	-	152-203	-	-	-
D41 MI pond	+	203	-	-	-	-
Age II						
C258 SA	7	81	71-	-	-	-
B20 MN Superior N. F.	+	89	51-165	-	-	-
M215 KS Big Blue R.	6	99	-	-	-	-
S91, E9 NY ponds	+	-	102-178	-	-	-
R97 NT Great Slave L.	2	165	-	-	-	-
L110 IL creek	5	170	-	5	54	-
S89 WI Muskellunge L.	25	175	-	-	-	-
K34 MT Missouri R.						
C. c. suckleyi	22	188	-	-	-	-
B192 OH creek	+	193	-	-	-	-
H258 WY Pathfinder L.	6	208	-	6	136	-
B222 NY Wolf L. *C. c. utawana*	1	218	-	1	119	-

Age II (cont.)	No.	TL	Range	No.	Weight	Range
W98 WY	+	-	218-267	-	-	-
L129 IL	+	246	-	-	-	-
C35, E3 MN	42	318	211-480	9	380	170-680
B150 IA Little Wall L.	1	460	-	1	1134	-
Age III						
B20 MN Superior N. F.	+	89	76-267	-	-	-
S91 NY	+	-	114-152	-	-	-
C258 SA	6	142	124-	-	-	-
H143 CO *C. c. suckleyi*	4	180	109-272	4	116	20-266
M215 KS Big Blue R.	1	208	-	-	-	-
B222 NY Catlin L.	5	211	206-218	5	111	-
B222 NY *C. c. utawana*	14	236	-241	14	139	-156
L110 IL creek	4	213	-	4	119	-
S89 WI Muskellunge L.	69	218	-	-	-	-
F113 NY pond	+	-	178-330	-	-	-
B192 OH	+	257	-	-	-	-
W98 WY	+	-	272-305	-	-	-
H258 WY Pathfinder L.	6	307	-	6	345	-
L129 IL	+	307	-	-	-	-
K34 MT Missouri R.						
C. c. suckleyi	62	277	-	-	-	-
T76 WI lake	4	391	-399	4	567	524-606
C35, E3 MN	323	404	269-572	23	791	425-1066
C13 MI Deep L.	1	396	-	-	-	-
Age IV						
S91 NY	+	-	152-203	-	-	-
B20 MN Superior N. F.	+	190	102-318	-	-	-
H143 CO *C. c. suckleyi*	5	211	196-231	5	130	119-164
C258 SA	12	216	201-	12	91	-
B222 NY Catlin L.	15	229	208-	15	136	-
Wolf L.	40	234	-262	40	145	-
Wolf L., *C. c. utawana* males	10	221	-	10	122	116-
females	6	264	-279	6	196	-213
S156 NK Gibson L.	38	231	-	-	-	-
S89 WI Muskellunge L.	126	246	-	-	-	-
R97 NT Great Slave L.	1	262	-	-	-	-
L110 IL creek	3	267	-	3	221	-
B192 OH	+	295	-	-	-	-
C215 MI lake	8	302	-	-	-	-
K34 MT Missouri R.						
C. c. suckleyi	42	302	-	-	-	-
W98 WY	+	-	307-328	-	-	-
L129 IL	+	351	-	-	-	-
H258 WY Pathfinder L.	6	386	-	6	638	-
R14 NY Skaneateles males	+	378	-	-	-	-
females	+	421	-	-	-	-
C35 MN L. Vermilion	27	421	348-493	23	921	510-1191
E3 MN	484	439	300-599	-	-	-
Age V						
B20 MN Superior N. F.	+	229	140-356	-	-	-
S156 NK Gibson L. males	65	251	-	-	-	-
females	11	269	-	-	-	-

	No.	TL	Range	No.	Weight	Range
C258 SA	10	254	-	10	167	-
S89 WI Muskellunge L.	193	267	-	-	-	-
B222 NY lakes	31	267	224-305	31	204	-
S91 NY	+	-	203-343	-	-	-
L110 IL creek	2	302	-	2	298	-
R97 NT Great Slave L.	6	315	-	-	-	-
K34 MT Missouri R.						
C. c. suckleyi	40	340	-	-	-	-
H258 WY Pathfinder L.	3	409	-	6	720	-
R14 NY Skaneateles males	+	394	-	-	-	-
females	+	437	-	-	-	-
C215 MI	11	434	-	-	-	-
C35 MN L. Vermilion	14	437	406-475	12	1012	737-1162
E3 MN	398	450	330-660	-	-	-
L129 IL	+	455	-	-	-	-
Age VI						
B20 MN Superior N. F.	+	279	165-356	-	-	-
S156 NB Gibson L. males	60	272	-	-	-	-
females	14	292	-	-	-	-
S89 WI Muskellunge L.	139	290	-	-	-	-
B222 NY lakes	10	292	267-320	10	278	-
S91 NY	+	-	343-394	-	-	-
C258 SA	27	358	-432	27	533	-
K34 MT Missouri R.						
C. c. suckleyi	32	381	-	-	-	-
R97 NT Great Slave L.	8	386	-	-	-	-
C215 MI	4	455	-	-	-	-
E3, C35 MN	115	462	361-660	9	1540	879-2296
R14 NY Skaneateles males	+	439	-	-	-	-
females	+	472	-	-	-	-
W13 CT	1	635	-	1	3147	-
Age VII						
S156 NK Gibson L. males	19	287	-	-	-	-
females	13	320	-	-	-	-
B20 MN Superior N. F.	+	305	203-394	-	-	-
S89 WI Muskellunge L.	68	307	-	-	-	-
B222 NY lakes	8	338	-	8	411	-
K34 MT Missouri R.						
C. c. suckleyi	11	378	-	-	-	-
R97 NT Great Slave L.	2	445	-	-	-	-
C258 SA	23	445	-490	23	1015	-
R14 NY Skaneateles males	+	437	-	-	-	-
females	+	486	-	-	-	-
C215 MI	12	467	-	-	-	-
C35, E3 MN	40	467	361-541	1	1820	-
Age VIII						
S156 NK Gibson L. males	4	302	-	-	-	-
females	7	358	-	-	-	-
S89 WI Muskellunge L.	15	340	-	-	-	-
B20 MN Superior N. F.	+	343	254-406	-	-	-
K34 MT Missouri R.						
C. c. suckleyi	5	406	-	-	-	-

Age VIII (cont.)	No.	TL	Range	No.	Weight	Range
R14 NY Skaneateles males	+	450	-	-	-	-
females	+	490	-	-	-	-
R97 NT Great Slave L.	1	465	-	-	-	-
C258 SA	20	498	450-531	20	1446	-
C35, E3 MN	11	518	450-572	1	1540	-
D196 MD Deep Creek L.	5	-	508-584	5	-	1361-1814
Age IX						
S89 WI Muskellunge L.	3	361	-	-	-	-
B20 MN Superior N. F.	+	368	-	-	-	-
R97 NT Great Slave L.	3	505	498-	-	-	-
E3 MN	5	528	511-538	-	-	-
C258 SA	8	544	508-589	8	1902	-
Age X						
B20 MN Superior N. F.	+	406	343-470	-	-	-
R97 NT Great Slave L.	3	544	516-577	-	-	-
Age XI						
R97 NT Great Slave L.	2	589	554-625	-	-	-
Age XII						
R97 NT Great Slave L.	1	599	-	-	-	-

Average TL at each annulus

	No.	1	2	3	4	5	6	7	8	9	10
E3, S62 MN	2655+	109	206	295	351	401	424	437	490	495	
K33, M258 MN	1065	94	180	259	333	378	424	460	495		
C35 MN L. Vermilion	94	142	234	312	366	411	465	478	472		
R51 OH	+	64	229	318	384	432	457	470			
J42 OK Ft. Gibson	3	69	178	274	333						
B134 CO Loveland L.											
C. c. suckleyi	5	145	257	328							
R136 SA Churchill L.	15	155	213	269	320	396	478	531			
P112 MO Salt R.	59	97	173	229	297						
H268 CO Shadow Mt. L.											
C. c. suckleyi males	56	69	135	196	249	267					
females	66	76	155	229	295	328					
M453, P159 MT streams	163	53	135	198	254	277	302	328	333	378	394
lakes	282	53	147	229	295	328	378	391	417	434	
P122 WI Flora L.	130	147	224	290	366	401					
After thinning	250	160	257	338	399	429					
M371 ON pond	+	61	190	325	399	445	483				
J101 OK L. Eucha	19	142	315	394	429						

Mean TL and increments at each year of life

	1	2	3	4	5	6	7	8	9	10
S89 WI Muskellunge L.	71	117	163	203	231	262	290	310	345	335
Increment	71	46	46	41	36	36	30	28	25	23
Number	2990	2954	2739	2301	1820	1092	438	73	14	2
L110 IL Hutchins C.,										
Clear C.	51	102	160	224	279					
Increment	51	61	61	41	33					
Number	34	14	9	5	2					

Mean TL and increments at each year of life

	1	2	3	4	5	6	7	8	9	10
H258 WY Pathfinder L.	71	157	251	330	376					
Increment	71	89	94	81	64					
Number	22	21	15	9	3					
K34 MT Missouri R.										
C. c. *suckleyi*	66	135	208	257	310	348	358	386		
Increment	66	71	69	61	53	43	18	28		
Number	223	214	192	130	88	48	16	5		

Weight at each year of life

	No.	1	2	3	4	5	6	7	8	9
S89 WI Muskellunge L.	2990	3	20	48	79	130	179	244	320	403
R136 SA Churchill L.	15	28	108	198	340	737	1247	1843		

Annual increment as percentage of length at beginning of year

TL	S89 WI Muskellunge L.	L110 IL creeks	K34 MT Missouri R. C. c. *suckleyi*	Range
25	102	158	-	88-196
51	63	-	98	47-132
76	63	83	97	31-97
102	37	-	48	20-55
127	28	64	47	24-61
152	27	-	30	19-40
178	19	31	34	13-35
203	17	23	25	10-30
229	16	-	27	10-31
254	12	-	-	7-22
279	10	-	15	7-17
305	8	-	13	5-13
370	8	-	2	2-11
356	-	-	8	-

Ages of older suckers may be underestimated from scale readings since marked spawning-run suckers had to be older on recapture than scales indicated (G154). Scale readings are probably valid for the first 5 or 6 years, however (G154).

In most studies growth was computed from scale measurements by direct proportion but H268 and P122 made Fraser-type corrections. Scales were found to form when the suckers were 23-25 mm SL but the growth is rapid early after scale formation and no Fraser-type modification was believed necessary by S89. The dorsal-ventral diameter was found to give better data than the anterior-posterior diameter or radius by S89. Scale description and evidence of the validity of scale readings were given in S89. In Waskisiu Lake, Saskatchewan (C258), scales of 105 of 225 suckers were not considered readable. Most of these were older suckers. Pectoral fin-ray sections were easier to read than scales (S124). Of 78 suckers in a Kentucky stream, 11.5% had aberrant scalation (M407). Lee's phenomenon of apparent change of growth was evident in the Muskellunge Lake, Wisconsin, study (S89) and was believed to be the result of selection of the larger fish in the younger age groups.

There is little evidence of regional differences in growth in the data tabulated above. Growth is very slow in some waters and rapid in others. In a series of Minnesota lakes growth rate was found to be correlated with total dissolved solids, total carbonates, hydrogen-ion concentration, plankton abundance, bottom fauna and length of growing season (E94).

Thinning of the population increased the growth of suckers in Flora Lake, Wisconsin (P122). In New York ponds the mean total length in fall was negatively correlated with population density, P, recorded numbers of fish per acre (F66):

$$\log TL = 2.1272 - 0.3762 \log P$$

In Minnesota ponds the average total lengths at 60 days decreased as the population density increased (D81):

if 4000 suckers per acre	76 mm at 60 days
8000	64 mm
19000	51 mm

The average daily increments of young suckers in Waskisiu Lake, Saskatchewan (C258), were 0.1 mm FL from June 5-15 and 0.4 mm from June 15-19.

Female suckers grew faster than males in B222, S156, R14, H268, S89. Females averaged larger than males in spawning runs in British Columbia (G154) and continued to return in subsequent years more than males, suggesting a greater longevity. Spawning mortality was estimated at 16-20% (G154).

In Muskellunge Lake, Wisconsin (S89), seasonal growth was completed by mid-September and most of it was over by mid-August. There did not seem to be differences between the end of the growing season for various age groups.

Annual growth of adult suckers in a British Columbia lake was 15-20 mm FL according to tag returns and length frequencies (G154). White suckers disappeared from Lewis and Clark Lake, South Dakota, within 3 years of impoundment (W192).

Suckers matured at 200 mm, in their third year in an Ontario pond (M371). In Muskellunge Lake, Wisconsin (S89), a few females matured at age III but no males until age IV. In Waskisiu Lake, Saskatchewan (C258), males matured at ages VI and VII and females at ages VI to IX. *C. c. suckleyi* males matured at age II and at 150 mm TL but females at age IV and 267 mm TL in Colorado (H268). In a British Columbia lake, white suckers first matured at age VI and at 230 mm FL for males and 260 mm FL for females (G154).

	TL	Weight	No. of females	No. of eggs per female Mean	Range
S91 NY	305	-	1	18000	-
	381	-	1	31200	-
D10 NY *C. c. utawana*	120	-	1	775	-
	127	-	1	1000	-
	152	-	-	1300	-
	178	-	-	1500	-
L14 OH	343	-	1	24600	-
R14 NY	406	-	1	21800	-
	432	-	3	28400	26700-30400
	457-737	-	4	40525	36100-47800
V21 MN	178-203	-	3	42000	36000-46000
	356-405	-	9	67000	46000-81000
	406-431	-	12	72000	56000-114,000
	432-482	-	50	92000	66000-134,000
	483-510	-	8	111,000	93000-139,000

	TL	Weight	No. of females	No. of eggs per female Mean	Range
E4 MN	-	-	-	-	36000-140,000
D81, D41 MN, stripping	-	-	-	-	-47800
C258 SA	-	936	1	23000	-
	-	1418	1	42300	-
H191 CO *C. c. suckleyi*	348	454	1	24551	-
	320	312	1	11830	-
H268 CO *C. c. suckleyi*	368	567	1	21640	-
	427	794	1	18840	-

Suckers spawn in late March in Illinois (H8), in late April to early May in Michigan (R260), in late May to mid-June in Wisconsin (S89), in mid-May in Saskatchewan (C258), and in Colorado, *C. c. suckleyi,* from late May to August 6 (H268). They spawn in riffles over gravel with several males per female (R260). In Colorado some suckers spawned in riffles in streams where the gravel was scoured by the current, but others spawning in a lake were observed to clean the gravel before spawning (H268). Males have pearl organs on the lower part of the caudal, anal, and pelvic fins whereas the females do not. Pearl organs are shed shortly after spawning (R260). Artificial propagation methods and stripping are outlined in J62 and H195. Males had an average 8% weight loss on spawning, and females 14% (B226).

The incubation period was 5 days at 18C, 7 days at 15.5-16.1C, and 11 days at 13.6C (B226). Fry were 12-14 mm at first downstream movement (G154).

	Lethal temperature		
When acclimated at	Low	High	Citation
5C	-	26-26.3	H253, S333
10	-	27.7-28	H253, S333
15	-	29.3	S333
20	2.5-3.0	29-29.3	H253, S333
25	6	29.3	S333
25-6	-	31.2	B201

When suckers at 7C were frightened into warmer water, they died (T102). Suckers were most abundant at 19-21C in a Colorado reservoir and they went deeper to stay in these temperatures as the summer progressed (H338).

H143 CO   9 suckers, 109-272 mm TL, 1,204,000 red blood cells per cu. mm, range 790,000-1,666,000; hemoglobin 10.9.

Young suckers feed predominantly on entomostracans, small insects, rotifers, and algae (N104, N117, D168, D183, F148, H268). From 13-23 mm they were plankton feeders, and from 23-89 mm they grazed on weeds or solid surfaces but not mud (H268). Larger and adult suckers feed on chironomids, entomostraca, amphipods, fingernail clams, snails, and detritus (H268, C258, D183, S357, M404, B226). No evidence of egg depredation was reported by C258 or by E93 on suckers taken from salmon-spawning areas.

In Flora Lake, Wisconsin (P122), the number of suckers over 140 mm was estimated at 39.0, 33.6, 38.8, 46.4, and 48.7 per hectare in 1952 to 1956, respectively, giving 7.4 to 14.5 kg/ha. In Shadow Mountain Reservoir, Colorado, the sucker population, over 254 mm long, was estimated at 91.7 kg/ha. The angler harvest in Quabbin Reservoir, Massachusetts (M296), was estimated at .22 and .07 kg/ha in 1954 and 1955.

The sizes taken in gill nets were given as:

Mesh size stretch		51	76	102	127	140			
R97 NT									
Mean FL		224	318	401	467	465			
Mean weight		186	680	1225	1950	1905			
Mesh size	38	51	64	73	76	83	86	95	102
C17 MN L. of Woods									
Number	36	26	58	44	86	101	170	55	888
Mean FL	155	201	244	323	305	312	358	363	396
Range of center 50%	142-157	188-203	229-254	290-353	277-330	290-318	333-376	345-376	368-414

	South Bay B251			Georgian Bay B251		
Nylon mesh size	No.	Mean FL	Range	No.	Mean FL	Range
38	6	254	241-264	15	198	165-310
44	3	234	-	68	203	178-361
51	48	264	193-399	123	241	190-404
57	57	274	221-366	146	272	208-417
64	108	295	241-442	202	287	241-411
70	120	318	244-417	240	320	254-475
76	117	343	267-429	179	338	254-437
89	263	363	305-429	322	378	257-475
102	131	388	330-488	243	406	345-488
114	51	404	348-488	89	432	371-493
127	8	457	432-480	-	-	-

B251 South Bay	308 fish	girth = -0.414 mm + .593 FL
B251 Georgian Bay	257 fish	girth = -0.605 mm + .617 FL

BLUEHEAD SUCKER, *Catostomus discobolus* Cope
   Colorado River drainage above the mouth of the Grand Canyon; Snake River, Idaho and Wyoming; and Bear River drainage and Weber River drainage of the Bonneville Basin (S507). This species includes most of the specimens previously known as *Pantosteus delphinus* (Cope).
   In Glen Canyon, Utah, bluehead suckers fed mostly on algae and to a limited extent on insects (M340).

FLANNELMOUTH SUCKER, *Catostomus latipinnis* Baird and Girard
   Found in the Colorado River drainage (E119).

B240 WY	513 mm	1332 g	
M340 UT Green R.			

SL	No.	Weight	K
86	1	16	2.51
182	3	81	1.34
221	3	156	1.45
250	7	225	1.44
272	4	291	1.43
301	6	404	1.48
322	4	475	1.46
354	1	620	1.40
374	10	694	1.32
396	5	878	1.41
481	1	1610	1.45

M340 UT Green R.

	No.	Mean TL	Range
Age I	1	104	-
II	3	165	142-183
III	2	295	284-305
IV	2	319	307-330
V	2	356	348-363
VI	2	363	-

The food of 4 flannelmouth suckers was found to be algae and unidentified material (M340).

LARGESCALE SUCKER, *Catostomus macrocheilus* Girard
Found in the Columbia River (E119).

M303 ID	113 suckers	12-20 mm	$\log W = -2.517 + 4.9125 \log FL$
	76 suckers	20-38 mm	$\log W = -2.048 + 3.350 \log FL$

A major change in growth was noted at 20 mm FL.
In Nicola Lake, British Columbia, 15 young suckers in July were from 20-25 mm and averaged 23 mm FL (N82).

Average calculated FL at each annulus

	1	2	3	4	5	6	7	8	9	10	11
C53 BC Okanagan	46	76	124	170	190	229	284	305	340	356	376

Average calculated TL at each annulus

	No.	1	2	3	4	5	6	7	8
P159 MT lakes	205	46	84	137	188	262	328	399	427

The spawning run in Oregon is in April and May (L193).
The lethal temperature determined for largescale suckers acclimated at 19C was 29.4C (B200).
Young suckers, 12-15.9 mm FL, fed at the surface on diatoms, rotifers, protozoans, and cladocerans and by 24 mm started feeding on the bottom on the same organisms (M303), including some sand and filamentous algae. Chironomids and annelids were the principal foods of adults (M325).

SACRAMENTO SUCKER, *Catostomus occidentalis* Ayers
Found in central California (E119, B345).

B345 CA Merced R.	Annulus	1	2	3	4	5
	Mean calculated FL	84	142	211	277	330
	Weight	11	42	125	264	428

They spawn in spring or early summer and 280-380 mm FL females contained 4720-10932 eggs (B345).
Newly hatched young feed on plankton near the surface and older suckers feed on insects, molluscs, algae, and higher plants on the bottom (B345).

MOUNTAIN SUCKER, *Catostomus platyrhynchus* (Cope)
Streams of the Great Basin in Utah, Nevada and California; headwaters of the Green River, Utah, Colorado, and Wyoming and east into South Dakota and

northwestern Nebraska; north into Saskatchewan, Alberta, and British Columbia (S507). This species also includes the Lahontan sucker, *Pantosteus lahontan* (Rutter).

Lahontan suckers were found with ripe eggs in mid to late July in Nevada
(L193).

TAHOE SUCKER, *Catostomus tahoensis* Gill and Jordan
Lake Tahoe, California, and Nevada (E119).

These suckers spawn from late April to early June at water temperatures
of 12-14C, over coarse sand and gravel in less than 305 mm of water (L193).
They are largely herbivorous but also eat some *Gammarus* and chironomids
(L193).

CUI-UI, *Chasmistes cujus* Cope
Pyramid Lake, Nevada (E119, L193). The largest male was 540 mm,
1588 g and the largest female was 629 mm, 2722 g (L193). In April 15-May 16
they run into tributary streams to spawn. Cui-ui feed on zooplankton (L193).

BLUE SUCKER, *Cycleptus elongatus* (LeSueur)
In large rivers from South Dakota to Pennsylvania south to Tennessee and
Mexico (E119, T113).

F100,124,125 SD Oahe

TL	318	368	483	521	546	572	597	622	648	927
Weight	186	290	840	998	1188	1225	1438	1665	1905	2504

S472 AL

TL	457	457	508	508	533	635
Weight	703	907	1043	1361	1451	2246

$$\log W = -5.57 + 3.19 \log TL$$

	No.	TL	K(TL)
F100,105,124,125 SD Oahe	14	483-635	0.67
S276, N115 SD Gavins Point	12	432-686	0.72
S226,228,229 SD Ft. Randall	20	457-660	0.76
S472 AL	6	457-635	0.89

	No.	SL	TL	Range	No.	Weight	Range
Age 0-June							
M133 OK Grand L.	2	43	-	33-50	-	-	-
Age II							
M133 OK L. Texoma	3	374	-	365-387	3	720	709-732
S228 SD Ft. Randall	+	-	356	-	-	-	-
Age IV							
J31 OK Grand L.	2	-	630	605-655	2	2200	1973-2433
S228 SD Ft. Randall	+	-	559	-	-	-	-

Average TL and increments at each annulus

	No.	1	2	3	4	5	6	7
J31 OK Grand L.	2	178	328	434	508	579	630	
S229, S276 SD Gavins Point	-	140	295	424	478	544	617	
Increment	-	140	155	130	79	91	56	
Number	-	16	16	14	9	4	2	
F125 SD Oahe	-	109	196	290	411	462	511	582
Increment	-	109	86	94	122	36	74	71
Number	-	5	5	5	5	4	2	2
S226 SD Ft. Randall	-	117	257	358	478	521		
Increment	-	117	140	135	119	104		
Number	-	10	10	8	8	2		

CREEK CHUBSUCKER, *Erimyzon oblongus* (Mitchill)
Wisconsin to New England and south to Alabama and Texas (E119).

L110 IL	54 chubsuckers	47-168 mm SL	TL = 1.19 SL

C261 NC

SL	148	165	187	200	220	240	260	280	300	320	335
Males, Wt.	80	122	-	182	214	454	584	721	826	931	-
No.	2	1	-	1	5	12	21	29	13	3	-
Females, Wt.	-	111	189	270	312	427	510	619	817	1101	1174
No.	-	1	5	5	26	41	37	16	12	11	2
Mean K(SL)	2.64	2.71	2.79	2.85	2.91	2.97	3.02	3.07	3.12	3.17	3.21

C261 NC	males	$\log W = -5.0507 + 3.219 \log SL$
	females	$\log W = -5.00977 + 3.202 \log SL$
L110 IL	56 chubsuckers 65-168 mm	$\log W = -4.83 + 3.08 \log SL$
	average K = 2.12	$r_{xy} = 0.98$
E105 MD	419 mm TL	1814 g

	No.	TL	Range	No.	Weight
Age 0-Sept.					
F66 NY pond	+	66	-	-	-
Age I					
L110 IL Hutchins C., Clear C.	29	102	-	29	14
C261 NC pond	14	155	130-180	-	-
Age II					
L110 IL Hutchins C., Clear C.	22	145	-	22	40
C261 NC	131	300	183-378	-	-
Age III					
C261 NC	82	318	234-366	-	-
Age IV					
C261 NC	17	351	305-384	-	-
Age V					
C261 NC	16	373	323-411	-	-
Age VI					
C261 NC	4	401	391-404	-	-

TL and increments at each annulus

	1	2	3	4	5	6
L110 IL Hutchins C., Clear C.	48	104				
Increment	48	48				
Number	51	22				
C261 NC Raleigh L. females	124	241	307	351	378	399
Increment	124	137	76	43	25	15
Number	170	162	80	30	19	4
males	147	251	315	353	366	
Increment	147	84	97	33	23	
Number	94	88	39	7	1	

The annulus forms in March-April in North Carolina (C261). Growth was computed on a direct-proportion basis in both studies.

They spawn March 19-April 22 in North Carolina (C261) and the number of eggs per female were reported as:

		Eggs per female	
Weight	No.	Mean	Range
111-196	3	11329	8694-16548
262-302	5	18253	11774-24605
371-438	3	25956	25001-27533
484-785	3	40519	32998-50413
956-970	3	59565	35505-72360

Entomostraca, rotifers, small insects and algae constitute the food of creek chubsuckers (F148).

LAKE CHUBSUCKER, *Erimyzon sucetta* (Lacépède)
Southeastern Minnesota to New York and south to Florida and Texas (E119, T113).

A26 FL Silver Springs   387 mm chubsucker weighed 907 g

M242 FL L. Panasoffkee

TL	241	254	279-92	305-18	330-43	356-68	381-94
Weight	181	227	363	454	635	771	953
Range	-	-	272-499	408-544	454-771	726-816	907-998
Number	1	1	10	15	14	4	2

B329 WI   3 chubsuckers   208-211 mm TL   K(TL) = 1.51   range 1.38-1.57

Growth during the first month was 0.5 mm per day in Portage Lake, Michigan (H86).

	No.	SL	FL	TL	Range	No.	Weight
Age 0-May-June							
R27, H8 IL	+	-	-	X	18-33	-	-
Age II							
B77 MI Third Sister L.	+	-	-	58	41-79	-	-
R34 IN Shoe L.	19	-	168	-	155-185	-	-
U3 NY Grovers L.	2	193	-	-	-	-	-

	No.	SL	FL	TL	Range	No.	Weight
Age III							
C215 MI Whitmore L.	17	-	-	180	-	-	-
R34 IN Shoe L.	81	-	190	-	165-231	-	-
B329 WI	1	-	-	208	-	1	125
U3 NY Grovers L.	1	219	-	-	-	-	-
Age IV							
B329 WI	2	-	-	211	-	2	142
C215 MI Whitmore L.	13	-	-	216	-	-	-
R34 IN Shoe L.	16	-	231	-	216-254	-	-
U3 NY Grovers L.	51	255	-	-	-	-	-
Age V							
C215 MI Whitmore L.	5	-	-	231	-	-	-
C58 MI	+	-	-	254	-	-	-
U3 NY Grovers L.	59	268	-	-	-	-	-
W13 CT	1	300	-	353	-	1	765
Age VI							
C215 MI Whitmore L.	2	-	-	244	-	-	-
U3 NY Grovers L.	67	270	-	-	-	-	-
Age VII							
U3 NY Grovers L.	14	268	-	-	-	-	-
Age VIII							
U3 NY Grovers L.	6	266	-	-	-	-	-

		Average calculated TL at each annulus					
	No.	1	2	3	4	5	6
L157 NC Black L.	9	53	137	190	224	244	-
Jones L.	9	53	135	193	244	284	305
Salters L.	13	56	135	198	244	254	259
Singletown L.	9	66	152	213	244	264	-
White L.	4	104	203	-	-	-	-

Lake chubsuckers spawn in March-April in Illinois in streams (B307, R27). The male cleans an area among gravel for a nest. Females produced 3000 to 20000 eggs per female in Michigan (C56).

Food of small chubsuckers, 30-42 mm, was copepods, cladocerans, and chironomids (E121). Chubsuckers are suggested as being ideal forage fish for bass (B307). Ridge Lake, Illinois, had a population of 187 chubsuckers per acre (B307).

		Sizes in gill net meshes		
Mesh size, bar		38	51	76
L216 LA	No.	6	1	1
	mean TL	295	300	224
	range	229-432	-	-

ALABAMA HOG SUCKER, *Hypentelium etowanum* (Jordan)
Alabama River system (E119).

			Weight		
S472 AL	TL	No.	Mean	Range	K(TL)
	51	17	2.4	1.8-	1.83
	76	42	3.8	2.7-6.0	.86
	102	3	9	6-	.89
	127	4	23	16-	1.11
	152	8	32	-	.89
	178	5	59	50-64	1.05
	203	1	95	-	1.14

Because the weights of the 51 mm fish were too high, the regression given in S472 is not valid.

NORTHERN HOG SUCKER, *Hypentilium nigricans* (LeSueur)
Minnesota to New York, south to northern Alabama, southern Louisiana and Oklahoma (E119, T113).

TL	No.	Weight	Range	Citation
76	1	4	-	S472 AL
102	1	8	-	S472
127	5	23	20-26	S472
152	+	43	-	P113 MO
	4	45	-	S472
178	2	48	45-51	S472
	+	62	-	P113
203	+	45	-	W132 TN
	2	79	71-85	S472
	+	99	-	P113
229	1	122	-	S472
	+	136	-	W132
	+	142	-	P113
254	2	147	-	S472
	+	193	-	P113
279	+	252	-	P113
	1	286	-	S472
305	+	326	-	P113
330	+	408	-	W132
	+	411	-	P113
356	+	505	-	P113
	1	680	-	E105 MD
381	+	621	-	P113
559	1	1889	-	T113 L. Erie

P113 MO			$\log W = -4.697 + 2.902 \log TL$
S472 AL			$\log W = -4.96 + 2.98 \log TL$
S472 AL	19 fish	76-280 mm	K(TL) = 1.05 range 0.86-1.30

	No.	TL	Range	No.	Weight	Range
Age 0-Oct.						
T113 OH	+	-	51-89	-	-	-
Age I						
R11 NY Genesee	3	43	41-46	-	-	-
R11 NY Catatonk	52	79	41-109	-	-	-

	No.	TL	Range	No.	Weight	Range
R91 VA Roanoke R.	26	79	61-104	-	-	-
B192 OH Massie C.	4	-	104-117	-	-	-
J25 OK Illinois R.	73	196	140-262	73	82	-
Age II						
R11 NY Genesee	23	99	84-122	-	-	-
R11 NY Catatonk	34	122	68-140	-	-	-
R91 VA Roanoke R.	5	145	130-160	-	-	-
B192 OH Massie C.	+	193	-	-	-	-
J25 OK Illinois R.	12	287	267-310	12	247	-
Age III						
R11 NY Genesee	35	142	102-183	-	-	-
R11 NY Catatonk	37	178	109-211	-	-	-
J42 OK Ft. Gibson	1	333	-	-	-	-
J25 OK Illinois R.	18	353	330-371	18	490	-
Age IV						
R11 NY Genesee	16	190	150-224	-	-	-
R11 NY Catatonk	16	213	180-251	-	-	-
J25 OK Illinois R.	5	373	361-388	5	606	-
Age V						
R91 VA Roanoke R.	1	216	-	-	-	-
R11 NY Genesee	58	239	196-292	-	-	-
Age VI						
R91 VA Roanoke R. male	1	241	-	-	-	-
R11 NY Genesee	13	251	213-297	-	-	-
Age VII						
R11 NY Genesee	21	274	231-318	-	-	-
R91 NY Roanoke R.	1	279	-	-	-	-
Age VIII						
R11 NY Genesee	7	287	257-320	-	-	-
Age IX						
R11 NY Genesee	1	351	-	-	-	-

		Average calculated SL		
	No.	1	2	3
E3 MN	43	95	167	232

	No.	1	2	3	4	5	6	7	8	9
		Average calculated TL at each annulus								
P113 MO	1213	86	165	246	300	330	356	376	388	399
poorest station		69	127	211	264	277	340			
best station		104	224	338	381	391	388			
J42 OK Ft. Gibson	6	71	188	274						
H150 OK Illinois R.	243	89	188	300	358					
J25 OK Illinois R.	108	94	213	312	358					
J101 OK L. Eucha	8	109	221	325	376					
J101 OK Spavinaw L.	27	119	277	351	356					

Of 130 hog suckers examined in a Kentucky stream, 15.4% had scale aberrations, mostly above the lateral line (M407). Stunting is common in small streams with steep gradients (T113). Males grow more rapidly through the first 4 years (R11, R91) but in older age groups females averaged larger than the males.

Most males matured at age II with a few at age I, 71 mm TL, and most females matured at age III in the Roanoke River, Virginia (R91). In Illinois they spawn in April (H8); in Ohio in late March to early June (T113); in New York in April-May (H8); in Michigan, in May (R260). They spawn in rapids and both males and females have pearl organs though the pearl organs are smaller on females. The anal fin of the male is longer than that of the female (R260).

Hog suckers feed on the bottom fauna (M404, R260) and *Notropis* and *Campostoma* minnows may follow to pick up food items dislodged by hog suckers (R260).

ROANOKE HOG SUCKER, *Hypentelium roanokense* Raney and Lachner
Headwaters of Roanoke River, Virginia (E113).

R91 VA Roanoke R.			TL = 1.23 SL

R91 VA Roanoke R. Age	No.	Mean TL	Range
I females	17	64	56-71
males	16	69	64-81
II males	25	81	66-104
females	26	89	74-120
III males	1	94	-
females	4	119	112-124
IV males	3	97	89-99
females	9	122	97-140
V females	2	127	74-127

SMALLMOUTH BUFFALO, *Ictiobus bubalus* (Rafinesque)
Mississippi River drainage from Pennsylvania to South Dakota south to Louisiana, and through Texas to Mexico (E119, T113).

S17 TN Reelfoot L.	107 buffalo	190-750 mm SL	TL = 1.213 SL
C30, M302 IA	6 buffalo		TL = 1.21 SL

		Weight			
		Mean	Central		Citation of
TL	No.	of means	50%	Range	extremes
51-75	+	9	-	-	Oahe
76-101	14+	10	-	6-18	S472, Oahe
102-126	39+	20	-	9-27	S472, Oahe
127-151	127+	36	-	22-56	S472, S472
152-177	237+	54	45-68	36-68	S472, Oahe
178-202	128+	91	86-100	77-113	J31, Oahe
203-228	43+	136	113-145	91-181	S472, S472
229-253	111+	186	177-181	23-254	S472, S472
254-278	266+	259	240-268	36-322	S472, Oahe
279-304	436+	345	331-340	50-477	S472, S472
305-329	388+	445	426-445	295-680	S472, S472
330-355	520+	571	539-562	381-740	S472, Oahe
356-380	640+	712	666-712	590-944	S472, S472
381-405	479+	843	794-848	626-1134	S472, S472
406-431	452+	1003	939-1003	816-1474	S472, S472
432-456	301+	1234	1188-1225	967-1588	S472, S472

TL	No.	Weight Mean of means	Central 50%	Range	Citation of extremes
457-482	186+	1478	1438-1547	1003-2495	S472, S472
483-507	126+	1719	1710-1760	1152-2132	S472, S472
508-532	77+	2159	2100-2177	1724-2576	S472, S472
533-558	56+	2527	2478-2736	2059-2858	S472, S472
559-583	52+	2989	2826-3098	2087-3243	S472, S472
584-609	23+	3407	3220-3502	2912-3856	S472, S472
610-634	9+	3643	-	2722-4151	W132, Oahe
635-659	8+	4708	-	3629-4953	S472, Oahe
660-685	6+	5389	-	4377-6083	S472, Oahe
686-710	2+	6115	-	4900-7009	S472, Oahe
711-736	2+	6881	-	5443-7983	S472, Oahe
737-761	2	7371	-	6804-7938	S472
762-786	7	8310	-	6532-9185	S472, M302
838-863	1	11567	-	-	S17
864-888	2	10465	-	-	J31

The above data came from Alabama A39, S472; Iowa M302; Oklahoma J31; Tennessee W132; and from the following reservoirs in South Dakota:

Gavins Point Reservoir N94, 115, S229, 276
Oahe Reservoir F100, 105, 124, 125
Ft. Randall Reservoir S225, 226, 228, 230, 330

The numbers of buffalo at each length were not indicated for the reservoirs but the totals were for 424, 359+, and 638 respectively. The central 50% does not include the mean of the means in many cases because the Oahe figures were much higher than the other data.

				r
S472 AL	3830 fish	76-762 mm	$\log W = -4.91 + 3.04 \log TL$	
M505 TN	655 fish	290-695 mm TL	$\log W = -3.046 + 3.0 \log TL$	
H420 TN Watts Bar L.	gravid females		$\log W = -4.826 + 3.010 \log TL$	.91
	ripe males		$\log W = -4.154 + 2.747 \log TL$	.90
	spent females		$\log W = -3.488 + 2.494 \log TL$	.85
	spent males		$\log W = -5.048 + 3.075 \log TL$	.99
	ripe males and females		$\log W = -4.704 + 2.959 \log TL$	.91
	spent males and females		$\log W = -4.111 + 2.726 \log TL$	.91
J31 OK Grand L.	537 buffalo		$\log W = -4.749 + 3.212 \log TL$	
S225 SD Ft. Randall			$\log W = -4.239 + 2.736 \log TL$	

		No.	K(SL)	Range
E28 TN Chickamauga L.		596	3.1	-
M302 IA	760-790 mm	5	3.22	-
P179 TX	82-605 mm	19	3.80	3.14-4.66
			K(TL)	
N94, 115; S227, 229, 276 SD Gavins Point L.		430	1.29	-
S222 SD Ft. Randall tailwaters		2	1.36	-
S225, 226, 228, 230, 330, 393 SD Ft. Randall L.		381	1.45	-
F100, 105, 124, 125 SD Oahe L.		544	1.53	-
S472 AL	76-762 mm	3812	1.53	1.33-1.83

No trend was evident in condition factors in the South Dakota reservoirs in the first years after impoundment. No sexual difference in condition factors was reported by S276, S229, S228, S226 but in Oahe (F105, F100, F125) males had slightly higher average K(TL) values than females. No trend in condition factor with increase in length was reported by S225, S228, and S229 but in S276 and S472 a tendency for K(TL) to increase with length was noted.

	No.	TL	Range	No.	Weight	Range
Age 0-May						
W51 LA ponds	+	51	-	-	-	-
Age 0-June						
S230 SD Ft. Randall	+	28	-	-	-	-
W51 LA ponds	+	76	-	+	5	-
Age 0-July						
S228 SD Ft. Randall	+	48	43-53	-	-	-
Age 0-Aug.						
S228 SD Ft. Randall	+	53	-	-	-	-
S207 KS Fall River L.	2	112	104-122	2	18	-
Age 0-Oct.						
H150 OK Tenkiller L.	4	130	-	-	-	-
Age I						
M215 KS Big Blue R.	11	61	-	-	-	-
H127 OK Wister L.	1	127	-	-	-	-
S207 KS Fall River L.	2	145	-	2	41	-
J42 OK Ft. Gibson L.	3	216	-	-	-	-
T44 OK Grand L.	3	218	-	3	91	-
U11 SD Oahe L.	32	241	-	32	227	-
S17 TN Reelfoot L.	6	284	229-356	6	408	181-907
Age II						
M215 KS Big Blue R.	14	112	-	-	-	-
S207 KS Fall River L.	1	180	-	1	73	-
E28 TN Chickamauga L.	2182	183	124-262	-	-	-
H420 TN Watts Bar L.	1	188	-	-	-	-
T44 OK Grand L.	28	257	-	28	227	-
U11 SD Oahe L.	37	297	259-378	37	422	272-907
J42 OK Ft. Gibson L.	5	310	-	-	-	-
H127 OK Wister L.	8	343	-	-	-	-
S17 TN Reelfoot L.	14	388	338-424	14	998	816-1497
Age III						
M215 KS Big Blue R.	5	168	-	-	-	-
S207 KS Fall River L.	62	239	211-267	62	186	-
H420 TN Watts Bar L.	1	292	-	-	-	-
T44 OK Grand L.	271	305	-	271	381	-
I10 IA	1	394	-	1	794	-
U11 SD Oahe L.	4	401	-	4	998	-
H127 OK Wister L.	20	409	-	-	-	-
S17 TN Reelfoot L.	20	439	376-486	20	1406	1043-2268
M505 TN Watts Bar L.	88	450	391-521	-	-	-
Age IV						
W192 SD Lewis & Clark L.	3	259	-	-	-	-
T44 OK Grand L.	13	340	-	13	481	-
M505 TN Watts Bar L.	392	462	419-544	-	-	-
U11 SD Oahe L.	44	465	411-569	44	1656	1089-1724
S17 TN Reelfoot L.	28	467	411-533	28	1588	907-2449
H127 OK Wister L.	2	488	-	-	-	-

	No.	TL	Range	No.	Weight	Range
Age V						
W192 SD Lewis & Clark L.	34	300	-	-	-	-
T44 OK Grand L.	1	371	-	1	1506	-
M505 TN Watts Bar L.	458	470	401-594	-	-	-
U11 SD Oahe L.	65	490	457-615	65	1937	1769-3130
H127 OK Wister L.	6	521	-	-	-	-
S17 TN Reelfoot L.	19	544	460-643	19	3175	1406-4580
Age VI						
W192 SD Lewis & Clark L.	61	330	-	-	-	-
H420 TN Watts Bar L.	6	414	-	-	-	-
M505 TN Watts Bar L.	225	480	429-594	-	-	-
H127 OK Wister L.	1	572	-	-	-	-
U11 SD Oahe L.	6	592	513-640	6	3298	2812-4082
S17 TN Reelfoot L.	5	594	546-655	5	3629	3175-4580
Age VII						
H420 TN Watts Bar L.	19	427	-	-	-	-
W192 SD Lewis & Clark L.	2	455	-	-	-	-
M505 TN Watts Bar L.	79	503	450-599	-	-	-
U11 SD Oahe L.	10	620	559-691	10	4210	3402-7985
S17 TN Reelfoot L.	4	648	630-668	4	4491	4037-4808
Age VIII						
H420 TN Watts Bar L.	25	460	-	-	-	-
M505 TN Watts Bar L.	20	536	465-650	-	-	-
U11 SD Oahe L.	3	658	559-	3	5216	-
Age IX						
H420 TN Watts Bar L.	21	475	-	-	-	-
M505 TN Watts Bar L.	4	620	526-805	-	-	-
U11 SD Oahe L.	1	681	-	1	6441	-
S17 TN Reelfoot L.	7	782	752-838	7	8983	4536-9300
Age X						
H420 TN Watts Bar L.	18	475	-	-	-	-
M505 TN Watts Bar L.	3	594	500-737	-	-	-
Age XI						
H420 TN Watts Bar L.	14	493	-	-	-	-
Age XII						
H420 TN Watts Bar L.	11	503	-	-	-	-
S17 TN Reelfoot L.	4	837	787-909	4	11567	10570-12928
Age XIII						
H420 TN Watts Bar L.	3	536	-	-	-	-

							TL at each annulus							
	No.	1	2	3	4	5	6	7	8	9	10	11	12	13	14
S225 SD Ft. Randall before impoundment	1	56	132												
E28 TN Chickamauga L.	596	97	163												
J42 OK Ft. Gibson L.	220	89	178	211	277	378	447								
S391 OK Rock C.	14	104	163	203	244	290	330	366	419						
I10 IA Desota Bend	1	89	185	318											
P113 MO St. Francis R.	50	127	201	254	295	330	363	396	460	490	526	592	615	615	632
L195 OK Arkansas R.	16	152	239	269	310	361	384	361							
J31 OK Grand L.	473	114	188	259	315	361	396								
P112 MO Salt R.	71	130	244	325	381	429	493	521	554						
C270 MN Mississippi R.	8	147													
J45 OK Verdigris R.	17	160	267	363	467	498	544								
J31 OK Neosho R.	92	188	290	384	442	500	546	587	650	782	876				
H303 OK Lawtonka L.	1	183	371	411	488										

TL and average increments at each annulus

	1	2	3	4	5	6	7	8	9	10	11	12	13	14
T44 OK Grand L.	155	216	264	302	338									
Increment	155	61	51	36	46									
Number	316	316	313	285	14									
S229, S276, N94 SD Gavins	104	175	239	274	378	427	470	505	544					
Increment	104	76	71	33	51	48	43	30	33					
Number	304	207	88	82	3	3	3	2	1					
F100, 105, 124, 125 SD Oahe	104	193	315	368	419	478	508	584	625					
Increment	104	94	94	58	53	74	46	58	41					
Number	321	261	66	29	19	9	6	1	1					
S225, 226, 228, 393 SD Ft. Randall	132	262	345	401										
Increment	132	135	91	66										
Number	289	214	48	4										
H420 TN Watts Bar L.	91	150	196	239	279	320	353	386	409	429	450	478	508	
Increment	91	58	46	43	41	33	33	36	36	30	28	28	30	
Number	100	100	99	98	98	98	98	92	67	46	28	14	3	
M505 TN Watts Bar L.	112	257	353	414	442	462	490	528	582	594	645	668	732	742
Increment	112	145	96	64	43	36	30	30	36	36	23	20	18	10
Number	1271	1271	1271	1183	791	333	108	29	9	5	2	2	1	1

Average annual increment as percentage of TL at beginning of year and instantaneous growth rates M505

TL	76-101	102-126	127-151	152-177	203-228	229-253	254-278
% increments	81-111	82-143	126	44	36-40	36-38	37-38
Instantaneous	.59-.75	.60-.88	.82	.37	.31-.34	.31-.32	.32
TL	279-304	305-329	330-355	356-380	381-405	406-431	432-456
% increments	18-47	18-22	16-18	13	11-15	7-11	7-12
Instantaneous	.17-.39	.17-.20	.15-.17	.12	.10-.14	.068-.113	.068-.113
TL	457-482	483-507	508-532	533-558	559-583	584-609	610-634
% increments	6-8	6-9	6.0-6.8	7.2	5.5-6.6	4.2	6.0
Instantaneous	.058-.077	.058-.086	.058-.068	.068	.058-.068	.039	.058
TL	635-659	660-685	686-710	711-736	-	-	-
% increments	3.3	2.1	2.1	1.0-2.0	-	-	-
Instantaneous	.03	.02	.02	.01-.02	-	-	-

Average annual increments as percentages and instantaneous growth rates by year of life M505

Year of life	2	3	4	5	6	7	8	9	10	11	12-14
% increments	81-143	36-47	15-38	10-21	8-12	6-9	6-8	6.0-6.6	3.3-7.2	2.1-5.5	1.0-4.2
Instantaneous	.59-.89	.31-.39	.14-.32	.10-.19	.08-.11	.06-.08	.06-.08	.058-.068	.03-.068	.02-.058	.01-.039

No regional trends are evident in the growth data. Seasonal growth in South Dakota is reported to start in late May (S226) to the first week of June (S276, S229, S228) when water temperatures reach 18C (S228). Age I buffalo grew 25 mm per 27 days (S229); age II, 25 mm per 33 days (S276), and age III, 25 mm per 41 days (S228) during the summer in South Dakota. In Alabama 9 buffalo increased from an average of 227 to 2120 g in one year, and 3 buffalo increased from an average of 1710 to 1964 g from October to July (H199). Lee's phenomenon was pronounced in the M505 data. Growth was not significantly affected by radioactivity levels adequate to mark scales (M505).

No scale aberration was noted on 13 smallmouth buffalo in a Kentucky stream where many fish showed aberrations (M407).

Smallmouth buffalo first mature at age III (S228, S230) in Ft. Randall Reservoir but not until age IV in Gavins Point and only 50% were mature at age IV (N94).

They spawn in late May (S227) to early June (S393, S228, W192, N94) in South Dakota.

Smallmouth buffalo spawn from March through June at water temperatures about 17C (K118). A good year class formed the first year of impoundment but spawning success was poor in later years in Lewis and Clark Lake (W192).

Young buffalo feed mostly on copepods and cladocerans and these organisms continue to be an important food of adults, which also eat algae, duckweed, and insects (M506). The smallmouth buffalo feeds on the bottom, in contrast to the largemouth which is a plankton feeder. The absence of *Hexagenia* mayfly nymphs in the food indicated that the smallmouth buffalo feed near the shore rather than over deeper silt bottoms (M506).

L216 LA			Sizes in gill net mesh					
Mesh size (square)	25	38	51	64	76	89	102	114
Number	8	32	34	10	29	2	115	21
Mean TL	381	312	340	472	475	534	579	605
Range	236-541	234-419	249-470	401-577	-	521-546	-	-
Mean weight	1633	771	862	1996	2404	-	-	-
Range	227-2948	272-1588	227-2449	1451-2087	-	-	-	-

Linen caught larger buffalo than equivalent mesh of cotton; nylon caught larger yet (L216).

The red blood cell count for 4 smallmouth buffalo was 1,332,000 ± 88988 per cu. mm and the hemoglobin reading 40 ± 6 (S151).

Smallmouth buffalo remained fairly sedentary for two years after hatching and then moved upstream in Watts Bar Lake (M505).

BIGMOUTH BUFFALO, *Ictiobus cyprinellus* (Valenciennes)

Southern Alberta and Manitoba through southeastern Wisconsin to Lake Erie south to northern Alabama, and in the Mississippi to southern Louisiana (J101, E119). Introduced in Arizona and southern California (J101).

		No.	FL/SL	TL/SL
C30 IA		2	1.13	-
		20	-	1.25
C89 IA	300-592 mm SL	33	1.110	1.211
S16 TN	220-600 mm SL	118	-	1.224
B346 AZ		60	-	1.2
J90 SA	254-787 mm FL	359	1.13	-
M302 IA lakes	28-279 mm TL	22	-	1.28
	280-459 mm TL	140	-	1.25
	460-720 mm TL	78	-	1.23

In those cases where data were given in SL, they are converted to TL using the conversion factors from M302.

		Weight			
TL	No.	Mean of means	Central 50%	Range	Citation of extremes
25-50	5	1	-	-	M302
51-75	386	4	-	3-4	M302, S472
76-101	19624	9	-	6-14	S472, M302
102-126	25908+	20	-	14-27	S472, Oahe
127-151	4530+	32	27-32	23-50	S472, Oahe
152-177	810+	59	50-68	45-73	S472, Oahe
178-202	126+	104	100-113	73-172	S472, S472

		Weight			
		Mean of	Central		Citation of
TL	No.	means	50%	Range	extremes
203-228	186+	154	145-168	122-286	S472, S472
229-253	11+	249	222-336	191-363	Gavins, S472
254-278	3+	290	-	268-308	Gavins, Ft. Randall
279-304	26+	376	376-426	272-426	S472, M302
305-329	50+	494	435-522	318-390	S472, S472
330-355	344+	617	567-658	531-707	S472, M302
356-380	172+	762	707-839	622-1125	S472, S472
381-405	151+	976	948-1007	816-1043	S472, S472
406-431	169+	1161	1043-1211	771-1333	S472, Gavins
432-456	120+	1366	1265-1460	1125-1615	S472, Gavins
457-482	178+	1647	1628-1701	1248-1932	S472, Gavins
483-507	148+	1932	1783-2041	1497-2246	S472, Gavins
508-532	55+	2246	2232-2377	1919-2580	S472, Gavins
533-558	37+	2550	2440-2631	2014-2885	S472, Gavins
559-583	45+	2940	2812-3034	2404-3629	S472, S472
584-609	45	3338	3153-3362	3062-3910	S472, M302
610-634	35	3842	3714-4055	3371-4055	S472, J31
635-659	36	4468	4427-4900	4082-4900	S472, S472
660-685	32	5162	4962-5333	4935-5333	M302, J31
686-710	14	5620	-	3796-5965	M302, S472
711-736	6	6472	-	5443-7000	C30, J31
737-761	1	7920	-	-	J31
762-786	1	8990	-	-	J31
862	1	11567	-	-	J90
914-939	1	13862	-	-	C30

The above data came from Iowa C30, 89, 100, M302; Oklahoma J31; Manitoba J90; and from the following South Dakota reservoirs:

Ft. Randall S225, 226, 228, 230, 330	570 buffalo
Gavins Point N94, 115, S227, 229, 276	255 buffalo
Oahe F100, 124, 125	804 buffalo

M507 IA Coralville			$\log W = -5.028 + 3.193 \log SL$
J90 SA	827 buffalo	254-787 mm FL	$\log W = -4.445 + 2.91 \log FL$
S225 SD Ft. Randall			$\log W = -5.0173 + 3.085 \log TL$
J31 OK Grand L.			$\log W = -5.1755 + 3.403 \log TL$
S472 AL	52163 buffalo	51-419 mm	$\log W = -4.79 + 3.00 \log TL$
	549 buffalo	420-700 mm	$\log W = -5.62 + 3.30 \log TL$

	Length	No.	Mean	Range	Type
B346 AZ Roosevelt L.	380-679	956	2.40	-	K(SL)
M302 IA	30-560 mm	287	3.20	1.78-4.68	K(SL)
C36 IA	240-940 mm	47	3.32	2.31-4.68	K(SL)
S472 AL	51-139	50416	1.39	1.25-2.74	K(TL)
	140-700	2293	1.55	1.25-2.80	K(TL)
C89 IA Ruth's P.	356-740	118	1.50	1.08-1.86	K(TL)
S225,226,228,230,330,					
393 SD Ft. Randall	178-640	580+	1.61	1.39-1.75	K(TL)
N94,115, S227,229,					
276 SD Gavins	127-550	255	1.50	1.41-1.88	K(TL)
F100,105,124,125,					
SD Oahe	102-700	1015	1.66	1.52-1.77	K(TL)

The condition factors improved somewhat in Gavins Point Reservoir, South Dakota, being about 1.41 in the first 2 years of impoundment and 1.69-1.88 in the 4th and 5th year. An increase in the average lengths also occurred in the time as most of the bigmouth buffalo were only 129-203 mm long the first year. General increase in condition factor with increase in length was reported by S230, S276, S226 but was not evident in S229, S228, and M302. The computed length-weight relationship suggests an increase in condition with length in J31 but not in J90 and S225. No sex differences in condition factors were noted in S228, S229, S276, and S226. Females had higher values in F100, F124 and S230 but males had higher values in F105 and F125 and probably none of these differences is statistically significant.

	No.	TL	Range	No.	Weight	Range
Age 0-June						
J90 SA Pasque L.	+	-	13-28	-	-	-
S227 SD Gavins Point	75	30	-	-	-	-
H398 IA Coralville L.	+	58	-	+	2	-
Age 0-July						
J90 SA Pasque L.	+	-	20-46	-	-	-
B2, B3 IA lakes	27	46	25-64	-	-	-
Age 0-Aug.						
J90 SA Pasque L.	+	-	36-69	-	-	-
Qu'Appele R.	13	76	-	-	-	-
B2, B3 IA lakes	51	76	61-91	-	-	-
S227 SD Gavins Point	50	-	135-168	-	-	-
Age 0-Sept.						
F105 SD Oahe	9	168	-	-	-	-
Age 0-Oct.						
F35 IA Fairport	+	-	64-127	-	-	-
H398 IA Coralville L.	+	-	51-152	-	-	-
S311 AL ponds	+	-	102-127	-	-	-
H150 OK Tenkiller L.	3	287	-	-	-	-
Age 0-Nov.						
Y7 Israel	-	-	-	+	41	-
F32 IA Fairport	+	-	-152	-	-	-
B237, J65 AR ponds	+	-	102-203	+	-	-680
Age 0-Dec.						
G74 KS El Dorado L.	+	246	-	+	213	-
Age I						
E3 MN	10	249	193-290	-	-	-
M503 IA Coralville L.	1	262	-	1	254	-
U11 SD Oahe L.	54	279	-	54	363	-
Y7 Israel	-	-	-	+	576	-
S504 AR surface water	+	282	-	+	644	-
S311 AL ponds	-	-	-	+	604	-
S504 AR well water	+	297	-	+	848	-
S16 TN Reelfoot L.	6	335	279-368	6	698	454-853
S504 AR ponds Dec.	+	386	-	+	1533	-
B237 AR ponds	-	-	-	+	-	-1361
Age II						
U11 SD Oahe L.	180	348	290-470	180	721	408-1814
E3 MN	10	358	290-411	-	-	-
S16 TN Reelfoot L.	18	386	351-429	18	924	707-1420

	No.	TL	Range	No.	Weight	Range
I10 IA Snyder Bend	1	394	-	1	481	-
S504 AR surface water	+	401	-	+	1216	-
G74 KS El Dorado L.	+	401	-	+	924	-
M503 IA Coralville L.	15	411	373-480	15	1084	771-1800
S311 AL ponds	-	-	-	+	-	1075-1642
B237, J65 AR ponds	+	-	353-486	+	-	726-2268
S504 AR well water	+	467	-	+	1945	-
Age III						
J90 SA lakes	7	-	155-229	-	-	-
H150, J25 OK Illinois R.	9	287	246-330	9	313	-
S16 TN Reelfoot L.	34	424	284-483	34	1139	626-2041
I10 IA Missouri R.	2	432	381-483	2	1315	907-1729
J42 OK Ft. Gibson L.	1	452	-	-	-	-
M503 IA Coralville L.	3	460	432-505	3	1729	1225-2105
E3 MN	11	462	424-531	-	-	-
U11 SD Oahe L.	30	483	394-538	30	2068	1043-2268
G74 KS El Dorado L.	+	508	-	+	2327	-
B237 AR ponds	+	-	498-592	+	-	1846-3516
Age IV						
W192 SD Lewis & Clark L	+	241	-	-	-	-
J90 SA lakes	13	-	279-386	-	-	-
H150, J25 OK Illinois R.	29	447	394-531	29	1438	-
S16 TN Reelfoot L.	20	455	411-495	20	1415	1248-2041
I10 IA Missouri R.	1	470	-	1	1615	-
M503 IA Coralville L.	18	505	460-584	18	2014	1361-2908
E3 MN	8	516	472-594	-	-	-
U11 SD Oahe L.	163	526	475-632	163	2636	2404-2812
G74 KS El Dorado R.	+	579	-	+	3316	-
G40 LA Cross L.	1	584	-	1	3629	-
Age V						
J90 SA lakes	29	-	259-434	-	-	-
W192 SD Lewis & Clark L.	+	356	-	-	-	-
I10 IA Missouri R.	1	462	-	1	934	-
H150, J25 OK Illinois R.	11	513	478-541	11	2164	-
S16 TN Reelfoot L.	25	528	460-584	25	2391	1502-3742
U11 SD Oahe	165	554	500-653	165	2989	2540-3719
M503 IA Coralville L.	5	582	533-663	5	2435	2313-5534
E3 MN	6	587	531-655	-	-	-
G74 KS El Dorado L.	+	638	-	+	3883	-
Age VI						
J90 SA Pasque L.	1	315	-	-	-	-
W192 SD Lewis & Clark L.	+	417	-	-	-	-
H150 OK Illinois R.	+	526	-	-	-	-
S16 TN Reelfoot L.	12	597	566-640	12	3334	3175-4450
U11 SD Oahe L.	14	615	538-678	14	4482	3175-4990
E3 MN	17	653	556-719	-	-	-
Age VII						
J90 SA lakes	26	-	325-467	-	-	-
W192 SD Lewis & Clark L.	+	483	-	-	-	-
S16 TN Reelfoot L.	3	668	640-688	3	6241	6124-6350
U11 SD Oahe L.	13	681	627-701	13	6033	5806-6124
E3 MN	9	688	655-749	-	-	-
G40 LA Cross L.	1	762	-	1	6804	-

	No.	TL	Range	No.	Weight	Range
Age VIII						
J90 SA lakes	15	-	358-526	-	-	-
E3 MN	9	696	655-780	-	-	-
G74 KS El Dorado L.	1	716	-	-	-	-
Age IX						
J90 SA lakes	10	-	409-516	-	-	-
U11 SD Oahe L.	5	744	732-754	5	8237	7800-8528
E3 MN	4	757	719-780	-	-	-
G40 LA Cross L.	2	787	-	2	7938	7711-8165
Age X						
J90 SA lakes	12	-	429-503	-	-	-
E3 MN	4	882	811-967	-	-	-
Age XI						
J90 SA Pasque L.	1	467	-	-	-	-
E3 MN	9	884	811-1001	-	-	-
Age XII						
J90 SA lakes	5	-	566-681	-	-	-
G40 LA Cross L.	1	914	-	1	21320	-
E3 MN	3	955	939-967	-	-	-
Age XIII						
J90 SA lakes	2	615	607-622	-	-	-
Age XIV						
J90 SA lakes	2	660	607-719	-	-	-
G40 LA Cross L.	2	978	940-1016	2	23587	22680-24494
Age XV						
J90 SA lakes	2	630	599-663	-	-	-
Age XVII						
J90 SA lakes	2	724	711-737	-	-	-
Age XVIII						
J90 SA lakes	2	688	676-703	-	-	-
Age XIX						
J90 SA lakes	2	734	726-742	-	-	-
Age XX						
J90 SA lakes	1	757	-	-	-	-

					Average TL at each annulus							
	No.	1	2	3	4	5	6	7	8	9	10	11	12
F13 WI	1130	147	297	388	467	526	577	579					
E3, E2 MN	199	147	267	373	447	523	587	640	686	739	793	831	900
J25 OK Illinois R.	47	150	246	328	414	478							
J42 OK Ft. Gibson	20	135	269										
J31 OK Grand L.	107	203	305	394	470	533	587	655	711				
J45 OK Verdigris P.	12	244	333	414	478	531	574	645					
C270 MN Mississippi R.	16	130											

				Average TL and increment at each annulus						
	1	2	3	4	5	6	7	8	9	10
I10 IA Missouri R.	135	244	330	368	388					
Increment	135	109	109	86	74					
Number	5	5	4	2	1					
S225, 226, 228, 230, 393										
SD Ft. Randall L.	150	290	371	442	472	488	531	556		
Increment	150	145	99	79	53	53	51	53		
Number	325	265	65	17	11	4	3	2		

	Average TL and increment at each annulus									
	1	2	3	4	5	6	7	8	9	10
S229, S276, N94, N115										
SD Gavins	130	231	312	378	442	538				
Increment	130	104	89	64	61	43				
Number	176	122	54	46	29	5				
F100, 105, 124, 125										
SD Oahe	157	257	333	373	442	480				
Increment	157	119	86	81	76	107				
Number	403	131	47	7	4	2				
M507 IA Coralville	175	328	388	432	467	513	584	678	688	703
Increment	175	140	58	43	33	41	33	20	13	18
Number	236	173	163	152	120	61	15	5	2	1
B346 AZ Roosevelt L.	208	361	455	503	538	569	597	582		
Increment	208	152	94	48	36	30	28	18		
Number	490	490	490	490	488	412	230	76		

The growth of bigmouth buffalo in Saskatchewan lakes appears to be slower than in more southern parts of the range, but other regional differences are not clear. Growth was computed on the basis of direct proportion between scale and body growth. Characteristics of the scales and annuli were described in J90. Annuli form from June 12-19 in Saskatchewan (J90), from June 4-17 in Iowa (M507), and from the last week of May through the first week of June in South Dakota (S229, S228, S276, S225, S226). Growth was slow in a very abundant year class (J90). Growth increased following a winter kill (M507). About 25% of the annual growth was completed by June 27; 50% by July 18; 75% by August 25; and 100% by September 16 in Iowa (M507). The females grew slightly faster and lived longer than the males (U11). Most of the large buffalo were females (J90). Growth was better in ponds with well water (with higher mineral content) than in surface water in Arkansas (S504).

Buffalo of 28-85 g increased 0.22 to 0.5 g per day from November to March in Israel and 1.4 g per day from March to September (Y7). Age I buffalo increased 25 mm per 16-20 days (S225, S229) and age III buffalo increased 25 mm per 24 days (S228) in the summer in South Dakota.

Bigmouth buffalo mature at age III in Gavins Point Reservoir, South Dakota (N94) and at age III to age IV in Ft. Randall (S228, S230). They spawn late May and early June in South Dakota (F105, S393, S228, S227).

In Saskatchewan (J90) the smallest mature male was 305 mm FL and female was 318 mm and the largest immature male was 445 mm and female, 470 mm. Bigmouth buffalo spawn from May 25 to June 15 in Saskatchewan.

Bigmouth buffalo spawn in shallow water with much splashing. Most of the eggs adhere to vegetation. Spawning may start at 14.4C but most of it occurs at 15.6 to 18.3C (C326, S390). Eggs spawned at 26.7C were still satisfactory (S390). Spawning is stimulated by the addition of fresh-water (C326, S390, J90) perhaps due to dilution of waste products or of hormone-like secretions which repress reproduction (S390). A 665 mm female, 6532 g, had 750,000 eggs (J90), and 400,000 eggs have been reported as the average per female (K118, M426).

Bigmouth buffalo feed mostly upon bottom organisms and zooplankton, on cladocerans, copepods, and chironomids (D161, M425, M506, S357, J90). Algae rarely contribute significantly to the food. The gill rakers are well adapted to retain entomostracans (J90). Some adults had 1,250,000 entomostracans per stomach (M425). Feeding occurred through the winter under the ice in Iowa but the volumes of food were smaller than in summer (M425).

Mesh size (bar)	25	38	51	64	76	102	114
			Sizes in gill net mesh				
L216 LA							
No.	3	9	15	10	6	272	17
Mean TL	267	257	538	445	493	599	635
Weight	1406	318	3357	1089	2313	4128	5352

Nylon caught larger buffalo than cotton or linen gill nets (L126).

Bigmouth buffalo are poor bioassay animals because they are susceptible to parasites, hard to hold in tanks, grow too fast to be of proper size much of the year, and are too excitable (G156).

BLACK BUFFALO, *Ictiobus niger* (Rafinesque)

Southern Minnesota to Michigan and south to Mexico (E119).

C30, M302 IA 3 fish TL = 1.18 SL

C30 IA		Standard length-weight relationship				
SL	380	380	390	420	550	600
Weight	1673	1701	1701	2268	5557	6747

			Weight		K(TL)	
	TL	No.	Mean	Range	Mean	Range
H156 OK Canton L.	394	7	871	766-989	1.44	1.33-1.55
	421	4	1048	980-1148	1.41	1.36-1.50
	445	12	1288	1034-1492	1.50	1.27-1.75
	465	10	1483	1310-1651	1.50	1.36-1.61
	490	5	1855	1673-2001	1.52	1.50-1.66
	521	3	2172	2068-2209	1.55	1.44-1.61
	554	1	2327	-	1.36	-
	599	1	3565	-	1.66	-
T113 OH	711	1	7031	-	1.96	-
M302 IA	929	2	14601	-	1.82	-

H156 log W = -5.0997 + 2.80 log TL (on the basis of the K(TL) values it would appear that the slope should be over 3.0)

E28 TN Chickamauga 47 buffalo average K(SL) 2.8

C36, M302 IA 9 buffalo 380-750 mm SL average K(SL) 3.03 range 2.80-3.34

	No.	TL	Range	No.	Weight	Range
Age 0-Sept.						
T113 OH	+	-	38-102	-	-	-
Age 0-Oct.						
H150 OK Tenkiller R.	6	302	-	-	-	-
Age I						
J25 OK Illinois R.	1	140	-	1	42	-
G74 KS El Dorado L.	+	310	-	+	397	-
Age II						
J25 OK Illinois R.	47	213	183-251	47	130	-
E28 TN Chickamauga L.	43	229	-	-	-	-
J31 OK Ft. Gibson L.	11	312	-	-	-	-
H127 OK Wister L.	25	348	-	-	-	-
G74 KS El Dorado L.	+	368	-	+	794	-
H156 OK Canton L.	18	427	-	-	-	-

	No.	TL	Range	No.	Weight	Range
Age III						
J25 OK Illinois R.	2	292	274-307	2	312	-
H127 OK Wister L.	56	429	-	-	-	-
G74 KS El Dorado L.	+	447	-	+	1247	-
H156 OK Canton L.	24	475	-	-	-	-
Age IV						
J25 OK Illinois R.	7	384	328-424	7	536	-
H127 OK Wister L.	81	488	-	-	-	-
H156 OK Canton L.	1	523	-	-	-	-
G74 KS El Dorado L.	+	544	-	+	2325	-
Age V						
H127 OK Wister L.	12	551	-	-	-	-
G74 KS El Dorado L.	+	627	-	-	-	-
Age VI						
J25 OK Illinois R.	7	483	452-549	7	2038	-
H127 OK Wister L.	1	635	-	-	-	-
Age VII						
J25 OK Illinois R.	3	602	584-630	3	3175	-
H127 OK Wister L.	1	678	-	-	-	-
Age VIII						
J25 OK Illinois R.	2	635	-	2	4323	-
Age XXIV						
A2 IL	-	-	-	1	15422	-

				Average TL at each annulus					
	No.	1	2	3	4	5	6	7	8
E28 TN Chickamauga L.	43	135	211						
J25 OK Illinois R.	69	91	201	284	353	427	483	544	589
Poteau R.	+	185	292	381	447				
Wister L.	+	236	340						
H150 OK Illinois R.	92	84	198	325	381	452	493	577	
J31 OK Ft. Gibson L.	7	122	208						
E61 OK Salt Creek	3	109	193						
J45 OK Verdigris R.	34	152	257	351	452	526	574		
tributaries	6	135	193	229	269				
B183 OK Canton L.	66	190	277	315					
H156 OK Canton L.	43	208	330	432	480	523			

L216 LA			Sizes in gill nets			
Mesh size (bar)	25	51	64	76	102	114
No.	3	14	5	3	43	14
Mean TL	366	401	457	533	632	665
Weight	907	816	1724	3175	4944	6124

HYBRID BUFFALO

Crosses between *I. niger* X *I. cyprinellus*, *I. niger* X *I. bubalus*, and *I. bubalus* X *I. cyprinellus* were secured by hormone injections, hand-stripping, and artificial hatching methods (S504). Natural spawning after hormone treatment also produced *I. niger* X *I. cyprinellus* offspring. All crosses showed better average growth for the first 220 days than did control lots of smallmouth and of bigmouth buffalo. Natural hybrids between smallmouth and bigmouth buffalo are also reported from Ohio (T113).

SPOTTED SUCKER, *Minytrema melanops* (Rafinesque)
Larger streams, southern Minnesota to Pennsylvania and south to Texas and Florida (E119). Intolerant of turbid waters and industrial pollutants (T113).

TL	No.	Weight Mean	Range	Citation
76	422	5	4-9	S472 AL
102	161	9	6-25	S472
127	26	21	11-45	S472
152	24	43	15-82	S472
178	97	64	41-91	S472
178-202	5	64	64-68	V44 LA
203	277	82	43-141	S472
203-228	3	82	77-82	V44
229	296	127	77-227	S472
229-253	1	132	-	V44
254	203	177	113-268	S472
279	182	227	145-367	S472
305	246	322	227-454	S472
330	134	395	308-530	S472
356	88	472	340-590	S472
381	45	571	331-680	S472
406	10	731	544-766	S472
432	9	943	635-1179	S472
462	1	1361	-	J28 OK

	No.		K(TL) Mean	Range
J28 OK Lower Spavinaw L.	388	log W = -5.753 + 3.341 log TL	1.14	-
Upper Spavinaw L.	64	log W = -5.408 + 3.339 log TL	1.24	-

(The intercepts reported in J28 were assumed to be -3.xxx rather than -8.xxx)

	No.		K(TL) Mean	Range
S472 AL 51-150 mm	610	log W = -4.99 + 3.00 log TL	1.01	-
151-450 mm	1611	log W = -4.93 + 2.99 log TL	1.07	1.00-1.22

W142 IL Little Grassy L.   35 182-252 mm   mean K(SL) = 2.00, the value decreasing with age and length.

	Age 0	No.	
H76 WV Back C.		17	41 mm FL

	No.	TL	Range	No.	Weight	Range
Age 0-Sept.						
T113 OH	+	-	51-102	-	-	-
Age 0-Oct.						
H150 OK Tenkiller L.	12	170	-	-	-	-
V44 LA Springhill L.	9	201	185-229	9	77	62-130
Age I						
H150 OK Tenkiller L.	1	137	-	-	-	-
C270 MN Mississippi R.	38	157	119-213	-	-	-
J42 OK Ft. Gibson L.	6	246	-	-	-	-
Age II						
H150 OK Tenkiller L.	1	206	-	-	-	-
J42 OK Ft. Gibson L.	1	292	-	-	-	-
C270 MN Mississippi R.	31	307	257-335	-	-	-

	No.	TL	Range	No.	Weight	Range
Age III						
H150 OK Tenkiller L.	2	277	-	-	-	-
C270 MN Mississippi R.	4	371	343-391	-	-	-
Age IV						
C270 MN Mississippi R.	2	411	399-421	-	-	-
Age V						
C270 MN Mississippi R.	8	427	396-445	-	-	-
Age VI						
C270 MN Mississippi R.	2	432	419-442	-	-	-

		Average calculated SL at each annulus		
	No.	1	2	3
E3 MN	1	65	192	
W142 IL Little Grassy L.	35	86	190	227
Increments		86	104	35
Number		35	31	23

		Average calculated TL at each annulus					
	No.	1	2	3	4	5	6
J42 OK Ft. Gibson	2	51	147				
C270 MN Miller L.	76	48	170	297			
E61 OK Salt Creek	3	112	206				
J45 OK Verdigris R.	32	142	206	254	295		
J27 OK Lower Spavinaw L.	45	107	234	282	320	368	
J28 OK Lower Spavinaw L.	388	130	259	323	373	419	
Upper Spavinaw L.	64	155	287	386	414	439	
J101 OK L. Eucha	710	152	274	348	388	409	
J101 OK Spavinaw L.	762	150	272	328	351	373	
C270 MN Mississippi R.	85	61	203	284	351	396	414
Increments		61	135	112	79	51	33
Number		85	47	16	12	10	2

Growth rates were calculated on a direct-proportion basis except in J28 when a Fraser-type modification was used. Of 139 spotted suckers in a Kentucky stream (M407) 15.1% had scale aberrations, mostly on the right side and above the lateral line.

Dwarf males may breed at 150 mm (T113). In Oklahoma (J28) spotted suckers spawned in late April and May at 15-18C on riffle areas above pools. Eggs hatched in 7 to 12 days.

The red blood cell count of two spotted suckers was 1,220,000 ± 240,416 cells per cu. mm (S151).

L216 LA		Sizes in gill net meshes		
Mesh size (bar)	25	38	51	64
No.	2	11	16	19
Mean TL	445	455	447	460
Range	432-445	394-495	406-470	432-495

SILVER REDHORSE, *Moxostoma anisurum* (Rafinesque)

Eastern Manitoba to St. Lawrence River in Quebec and south to northern Alabama and Missouri (E119, T113).

	SL range	Number	FL/SL	TL/SL	TL/FL
M394 IA	100-199	5	1.177	1.312	1.114
	200-299	21	1.169	1.319	1.129
	300+	15	1.151	1.291	1.122
C22 MN	300-399	4	1.120	-	-
	400-449	8	1.108	-	-
	450-570	4	1.102	-	-
	300-570	12	-	1.140	-

M395 IA   190 redhorse   76-559 mm $\log W = -4.263 + 3.124 \log TL$
Mean K(TL) ranged from 1.08 in June to 1.43 in May.

M395 IA Des Moines R.

Age group	I	II	III	IV	V	VI	VII	VIII	IX
Mean TL	109	168	302	348	391	452	498	513	518
Number	7	31	41	71	21	12	5	7	1

	No.	\multicolumn{10}{c}{Mean TL at each annulus}								
		1	2	3	4	5	6	7	8	9	10
P113 MO St. Francis R.	37	114	201	269	312	351	386	414	437	455	462
Meramec R.	32	94	178	269	338	399	439	467			
Gasconade R.	41	94	170	262	333	406	455	457			
M395 IA Des Moines R.	196	109	196	272	328	378	432	470	505	511	
Increment		109	84	69	64	58	51	43	43	46	
Number		196	189	158	117	46	25	13	8	1	
Weight		13	80	209	413	644	967	1261	1579	1656	
Increment, g		13	67	118	200	263	308	331	381	417	

In a Kentucky stream 2 of 4 silver redhorse had aberrant scalation (M407). Growth of silver redhorse in M395 was computed with a 32 mm intercept correction.

Silver redhorse mature at age V (M395) and spawn in early May when water temperatures are about 13.5C (M395).

M395 IA   3 females   338-490 mm TL   14910-36340 eggs

Chironomids, mayflies and caddisflies were the principal foods (M395).

BLACKFIN SUCKER, *Moxostoma atripinne* Bailey

		\multicolumn{3}{c}{SL in April-May}		
		No.	Mean	Range
B126 KY Green R.				
Age I		20	55	40-67
Age II Males		8	81	72-88
Females		1	85	-
Age III Males		3	110	105-115
Females		8	122	106-128

In Kentucky, blackfin suckers spawn in April at temperatures of 12-18C on gravel riffle areas (B126). Males mature at age II, females at III.

SHORTHEAD REDHORSE, *Moxostoma breviceps* (Cope)
  Ohio River drainage in Ohio, Pennsylvania, to northern Alabama (T113, E119). Particularly susceptible to industrial pollutants (T113).

W132 TN						
TL	203	229	254	279	330	406
Mean weight	91	136	181	204	303	545
Number	3	1	5	2	3	1
T113 OH	51–102 mm in Oct. of first year					

  Of 56 shorthead redhorse in a Kentucky stream, 51.8% had aberrant scalation (M407).
  Diptera larvae, Cladophora, and detritus made up the food of 5 shorthead redhorse in a Kentucky stream (M404).

RIVER REDHORSE, *Moxostoma carinatum* (Cope)
  Nebraska to St. Lawrence River and south to northern Georgia, Alabama and Oklahoma (E119, T113).

TL	No.	Weight Mean	Weight Range	Citation
127–151	4	32	–	W132 TN
152–177	+	41	–	P113 MO
	3	45	–	W132
178–202	+	64	–	P113
	1	136	–	W132
203–228	+	95	–	P113
229–253	+	136	–	P113
254–278	1	136	–	S472 AL
	+	186	–	P113
279–304	+	245	–	P113
305–329	2	318	–	S472
	+	313	–	P113
	1	454	–	W132
330–355	+	399	–	P113
	3	408	–	S472
356–380	5	472	454–499	S472
	+	494	–	P113
381–405	5	590	576–635	S472
	+	609	–	P113
406–431	+	735	–	P113
	3	816	–	S472
432–456	+	880	–	P113
	5	898	862–953	S472
457–482	1	816	–	W132
	4	953	862–985	S472
	+	1038	–	P113
483–507	3	1148	–	S472
	+	1174	–	P113
508–532	+	1420	–	P113
	7	1433	1406–1460	S472
533–558	6	1620	1533–1814	S472
	+	1638	–	P113
559–583	1	1814	–	S472
	+	1878	–	P113

TL	No.	Weight Mean	Weight Range	Citation
		Weight		
584-609	+	2141	-	P113
610-634	4	2291	2268-2300	S472
	+	2427	-	P113
635-659	+	2740	-	P113
	2	2994	2631-3311	S472
660-685	+	3071	-	P113
737	1	4763	-	T113 OH

S472 AL	52 redhorse	250-640 mm	log W = -4.72 + 2.88 log TL
P113 MO			log W = -4.829 + 2.949 log TL
S472 AL	51 redhorse	250-640 mm	K(TL) = 1.07, range 0.83-1.22

TL at capture

Age Group	II	III	IV	V
H150 OK Illinois R.	279	340	394	500
J42 OK Ft. Gibson			391 (2 fish)	

Calculated TL at each annulus

	No.	1	2	3	4	5	6	7	8	9	10	11	12
P113 MO													
Gasconade R.	85	43	122	193	274	315	386						
St. Francis R.	21	61	135	206	264	297	351	414	452	503	531	559	597
Meramec R.	54	48	142	241	345	424	493	528	597	612	635		
E61 OK Salt C.	4	137	203	274	335								
J27 OK													
L. Spavinaw	1	64	254	376									

GRAY REDHORSE, *Moxostoma congestum* (Baird and Girard)
Rio Grande valley and vicinity, Texas (E119).

D179 TX   290 redhorse   191-514 mm SL   K(SL) 2.23   range 1.47-3.80
After thinning shad population, 7 redhorse, 280-336 mm K(SL) 2.52
range 1.91-2.82

BLACK REDHORSE, *Moxostoma duquesnei* (LeSueur)
Southern Minnesota to St. Lawrence near Montreal and south to northern
Alabama and Oklahoma (E119, T113).

B242 MO Big Piney R.   200 redhorse   TL = 1.227 SL, no trend with length
        Niangua R.   102 redhorse   TL = 1.234 SL, no trend with length

W132 TN											
TL	356	457	483	508	559	584	610	686			
Mean weight	318	680	907	907	1255	1701	1780	2404			
Number	1	1	1	2	3	2	4	1			
P113 MO											
TL	152	178	203	229	254	279	305	310	356	381	406
Mean weight	40	60	88	125	170	227	295	378	471	578	703

B242 MO Niangua R. 1619 redhorse, 60–290 mm $\log W = -4.58 + 2.94 \log SL$
   Big Piney R.   1775 redhorse, 75–330 mm $\log W = -4.59 + 2.95 \log SL$
B239 MO Niangua R.   599 redhorse $\log W = -4.58 + 2.8 \log TL$
   Big Piney R.   538 redhorse $\log W = -4.24 + 2.7 \log TL$
P113 MO $\log W = -4.977 + 2.998 \log TL$

TL at capture

| | B242 MO (superseding B239 data) | | | | H150 OK |
| | Niangua R. | | Big Piney R. | | Illinois R. |
Age group	No.	Mean	No.	Mean	Mean
I	105	117	115	147	-
II	285	157	202	196	-
III	298	198	317	249	325
IV	360	231	402	279	363
V	259	262	375	302	-
VI	192	287	267	323	-
VII	92	307	73	345	-
VIII	27	307	18	363	-
IX	6	305	1	356	-
X	1	323	-	-	-

Calculated TL at each annulus

	No.	1	2	3	4	5	6	7	8	9	10
P113 MO statewide	3078	89	165	236	279	307	325				
headwaters	1596	89	152	218	262	295	323				
middle	1364	89	165	236	282	315	325				
lower	115	97	198	284	315						
poorest station	-	66	122	190	234	259	279	300	302	328	345
best station	-	99	234	373	394	343	353	381	394	394	
J101 OK L. Eucha	109	119	284	384	484	480					
B242 MO Niangua R.	1625	84	127	173	213	244	269	287	297	302	325
Increment		84	43	46	43	30	25	23	25	5	13
Number	1625	1520	1235	937	577	318	126	34	7	1	
MO Big Piney R.	1775	94	168	229	277	290	310	320	325	338	
Increment		94	74	64	48	15	18	13	5	10	
Number	1770	1655	1453	1136	734	359	92	19	1		
P71 MO Black R.											
Increments		89	86	61							
Clearwater L.											
1948 increment		91	147	114	81						
Increments (1953)		-	99	58	41						

The validity of annuli was demonstrated by B242 through the agreement in growth and abundance of year classes in different years of collection. A true annulus was characterized by a break in uniform pattern of circuli and a slight distortion of the radii on the anterior field, cutting over on the lateral fields, and a more or less translucent band on the posterior field. False annuli, with cutting over on one side only, were found on 2.6% of the scales from Niangua River and on 1.6% from Big Piney River (B242). Of 24 black redhorse taken in a Kentucky stream 25% had aberrant scalation (M407). In Missouri the first two annuli formed in April or May but those of older fish formed somewhat later (B242). Two lines were needed to describe the body-scale relationship

(B239, B242), and lengths at various annuli in these two papers were computed using the observed body-scale relationships.

Growth appears to be faster downstream than in the headwaters (P113). Females grew slightly faster than males in Big Piney River but not in the Niangua (B242). There was no evidence of growth compensation (B242).

Both males and females matured at age III with minimum lengths for males at maturity of 150 mm and 200 mm SL in Niangua and Big Piney Rivers respectively and for females, 170 mm and 210 mm (B242). They spawn in late April or early May at surface temperatures of 13-23C (B239, B242). A correlation between spawning time and full moon in April was reported (B242). Males defend territories on the riffle areas in 150 to 600 mm of water and usually two males accompany each female at spawning (B239, B242).

B242 MO	4 females	170-180 mm SL	1343 eggs per female
	12 females	200-230 mm SL	1661 eggs per female
	9 females	240-260 mm SL	2402 eggs per female
	3 females	270-280 mm	3890 eggs per female
	2 females	310-320 mm	5666-6005

The young usually remain in relatively quiet pools; years with high water in May usually produce a year class of low abundance (B242).

Black redhorse usually feed in schools, with less schooling at about 150 mm SL than when smaller or larger (B242). Most feeding is in the early hours of night and none occurs during spawning (B242). Small black redhorse (under 75 mm) feed mostly on phytoplankton and some zooplankton, but larger redhorse feed mostly on Diptera, cladocerans, and copepods and a few other insects and invertebrates (B239, B242).

GOLDEN REDHORSE, *Moxostoma erythrurum* (Rafinesque)

Minnesota to southern Ontario south to northern Georgia and northeast Texas (E119, T113).

	SL	No.	FL/SL	TL/SL	TL/FL
M395 IA	0-99	13	1.165	1.286	1.104
	100-199	27	1.160	1.290	1.112
	200-299	70	1.149	1.279	1.113
	over 300	3	1.127	1.249	1.109
L110 IL	74-272	72	-	1.22	-

	W132 TN		P113 MO	S472 AL		
TL	No.	Mean	Mean	No.	Mean	Range
51	-	-	-	3	0.3	.03-0.6
76	-	-	-	99	8	0.1-11
102	-	-	-	165	14	8-29
127	-	-	-	79	26	1-57
152	-	-	42	33	54	34-113
178	5	62	65	41	77	42-136
203	7	116	96	59	99	57-159
229	10	153	139	79	136	76-198
254	7	196	190	58	198	136-286
279	1	227	250	77	250	181-362
305	9	307	326	121	337	227-471

Weight (header spanning Mean / No. Mean Range columns)

Weight

	W132 TN		P113 MO		S472 AL	
TL	No.	Mean	Mean	No.	Mean	Range
330	7	389	411	126	408	272-524
356	15	499	513	223	507	397-680
381	3	621	630	246	567	-876
406	1	589	763	194	723	576-907
432	-	-	913	111	808	536-953
457	-	-	-	23	984	548-1140
483	-	-	-	10	1140	953-1273
508	1	1270	-	13	1290	1128-
533	1	1814	-	6	1602	1315-

M505 TN   2 redhorse 584-605 mm   2000-2124 g
T113 OH A 660 mm fish weighed 2041 g
L110 IL creeks   72 redhorse 74-272 mm log W = -4.85 + 3.07 log SL
P113 MO   log W = -4.881 + 2.975 log TL
M395 IA Des
 Moines R.   494 redhorse 53-513 mm log W = -4.202 + 3.098 log TL
L110 IL Hutchins C.,
 Clear C.   72 redhorse 74-272 mm   average K(SL) 2.01
E23 TN Chickamauga   21 fish   average K(SL) 2.3
R55 OH   average K(TL) 1.39
S472 AL   420 redhorse 51-200 mm TL K(TL) 1.48 range 1.25-2.44
                    1346 redhorse 203-533 mm TL K(TL) 1.10 range 0.97-1.22
M395 IA Des Moines R.   monthly mean K(TL) from 1.02 in June to 1.19 in May

		TL			
	No.	Mean	Range	No.	Weight
Age 0-Aug.					
L108 IL Big C.	1	99	-	-	-
Age I					
L108 IL Big C.	1	97	-	-	-
L110 IL Hutchins C., Clear C.	29	114	-	29	17
B192 OH Massie C.	+	117	-	-	-
M395 IA Des Moines R.	41	124	-	-	-
J25 OK Illinois R.	250	175	102-302	250	79
D166 KS	29	188	157-218	-	-
Age II					
L108 IL Big C.	40	152	-	-	-
L110 IL Hutchins C., Clear C.	26	180	-	26	74
B192 OH Massie C.	+	180	-	-	-
M395 IA Des Moines R.	30	188	-	-	-
E28 TN Chickamauga L.	41	244	-	-	-
J25 OK Illinois R.	92	272	201-351	92	258
D166 KS	26	277	236-343	-	-
G74 KS El Dorado L.	+	325	-	+	373
Age III					
L108 IL Big C.	11	196	-	-	-
B192 OH Massie C.	+	257	-	-	-
L110 IL Hutchins C., Clear C.	5	259	-	5	201
M395 IA Des Moines R.	254	264	-	-	-
G74 KS El Dorado L.	+	340	-	+	454
J25 OK Illinois R.	111	356	249-419	111	463

	No.	TL Mean	Range	No.	Weight
Age IV					
L108 IL Big C.	2	264	-	-	-
M395 IA Des Moines R.	131	290	-	-	-
B192 OH Massie C.	+	295	-	-	-
J25 OK Illinois R.	65	414	300-455	65	765
Age V					
L108 IL Big C.	14	282	-	-	-
B192 OH Massie C.	+	320	-	-	-
M395 IA Des Moines R.	27	325	-	-	-
L110 IL Hutchin C.	2	345	-	2	748
J25 OK Illinois R.	18	490	414-538	18	1282
Age VI					
L108 IL Big C.	2	295	-	-	-
M395 IA Des Moines R.	3	472	-	-	-
J25 OK Illinois R.	10	566	541-587	10	1814
Age VII					
L108 IL Big C.	1	432	-	-	-
M395 IA Des Moines R.	1	488	-	-	-
J25 OK Illinois R.	3	610	587-625	3	2211

Calculated TL at each annulus

	No.	1	2	3	4	5	6	7	8	9	10	11
P113 MO statewide	2549	79	150	218	272	310	333					
headwaters	682	76	147	190	236	274	302					
middle	1602	79	147	218	274	307	330					
lower stretches	265	79	168	262	333	371	399					
poorest station	-	61	119	165	208	239	274	300	307	335	333	356
best station	-	86	208	333	396	419	434	445	457	376	404	366
R55 OH	-	97	251	330	391	434						
E28 TN Chicka-mauga L.	41	160	216									
J25 OK Illinois R.	549	94	218	305	378	445	521	579	625			
E61 OK Salt C.	5	147	208	269								
J27 OK L. Spavinaw	8	86	262	384	434	472	528	584				
J45 OK Verdigris R.	7	135	213	274	312							
tributaries	48	130	206	279	325							
K71 TN White Oak L.	74	140	226	279	325							
J101 OK L. Eucha	89	145	302	391	434	490						
J101 OK Spavinaw L.	162	122	297	361	427	551						

Calculated weight at each annulus

	1	2	3	4	5	6	7
M395 IA Des Moines R.	5	56	150	240	386	1057	1352
Increments	5	51	91	118	154	354	150
Number	487	446	416	162	31	4	1
L110 IL Hutchins C., Clear C.							
Increments	1.4	16	86	127	163		
Number	62	33	7	2	2		

Calculated TL and increments at each annulus

	1	2	3	4	5	6	7
L108 IL Big C.	51	104	165	224	254	318	406
Increments	51	58	79	51	38	28	25
Number	128	70	30	19	17	3	1
L110 IL Hutchins C., Clear C.	56	119	208	262	315		
Increments	56	64	86	61	53		
Number	62	33	7	2	2		
M395 IA Des Moines R.	84	175	241	279	325	450	488
Increments	84	91	64	56	48	56	18
Number	487	446	416	162	31	4	1
H303 OK Lawtonka L.	160	297	361				
Increments	160	127	48				
Number	12	11	2				
P71 MO Black R. increments	61	94	102	66			
Clearwater L. 1948 increments	76	173	124	84			
1953 increments	-	109	94	38			
H420 TN Watts Bar L.	157	284	368	455	500		
Increments	157	127	89	76	46		
Number	17	17	6	1	1		

Growth of golden redhorse was usually calculated on a direct-proportion basis but B395 used a 25 mm intercept correction. Of 103 golden redhorse in a Kentucky stream, 22.3% had aberrant scalation (M407).

Growth appears to be faster in the few reservoirs in which golden redhorse have been studied (E28, G74, J27) than in most rivers and streams. Growth seems to be slower in headwater streams (P113, L108, L110) than in larger rivers.

In Iowa, males matured at age III and females at age IV (M395). They spawned in mid-May at water temperatures of about 15.5C.

B395 IA Des Moines R.	No. eggs per female = (-33.1 + 0.136 TL) 1000	
19 females 292-399 mm TL	6100-25350 eggs per female	
R55 OH	5000 g female	35000 eggs

Chironomids, mayflies, caddisflies, oligochaetes, and fingernail clams were the principal food items (M404, M395).

NORTHERN REDHORSE, *Moxostoma macrolepidotum* (LeSueur)
Alberta to Hudson Bay to Lake Champlain south to Oklahoma and Arkansas. Not in much of Ohio drainage (E119, T113). Also streams of Delaware and the Chesapeake Bay.

	No.	SL range	FL/SL	TL/SL	TL/FL
C27 MN L. of Woods	9	-99	1.124	1.240	-
	7	200-299	1.114	1.240	-
	163	300-499	1.097	1.184	-
C30 IA	1	250	1.16	1.28	-
M395 IA Des Moines R.	10	-99	1.191	1.309	1.098
	44	100-199	1.154	1.292	1.119
	28	200-299	1.144	1.275	1.114
	10	300+	1.135	1.256	1.107

| | | Weight | | |
TL	No.	Mean of means	Range	Citations
51-75	+	14	-	F105
76-101	+	23	-	F105
102-126	+	27	-	F105
127-151	+	36	-	F105
152-177	+	45	36-54	P112, F100, F105
178-202	+	73	59-91	P112, S227, F100, F105
203-228	+	95	86-113	P112, Oahe, Gavins
229-253	+	141	127-163	P112, Oahe, Gavins
254-278	+	191	168-227	P112, Oahe, Gavins
279-304	1+	249	222-277	P112, C20, Oahe, Gavins
305-329	1+	331	268-431	P112, Gavins, Oahe, C20
330-355	8+	372	349-495	P112, Oahe, Gavins, G135
356-380	25+	530	449-626	T44, P112, Oahe, Gavins, G135, C20
381-405	50+	644	562-762	P112, C20, G135, Oahe
406-431	97+	821	785-912	G135, T44, C20, F105
432-456	126	953	893-1016	G135, C20
457-482	59	1188	1103-1361	G135, C20
483-507	23	1388	1248-1729	E105, G135, C20
508-532	5	1570	1479-1701	G135, C20
533-558	2	2228	-	C20
559-583	2	2554	-	C20
620	1	1871	-	T113

The above data came from Maryland, E105; Minnesota, C20; Mississippi River, G135; Missouri, P112; Ohio, T113; Oklahoma, T44; and the following reservoirs in South Dakota: Oahe, F100, 105, 124, 125; Gavins Point, N94, S227, 229, 276.

	TL	No.	
G42 Upper Mississippi R.	254-546	181	log W = -3.20 + 2.83 log TL
P112 MO	-	-	log W = -4.887 + 2.958 log TL
M395 IA Des Moines R.	91-655	318	log W = -4.042 + 3.021 log TL

	SL	No.	K(SL)
C23 MN 1941-43	200-550	37	1.96
E3 MN 1936-41	229-507	194	2.06
C22 MN L. of Woods	-	201	2.23
C36 IA	250	1	2.26

	TL	No.	K(TL)	Range
M395 IA Des Moines R.	-	318	-	0.91-1.11
N94, 115, S227, 276 SD				
Gavins L.	178-360	282	1.02	0.98-1.11
F100, 105, 124, 125 SD				
Oahe L.	51-406	40	1.09	0.99-1.35
S228 SD Ft. Randall L.	279-406	12	1.27	-

The mean condition factor decreased from 1.11 in 1957 to 1.01 in 1959 in Gavins Point Reservoir and from 1.27 to 0.99 in Oahe Reservoir in the first few years after the reservoirs were impounded. In the Des Moines River, Iowa (M395), the average K factor was low, 0.91, in June and rose to 1.11 by October. No sex or size differences in condition factors were noted at Gavins Point, South Dakota (S276).

	No.	TL	Range	No.	Weight
Age 0-July					
F5 L. Erie	1	-	-	-	-
J42 OK Ft. Gibson L.	1	61	-	-	-
M133 OK Chikaski R.	8	-	38-56	-	-
Age 0-Oct.					
H150 OK Tenkiller L. *M. m. pisolabrum*	+	157	-	-	-
Age I					
J42 OK Ft. Gibson L.	33	155	-	-	-
M395 IA Des Moines R.	154	157	-	-	-
H150 OK Tenkiller L.	1	180	-	-	-
T44 OK Grand L.	6	206	-	6	91
Age II					
R230 Saskatchewan R.	-	91	-	-	-
J42 OK Ft. Gibson L.	6	241	-	-	-
T44 OK Grand L.	3	249	-	3	150
M395 IA Des Moines R.	70	259	-	-	-
E3 MN	3	315	-	-	-
H150 OK Tenkiller L. *M. m. pisolabrum*	+	384	-	-	-
Age III					
R230 Saskatchewan R.	+	122	-	+	57
J42 OK Ft. Gibson L.	7	295	-	-	-
T44 OK Grand L.	5	300	-	5	278
M395 IA Des Moines R.	58	330	-	-	-
E3 MN	10	376	345-391	-	-
H150 OK Tenkiller L. *M. m. pisolabrum*	-	427	-	-	-
Age IV					
R230 Saskatchewan R.	+	297	-	+	284
M395 IA Des Moines R.	27	345	-	-	-
J42 OK Ft. Gibson L.	5	361	-	-	-
T44 OK Grand L.	1	363	-	1	454
E3 MN	19	388	345-452	-	-
Age V					
R230 Saskatchewan R.	+	335	-	-	340
T44 OK Grand L.	2	384	-	2	638
M395 IA Des Moines R.	7	404	-	-	-
J42 OK Ft. Gibson L.	1	406	-	-	-
E3, C270 MN	12	414	381-480	-	-
Age VI					
T44 OK Grand L.	1	411	-	1	794
E3 MN	4	424	394-450	-	-
M395 IA Des Moines R.	1	432	-	-	-
Age VII					
R230 Saskatchewan R.	+	368	-	+	482
E3 MN	2	511	-	-	-
M395 IA Des Moines R.	1	536	-	-	-
Age VIII					
R230 Saskatchewan R.	+	391	-	+	624
E3 MN	2	617	572-660	-	-
M395 IA Des Moines R.	1	655	-	-	-
Age IX					
R230 Saskatchewan R.	+	411	-	+	680
E3 MN	1	660	-	-	-

	No.	TL	Range	No.	Weight
Age X					
R230 Saskatchewan R.	+	429	-	+	851
Age XI					
R230 Saskatchewan R.	+	432	-	+	879
Age XII					
R230 Saskatchewan R.	+	462	-	+	1134

Average calculated TL at each annulus

	No.	1	2	3	4	5	6	7	8	9
E3 MN	101	107	190	264	330	381	427	483	579	615
K33 MN	53	97	173	251	328	381	406	442	503	
S192 MD Potomac R.	+	86	254	378	439					
J42 OK Ft. Gibson L.	27	69	203	315	371	445	480			
H150 OK Illinois R.	479	79	188	328	384	447	538			
P112 MO Salt L.	75	91	173	246	290	343	378			
P113 MO statewide	744	107	193	264	305	335	371			
headwaters	38	107	201	274	318					
middle	479	109	190	259	300	330	361			
lower stretches	227	107	196	267	310	353	391			
poorest mean	+	81	168	236	287	310	343	388	462	503
best mean	+	124	211	279	320	373	404	421		
W192 SD Lewis & Clark L.	+	66	173	259	320	348	406			
P159 MT streams	15	36	94	155	231	279	333	368		
H150 OK Illinois R.										
M. m. pisolabrum	84	94	244	384						

Average calculated TL and increment
at each annulus

	1	2	3	4	5	6	7	8
M395 IA Des Moines R.	117	224	292	343	417	521	589	655
Increments	117	114	74	69	51	51	25	13
Number	319	165	95	37	10	3	2	1
S225, S226, S228 SD Ft. Randall	84	216	264	328				
Increments	84	127	97	112				
Number	55	51	29	7				
S227, S229, S276, N94, N115 SD Gavins	81	224	287	325	356			
Increments	81	145	74	43	25			
Number	256	225	79	48	5			
F104, F105 SD Oahe	71	203	307					
Increments	71	132	140					
Number	11	11	1					

Average calculated weight and increment
at each annulus

	1	2	3	4	5	6	7	8
M395 IA Des Moines R.	15	116	262	422	771	1515	2214	3062
Increment	15	103	154	209	249	408	268	181
Number	319	165	95	37	10	3	2	1

The annulus formed May 16 (S276) to June 1 (S279) in South Dakota.

Growth was computed on a direct-proportion basis except in P112 and M395 which used intercept corrections.

Growth of northern redhorse appears to be lower in the Saskatchewan River (R230) and in Montana streams (P159) than elsewhere. The small amount of data on the shorthead (pealip) redhorse, M. m. pisolabrum, (H150) suggests that is is a rapid-growing subspecies. Growth immediately after impoundment was better than prior to impoundment at Ft. Randall, South Dakota (S225). Increments for the first through third year prior to impoundment were 53, 74, and 58 mm respectively and for the second through fourth year after impoundment were 147, 99, and 109 mm respectively. Age I redhorse grew 25 mm per 16 days (S227) and age II redhorse 25 mm per 39 days (S229, S276) in the growing season in South Dakota.

Some males and females mature at age II (S229, S276) but in other populations not until age III (N94, M395). They spawn in late April in the Des Moines River, Iowa (M395), and in May in Illinois (H8) and Michigan (R260). They spawn on riffles (H8, R260) but on the quieter upper parts than common suckers (R260). The Michigan streams in which they spawn are at least 10 m wide (R260). The fins of females are shorter than those of mature males (R260). Occasional females have small pearl organs (R260).

M133 OK   293 mm SL female   29732 eggs
M395 IA Des Moines R. 9 females 325-460 mm TL 13500-27150 eggs per female
              No. of eggs per female = 1000 (-19.7 + 0.1 TL)

Small redhorse, 90-125 mm, fed on Cladocera and chironomids (N117), large redhorse on benthic insects (M395, R230).

		Sizes in gill net meshes								
Mesh size (stretch)	38	51	64	73	76	83	86	89	95	102
C22 MN L. of Woods No.	2	1	3	2	7	3	5	2	22	427
mean FL	292	419	307	279	351	343	343	406	384	396

=====

BLACKTAIL REDHORSE, Moxostoma poecilurum (Jordan)
Coastal streams of the Gulf of Mexico (E119).

	A40 AL L. Martin		S472 AL Weight			
TL	No.	Weight	No.	Mean	Range	K(TL)
102	-	-	35	15	5-45	1.44
127	-	-	45	21	18-45	1.02
152	-	-	75	45	23-77	1.30
178	4	86	92	59	45-91	1.02
203	12	86	115	86	68-136	1.02
229	14	122	74	122	68-204	1.05
254	2	220	117	172	109-227	1.05
279	-	-	113	249	181-404	1.14
305	1	318	71	295	227-499	1.05
330	-	-	52	390	272-499	1.08
356	4	499	46	513	376-567	1.14
381	3	590	15	635	544-703	1.16
406	-	-	13	780	712-998	1.16
432	-	-	12	930	771-1043	1.16
457	-	-	5	1034	953-1225	1.08
483	-	-	4	1202	1157-1216	1.08
508	-	-	2	1384	1361-1406	1.05

S472 AL   906 redhorse   102-508 mm   log W = -4.90 + 2.97 log TL

TORRENT SUCKER, *Moxostoma rhothoecum* (Thoburn)
Headwaters of the James and Kanawha rivers and, possibly, Potomac River, Virginia (E119).

R12 NY

Age		No.	TL	Range
0-Oct.		9	38	30-46
I April		10	48	41-51
II April,	females	50	69	56-81
	males	54	74	51-91
III April,	males	20	102	79-124
	females	22	112	89-130
IV April,	males	7	119	109-124
	females	26	127	109-145
V April,	males	2	140	-
	females	9	147	130-160
VI April,	females	2	157	145-160
VII April,	females	3	165	-

Three females, 107, 112, and 145 mm TL had 742, 1190, and 1749 eggs respectively (R12).
Plant material constituted 80% of the food with insects also taken (F148).

HUMPBACK SUCKER, *Xyrauchen texanus* (Abbott)
Lower Colorado River (E119).

M340 UT Green R.   419 mm SL (= 488 mm TL)   1000 g   K(SL) 1.36
S146 WY   521 mm TL   weighed 1390 g
D42 CA Colorado R.   4 suckers, 538-587 mm FL, weighed 2177-2540 g
L193 NV   spawning males   538-541 mm FL   2177-2495 g

Larvae 10 mm TL were collected March 16 in Nevada (L193).
In Lake Mead, Nevada, humpback suckers spawn March 1 to April 15, at water temperatures of 12-18C and at depths of 0.6-5 m (M340). In Lake Havasu they spawn in early March at temperatures of 17-22C (D177). Several males herd a female into spawning position near the bottom in water 0.4-2 m deep and with silt over sand bottom. The eggs are then broadcast. Males have pronounced tubercles on the caudal, anal and pelvic fins and on the frontal bone (D42).
Diatoms and filamentous algae constitute the principal foods of humpback suckers (D13, M340).

WHITE CATFISH, *Ictalurus catus* (Linnaeus)
Coastal streams from New Jersey to Florida (H382, T113). Widely introduced in California (M508) and elsewhere (T113). They live in fresh to slightly brackish waters. In streams they are most abundant in areas intermediate between the faster waters preferred by channel catfish and the slow waters with silt bottom preferred by bullheads (T113).

FL	TL	No.	Weight Mean of means	Range	Citations
-	25	2	0.2	-	S472
-	51	77	1.9	1.6-2.2	S472
-	76	25	4.4	2.3-5.4	S472
-	102	9	13	-	S472
102	-	+	14	-	M508
-	127	16	24	14-24	S472
127	-	+	27	-	M508
-	152-177	17	59	45-141	S472, S365
152	-	+	50	-	M508
-	178-202	25+	68	50-95	M38, S365, S472
178-180	-	1+	82	77-82	M140, M508
-	203-228	19+	95	68-136	M38, S365
203-228	-	3+	141	122-177	M508, M140
-	229-253	24+	141	91-181	M38, S365
229-253	-	31+	209	127-277	M508, M140
-	254-278	35+	240	181-322	S365, M38, M242
254-278	-	36+	268	200-372	M508, M140
-	279-304	71+	318	231-454	S472, S365, M38, M242
279-304	-	64+	354	268-504	M508, M140
-	305-329	58	413	240-771	S365, M242
305-329	-	43+	449	349-698	M508, M140
-	330-355	62+	531	272-1043	M38, S365, M242
330-355	-	30+	567	477-699	M508, M140
-	356-380	46+	703	363-953	S365, M38, M242
356-380	-	5+	658	600-731	M508, M140
-	381-405	42+	1007	635-1225	S365, M242, M38
381-405	-	1+	798	794-803	M508, M140
-	406-431	50	1057	680-1497	S472, M242, S365
406-431	-	2+	1057	967-1202	M508, M140
-	432-456	22	1324	1179-1724	S509, M242, S365
432	-	+	1161	-	M508
-	457-482	19	1588	1043-2177	S365
457	-	+	1379	-	M508
-	483-507	11	1724	1406-3538	S365
483-507	-	2+	1941	1620-2500	M508, M140
-	508-532	7	2177	1860-2631	M242, S365
-	533-558	11	2449	1860-3175	S365
-	559	4	3084	2495-3538	S365
-	592	2	2767	-	S365

The above data came from Alabama S472; California M140, M508; Florida M242; Maryland S509; South Carolina S365; and Virginia M38.

M508 CA Pine Flat L.	100-483 mm FL	$\log W = -4.859 + 3.006 \log FL$	
S509 MD	124-440 mm TL	$\log TL = -1.979 + .1689 \log W$	
S472 AL 137 catfish	25-406 mm TL	$\log W = -4.93 + 3.01 \log TL$	
S472 AL 137 catfish	25-406 mm TL	$K(TL)$ 1.30, range 0.97-1.41	

CA Clear L.

Age group	No.	M140 - from preopercles FL Mean	Range	M335 - approximate FL Range
I	2	178	-	-
II	2	226	208-244	152-201
III	15	251	221-323	203-251
IV	22	290	226-353	254-302
V	10	320	267-353	305-353
VI	3	325	323-328	-

S504 AR   Yearling white catfish stocked in May at 178 mm TL, 36 g, reached 259 mm and 195 g in six months.

S365 SC Santee-Cooper L.

Age	0	I	II	III	IV	V	VI	VII	VIII	IX	X	XI
Mean TL	66	124	168	241	310	378	414	434	480	500	462	559
Number	11	6	15	13	30	24	27	34	12	5	1	1

P127 AL   Age II males   998-1270 g   females 816-907 g
S509 MD   a XIV male was 574 mm TL

Calculated FL at each annulus

	1	2	3	4	5	6	7	8	9
M508 CA Clear L.	107	168	226	277	320	348	371	381	401
Sacramento R.	79	132	175	213	251	292	328	351	

Calculated TL and increment at each annulus

	1	2	3	4	5	6	7	8	9	10	11	12	13	14
S365 SC Santee-Cooper	81	137	206	272	325	366	399	437	470	472	559			
Increments	81	56	69	66	56	43	41	36	38	33	64			
Number	168	162	147	134	104	80	53	19	7	2	1			
S509 MD Patuxent R.	117	150	185	211	244	264	300	338	378	424	445	508		
Increments	117	33	36	23	56	20	25	30	41	46	41	38		
Number	469	470	469	306	258	180	14	4	4	4	2	2		
MD millpond (1 fish)	112	127	165	198	239	282	307	335	353	381	424	460	486	516

Ages were determined from rings on preopercles (M140), on dorsal spines (S365) and on vertebrae (S509). Back calculations from the vertebrae were computed on the basis of a curvilinear vertebrae-fish length relationship (S509). Those from spines (S365) assumed a direct-proportion relationship.

The smallest mature female reported is 208 mm TL in South Carolina (S365) and 211 mm FL in California (M140). Age II white catfish produced 2500 to 4000 fry per spawn (P127). In Virginia 2 females, 300-320 mm TL, had 3200 and 3500 eggs (M38). Spawning reaches its peak in South Carolina in June (S365), in the District of Columbia in July (G133), and in California in June and July (M508). The male guards and aerates the eggs (G133) and cares for the young (M508). Eggs hatch in 6-7 days at 24-29C (P127).

Catfish tagged as age II, III, and IV were recaptured as much as 7 years later; age V as much as 6 years later; and older fish as much as 5 years later, indicating a life span of at least 11 years (M335). Two staple-tagged white catfish were caught 10 years, 8 months later (L214). Returns from hydrostatic tags were a bit lower than those from staple and disk danglers, but the results were not substantially different (M335, P80). The rate of exploitation, u, was estimated at 0.05-.33; mean annual survival at 0.43-0.81; mean annual total mortality at 0.19-0.57; instantaneous mortality rate 0.21; and instantaneous fishing mortality rate 0.11 (M335, R280).

White catfish were found to feed on pondweeds, aquatic insects, and predominantly on fish, shad, alewives, etc. (G97, S365, M508). Fish eggs are also eaten in quantities when available (M508).

Summary and bibliography: M508.

BLUE CATFISH, *Ictalurus furcatus* (LeSueur)

Large rivers from South Dakota to Ohio and south into Mexico (E113, T119).

TL	Number	Mean	Weight Range	K(TL)	Citations
25	1313	0.4	0.2-0.5	2.57	S472 AL
-	1	0.5	-	-	A39 AL
51	6	0.9	-	-	A39
-	6132	1.5	0.1-3.2	1.13	S472
76	6121	3.5	0.8-1.5	0.80	S472
102	1970	8.2	3.6-20	0.77	S472
127	1150	19	6-59	0.91	S472
152	2358	28	20-68	0.79	S472
-	1	45	-	-	A39
178	1717	45	19-86	0.81	S472
203	7	68	27-91	-	A38 AL
-	1402	68	30-127	0.82	S472
229	15	95	91-109	-	A38, A39, J36 OK
-	1361	109	64-200	0.91	S472
254	15	136	91-141	-	A38, A39, J36
-	745	145	86-231	0.87	S472
279	18	191	181-227	-	A38, A39, J36
-	606	181	136-340	0.84	S472
305	19	236	181-272	-	A38, A39, J36
-	473	240	200-490	0.85	S472
330	24	318	272-363	-	A38, A39, J36
-	464	304	-454	0.84	S472
356	14	386	272-454	-	A38, A39, J36
-	390	399	286-526	0.88	S472
381	8	508	499-513	-	A39, J36
-	280	494	431-662	0.90	S472
406	10	580	-	-	J36
-	265	617	472-	0.92	S472
432	17	731	680-735	-	A38, A39, J36
-	191	750	544-885	0.93	S472
457	15	871	848-953	-	A39, J36
-	133	902	477-1179	0.95	S472
483	10	1089	1034-1315	-	A39, J36
-	111	1112	653-1565	0.99	S472
508	6	1270	1225-1361	-	A38, A39, J36
-	88	1397	1066-1900	1.07	S472
533	10	1538	1511-1588	-	A39, J36
-	75	1524	1301-1950	1.01	S472
559	11	1800	1588-2041	-	A39, J36

TL	Number	Weight Mean	Weight Range	K(TL)	Citations
-	61	1828	1588-2200	1.05	S472
584	14	2041	2005-2132	-	A39, J36
-	65	2096	1968-2608	1.08	S472
610	5	2467	2336-2585	-	A38, A39, J36
-	28	2513	2327-2872	1.11	S472
635	6	2758	-	-	J36
-	33	2912	2540-3175	1.14	S472
660	5	3166	3017-3266	-	A39, J36
-	29	3320	2890-3969	1.15	S472
686	8	3529	-	-	J36
-	18	3783	3261-4082	1.17	S472
711	9	4314	4082-4400	-	A38, A39, J36
-	21	4372	4082-4958	1.22	S472
737	2	4808	-	-	J36
-	12	4606	4173-5851	1.15	S472
762	3	5883	-	-	J36
-	11	5720	4763-7076	1.29	S472
787	4	6650	6405-6895	-	A39, J36
-	9	6527	4990-7167	1.34	S472
813	2	6876	6491-7258	-	J36, E105 MD
-	7	7244	6523-8210	1.35	S472
838	5	7675	7666-7711	-	A39, J36
-	6	7680	6420-8709	1.30	S472
864	1	8618	-	-	J36
-	8	8523	7668-8959	1.32	S472
889	1	9971	-	-	J36
940	1	12588	-	1.50	S472
965	3	12823	12247-13132	1.43	S472
1016	4	17523	13608-18824	1.67	S472
-	1	18144	-	-	A39
1067	2	20158	20131-20185	1.66	S472
1092	1	20140	-	1.53	S472
1118	1	20866	-	-	J36
1143	1	22635	-	1.52	S472
1194	2	23134	22680-23587	1.36	S472

T113 Ohio reports a 41730 g specimen about 1370 mm long.
S227 SD Gavins Point L.   11 catfish   330-508 mm TL   K(TL) 0.97

	No.	TL	log TL
J36 OK L. Texoma	148	229-690	= -5.118 + 3.399
	27	691-1144	= -5.961 + 3.706
S472 AL	24035	51-305	= -4.91 + 2.94
	2123	330-610	= -5.97 + 3.36
	197	630-1200	= -6.49 + 3.55
H420 TN Watts Bar L.	-	170-660	= -6.226 + 3.437 r = .91
C347 TN Kentucky L.	134	127-838	= -6.657 + 3.196

K133 LA   Length-weight relationship described with a cubic equation showing
rapid increase in weight from 711-813 mm with slower increase
thereafter.

	No.	TL	No.	Weight
Age I				
G155 AR April	-	-	+	32
M314 TX Nov.	-	-	+	68
H420 TN Chickamauga L.	4	226	-	-
H420 TN Watts Bar L.	8	264	-	-
G155 AR Oct.	-	-	500	322
Age II				
C347 TN Kentucky L.	4	229	-	-
H420 TN Chickamauga L.	21	254	-	-
H420 TN Watts Bar L.	7	338	-	-
G155 AR April	-	-	+	313
Aug.	-	-	75	948
Age III				
C347 TN Kentucky L.	16	269	-	-
H420 TN Chickamauga L.	9	287	-	-
H420 TN Watts Bar L.	2	376	-	-
Age IV				
C347 TN Kentucky L.	14	310	-	-
H420 TN Chickamauga L.	4	302	-	-
H420 TN Watts Bar L.	1	437	-	-
Age V				
H420 TN Chickamauga L.	3	348	-	-
C347 TN Kentucky L.	17	366	-	-
H420 TN Watts Bar L.	1	470	-	-
Age VI				
H420 TN Chickamauga L.	4	353	-	-
C347 TN Kentucky L.	9	434	-	-
H420 TN Watts Bar L.	2	516	-	-
Age VII				
H420 TN Chickamauga L.	2	427	-	-
C347 TN Kentucky L.	22	498	-	-
H420 TN Watts Bar L.	1	594	-	-
Age VIII				
C347 TN Kentucky L.	29	617	-	-
H420 TN Watts Bar L.	1	653	-	-
Age IX				
C347 TN Kentucky L.	19	698	-	-
H420 TN Watts Bar L.	1	773	-	-
Age X				
C347 TN Kentucky L.	4	862	-	-

Calculated TL at each annulus from spines

	1	2	3	4	5	6	7	8	9	10	11
J36 OK L. Texoma	145	254	351	442	533	655	769	871	1026	1067	1118
Number	190	189	142	112	68	37	16	4	1	1	1

	1	2	3	4	5	6	7	8	9	10
H420 TN Watts Bar L.	168	239	290	356	409	462	531	587	663	
Increments	168	79	61	64	53	53	71	66	79	
Number	24	16	9	7	6	5	3	2	1	
Chickamauga L.	119	198	239	272	300	330	386			
Increments	119	79	43	33	28	36	43			
Number	47	43	22	13	9	6	2			

	1	2	3	4	5	6	7	8	9	10
C347 TN Kentucky L.	135	198	251	297	356	429	513	582	698	846
Increments	135	74	64	46	58	71	84	53	117	79
Number	134	134	130	114	100	83	74	52	23	4
K133 LA Delta Refuge	190	386	508	638	749	849				
Increments	190	196	114	147	102	94				
Number	57	54	28	24	15	3				

Blue catfish were taken for only 2 years after impoundment in Lewis and Clark Lake and disappeared from other South Dakota reservoirs (W192). An intercept of -269 mm was recorded for the body-pectoral spine relationship (C347).

Blue catfish were found to be able to adjust to rapid transfer from fresh water to 0.69% saline (G86).

Zooplankton was the principal food of blue catfish under 125 mm (D57); immature bottom insects, much organic detritus and a few fish were found in larger blue catfish in the brackish waters in Louisiana (D57, L140). In the Tombigbee River most of the larger blue catfish had eaten fish, but in the Tensaw and Alabama rivers catfish of the same size had eaten mostly invertebrates, particularly crayfish and mayflies (B347).

A 511 mm blue catfish was taken in 64 mm mesh gill net (L216).

BLACK BULLHEAD, *Ictalurus melas* (Rafinesque)
Montana to western New York south to Alabama and Texas (E119, T113).

	Number	Length range SL	TL/SL
L29 IA East L.	13	134-321	1.173
C30, C33 IA	140	40-209	1.179
M302 IA lakes	325	110-407	1.186
J25 OK	128	127-202	1.218
C157 IA Little Wall L.	72	under 100	1.227
	197	101-130	1.212
	149	131-160	1.208
	22	over 161	1.192
F65 IA Clear L.	138	136-155	1.211
	318	156-180	1.203
	142	181-200	1.193
	55	201-250	1.186

TL	SL	No.	Mean of means	Weight Central 50%	Range	Citations of extremes
25-51	21-41	215	0.7	0.3-0.9	0.3-2.0	S472, H421
52-75	42-62	144+	3.0	2.0-4.3	1.0-5.7	S472, H421
76-101	63-83	1096+	9.6	8.5-11.1	5.7-14.2	H421, F105
102-126	84-104	1511+	21	19-26	11-43	C100, L12
127-151	105-125	1219+	35	31-39	14-79	C30, C33
152-177	126-146	958+	62	56-63	25-142	C30, C33
178-202	147-167	1509+	96	95-99	40-213	C30, B150
203-228	168-189	1667+	145	133-156	71-227	B150, C33
229-253	190-211	1394+	207	190-227	133-300	H156, M508
254-278	212-233	1385+	281	252-315	184-454	H156, C33
279-304	234-255	1250+	383	318-425	269-680	C30, C89

| | | | Weight | | | |
TL	SL	No.	Mean of means	Central 50%	Range	Citations of extremes
305-329	256-277	153+	493	463-527	326-630	M302, M302
330-355	278-299	147	613	598-652	488-771	M302, M302
356-380	300-320	108	762	-	680-907	M302, H421
381-405	321-341	34	857	-	680-992	C30, H421
406-431	-	4	1049	-	1021-1077	M302
434	-	1	1106	-	-	M302
467	-	1	1474	-	-	M302

The above data came from Alabama S472; California M509; Colorado C208, L126; Illinois M499; Iowa B150, C30,33,38,39,100-107,157, F31, M130, 246; Minnesota M136; Ohio L12; Oklahoma H156, H21; South Dakota F100,105, 124,125, S225-229; Texas D179; Wisconsin P188.

		No.	Length	
F65 IA Clear L.	1950	66	-	$\log W = -4.258 + 2.86 \log SL$
	1951	559	-	$\log W = -4.049 + 2.801 \log SL$
	1952	257	-	$\log W = -2.848 + 2.285 \log SL$
West end of Clear L.	1952	56	-	$\log W = -3.676 + 2.620 \log SL$
Ventura marsh males		41	-	$\log W = -2.591 + 2.156 \log SL$
females		53	-	$\log W = -2.188 + 1.970 \log SL$
C157 IA Little Wall L.		441	-	$\log W = -4.439 + 2.935 \log SL$
S225 SD Ft. Randall		-	51-305	$\log W = -4.613 + 2.887 \log TL$
P188 WI L. Butte des Morts		-	127-305	$\log W = -4.710 + 2.924 \log TL$
S472 AL		2049	25-127	$\log W = -5.18 + 3.14 \log TL$
H419, 421 OK		-	46-394	$\log W = -5.267 + 3.180 \log TL$
C100 IA Williams L.		123	76-229	$\log W = -5.552 + 3.326 \log TL$
C106 IA ponds		244	76-254	$\log W = -5.791 + 3.414 \log TL$
F31 IA ponds		55	102-229	$\log W = -5.437 + 2.722 \log TL$
J25 OK		128	155-246	$\log W = -4.798 + 2.977 \log TL$
H156 OK Canton L.		802	127-305	$\log W = -5.523 + 2.892 \log TL$
M246 IA ponds		243	76-279	$\log W = -4.995 + 3.060 \log TL$

			K(SL)	
	SL	No.	Mean	Range
C33 IA ponds	40-209	58	1.77	-
C36 IA Clear L.	150-170	3	1.97	1.61-2.19
C38, C157 IA Little Wall L.	71-125	621	2.25	-
L39 IA Red Haw L.	-	13	2.34	-
M302 IA lakes	127-450	325	2.45	1.66-3.13
D179 TX	126-212	66	2.57	2.06-3.45
C36 IA Ahquabi L.	211-279	3	2.78	2.71-2.83
M189 IA East Okoboji normal	195-254	15	2.81	2.46-3.43
blind	168-216	14	2.81	1.96-3.35
R115 IA Lost Island L. 1946-48	120-230	237	2.40	-
1949	180-250	107	3.07	-
1950	-	48	2.76	-
F65 IA Clear L.	-	1214	3.01	-

	TL	No.	K(TL) Mean	Range
C100,105,106 IA ponds	76-126	148	1.25	0.86-2.03
	127-151	193	1.11	0.75-1.77
	152-177	68	1.28	-1.58
	178-202	141	1.36	-1.91
	203-228	42	1.41	-1.97
	229-264	11	1.58	1.44-1.77
P188 WI L. Butte des Morts	140-302	2029	1.29	1.19-1.50
H156 OK Canton L.	127-305	802	1.30	1.02-1.66
S225,226,228 SD Ft. Randall L.	51-305	1122+	1.38	-
N94, S227,229,276 SD Gavins L.	102-254	134	1.38	-
B150 IA Little Wall L.	76-279	361	1.39	-
S472 AL	25-127	2049	1.41	-
J25 OK	152-246	128	1.41	-
M499 IL	114-318	416	1.44	1.33-1.80
M246 IA ponds	76-279	243	1.44	-
F100,105,124,125 SD Oahe L.	127-305	343	1.66	-
J96 OK Rod and Gun Club L.	-	-		1.66-1.77
H424 OK South Rod and Gun Club L.	-	-	1.30	-
after thinning	-	-	1.39	-

No obvious change in K with change in length was noted in B150, H156, M246, M302, P188, and S225 but K increased with length in Iowa ponds (C100, 105,106) and in Illinois (M499). The very low slopes of the Ventura marsh fish (F65), 2.156 and 1.970, were associated with a pronounced decline in condition from mid-July to the end of August. No sex differences in condition factor were noted by F65, S228, and S229. Condition-factor seemed to be lower in mid-July, toward the end of the spawning season, than earlier in Gavins Point Lake, South Dakota (S229), but no such seasonal change was noted at Ft. Randall (S228). Differences in ponderal indexes in different parts of Clear Lake suggested the presence of several populations (F65).

	No.	Mean TL	Range	No.	Weight	Range
Age 0-June						
C134 OK Stillwater C.	1	18	-	-	-	-
C284 OH	+	-	19-25	-	-	-
Age 0-July						
C134 OK Stillwater C.	22	-	20-33	-	-	-
R38 IN	+	33	-	-	-	-
B3, C102, C109, F65 IA Clear L.	11039	36	13-58	-	-	-
B150, C157 IA Little Wall L.	158+	43	20-91	-	-	-
Age 0-Aug.						
C284 OH	+	38	-	-	-	-
B159 OR	11	-	41-53	-	-	-
B2 IA L. Okoboji	23	46	33-58	-	-	-
B3, C102, C109, F65 IA Clear L.	8335	46	15-99	-	-	-
L27 IA pond	226	69	33-89	-	-	-
S226 SD Ft. Randall	+	76	-	-	-	-
Age 0-Sept.						
C102 IA Clear L.	261	46	33-58	-	-	-
Age 0-Oct.						
B4 IA Clear L.	55	79	41-114	-	-	-

	No.	Mean TL	Range	No.	Weight	Range
Age 0-Dec.						
J34 OK Ardmore L.	+	168	-	-	-	-
Age 0						
M448 IA Chariton R.	35	25	-	-	-	-
H332 KS turbid ponds	1	33	-	-	-	-
R115 IA Lost Island L.	+	38	25-51	+	3	-
H398 IA Coralville L.	+	-	51-102	-	-	-
R31 IN	+	84	-	-	-	-
Age I						
D166 KS Neosho R.	+	51	38-97	-	-	-
R115 IA Lost Island L.	+	56	51-84	+	6	-
B97 IL Gale L.	+	71	64-76	-	-	-
R38 IN Sweet Gum P.	8	81	69-91	8	9	6-14
L27 IA ponds	58	86	58-119	-	-	-
C157 IA Little Wall L.	61	86	38-147	24	26	9-43
L27, F65 IA Clear L.	432	97	46-165	-	-	-
M448 IA Chariton R.	74	112	-	74	28	-
S349 ND	3	114	109-117	-	-	-
D166 KS Marais des Cygnes R.	60	117	89-168	540	20	-
M246, M295 IA Silver L.	+	127	-	-	-	-
R31 IN gravel pit	+	130	-	-	-	-
R38 IN Grassy P.	6	132	114-145	6	40	23-60
B150 IA Little Wall L.	128	135	112-150	-	-	-
H332 KS turbid ponds	45	152	-	-	-	-
C208, L126 CO	7	145	127-152	-	-	-
S207 KS Fall River L.	4	170	130-196	4	76	34-113
G74 KS El Dorado L.	+	180	-	-	-	-
B131 MI ponds	519	180	-	519	82	-
L27 IA Von Zom. pond	2	239	234-241	-	-	-
C346 OR Empire L.	2	251	-	-	-	-
Age II						
M232 IA Center L.	+	94	-	+	6	-
L27 IA ponds	58	130	99-188	-	-	-
M295 IA Silver L.	+	132	-	-	-	-
M295 IA E. Okoboji L.	+	137	-	+	28	-
C157 IA Little Wall L.	174	140	99-175	169	37	11-74
M448 IA Chariton R.	101	145	-	101	43	-
S349 ND	14	145	114-170	-	-	-
M295 IA Lost Island L.	+	155	-	-	-	-
O34 OR Lake of Woods	+	155	-	-	-	-
H332 KS turbid ponds	90	157	-	-	-	-
F65 IA Clear L.	551	163	122-218	-	-	-
B150 IA Little Wall L.	25	178	152-201	-	-	-
H150 OK Tenkiller L.	4	180	170-198	4	85	-
L126, C208 CO	121	188	165-279	-	-	-
G74 KS El Dorado L.	+	193	-	+	85	-
P188 WI L. Butte des Morts	5	198	157-216	-	-	-
H156 OK Canton L.	35	208	-	-	-	-
S207 KS Fall River L.	9	218	185-251	9	150	111-301
T16 IL Horseshoe L.	+	229	-	-	-	-
H150 OK Tenkiller L.	4	279	264-318	4	386	-
C346 OR Empire L.	25	287	-	-	-	-
H332 KS clear ponds	1	302	-	-	-	-

	No.	Mean TL	Range	No.	Weight	Range
Age III						
C157 IA Little Wall L.	157	152	119-188	157	48	20-82
L27 IA ponds	60	165	102-208	-	-	-
M295, R115 IA Lost Island L.	+	168	152-185	+	-	57-142
M448 IA Chariton R.	19	173	-	19	77	-
O34 OR Lake of Woods	+	175	-	-	-	-
M295 IA E. Okoboji L.	+	183	-	+	85	-
J25 OK Illinois R.	17	185	168-201	-	-	-
H332 KS turbid ponds	38	190	-	-	-	-
M295 IA Silver L.	+	201	-	+	122	-
L29 IA Red Haw L.	5	198	163-241	5	116	54-190
H150 OK Tenkiller L.	4	206	203-211	-	-	-
P188 WI L. Butte des Morts	41	211	150-267	-	-	-
F65 IA Clear L.	663	213	178-267	-	-	-
C208, L126 CO	62	221	157-297	-	-	-
S349 ND	98	221	188-259	-	-	-
B150 IA Little Wall L.	7	221	203-251	-	-	-
H156 OK Canton L.	123	234	-	-	-	-
H332 KS clear ponds	1	262	-	-	-	-
H150 OK Tenkiller L.	6	292	244-368	6	490	-
C346 OR Empire L.	25	300	-	-	-	-
Age IV						
R115 IA Lost Island L.	+	152	-	+	71	-
C157 IA Little Wall L.	81	170	140-216	63	65	34-122
L27 IA ponds	64	175	135-206	-	-	-
F65 IA Clear L.	341	201	157-246	-	-	-
M448 IA Chariton R.	4	206	-	4	122	-
M295 IA Lost Island L.	+	208	-	-	-	-
H332 KS turbid ponds	7	211	-	-	-	-
O34 OR Lake of Woods	+	216	-	-	-	-
H156 OK Canton L.	3	249	-	-	-	-
L126, C208 CO	11	251	193-351	-	-	-
L29 IA Red Haw L.	2	259	244-302	2	284	196-369
P188 WI L. Butte des Morts	12	267	239-295	-	-	-
S349 ND	11	284	257-315	-	-	-
H150 OK Tenkiller L.	3	295	236-378	3	490	284-907
C346 OR Empire L.	14	307	-	-	-	-
Age V						
C157 IA Little Wall L.	33	188	152-229	18	82	42-196
L27 IA ponds	59	190	157-269	-	-	-
R115, M295 IA Lost Island L.	+	198	183-213	+	94	-
O34 OR Lake of Woods	+	224	-	-	-	-
H156 OK Canton L.	1	226	-	-	-	-
F65 IA Clear L.	1135	246	203-310	-	-	-
C208 CO	1	262	-	-	-	-
H332 KS turbid ponds	7	262	-	-	-	-
H332 KS clear ponds	3	318	-	-	-	-
C346 OR Empire L.	11	320	-	-	-	-
Age VI						
L27 IA ponds	15	196	170-211	-	-	-
R115 IA Lost Island L.	+	203	-	+	113	-
O34 OR Lake of Woods	+	229	-	-	-	-
H332 KS turbid ponds	1	239	-	-	-	-

Age VI (cont.)	No.	Mean TL	Range	No.	Weight	Range
C157 IA Little Wall L.	9	246	206-287	5	210	176-238
C208 CO	1	262	-	-	-	-
F65 IA Clear L.	413	292	231-338	-	-	-
H332 KS clear ponds	3	320	-	-	-	-
C346 OR Empire L.	3	338	-	-	-	-
L29 IA Red Haw L.	1	376	-	1	397	-
Age VII						
L27 IA Barringer Slough	4	211	201-229	-	-	-
R115 IA Lost Island L.	+	216	190-234	+	136	-
F65 IA Clear L.	102	302	259-345	-	-	-
C346 OR Empire L.	3	351	-	-	-	-
Age VIII						
C157 IA Little Wall L.	1	213	-	-	-	-
R115 IA Lost Island L.	+	246	-	+	266	-
F65 IA Clear L.	96	310	277-343	-	-	-
Age IX						
M295, R115 IA Lost Island L.	+	264	-	+	287	-

		Calculated TL at each annulus					
	No.	1	2	3	4	5	6
H421,419 OK state average	-	94	170	229	274	312	351
slowest	-	25	56	102	152	216	310
fastest	-	257	391	366	343	437	366
H421 OK 19 streams	89	81	140	196	208	234	
25 ponds, 0-10 acres	187	97	170	236	282	351	
25 lakes, 10-150 acres	253	97	175	236	302	338	351
8 lakes, 150-500 acres	186	107	188	239	307	335	
16 reservoirs, over 500 acres	458	99	180	241	269	279	351
H150 OK Illinois R.	21	89	175	234	353		
F64 OK Little R.	15	94	173	277			
E126 OK Rocket Plant L.	3	137	226	320	345		
H424 OK S. Rod and Gun Club L.	99	109	203	244	302	328	366
J101 OK L. Eucha	164	178	254	315			

Since the data from E61, H156, J25,34,42,45, and S391 are included in H421 they are not repeated here.

		Calculated TL and increments for each year of life		
	1	2	3	4
S349 ND Heart Butte L.	53	122	185	226
Increment	53	69	64	41
Number	26	23	10	3
S349 ND L. Ashtabula	61	135	185	
Increment	61	48	51	
Number	58	58	58	
S349 ND L. Darling	66	140	201	257
Increment	66	76	58	61
Number	43	43	42	9

Age was determined from rings on the vertebrae (L27, L29, F65, C157), on sections of dried pelvic fins (S349), or on spines (most of the papers). A razor blade was satisfactory for sectioning the dried pelvic fins and xylol aided in making the sections translucent (S349). Annuli on spines usually formed the second week of May in Oklahoma (H421). Growth tended to be faster in clear than in turbid water (H332). Bullheads are quite subject to overcrowding and slow growth. In smaller ponds they rarely maintain suitable growth (H421). Bullheads transferred from a crowded population in Lost Island Lake, Iowa (M295), at age V grew from an average of 100 g to average 254 g in 3.5 months under uncrowded conditions. In Silver Lake, Iowa (M295), bullheads averaged 127 mm in September at age I and only 132 mm and 41 g in September at age II, but after a heavy winter kill averaged 201 mm and 122 g at age III. After angler removal of 185 kg/ha in Lost Island Lake, age IV bullheads at 208 mm increased only to 213 mm at age V (M295). In East Okoboji Lake, Iowa, 574 kg/ha were removed in one year (M295). Rapid growth in the Empire Lakes, Oregon (C346) may have been the result of recent introduction and low population density.

Growth in streams was in general slower than in ponds and reservoirs (H421). "Although the black bullhead catfish demonstrate a tendency toward becoming a dominant species in ponds, they are apparently unable to withstand competition in reservoir populations. In newly impounded reservoirs they frequently develop significantly large populations which flourish for a short time only to begin an apparently irreversible decline until within a few short years they become only an incidental species" (H421). Bullheads disappeared from the catch in Lewis and Clark Lake and in Ft. Randall Reservoir soon after impoundment (W192). After the Humboldt Impoundment on the Des Moines River was treated with rotenone in 1956, bullheads became very abundant but by 1962 they were almost nonexistent in the area (H422).

Black bullheads, introduced during a flood, eliminated the crayfish, frogs, and topminnows from ponds in Louisiana, except for 2 sand-bottomed ponds where few bullheads remained (V23).

Black bullheads were mature at 254 mm and at age III in the Mississippi River in Illinois (B94).

C108 IA	No.	TL	Weight	Eggs/Female	Range
	3	183-193	91-116	2155	1638-2430
	3	208-211	136-170	2734	2477-3128
	1	224	-	6820	-
S151 IL	14 bullheads, 1,610,000 ± 475,426 red blood cells per cu. mm				
	55.6 ± 8.9 hemoglobin				

Black bullheads spawn from late June to late July in South Dakota (S227, S228, S229).

When acclimated at 23C, the upper lethal temperature for black bullheads was 35C (B200).

Small bullheads (36-76 mm TL) fed mostly on *Hyalella*, entomostracans, and small insect larvae (E121, D170). Little feeding occurred at midday or midnight; the peak feeding was just before dawn (D170, D197). Rates at which *Hyalella* were digested by small bullheads were determined by D170.

Young bullheads school during the day (D197), their social appetite stimulated by tactile and gustatory senses (B348). In late afternoon, as the light intensity decreases, the visual stimuli of the other fish diminish and the hungry fish begin feeding again. After dark the young bullheads are relatively inactive and do not feed again until it begins to get light (D197). Adult bullheads tend to be active all night but are inactive during the day, apparently resting in dense vegetation (D197).

Insects, entomostracans, plant debris, fish, and a few frogs were foods of larger bullheads (R115, K73, K74).

Black bullheads adjust well to captivity except that the spines catch in nets and often become injured. They were more resistant to oil refinery effluents than other species of fish tested (W214).

YELLOW BULLHEAD, *Ictalurus natalis* (LeSueur)

North Dakota to Hudson River south to Florida and Mexico (E113, T119).

C30 IA   2 bullheads   TL = 1.17 SL

TL	No.	Mean of means	Weight Central 50%	Range	Citations
25-51	53	1.1	-	-	S472, L12
52-75	73+	4	2.3-6.0	0.4-1.7	S472, K71, L12
76-101	160+	7.7	4.5-11.1	1.4-8.5	S472, K71, L12
102-126	40+	15.9	13.6-18.1	3.7-11.3	S472, K71, L12
127-151	51+	26	-	9.1-4.0	S472, K71
152-177	49+	46	41-57	14-34	S472, W132, K71
178-202	232+	97	82-102	34-77 48-159	S472, C30, K71, P188, L12, D179, W132
203-228	287+	133	125-136	91-227	S472, L12, K71, P188, W132
229-253	418+	196	193-207	136-238	S472, L12, P188, K71, W132, E32
254-278	552+	272	227-290	181-369	S472, W132, P188, L12, K71, E32
279-304	361+	360	304-414	272-482	S472, P188, L12, W132, K71, E32
305-329	51+	488	406-493	394-652	S472, E32, L12, K71
330-355	22+	612	504-700	445-765	S472, N52, E32, C30, L12, K71
356-380	7+	779	-	624-901	S472, E32, K71, L12
381-405	1	794	-	-	L12

The above data came from Alabama S472; Indiana E32; Iowa C30; Ohio L12; Tennessee K71, W132; Texas D179; Wisconsin N52, P188. The weights from S472 were usually below the others and those of L12 were usually high.

S472 AL   371 bullheads   25-356 mm   log W = -4.11 + 2.66 log TL

The slope should probably be greater because the 25 and 51 mm fish had obviously high weights.

P188 WI L. Butte des Morts   log W = -4.792 + 2.973 log TL

W105 IL Crab Orchard L.	6 bullheads	108-385 g	avg. K(SL) 2.80 range 2.05-4.45

D179 TX   1 bullhead   160 mm SL   K(SL) 2.69

S472 AL   317 bullheads   76-356 mm   K(TL) 1.21   1.05-1.44

The trend over this range would suggest a slope of over 3.0.

P188 WI L. Butte des Morts 1694 bullheads 178-279 mm K(TL) 1.43   1.36-1.66

	No.	Mean TL	Range	No.	Weight
Age 0-first month					
S59 VA	+	-	5-8	-	-
Age 0-two months					
S59 VA	+	51	-	-	-
Age 0-July					
J42 OK Ft. Gibson	3	74	-	-	-
Age 0-Aug.					
W105 IL Crab Orchard L.	1	33	-	-	-
Age 0-Oct.					
B4 IA Clear L.	2	66	61-71	-	-
Age I					
B77 MI	411	53	30-66	411	1.7
S59 VA January	+	-	56-64	-	-
J25 OK Illinois R.	10	122	99-135	10	23
J42 OK Ft. Gibson L.	6	130	-	-	-
B86 IL Onized L.	333	-	102-178	-	-
L29 IA Red Haw L.	1	180	-	1	82
S201 TN Reelfoot L.	1	183	-	1	57
Age II					
B77 MI	60	124	89-150	60	26
J25 OK Illinois R.	26	180	150-229	26	88
J42 OK Ft. Gibson L.	2	180	-	-	-
P188 WI L. Butte des Morts	4	206	180-244	-	-
B86 IL Onized L.	12	-	216-267	-	-
H150 OK Tenkiller L.	1	284	-	-	-
S201 TN Reelfoot L.	57	333	267-394	57	516
Age III					
J25 OK Illinois R.	14	218	190-264	14	133
P188 WI L. Butte des Morts	18	226	193-274	-	-
B86 IL Onized L.	2	330	305-356	-	-
S201 TN Reelfoot L.	250	371	307-429	250	720
Age IV					
P188 WI L. Butte des Morts	7	272	226-295	-	-
J25 OK Illinois R.	4	272	246-305	4	278
L29 IA Red Haw L.	1	333	-	1	638
S201 TN Reelfoot L.	130	401	318-450	130	901
Age V					
S201 TN Reelfoot L.	69	429	384-470	69	1085

	No.	Average calculated TL at each annulus				
		1	2	3	4	5
S391 OK Rock Creek	8	64	114	188	218	
J45 OK Verdigris R.	16	66	140	190	221	
H150, J25 OK Illinois R.	61	64	145	185	244	323
F64 OK Little R.	166	76	152	203	272	396
E61 OK Salt C.	3	76	150	221		

During the first month of life, yellow bullheads grew 0.6 mm per day in Michigan (H86).

Yellow bullheads spawn in May-June in Illinois (R27).

	Weight of females	No.	Eggs per female	
			Mean	Range
U1 IN	170	2	1847	1652-2042
	284	7	2334	2028-2977
	397	4	3607	2715-4452
	680	1	6660	-
E4, V21 MN	397-567	3	4157	3950-4270

Annual mortality rates were estimated at 83% in Spear Lake (R152), at 80% in Shoe Lake and at 90% in Muskellunge Lake (R34). No homing nor territorial tendency was noted in marked yellow bullheads in 16-hectare Wintergreen Lake (F97).

S151 IL   11 bullheads   1,490,000 ± 488,000 red blood cells per cu. mm
                                    55.6 ± 9.3 hemoglobin

Yellow bullheads in a Texas lake fed mostly on fish (M314), and in Maryland mostly on crustacea and insects (F148).

		Sizes taken in gill net meshes			
Mesh size		25	38	51	64
L216 LA	No.	6	20	7	3
	Mean TL	236	307	345	325
	Range	165-356	241-406	318-381	254-394

BROWN BULLHEAD, *Ictalurus nebulosus* (LeSueur)
    Southern Manitoba, Great Lakes to Maine south to Florida and northern Mississippi (T119). Introduced elsewhere.

C37 MN	785 bullheads	100-350 mm	TL = 1.135 SL
C30 IA	3 bullheads	-	TL = 1.16 SL

		Mean of	Weight Central		
TL	No.	means	50%	Range	Citations
25-50	160	1.3	-	-	L12
51-75	205	3.1	-	-	L12
76-101	2929	8.8	-	0.6-10.8	F71, L12, S472
102-126	231	27	-	8.5-51	F71, L12, S472
127-151	276	34	-	14-57	Below
152-177	443+	57	45-64	31-85	Below
178-202	342+	88	82-94	37-136	Below
203-228	623+	136	116-147	82-198	Below
229-253	1888+	213	173-233	113-233	Below
254-278	1839+	272	241-306	198-369	Below
279-304	878+	343	309-357	235-548	Below
305-329	777+	460	440-500	284-548	Below
330-355	261+	606	589-646	408-862	Below
356-380	250+	740	589-811	539-954	Below
381-405	278	868	-	612-1247	Below
406-431	177	1021	-	819-1455	S472, L12, M242
432-456	101	1324	-	901-1950	S472, M242
457-482	100	1497	-	1315-1950	S472, E105, M242
483-507	21	1823	-	1671-2234	M242
508-532	1	2087	-	-	M242

The above data came from Alabama S472; California E134; Florida M242; Idaho S147; Illinois M499; Iowa C30; Maryland E105; Michigan R24; Minnesota C22, S62; Ohio L12; Wisconsin M20, P188; and Czechoslovakia F71.

	Length	No.	
S65 NS L. Jesse	-	-	$\log W = -5.166 + 3.125 \log FL$
P188 WI L. Butte des Morts	152-292	1634	$\log W = -5.061 + 3.065 \log TL$
S472 AL	51-483	66443	$\log W = -4.31 + 2.76 \log TL$

This regression (S472) is influenced by large numbers of 25-102 mm bullheads which are obviously recorded at too high a weight. The slope for the larger bullheads should be over 3.0.

	Size	No.	K(SL)
C36 IA	222 mm	1	2.14
W105 IL	505 g	1	2.25
C22,23,24, E3 MN	-	1159	2.37

			K(TL)	
	Length	No.	Mean	Range
P188 WI L. Butte des Morts	152-292	1634	1.26	1.17-1.34
S472 AL	102-457	4602	1.27	1.06-1.50
M499 IL	140-318	70	1.40	1.16-1.80
M259 MN L. Traverse	-	272	-	1.38-1.52
C23 MN	-	387	1.63	-

In general, condition increased with length in P188 and in S472 from 127 to 457 mm.

	No.	Mean TL	Range	No.	Mean Weight
Age 0-May					
S57 NC	+	-	5-18	-	-
Age 0-June					
R13 NY	197	18	13-23	-	-
C284 OH St. Mary's	+	25	-	-	-
S57 NC	+	51	-	-	-
Age 0-July					
M243 NY ponds	30	23	20-28	-	-
F71 Czech.	292	33	20-43	-	-
R13 NY	265	33	20-43	-	-
H76 WV	1	33	-	-	-
C284 OH St. Mary's	+	43	-	-	-
R27 IL	+	46	-	-	-
Age 0-Aug.					
M243 NY L. George	15	23	-	-	-
F71 Czech.	32	41	33-53	-	-
C284 OH St. Mary's	+	76	-	-	-
Age 0-Sept.					
M243, R13 NY ponds	43	58	43-74	-	-
Age 0-Oct.					
R13 NY	53	69	66-89	-	-
C284 OH St. Mary's	+	71	-	-	-
T113 OH	+	-	51-124	-	-

	No.	Mean TL	Range	No.	Mean Weight
Age 0-Dec.					
E32 IN	1	53	-	-	-
Age 0					
E9 NY	+	64	-	-	-
F71 Czech.	3	86	76-97	-	-
Age I					
E9, E11 NY	150+	-	76-102	-	-
T113 OH	+	-	69-152	-	-
F71 Czech.	98	127	94-165	-	-
O24, G75 OR Ten Mile L.	2	112	76-145	-	-
C279 OR Empire L.	+	234	-	-	-
Age II					
E9 NY	+	-	127-152	-	-
K61 CA Pardee L.	4	150	140-157	-	-
P188 WI L. Butte des Morts	1	152	-	-	-
F71 Czech.	165	163	122-208	-	-
E134 CA L. Almanor	+	-	130-198	-	-
O24, G75 OR Ten Mile L.	197	206	170-259	-	-
C279 OR Empire L.	+	282	-	-	-
L29 IA Red Haw L.	1	216	-	1	147
Age III					
F71 Czech.	47	185	160-229	-	-
P188 WI L. Butte des Morts	30	193	157-246	-	-
K61 CA Pardee L.	8	196	170-221	-	-
E134 CA L. Almanor	+	-	198-254	-	-
O35 OR Eel L.	6	236	216-259	-	-
O24, G75 OR Ten Mile L.	93	251	180-320	-	-
L29 IA Red Haw L.	1	246	-	1	201
C279 OR Empire L.	+	287	-	-	-
Age IV					
K61 CA Pardee L.	1	173	-	-	-
F71 Czech.	10	231	190-300	-	-
P188 WI L. Butte des Morts	20	241	203-290	-	-
M259 MN L. Traverse	272	262	-	272	269
E134 CA L. Almanor	+	-	254-284	-	-
O35 OR Eel L.	1	267	-	-	-
O24, G75 OR Ten Mile L.	114	272	236-366	-	-
Age V					
P188 WI L. Butte des Morts	5	267	257-290	-	-
O35 OR Eel L.	2	290	284-297	-	-
F71 Czech.	3	318	310-325	-	-
O24 OR Ten Mile L.	37	277	259-320	-	-
Age VI					
O24 OR Ten Mile L.	1	290	-	-	-

		Mean Calculated TL at each annulus				
	No.	1	2	3	4	5
F71 Czech. R. Elbe	101	81	119	160	201	
pond	222	81	124	160	211	272
F64 OK Little R.	41	91	180	259	312	361
S192 MD Monocacy R.	+	147	203	239		

Age was determined from rings on the vertebrae (F71, M259, L29) or on the spines (F64, S192). Direct proportion was used in computing lengths at previous annuli by F64 and S192, but F71 made a Fraser-type correction.

When introduced into mountain lakes better suited for trout, brown bullheads stunt (E134).

Brown bullheads spawn March to May in Florida (M242), May to June in Maine (E93) and Illinois (R27), and in June in Wisconsin and Michigan (G133). Spawning may continue through September in Alabama (S310). The male clears the nest and guards the eggs and nest (G133). Both parents may be involved in nest building and care of young (E134). The first hatch occurred when the water reached 27C, 150 mm from the surface (S334). Incubation takes 5-14 days (E134).

	Size of female	No.	No. of eggs per female Mean	Range
S123, N25, K118, M426	-	-	-	2000-13000
M38 VA	218 mm	1	2400	-
T27 NY	454 g	1	2188	-
E4, V21 MN	267-330 mm	5	10043	6180-13800

When acclimated at	Lethal temperatures Minimum	Maximum	Citations
5-6C	-	28.6-29	S333, B201, B202
10	-	30-30.2	B202, S333
20	-1.0-1.0	33-33.4	B201, B202, S333
25	1.3	35-35.5	B201, B202, S333
30	3.7	36.5-37	B201, B202, S333
36 (ultimate)	7.0	37.5	B201, S333

Brown bullheads were observed swimming in 37-38C and would enter 40C water for worms (T102). The LD-50 temperature for these bullheads appeared to be 37.2C.

Fry and fingerlings to 75 mm eat zooplankton and chironomids (K125, N117, S372); adults feed on insects, fish, fish eggs, molluscs, and plants (S372, F148, K125, J53). Over half the bullheads collected during the spawning time of lake trout had eaten trout eggs (E93).

Brown bullheads bury themselves in the bottom muds under some conditions at temperatures of 0-18C, more commonly at the lower temperatures (L215). They may remain inactive under the mud long enough to escape toxic substances used for their eradication.

In fertilized ponds in Alabama (S325), 168-280 kg per ha of brown bullheads were raised, but in ponds where the bullheads were fed, 1000-1345 kg per ha were raised. Standing crops ranged from 0.9 to 148 kg per ha, and annual yields from 0.3 to 121 kg per ha (E134). Angler harvest per hectare from Quabbin Reservoir, Massachusetts (M296), was 0.40, 0.67, 0.72, 0.73, and 0.76 kg in various years.

In Clear Lake, California (L214), a tagged bullhead was caught 8 years later, although it was 251 mm FL at tagging. Mean annual survival was estimated at 76% and rate of exploitation at 7% in Clear Lake (M335) and at 52% and 14% in Folsom Lake, California (R280).

Brown bullheads hybridize with black bullheads (T113).

A summary with bibliography E134.

FLAT BULLHEAD, *Ictalurus platycephalus* (Girard)
Coastal streams North Carolina to northern Florida (E119).

S472 AL	TL	No.	Weight Mean	Weight Range	K(TL)
	51	14	1.3	0.9-1.4	0.96
	76	22	5.4	4.4-8.2	1.23
	102	7	12.2	10-23	1.16
	127	9	27	21-27	1.33
	152	4	45	-	1.31
	178	8	64	59-91	1.13
	203	5	95	-	1.11
	229	1	150	-	1.26

S472 AL	70 fish	51-229 mm	log W = -5.34 + 3.197 log TL

CHANNEL CATFISH, *Ictalurus punctatus* (Rafinesque)
Montana, southern Manitoba, to southern Quebec, south, west of the Appalachians, to Florida and Mexico (E119, T113). Introduced elsewhere.

C30, C89, L29 IA	30 catfish	-	FL = 1.091 SL; TL = 1.201 SL
F63 OK	697	51-216 mm SL	TL = -0.08 + 1.288 SL
	189	217-559 mm SL	TL = 1.237 + 1.138 SL
M279 IA	421	76-584 mm (no trend)	FL = 1.14 SL; TL = 1.36 SL
M513 QU	2566	200-550 mm FL	TL = 17 mm + 1.056 FL

When data have been recorded as SL or FL they are converted to TL on the basis indicated in the length-weight summary. The lengths indicated are the lower limits of the size class, not the class centers.

SL	FL	TL	No.	Mean of means	Weight Central 50%	Range	Citations
21	23	25	4869+	0.5	-	0.05-20	A39, S345, S472 AL
42	46	51	19462+	1.5	-	1.4-3.2	A39,40, S345,472 AL
			+	14	-	-	F105 SD
62	69	76	7835+	3.1	-	0.4-14	A39,40, S345,472 AL
			8	4.5	-	-	W132 TN
			4	6.8	-	-	M279 Des Moines R.
			+	18	-	-	F105 SD
85	91	102	2452	9	-	3-30	S472 AL
			29+	15	9-23	5-27	Combined others
106	114	127	2898	15	-	6-38	S472 AL
			27	23	-	-	H170, M279 Des Moines
			+	18	-	18-36	SD
			61+	23	18-27	14-41	Combined, except S472
127	140	152	2582	25	-	12-59	S472 AL
			31	27	-	-	W132 TN
			10	32	-	-	H156 OK
			48	36	-	-	H168,170, M279
			+	36	-	23-50	SD
			119+	36	27-41	23-50	Combined, except S472
148	163	178	1734	43	-	23-100	S472 AL
			28	45	-	-	W132 TN

SL	FL	TL	No.	Mean of means	Weight Central 50%	Range	Citations
			721	50	-	-	H140,152,156, F63 OK
			+	54	-	41-68	SD
			22	59	-	-	S365 SC
			217	68	-	-	H168,170, M279 D.M.R.
			1044+	54	45-59	27-91	Combined, except S472
170	185	203	1583	64	-	36-113	S472 AL
			67	68	-	-	W132 TN
			674	73	-	-	OK
			+	77	-	64-86	SD
			12	82	-	50-141	SC
			222	86	-	77-95	Des Moines R.
			1044+	77	68-82	50-141	Combined, except S472
192	208	229	1211	86	-	-141	S472 AL
			64	91	-	-	W132 TN
			57	91	-	-	A39,40 AL
			959	104	-	-	OK
			+	109	-	95-118	SD
			269	122	-	104-136	Des Moines R.
			7	141	-	104-367	S365 SC
			1376+	104	95-113	64-367	Combined, except S472
215	231	254	1057	127	-	43-240	S472 AL
			32	132	-	-	W132 TN
			18	136	-	-	A32 Miss. R.
			17	136	-	-	A39,40 AL
			795	141	-	-	OK
			+	141	-	122-181	SD
			172	159	-	-	Des Moines R.
			1073+	150	136-168	95-227	Combined, except S472
237	254	279	953	172	-	104-295	S472 AL
			34	186	-	-	W132 TN
			53	191	-	-	A32 Miss. R.
			683	191	-	-	OK
			+	191	-	163-245	SD
			96	195	-	-	Des Moines R.
			933+	191	172-200	91-254	Combined, except S472
260	277	305	750	227	-	150-358	S472 AL
			73	245	-	-	W132 TN
			+	245	-	218-318	SD
			51	254	-	-	A32 Miss. R.
			684	254	-	-	OK
			67	259	-	-	Des Moines R.
			916+	249	231-263	168-327	Combined, except S472
284	300	330	594	308	-	227-	S472 AL
			29	295	-	-	A32 Miss. R.
			+	308	-	290-431	SD
			770	318	-	-	OK
			78	340	-	-	Des Moines R.
			45	404	-	-	W132 TN
			962+	331	295-345	181-468	Combined, except S472
306	323	356	434	376	-	227-862	S472 AL
			55	376	-	-	W132 TN
			33	381	-	-	Des Moines R.
			+	395	-	363-522	SD
			362	404	-	-	OK
			535+	413	376-440	254-680	Combined, except S472
328	345	381	341	463	-	249-626	S472 AL
			47	490	-	-	W132 TN

SL	FL	TL	No.	Mean of means	Weight Central 50%	Range	Citations
			325	504	-	-	OK
			+	504	-	459-672	SD
			20	513	-	-	Des Moines R.
			53	526	-	-	A32 Miss. R.
			23	571	-	-	C150 OR
			514+	522	481-567	363-798	Combined, except S472
350	368	406	248	580	-	227-862	S472 AL
			+	626	-	567-825	SD
			48	631	-	-	W132 TN
			464	635	-	-	OK
			18	649	-	-	Des Moines R.
			36	667	-	-	C150 OR
			55	731	-	-	A32 Miss. R.
			634+	653	608-698	445-962	Combined, except S472
372	391	432	149	712	-	331-934	S472 AL
			303	776	-	-	OK
			+	776	-	703-998	SD
			25	785	-	-	W132 TN
			30	816	-	-	C150 OR
			75	821	-	-	A32 Miss. R.
			450+	789	753-821	499-1089	Combined, except S472
394	417	457	92	862	-	544-1089	S472 AL
			20	889	-	-	W132 TN
			24	912	-	-	Des Moines R.
			+	948	-	839-1197	SD
			243	953	-	-	OK
			60	985	-	-	A32 Miss. R.
			17	1021	-	-	C150 OR
			374+	953	889-1043	508-1950	Combined, except S472
416	439	483	57	1007	-	771-1225	S472 AL
			18	1016	-	-	W132 TN
			181	1139	-	-	OK
			10	1139	-	-	Des Moines R.
			+	1184	-	998-1429	SD
			50	1193	-	-	A32 Miss. R.
			282+	1171	1098-1193	958-1497	Combined, except S472
438	462	508	29	1310	-	1112-1701	S472 AL
			119	1366	-	-	OK
			+	1375	-	1239-1673	SD
			20	1469	-	-	C150 OR
			15	1474	-	-	Des Moines R.
			38	1511	-	-	A32 Miss. R.
			235+	1433	1310-1511	1112-2313	Combined, except S472
460	486	533	16	1492	-	1361-1702	S472 AL
			+	1633	-	1469-1928	SD
			12	1628	-	-	C150 OR
			121	1641	-	-	OK
			31	1769	-	-	A32 Miss. R.
			202+	1660	1497-1769	1243-2812	Combined
482	508	559	12	1574	-	907-2041	S472
			+	1855	-	1738-2204	SD
			12	1878	-	-	C150 OR
			64	1887	-	-	OK
			38	2164	-	-	A32 Miss. R.
			141+	1910	1814-1996	907-2858	Combined
504	531	584	+	2141	-	2014-2550	SD
			49	2191	-	-	OK
			15	2490	-	-	A32 Miss. R.
			86+	2304	2137-2490	1134-3402	Combined

SL	FL	TL	No.	Weight Mean of means	Weight Central 50%	Weight Range	Citations
526	554	610	29	2475	-	-	OK
			+	2495	-	2390-2926	SD
			21	2740	-	-	A32 Miss. R.
			15	3220	-	2449-3765	S365 Santee R.
			76+	2580	2404-2740	2268-3765	Combined
548	577	635	28	2858	-	-	OK
			+	2971	-	2876-3266	SD
			21	3856	-	3175-4536	S365 Santee R.
			62+	3111	2449-3229	2722-4536	Combined
560	599	660	17	3225	-	-	OK
			14	4264	-	3583-4990	S365 Santee R.
			42+	3522	3202-3788	2495-4990	Combined
582	622	686	14	3683	-	-	F63 OK
			26	4944	-	3992-5897	S365 Santee R.
			58+	4100	3856-4241	2948-5897	Combined
604	648	711	29	5670	-	4445-7804	S365 Santee R.
			43+	4712	4264-5239	3833-7804	Combined
626	671	737	23	6214	-	5216-7394	S365 Santee R.
			35+	5271	4622-6214	4082-7394	Combined
648	693	762	22	7076	-	5897-8165	S365 Santee R.
			33	6169	5529-7076	5171-8165	Combined
670	716	787	24	7938	-	6350-9662	S365 Santee R.
			31	7258	-	6350-9662	Combined
692	739	813	20	8710	-	7756-10025	S365 Santee R.
			25	8165	-	7031-10025	Combined
714	762	838	22	9300	-	8346-10886	S365 Santee R.
			28+	7847	-	6124-10086	Combined
736	785	864	10	10342	-	9616-11340	S365 Santee R.
			17	9253	-	7031-12247	Combined
758	807	889	+	8165	-	-	R46 OH
			12	11113	-	9979-12066	S365 Santee R.
780	831	914	1	9526	-	-	M242 FL
			4	12746	-	12066-13381	S365 Santee R.
802	853	940	1	9526	-	-	M242 FL
			9	14923	-	13608-17690	S365 Santee R.
824	876	965	9	15967	-	13790-17509	S365 Santee R.
846	902	991	2	15513	-	15332-15695	S365 Santee R.
858	925	1016	6	18462	-	16556-21909	S365 Santee R.
880	947	1041	2	21138	-	20911-21319	S365 Santee R.
902	971	1067	1	19732	-	-	S365 Santee R.
1078	1156	1270	1	24948	-	-	G36 SD

The South Dakota reservoir data in the above table came from F100, F105, F124, F125, N94, N115, S225, S226, S227, S228, S229, S276, and S393 and included data on 1726 catfish. The Oklahoma reservoir data came from F63, H140, H152, H156; the Des Moines River, Iowa, data came from H168, H170 and M279. Additional data included in the combined came from Alabama S345; California M328; Colorado L126; Georgia M186, M187; Illinois M499; Iowa C30, C89, L29; Kansas D97; Maryland E105; Missouri P113; Nevada W52; New York R240; Ohio O7, R46; Oregon B153, C150; Tennessee S200; Virginia M38; and Lake Huron S164.

The Santee tailrace catfish were reported to be heavier for their lengths than the reservoir fish (S365). Since a high proportion of the large catfish in this compilation came from the Santee-Cooper Reservoir, the length-weight relationship may be biased. These fish tended to be heavier than fish from other waters at the same lengths.

	TL	No.	
M513 QU St. Lawrence R.	-	-	log W = -4.012 + 3.039 log FL
D198 L. Erie	-	145	log W = -4.910 + 2.956 log TL
H156 OK Canton L.	-	154	log W = -4.667 + 3.056 log TL
M187 GA	178-559	988	log W = -5.375 + 3.132 log TL
M279 IA Des Moines R.	76-660	558	log W = -5.401 + 3.133 log TL
M186 GA	178-610	278	log W = -5.791 + 3.256 log TL
P112,113 MO Salt R.	203-559	-	log W = -5.839 + 3.30 log TL
H140,152,419 OK	178-762	3091	log W = -5.289 + 3.364 log TL
F63 OK	178-864	4617	log W = -5.385 + 3.407 log TL
C347 TN Kentucky L.	102-635	85	log W = -5.587 + 3.516 log TL
A32 IA Mississippi R.	-	-	log W = -6.759 + 3.66 log TL
H420 TN Watts Bar L. females (r = 0.96)	-	-	log W = -6.786 + 3.658 log TL
males (r = 0.94)	-	-	log W = -6.936 + 3.739 log TL

The slope, 3.133, determined for the Des Moines River fish (M279) was found to be significantly greater than 3.0, and the correlation coefficient for the log weight-log length relationship was 0.993. The slope for 7791 catfish receiving supplemental feeds was 3.099 and that of 864 catfish from Kansas reservoirs was 3.304 but throughout the range of sizes the fed fish were heavier at each length (S510).

				K(SL)
	SL	No.	Mean	Range
C36, L29 IA	348-736	13	1.95	-
C179 TX Medina L.	124-505	836	1.68	1.01-2.78
after thinning shad population	135-520	490	1.96	1.33-2.89
2 years after thinning	152-487	274	1.68	1.14-2.51

				K(TL)
	TL	No.	Mean	Range
C89 IA Ruth's P.	229-406	20	0.75	-1.16
H156 OK Canton L.	102-635	154	0.75	0.50-1.19
S225,226,228,393 SD Ft. Randall L.	-	528	0.75	0.66-0.78
M279 IA Des Moines R.	76-660	558	0.83	-1.33
N94,115, S227,229,276 SD Gavins L.	-	1202	0.89	-1.00
B150 IA Little Wall L.	356-457	2	0.97	0.88-1.05
F100,105,124,125 SD Oahe L.	-	370	0.97	0.80-1.05
S510 KS ponds Age II	-	-	-	0.91-1.22
fingerlings	-	-	-	0.86-1.11
S504 AR ponds Age II	-	-	1.12	1.05-1.20

Condition factors increased with increase in length in M279, S225, S229, S276, S228, S226, but not in S227 or S393. In most length-weight regressions the slope was greater than 3.0 indicating an increase in condition with increase in length. No differences were noted in length-weight regressions of male and female catfish if taken outside the spawning season (A32, D198). No sex differences were reported in S226, S227, S228, S229, and S393, but males tended to have higher K values in F100, F105, and F125. In Gavins Lake, a decrease in condition as the season progressed was evident one year (S276). In Kansas ponds condition decreased in September and October even when the fish were fed (S510). In Kansas reservoirs condition improved from May through October (S510) but was significantly lower than that of channel catfish being fed in ponds, except in October when there was little difference.

In general, condition decreased as standing crop increased, as alkalinity decreased, and as oxygen level became more variable within each 24-hour cycle (S510). When catfish were not fed, the condition index was highly correlated (r = .75 and .43 respectively) with alkalinity and diurnal changes in alkalinity of the ponds (S510) and with afternoon temperatures (r = .66) and diurnal changes in temperature (r = .60). Little or no difference was found in the condition of catfish living in aerated and nonaerated ponds (S510).

Reduction of the shad population in Medina Lake, Texas (D179), was followed by an increase in condition of the channel catfish but this increased condition was lost within 2 years.

		TL					
	No.	Mean of means	Central 50%	Range	No.	Mean Weight	Range
Age 0-June							
C134 OK	+	-	-	13-18	-	-	-
T111 KS	-	-	-	-	+	-	5-14
Age 0-July							
S41 IA; M74 MO; G102, T111, D166 KS; G32 OH; C134 OK; T56 TX	58+	30	-	8-48	+	-	0.1-41
O37 OK	40	56	-	38-71	40	1.8	-
N115 SD	+	71	-	-	-	-	-
N85 AR albino	+	89	-	-	-	-	-
N69 AR	+	127	-	-	+	18	-
Age 0-Aug.							
C284, G32, L80 OH	+	-	-	43-64	-	-	-
T56 TX; S345 AL; G102, T111 KS	-	-	-	-	+	-	0.5-82
M215, D166 KS	265+	64	-	33-74	-	-	-
F33, S41 IA	55+	76	-	51-104	-	-	-
Age 0-Sept.							
G32 OH; R240 NY	+	71	-	69-74	-	-	-
T56 TX; G102, T111 KS	-	-	-	-	+	-	1.4-154
D166 KS river	35	86	-	38-102	-	-	-
S41 IA	15	94	-	79-112	-	-	-
D166 KS river	633	119	-	51-152	-	-	-
Age 0-Oct.-Dec.							
T78,79,95 KS crowded ponds	+	-	-	51-89	25+	5	-
D166 KS rivers	173	74	-	51-94	-	-	-
B6 IA; M74 MO	+	-	-	64-140	-	-	-
D26, D27 KS	+	-	-	76-178	-	-	-
T56 TX Oct.	-	-	-	-	+	9	5-14
T56 TX Nov.-Dec.	-	-	-	-	+	15	9-18
M38 VA	+	152	-	-	-	-	-
S504 AR	+	190	-	-	+	45	-
A9, A10 OK	+	206	-	-	+	-	1-43
Age 0							
M448,503 IA	187	56	-	53-66	-	-	-
J26,41, S159, T44 OK	82+	71	-	38-94	72	3	-
H332 KS pond	13	58	-	-	-	-	-
D97 KS rivers	321	94	-	-	321	9	-
S365 SC Moultree L.	3	104	-	58-122	-	-	-
C271 OK Canton L.	9	193	-	17-249	-	-	-
Age I							
M183 MO L. of Ozarks	53	64	-	-	-	-	-
W192 SD Lewis & Clark L.	2	109	-	-	-	-	-
H172,176,174 IA Des Moines R.	4+	86	-	53-91	-	-	-
T78,79,105 KS crowded ponds	792+	117	-	76-221	2453	-	5-28
B183, H127,140,150,152,156, J25, 26,41,42, O37, S159,470, T44 OK	1091	160	127-175	91-265	172	-	11-32
D97,166, M215, S207 KS rivers	473	137	114-160	64-190	60	36	14-41
H423, M503 IA Coralville L.	208	147	-	102-175	-	-	-
G74, B81, S340 KS lakes	+	-	-	137-356	+	18	-

		TL					
Age I (cont.)	No.	Mean of means	Central 50%	Range	No.	Mean Weight	Range
S365 SC Moultree L.	20	175	-	145-221	20	-	45-91
R240 NY 7 ponds	-	-	-	-	+	-	9-82
S512, M448, H174, B6, I10, H176, U4 IL and IA rivers	62+	163	137-196	71-216	26	36	23-82
T78,79,95, H332 KS good ponds	75+	234	221-231	211-305	40+	100	45-127
J25, C271 OK 2 reservoirs	134+	262	-	145-394	-	-	-
M314, P126, S345 TX and AL ponds	-	-	-	-	4260+	-	68-758
M38 VA	+	-	-	254-279	-	-	-
S514,200,394 TN Reelfoot L.	141	307	-	165-368	125	249	73-268
G155, N69, S504 AR ponds	+	279	-	203-356	+	408	141-907
Age II							
M183 MO L. of Ozarks	50	132	-	-	-	-	-
T78,79,95 KS crowded ponds	103+	140	-	127-155	103+	18	-
H170,172,176,174 IA Des Moines R.	161+	147	-	84-203	-	-	-
W192 SD Lewis & Clark L.	17	157	-	-	-	-	-
S365 SC Moultree L.	25	201	-	145-221	25	-	45-272
B236 OR	51	185	-	140-236	51	64	23-150
A86, B6, H176, L29, M195, M448 IA; L38 IL	55+	206	190-229	135-302	51	227	54-907
H423, M503 IA Coralville L.	81	218	-	170-246	-	-	-
K64, L126 CO	57	234	-	168-381	-	-	-
D97,166, G74, M215, S207 KS rivers	360+	234	188-246	152-330	99+	141	45-213
O37, H152,156, B183, W7,S470 T44, S159, H127, J26,25, W53, H140, J41,42 OK	1621+	224	196-254	127-480	222	73	41-154
A32, I10, H174, U4,5, S512 IA and IL Miss. and Mo. R.	357+	241	213-257	178-302	8	91	77-113
R240 NY 8 ponds	-	-	-	-	+	-	45-363
C149,150 OR	44	305	-	178-419	-	-	-
H332, T78,79,95 KS ponds, good growth	50+	330	295-343	249-498	49+	363	186-1080
S514,200,394 TN Reelfoot L.	478	348	-	282-419	478	395	-
C271, J25, H150 OK Canton L., Tenkiller L.	102+	368	-	206-610	67	381	-
G156, S504 AR ponds	+	399	-	366-434	+	749	354-1158
N85 AR albino	-	-	-	-	+	794	-907
K41 IN ponds	+	442	-	-	+	-	-1080
B109 TX ponds	+	447	-	404-486	+	821	626-1080
Age III							
T95 KS ponds, stunted	115	152	-	-	115	23	-
M183 MO L. of Ozarks	14	193	-	-	-	-	-
W192 SD Lewis & Clark L.	6	206	-	-	-	-	-
H170,172,176,174 IA Des Moines R.	332	213	203-234	147-295	-	-	-
H423, M503 IA Coralville L.	68	249	-	183-284	-	-	-
A86, H176, M195, L38, G74, S207, M448 IL, IA, KS	28+	282	244-330	190-470	23+	168	59-399
K64, L126 CO	60	297	-	193-294	-	-	-
R240 NY 6 ponds	-	-	-	-	+	-	181-907
H127,140,152, J26,25,41,42, O37, S159,470, T44 OK	1569	282	244-323	193-605	221	154	95-299
S365 SC Moultree L.	4	312	-	226-447	4	-	363-1043
A32, H174, I10, U4,5, S511,512, 513 IA, IL Miss. & Mo. R.	1034+	315	297-343	241-401	1	227	-
D97, H332, T79,95 KS ponds, good growth	100	335	297-330	295-495	97	331	168-1048
B236, C149, C150 OR	79	373	-	279-495	8	381	313-513
S514,200,394 TN Reelfoot L.	360	404	-	333-500	360	531	413-671
B183, C271, H150,156, J25 OK Tenkiller L., Canton L.	29+	424	348-486	300-493	7	753	-

	No.	TL Mean of means	Central 50%	Range	No.	Mean Weight	Range
Age IV							
T95 KS pond, stunted	78	183	-	-	42	45	-
M183 MO L. of Ozarks	66	236	-	-	-	-	-
W192 SD Lewis & Clark L.	4	249	-	-	-	-	-
H170,172,176,174 IA Des Moines R.	297+	249	224-269	178-368	-	-	-
A86, B150, C82, G74, H176,423, M195,448,503,215 IA, KS	88+	297	269-315	229-384	11+	200	100-381
K64, L126 CO	40	333	-	239-429	-	-	-
H127,140,152, J26,41, S470, O37, T44, S159 OK	1208+	333	284-378	229-559	70	295	159-522
A32, H174, I10, S511,513, U4,5 IA, IL Miss. & Mo. R.	1153+	368	345-386	284-427	7	608	172-816
H332, T95 KS ponds	88	391	-	305-518	84	671	227-1066
S365 SC Moultree L.	4	391	-	259-544	4	-	363-1547
L38 IL Kaskaskia R.	+	442	-	-	-	-	-
S514,200,394 TN Reelfoot L.	256	450	-	358-546	125	925	-1089
C149,150 OR	23	455	-	381-546	-	-	-
R240 NY 4 ponds	-	-	-	-	+	-	590-2132
B183, C271, H150,156, J25 OK Tenkiller L., Canton L.	30+	472	467-486	399-643	14	867	717-1021
M510,353 CA	+	-	-	457-531	-	-	-
Age V							
T95 KS ponds, stunted	343	196	-	-	514	64	-
M183 MO L. of Ozarks	17	277	-	-	-	-	-
W192 SD Lewis & Clark L.	6	284	-	-	-	-	-
H170,172,176,174 IA Des Moines R.	346+	300	279-343	130-462	-	-	-
H423, C281, H176, G74, M215, 448,503 IA, KS	93+	343	307-373	267-483	+	268	213-399
K64, L126 CO	23	376	-	297-396	-	-	-
H140,127,152, J26, O37, S470, 158, T21,44, W7,53 OK	870	381	312-417	297-754	32	481	249-1270
H332, T95 KS ponds	40	386	-	274-531	33	617	218-689
A32, I10, U4,5, H174, S511,513 Miss. R., Mo. R.	203+	437	399-452	330-615	6	1397	499-2585
H176, B150 IA lakes	2	455	-	450-460	1	976	-
H150, J25 OK Tenkiller L.	11+	488	-	406-541	4	1310	-
C149,150 OR	6	511	-	445-572	-	-	-
S200,514 TN Reelfoot L.	170	500	-	419-610	170	1270	1171-1411
R240 NY 4 ponds	-	-	-	-	+	-	907-2449
S365 SC Moultree L.	4	521	-	419-587	4	-	1089-3538
Age VI							
M183 MO L. of Ozarks	102	318	-	-	-	-	-
W192 SD Lewis & Clark L.	31	328	-	-	-	-	-
H170,172,176,174 IA Des Moines R.	117+	333	284-386	239-559	-	-	-
A86, H423,176, M215,448 KS rivers	52	401	363-394	340-495	3	1188	-
H127,152,140, J26,41, S470, O37, S159, T44 OK	236+	460	442-452	343-810	9	544	404-595
H332 KS ponds	6	429	-	-	-	-	-
H174, A32, I10, U4,5, S511 Miss. & Mo. rivers	200+	472	460-511	318-622	2	1252	953-1547
C271, H150, J25 OK Tenkiller L.	4+	536	-	531-541	3	1764	-
G74, H176, L29 KS lakes	4	544	-	503-627	1	1588	-
K64, L126 CO	8	538	-	384-686	-	-	-
R240 NY 4 ponds	-	-	-	-	+	-	1451-2631
S200 TN Reelfoot L.	17	569	-	521-610	17	1865	-
S365 SC Moultree L.	7	592	-	483-744	7	-	1633-6804
C149,150 OR	8	599	-	546-724	-	-	-
Age VII							
M513 QU St. Lawrence R.	2	338	-	330-345	-	-	-
M183 MO L. of Ozarks	64	345	-	-	-	-	-

	No.	Mean of means	TL Central 50%	Range	No.	Mean Weight	Range
Age VII (cont.)							
H170,172,176,174 IA Des Moines R.	64+	363	315-404	297-584	-	-	-
W192 SD Lewis & Clark L.	25	378	-	-	-	-	-
A86, H423,176 IA other rivers	35+	457	406-554	363-671	-	-	-
G74, M215 KS	8	483	-	457-505	-	-	-
H127,140,152,156, J25,42, O37, S159, T44, W7,53 OK	99	521	472-561	391-744	7	1021	821-1700
S511, A32, U4,5 Miss. & Mo. R.	131+	533	518-536	406-698	-	-	-
K64 CO	7	541	-	-	-	-	-
H176, L29 IA lakes	2	605	-	561-648	1	2327	-
R240 NY 2 ponds	-	-	-	-	+	-	2359-2449
C150 OR	5	607	-	508-800	-	-	-
S200 TN Reelfoot L.	5	617	-	587-635	5	2540	-
S365 SC Moultree L.	18	678	-	599-775	18	-	2540-7167
Age VIII							
M513 QU St. Lawrence R.	3	361	-	351-368	-	-	-
M183 MO L. of Ozarks	52	366	-	-	-	-	-
H170,172,176 IA Des Moines R.	17+	376	-	297-640	-	-	-
W192 SD Lewis & Clark L.	7	447	-	-	-	-	-
D97, G74, M215 KS	62	511	-	483-556	56	953	-
A86, H423 IA	19	457	-	445-503	-	-	-
H140,152, J26, S159, T44, W7,53 OK	59	536	523-559	450-754	11	1511	1474-1520
A32, I10, S511, U5 Miss. & Mo. R.	39+	561	541-566	483-653	1	2041	-
R240 NY one pond	-	-	-	-	+	2722	-
S200 TN Reelfoot L.	4	663	-	622-686	4	3647	-
S365 SC Moultree L.	34	714	-	584-851	34	-	3175-10206
C150 OR	3	716	-	673-775	-	-	-
L29 IA East L.	2	810	-	762-858	3	6074	5529-6637
Age IX							
H170,172,176 IA Des Moines R.	7+	363	-	287-648	-	-	-
M183 MO L. of Ozarks	16	376	-	-	-	-	-
M513 QU St. Lawrence R.	1	386	-	-	-	-	-
A86, H423 IA	10	480	-	419-500	-	-	-
W192 SC Lewis & Clark L.	2	505	-	-	-	-	-
S511, A32, U5 Mississippi R.	12+	607	-	526-673	-	-	-
D97 KS ponds	5	-	-	610-660	-	-	-
H140, T44, H152 OK	21	625	-	615-663	-	-	-
C150 OR	4	693	-	622-749	-	-	-
S365 SC Moultree L.	34	769	-	617-1000	34	-	3311-16556
Age X							
C135 ON Ottawa R.	1	363	-	-	1	426	-
H170,172,176 IA Des Moines R.	4+	371	-	325-488	-	-	-
M513 QU St. Lawrence R.	4	401	-	351-439	-	-	-
A86 IA Red Rock area	2	602	-	-	-	-	-
A32, U5 Mississippi R.	1	711	-	-	-	-	-
H140,152 OK	5	716	-	-	-	-	-
C150 OR	1	800	-	-	-	-	-
S365 SC Moultree L.	24	818	-	650-978	24	-	3992-18598
Age XI							
M513 QU St. Lawrence R.	2	406	-	384-432	-	-	-
H170,176 IA Des Moines R.	2	427	-	406-447	-	-	-
A32, U5 Mississippi R.	2	665	-	622-686	-	-	-
L29 IA East L.	-	-	-	-	1	4536	-
C150 OR	1	775	-	-	-	-	-
S365 SC Moultree L.	19	851	-	683-985	19	-	4128-15650
Age XII							
M513 QU St. Lawrence R.	3	432	-	388-472	-	-	-
H172 IA Des Moines R.	1	437	-	-	-	-	-
A32, U5 Mississippi R.	2	724	-	719-732	-	-	-
S365 SC Moultree R.	3	862	-	787-869	3	-	7530-15513
Age XIII							
M513 QU St. Lawrence R.	2	447	-	442-452	-	-	-
H172 IA Des Moines R.	1	592	-	-	-	-	-
S365 SC Moultree L.	8	828	-	775-1039	8	-	9934-19732
C150 OR	1	851	-	-	-	-	-
L29 IA East L.	1	887	-	-	1	8392	-

	No.	TL Mean of means	Central 50%	Range	No.	Mean Weight	Range
Age XIV							
M513 QU St. Lawrence R.	2	442	-	419-462	-	-	-
S365 SC Moultree L.	3	929	-	851-976	3	-	8664-17237
Age XV							
M513 QU St. Lawrence R.	1	447	-	-	-	-	-
Age XVI							
M513 QU St. Lawrence R.	2	486	-	478-490	-	-	-
Age XVII							
M513 QU St. Lawrence R.	1	495	-	-	-	-	-
Age XVIII							
M513 QU St. Lawrence R.	3	486	-	462-523	-	-	-
Age XIX							
M513 QU St. Lawrence R.	2	488	-	478-495	-	-	-
Age XXII							
M513 QU St. Lawrence R.	1	544	-	-	-	-	-
About Age XL							
M513 QU St. Lawrence R.	1	807	-	-	-	-	-

Mean TL (at capture, H152, or at last annulus, D97)
in various age groups

	No.	I	II	III	IV	V	VI	VII	VIII	IX	X	XI	XII	XIII	XIV
H152 OK summary															
5 stunted ponds	1149	107	155	201	251	279	356	445	483	602					
4 ponds with repro- duction	142	107	175	229	257	292	345	381							
5 ponds without	23	-	-	406	488	475	518	-	533	602					
lakes, 5-100 acres															
6 with reproduction	159	137	206	269	366	457	396	447	546	625	648				
10 without	174	249	394	493	551	544	536	546	-	-	770				
4 lakes, 100-500 acres	460	150	239	272	312	384	401	508	554						
10 lakes, over 500	1304	173	234	312	376	439	498	625	-	561					
3 rivers	294	142	216	292	388	439	505	574	650	686	724				
3 new impoundments	349	290	361	421	480	513	569	589	627						
D97 KS summary															
15 state lakes	689	74	213	254	333	338	373	434	434	480	645	381	526	-	671
slowest	-	41	152	183	249	272	287	356	315	429	511	-	480		
fastest	-	91	257	323	434	401	457	538	589	556	780	-	569		
11 city-county lakes	538	155	216	267	262	318	356	411	429	503	521	625	-	762	691
slowest	-	-	135	173	208	221	254	305	348	419	505				
fastest	-	-	330	411	335	490	556	617	556	701	551				
Cedar Bluff L.	84	-	-	272	201	236	338	668							
Kanopolis L.	165	-	160	203	241	282	320	292	368	505					
6 streams	91	66	180	272	284	335	320	335	544						
slowest	-	-	132	160	211	246	-	-	373						
fastest	-	-	246	328	421	447	-	-	714						

	No.	1	2	3	4	5	6	7	8	9	10	11	12	13	14
						Average calculated TL at each annulus										
H419, F63 OK															
summary	7717	102	216	302	368	409	452	505	556	607	630	645	648	655	732
slowest	-	48	102	147	160	196	213	246	264	300	330	414	508	554	701
fastest	-	239	460	554	660	714	770	795	787	833	858	882	813	749	762
6 streams	1042	107	196	279	348	409	472	495							
reservoirs over 500 acres															
7 new	700	117	274	335	445	526	556	523	643						
16 old	3291	91	178	249	305	363	417	472	531	577	602	589	572	658	701
10 clear	-	97	190	259	318	384	442	503	561	658	706	747			
5 turbid	-	76	152	213	262	310	358	421	480	467	498	511			
lakes, 111-500 acres															
15 lakes	1672	91	183	251	307	368	414	478	483	528	541	569	632	653	762
5 clear	-	104	221	312	371	421	483	538	521	559	559	544			
10 turbid	-	84	163	224	277	340	381	450	472	518	531	577			

	No.	1	2	3	4	5	6	7	8	9	10	11	12	13	14
								Average calculated TL at each annulus							
lakes, 5-110 acres															
39 lakes	789	112	234	340	417	472	513	587	680	732	759	782	673		
30 clear	-	117	244	414	442	500	546	602	688	734	762				
7 turbid	-	89	196	267	325	363	391	462	551	711	749				
Weighted average based on acreage	-	109	206	274	335	371	414	465	533	592	627	640	691	658	709

The above is believed to include data from E61, J27, J31, H150, J47, H152, J25, H140, S159, but not the following later papers:

	No.	1	2	3	4	5	6	7	8	9	10	11	12	13	14
K65,66 OK Hiwasse L.	250	107	206	262	318	394	465	465							
J34 OK Ardmore	94	117	257	356	384	414	442	472	498						
F64 OK Little R.	105	104	216	305	366	478	526	610							
F62 Sub Prison	62	58	119	173	216	246	287	361	424						
after reclam.	83	84	188	323											
H220 Ft. Gibson	1450	81	175	249	335	419	495	572	602	660	·762				
J45 Verdigris R.	38	84	140	198	246	302	351	406	455						
tributaries	35	89	155	213	264	302	351	406	455						
S391 OK Rock C.	23	81	145	229	305	406									
H303 OK Lawtonka	65	104	196	284	351	417	467	521	589	645					
F87 Little R.	58	94	198	277	330	452									
cutoff lakes	47	109	221	315	381	483	526	610							
K64 CA L. Havasu	52	76	137	168	196	236	284	351	388	455	503	475	541	572	597
Colorado R.	137	71	175	236	305	386	467	531							
A32 Miss. R.	535	76	160	231	297	361	421	486	536	610	676	658	711		
D198 L. Erie	665	64	165	226	269	297	330	363							
T127 Kentucky L.	116	109	173	224	264	310	363	424	495						
T127 East TN reservoirs	261	114	203	295	358	419	480	518	541	574					
J101 OK L. Eucha	118	142	279	373	439	478	531								
J101 OK Spavinaw L.	306	127	246	323	388	452	513	569	650	693					
S192 MD Potomac R.	+	196	267	330	399	460	521								

	1	2	3	4	5	6	7	8	9	10	11	12	13	14
				Average calculated total lengths and increments										
S365 SC Moultree L.	86	185	284	368	442	531	602	665	·726	773	807	853	916	904
Increments	86	99	99	84	74	86	74	71	69	58	51	61	56	41
Number	207	187	162	158	154	150	143	125	91	57	33	14	11	3
F124, F125 SD Oahe L.	122	201	267	348	437	513	549	602	663					
Increments	122	79	79	81	74	86	36	43	36					
Number	60	57	25	15	10	4	4	3	1					
C347 TN Kentucky L.	109	170	221	262	307	363	424	495	566	640				
Increments	109	61	51	46	51	58	64	64	71	74				
Number	94	94	86	76	64	52	47	36	22	10				
H423 IA Coralville L.	84	157	211	267	320	351	384	411	381					
Increments	84	74	58	56	53	33	33	18	25					
Number	265	219	161	118	89	34	11	2	1					
H420 TN Watts Bar L.	163	239	290	333	366	404	427	462	488	523				
Increments	163	76	53	46	38	41	41	36	33	36				
Number	55	55	50	36	28	18	7	6	1	1				
M183 MO L. of Ozarks	53	117	168	206	241	269	295	325						
Increments	53	66	53	41	38	36	38	43						
Number	381	331	317	251	234	132	68	16						
P112 MO Salt R., upper	69	119	173	218	269	345	404	450						
Increments	69	51	51	56	61	71	56	46						
Salt R., middle	61	112	170	226	272	318	356	401	505					
Increments	61	51	58	61	61	56	56	64	76					
Salt R., lower	66	135	206	259	297	340	399							
Increments	66	66	74	66	61	51	46							
H174 IA Mississippi R.	66	150	211	254	274	315								
Increments	66	81	64	46	28	30								
Number	91	85	72	48	15	5								
transplanted														
Increments	-	152	109	97	84	114	107							
Number	-	6	13	24	33	10	5							
M279 IA Des Moines R.	46	124	196	257	312	381	442	490	546	617	645	640	676	
Increments	46	81	74	69	66	69	61	48	46	23	36	36	36	
Number	400	477	285	181	124	104	77	42	11	3	2	1	1	

					Average calculated total lengths and increments								
	1	2	3	4	5	6	7	8	9	10	11	12	13	14
I10 IA Missouri R. lakes	61	145	239	330	432	457	457	513						
Increments	61	81	99	89	76	48	53	56						
Number	37	28	19	18	9	3	1	1						
H170 IA Des Moines R.	84	150	190	224	267	279	305	310	333	343	406			
Increments	84	64	38	28	28	25	28	30	46	30	46			
H156 OK Canton L.	94	208	305	386	442	536	594							
Increments	94	107	81	91	89	94	58							
Number	107	97	51	9	1	1	1							
O37 OK Heyburn L.	86	165	224	305	394	472	561							
Increments	86	102	64	89	107	94	99							
Number	206	186	138	23	3	2	1							

		Average calculated weight at each annulus						
	No.	1	2	3	4	5	6	7
D198 L. Erie	665	2	45	113	186	249	345	445

Annual increments as percentages of TL at beginning of year. Number of specimens in parentheses.

	M183 MO L. of Ozarks		H156 OK Canton L.	
TL	Mean	Range	Mean	Range
25-50	116 (102)			
51-75	125 (428)	95(16)-150(14)		
76-101	46 (16)			
102-126	47 (552)	32(52)-51(102)	64 (458)	51(188)-91(107)
127-151	34 (214)	26(16)-42(66)		
152-177	26 (370)	22(16)-31(102)	37 (663)	31(348)-43(315)
178-202	20 (184)	15(52)-23(52)	36 (351)	34(163)-38(188)
203-228	21 (352)	14(16)-28(102)	29 (770)	14(348)-46(107)
229-253	17 (132)	13(52)-20(64)	19 (1071)	17(348)-20(188)
254-278	16 (168)	16(52)-61(64)	34 (163)	
279-304	13 (32)	11(16)-15(16)	15 (1234)	10(348)-23(163)
305-329			26 (770)	15(348)-37(315)
330-355			19 (1049)	15(535)-27(188)
381-405			15 (642)	14(107)-15(535)
406-431			14 (829)	8(351)-21(163)
432-456			22 (270)	21(107)-22(163)
457-482			18 (723)	16(535)-21(188)
483-507			14 (478)	12(315)-18(163)
533-558			7 (957)	6(535)-11(107)
559-583			4 (163)	
584-609			2 (163)	
610-634			2 (163)	

In most cases age and growth have been determined from rings on the pectoral spines (e. g., S159, H220, M279), but in others from rings on the centra of vertebrae (F63, M183). Growth was determined from marked specimens and used to estimate the ages given in M513. Some of the tagged catfish were out 11 years before recapture. A53 found the centra of vertebrae to be better than the cliethrum, operculum, or otolith, but indicated that these other bones may also be used in aging catfish. D97 found that ages as determined from spines compared well with known-age fish of two year classes. Growth was usually calculated on direct proportion, but S470 used a Fraser-modification with 33 mm TL as intercept. The body-spine relationship was found to be a straight line on a log-log plot when the spine was cut at the same point on all fish, but was curvilinear if cut as had been done in some earlier work (D198). Lee's phenomenon was evident in the data in S470. The annulus was formed in April and early May in the Des Moines River, Iowa (H174), and in June in Canton Reservoir, Oklahoma (C271). D166 followed the growth of a dominant year class. In the Des Moines River, Iowa (H169), tagged channel catfish added 75 to 100 mm in length per year when under 305 mm TL, but grew more slowly thereafter. In this river growth tends to be slow in June, rapid in July, and then slower in August (H174). A tagged channel catfish in Sacramento Valley, California (M328) grew 152 mm in 2 years, reaching 533 mm TL. A catfish tagged at 302 mm was recaptured after 8 years, 2 months (L214).

The tabulated data give little evidence of regional differences in growth rate though more of the faster-growing catfish reported have come from the southern part of the range. There is little evidence of any consistent difference in the growth of catfish in lakes compared to streams (see particularly H152, D97, and F63), nor evidence that size of lake has any significant effect on average growth. In Salt River, Missouri (P112), there was little evidence of difference in average growth in upper, middle, and lower sections of the river. After age V growth was faster in Lake Moultree, South Carolina, than in the tailrace (S365). Growth averaged faster in clear than turbid waters (F63).

Channel catfish under 2 years old in unchanneled portions of the Missouri River were larger than those in channelized portions of the river but the reverse was true of older catfish (R281). Availability of suitable food and population density may have been the major causes of the growth differences. Growth was faster in ponds with well-water than with surface water in Arkansas (S504) probably because the well-water had more minerals. Growth was even better when the ponds were fertilized.

In Kansas ponds, D97 suggested that the slower growth in turbid waters might be partly due to overpopulation. Growth was also slower when drought reduced the water level (D97). In another study (H332) growth was as rapid in turbid as in clear ponds. Lower water levels were associated with slower growth of channel catfish in 1955 than in 1956 in the Des Moines River, Iowa (M279). Normal flow in 1957 after a drought in Blue River, Kansas (M215), resulted in good growth and reproduction, but competition of the 1957 year class was therefore greater and at mid-age I the average length was less than of the 1956 year class.

In New York ponds, average growth rate was roughly inversely proportional to population density (R240), and in Alabama age I-Sept. catfish averaged 454 g when stocked at 1000 per acre compared to 318 g when stocked at 2000 per acre, even though the catfish were fed in both ponds (P126). Slow growth in Grand Lake, Oklahoma, was blamed on overcrowding (J31), and also in many Kansas ponds (T78, T79, T95, T105). Catfish from slow-growing populations grew rapidly when transplanted to less crowded ponds (T78, T95, D97). These fish, transplanted when 200-300 mm TL, grew 25-104 mm in 10.5 months and 107-224 mm in 22.5 months (D97). Those stocked in Prairie Lake in August had increased as follows:

by	May	June	July	August	September	October	22-24 months
Increment	43	51	99	109	130	127	196
Number	19	7	32	66	13	6	60

Elimination of the population and restocking resulted in increased growth rates in Sub Prison Lake, Oklahoma (F62). Angling success also improved after an overcrowded population was eliminated and catfish were restocked in the Des

Moines River, Iowa (H422). Growth is rapid in new impoundments (H152, J47, F63, H220) but decreases as the population density increases (H220). Slow growth of channel catfish in the Humboldt area of the Des Moines River, Iowa, was associated with a high population density (H170, H172, H176) and growth was more rapid after the population was reduced (H174). Growth was faster in ponds and lakes where there was no reproduction, in Oklahoma (H152, F63).

Average growth decreased as the stocking rate increased in Kansas ponds, although total production increased (S510). Increased water flow through tanks increased the growth of otherwise crowded catfish but even with continuous flow population density reduced average growth rate. However, the effects were relatively small and it was suggested that catfish might be held in a confined area of a large body of water to simplify feeding and harvest.

Growth of albino catfish was the same as that of normal catfish, but when stocked at 75-150 mm the survival rate in ponds was only 79.8-82.5% compared to 93.5-94.9% for the normal (P190).

Growth improved as the protein content of the diet was increased from 6.6% to 15.8% and then to 25.3%, but showed no further increase when raised to 34.8% (N134, S431). Other studies (S510) showed increases up to 40% but little additional increase with 60% protein (S510). Casein and a mixture of casein and soybean protein produced better growth than soybean protein alone (K136).

Growth slowed and the fish actually lost weight after mid-September in Kansas ponds (T105). In two other ponds growth ceased about July 8, with stocking crops of 230-1350 kg per ha even though the fish were being fed (S510). Removal of half the fish from one pond on August 8 resulted in renewed growth and the poundage by October 4 was again 1230 kg. Mechanical aeration of the other pond also resulted in increased growth and the standing crop reached 2480 kg per ha by November 1. Aeration failed to have beneficial effects on growth in other experiments however and probably is effective only when growth is inhibited by low dissolved oxygen or by oxidizable substances. Growth rates were positively correlated with total alkalinity and with the extent of diurnal change in alkalinity in ponds where catfish were not fed, but feeding largely masked any relation to alkalinity or other chemical factors (S510). Stresses caused by oxygen levels less than 3 ppm retarded growth of channel catfish although the catfish showed distress, by surfacing, only when oxygen levels were below 1 ppm (S510).

In Texas ponds, channel catfish will mature 18 months after hatching (M314). Albino catfish matured at age II and at 907 g in Arkansas (N85). In Oklahoma (C271) female catfish mature at age II and the smallest was 358 mm. Females were mature at 267-279 mm (one at 178 mm) and males at 305-318 mm (one at 216 mm) in Kansas ponds (D97, D99). In Kansas rivers catfish mature at 305-381 mm (D166). In South Dakota reservoirs a few mature at 279-305 mm (S227, N115), others at 330-356 mm (S227, S229), but most not until 380 mm (S229, S228, S393). The smallest mature female found in Moultree Lake, South Carolina, was 615 mm TL and weighed 2540 g, but few smaller fish were seen, mature or otherwise. A summary of the literature (K118) suggested that channel catfish mature at 4-6 years and at 330-560 mm. In the Mississippi River the smallest mature males were 292 mm and females 318 mm (G43) but most mature at 305-381 mm or age IV to V (B94). In California, a 216 mm female was mature (D13). In Lake Erie about one half the females were mature at 250-280 mm and males at 280-310 mm (D198). Males and females first matured at age IV, about 272 mm, in the Des Moines River, Iowa (A86).

In South Carolina catfish spawn in March and April, but mostly in June-July (S365); in Oklahoma, in June-July (C271); in South Dakota, they spawned prior to mid-June (S227, S228, S393).

	Size of females	No. of females	No. of eggs per female Mean	Range
C82 IA	254 mm TL	1	2000	-
	324 mm, 397 g	1	3100	-
	680 g	1	4000	-
	1814 g	1	8000	-
L13 OH	413 mm	1	8110	-
D13 CA Colorado R.	216 mm	1	1600	-
	660 mm	1	34500	-
M38 VA	406-508 mm	10	7430	4200-10600
B81, D26 KS	-	+	-	2500-70000
D27 KS	1814 g	1	10500	-
M426 literature review	-	-	-	3000-20000
K118 literature review	-	-	-	2000-70000

There are probably two spawning periods in Kansas rivers, at least some years (D166).

Males are darker and have broader heads than females, particularly during the breeding season. In males there is a single urinogenital pore, but in females the genital and urinary pores are separate (M234). The male urinogenital pore is toward the tail and pointed; the female genital pore is closer to the vent, a slit-like pore with a flap (C292). As spawning time approaches the female genital area may be seen to pulsate. Spawning usually occurs at about 23.9C (K118). Spawning behavior is described in C194. Spawning can be delayed by holding brood stock in water 16.7-18.4C, and then transferring the stock to warmer water than that in which spawning is desired (C292). Males may be used for several spawnings (C292). In Texas channel catfish were observed to spawn only between 21-27C, except when some spawning occurred at 15.5C after the temperature had been over 21C. Most spawning was at 21.7C (M314). Spawning and propagation methods are described in G109, S337, S420. Pituitary is used in promoting spawning (S337) in many situations. Embryology is described in S377.

In Alabama ponds, stocking with 7410 fingerlings per hectare (3000 per acre) gave the best production (S345):

No. stocked per hectare	2470	4940	7410	9880	12350
Kg/ha harvested	1072	1699	1803 2632	2016	2246

With supplemental feeding and fertilizing, production of channel catfish in the ponds one year averaged about 1680 kg/ha and ranged from 866 to 2649 (S309, S345, S325), from 1031 to 2029 kg/ha (S420), and in Texas 168 to 785 kg/ha with an average for 5 ponds of 437 kg (M314). In the pond with 168 kg, an additional 717 kg of other fish were produced, and in the pond with 785 kg of catfish there were 376 kg of other fish, but the other 3 ponds had only catfish. Continued feeding after a pond reaches carrying capacity is inefficient and potentially harmful through pollution (S510).

In six New York ponds, the standing crop of catfish averaged 175 kg/ha and ranged from 90-291 (R240). A Kansas pond had 443 kg/ha, but if the water level had been normal there would have been only 72 kg/ha (D97). Marking and recovery estimates of the number of channel catfish over 147 mm TL in Lake Lawtonka, Oklahoma (H301) were 479 catfish or 12.2 kg/ha. It was necessary to correct for homing in making the estimate.

In Kansas, ponds produced 110-182 kg/ha, average 147, at the end of one growing season and 151-205 kg/ha, average 188, at the end of two growing seasons without supplemental feeding (S510). Natural fertility and fertilization affected the standing crop. Standing crop of channel catfish did not seem to be affected by presence or absence of blue gills or bass (S510).

In the Des Moines River 6.6% of the channel catfish tagged at over 229 mm were recaptured within 4 years (B351). Of 797 channel catfish tagged in the Sacramento Valley, California (M328), 18.3% were recaptured the first year, 6.8% the 2nd, 2.5% the 3rd, and 0.1% the 4th. Annual mortality was calculated at 56% and fishing mortality at 30% and the values for Ricker's (R283) vital statistics were as follows:

s	p	q	u	v	a	i	m	n
.440	.445	.376	.304	.256	.560	.821	.359	.314

In the Mississippi River tagged channel catfish were caught at 214 miles upstream after 33 months, and at 111 miles downstream in 36 days (H333). In a Kansas pond, there was 94.3% survival from 76 to 226 m (T105) and the production was 1806 kg/ha. Of the total annual increase in weight, 43.6% occurred in August, 29.4% in July, 18.2% in June, 9.5% in September, 2.3% in May, and in October there was a decrease of 3.0%.

The catch per hour in Kanopolis Reservoir, Kansas, showed the following trend, when the size limit was removed in 1954 (D97):

Year	1953	1954	1955	1956	1957
Catch per hour	0.1	0.25	0.1	0.04	0.04

The ultimate maximum temperature for channel catfish was reported to be 33.5C (S333):

When acclimated at	Lethal temperatures	
15	0	30.3
20	2.5	32.8
25	6.0	33.5

Toleration of minimum oxygen concentration increases with temperature (M351):

C	Shock minimum	Acclimation
25	0.95 ppm	0.70-0.97 ppm
30	1.03	0.85-0.96
35	1.05-1.09	0.83-1.12

The channel catfish used by M351 apparently could stand temperatures of 35C, which is beyond that reported by S333. Channel catfish are most active just preceding sunrise and sunset (S427).

Young channel catfish tend to feed primarily on aquatic insects (B6, E121) or on bottom arthropods (D56) but beyond 100 mm TL they are usually omnivorous (B6, D56, M340, H243, C271, D161) or piscivorous (D163, R281, S365). Among the more unusual food items listed were a snake skin, an adult bobwhite (C271), and hydroids (D56). At one season the main food items in the Des Moines River were elm seeds (B6). S365 mentions that channel catfish in Lake Moultree, South Carolina, fed mostly on herring, bluegills, and small catfish, but rarely ate shad which were heavily fed upon by *Ictalurus catus*. In Ouachita River, Louisiana, also, channel catfish ate no shad (D161). In Norris

Lake, Tennessee (D163) and in Canton Reservoir, Oklahoma (C271), gizzard shad were important in the food of channel catfish. *Hexagenia* mayflies were most important in the food of channel catfish from 241-546 mm TL in one study (H243) and caddisflies in another (B317).

In food preference studies (L217) channel catfish showed a pronounced preference for crayfish of the right size over fathead minnows and these over fingerling bluegills, green sunfish, and golden shiners. Fingerling carp and bullheads were poorly utilized and tadpoles were killed but not eaten.

Channel catfish apparently reduced the growth and condition of bluegill and redear sunfish, probably by competing with them for chironomids (B350). Resulting reduction of bluegill reproduction also limited bass growth.

Calories required to produce a kg of catfish fingerlings varied from 1942 to 2370 in four diets (D199). Artificial feeds are used extensively in catfish farming (S510, D199, H425, H426, N134).

In Louisiana (L216) the following TL of channel catfish were taken:

in 38 mm bar mesh gill net	1	401	
in 51 mm bar mesh gill net	1	587	
in 64 mm gar mesh gill net	3	498	457-521

A summary and bibliography are given in M510.

HYBRID CATFISH

Attempts were made to artificially produce nine types (including reciprocal matings) of catfish using channel, blue, white, and flathead catfish (S517). Little success resulted when the flathead female was used, but the six other crosses resulted in offspring. Hybrid vigor may be present in most crosses, but was most evident in the white catfish male X blue catfish female.

Offspring from a blue catfish male X channel catfish female grew more rapidly than the parent species under the same conditions (G155).

	I April	I October	II April	II August	
Blue catfish	32	322	313	948	grams
Channel catfish	23	381	386	1157	
Hybrid	14	426	499	1451	

At 8 weeks the hybrid fingerlings averaged 43 mm and 0.9 g and survival was 61% to this age.

MOUNTAIN MADTOM, *Noturus eleutherus* Jordan

Ohio River drainage from Pennsylvania to northern Alabama (T113). Young of the year in October are 25-58 mm and yearlings 36-64 mm in Ohio (T113). The largest was 127 mm.

SLENDER MADTOM, *Noturus exilis* Nelson

Southern Minnesota to West Virginia, south into Oklahoma (E113). The maximum length in the Tennessee River was 108 mm SL (H376). These madtoms did not eat prepared foods and required special attention in bioassay tests (B349).

STONE CAT, *Noturus flavus* Rafinesque

Montana to Quebec and south to northern Alabama and northeastern Oklahoma (T113).

G91 OH   log W = -4.426 + 2.841 log SL

F100,105,124,125 SD Oahe L.   76 stone cats

TL	64	89	114	140	165	190	216	241	267
Weight	5	9	23	36	59	82	122	163	204

N94, F100,105,124,125 SD Oahe L.   119 stone cats
                    Average K(TL) 1.24   range 1.04-1.71

T113 OH 0-Oct., 30-81 mm; age I, 56-102 mm;   maximum 312 mm, 482 g

Age	I	II	III	IV	V	VI	VII	VIII	IX
G91 OH streams									
Mean SL	53	77	89	98	107	126			
Number	13	5	6	7	6	10			
G91 L. Erie									
SL						207	209	-	233
Number						6	14		1

	Calculated SL and increments at each annulus								
	1	2	3	4	5	6	7	8	9
G91 OH streams	54	73	89	104	116	129			
Increments	54	19	16	14	11	8			
Number	47	34	29	23	16	10			
G91 L. Erie	68	121	162	181	195	203	208	224	237
Increments	68	53	41	19	14	8	7	12	13
Number	21	21	21	21	21	21	14	1	1

Growth was calculated using the spine-body length relationship determined from the data (G91). The cleithrum, operculum, or otolith may be used for age determination but the centrum of vertebrae was better (A53).

No sex difference in growth was noted (G91). Growth compensation was demonstrated. The more rapid growth in Lake Erie was believed to be due to the abundance of mayfly naiads as food.

In Ohio stone cats spawn early June to late August, with the peak in the last half of June (G91). The eggs are placed under stones and guarded by both parents. They may spawn in streams or in shallow rocky areas of the lake. The average number of eggs per female was 973, with a range of 767-1205 (L80).

In Lake Erie, mayfly naiads were the principal food (G91).

NORTHERN MADTOM, *Noturus furiosus* Jordan and Meek
    Southeastern Michigan, Ohio, Indiana and northern Kentucky (T113). Young of the year in October were 25-58 mm, yearlings 36-64 mm and the maximum 132 mm (T113).

TADPOLE MADTOM, *Noturus gyrinus* (Mitchill)
    Southern Manitoba to Quebec south to Florida and Texas, but not in the Appalachian or Ozark areas (E119, T113). Introduced into Columbia River drainage (E119).

				SL	
	Age	No.	Mean	Range	Weight
B159 OR	0-Aug.	15	-	9-11	-
H80 MN Demming L.	0	2176	26	15-35	0.3
	I	1978	62	48-85	4.2
	II	52	89	78-104	16.3

	Age	TL
G130 IA Clear L.	0-July	23-25
	0-Aug.	18-51
	0-Sept.	18-56
T113 OH	0-Oct.	20-51
	I	30-64
	Largest	112

E32 IN   average, 50 eggs per female. One 50 mm female had 93 eggs.
    Females with eggs were taken on July 1 in Illinois (R27), and July 15-17 in Iowa (G130). In an Ohio lake, these madtoms fed on Cladocera, ostracods, *Hyalella*, chironomids, and debris (E121).
    Hybrids with brindled madtoms, *N. miurus* are known (T113).

MARGINED MADTOM, *Noturus insignis* (Richardson)
    New York to Georgia, mostly east of the Appalachians (E119). There are some reports of margined madtoms up to 305 mm TL (H376).

C250 PA   250 madtoms $SL = -0.65 \, mm + 0.851 \, TL$

Mean TL	19	24	35	43	66	75	85	95	105	115	125	135	145
Mean Wt.	0.1	0.3	0.5	0.9	3.0	4.1	6.7	9.2	12.1	16.1	21.9	26.0	35.8
Number	9	17	12	6	16	30	17	36	24	29	23	19	12

$$\log W = -4.748 + 2.89 \log TL$$

C250 PA Bald Eagle C.

						I-Apr.	July		II-Apr.			
Age	0-July	Aug.	Sept.	Oct.	Nov.	June	Aug.	Nov.	Aug.	Nov.	III	IV
TL	21	34	42	46	53	58	73	89	103	115	118	134
Number	31	13	12	9	4	62	118	35	251	44	145	40

    Age was determined from spine sections. After maturity most of the growth occurred in late summer (C250). Males may mature in the 2nd summer, but most did in the third, when females also matured. A 122 mm female had 107 eggs.
    Three specimens were found to be eating insects and unidentified fish (F148).

SPECKLED MADTOM, *Noturus leptacanthus* Jordan
    South Carolina to Louisiana (E119).

S472 AL	TL	No.	Weight	K(TL)
	25	4	0.3	2.08
	51	47	2.6	2.02
	76	42	5.0	1.08
	102	22	8.6	0.83
	127	1	18	0.89

BRINDLED MADTOM, *Noturus miurus* Jordan
Western New York and southern Ontario to Tennessee, Louisiana, and southeastern Kansas (E119, T113). Young of the year are 25-56 mm in October, yearlings 36-114 mm, and the maximum 132 mm, in Ohio (T113).

FRECKLED MADTOM, *Noturus nocturnus* Jordan and Gilbert
Indiana to Oklahoma (E119) and Alabama (S472).

S472 AL

TL	No.	Weight Mean	Range	K(TL)
25	17	0.5	0.4-	4.04
51	95	1.3	1.2-1.7	0.97
76	6	3.8	3.0-	0.86
102	12	10	5-14	0.94

Because the 25 mm fish are obviously too heavy, the length-weight regression given in S472 is not valid.

FLATHEAD CATFISH, *Pylodictus olivaris* (Rafinesque)
Large rivers, Mississippi Valley and into Mexico (E119). Also Lake Erie (T113).

TL	No.	Mean of means	Weight Central 50%	Range	Citation of extremes
25-75	1410	1.4	-	0.2-5.4	S472, A39
76-126	1627	8.2	5-14	2.2-17	S472, M153
127-177	369	29	23-36	5-59	S472, S472
178-228	415	73	64-82	41-141	S472, S472
229-278	255	145	136-159	45-272	S472, S472
279-329	247+	272	254-277	172-399	S472, S472
330-380	173+	440	413-449	272-635	S472, S472
381-431	107+	644	621-707	408-907	S472, S472
432-482	83+	1034	880-1070	680-1134	S472, J24
483-532	49+	1529	1384-1588	1029-1740	S472, M511
533-583	30+	1923	1747-1928	1361-2268	S472, A38
584-634	33+	2531	2418-2690	1814-3402	S472, S472
635-685	33+	3520	3220-3901	2676-3901	S472, W132
686-736	41	4500	4359-4613	3719-5670	S472, S472
737-786	26+	5484	5539-5661	4627-6124	S472, S472
787-837	17+	6904	6804-7076	6124-7666	M511, S472
838-888	15+	7729	-	6713-8301	M511, M153
889-939	20	10433	-	8936-11022	M511, S472
940-990	17+	13336	-	11748-16783	M153, J13
991-1040	10+	13700	-	12565-14697	W132, W132
1041-1091	10+	16065	-	15876-20866	W74, J13
1092-1142	11+	18326	-	-	M153, J24
1142-1192	2	21546	-	20412-22680	W74, M153
1346	1	37194	-	-	T113
1410	1	43090	-	-	M153

The above data came from Alabama A38,39,40, S472; Iowa M511; Ohio T113; Oklahoma J13,24, M153; Tennessee W132; West Virginia W64.

	No.	TL	
H420 TN Watts Bar L.	-	-	$r = .966$ $\log W = -6.080 + 3.421 \log TL$
M511 IA Des Moines R.	59	102-991 mm	$= -5.334 + 3.138$
J24 OK Grand L.	-	-	$= -4.917 + 3.233$
H419, M153 OK	-	-	$= -4.974 + 3.255$
M291 KS	62	-	$= -5.387 + 3.099$
S472 AL	3564	51-305	$= -4.75 + 2.89$
	297	330-610	$= -5.45 + 3.18$
	26	635-914	$= -6.15 + 3.44$

The slope of the 51-305 mm catfish is probably unduly affected by the 51 mm catfish which had too high reported weights.

	Length	No.	Mean K(SL)	Range
D179 TX	225-3771 mm	52	1.96	1.24-2.83
			Mean K(TL)	
M511 IA Des Moines R.	102-987 mm	59	1.05	0.75-1.41
S472 AL	76-914 mm	2713	0.97	0.72-1.44
F105, N94,115, S228,229, 276,393 SD	-	43	1.12	1.00-1.39

The condition indexes for M511 and S472 show an increase with increase in length, which is also indicated by the slope of the regression lines.

	TL (except when marked SL)			Weight		
	No.	Mean	Range	No.	Mean	Range
At hatching						
S271 AL	+	13	-	-	-	-
Age 0-June						
S271 AL (22 days old)	+	-	20-36	+	0.5	-
Age 0-July						
M291 KS rivers	21	33	18-56	-	-	-
Age 0-Aug.						
M291 KS rivers	104	69	30-104	-	-	-
M511 IA Des Moines R.	1	104	-	-	-	-
Age 0-Sept.						
H427 TX (total survival in one pond)	8	229	-	8	227	-
TX	+	-	89-102	+	3	-
Age 0-Oct.						
D166 KS	+	102	81-122	-	-	-
M511 IA Des Moines R.	13	107	76-135	-	-	-
F8 TX	+	203	-	+	0.9	-
Age 0						
O37, T44 OK	4	71	51-86	4	5	-
Age I						
J24, O37, T44 OK	28	132	84-126	25	32	14-59
M511 IA Des Moines R.	11	185	104-249	-	-	-
C312, D166, S207 KS	5+	201	152-241	4	54	45-59
B94, U4 Mississippi R.	+	193	178-203	-	-	-

	No.	TL (except when marked SL) Mean	Range	No.	Weight Mean	Range
Age II						
L38 IL Kaskaskia R.	+	SL 203	-	-	-	-
J24, T44 OK Grand L.	67	208	145-366	67	100	54-181
M511 IA Des Moines R.	23	269	231-310	-	-	-
C312, D166, S207 KS	15+	287	221-378	11	345	186-486
B24, U4 Mississippi R.	+	297	254-330	-	-	-
O37 OK Heyburn L.	6	439	305-559	6	590	-
W7 OK Duncan L.	1	SL 381	-	1	907	-
Age III						
L38 IL Kaskaskia R.	+	SL 279	-	-	-	-
J24, T44 OK Lower Grand L.	38	246	203-267	38	150	
M511 IA Des Moines R.	6	343	315-366	-	-	-
J24 OK Upper Grand L.	13	351	224-409	13	431	-
B94, U4 Mississippi R.	+	373	305-414	-	-	-
C312, S207 KS	8	401	279-553	1	680	-
M153 OK Boomer L.	4	467	429-508	-	-	-
H420 TN Watts Bar L.	1	663	-	-	-	-
B181, O37 OK Heyburn L.	12+	665	572-737	12	3375	-
Age IV						
J24, T44 OK Lower Grand L.	36	305	236-434	36	327	-
L38 IL Kaskaskia R.	+	SL 356	-	-	-	-
B94, U4 Mississippi R.	+	429	386-455	-	-	-
J24, OK Upper Grand L. and Neosho R.	14	472	335-587	14	1248	-
M153 OK Boomer L.	3	582	551-597	-	-	-
C312 KS	4	622	602-648	-	-	-
H420 TN Watts Bar L.	9	732	-	-	-	-
Age V						
J24, T44 OK Grand L.	29	429	246-561	29	893	726-953
B94, U4 Mississippi R.	+	490	445-546	-	-	-
L38 IL Kaskaskia R.	+	SL 432	-	-	-	-
C312 KS	2	648	584-711	-	-	-
J24 OK Neosho R.	3	615	589-635	3	2903	-
H420 TN Watts Bar L.	6	625	-	-	-	-
M153 OK Boomer L.	17	698	650-729	-	-	-
Age VI						
J24, T44 OK Grand L.	9	498	381-587	9	1315	590-1633
B94, U4 Mississippi R.	+	561	559-594	-	-	-
J42 OK Ft. Gibson	1	630	-	-	-	-
J24 OK Neosho R.	15	660	602-729	15	3629	-
H420 TN Watts Bar L.	4	782	-	-	-	-
M153 OK Boomer L.	4	818	775-871	-	-	-
C312 KS	5	818	749-909	-	-	-
Age VII						
J24 OK Lower Grand L.	2	498	-	2	1547	-
J24, T44 OK Upper Grand L., Neosho R.	15	686	599-754	15	4354	-

		TL (except when marked SL)				
Age VII (cont.)	No.	Mean	Range	No.	Mean	Range
B94, U4 Mississippi R.	+	608	610–698	-	-	-
C312 KS	2	851	813–889	-	-	-
M153 OK Boomer L.	4	895	802–978	-	-	-
H420 TN Watts Bar L.	1	895	-	-	-	-
Age VIII						
J24 OK Lower Grand L.	1	452	-	1	1134	-
J24 OK Neosho R.	24	764	615–862	24	5806	-
B94, U4 Mississippi R.	+	795	660–864	-	-	-
Age IX						
B94 IA Mississippi R.	+	711	-	-	-	-
J24 OK Neosho R.	7	844	759–953	7	8029	-
B94, U4 IL, MO Mississippi R.	+	895	889–902	-	-	-
C312 KS	2	1031	902–1135	-	-	-
M153 OK Boomer L.	2	1029	1012–1044	-	-	-
H420 TN Watts Bar L.	1	1077	-	-	-	-
Age X						
B94 IA Mississippi R.	+	838	-	-	-	-
J24 OK Neosho R.	8	920	780–1046	8	10705	-
U4 IL Mississippi R.	+	953	-	-	-	-
M153 OK Boomer L.	3	965	851–991	-	-	-
C312 KS	1	1118	-	-	-	-
Age XI						
B94 IA Mississippi R.	+	902	-	-	-	-
J24 OK Neosho R.	8	1012	866–1102	8	14470	-
M153 OK Boomer L.	6	1002	914–1167	-	-	-
H420 TN Watts Bar L.	2	1148	-	-	-	-
Age XII						
C312 KS	1	889	-	-	-	-
B94 MO Mississippi R.	+	965	-	-	-	-
J24 OK Neosho R.	4	1092	1062–1128	4	16828	-
M153 OK Boomer L.	4	1161	1092–1219	-	-	-
Age XIII						
B94 IA Mississippi R.	+	940	-	-	-	-
M153 OK Boomer L.	9	1123	1046–1384	-	-	-
Age XIV						
B94 IA Mississippi R.	+	978	-	-	-	-
M153 OK Boomer L.	6	1133	1095–1186	-	-	-
Age XV						
B94 MO, IA Mississippi R.	+	1016	991–1041	-	-	-
M153 OK Boomer L.	5	1176	1123–1410	-	-	-
Age XVI						
M153 OK Boomer L.	4	1110	945–1283	-	-	-
Age XVIII						
M153 OK Boomer L.	4	1097	1067–1130	-	-	-
Age XIX						
M153 OK Boomer L.	1	1067	-	-	-	-

Average calculated TL at each annulus

	No.	1	2	3	4	5	6	7	8	9	10	11	12	13	14	15	16	17	18	19
J24 OK Lower Grand	59	64	127	185	259	340	386	455	442											
J45 OK Verdigris R.	28	91	155	206	274	320	373	419	523	584	615									
tributaries	4	81	170	249																
S391 OK Rock C.	6	102	178	224	282															
P112 MO Salt R.	52	76	155	231	300	348	421	452	503	599										
J31 OK Grand L.	221	86	160	241	323	384	439	486	518											
J25 OK Quall's L.	23	71	178	277	351	406	528	572												
E61 OK Salt C.	9	107	203	277	427															
J24 OK Upper Grand L.	61	86	175	287	411	465	544	625												
M291 KS Neosho R.	79	97	226	320	391	437														
J25 OK Tenkiller L.	19	84	201	330	437	516														
H150 OK Illinois R.	46	-	257	333	442	518														
J27 OK Spavinaw L.	1	86	246	356	414															
M291 KS Big Blue R.	74	142	262	366	483	630	701	773												
H220 OK Ft. Gibson	314	132	264	378	508	648	737	831	896											
J24 OK Neosho R.	86	130	259	381	490	584	655	719	785	878	945	1014	1074							
M153 OK	723	117	246	386	508	594	658	734	824	891	973	991	1054	1087						
J25, J41 OK Ft. Gibson	31	89	229	394	493	627	638													
H419 OK	+	114	234	358	486	579	645	686	764	777	849	904	973	1005	991	991	914	950	1085	1067
slowest	-	38	122	173	231	292	376	421	465	538	559	592	607	627	650	674	693	826	1057	
fastest	-	254	528	691	742	916	940	1031	1102	1057	1105	1113	1168	1194	1163	1179	1107	1095	1097	
J101 OK L. Eucha	16	170	287	394																
J101 OK Spavinaw L.	29	130	221	300	399	490														
L195 OK Arkansas R.	24	168	320	450	579	587														
L195 OK Cimarron R.	16	135	290	371	452	493	579	660	703											

Average weight at each annulus

	1	2	3	4	5	6	7	8	9	10	11	12	13	14	15	16	17	18	19
H419 OK	14	118	468	1284	2282	3243	3955	5625	5942	7893	9707	12338	13744	13109	13109	10070	11431	17608	16647
slowest	0.5	14	45	113	245	558	811	1116	1800	2032	2449	2658	2962	3325	3674	4100	7212	16148	
fastest	154	1696	4051	5080	10151	11022	14923	18552	16148	18688	19097	22408	24041	22090	23043	18824	18144	18280	
M511 IA Des Moines R.	5	41	132	381	912	1602	2545	3511	4808	6214	7530	8709	9480	8437	6305				
Increments	5	36	95	254	495	689	848	967	1170	1456	1429	1243	1094	731	513				

FLATHEAD CATFISH

	Average calculated TL and increments at each annulus												
	1	2	3	4	5	6	7	8	9	10	11	12	13
H420 TN Watts Bar L.	239	353	475	579	643	780	907	985	1021	1074	1100		
Increments	239	114	122	107	91	79	64	61	36	53	26		
Number	24	24	24	23	14	8	4	3	3	2	2		
M511 IA Des Moines R.	76	163	236	333	439	526	612	676	747	811	862	902	927
Increments	76	91	79	99	97	86	76	64	64	66	53	43	36
Number	61	57	49	33	22	20	17	15	11	10	9	7	5
H303 OK Lawtonka L.	124	307	541	665									
Increments	124	196	231	152									
Number	8	5	3	1									
O37 OK Heyburn L.	160	340	564										
Increments	160	185	274										
Number	25	18	12										

Age and growth were determined in most studies by studying the rings on sections of the pectoral spines. One to three early annuli were missing on some of the spines from larger fish (M511). Growth of tagged fish suggested the growth as determined by spines was of the right magnitude (M511). Slow growth in Grand Lake, Oklahoma, was believed to be due to overpopulation (J31) and growth was more rapid on the shallow mud flats than in the clear rocky areas. In Salt River, Missouri, growth was more rapid in the downstream sections (P112). Growth in the Mississippi River was faster in Missouri and in Illinois than in Iowa, farther north (B94, U4). Faster growth in the Blue River, Kansas, was associated with use of fish as food earlier than in the Neosho River (M291). Young flathead catfish grew more slowly in Fort Gibson Reservoir, Oklahoma (H220), after impoundment than before.

Yearling flathead catfish stocked in May at 178 mm and 45 g averaged 384 mm and 635 g 6 months later (S504). Yearlings stocked at 50 per acre with an abundance of forage fish averaged 862 g, with 92% survival, at the end of the season, compared to 680 g and 86% survival when stocked with other fish (S504). Flathead catfish were larger at all ages in channeled than in unchanneled areas of the Missouri River (L219), probably because of greater abundance and aggregation of forage fish around pilings in the channeled area.

In Kansas, male flatheads mature at 380-460 mm or at 3-5 years, and females at 460-510 mm or 4-6 years (D166, M291) or at 4 years (H427). In the Mississippi River, a few females were mature at 380 mm, but most not until 457 mm, at age IV-V (B94). Loss of a light patch at the tip of the upper lobe of the caudal fin may indicate sexual maturity (M291). Spawning is in early June to late July in Kansas (M291, D166) and in May in Texas (H427). The male guards the nest and young (K118).

M291 KS   3 females   305-610 mm TL   6900-11300 eggs per female

Eggs hatched in 6-7 days at 75-82F (G157). At 88 days the young fish were 48 mm and 1.7 g, with 60% survival (G157). At 110 days, young were 75-102 mm in troughs (S518). Flathead catfish may spawn in brood pens without (H427) or with gonadotropin injections (G157, S518). In lightly stocked troughs the young fish developed and defended territories, a condition not observed in heavily stocked troughs (S518). Propagation and rearing methods are given by G157, H427, S518, S271.

In plastic-lined tanks flathead catfish ate largemouth bass, white catfish, green sunfish, and goldfish, in that order of selectivity, but in earthen ponds white catfish were more vulnerable than the largemouth bass (H428). Fifty large flathead catfish per acre did not completely correct the balance in a population of stunted bluegills in 320 days (H428).

Sizes taken in gill net meshes

	Bar measure	No.	TL	Range
L216 LA	25	1	358	
	38	1	297	
	64	2	503	457-546
	76	3	638	572-711
	89	3	630	572-660

Flathead catfish under 100 mm TL were usually in riffles and fed on insect larvae, mostly Ephemeroptera and Trichoptera (D166, L219, M291). When 100 to 250 mm, they tended to be dispersed and feed on insect larvae, fish and crayfish (L219, D166, M291, B347). Those from 300-400 mm usually remained in or near brushpiles, and larger catfish tended to be solitary and in deeper pools (M291).

AMERICAN EEL, *Anguilla rostrata* (LeSueur)

The eel is a catadromous fish spawning in the deep Atlantic near Bermuda, but living most of its life in fresh water. They enter rivers from Labrador to Mexico and may go almost to the headwaters though not extending further up the Missouri River than South Dakota (E119, T113). Only the females go far above the tidal zone of rivers (E119).

		Weight		
TL	No.	Mean	Range	Citation
51	1	0.9	-	H61 Chesapeake Bay
76	2	1.7	1.1-2.0	H61
127	15	2.3	2.3-4.5	A39, S472 AL
152	13	2.6	1.4-	A39, S472
178	26	10	4-11	A39, H61, S472
203	6	18	-	S472
229	11	33	-	S472
254	12	23	15-45	S472
278	4	31	12-45	S472
305	10	49	28-68	H61, S472
330	17	79	60-96	H61, S472
381	7	91	68-96	H61, S472
406	7	119	82-156	H61, S472
432	5	150	136-159	H61, S472
457	1	181	-	S472
483	6	224	213-233	H61, S472
508	4	267	227-315	H61
533	10	292	227-340	H61, S472
559	11	372	290-417	H61, S472
584	9	439	340-567	H61, S472
610	8	553	408-635	H61, S472
762	2	1097	959-1261	H61
864	1	1701	-	F10 IL
1066	1	2552	-	E105 MD
1092	1	2948	-	J13
1321	1	3400	-	T113 OH
1524	1	6800	-	S386 MD

S256 NK $\log W = -5.748 + 3.035 \log TL$
S472 AL 144 eels 127–610 mm $= -6.94 + 3.47$ $r = .955$

S65 NS   149   91–750   average K(SL) 0.19 range 0.13–0.25
S472 AL 143   127–610   K(TL) 0.18 range 0.08–0.28

			TL		Weight		
		No.	Mean	Range	Mean	Range	Citation
Age	I	11	-	58–99	-	-	H61 Chesapeake
	II	13	-	112–175	-	-	H61
	III	18	241	216–259	26	17–34	S256 NK
	IV	99	292	236–302	48	28–51	S256
	V	223	348	312–356	82	62–91	S256
	VI	556	368	340–411	102	82–139	S256
	VII	491	386	358–478	128	99–235	S256
	VIII	206	462	429–549	227	184–366	S256
	IX	101	500	472–566	286	252–403	S256
	X	29	549	488–607	383	272–499	S256
	XI and older	7	671	612–744	794	542–1077	S256

Ages of the younger fish were determined by length-frequencies (H61) and those of the older fish by scale examination (S256). B360, however, questioned the scale data on the European eel and said that growth is extremely variable (G111). Smaller eels were found in the stream with the higher population density in one study (G111). Two eels tagged at 360 mm grew at least 138 and 325 mm per year (G111).

Males mature at 279–305 mm and females at 457 mm, and at 5 to 20 years of age (K118). They spawn in the Sargasso Sea, in the winter, and die after spawning (K118). The number of eggs per female has been reported at 5–20 million (M426), 9 million (B23), and 10.7 million in a 813 mm female (F10).

Eels were found living in water up to 33 and 35C (T102).

Insects were the principal foods in a New Brunswick stream, but crayfish and fish, about 14% of which were salmon fry, were also important (G63).

It was estimated that the annual run from Crecy Lake, New Brunswick, amounted to an annual removal of 1.5–5.2 kg/ha of fish flesh (S519).

WHITE RIVER KILLIFISH, *Crenichthys baileyi* (Gilbert)
  Moapa River, California, and White River and Pahranagat Valleys, Nevada (E119).

   K17 Nevada   Age group   0–15 days   7.3 mm SL

  They spawn at 32C at night, laying one egg at a time, adhesive to vegetation (K17).

DEVILS HOLE PUPFISH, *Cyprinodon diabolis* Wales
  Devils Hole, Ashe Meadows, Nye County, Nevada (E119). With the constant temperature of the spring, 33–34C, breeding appears to continue throughout the year.

DESERT PUPFISH, *Cyprinodon macularius* Baird and Girard
  Desert streams of southern Nevada and Utah and southeastern California (E119).

No differences in length-weight relationships were noted at different temperatures and salinity in the Salton Sea (K92).

At hatching, desert pupfish average 5.3 mm TL (C68, K92), but at temperatures above 30C and 4.5% salinity the average size at hatching may be only 4 mm.

K92 CA Salton Sea	Age in weeks	Length
	4	7.6-11.8
	8	9.5-16.9
	16	12.6-24.0
	24	15.1-28.2

In less than a year they may reach 45 mm SL (W147). Best growth is at 30C, next at 25C, next 35C, then 20C and 15C. After 12 weeks, the pupfish kept at 25C, 30C and 35C were about the same size and by 35 weeks, the 20C fish had reached this length. By this time most of the 30C fish were dead. Growth was also best at 3.5% salinity, next at 1.3%, then 5.5%, and then in fresh water (K92). They will grow in up to 7% salinity (W147). Growth rate was reduced in crowded populations (K92). Maximum food intake was at 30C, 20 worms per day; next at 35C, 18 per day; then 25C, 20C, and 15C, 2 worms per day (K92). Food conversion was maximal at 20C, then 15C, 25C, 30C, and 35C. Food conversion was better at 1.5% salinity than at 3.5% or fresh water (K92).

Desert pupfish may mature and spawn the same season if hatched early (W147), and sexual maturity is reached earlier in fast-growing than slow-growing individuals (K92). They spawn only above 20C, with maximum spawning at 28-32C (K92).

VARIEGATED MINNOW, *Cyprinodon variegatus* Lacépède
Fresh and brackish water along the coast from Cape Cod to the Rio Grande (E119).

At hatching, variegated minnows averaged 4 mm (K28). In August they were 32 mm, in Maryland (H60). This species is found in 0.2-3.57% saline waters (G86). They feed mostly on algae and mosquito larvae and pupae (H305). They spawn from April to late summer (K118).

MARSH KILLIFISH, *Fundulus confluentus* Goode and Bean
Coastal swamps Maryland to Florida and west to Louisiana (E119).

This killifish was feeding primarily on mosquito larvae and pupae, but also ate some Palaemonetes shrimp and a few small fish (H305).

BANDED KILLIFISH, *Fundulus diaphanus* (LeSueur)
North Dakota and Iowa to New York, and Quebec to South Carolina (E119).

L132 MN Linwood L.	Age 0-July	36	25 TL	
	0-Aug.	101	41	
	0-Sept.	124	46	
	0-Oct.	20	48	
T113 OH	0-Oct.	+	20-58 TL	
S156 NK Gibson L.	I	+	51	33-64
T113 OH	I	+	38-71	

566 BANDED KILLIFISH

Scott (S520) discussed scale regeneration in this species.

Banded killifish mature at 64 mm and one female had 252 eggs (K118). They spawn April to September (K118), in Quebec in May (R259) and in Illinois in May (R27). The temperature at spawning was recorded at 21C (R259). The male courts and selects the female and the eggs are released in clusters, with filaments which attach to plants (R259).

In Telford Lake, Nova Scotia, bottom organisms constituted the food and no pelagic microcrustacea were taken (S70). In Ontario banded killifish were versatile feeders taking aquatic and small flying insects, planktonic crustacea, some plant seeds, and fine algae material (K125, 129). They were the only species eating ostracods (K125). They have been found in water up to 38.3C (T102).

MUMMICHOG, *Fundulus heteroclitus* (Linnaeus)
Brackish and fresh water from Maine to Texas (E119).

R262 CT		Quinnipiac R. salinity 1.6-3.6 ppt		TL	Harbor 20-30 ppt	
Age	No.	Mean	Range	No.	Mean	Range
0-June	24	12.1	7-25	14	21.6	14-28
0-July	50	23.5	10-40	62	25.6	13-47
0-Aug.	24	29.8	19-52	12	36.7	22-46

Sexual maturity of male mummichog kept in fresh water was less advanced than those kept in brackish water (B352). This species has been used for much research on hormones and other physiological problems. It has also been widely used in embryological experimentation (B353, O46, O47, S521, R262). Eggs of this species in North Carolina lacked fibrils on the chorion, characteristic of eggs at Woods Hole, Massachusetts (B353). Spawning season is from April to September in North Carolina (B353, K118) but only in late June and early July at Woods Hole (B353). One female had 460 eggs (K118).

PLAINS KILLIFISH, *Fundulus kansae* Garman
South Dakota and Wyoming south to Texas (E119).
The larval development to 18 mm is described by K117.

C134 OK Stillwater C.   age 0-June   12-19 mm SL

Plains killifish are usually in shallow streams with sandy bottoms, highly alkaline or saline (C315). Spawning is in June and July in Oklahoma (H429), in July in Wyoming (S146), and in August in New Mexico (K117) at 28C. They spawn in shallow pools, 50-100 mm deep over gravel and sand bottom. Males isolate themselves over the spawning grounds but do not have real territories nor nests (K117). Courtship involves circling and sinuous swimming. Males have orange on the fins and broader and fewer vertical bars than females (C315). The female has a membraneous fold ensheathing the anal fin base.

Surface insects and floating material constitute the principal foods, but they may also take bottom forms stirred up by quick sidewise darting motions (S146). They make good aquarium fishes, but did not hold well enough in the laboratory to rate highly as bioassay animals (G156).

STRIPED KILLIFISH, *Fundulus majalis* (Walbaum)
Brackish water, sometimes entering fresh water, Cape Cod to Florida (E119).

C306 MD Hungerford C.　　　males　　　$\log TL = 1.557 + 0.338 \log W$
　　　　　　　　　　　　　　females　　　$\log TL = 1.620 + 0.330 \log W$

R262 CT	New Haven Harbor, salinity 20-30 ppt		
Age	No.	TL	Range
0-June	10	12.4	9-15
0-July	57	28.4	16-45
0-Aug.	12	39.7	27-51

	Average calculated TL at each annulus						
	1	2	3	4	5	6	7
C306 MD Males	65	86	103	115	126	130	
Increments	65	22	16	11	10	9	
Number	158	135	79	43	14	2	
Females	61	89	106	120	128	138	150
Increments	61	21	17	13	10	11	10
Number	156	138	116	79	44	15	4

These killifish spawn April to September and males mature at 64 mm and females at 76 mm (K118). A female produced 540 eggs (K118). Embryology is described (R262).

BLACKSTRIPE TOPMINNOW, *Fundulus notatus* (Rafinesque)
Iowa to Ohio, south to Mississippi and east Texas (E119, T113).

T113 OH　　age 0-Oct., 25-51 mm　　　age I, 28-64 mm　　　largest, 76 mm

This species was spawning in Illinois in late May (R27).

BLACKSPOTTED TOPMINNOW, *Fundulus olivaceous* (Storer)
Oklahoma and Texas to western Florida (E119).

				Weight		
TL	Citation	No.	Mean	Range	K(TL)	
25	S472 AL	43	0.45	0.35-0.50	2.83	
	A39, 40 AL	4	1.1	-	-	
51	S472 AL	108	1.2	0.5-1.4	0.93	
	A39, 40 AL	19	0.9	-	-	
76	S472 AL	66	3.8	2.3-9	0.86	
	A39, 40 AL	2	5.9	-	-	

Since the recorded weights for the 25 mm fish are obviously too high, the length-weight regression in S472 is erroneous.

This species was rated as a fair bioassay animal (B349), because they ate well and could be kept easily in the laboratory, but these topminnows were initially overly excitable in test chambers.

RAINWATER KILLIFISH, *Luciana parva* (Baird and Girard)
Swamps and brackish waters along the coast from Cape Cod to Mexico (E119). Copepods and mosquito larvae were the foods of this species in a Florida marsh (H305).

RIVULUS, *Rivulus marmortus* Poey

H160 FL Indian R. age 0-Aug. 22 fish　　7.5-11.1 mm SL　　10.0-14.2 mm TL

The caudal fin is shorter proportionately on longer fish, so that at 7 mm, TL = 1.03 SL, and at 30 mm, TL = 1.022 SL (H160).

MOSQUITOFISH, *Gambusia affinis* (Baird and Girard)
Southern Indiana and Illinois south to Mexico and Florida and north up the coast to New Jersey (E119, T113). Also widely introduced in warm parts of the world for mosquito control.

H60 U.S.	age 0-June	8-10 mm SL		age 0-Sept.	13-25 mm SL	

K22 IL		Females			Males	
Age	No.	Mean TL	Range	No.	Mean TL	Range
0-June	646	16	10-20	1321	17	10-23
0-July	1413	23	14-29	3075	26	11-45
0-Aug.	2309	23	11-31	4067	26	10-49
0-Sept.	359	21	15-34	3154	26	13-49
0-Oct.	253	19	14-28	4545	28	11-49
I-May	109	27	22-29	41	38	33-45
I-June	315	28	23-31	145	43	32-59
I-July	192	28	24-30	540	45	33-58
I-Aug.	126	28	24-31	921	40	34-55
I-Sept.	-	-	-	398	40	36-47
I-Oct.	-	-	-	196	40	36-45

In a spring with water temperatures of 22-24C year-round in Montana, seasonal changes in size gave no indication of growth, and spawning occurred from March through October (B354). The smallest female with yolk-sac fry was 29 mm in Montana (B354), compared to 23 mm in North Carolina (H60).

Mosquitofish spawn from May to September, and mature at 30-36 mm (K118). Eggs per female range from 1-315 (K22), 76 eggs (K27), 100 eggs (S58), or average 40, with a maximum of 63 (H60), or 211 for *Gambusia affinis holbrooki* (H61).

The ultimate maximum temperature was reported at 37.3C (S333).

Acclimated at	Lethal temperatures	
15C	1.5	35.4
20	5.5	37.3
35	14.5	37.3

The principal foods were mosquito larvae and pupae, a few copepods, algae and small fish (H305, F148, B355).

Gambusia alvarezi Hubbs and Springer (Common name not designated.)
The maximum length was 42 mm, a female, in a Texas spring (H369).

BIG BEND GAMBUSIA, *Gambusia gaigei* Hubbs
In Texas springs at 25-32.5C, females reached a maximum of 28 mm and males of 21 mm, but in the laboratory at cooler temperatures females reached 45 and males 30 mm (H369).

LARGESPRING GAMBUSIA, *Gambusia geiseri* Hubbs and Hubbs
In Texas springs at 19-26C, the maximum length was 37 mm, a female (H369).

Gambusia hurtadoi Hubbs and Springer (Common name not designated.)
    In Texas springs at 29-33C, the maximum length of females was 36 mm
and of males 22 mm, but in the laboratory at cooler temperatures, females
reached 45 mm and males 31 mm (H369).

PECOS GAMBUSIA, *Gambusia nobilis* (Baird and Girard)
    In Texas springs, at 21-26C, high in carbonates, females reached a maxi-
mum length of 40 mm (H369).

BLOTCHED GAMBUSIA, *Gambusia senilis* Girard
    In Texas ditches at 12-30C, females reached a maximum length of 46 mm
(H369).

SAILFIN MOLLY, *Molliensia latipinna* LeSueur
    Common near coast and in fresh water from South Carolina into Mexico
(E119). These fish feed mostly on vascular plants and on mosquito larvae in a
Florida marsh (H305).

OZARK CAVEFISH, *Amblyopsis rosae* (Eigenmann)
    In small streams in caves in southeastern Missouri and Arkansas (E119,
P169).

P169	Age	No.	SL	Range
Free swimming fry	5-6 months	-	-	-
First scales	10-12 months	-	-	12-14
Vent migration complete	12-14 months	-	-	12-18
First annulus	19-21 months	-	-	16-24
First reproduction	36-48 months	-	-	36-42
Oldest individuals	45-57 months	-	-	44-52
just under 1 year		2	10	-
2 years		3	17	16-19
2.9-3.9 years		16	30	26-36
3.9-4.7 years		15	40	35-47
4.9-5.3 years		4	46	43-50

    The slopes of the growth curves were 1.77, 0.60, and 1.03 for the 1st, 2nd,
and later years respectively (P169). The average number of ova in 8 mature
females was 23. It was estimated that 20% of the adult females breed per year
and the compound interest population growth rate is 0.002 per year (P169).
The annual breeding cycle is probably the same as in the northern cavefish.
    Ozark cavefish feed on the substrate and under rocks, eating copepods and
a few small salamanders, crayfish, isopods, amphipods, and their own young
(P169). Metabolic rates are low and this species is probably most adapted of
the cavefish to living in complete darkness.

NORTHERN CAVEFISH, *Amblyopsis spelaea* DeKay
    In subterranean streams of the cave regions of Kentucky and southern
Indiana (E119).

P169		No.	SL	Range
Free swimming fry	5-6 months	-	-	8-10
First scales	12-13 months	-	-	15-17
Vent migration complete	17-20 months	-	-	16-28
First annulus	19-21 months	-	-	18-26
First reproduction	36-48 months	-	-	45-56
Oldest individuals	73-84 months	-	-	76-85
	just under 1 year	3	8	8-9
	1.9-2.4 years	3	16	15-17
	2.9-3.6 years	6	30	26-37
	3.8-4.5 years	7	41	34-51
	4.7-5.2 years	8	56	50-64
	5.6-6.2 years	11	66	56-77
	6.6-7.1 years	6	75	73-78

The slopes of the growth curves were 1.40, 0.65, and 1.30 for the 1st, 2nd, and later years respectively (P169). The average number of ova in 7 mature females was 69.5. It was estimated that 10% of the adult females breed each year and the compound interest population growth was .004 per year (P169). Breeding occurs during high water from February through April. The females carry the eggs in their gill cavities until the yolk sacs are lost by the fry, which thus become free-swimming in late summer or early fall.

Cavefish smaller than 45 mm mainly eat copepods, and larger cavefish eat *Asellus, Crangonyx*, small crayfish, and their own young (P169). Metabolic rates are low, and show a greater decline from active to resting stage than non-cavefish (P169).

SPRING CAVEFISH, *Chologaster agassizi* Putnam

Springs in southern Illinois, Kentucky, and Tennessee (E119). Sometimes in caves or wells (P169). In the springs, they usually retreat underground during the day, emerging into the spring at night (P169).

P169		No.	SL	Range
Free swimming fry	1 month	-	-	-
First scales	2.5 months	-	-	15-19
Vent migration complete	4-5 months	-	-	15-32
First annulus	8-10 months	-	-	25-36
First reproduction	11-12 months	-	-	34-44
Oldest individuals	14-28 months	-	-	38-61
Age 0 to 0.8, Mammoth Cave		3	20	19-21
springs		19	22	15-29
Age 0.9-1.8, Mammoth Cave		4	34	30-37
springs		15	44	34-53
Age 2, Mammoth Cave		1	45	-
springs		5	48	46-51
Age 2.7-3.0, springs		2	59	58-60

The slopes of the growth curves in the springs were 7.28 and 1.42 in the first half year and later years respectively (P169). Spawning is underground in February when water levels are at the year's maximum. Fry appear and adults reappear in the open springs in early March. The average number of ova in 7 adult females was 152. It was estimated that all adult females breed each year and the compound interest population growth rate is 1.00 per year (P169).

W94 IL	3 females, 53-59 mm	80-101 eggs per female
	2 females, 60-61 mm	202-285 eggs per female

Food is more available in springs than in caves, which probably accounts for the slower growth in Mammoth Cave (P169). Activity is less than in the swampfish.

SWAMPFISH, *Chologaster cornuta* Agassiz
Lowland swamps from southern Virginia to Florida (E119).

P169		No.	SL	Range
Free swimming fry	0.6 month	-	-	-
First scales	1.5 months	-	-	7-18
Vent migration complete	3-4 months	-	-	16-30
First annulus	7-10 months	-	-	21-48
First reproduction	11-12 months	-	-	23-55
Oldest individuals	14-15 months	-	-	25-57
	0.4 year	7	16	14-17
	0.9-1.3 years	40	-	24-44

The slopes of the growth curves were 6.25 and 1.97 for the first half year and the rest of life respectively (P169). Swampfish live in cypress swamps and small shaded streams but come to open streams in early April to spawn (P169). The average number of mature ova in 13 adult females was 98 and the compound interest population growth is estimated at 1.00 per year (P169).

The swampfish is nocturnal and hides under debris with only the head exposed (P169). They feed on small crustacea and insects.

SOUTHERN CAVEFISH, *Typhlichthys subterraneus* Girard
Streams in cave regions of Kentucky, Tennessee, Alabama, and Missouri (E119). They live in caves near the water table where conditions are more uniform than for other cavefish, but where food is scarce (P169).

P169		No.	SL	Range
Free swimming fry	2-3 months	-	-	-
First scales	3 months	-	-	9-10
Vent migration complete	5 months	-	-	10-23
First annulus	7-10 months	-	-	21-30
First reproduction	22-24 months	-	-	33-40
Oldest individuals	38-50 months	-	-	45-62
	0.4-0.6 year	3	13	12-13
	0.8-1.3 years	12	23	18-29
	1.8-2.8 years	24	36	28-47
	2.9-3.5 years	7	45	43-48
	4.1-4.4 years	2	58	55-60

The slopes of the growth curves were 3.50 and 1.08 for the first and later years respectively. Breeding is probably in spring as the females are spent in June and July. The average number of mature ova in 10 adult females was 49.8. It was estimated that about one half the adult females spawn each year and that the compound interest growth rate is 0.01 per year (P169).

Copepods were the principal foods but stomachs also contained larvae of caddisflies and midges, Cladocera, isopods and crayfish (P169).

BURBOT, *Lota lota* (Linnaeus)

Fresh and brackish water in Europe, northern Asia, and North America (S471). In North America it is found as far south as Oregon, Missouri, and Pennsylvania (T113).

V18 L. Michigan	TL = 1.071 SL	C24 MN	50 fish	TL = 1.070 SL
C22 MN L. of Woods	TL = 1.071 SL	C35 MN L. Vermilion		TL = 1.085 SL

In the tabulations which follow SL has been converted to TL on the basis of TL = 1.071 SL, and FL has been considered equal to TL

		Weight		
TL	No.	Mean	Range	Citations
254-278	+	154	-	F124 SD Oahe
279-304	8+	231	191-259	F124, C125 L. Erie
305-329	33+	249	222-272	F124, F125 SD Oahe, C125
330-355	71+	286	227-318	F124, F125, C125, M157 ON L. Simcoe
356-380	138+	354	318-372	F124, F125, C125, M157
381-405	197+	386	227-426	F125, C125, M157, S164 L. Huron
406-431	416+	531	499-549	F125, C125, M157
432-456	206+	635	595-649	F125, C125, M157
457-482	166+	762	726-816	F125, C125, M157
483-507	142+	971	889-1179	C125, M157
508-532	155+	1003	771-1134	C125, M157, S164
533-558	335+	1306	626-1361	C125, M157
559-583	173+	1588	1474-2101	C125, M157, H26 Can.
584-609	110+	1686	1620-1860	C125, M157
610-634	118+	2005	1794-2608	C125, M157, H26
635-659	76+	1928	1588-2223	C125, M157, S164
660-685	66+	2418	2172-3742	C125, M157, H26
686-710	7+	2636	1928-3121	C125, M157, H26, C126 BC
711-736	2+	3084	2994-3175	M157, H26
737-761	4+	3674	3266-4109	M157, H26
762-786	+	2722	-	S164 L. Huron
787-812	+	3810	-	M157
813-837	+	3992	-	M157
838-863	+	4717	-	M157
864-888	2	6350	4990-7738	H26, R97 NT Great Slave L.
938	1	8392	-	K93 NT Great Slave L.
940	1	6804	-	H26
965	1	5302	-	H26
991	1	11622	-	H26
1016	1	11340	-	H26
1041	1	11794	-	H26

The largest reported for North America was 27216 g in Alaska by Dall 1870, and for Europe, 34020 g by Wynne-Edwards 1951 (K93, K98).

C22, 23, 34, 35 MN	67 burbot	Average K(SL) 0.87	0.74-1.10
S225, 276, F105, 124, 125 SD	20+ burbot	Average K(TL) 0.67	0.63-1.22

H205 MB L. Winnipeg	$\log W = -3.071 + 2.65 \log TL$
Mukutawa R.	$\log W = -3.770 + 3.18 \log TL$

The slopes were statistically different.

	No.	TL Mean	Range	No.	Weight Mean	Range
Age 0-June						
F4, F5 L. Erie	97	-	3-13	-	-	-
Age 0-July						
F4, F5, C125 L. Erie	23+	-	8-33	-	-	-
C53 BC Okanagan	3	28	28-30	-	-	-
Age 0-Aug.						
F4, F5, C125 L. Erie	4+	-	30-64	-	-	-
C53 BC Okanagan	1	38	-	-	-	-
T113 OH	+	-	25-102	-	-	-
Age 0-Sept.						
C125 L. Erie	+	74	-	-	-	-
Age 0-Oct.						
C125 L. Erie	+	99	-	-	-	-
Age 0						
R160 NY Susquehanna R.	5	81	-	-	-	-
M277 Ger.	+	-	109-211	-	-	-
Age I						
M28 ON	3	-	119-160	-	-	-
T113 OH	+	-	152-203	-	-	-
C125 L. Erie	11	226	-	-	-	-
M277 Ger.	+	-	180-391	-	-	-
Age II						
R160 NY Susquehanna R.	21	208	-	-	-	-
M28 ON	2	254	249-259	-	-	-
C125 L. Erie	211	345	257-470	-	-	-
M277 Ger.	+	-	221-511	-	-	-
Age III						
C35 MN L. Vermilion	5	351	312-386	5	313	200-459
M28 ON	3	-	279-381	-	-	-
R160 NY Susquehanna	72	251	-	-	-	-
H205 MB Mukutawa	5	373	-	5	318	-
C125 L. Erie	488	404	300-516	-	-	-
M277 Ger.	+	-	279-551	-	-	-
H205 MB Winnipeg L.	4	411	-	4	590	-
H156, H157 ON L. Simcoe	16+	450	419-495	16	735	567-907
Age IV						
R160 NY Susquehanna	59	279	-	-	-	-
H28 ON	5	-	300-391	-	-	-
C35 MN L. Vermilion	5	427	333-503	5	1034	281-993
C125 L. Erie	470	455	343-579	-	-	-
H205 MB Muktuawa R.	29	460	-	29	816	-
M156 ON Simcoe L.	13	465	394-533	13	1275	794-1615
H205 ON L. Winnipeg	16	490	-	16	998	-
M277 Ger.	+	-	300-569	-	-	-
Age V						
R160 NY Susquehanna	7	320	-	-	-	-
M28 ON	2	381	371-391	-	-	-
C35 MN L. Vermilion	3	503	439-559	3	912	539-1161
H205 MB	46	516	511-526	46	1120	1089-1134
C125 L. Erie	615	528	386-645	-	-	-
M156 ON L. Simcoe	10	556	457-648	10	1502	967-1729
M277 Ger.	+	-	371-660	-	-	-

	No.	TL Mean	Range	No.	Weight Mean	Range
Age VI						
R160 NY Susquehanna R.	5	328	-	-	-	-
M28 ON	2	384	351–439	-	-	-
R97 NT Great Slave L.	1	427	-	1	454	-
C35 MN L. Vermilion	3	523	503–536	3	989	821–1120
H205 MB	71	551	531–566	71	1393	1179–1547
M156, M157 ON L. Simcoe	15+	577	521–635	15	1588	1361–1814
C125 L. Erie	204	579	470–706	-	-	-
M277 Ger.	+	-	681–701	-	-	-
Age VII						
R160 NY Susquehanna R.	4	366	-	-	-	-
M28 ON	3	-	455–490	-	-	-
H205 MB L. Winnipeg	54	554	-	54	1315	-
C125 L. Erie	104	597	470–686	-	-	-
M277 Ger.	+	762	-	-	-	-
M156, M157 ON L. Simcoe	22+	645	622–660	22	2001	1701–2722
H205 MB Mukutawa R.	15	615	-	15	2087	-
Age VIII						
M28 ON	4	-	419–500	-	-	-
R97 NT Great Slave L.	1	508	-	1	1361	-
H205 MB L. Winnipeg	50	587	-	50	1678	-
C125 L. Erie	103	620	513–706	-	-	-
M156, M157 ON L. Simcoe	28+	681	635–749	28	2467	1814–2948
H205 MB Mukutawa R.	6	676	-	6	2722	-
Age IX						
M28 ON	3	-	419–480	-	-	-
R97 NT Great Slave L.	2	579	564–592	2	1120	934–1306
H205 MB L. Winnipeg	57	632	-	57	2223	-
C125 L. Erie	62	635	536–773	-	-	-
H205 MB Mukutawa R.	1	719	-	1	3901	-
M156, M157 ON L. Simcoe	10+	739	724–762	10	2745	2608–2948
Age X						
C125 L. Erie	35	660	579–749	-	-	-
H205 MB L. Winnipeg	20	668	-	20	2223	-
H205 MB Mukutawa R.	4	719	-	4	3810	-
M156, M157 ON L. Simcoe	1+	789	-	1	3175	-
Age XI						
R97 NT Great Slave L.	1	711	-	1	3289	-
H205 MB L. Winnipeg	22	921	-	1	2994	-
M156, M157 ON L. Simcoe	4	795	762–800	4	3121	2894–3175
Age XII						
M28 ON	1	541	-	-	-	-
H205 MB L. Winnipeg	18	782	-	18	3538	-
H205 MB Mukutawa R.	1	820	-	1	4627	-
M156, M157 ON L. Simcoe	1+	833	-	1	4309	-
Age XIII						
M28 ON	1	531	-	-	-	-
H205 MB L. Winnipeg	11	795	-	11	3810	-
M157 ON L. Simcoe	+	838	-	-	-	-
Age XIV						
H205 MB L. Winnipeg	6	787	-	6	3357	-
Age XVI						
R97 NT Great Slave L.	1	775	-	1	2722	-

The ages were determined by length frequencies and by scale analysis, by C135, but most other studies used otoliths (M156, M157, R160, H205, M277). Heating glycerine-stored otoliths at 170-180C for 40-60 minutes brought out the annuli (L143).

Growth was more rapid in brackish than in fresh water (M277). No sex differences in growth rate were noted except that all burbot over age VIII were females (M156). Disappearance of burbot after 3 years of impoundment in Lewis and Clark Lake, South Dakota, suggests that ecological conditions in the impoundment are unfavorable (W192).

Male burbot matured at age III and at 343 mm or 255 g in Lake Simcoe, Ontario, females at age III, 419 mm or 680 g (M157). Males appear on the spawning areas first with spawning reaching its peak January 31-February 5 in Lake Simcoe (M156), and February 3-17 in Poland (B285). Burbot did not feed during spawning but ran to tributaries to feed shortly after spawning (M156). One tagged burbot travelled 43 kilometers in 5 months and 6.2% of the tagged burbot were recaptured (S76).

	Size of females	No.	No. of eggs per female Mean	Range
B55 WY	305 mm	1	64498	-
	572 mm	1	451,473	-
	625 mm	1	522,385	-
	838 mm	1	1,444,122	-
C2, S146 WI	698 mm	1	1,153,144	-
D12			800,000	-
B23 Europe			160,000	-
M73			-	160,000-670,000
M157 ON L. Simcoe	419 mm, 680 g	1	45600	-
	711 mm	1	1,018,050	-

Small burbot, 80-123 mm, fed on insects and *Cottus* spp. (N117). Of 136 burbot collected in Lake Mille Lac, Minnesota (B231), 120 had eaten fish, mostly perch, and some had also eaten *Hexagenia* mayfly naiads. In Cree Lake, burbot had eaten only fish: cisco, whitefish and burbot (R138). Fish and crayfish were the main foods of burbot in Lake Vermilion (D183).

The effect of gill net selectivity on average sizes at various ages is evident in the following data from Lake Winnipeg (H205):

Age	II	III	IV	V	VI	VII	VIII	IX	X
In otter-trawls									
Mean weight	91	363	499	635	-	-	-	1855	-
Number	2	7	9	8	-	-	-	3	-
In gill nets									
Mean weight	363	454	680	771	771	726	862	953	1588
Number	1	9	18	38	29	25	14	1	2

			Sizes of burbot taken in gill net meshes					
Mesh Size	38	51	64	76	83	95	102	127	140
R97 NT									
Number	6	17	-	16	-	-	35	16	28
Mean TL	414	429	-	518	-	-	546	620	615
Mean Weight	907	771	-	1588	-	-	1406	2495	2313
C22 MN L. of Woods									
Number	-	2	4	7	7	3	111	-	-
Mean TL	-	267	361	531	457	732	589	-	-

FOURSPINE STICKLEBACK, *Apeltes quadracus* (Mitchill)

Mostly marine along Atlantic coast from Virginia northward, but sometimes in coastal streams (E119).

The largest fourspine stickleback from Maryland was 51 mm (S386). Males in breeding colors were reported as early as April 16 and gravid females as late as July 24 in Massachusetts (E97).

BROOK STICKLEBACK, *Eucalia inconstans* (Kirtland)

Eastern British Columbia to Quebec south to Kansas and Ohio (E119, T113). In clear, cold, heavily vegetated streams and lakes (W217).

T113 OH	age 0-Oct.	30-51 mm;	largest	69 mm

Size distribution in some ponds suggested that a few sticklebacks might be age III (W217). Brook stickleback mature at age I (W217). They spawn in late April to mid-June in southern Michigan and one month later in northern; in New York in April and May; and in Minnesota in April to June (J102), at temperatures of 4.5-21C. The male builds a globular nest and maintains a territory. Females produce up to 250 eggs at a time. The eggs are 1.3 mm in diameter and adhesive (W217). The eggs hatch in 8 days at 18C, and in 203-232 hours at 15-17C. The fry are 5 mm in length at hatching (W217).

Brook sticklebacks eat insects, crustacea, snails, fish eggs, oligochaetes, sponges, and algae (W217, N104). When acclimated at 25-26C the lethal temperature was 30.6C (B201).

THREESPINE STICKLEBACK, *Gasterosteus aculeatus* Linnaeus

Fresh water and marine in northern hemisphere. In the U.S. it occurs along the Pacific and Atlantic coasts and in coastal streams (E119). Marine forms have many (30-35) long plates whereas the freshwater populations usually have 5-7 in their armour (G71).

J80 England	SL = 0.90 FL with no evidence of change with length			
G71 AK	FL = 1.17 SL		TL = 1.21 SL	

R223 AK L. Aleknagik

FL	52	55	60	64	67	71	75	79	84
Grams	1.14	1.43	1.81	2.28	2.73	3.17	3.73	3.77	6.38
Number	8	20	27	50	34	31	7	3	2

	No.	Mean FL	Range
At hatching			
C153 BC Cowichan L.	+	7	-
Age 1 month			
C153 BC Cowichan L.	+	16	-
Age 2 months			
C153 BC Cowichan L.	+	22	-
Age 86 days			
V33 CA freshwater	+	16	-
salt water	+	20	-
Age 0-May			
G71 AK Bare L.	+	28	-
Age 0-June			
G71 AK Bare L.	+	29	26-32

	No.	Mean FL	Range
Age 0-July			
G71 AK Karluk L.	+	28	-
G71 AK Bare L.	+	34	32-37
Age 0-May-July			
J80 England freshwater	310	15	6-30
R223 AK Aleknagik L.	+	12	-
Age 3 months			
C153 BC Cowichan L.	+	33	-
Age 4 months			
C153 BC Cowichan L.	+	45	-
Age 0-Aug.			
G71 AK Bare L.	+	40	25-48
G71 AK Karluk L.	+	32	-
Age 0-Sept.			
G71 AK Karluk L.	+	39	-
Age 5 months			
B356 Britain	+	-	26-31
Age 6 months			
W215 aquarium, well fed	+	-	-70
Age 0-Aug.-Oct.			
J80 England freshwater	454	27	14-44
Age 0-Nov. to I-Jan.			
J80 England freshwater	145	31	18-45
J80 England Windermere	293	33	21-40
Age 10 months			
C153 BC Cowichan L.	+	64	-
Age I			
R223 AK L. Aleknagik	+	30	-
B357, W215 Ger.	+	-	30-40
J80 England L. Windermere	231	40	30-49
England Birket	348	41	18-54
England Esdale, marine	7	49	45-53
G71 AK Bare L.	+	45	40-50
G71 AK Karluk L.	+	55	-
Age II			
G71 AK Bare L.	+	50	47-56
R223 AK L. Aleknagik	+	53	-
B357, W215 Ger.	+	-	45-50
J80 England L. Windermere	58	47	42-53
England Birket	254	47	31-62
England Esdale, marine	14	67	57-74
G71 AK Karluk L.	+	73	-91
Age III			
R223 AK L. Aleknagik	+	62	-
J80 England Birket	99	54	35-73
England Esdale, marine	6	86	72-96
B357, W215 Ger.	+	-	50-70
Age IV			
R223 AK L. Aleknagik	+	75	-

Growth is apparently faster in marine than fresh water (J80). Long day length and high temperature are reported to inhibit growth during the first year (H390). Only females reached age IV (R223).

In the first year of life sticklebacks grow less in Karluk Lake than in Bare Lake because of slower warming of the water and later breeding (G71), but in later years growth is more rapid in Karluk Lake. Otoliths were used in aging sticklebacks (J80, and some G71) but most fish in G71 were aged by length frequencies. Females attain a somewhat larger size than males (G71).

Threespine sticklebacks may spawn before a year old (C153) at age I (H390, J80), but in Lake Aleknagik, Alaska, do not mature until age II, at about 60 mm (R223). Age II was also reported as youngest maturity by B356, W215. In Bare and Karluk Lakes sticklebacks spawn at age I or II and most die shortly after spawning although some age I fish may survive (G71). The smallest mature males and females were 28 mm (J80). They spawn in March in Italy and Spain; March-April in France and Belgium; April-May in Germany and Holland; May-June in Great Britain; June in southern Scandinavia; and June-July in Greenland (H390) and in Alaska (R223). They spawn at temperatures of 5-20C (H390) and the male nests and guards the territory. A male may breed 5 times in a season (H390). In fresh water, females produce 100-150 eggs per spawning, and in marine habitats 250-350 (H390). Embryology is described (S522). This species has been widely used in behavioral and physiological research (B357, W215, V33, W216).

R223 AK L. Aleknagik   No. of mature eggs = 105 + 2.80 (FL - 64 mm)

In the southern part of the range, the species is usually fresh water, but marine to the north (H390). Individuals may be acclimated from fresh water to 3.5% salt gradually (K79). Body temperature of -0.7C was fatal (B202).

Insects, mostly adult chironomids, small invertebrates and some algae constituted the food of sticklebacks from the Sacramento River (M419), and in Alaska (G71).

The sizes taken in gill nets in Alaska were (H313)

Mesh size (stretch)	No. of fish	Mean FL	Range
10 mm	44	43	39-47
13	293	57	45-67
19	12	59	53-70

NINESPINE STICKLEBACK, *Pungitius pungitius* (Linnaeus)
Fresh and brackish waters of the northern hemisphere. In North America it gets as far south as northern Minnesota, the Great Lakes except Lake Erie, and New Jersey (E119, S471).

J80 England   SL = 0.88 FL, with no change with length.

J80 England, Birket	No.	Mean FL	Range
Age 0-May-July	26	22	15-32
0-Aug.-Oct.	165	29	14-39
0-Nov.-Jan.	3	35	33-37
I	197	37	28-48
II	179	41	32-50
III	65	46	34-55

From 32-174 eggs were stripped per female (L220). Optimum survival of eggs was secured at 19-24C, but some survived at 16-26C (L220). Higher vertebral counts resulted when the eggs and fry were raised at higher temperatures (L220).

Both males and females mature at age I, the smallest mature male being 32 mm and female, 34 mm (J80).

H313 AK		Gill nets	
Mesh size (stretch)	Number of sticklebacks	Mean FL	Range
10	147	47	37-62
13	23	55	46-60

TROUT-PERCH, *Percopsis omiscomaycus* (Walbaum)
Eastern British Columbia to Labrador south to Missouri, Kentucky, and New Jersey (E119, T113). Requires cool, clear water.

K85, K119 L. Erie	272 trout-perch	TL = 2.37 mm + 1.215 SL or TL = 1.25 SL
K85 L. Erie	272 trout-perch	log W = -5.0321 + 3.08 log SL

	No.	Mean TL	Range
Hatching			
M324 MN	+	5.25	-
F5 L. Erie	+	5.8	-
Yolk sac, absorbed			
M324 MN	+	6.25	-
Age 0-June			
F5 L. Erie	+	-	6-8
M324 MN Lower Red L.	40	13.6	8-18
P145 WI L. Winnebago	+	-	25-30
Age 0-July			
F5 L. Erie	+	-	6-10
M324 MN Lower Red L.	629	19.2	9-41
Age 0-Aug.			
F5 L. Erie	+	-	16-35
M324 MN Lower Red L.	4153	28.6	10-52
M324 WI L. Winnebago	246	60.5	-
K85 L. Erie	701	-	16-53
Age 0-Sept.			
K85 L. Erie	500	-	21-63
Age 0-Oct.			
T113 OH	+	-	38-76
K85 L. Erie	169	-	40-68
M324 WI L. Winnebago	759	84.0	-
C275 L. Erie	+	84.0	-
Age I			
K85 L. Erie, March	204	-	65-93
P145 WI L. Winnebago, fall	+	-	81-91
T113 OH	+	-	64-114
M512 MN Lower Red L.	239	67	58-79
Age II			
M512 MN Lower Red L. males	60	88	80-97
females	49	90	80-99
Age III			
M512 MN Lower Red L. males	9	97	94-102
females	49	104	94-115

	No.	Mean TL	Range
Adults			
K85 L. Erie (Age II+) March	+	-	95-113
L. Erie (Age I+) Sept. males	+	-	70-105
females	+	-	70-123

	Average TL in mm at each annulus			
	1	2	3	4
M324, M512 MN Lower Red L.				
males, mean TL	50.8	85.4	97.8	
Increment	50.8	37.4	15.3	
Number	941	536	35	
females, mean TL	51.4	88.6	102.8	112.0
Increment	51.4	40.8	16.1	6.4
Number	572	485	131	4

K85 was unable to read the scales and used length frequencies and probability paper. M512 (and M324) were able to distinguish annuli on the scales as zones of closely spaced circuli followed by widely spaced circuli on the anterior field and by anastomosis in the lateral fields. False annuli were fairly common in males in the second summer. In midsummer both age I and II had only one annulus because annulus II did not form till late in July or early August, but the two age groups could usually be separated by the amount of scale growth beyond the annulus. Growth was calculated using a correction factor of 8.3 mm. Lee's phenomenon was apparent in the data, probably due to selective sampling. Growth compensation also occurred but was not complete so that the initial ranking was not changed. M512 and M324 give a good discussion of growth compensation.

Growth of young of the year ends October 15-30, at 15.5C in Lake Erie (C275). Growth of older fish was delayed until late July or August in Lower Red Lake, perhaps because of the spawning season (M512). More rapid growth in Lake Winnebago than in Red Lake was believed to be due to the longer growing season (M512). No sex difference in growth was reported for Lake Erie by K85, although adult females were larger than the males, probably because of a longer life. In Red Lake, females grew faster and lived longer than the males (M512), but the difference was not evident until the second year, when males first matured. In Red Lake (M512) mean July through September temperatures were correlated with first year growth (r = .84) and population density had little effect on first year growth. In the second year temperature also affected growth but the effect was modified by growth compensation (M512).

Most trout-perch spawn at age I in Lake Erie (K85) and few males or females live to spawn at age II. In Red Lake, Minnesota (M512), many males and a few females spawn at age I but most females spawn at age II. In Lake Winnebago (P145) they spawn at 16-20C, in May to middle June, on sand bars or among rocks in shallow water, with no care of the young. At Heming Lake, Manitoba (L96), they ascend small tributaries to spawn on silt and boulder bottoms at 6-11C in May. In Lake Erie they spawn June 1-15, at 20C (C275), or at 19-21.4C (K119) over sand, gravel or rocks (K119, K85). In Red Lake there are two or more peaks of spawning in June and July, with some runs in early May to early September. They spawn along the beaches or in streams at less than 1 meter depth, on hard gravel bottoms or over vegetation (M512). Spawning is in the dark (M512). Frequently there is high mortality after spawning (P145, K119). In male trout-perch the urinogenital opening is small

and cannot be probed with a 0.5 mm diameter probe, but the urinogenital opening of females will accept a 1 mm probe (K85).

	Size of females	No. of females	No. of eggs per female Mean	Range
L80 L. Erie	61-71 mm, 3.1-4.9 g	9	293	210-350
	76 mm 7.3-8.1 g	2	602	475-728
K85 L. Erie	62 mm	1	238	-
	75-78	4	433	230-674
	81-88	4	715	509-1042
	93	1	668	-
M324 MN Red L.	67-79 mm	7	-	190-310
	80-89	6	-	270-420
	90-99	5	-	400-580
	100-109	5	-	520-860
	110-119	5	-	710-1280
	125	1	1330	-
L96 MB Heming L.	3.1-4.9 g	9	293	210-350
	7.3-8.1 g	2	601	475-728
M512 MN Red L.	log No. of eggs = -3.247 + 3.029 log TL			

Incubation takes 8 days (E119) or 250 degree-days (above 0C) (M324). In
Red Lake there was an alternation of year classes 1952-57 with good brood
stock producing good year classes, but multiple-variance analysis indicated
that the abundance of year classes was related to temperature and wind rather
than to brood stock (M512). Warm summers resulted in bigger year classes.

Young trout-perch feed mostly on ostracods, *Gammarus*, *Leptodora*, and
chironomids (K85) or on zooplankton (P145), and larger trout-perch on chironomids (N117), mayflies and other insects (P145, D183, K85). Occasional small
minnows and darters were eaten (K85).

SAND ROLLER, *Percopsis transmontana* (Eigenmann and Eigenmann)
Lower Columbia River drainage (E119). A female had 4748 eggs (M426).

PIRATE PERCH, *Aphredoderus sayanus* (Gilliams)
From southern Minnesota and western Ohio south to Texas and Florida
and in the coastal plain to New York (E119, T113).

A39 AL	a 51 mm TL pirate perch weighed 3.4 g
E105 MD	a 76 mm TL pirate perch weighed 5.7 g

S472 AL	TL	No.	Weight Mean	Range	K(TL)
	25	6	0.7	0.3-0.9	4.49
	51	23	2.8	1.4-3.6	1.52
	76	15	3.3	2.1-10.9	0.75

Since the recorded weights of the 25 and probably 51 mm fish are too high,
the regression given in S472 is not valid.

H151 OK Sub Prison L.	No.	Mean TL	Range
Age I	58	84	66-97
II	15	99	81-109
III	9	117	104-127

		Average calculated TL at each annulus			
	No.	1	2	3	4
F62, H151 OK	82	56	84	102	117

Early relative growth and changes in morphology including the shift in the position of the vent is described in M403.

Insects are the main food (F148).

Conversion Tables

INCHES (BY TENTHS) TO NEAREST MILLIMETER

	0	0.1	0.2	0.3	0.4	0.5	0.6	0.7	0.8	0.9
0	0	3	5	8	10	13	15	18	20	23
1	25	28	30	33	36	38	41	43	46	48
2	51	53	56	58	61	64	66	69	71	74
3	76	79	81	84	86	89	91	94	97	99
4	102	104	107	109	112	114	117	119	122	124
5	127	130	132	135	137	140	142	145	147	150
6	152	155	157	160	163	165	168	170	173	175
7	178	180	183	185	188	190	193	196	198	201
8	203	206	208	211	213	216	218	221	224	226
9	229	231	234	236	239	241	244	246	249	251
10	254	257	259	262	264	267	269	272	274	277
11	279	282	284	287	290	292	295	297	300	302
12	305	307	310	312	315	318	320	323	325	328
13	330	333	335	338	340	343	345	348	351	353
14	356	358	361	363	366	368	371	373	376	378
15	381	384	386	388	391	394	396	399	401	404
16	406	409	411	414	417	419	421	424	427	429
17	432	434	437	439	442	445	447	450	452	455
18	457	460	462	465	467	470	472	475	478	480
19	483	486	488	490	493	495	498	500	503	505
20	508	511	513	516	518	521	523	526	528	531
21	533	536	538	541	544	546	549	551	554	556
22	559	561	564	566	569	572	574	577	579	582
23	584	587	589	592	594	597	599	602	605	607
24	610	612	615	617	620	622	625	627	630	632
25	635	638	640	643	645	648	650	653	655	658
26	660	663	665	668	671	673	676	678	681	683
27	686	688	691	693	696	698	701	703	706	709
28	711	714	716	719	721	724	726	729	732	734
29	737	739	742	744	747	749	752	754	757	759
30	762	765	767	770	772	775	777	780	782	785
31	787	790	792	795	798	800	803	805	808	810

CONVERSION TABLES

INCHES (BY TENTHS) TO NEAREST MILLIMETER (continued)

	0	0.1	0.2	0.3	0.4	0.5	0.6	0.7	0.8	0.9
32	813	815	818	820	823	826	828	831	833	836
33	838	841	843	846	848	851	853	856	859	861
34	864	866	869	871	874	876	879	881	884	886
35	889	892	894	897	899	902	904	907	909	912
36	914	917	919	922	925	927	930	932	935	937
37	940	942	945	947	950	953	955	958	960	963
38	965	968	970	973	975	978	980	983	986	988
39	991	993	996	998	1001	1003	1006	1008	1011	1013
40	1016	1019	1021	1024	1026	1029	1031	1034	1036	1039
41	1041	1044	1046	1049	1052	1054	1057	1059	1062	1064
42	1067	1069	1072	1074	1077	1080	1082	1085	1087	1090
43	1092	1095	1097	1100	1102	1105	1107	1110	1113	1115
44	1118	1120	1123	1125	1128	1130	1133	1135	1138	1140
45	1143	1146	1148	1151	1153	1156	1158	1161	1163	1166
46	1168	1171	1173	1176	1179	1181	1184	1186	1189	1191
47	1194	1196	1199	1201	1204	1207	1209	1212	1214	1217
48	1219	1222	1224	1227	1229	1232	1234	1237	1240	1242
49	1245	1247	1250	1252	1255	1257	1260	1262	1265	1267

FEET TO METERS

	0	1	2	3	4	5	6	7	8	9
		0.30	0.61	0.91	1.22	1.52	1.83	2.13	2.44	2.74
10	3.05	3.35	3.66	3.96	4.27	4.57	4.88	5.18	5.49	5.79
20	6.10	6.40	6.71	7.01	7.32	7.62	7.92	8.23	8.53	8.84
30	9.14	9.45	9.75	10.05	10.36	10.66	10.97	11.28	11.58	11.89
40	12.19	12.50	12.80	13.11	13.41	13.72	14.02	14.33	14.63	14.94
50	15.24	15.54	15.85	16.15	16.46	16.76	17.07	17.37	17.68	17.98
60	18.29	18.59	18.90	19.20	19.51	19.81	20.12	20.42	20.73	21.03
70	21.34	21.64	21.95	22.25	22.56	22.86	23.16	23.47	23.77	24.08
80	24.38	24.69	24.99	25.30	25.60	25.91	26.21	26.52	26.82	27.13
90	27.43	27.74	28.04	28.35	28.65	28.96	29.26	29.57	29.87	30.18
100	30.48	30.78	31.09	31.39	31.70	32.00	32.31	32.61	32.92	32.22

MILES TO KILOMETERS

	0	1	2	3	4	5	6	7	8	9
		1.6	3.2	4.8	6.5	8.1	9.7	11.3	12.9	14.5
10	16.1	17.8	19.4	21.0	22.6	24.2	25.8	27.4	29.1	30.7
20	32.3	33.9	35.5	37.1	38.7	40.4	42.0	43.6	45.2	46.8
30	48.4	50.0	51.7	53.3	54.9	56.5	58.1	59.7	61.3	63.0
40	64.6	66.2	67.8	69.4	71.0	72.6	74.3	75.9	77.5	79.1
50	80.7	82.3	83.9	85.6	87.2	88.8	90.4	92.0	93.6	95.2
60	96.9	98.5	100.1	101.7	103.3	104.9	106.5	108.2	109.8	111.4
70	113.0	114.6	116.2	117.8	119.5	121.1	122.7	124.3	125.9	127.5
80	129.1	130.8	132.4	134.0	135.6	137.2	138.8	140.4	142.1	143.7
90	145.3	146.9	148.5	150.1	151.7	153.4	155.0	156.6	158.2	159.8
100	161.4	163.0	164.8	166.3	167.9	169.5	171.1	172.7	174.3	176.0

OUNCES TO GRAMS

Pounds

Ounces	0	1	2	3	4	5	6
0	0	454	907	1361	1814	2268	2722
1	28	482	936	1389	1843	2296	2750
2	57	510	964	1418	1871	2325	2778
3	85	539	992	1446	1899	2353	2807
4	113	567	1021	1474	1928	2381	2835
5	142	595	1049	1503	1956	2410	2863
6	170	624	1077	1531	1985	2438	2892
7	198	652	1106	1559	2013	2466	2920
8	227	680	1134	1588	2041	2495	2948
9	255	709	1162	1616	2070	2523	2977
10	284	737	1191	1644	2098	2552	3005
11	312	765	1219	1673	2126	2580	3033
12	340	794	1247	1701	2155	2608	3062
13	369	822	1276	1729	2183	2637	3090
14	397	851	1304	1758	2211	2665	3119
15	425	879	1332	1786	2240	2693	3147

HUNDREDTHS OF POUND TO NEAREST GRAM

	0	0.01	0.02	0.03	0.04	0.05	0.06	0.07	0.08	0.09
0.0		5	9	14	18	23	27	32	36	41
0.1	45	50	54	59	64	68	73	77	82	86
0.2	91	95	100	104	109	113	118	122	127	132
0.3	136	141	145	150	154	159	163	168	172	177
0.4	181	186	191	195	200	204	209	213	218	222
0.5	227	231	236	240	245	249	254	259	263	268
0.6	272	277	281	286	290	295	299	304	308	313
0.7	318	322	327	331	336	340	345	349	354	358
0.8	363	367	372	376	381	386	390	395	399	404
0.9	408	413	417	422	426	431	435	440	445	449

POUNDS BY TENTHS TO NEAREST GRAM

	0.0	0.1	0.2	0.3	0.4	0.5	0.6	0.7	0.8	0.9
0		45	91	136	181	227	272	318	363	408
1	454	499	544	590	635	680	726	771	816	862
2	907	953	998	1043	1089	1134	1179	1225	1270	1315
3	1361	1406	1451	1497	1547	1588	1633	1678	1724	1769
4	1814	1860	1905	1950	1996	2041	2087	2132	2177	2223
5	2268	2313	2359	2404	2449	2495	2540	2585	2631	2676
6	2722	2767	2812	2858	2903	2948	2994	3039	3084	3130
7	3175	3220	3266	3311	3357	3402	3447	3443	3538	3583
8	3629	3674	3719	3765	3810	3856	3901	3946	3992	4037
9	4082	4128	4173	4218	4264	4309	4354	4400	4445	4491
10	4536	4581	4627	4672	4717	4763	4808	4853	4899	4944
11	4989	5035	5080	5125	5171	5216	5262	5307	5352	5398

CONVERSION TABLES

POUNDS BY TENTHS TO NEAREST GRAM (continued)

	0.0	0.1	0.2	0.3	0.4	0.5	0.6	0.7	0.8	0.9
12	5443	5488	5534	5579	5625	5670	5715	5761	5806	5851
13	5897	5942	5987	6033	6078	6123	6169	6214	6260	6305
14	6350	6396	6441	6486	6532	6577	6622	6668	6713	6758
15	6804	6849	6895	6940	6985	7031	7076	7121	7167	7212
16	7257	7303	7348	7394	7439	7484	7530	7575	7620	7666
17	7711	7756	7802	7847	7892	7938	7983	8029	8074	8119
18	8165	8210	8255	8301	8346	8391	8437	8482	8527	8573
19	8618	8663	8709	8754	8800	8845	8890	8936	8981	9026
20	9072	9117	9163	9208	9253	9299	9344	9389	9435	9480
21	9525	9571	9616	9661	9707	9752	9798	9843	9888	9934
22	9979	10024	10070	10115	10160	10206	10251	10296	10342	10387
23	10433	10478	10523	10569	10614	10659	10705	10750	10795	10841
24	10886	10932	10977	11022	11068	11113	11158	11204	11249	11294
25	11340	11385	11430	11476	11521	11567	11612	11657	11703	11748
26	11793	11839	11884	11929	11975	12020	12065	12111	12156	12202
27	12247	12292	12338	12383	12428	12474	12519	12564	12610	12655
28	12701	12746	12791	12837	12882	12927	12973	13018	13063	13109
29	13154	13200	13245	13290	13336	13381	13426	13472	13517	13562
30	13608	14061	14515	14968	15422	15876	16329	16783	17236	17690
40	18144	18597	19051	19504	19958	20412	20865	21319	21772	22226
50	22680	23133	23587	24040	24494	24947	25401	25855	26308	26762
60	27215	27669	28123	28576	29030	29483	29937	30391	30844	31298
70	31751	32205	32658	33112	33566	34019	34473	34926	35380	35834
80	36287	36744	37195	37648	38102	38555	30990	39463	39916	40370
90	40823	41277	41731	42184	42638	43091	43545	43998	44452	44906

ACRES TO HECTARES

	0	1	2	3	4	5	6	7	8	9
		0.40	0.81	1.21	1.62	2.02	2.43	2.83	3.24	3.64
10	4.05	4.45	4.86	5.26	5.67	6.07	6.48	6.88	7.28	7.69
20	8.09	8.50	8.90	9.31	9.71	10.01	10.52	10.93	11.33	11.74
30	12.14	12.55	12.95	13.36	13.76	14.16	14.57	14.97	15.38	15.78
40	16.19	16.59	17.00	17.40	17.81	18.21	18.62	19.02	19.43	19.83
50	20.24	20.64	21.04	21.45	21.85	22.26	22.66	23.07	23.47	23.88
60	24.28	24.69	25.10	25.50	25.90	26.31	26.71	27.11	27.52	27.92
70	28.33	28.73	29.14	29.54	29.95	30.35	30.76	31.16	31.57	31.97
80	32.38	32.78	33.19	33.59	33.99	34.40	34.80	35.21	35.61	36.02
90	36.42	36.83	37.23	37.64	38.04	38.45	38.85	39.26	39.66	40.07

POUNDS/ACRE TO KILOGRAMS/HECTARE

1 pound/acre = 1.12085 kilo/hectare

	0	1	2	3	4	5	6	7	8	9
		1.1	2.2	3.4	4.5	5.6	6.7	7.8	9.0	10.1
10	11.2	12.3	13.5	14.6	15.7	16.8	17.9	19.1	20.2	21.3
20	22.4	23.5	24.7	25.8	26.9	28.0	29.1	30.3	31.4	32.5
30	33.6	34.7	35.9	37.0	38.1	39.2	40.4	41.5	42.6	43.7
40	44.8	46.0	47.1	48.2	49.3	50.4	51.6	52.7	53.8	54.9
50	56.0	57.2	58.3	59.4	60.5	61.6	62.8	63.9	65.0	66.1
60	67.3	68.4	69.5	70.6	71.7	72.9	74.0	75.1	76.2	77.3
70	78.5	79.6	80.7	81.8	82.9	84.1	85.2	86.3	87.4	88.5
80	89.7	90.8	91.9	93.0	94.2	95.3	96.4	97.5	98.6	99.8
90	100.9	102.0	103.1	104.2	105.4	106.5	107.6	108.7	109.8	111.0

CONDITION FACTOR C TO K

	0	1	2	3	4	5	6	7	8	9
10	0.28	0.30	0.33	0.36	0.39	0.42	0.44	0.47	0.50	0.53
20	0.55	0.58	0.61	0.64	0.66	0.69	0.72	0.75	0.78	0.80
30	0.83	0.86	0.89	0.91	0.94	0.97	1.00	1.02	1.05	1.08
40	1.11	1.14	1.16	1.19	1.22	1.25	1.27	1.30	1.33	1.36
50	1.39	1.41	1.44	1.47	1.50	1.52	1.55	1.58	1.61	1.63
60	1.66	1.69	1.72	1.75	1.77	1.80	1.83	1.86	1.88	1.91
70	1.94	1.97	1.99	2.02	2.05	2.08	2.11	2.13	2.16	2.19
80	2.22	2.24	2.27	2.30	2.33	2.35	2.38	2.41	2.44	2.47
90	2.49	2.52	2.55	2.58	2.60	2.63	2.66	2.69	2.71	2.74

TEMPERATURE—FAHRENHEIT TO CENTIGRADE

	0	1	2	3	4	5	6	7	8	9
30	-1.1	-0.6	0	0.6	1.1	1.7	2.2	2.8	3.3	3.9
40	4.4	5.0	5.6	6.1	6.7	7.2	7.8	8.3	8.9	9.4
50	10.0	10.6	11.1	11.7	12.2	12.8	13.3	13.9	14.4	15.0
60	15.6	16.1	16.7	17.2	17.8	18.3	18.9	19.4	20.0	20.6
70	21.1	21.7	22.2	22.8	23.3	23.9	24.4	25.0	25.6	26.1
80	26.7	27.2	27.8	28.3	28.9	29.4	30.0	30.6	31.1	31.7
90	32.2	32.8	33.3	33.9	34.4	35.0	35.6	36.1	36.7	37.2
100	37.8	38.3	38.9	39.4	40.0					

Citations

A1 Adams, C. C. and T. L. Hankinson, 1928. The ecology and economics of Oneida Lake fish. Roosevelt Wildl. Ann., 1(3&4): 241-358.

A2 Adams, L. A., 1931. Determination of age in fishes. Trans. Ill. St. Acad. Sci., 23(3): 219-26.

A3 _____, 1942. Age determination and rate of growth of *Polydon spathula* by means of the growth rings of the otoliths and dentary bone. Amer. Midland Nat., 28(3): 617-30.

A5 Agassiz, A., 1878. The development of *Lepidosteus*. Part I. Proc. Amer. Acad. Arts Sci., 13: 65-76.

A9 Aldrich, A. D., 1948. Production of fathead minnows at Holdenville State Fish Hatchery. Okla. Game Fish News, 1948(Feb.): 10-11.

A10 _____, 1949. Progress report: improvement of fish-cultural practices in Oklahoma. Progr. Fish Cult., 11(1): 25-30.

A11 Allen, C. R. K., 1932. Physical changes during the early development of the salmon. Proc. Nova Scotian Inst. Sci., 18: 34-49.

A12 Allen, K. R., 1941. Studies on the biology of the early stages of the salmon *(Salmo salar)*. 3. Growth in the Thurso River system, Caithness. J. Anim. Ecol., 10: 273-95.

A13 American Fisheries Society, 1960. A list of common and scientific names of the better known fishes of the United States and Canada. Second edition. Amer. Fish. Soc. Spec. Publ., 2: 102 p.

A14 Applegate, V. C., 1943. Partial analysis of growth in a population of mudminnows, *Umbra limi* (Kirtland). Copeia, 1943(2): 92-6.

A15 _____, 1947. The menace of the sea lamprey. Mich. Conserv., 16(4): 6,7,10.

A16 _____, 1947. Growth of some lake trout, *Cristivomer n. namaycush,* of known age in inland Michigan lakes. Copeia, 1947(4): 237-41.

A17 Atkins, C. G., 1874. On the salmon of eastern North America, and its artificial culture. Rep. U. S. Comm. Fish Fish., 1871-72; 226-335.

A18 _____, 1906. Experiments in fasting of fry. Trans. Amer. Fish. Soc., 35: 123-42.

A19 _____, 1910. Foods for young salmonoid fishes. Proc. 4th Int. Fish. Cong., Bull. U. S. Fish Comm., 28: 839-51.

A20 _____, 1911. Notes on foreign fish culture and fisheries. Trans. Amer. Fish. Soc., 41: 422-38.

A21   Atkinson, N. J., 1932. A study of comparative results from stocking barren lakes with rainbow trout. Trans. Amer. Fish. Soc., 62: 197-201.

A22   Armstrong, G. C., 1949. Mortality, rate of growth, and fin regeneration of marked and unmarked lake trout fingerlings at the Provincial Fish Hatchery, Port Arthur, Ontario. Trans. Amer. Fish. Soc., 77: 129-31.

A23   Alm, G., 1949. Influence of heredity and environment on various forms of trout. Rep. Inst. Freshw. Res. Drottning., 29: 29-34.

A24   Arthur, W., 1878. Brown trout introduced into Otago. Trans. N. Z. Inst., 11.

A25   _____, 1883. Brown trout introduced into Otago II. Trans. N. Z. Inst., 16.

A26   Allen, E. R., 1946. Fishes of Silver Springs, Florida. Silver Springs, Fla., 36 p.

A27   Allen, K. R., 1951. The Horokiwi stream; a study of a trout population. N. Z. Mar. Dept. Fish. Bull., 10: 1-231.

A28   Alm, G., 1950. The sea-trout population in the Ava Stream. Rep. Inst. Freshw. Res. Drottning., 31: 26-56.

A30   Anderson, L. R., 1948. Unusual items in the diet of the northern muskellunge. Copeia, 1948(1): 63.

A32   Appelget, J. and L. L. Smith, Jr., 1951. The determination of age and rate of growth from vertebrae of the channel catfish, *Ictaluris lacustris punctatus*. Trans. Amer. Fish. Soc., 80: 119-39.

A33   Applegate, V. C., 1950. Natural history of the sea lamprey, *Petromyzon marinus*, in Michigan. Spec. Sci. Rep. U. S. Fish Wildl. Serv., 55: 1-237.

A34   _____, 1951. Sea lamprey investigations. II. Egg development, maturity, egg production and percentage of unspawned eggs of sea lampreys, *Petromyzon marinus*, captured in several Lake Huron tributaries. Pap. Mich. Acad. Sci. Arts Lett., 35: 71-90.

A38   Ala. Dept. of Conserv., 1954. Report for fiscal year, October 1, 1953-September 30, 1954. 222 p.

A39   _____, 1957. Report for fiscal year, October 1, 1955-September 30, 1956. 205 p.

A40   _____, 1958. Report for fiscal year, October 1, 1956-September 30, 1957. 200 p.

A43   Allen, G. H., 1956. Age and growth of the brook trout in a Wyoming beaver pond. Copeia, 1956(1): 1-9.

A44   Allen, K. R., 1938. Deterioration of Windermere trout. Salm. Trout Mag., 91: 152-56.

A45   Alm, G., 1955. Artificial hybridization between different species of the salmon family. Rep. Inst. Freshw. Res. Drottning., 36: 13-56.

A46   Alvord, W., 1954. Validity of age determinations from scales of brown trout, rainbow trout, and brook trout. Trans. Amer. Fish. Soc., 83: 91-103.

A48   Andrews, C. W. and E. Lear, 1956. The biology of Arctic char (*Salvelinus alpinus* L.) in northern Labrador. J. Fish. Res. Bd. Can., 13(6): 843-60.

A50   Allen, G. H. and L. G. Claussen, 1960. Selectivity of food by brook trout in a Wyoming beaver pond. Trans. Amer. Fish. Soc., 89(1): 80-81.

A51   Alm, G., 1959. Connection between maturity, size, and age in fishes. Rep. Inst. Freshw. Res. Drottning., 40: 5-145.

A52   Allen, W. F., 1911. Notes on the breeding season and young of *Polydon spathula.* J. Wash. Acad. Sci., 1: 280-83.

A53   Archibald, K. D., 1934. Age determination in the catfish. MA thesis Ohio St. Univ. 22 p., 10 plates.

A54   Andrekson, A., 1949. A study of the biology of the cutthroat trout in the Sheep River with special reference to Gorge Creek. MS thesis Univ. Alta. 74 p.

A55   Allen, G. H. and G. A. Sanger, 1960. Fecundity of rainbow trout from actual count of eggs. Copeia, 1960(3): 260-61.

A58   Antosiak, B., 1961. Wzrost szczupaka (*Esox lucius* L.) w jexiorach okolic Wegorzewa. (The rate of growth of pike in Wegorzewa District Lakes. Eng. summary 600-602.) Roczniki Nauk Rolniczych (Warsaw), 77: 581-602.

A59   Angelovic, J. W., W. F. Sigler, and J. M. Neuhold, 1961. Temperature and fluorosis in rainbow trout. J. Water Pollution Control Feder., 33(4): 371-81.

A60   Ali, M. A., P. Copes, and W. R. Stevenson, 1961. Correlation of morphological and intra-ocular measurements in the Atlantic salmon *(Salmo salar)* yearling. J. Fish. Res. Bd. Can., 18(2): 259-72.

A61   Alm, G., 1961. Die Ergebnisse der Fishaussätze in den Kalarne-Seen (with English summary). Rep. Inst. Freshw. Res. Drottning., 42: 5-83.

A62   Alexander, G. R. and D. S. Shetter, 1961. Seasonal mortality and growth of hatchery-reared brook and rainbow trout in East Fish Lake, Montmorency County, Michigan, 1958-59. Pap. Mich. Acad. Sci. Arts Lett., 46: 317-28.

A63   Agersborg, H. P. K., 1934. When do the rainbow trout spawn? Trans. Amer. Fish. Soc., 64: 167-69.

A64   Anderson, R. B., 1962. A comparison of returns from fall- and spring-stocked hatchery-reared lake trout in Maine. Trans. Amer. Fish. Soc., 91(4): 425-27.

A66   Adirondack League Club, 1961. Fishery management report for 1961. 29 p. mimeo.

A67   Atton, F. M., 1959. The invasion of Manitoba and Saskatchewan by carp. Trans. Amer. Fish. Soc., 88(3): 203-05.

A68   Alikunhi, K. H., 1958. Observations on feeding habits of young carp fry. Indian J. Fish., 5(1): 95-106.

A69   _____ and S. Nayarajo Rao, 1948. Observations on the growth of *Cyprinus carpio* in tropical environment at the Chetput Fish Farm, Madras. Proc. Indian Sci. Cong., 35(Part III): 206.

A73   Applegate, V. C., 1951. The sea lamprey in the Great Lakes. Sci. Mon. N. Y., 72(5): 275-81.

A74   _____ and J. W. Moffett, 1955. Sea lamprey and lake trout. Twentieth-Century Bestiary, Simon and Schuster, N. Y.: 9-16.

A75   _____ and C. L. Brynildson, 1952. Downstream movement of recently transformed sea lampreys, *Petromyzon marinus,* in the Carp Lake River, Michigan. Trans. Amer. Fish. Soc., 81: 275-90.

A76   _____, 1961. Downstream movement of lamprey and fishes in the Carp Lake River, Michigan. Spec. Sci. Rep. U. S. Fish Wildl. Serv., 387: 1-71.

A77   Atkinson, C. E., 1951. Feeding habits of adult shad *(Alosa sapidissima)* in fresh-water. Ecology, 32(3): 556-57.

A78   Auburn University Fisheries Staff, 1960. Effect of threadfin shad on fish populations in Lake Martin. Auburn Univ. and Ala. Dept. Conserv. 75 p. mimeo.

A79   Averett, R. C., 1962. Studies of two races of cutthroat trout in north-
       ern Idaho. Idaho Dept. Fish Game Dingell-Johnson Completion Rep.,
       F-47-R-1: 57 p.

A80   Allen, K. R., 1962. The natural regulation of population in the salmo-
       nidae. N. Z. Sci. Rev., 20: 58-62.

A81   Armstrong, R. H. and R. F. Blackett, 1966. Digestion rate of the Dolly
       Varden. Trans. Amer. Fish. Soc., 95(4): 429-30.

A82   _____, 1965. Annotated bibliography on the Dolly Varden char. Alaska
       Dept. Fish Game Res. Rep., 4: 26 p.

A83   Antosiak, B., 1963. Pike food in some lakes of Wegorzewo District
       (Polish with Eng. summary). Polish Agric. Annu. Ser. B, 82(2): 295-
       317.

A84   Armbruster, D., 1966. Hybridization of the chain pickerel and northern
       pike. Progr. Fish Cult., 28(2): 76-8.

A85   Al-Rawi, T. R., 1964. Reading of scales of river carpsuckers, *Carpiodes
       carpio*. MS thesis Iowa St. Univ. Lib. 75 p.

A86   Ackerman, G. L., 1965. Age structure of spawning channel catfish.
       Iowa Conserv. Comm. Quart. Biol. Rep., 17(3): 52-8.

B1   Bailey, R. M., 1938. The fishes of the Merrimack watershed. N. H.
       Fish Game Dept. Surv. Rep., 3: 149-85.

B2   _____, 1943. Progress report — Fisheries Research in Spirit and Oko-
       boji Lakes. Quart. Rep. Iowa Coop. Wildl. Fish. Res. Units, 1943(July-
       Sept.): 11-13.

B3   _____, 1943. Fisheries research in Clear Lake, Iowa. Quart. Rep.
       Iowa Coop. Wildl. Fish. Res. Units, 1943(July-Sept.): 14-17.

B4   _____, 1943. Fisheries research in Clear Lake, Iowa. Quart. Rep.
       Iowa Coop. Wildl. Fish. Res. Units, 1943(Oct.-Dec.): 16-17.

B5   _____ and H. M. Harrison, Jr., 1945. The fishes of Clear Lake, Iowa.
       Iowa St. J. Sci., 20(1): 57-77.

B6   _____ and _____, 1948. Food habits of the southern channel catfish
       (Ictalurus lacustris punctatus) in the Des Moines River, Iowa. Trans.
       Amer. Fish. Soc., 75: 11-138.

B9   Baird, S. F., 1887. Report of the Commission. Rep. U. S. Fish Comm.
       1885, 13: XIX-CXII.

B10   Bajkov, A., 1927. Reports of the Jasper Park Lakes Investigations,
       1925-26. I. The fishes. Contr. Canad. Biol. Fish., NS 3(16): 379-404.

B11   _____, 1930. A study of the whitefish *(Coregonus clupeaformis)* in
       Manitoban lakes. Contr. Canad. Biol. Fish., NS 5(15): 442-55.

B12   _____, 1930. Fishing industry and fisheries investigations in the Prai-
       rie Provinces. Trans. Amer. Fish. Soc., 60: 215-37.

B15   Ball, R. C., 1948. Relationship between available fish food, feeding
       habits of fish and total fish production in a Michigan lake. Mich. St.
       Coll. Tech. Bull., 206: 1-59.

B16   _____, 1949. Experimental use of fertilizer in the production of fish-
       food organisms and fish. Mich. Agric. Exp. Sta. Tech. Bull., 210: 1-28.

B17   Barney, R. L., 1925. A confirmation of Borodin's scale method of age-
       determination of Connecticut River shad. Trans. Amer. Fish. Soc., 54:
       168-77.

B20   Bean, L. S., 1936. Fish yield on the National Forests. Region Nine.
       Proc. N. Amer. Wildl. Conf., 1: 301-4.

B21   Bean, T. H., 1887. Report on examination of clupeoids from carp ponds.
       Bull. U. S. Fish Comm., 6: 441-42.

B22   Bean, T. H., 1892. Observations upon fishes and fish-culture. Bull.
U. S. Fish Comm., 10: 49-61.

B23   _____, 1902. Food and game fishes of New York. N. Y. Forest Fish
Game Comm. Rep., 7: 251-460.

B24   _____, 1908. The muskalonge of the Ohio Basin. Trans. Amer. Fish.
Soc., 37: 145-51.

B27   Beckman, W. C., 1942. Length-weight relationship, age, sex ratio, and
food habits of the smelt (Osmerus mordax) from Crystal Lake, Benzie
County, Michigan. Copeia, 1942(2): 120-24.

B30   _____, 1948. The length-weight relationship, factors for conversions
between standard and total lengths, and coefficients of condition for
seven Michigan fishes. Trans. Amer. Fish. Soc., 75: 237-56.

B32   Belding, D. L., 1920. The preservation of the alewife. Trans. Amer.
Fish. Soc., 49(2): 92-104.

B33   _____, 1934. Improved technical methods for determining the annual
growth of salmon parr by scale measurements. Trans. Amer. Fish.
Soc., 64: 103-6.

B34   _____, 1935. Observations on the growth of Atlantic salmon parr.
Trans. Amer. Fish. Soc., 65: 157-60.

B35   _____, 1937. Atlantic salmon parr of the west coast rivers of Newfoundland. Trans. Amer. Fish. Soc., 66: 211-24.

B36   _____, 1938. The salmon of the Moisie River. Trans. Amer. Fish.
Soc., 67:195-206.

B37   _____ and M. P. Clark, 1938. Observations on the salmon parr of the
Margaree River. Trans. Amer. Fish. Soc., 67: 184-94.

B38   _____, M. J. Pender, and J. A. Rodd, 1932. The early growth of salmon
parr in Canadian hatcheries. Trans. Amer. Fish. Soc., 62: 211-23.

B43   Bennett, L. H., 1948. Pike culture at the New London, Minnesota, Station. Progr. Fish Cult., 10(2): 95-7.

B45   Berg, L. S., 1932. Les poissons des eaux douces de l R. S. S. et des
pays limitrophes. Third edition. Leningrad.

B46   Berry, J., 1937. Possible influence of iodine on growth of trout. Rep.
Avon Biol. Res. S'hampton (G. Brit.), 4: 59-64.

B48   Bhatia, D., 1932. Factors involved in production of annual zones on
scales on rainbow trout. II. J. Exp. Zool., 9: 6-11.

B49   Bigelow, N. K., 1923. The food of young suckers (Catostomus commer-
sonni) on Lake Nipigon. Publ. Ont. Fish Res. Lab., 21: 81-115.

B50   Bing, F. C., 1927. A progress report upon feeding experiments with
brook trout fingerlings at the Connecticut state fish hatchery, Burling-
ton, Conn. Trans. Amer. Fish. Soc., 57: 266-80.

B55   Bjorn, E. E., 1940. Preliminary observations and experimental study
of the ling, Lota maculosa (LeSueur), in Wyoming. Trans. Amer. Fish.
Soc., 69: 192-96.

B56   Black, J. D., 1945. Natural history of the northern mimic shiner, Notropis volucellus volucellus Cope. Invest. Ind. Lakes Streams, 2(18):
449-69.

B57   _____ and L. O. Williamson, 1947. Artificial hybrids between muskellunge and northern pike. Trans. Wis. Acad. Sci. Arts Lett., 38: 299-
314.

B58   Blair, A. A., 1937. The validity of age determination from the scales
of landlocked salmon. Science, 86: 519-20.

B59   _____, 1943. Atlantic salmon of the East Coast of Newfoundland and
Labrador, 1939. Res. Bull. Dept. Nat. Resources Newf., 13: 1-21.

B62   Bonham, K. and R. W. Williams, 1948. Effect of population pressure
       upon rate of growth and food conversion of fingerling cutthroat trout.
       Progr. Fish Cult., 10(1): 15-18.

B63   Borodin, N. A., 1925. Biological observations on the Atlantic sturgeon
       (Acipenser sturio). Trans. Amer. Fish. Soc., 55: 184-90.

B64   _____, 1925. Age of shad (*Alosa sapidissima* Wilson) as determined by
       scales. Trans. Amer. Fish. Soc., 54: 178-84.

B65   Bower, S., 1897. The propagation of smallmouth black bass. Trans.
       Amer. Fish. Soc., 25: 127-36.

B66   Brakeley, J. R., 1889. Rapid growth of carp due to abundance of food.
       Bull. U. S. Comm. Fish., 7: 20.

B67   Breder, C. M., Jr., 1920. Some notes on *Leuciscus vandoisulus* (Cuv.
       and Val.). Copeia, 1920: 35-8.

B68   _____, 1926. Fish notes for 1925 from Sandy Hook Bay. Copeia, 153:
       121-28.

B69   _____, 1936. Long-lived fishes in the aquarium. Bull. N. Y. Zool.
       Soc., 39: 116-17.

B70   _____ and D. R. Crawford, 1922. The food of certain minnows. Zoo-
       logica N. Y., 2(14): 287-327.

B71   _____ and R. F. Nigrelli, 1936. The winter movements of the land-
       locked alewife, *Pomolobus pseudoharengus* (Wilson). Zoologica N. Y.,
       21(3): 165-75.

B73   Brice, J. J., 1898. A manual of fish-culture, based on the methods of
       the United States Commission of Fish and Fisheries with chapters on
       the cultivation of oysters and frogs. Rep. U. S. Comm. Fish., 1897(Ap-
       pend.): 340 p.

B74   Brown, C. J. D., 1938. Observations on the life-history and breeding
       habits of the Montana grayling. Copeia, 1938(3): 132-36.

B76   _____, 1943. Age and growth of Montana grayling. J. Wildl. Mgmt.,
       7(4): 353-64.

B77   _____ and R. C. Ball, 1943. A fish population study of Third Sister
       Lake. Trans. Amer. Fish. Soc., 72: 177-85.

B78   _____ and C. Buck, Jr., 1939. When do trout and grayling fry begin to
       take food? J. Wildl. Mgmt., 3(2): 134-40.

B79   _____ and J. W. Moffett, 1942. Observations on the number of eggs and
       feeding habits of the cisco *(Leucichthys artedi)* in Swain's Lake, Jack-
       son County, Michigan. Copeia, 1942(3): 149-52.

B81   Brown, L., 1942. Propagation of the spotted channel catfish *(Ictalurus
       lacustris punctatus)*. Trans. Kans. Acad. Sci., 45: 311-14.

B82   Bund, J. W. W., 1912. Questions on the life history of the salmon aris-
       ing from a study of its scales. Trans. Worcester Nat. Club, 5: 311-24.

B84   Borovicka, R. L., 1949. East and Paulina Lake fishery. Ore. St. Game
       Comm. Bull., 4(4): 1,4,6-7.

B86   Bennett, G. W., 1945. Overfishing in a small artificial lake, Onized
       Lake near Alton, Illinois. Bull. Ill. Nat. Hist. Surv., 23(3): 372-406.

B89   Brown, M. E., 1946. The growth of brown trout (*Salmo trutta* Linn.) I.
       Factors influencing the growth of trout fry. J. Exp. Biol., 22: 118-29.

B90   Bajkov, A. D., 1949. A preliminary report on the Columbia River stur-
       geon. Ore. Fish Comm. Res. Briefs, 2(2): 3-10.

B91   Baldwin, N. S., 1950. The American smelt, (*Osmerus mordax* (Mitch-
       ill)), of South Bay, Manitoulin Island, Lake Huron. Trans. Amer. Fish
       Soc., 78(1948): 176-80.

B94   Barnickol, P. G. and W. C. Starrett, 1951. Commercial and sport fishery of the Mississippi River between Caruthersville, Missouri, and Dubuque, Iowa. Bull. Ill. Nat. Hist. Surv., 25(5): 267-350.

B95 Baten, W. D. and P. I. Tack, 1952. Relationships of weight and body measurements of adult smelt, *Osmerus mordax* (Mitchill). Progr. Fish Cult., 14(2): 50-55.

B97 Bennett, G. W., 1948. Winterkill of fishes in an Illinois lake. Ill. Nat. Hist. Surv. Biol. Notes, 19: 1-9.

B100 Bigelow, H. B. and W. C. Schroeder, 1948. Cyclostomes. Fishes of Western North Atlantic, mem. Sears Found. Mar. Res. New Haven, 1: 29-58.

B104 Bonham, K., 1950. Some tests with experimental groups of fingerling rainbow trout, *Salmo gairdnerii*, on uniformity and rate of growth, diet, and photographic size-recording. Trans. Amer. Fish. Soc., 79: 94-104.

B105 Bridge, T., 1943. Feeding experiments on Kamloops trout (*Salmo gairdnerii kamloops* Jordan). B. C. Game Comm. Rep., 1942: 13 p.

B106 Brown, C. J. D., 1951. The paddlefish in Fort Peck Reservoir, Montana. Copeia, 1951(3): 252.

B107 _____, 1952. Spawning habits and early development of the mountain whitefish, *Prosopium williamsoni*, in Montana. Copeia, 1952(2): 109-13.

B108 _____ and N. A. Thoreson, 1951. Ranch fish ponds in Montana; their construction and management. Mont. St. Coll. Agric. Exp. Sta. Bull., 480: 1-30.

B109 Brown, W. H., 1951. Results of stocking largemouth black bass and channel catfish in experimental Texas farm ponds. Trans. Amer. Fish. Soc., 80: 210-17.

B112 Bryant, M., Jr., 1950. A graphic record for hatcheries rearing artificially fed fish. Progr. Fish Cult., 12(2): 91-3.

B113 Brunson, R. B., 1952. Egg counts of *Salvelinus malma* from the Clark's Fork River, Montana. Copeia, 1952(3): 196-97.

B124 Brynildson, O. M., A. D. Hasler, and J. A. Larson, 1952. Bog lakes for trout. Wis. Conserv. Bull., 17(11): 11-13.

B126 Bailey, R. M., 1959. A new catostomid fish, *Moxostoma (Thoburnia) atripinne*, from the Green River drainage, Kentucky and Tennessee. Occ. Pap. Mus. Zool. Univ. Mich., 599: 1-19.

B127 _____ and F. B. Cross, 1954. River sturgeons of the American genus, *Scaphirynchus;* characters, distribution, and synonymy. Pap. Mich. Acad. Sci. Arts Lett., 39: 169-208.

B130 Ball, R. C., 1937. A seasonal study of the food of the common schiner (*Notropis cornutus* Mitchill). MS thesis Ohio St. Univ.

B131 _____ and J. R. Ford, 1953. Production of food-fish and minnows in Michigan ponds. Mich. Agric. Exp. Sta. Quart. Bull., 35(3): 384-91.

B134 Barnhart, R., 1955. Survey of Lake Loveland Reservoir, Larimer County, Colorado. Colo. Fish. Res. Unit Quart. Rep., 2(1&2): 31-40.

B141 Baxter, G. T., 1958. A study of the effect of fertilization on trout growth in an alpine lake in Wyoming. Wyo. Game Fish Dingell-Johnson Proj., F-7-R-4: 26 p.

B145 Benson, N. G., O. B. Cope, and R. V. Bulkley, 1959. Fishery management studies on Madison River system in Yellowstone National Park. Spec. Sci. Rep. U. S. Fish Wildl. Serv., 307: 1-29.

B146 Berry, F. H., 1957. Age and growth of the gizzard shad in Lake Newman, Florida. Proc. S. E. Assoc. Game Fish Comm., 11: 318-32.

B150 Birkenholz, D. and A. W. Fritz, 1956. Fishes of Little Wall Lake, Iowa. Ia. St. Univ. Coop. Fish. Unit. 21 p. typewritten.

B152   Bisbee, L. E., 1957. Southeast Oregon. Annu. Rep. Ore. St. Game
        Comm. Fish. Div., 1956: 92-105.

B153   _____, 1958. Southeast Oregon. Annu. Rep. Ore. St. Game Comm.
        Fish. Div., 1957: 106-20.

B154   _____, 1959. Southeast Oregon. Annu. Rep. Ore. St. Game Comm.
        Fish. Div., 1958: 101-28.

B155   Bishop, C. G., 1955. Age, growth and condition of trout in Prickley
        Pear Creek, Montana. Trans. Amer. Microscop. Soc., LXXIV(2): 134-
        45.

B156   Bjorn, T. C., 1957. A survey of the fishery resources of Priest and
        Upper Priest Lakes and their tributaries. Idaho Dept. Fish Game
        Dingell-Johnson Completion Rep., Proj. F-24-R: 176 p.

B158   Bodola, A., 1955. The life history of the gizzard shad, *Dorosoma cepe-
        dium* (LeSueur) in western Lake Erie. PhD disser. Ohio St. Univ.

B159   Bond, C. E. and L. E. Bisbee, 1955. Records of the tadpole madtom,
        Schilbeodes mollis, and the black bullhead, *Ameiurus melas,* from Ore-
        gon and Idaho. Copeia, 1955(1): 56.

B160   Bond, L. H., 1958. Rainbow trout, *Salmo gairdneri* Richardson. Fishes
        of Maine, 2nd ed., ed. W. H. Everhart: 38-41.

B162   Borovicka, R. L., 1956. Bend district. Annu. Rep. Ore. St. Game
        Comm. Fish. Div., 1955: 135-59.

B163   _____, 1957. Bend district. Annu. Rep. Ore. St. Game Comm. Fish.
        Div., 1956: 134-62.

B167   Brancamp, J. H., 1938. The chronological order of scale formation on
        the common shiner, *Notropis cornutus* (Mitchill). MS thesis Ohio St.
        Univ.

B168   Brasch, J., J. T. McFadden, and S. Kmiotek, 1958. The eastern brook
        trout, its life history, ecology and management. Wis. Conserv. Dept.
        Publ., 226: 1-11.

B169   Brezner, J., 1956. Some aspects in the life history of the northern
        river carpsucker, *Carpiodes carpio* (Rafinesque), in the Niangua Arm
        of the Lake of the Ozarks. MS thesis Univ. Mo. 79 p. typewritten.

B170   Bridges, C. H. and J. W. Mullan, 1958. A compendium of the life his-
        tory and ecology of the eastern brook trout, *Salvelinus fontinalis*
        (Mitchill). Mass. Fish. Bull., 23: 38 p.

B171   Briggs, J. C., 1953. The behavior and reproduction of salmonid fishes
        in a small coastal stream. Calif. Fish. Bull., 94: 1-62.

B172   Brown, C. J. D., 1955. A record-size pallid sturgeon, *Scaphirhynchus
        album,* from Montana. Copeia, 1955(1): 55.

B173   _____ and J. E. Bailey, 1952. Time and pattern of scale formation in
        Yellowstone cutthroat trout, *Salmo clarkii lewisii.* Trans. Amer.
        Microscop. Soc., 71(2): 120-24.

B174   _____ and R. J. Graham, 1954. Observations on the longnose sucker in
        Yellowstone Lake. Trans. Amer. Fish. Soc., 83: 38-46.

B175   _____ and G. D. Holton, 1953. Time of annulus formation on the scales
        of brook and rainbow trout. Trans. Amer. Microscop. Soc., 72(1): 47-8.

B178   Brunson, R. B., R. E. Pennington, and R. G. Bjorklund, 1952. On a fall
        collection of native trout *(Salmo clarkii)* from Flathead Lake, Montana.
        Proc. Mont. Acad. Sci., 12: 63-7.

B180   Buchholz, M. M., 1957. Age and growth of river carpsucker in Des
        Moines River, Iowa. Proc. Iowa Acad. Sci., 64: 589-600.

B183   Buck, D. H. and F. B. Cross, 1952. Early limnological and fish population conditions of Canton Reservoir, Oklahoma, and fishery management recommendations. Okla. A. M. Coll. Res. Found. 110 p.

B184   Burdick, M. E. and E. L. Cooper, 1956. Growth rate, survival and harvest of fingerling rainbow trout planted in Weber Lake, Wisconsin. J. Wildl. Mgmt., 20(3): 233-38.

B186   Buss, K. and J. E. Wright, 1958. Appearance and fertility of trout hybrids. Trans. Amer. Fish. Soc., 87: 172-81.

B187   Butler, G. E., 1952. Two poisoning projects in Manitoba. Canad. Fish Cult., 13: 10-14.

B192   Brown, E. H., Jr., 1960. Little Miami River headwater-stream investigations. Ohio Dept. Nat. Resources Div. Wildl. 1-143.

B193   Benson, N. G., 1960. Factors influencing production of immature cutthroat trout in Arnica Creek, Yellowstone Park. Trans. Amer. Fish. Soc., 89(2): 168-76.

B194   Burnet, A. M. R., 1959. Some observations on natural fluctuations of trout population numbers. N. Z. J. Sci., 2(3): 410-21.

B195   Buss, K. and R. McCreary, 1960. A comparison of egg production of hatchery-reared brook, brown, and rainbow trout. Progr. Fish Cult., 22(1): 7-10.

B196   Benson, N. G., 1960. Rocky Mountain sport fishery investigations. U. S. Fish Wildl. Serv. Circ., 81: 34-9.

B197   Budd, J. C. and F. E. J. Fry, 1960. Further observations on the survival of yearling lake trout planted in South Bay, Lake Huron. Canad. Fish Cult., 26: 7-13.

B200   Black, E. C., 1953. Upper lethal temperatures of some British Columbia freshwater fishes. J. Fish. Res. Bd. Can., 10(4): 196-210.

B201   Brett, J. R., 1944. Some lethal temperature relations of Algonquin Park fishes. Publ. Ont. Fish. Res. Lab., 63: 1-49.

B202   Bardach, J. E. and J. J. Bernstein, 1957. Extreme temperatures for growth and survival of fish. Table for Handbook of Biological Data. 7 p.

B203   Brett, J. R., 1941. Tempering versus acclimation in the planting of speckled trout. Trans. Amer. Fish. Soc., 70: 397-403.

B204   Black, V. S., 1951. Osmotic regulations in teleost fishes. Publ. Ont. Fish. Res. Lab., 71: 53-89.

B206   Bishai, H. M., 1960. Upper lethal temperatures for larval salmonids. J. Cons. Int. Explor. Mer., 25(2): 129-33.

B207   Battle, H. I. and W. M. Sprules, 1960. A description of the semibuoyant eggs and early developmental stages of the goldeye, *Hiodon alosoides* (Rafinesque). J. Fish. Res. Bd. Can., 17(2): 245-66.

R208   Burdick, M. E. and O. Brynildson, 1960. Fly fishing only. Wis. Conserv. Bull., 25(6): 11-14.

B209   Bersamin, S. V., 1958. A preliminary study of the nutritional ecology and food habits of the chubs (*Leucichthys* spp.) and their relation to the ecology of Lake Michigan. Pap. Mich. Acad. Sci. Arts Lett., 43: 107-18.

B210   Beyerle, G. B. and E. L. Cooper, 1960. Growth of brown trout in selected Pennsylvania streams. Trans. Amer. Fish. Soc., 89(3): 255-62.

B211   Budd, J. C., 1960. Survival and growth of tagged lake trout in South Bay, Lake Huron. Trans. Amer. Fish. Soc., 89(3): 308-9.

B212   _____, 1959. The use of the hybrid between eastern brook trout and lake trout in fishery management. Trans. N. E. Wildl. Conf., 10: 115-16.

B213 Buss, K. and R. McCreary, 1960. Competitive survival ability of brook, brown, and rainbow trout fingerlings in a stream-fed raceway. Progr. Fish Cult., 22(3): 99-102.

B214 Baxter, G. T., 1959. Experimental fertilization of an alpine lake in Wyoming. Univ. Wyo. Coop. Res. Prog., 3: 12 p. mimeo.

B215 Barbour, T., 1911. The smallest *Polyodon*. Biol. Bull. Wood's Hole, 21: 207-8.

B216 Bass, J. C. and C. D. Riggs, 1959. Age and growth of the river carpsucker, *Carpiodes carpio* (Rafinesque), of Lake Texoma. Proc. Okla. Acad. Sci., 39: 50-69.

B217 Bali, J. M., 1959. Scale analysis of steelhead trout, *Salmo gairdnerii gairdnerii* Richardson, from various coastal watersheds of Oregon. MS thesis Ore. St. Coll. 189 p.

B218 Bradley, M. C., 1948. An analysis of the growth and development of the fish in the freshwater impoundments at the Patuxent Research Refuge. MS thesis Univ. Md. 23 p.

B222 Bernhardt, R. W., 1957. Growth of fish in the waters of the Huntington Wildlife Forest. MS thesis Syracuse Univ. 93 p.

B223 Bulkley, R. V., 1957. The use of branchiostegal rays to determine age of lake trout, *Salvelinus namaycush* (Walbaum). MS thesis Utah St. Agric. Coll. 32 p.

B224 Blair, A. A., 1938. Factors affecting growth of the scales of salmon *(Salmo salar)*. PhD thesis Univ. Toronto. 227 p.

B225 Ball, J. N. and J. W. Jones, 1960. On the growth of the brown trout of Llyn Tegid. Proc. Zool. Soc. Lond., 134(1): 1-41.

B226 Bassett, H. M., 1957. Further life history studies of two species of suckers in Shadow Mountain Reservoir, Grand County, Colorado. MS thesis Colo. St. Univ. 112 p.

B227 Bjorklund, R. G., 1953. The lake whitefish, *Coregonus clupeaformis* (Mitchill), in Flathead Lake, Montana. MS thesis Mont. St. Univ. 144 p.

B228 Baldwin, N. S., 1948. A study of the speckled trout, *Salvelinus fontinalis* (Mitchill), in a pre-Cambrian lake. MA thesis Univ. Toronto. 55 p.

B229 Blair, A. A., 1932. Salmon *(Salmo salar)* of the Miramichi River System, 1931. MA thesis Univ. Toronto. 70 p.

B230 Bulkley, R. V., 1960. Use of branchiostegal rays to determine age of lake trout, *Salvelinus namaycush* (Walbaum). Trans. Amer. Fish. Soc., 89(4): 344-50.

B231 Bonde, T. and J. E. Maloney, 1960. Food habits of burbot. Trans. Amer. Fish. Soc., 89(4): 374-76.

B233 Bond, C. E., 1948. Fish management problems of Lake of the Woods, Oregon. MS thesis Ore. St. Coll. 109 p.

B235 Bauer, J. A. and W. O. Saltzman, 1960. Umpqua River. Annu. Rep., Ore. St. Game Comm. Fish. Div., 1959: 1-22.

B236 Bisbee, L. E., 1960. Southeast Oregon. Annu. Rep. Ore. St. Game Comm. Fish. Div., 1959: 109-31.

B237 Brady, L. and A. Hulsey, 1959. Propagation of buffalofishes, 1959. Proc. S. E. Assoc. Game Fish Comm., 13: 80-90.

B238 Bjorn, T. C., 1961. Harvest, age structure, and growth of game fish populations from Priest and Upper Priest Lakes. Trans. Amer. Fish. Soc., 90(1): 27-31.

B239 Bowman, M. L., 1954. Some aspects of the life history of the black redhorse (*Moxostoma duquesni* LeSueur), with reference to its association with the smallmouth bass (*Micropterus dolomieu* Lacépède) in two south central Missouri streams, the Niangua and the Big Piney. MS thesis Univ. Mo. 39 p.

B240 Bosley, C. E., 1960. Pre-impoundment study of the Flaming Gorge Reservoir. Wyo. Game Fish Comm. Fish Tech. Rep., 9: 81 p.

B242 Bowman, M. L., 1959. The life history of the black redhorse, *Moxostoma duquesni* (LeSueur), in Missouri. PhD thesis Univ. Mo. 144 p.

B245 Bouthillier, L. P., 1961. The influence of various diets on trout growth and pigmentation. Progr. Fish Cult., 23(4): 169-74.

B248 Bauer, J. A. and R. L. McDivitt, 1960. Umpqua River. Annu. Rep. Ore. St. Game Comm. Fish. Div., 1960: 1-21.

B249 Bisbee, L. E., 1960. Southeast Region. Annu. Rep. Ore. St. Game Comm. Fish. Div., 1960: 107-30.

B250 Buettner, H. J., 1961. Recoveries of tagged, hatchery-reared lake trout from Lake Superior. Trans. Amer. Fish. Soc., 90(4): 404-12.

B251 Berst, A. H., 1961. Selectivity and efficiency of experimental gill nets in South Bay and Georgian Bay of Lake Huron. Trans. Amer. Fish. Soc., 90(4): 413-18.

B254 Burkhard, W. T., 1961. Life history of the splake trout —Parvin Lake. Quart. Rep. Colo. Coop. Fish. Res. Unit, 7(Nov. 1960-Sept. 1961): 41-52.

B255 Buss, K., 1961. Record muskellunge—fact or fiction. Penn. Angler, June 1961. Reprint.

B256 _____ and J. G. Miller, 1961. Part VI — The age and growth of the northern pike in Pennsylvania. Penn. Angler, Mar. 1961. Reprint.

B257 _____ and _____, 1961. Part VII – The age and growth of the muskellunge in Pennsylvania. Penn. Angler, Apr. 1961. Reprint.

B259 Ball, O. P. and O. B. Cope, 1961. Mortality studies on cutthroat trout in Yellowstone Lake. U. S. Fish Wildl. Serv. Res. Rep., 55: 62 p.

B260 Blackett, R. F., 1962. Some phases in the life history of the Alaskan blackfish, *Dallia pectoralis*. Copeia, 1962(1): 124-30.

B261 Benson, N. G., 1961. Limnology of Yellowstone Lake in relation to the cutthroat trout. U. S. Fish Wildl. Serv. Res. Rep., 56: 33 p.

R262 Bulkley, R. V., 1961. Fluctuations in age composition and growth rate of cutthroat trout in Yellowstone Lake. U. S. Fish Wildl. Serv. Res. Rep., 54: 31 p.

B263 Brunson, R. B. and H. W. Newman, 1951. The summer food of *Coregonus clupeaformis* from Yellow Bay, Flathead Lake, Montana. Proc. Mont. Acad. Sci., 10: 5-7.

B264 _____, G. B. Castle, and R. B. Pirtle, 1952. Studies of *Oncorhynchus nerka* from Flathead Lake, Montana. Proc. Mont. Acad. Sci., 12: 35-43.

B265 Buss, K., 1957. The kokanee. Penn. Fish Comm. Spec. Purpose Rep. 13 p. mimeo.

B266 _____ and J. G. Miller, 1962. The age and growth of trout in Pennsylvania, (in) Age and growth of the fishes in Pennsylvania, Penn. Fish Comm.

B269 Blair, A. A., 1942. Regeneration of the scales of Atlantic salmon. J. Fish. Res. Bd. Can., 5(5): 440-47.

B270 Buss, K. and R. McCreary, 1962. A comparison of egg production of hatchery-reared brook, brown, and rainbow trout. Penn. Fish. Comm. (Bellefonte). 4 p. mimeo.

B271 Bulkley, R. V and N. G. Benson, 1962. Predicting year-class abundance of Yellowstone Lake cutthroat trout. U. S. Bur. Sport Fish. Wildl. Res. Rep., 59: 1-21.

B272 Buss, K., 1961. Pennsylvania's new muskellunge program. Penn. Angler, 30(8). Reprint.

B273 _____, 1961. Record northern pike—fact and fiction. Penn. Angler, 30(10): 6-7.

B274 Belding, D. L., 1940. The number of eggs and pyloric appendages as criteria of river varieties of the Atlantic salmon (Salmo salar). Trans. Amer. Fish. Soc., 69: 285-89.

B275 Brown, M. E., 1946. The growth of brown trout (Salmo trutta Linn.) II. The growth of two-year-old trout at a constant temperature of 11.5C. J. Exp. Biol., 22: 130-44.

B276 Benson, N. G., 1963. North Central reservoir investigations. U. S. Bur. Sport Fish. Wildl. Serv. Circ., 160: 94-7.

B277 Brown, C. J. D. and G. C. Kamp, 1942. Gonad measurements and egg counts of brown trout (Salmo trutta) from the Madison River, Montana. Trans. Amer. Fish. Soc., 71: 195-200.

B278 Brynildson, O. M., V. A. Hacker, and T. A. Klick, 1963. Brown trout: Its life history, ecology and management. Wis. Conserv. Dept. Publ., 234: 14 p.

B279 Buss, K., 1960. Data on known age hatchery trout. Personal communication. 3 p. mimeo. tables.

B280 Baldwin, N. S., 1957. Food consumption and growth of brook trout at different temperatures. Trans. Amer. Fish. Soc., 86: 323-28.

B281 Brett, J. R., 1956. Some principles in the thermal requirements of fishes. Quart. Rev. Biol., 31(2): 75-87.

B283 Bailey, M. M., 1963. Age, growth and maturity of round whitefish of the Apostle Islands and Isle Royale regions. U. S. Fish. Bull., 63(1): 63-75.

B284 Bjorn, T. C. and J. Mallet, 1964. Movements of planted and wild trout in Idaho river system. Trans. Amer. Fish. Soc., 93(1): 70-76.

B285 Bernatowicz, S., 1962. [Observations of the phenology of fish spawning] (in Polish with English summary). Roczniki Nauk Rolniczych, 81B(2): 307-33.

B287 Bailey, R. M. and T. Ugeno, 1964. Nomenclature of the blue chub and the tui chub, cyprinid fishes from western United States. Copeia, 1964(1): 238-39.

B288 Brofeldt, P., 1917. Bidrag till kannedomen om fiskbestandet i vara sjoar. Finlands Fiskerier Bd., 4: 172-212 not seen, quoted by W130.

B289 Buss, K., 1961. A literature survey of the life history and cultures of the northern pike. Penn. Fish. Comm. Benner Spring Fish Res. Sta. Special Purpose Rep. 58 p.

B290 Borzenko, M. P., 1926. Material for the biology of the carp, (Cyprinus carpio Linn.). Bull. Icthyol. Lab. Baku, 2(1): 1-67 original not seen, quoted in F98.

B292 Ball, R. C. and E. H. Bacon, 1951. Use of pituitary material in the propagation of minnows. Progr. Fish Cult., 16(3): 108-13.

B293 Black, J. D., 1946. Nature's own weed killer—the German carp. Wis. Conserv. Bull., 11(4): 3-7.

B294 _____, 1948. The spawning of carp in holding ponds. Wis. Conserv. Bull., 13(3): 6-7.

B295 Buck, D. H., M. A. Whitacre, and C. F. Thoits, III, 1960. Some experiments in the baiting of carp. J. Wildl. Mgmt., 24(4): 357-64.

B296   Burr, J. G., 1931. Electricity as a means of garfish and carp control. Trans. Amer. Fish. Soc., 61: 174-81.

B297   Brues, C. T., 1928. Studies on the fauna of hot springs in the western United States and the biology of thermophilous animals. Proc. Amer. Acad. Arts Sci., 63: 139-228.

B298   Bailey, R. M., 1954. Distribution of the American cyprinid fish, *Hybognathus hankinsoni* with comments on its original description. Copeia, 1954(4): 289-91.

B299   Britt, N. W., 1955. Stratification of western Lake Erie in summer of 1953: Effects on the *Hexagenia* (Ephemeroptera) population. Ecology, 36(2): 239-44.

B300   _____ and R. H. Gibbs, Jr., 1956. *Notropis callitaenia*, a new cyprinid fish from Alabama, Florida, and Georgia. Occ. Pap. Mus. Zool. Univ. Mich., 576: 1-15.

B301   _____ and R. D. Suttkus, 1952. *Notropis signipinnis*, a new cyprinid fish from southeastern United States. Occ. Pap. Mus. Zool. Univ. Mich., 542: 1-15.

B302   _____, 1959. Distribution of the American cyprinid fish *Notropis anogenus*. Copeia, 1959(2): 119-23.

B303   _____ and C. R. Gilbert, 1960. The American cyprinid fish *Notropis kanawha* identified as an interspecific hybrid. Copeia, 1960(4): 354-57.

B307   Bennett, G. W. and W. F. Childers, 1966. The lake chubsucker as a forage species. Progr. Fish Cult., 28(2): 89-92.

B308   Bean, T. H., 1903. Catalogue of the fishes of New York. N. Y. State Mus. Bull., 60: 784 p.

B309   Ballard, W. W. and R. G. Needham, 1964. Normal embryonic stages of *Polyodon spathula* (Walbaum). J. Morph., 114(3): 465-77.

B310   Bonham, K., 1941. Food of gars in Texas. Trans. Amer. Fish. Soc., 70: 356-62.

B311   Borodin, N. A., 1924. Age of shad estimated from examination of scales. Science, 60: 477.

B312   Bodola, A., 1966. Life history of the gizzard shad, *Dorosoma cepedianum* (LeSueur), in western Lake Erie. U. S. Fish Wildl. Serv. Fish. Bull., 65(2): 391-425.

B313   Berry, F. H., M. T. Huish, and H. L. Moody, 1956. Spawning mortality of the threadfin shad, *Dorosoma petenense* (Gunther) in Florida. Copeia, 1956(3): 192.

B314   Bigelow, H. B., 1963. Family Coregonidae. Sears Found. Mar. Res. Mem., 1(Part 3): 547-52.

B315   Beers, G. D. and W. J. McConnell, 1966. Some effects of threadfin shad introduction on black crappie diet and condition. J. Ariz. Acad. Sci., 4(2): 71-4.

B316   Benson, N. G. and R. V. Bulkley, 1963. Equilibrium yield and management of cutthroat trout in Yellowstone Lake. U. S. Fish Wildl. Serv. Res. Rep., 62: 44 p.

B317   Bisbee, L. E., 1962. Harney-Malheur district. Ann. Rep. Ore. St. Game Comm. Fish. Div., 1962: 135-63.

B318   Burns, J. W., 1966. Threadfin shad. Inland Fisheries Management, ed. A. Calhoun, Calif. Fish Game Dept.: 481-88.

B319   Bjorn, T. C., 1966. Salmon and steelhead investigations. Idaho Fish Game Dept. Dingell-Johnson Rep., Proj. F-49-R-4: (Job 3), 183 p.

B320   Bisbee, L. E., 1961. Harney and Malheur districts. Annu. Rep. Ore. St. Game Comm. Fish. Div., 1961: 114-27.

B321   Bauer, J. A. and R. L. McDivitt, 1961. Umpqua River district. Annu.
        Rep. Ore. St. Game Comm. Fish. Div., 1961: 1-22.
B322   Butler, R. L. and D. P. Borgeson, 1965. California "catchable" trout
        fisheries. Calif. Fish Bull., 127: 47 p.
B323   Balmain, K. H. and W. M. Shearer, 1956. Records of salmon and sea
        trout caught at sea. Scot. Freshw. Salm. Fish. Res., 11: 12 p.
B324   Boles, H. D. and D. P. Borgeson, 1966. Experimental brown trout
        management in Lower Sardine Lake, California. Calif. Fish Game,
        52(3): 166-72.
B325   Balon, E. K., 1964. On relative indexes for comparison of the growth
        of fishes. Acta Soc. Zool. Bohemoslov., 28(4): 369-79.
B326   Bigelow, H. B., 1963. Genus *Salvelinus* Richardson 1836. Sears
        Found. Mar. Res. Mem., 1(Part 3): 503-46.
B327   Bailey, M. M., 1964. Age, growth, maturity, and sex composition of
        the American smelt, *Osmerus mordax* (Mitchill), of western Lake Su-
        perior. Trans. Amer. Fish. Soc., 93(4): 382-95.
B328   Buss, K., 1962. A literature survey of the life history of the redfin and
        grass pickerels. Penn. Fish. Comm. Benner Spring Fish. Res. Sta.
        12 p. mimeo.
B329   Becker, G. C., 1964. The fishes of Pewaukee Lake. Trans. Wis. Acad.
        Sci. Arts Lett., 53: 19-27.
B330   _____, 1964. The fishes of Lakes Poygan and Winnebago. Trans. Wis.
        Acad. Sci. Arts Lett., 53: 29-52.
B331   Brown, E. H., Jr. and C. F. Clark, 1965. Length-weight relationship of
        northern pike, *Esox lucius* from East Harbor, Ohio. Trans. Amer.
        Fish. Soc., 94(4): 404-5.
B332   Balon, E. K., 1965. Wachstum des Hechtes (*Esox lucius* L.) in Orava-
        Stausee. Z. Fisch., NS 23: 113-58.
B333   Braum, E., 1964. Experimentelle untersuchungen zur ersten nährung-
        saufnahme und biologie an jungfischen von blaufelchen (*Coregonus wart-
        manni* Bloch), weissfelchen (*Coregonus fera* Jurine) und hechten (*Esox
        lucius* L.). Arch. Hydrobiol., 28(Suppl.): 183-244.
B334   Banks, J., 1965. The biology of pike from three lowland waters. Proc.
        Brit. Coarse Fish Conf. Univ. Liverpool, 2: 29-39.
B335   Barraclough, W. E., 1964. Contribution to the marine life history of
        the Eulachon *Thaleichthys pacificus*. J. Fish. Res. Bd. Can., 21(5):
        1333-37.
B336   Buss, K., 1960. The muskellunge. Penn. Fish Comm. Spec. Purpose
        Rep., 23 p. mimeo.
B337   Burns, J. W., 1966. Carp. Inland Fisheries Management, ed. A. Cal-
        houn, Calif. Dept. Fish Game: 510-15.
B338   _____, 1966. Tui chub. Inland Fisheries Management, ed. A. Calhoun,
        Calif. Dept. Fish Game: 528-30.
B339   _____, 1966. Hitch. Inland Fisheries Management, ed. A. Calhoun,
        Calif. Dept. Fish Game: 520-22.
B340   _____, 1966. Hardhead. Inland Fisheries Management, ed. A. Calhoun,
        Calif. Dept. Fish Game: 518-19.
B341   _____, 1966. Sacramento blackfish. Inland Fisheries Management, ed.
        A. Calhoun, Calif. Dept. Fish Game: 522-25.
B342   _____, 1966. Sacramento squawfish. Inland Fisheries Management,
        ed. A. Calhoun, Calif. Dept. Fish Game: 525-27.
B343   Behmer, D. J., 1965. Length-weight relationship and spawning of river

carpsuckers, *Carpiodes carpio*, in Des Moines River, Iowa. MS thesis Iowa St. Univ. Lib. 54 p.

B344   Behmer, D. J., 1966. Length-weight relationships as a measure of "condition" of river carpsuckers, *Carpiodes carpio*, in the Des Moines River. PhD disser. Iowa St. Univ. Lib. 154 p.

B345   Burns, J. W., 1966. Western sucker. Inland Fisheries Management, ed. A. Calhoun, Calif. Dept. Fish Game: 516-17.

B346   Beers, G. D., 1955. Effects of a threadfin shad introduction upon black crappie and buffalo fish populations in Roosevelt Lake. MS thesis Univ. Ariz. Lib. 57 p. not seen in entirety.

B347   Brown, B. E. and J. S. Dendy, 1961. Observations on the food habits of the flathead and blue catfish in Alabama. Proc. S. E. Assoc. Game Fish Comm., 15: 219-22.

B348   Bowen, E. S., 1931. The role of the sense organs in aggregations of *Ameiurus melas*. Ecol. Monogr., 1: 1-35.

B349   Bunting, D. L., II and W. H. Irwin, 1965. The relative resistances of seventeen species of fish to petroleum refinery effluents and a comparison of some possible methods of ranking resistances. Proc. S. E. Assoc. Game Fish Comm., 17: 293-307.

B350   Brown, B. E., 1965. Two-year study of a bass, sunfish, channel catfish population exposed to flooding and angling. Proc. S. E. Assoc. Game Fish Comm., 17: 367-72.

B351   Behmer, D. J., 1965. Movement and angler harvest of fishes in the Des Moines River, Boone County, Iowa. Proc. Iowa Acad. Sci., 71: 259-63.

B352   Ball, J. N. and G. E. Pickford, 1964. Pituitary cytology and freshwater adaptation in the killifish, *Fundulus heteroclitus*. Anat. Rec., 148(2): 358.

B353   Brummett, A. R., 1966. Observations on the eggs and breeding season of *Fundulus heteroclitus* at Beaufort, North Carolina. Copeia, 1966(3): 616-20.

B354   Brown, C. J. D. and A. C. Fox, 1966. Mosquito fish *(Gambusia affinis)* in a Montana pond. Copeia, 1966(3): 614-16.

B355   Barnickol, P. G., 1941. Food habits of *Gambusia affinis* from Reelfoot Lake, Tennessee, with special references to malaria control. J. Tenn. Acad. Sci., 6(1): 5-13.

B356   Bennett, M. A. Craig-, 1931. The reproductive cycle of the three-spined stickleback, *Gasterosteus aculeatus*, Linn. Philos. Trans. Roy. Soc. Lond., B219: 197-279.

B357   Bock, F., 1928. Die Hypophyse des Stichlings (*Gasterosteus aculeatus* L.) unter besonderer Berüchsichtigung der jahrescyklischen Veränderungen. J. Wiss. Zool., 131: 645-710.

B358   Breder, C. M., Jr. and D. E. Rosen, 1966. Modes of reproduction in fishes. Natural History Press, Garden City, N. Y. 941 p.

B359   Bridges, D. W. and J. M. Neuhold, 1966. Brown trout survival and movement in the Logan River. Proc. Utah Acad. Sci. Arts Lett., 43(1): 67-82.

B360   Bertin, L., 1956. Eels: a biological study. Cleaver-Hume, Lond. 192 p.

B361   Beverton, R. J. H. and S. J. Holt, 1957. On the dynamics of exploited fish populations. Fish. Invest. Lond. Ser., II(19): 533 p.

B362   Bergeron, J., 1962. Bibliographie du Saumon de L'Atlantique (*Salmo salar* L.). Que. Dept. Fish. Contr., 88: 64 p.

B363   Belding, D. L. and G. Préfontaine, 1961. A report on the salmon of the
        north shore of the Gulf of St. Lawrence and of the northeastern coast of
        Newfoundland. Que. Dept. Fish. Contr., 82: 104 p.

B364   Budd, J., 1957. Introduction of the hybrid between the eastern brook
        trout and lake trout into the Great Lakes. Canad. Fish Cult., 20: 25-8.

C1   Cahn, A. R., 1927. An ecological study of southern Wisconsin fishes,
        the brook silversides *(Labidesthes sicculus)* and the cisco *(Leucichthys
        artedi)* in their relations to the region. Ill. Biol. Monogr., 11(1): 1-151.

C2   _____, 1936. Observations on the breeding of the lawyer, *Lota macu-
        losa.* Copeia, 1936(3): 163-65.

C3   Calderwood, W. L., 1925. The relation of sea-growth and spawning
        frequency in *Salmo salar.* Proc. Roy. Soc. Edinb., 45: 142-48.

C4   _____, 1927. The salmon of the R. Grand Cascapedia, Canada. Proc.
        Roy. Soc. Edinb., 47: 142-47.

C5   Calhoun, A. J., 1944. Black-spotted trout in Blue Lake, California.
        Calif. Fish Game, 30(1): 22-42.

C7   Carbine, W. F., 1942. Observations on the life history of the northern
        pike, *Esox lucius* L., in Houghton Lake, Michigan. Trans. Amer. Fish.
        Soc., 71: 149-64.

C8   _____, 1943. The artificial propagation and growth of the common
        white sucker, *Catostomus c. commersonnii,* and its value as a bait and
        forage fish. Copeia, 1943(1): 48-9.

C9   _____, 1944. Egg production of the northern pike *Esox lucius* L., and
        the percentage survival of eggs and young on the spawning grounds.
        Pap. Mich. Acad. Sci. Arts Lett., 29: 123-38.

C10   _____, 1945. Growth potential of the northern pike *(Esox lucius).* Pap.
        Mich. Acad. Sci. Arts Lett., 30: 205-20.

C11   _____, 1947. The pike—a prized and spurned fish. Reprinted from
        Mich. Conserv. Aug. 1938, 7(12): 6-8, revised April 1947.

C12   _____ and V. C. Applegate, 1948. The movement and growth of marked
        northern pike *(Esox lucius* L.) in Houghton Lake and the Muskegon
        River. Pap. Mich. Acad. Sci. Arts Lett., 32: 213-38.

C13   _____ and _____, 1948. The fish population of Deep Lake, Michigan.
        Trans. Amer. Fish. Soc., 75: 200-237.

C14   Carl, G. C. and W. A. Clemens, 1948. The fresh-water fishes of Brit-
        ish Columbia. B. C. Prov. Mus. Handb., 5: 1-132.

C17   Carlander, K. D., 1942. An investigation of Lake of the Woods, Minne-
        sota, with particular reference to the commercial fisheries. Minn.
        Bur. Fish. Res. Invest. Rep., 42: 1-534 typewritten.

C19   _____, 1943. Growth rate of the spottailed minnow, *Notropis hudsonius*
        (Clinton), in Minnesota waters. Minn. Bur. Fish. Res. Invest. Rep., 50:
        3 p. typewritten.

C20   _____, 1943. Length-weight relationship of Minnesota fishes. Minn.
        Bur. Fish. Res. Invest. Rep., 17(revised): 23 p. typewritten.

C22   _____, 1944. Notes on the minor species of fish taken in the commer-
        cial fisheries at Lake of the Woods, 1939 to 1943. Minn. Bur. Fish.
        Res. Invest. Rep., 42(Suppl.2): 27 p. typewritten.

C23   _____, 1944. Notes on the coefficient of condition, K, of Minnesota
        fishes. Minn. Bur. Fish. Res. Invest. Rep., 41(revised): 40 p. type-
        written.

C24   _____, 1944. Relationship between standard, fork, and total lengths of
        some Minnesota fishes. Minn. Bur. Fish. Res. Invest. Rep., 19(re-
        vised): 55 p. typewritten.

C25   Carlander, K. D., 1944. The commercial fisheries of Lake of the
       Woods, 1942 and 1943. Minn. Bur. Fish. Res. Invest. Rep., 42(Suppl.1):
       37 p. typewritten.

C28   _____, 1945. Growth, length-weight relationship and population fluctu-
       ations of the tullibee, *Leucichthys artedi tullibee* (Richardson) with
       reference to the commercial fisheries, Lake of the Woods, Minnesota.
       Trans. Amer. Fish. Soc., 73: 125-36.

C30   _____, 1949. Project No. 39. Yellow pike-perch management. Progr.
       Rep. Iowa Coop. Wildl. Fish. Res. Units, 1949(Jan.-Mar.): 44-57.

C32   _____, 1950. Some considerations in the use of growth data derived
       from scale studies. Trans. Amer. Fish. Soc., 79: 187-94.

C33   _____ and R. A. Fredin, 1948. Project No. 37. Management of small
       ponds for fish production. Progr. Rep. Iowa Coop. Wildl. Fish. Res.
       Units, 1948(Oct.-Dec.): 101-5.

C34   _____ and L. E. Hiner, 1943. Preliminary report on fisheries investi-
       gations, Leech Lake, Cass County. Minn. Bur. Fish. Res. 18 p. type-
       written.

C35   _____ and _____, 1943. Fisheries investigation and management report
       for Lake Vermilion, St. Louis County. Minn. Bur. Fish. Res. Invest.
       Rep., 54: 1-175.

C36   _____ and J. W. Parsons, 1949. Project No. 39. Yellow pike-perch
       management. Progr. Rep. Iowa Coop. Wildl. Fish. Res. Units, 1949
       (April-June): 49-52.

C37   _____ and L. L. Smith, Jr., 1945. Some factors to consider in the
       choice between standard, fork, or total lengths in fishery investiga-
       tions. Copeia, 1945(1): 7-12.

C38   _____ and G. Sprugel, 1948. Project No. 42: Bullhead management:
       shallow lake investigations. Progr. Rep. Iowa Coop. Wildl. Fish. Res.
       Unit, 1948(Oct.-Dec.): 112-17.

C43   Catt, J. and A. W. H. Needler, 1946. Restoration of an abundant trout
       population by poisoning introduced yellow perch and restocking. Canad.
       Fish Cult., 1(1): 9-12.

C45   Chapman, P., 1884. Habits of the shad and herring as they appear in
       the Potomac River to one who has watched them for fifty years. Bull.
       U. S. Fish. Comm., 4: 61-4.

C47   Cheney, A. N., 1897. Concerning the work of the Fisheries, Game and
       Forest Commission of the State of New York. Trans. Amer. Fish.
       Soc., 25: 112-20.

C48   Churchill, W. S., 1947. The brook lamprey in the Brule River. Trans.
       Wis. Acad. Sci., 37: 337-46.

C49   Clark, C. F., 1943. Creek chub minnow propagation. Ohio Conserv.
       Bull., 7(6): 12-13.

C52   Clemens, W. A., 1922. A study of the ciscoes of Lake Erie. Publ. Ont.
       Fish. Res. Lab., 2: 27-37.

C53   _____, 1939. The fishes of Okanagan Lake and nearby waters. Bull.
       Fish. Res. Bd. Can., 56: 27-38.

C55   Cole, L. J., 1905. The German carp in the United States. Rep. U. S.
       Bur. Fish., 1904: 523-641.

C56   Cooper, G. P., 1935. Some results of forage fish investigations in
       Michigan. Trans. Amer. Fish. Soc., 65: 132-42.

C57   _____, 1936. Age and growth of the golden shiner *(Notemigonus chry-
       soleucas auratus)* and its suitability for propagation. Pap. Mich. Acad.
       Sci. Arts Lett., 21: 587-97.

C58   Cooper, G. P., 1936. Importance of forage fish. Proc. N. Am. Wildl. Conf., 1: 305-10.

C59   _____, 1937. Age, growth, and morphometry of the cisco, *Leucichthys artedi* (LeSueur), in Blind Lake, Washtenaw County, Michigan. Pap. Mich. Acad. Sci. Arts Lett., 22: 563-71.

C60   _____, 1937. Food habits, rate of growth and cannibalism of young largemouth bass *(Aplites salmoides)* in state-operated rearing ponds in Michigan during 1935. Trans. Amer. Fish. Soc., 66: 242-66.

C62   _____, 1940. A biological survey of the Rangeley Lakes with special reference to the trout and salmon. Maine Dept. Inl. Fish. Game Fish Surv. Rep., 3.

C64   _____ and J. L. Fuller, 1945. A biological survey of Moosehead Lake and Haymock Lake. Maine Dept. Inl. Fish. Game Fish Surv. Rep., 6: 1-160.

C66   Costen, H. E. T., F. T. K. Pendleton, and R. W. Butcher, 1936. River management: the making, care, and development of salmon and trout waters. Lippincott, Philadelphia. 263 p.

C67   Couch, J. H., 1922. Rate of growth of whitefish *(Coregonus albus)* in Lake Erie. Publ. Ont. Fish. Res. Lab., 7: 99-107.

C68   Cowles, R. B., 1934. Notes on the ecology and breeding habits of the desert minnow, *Cyprinodon macularis* Baird and Girard. Copeia, 1934(1): 40-42.

C69   Cramer, F. K., 1940. Notes on the natural spawning of cutthroat trout *(Salmo clarkii clarkii)* in Oregon. Proc. 6th Pacif. Sci. Cong., 3: 335-39.

C71   Creaser, C. W., 1925. The establishment of the Atlantic smelt in the upper waters of Great Lakes. Pap. Mich. Acad. Sci. Arts Lett., 5: 405-23.

C73   _____, 1929. The food of the yearling smelt from Michigan. Pap. Mich. Acad. Sci. Arts Lett., 10: 427-31.

C74   _____ and E. P. Creaser, 1934. The grayling in Michigan. Pap. Mich. Acad. Sci. Arts Lett., 20: 599-608.

C79   Curtis, B., 1934. The golden trout of Cottonwood Lakes (*Salmo aguabonita* Jordan). Trans. Amer. Fish. Soc., 64: 259-65.

C80   _____, 1935. The golden trout of Cottonwood Lakes. Calif. Fish Game, 21(2): 101-9.

C82   Canfield, H. L., 1947. Artificial propagation of those channel cats. Progr. Fish Cult., 9(1): 27-30.

C89   Carlander, K. D. and R. B. Moorman, 1949. Project 37. Management of small ponds for fish production. Quart. Rep. Iowa Coop. Wildl. Fish. Res. Units, 1949(July-Sept.): 54-83.

C91   _____ and J. W. Parsons, 1949. Project 39. Yellow pike-perch management. Quart. Rep. Iowa Coop. Wildl. Fish. Res. Units, 1949(Oct.-Dec.): 34-45.

C93   Cuerrier, J.-P., 1949. L'Esturgeon de Lac; Age-Croissance-Maturité. Chasse et Peche (Montreal), 1(6): 26.

C100   Carlander, K. D., 1952. Project 37. Farm ponds. Quart. Rep. Iowa Coop. Wildl. Fish. Res. Units, 17(4): 23-25.

C101   _____, T. S. English, and J. G. Erickson, 1950. Project 39. Yellow pike-perch management. Quart. Rep. Iowa Coop. Wildl. Fish. Res. Units, 16(1): 40-8.

C102   _____, J. L. Forney, and W. Pearcy, 1951. Project 39. Clear Lake Investigations. Quart. Rep. Iowa Coop. Wildl. Fish. Res. Units, 17(1): 37-43.

C103   Carlander, K. D. and R. C. Hennemuth, 1952. Artificial lakes. Quart.
        Rep. Iowa Coop. Wildl. Fish. Res. Units, 18(1): 32-35.

C104   ____, W. M. Lewis, C. E. Ruhr, and R. E. Cleary, 1953. Abundance,
        growth, and condition of yellow bass, *Morone interrupta* Gill, in Clear
        Lake, Iowa, 1944 to 1951. Trans. Amer. Fish. Soc., 82: 91-103.

C105   ____ and R. B. Moorman, 1950. Project 37. Management of small
        ponds for fish production. Quart. Rep. Iowa Coop. Wildl. Fish. Res.
        Units, 15(4): 36-40.

C106   ____ and ____, 1951. Project 37. Management of small ponds for
        fish production. Quart. Rep. Iowa Coop. Wildl. Fish. Res. Units, 16(3):
        43-5.

C107   ____ and ____, 1952. Project 37. Management of farm ponds for
        fish production. Quart. Rep. Iowa Coop. Wildl. Fish. Res. Units, 17(3):
        24-7.

C108   ____ and G. Sprugel, 1950. Project 42. Bullhead management. Quart.
        Rep. Iowa Coop. Wildl. Fish. Res. Units, 15(4): 44-5.

C109   ____ and R. R. Whitney, 1952. Clear Lake investigations. Quart. Rep.
        Iowa Coop. Wildl. Fish. Res. Units, 18(1): 27-30.

C117   Churchill, W. S., 1950. Experimental stocking of predators for popula-
        tion control. Wis. Div. Fish. Mgmt. Invest. Rep., 728: 1-8 mimeo.

C125   Clemens, H. P., 1951. The growth of the burbot, *Lota lota maculosa*
        (LeSueur) in Lake Erie. Trans. Amer. Fish. Soc., 80: 163-73.

C126   Clemens, W. A., R. V. Boughton, and J. A. Rattenbury, 1945. A pre-
        liminary report on a fishery survey of Teslin Lake, British Columbia.
        B. C. Fish. Dept. Rep., 1944: M70-5.

C128   Cooper, E. L., 1951. Validation of the use of scales of brook trout,
        Salvelinus fontinalis, for age determination. Copeia, 1951(2): 141-48.

C131   ____, 1952. Body-scale relationship of brook trout, *Salvelinus fonti-
        nalis,* in Michigan. Copeia, 1952(1): 1-4.

C132   ____ and N. G. Benson, 1951. The coefficient of condition of brook,
        brown, and rainbow trout in the Pigeon River, Otsego County, Michigan.
        Progr. Fish Cult., 13(4): 181-92.

C134   Cross, F. B., 1950. Effects of sewage and of a headwaters impound-
        ment on the fishes of Stillwater Creek in Payne County, Oklahoma.
        Amer. Midl. Nat., 43(1): 128-45.

C135   Cuerrier, J.-P., 1951. The use of pectoral fin rays for determining
        age of sturgeon and other species of fish. Canad. Fish Cult., 11:10-18.

C136   ____ and G. Roussow, 1951. Age and growth of lake sturgeon from
        Lake St. Francis, St. Lawrence River. Report on material collected in
        1947. Canad. Fish Cult., 10: 17-29.

C143   Cooper, E. L., 1952. Rate of exploitation of wild eastern brook trout
        and brown trout populations in the Pigeon River, Otsego County, Mich-
        igan. Trans. Amer. Fish. Soc., 81: 224-34.

C145   Cable, L. E., 1956. Validity of age determination from scales, and
        growth of marked Lake Michigan lake trout. U. S. Fish Wildl. Serv.
        Fish. Bull., 57(107): 1-59.

C146   Calderon-Andrese, E. G., 1955. Acclimation du brochet en Espagne.
        Proc. Int. Assoc. Theor. Appl. Limnol., 12: 536-42.

C147   Calderwood, W. L., 1927. Atlantic salmon in New Zealand. The salm
        salmon of Lake Te Anau. Salm. Trout Mag., 48: 241-52.

C148   Caldwell, D. K., H. T. Odum, T. R. Hellier, Jr., and F. H. Berry, 1957.
        Populations of spotted sunfish and Florida largemouth bass in a con-
        stant temperature spring. Trans. Amer. Fish. Soc., 85: 120-34.

C149   Campbell, H. J., 1956. Northeastern Oregon. Annu. Rep. Ore. St.
        Game Comm. Fish. Div., 1955: 73-83.
C150   _____, 1957. Northeastern Oregon. Annu. Rep. Ore. St. Game Comm.
        Fish. Div., 1956: 74-91.
C151   _____, 1958. Fishery Research. Annu. Rep. Ore. St. Game Comm.
        Fish. Div., 1957: 248-58.
C153   Carl, G. C., 1953. Limnobiology of Cowichan Lake, British Columbia.
        J. Fish. Res. Bd. Can., IX(9): 417-49.
C157   Carlander, K. D. and G. Sprugel, 1955. Fishes of Little Wall Lake,
        Iowa, prior to dredging. Proc. Iowa Acad. Sci., 62: 555-66.
C167   Cating, J. P., 1953. Determining age of Atlantic shad from their
        scales. Fish. Bull. U. S., 54(85): 187-99.
C173   Chaplin, B. R., 1954. Pittsfield record pickerel. Mass. Wildl., 1954
        (March): 9.
C174   Chapman, D. W., 1958. Studies on the life history of Alsea River steel-
        head. J. Wildl. Mgmt., 22(2): 123-33.
C185   Clark, C. F. and E. D. Now, 1954. Experimental propagation of north-
        ern pike at the St. Mary's fish farm. 16th Midw. Wildl. Conf. 9 p.
        mimeo.
C186   _____ and F. Steinbach, 1959. Observations on the age and growth of
        the northern pike, *Esox lucius* L., in East Harbor, Ohio. Ohio J. Sci.,
        59(3): 129-34.
C194   Clemens, H. P. and K. E. Sneed, 1957. Spawning behavior of the chan-
        nel catfish *Ictalurus punctatus*. U. S. Fish Wildl. Serv. Spec. Sci. Rep.
        Fish, 219: 1-11.
C197   Cochrun, K., 1956. Klamath District. Annu. Rep. Ore. St. Game
        Comm. Fish. Div., 1955: 160-67.
C198   _____, 1957. Klamath District. Annu. Rep. Ore. St. Game Comm.
        Fish. Div., 1956: 163-69.
C201   Cohen, D. M., 1954. Age and growth studies on two species of white-
        fishes from Point Barrow, Alaska. Stanf. Icthyol. Bull., 4(3): 168-88.
C205   Commercial Fisheries Review, 1954. Large sturgeon landed at Boston.
        Commer. Fish. Rev., 16(3): 56.
C206   _____, 1954. 152 year-old lake sturgeon caught in Ontario. Commer.
        Fish. Rev., 16(9): 28.
C207   Cook, C. H., 1910. The salmon as we know him. J. Salm. Trout
        Assoc., 1: 5-13.
C208   Cook, E. P., 1955. A lake inventory of Meredith Reservoir, Crowley
        County, Colorado. Colo. Dept. Game Fish Publ. 46 p.
C209   Cooper, E. L., 1953. Periodicity of growth and change of condition of
        brook trout *(Salvelinus fontinalis)* in three Michigan trout streams.
        Copeia, 1953(2): 107-13.
C210   _____, 1953. Returns from plantings of legal-sized brook, brown, and
        rainbow trout in the Pigeon River, Otsego County, Michigan. Trans.
        Amer. Fish. Soc., 82: 265-80.
C211   _____, 1953. Growth of brook trout *(Salvelinus fontinalis)* and brown
        trout *(Salmo trutta)* in the Pigeon River, Otsego County, Michigan.
        Pap. Mich. Acad. Sci. Arts Lett., 38: 151-61.
C212   _____, 1953. Mortality rates of brook trout and brown trout in the Pi-
        geon River, Otsego County, Michigan. Progr. Fish Cult., 15(4): 163-69.
C215   Cooper, G. P. and R. N. Schafer, 1954. Studies on the population of
        legal-size fish in Whitmore Lake, Washtenaw and Livingstone Counties,
        Michigan. Trans. N. Amer. Wildl. Conf., 19: 239-58.

C217 Cope, O. B., 1953. Length measurements of Lake Yellowstone trout. U. S. Fish Wildl. Serv. Spec. Sci. Rep. Fish., 103: 17 p.

C218 _____, 1956. The future of the cutthroat in Utah. Proc. Utah Acad., 32: 89-93.

C219 _____, 1956. The choice of spawning sites by cutthroat trout. Proc. Utah Acad., 32: 73-9.

C220 _____, 1956. Some migration patterns in cutthroat trout. Proc. Utah Acad., 33: 113-18.

C222 _____, 1958. Annotated bibliography on the cutthroat trout. U. S. Fish Wildl. Serv. Fish. Bull., 58(140): 417-42.

C223 _____, 1959. Rocky Mountain sport fishery investigations. U. S. Fish Wildl. Serv. Circ., 57: 20-28.

C224 Corbett, E. M., 1922. The length and weight of salmon. Salm. Trout Mag., 30: 206-14.

C227 Crossman, E. J., 1956. Growth, mortality and movements of a sanctuary population of maskinonge (*Esox Masquinongy* Mitchill). J. Fish. Res. Bd. Can., 13(5): 599-612.

C228 _____, 1959. Distribution and movements of a predator, the rainbow trout, and its prey, the redside shiner, in Paul Lake, British Columbia. J. Fish. Res. Bd. Can., 16(3): 247-67.

C229 _____, 1959. A predator-prey interaction in freshwater fish. J. Fish. Res. Bd. Can., (16): 269-81.

C230 _____ and P. A. Larkin, 1959. Yearling liberations and change of food as effecting [sic] rainbow trout yield in Paul Lake, British Columbia. Trans. Amer. Fish. Soc., 88(1): 36-45.

C233 Cuerrier, J. P., 1954. The history of Lake Minnewanka with reference to the reaction of lake trout to artificial changes in environment. Canad. Fish Cult., 15: 1-9.

C234 _____, 1949. Observations on the lake sturgeon (*Acipenser fulvescens* Raf.) in the Lake Saint Peter region during the spawning season. MS thesis Univ. Montreal 1949. 120 p. Rev. and supplemented manuscript.

C236 _____ and F. H. Schultz, 1957. Studies of lake trout and common whitefish in Watertown Lakes, Watertown Lakes National Park, Alberta. Canad. Wildl. Serv. Wildl. Mgmt. Bull., Ser. 3(5): 41 p.

C243 Cutting, R. E., 1958. Atlantic salmon, *Salmo salar* Linneaus. *In* Fishes of Maine, 2nd ed., ed. W. H. Everhart: 26-29.

C250 Clugston, J. P. and E. L. Cooper, 1960. Growth of the common eastern madtom, *Noturus insignis*, in central Pennsylvania. Copeia, 1960(1): 9-16.

C252 Christie, W. J., 1960. Variation in vertebral count in F^2 hybrids of *Salvelinus fontinalis* x *S. namaycush*. Canad. Fish Cult., 26: 15-21.

C254 Clemens, W. A. and G. V. Wilby, 1961. Fishes of the Pacific coast of Canada. Fish. Res. Bd. Can. Bull., 68(2nd ed.): 443 p.

C255 Caraway, P. A., 1951. The whitefish, *Coregonus clupeaformis* (Mitchill), of northern Lake Michigan, with special reference to age, growth, and certain morphometric characters. PhD thesis Mich. St. Coll. 140 p.

C258 Campbell, R. S., 1935. A study of the common sucker, *Catostomus commersoni* (Lacépède), of Waskesiu Lake. MA thesis Univ. Sask. 48 p.

C259 Carbine, W. F., 1936. The life history of the chub, *Trigoma atraria* Girard, of the Great Basin of Utah. MS thesis Univ. Utah. 107 p.

C261 Carnes, W. C., Jr., 1958. Contributions to the biology of the eastern chubsucker, *Erimyzon oblongus oblongus* (Mitchill). MS thesis N. C. St. Coll. 69 p.

C263   Corthell, R. A., 1960. Coos-Coquille District. Annu. Rep. Ore St. Game Comm. Fish. Div., 1959: 255-65.

C265   Calhoun, A. J., 1942. The biology of the black-spotted trout (*Salmo clarkii henshawi* Gill and Jordan) in two Sierra lakes. PhD disser. Stanford Univ. 218 p.

C270   Christenson, L. M., 1957. Some characteristics of the fish populations in backwater areas of the Upper Mississippi River. MS thesis Univ. Minn. 125 p.

C271   Cross, F. B., 1951. Early limnological and fish population conditions of Canton Reservoir, Oklahoma, with special reference to carp, channel catfish, largemouth bass, green sunfish and bluegill, and fishery management recommendations. PhD thesis Okla. A. M. Coll. 92 p.

C273   Chapoton, R. B., 1955. Growth characteristics of the northern redbelly dace, *Chrosomus eos* (Cope) in experimental ponds in northern Michigan. MS thesis Mich. St. Coll. 49 p.

C274   Conservationist (Wis.), 1961. Record splake. Conserv., 140: 5.

C275   Commercial Fisheries Review, 1961. Lake Erie fish population survey for 1961 season begins. Commer. Fish. Rev., 23(6): 23-4.

C276   Cuinat, R., 1960. Croissance et taille legale de la Truite fario dans quelques rivieres francaise. Ann. Sta. Centr. Hydrobiol. Appl., 8: 227-61.

C277   Campbell, R. N., 1961. The growth of brown trout in acid and alkaline waters. Salm. Trout Mag., Jan., 1961: 47-52.

C279   Corthell, R. A., 1960. Coos-Coquille District. Annu. Rep. Ore. St. Game Comm. Fish. Div., 1960: 278-90.

C281   Cooper, E. L., 1961. Growth of wild and hatchery strains of brook trout. Trans. Amer. Fish. Soc., 90(4): 424-38.

C282   Coble, D. W., 1961. Influence of water exchange and dissolved oxygen in redds on survival of steelhead trout embryos. Trans. Amer. Fish. Soc., 90(4): 469-74.

C283   Call, M. W., 1960. A study of the Pole Mountain beaver and their relation to the brook trout fishery. Wyo. Game Fish Comm., Univ. Wyo. Coop. Res. Proj., 2(MS thesis): 93-219.

C284   Clark, C. F., 1960. Lake St. Mary's and its management. Ohio Dept. Nat. Resources Div. Wildl. Publ., W-324: 107 p.

C286   Cooper, E. L., J. A. Boccardy, and J. K. Andersen, 1962. Growth rate of brook trout at different population densities in a small infertile stream. Progr. Fish Cult., 24(2): 74-80.

C287   Chicewicz, M., 1961. [Embryonic development of lake trout (*Salmo trutta morpha lacustris* L.) from Wdzydze Lake.] (English summary.) Rocz. Nauk Rolniczych, 93: 557-94.

C290   Crossman, E. J., 1962. The redfin pickerel, *Esox a. americanus,* in North Carolina. Copeia, 1962(1): 114-23.

C292   Crawford, B., 1960 (revised). Propagation of channel catfish *(Ictalurus lacustris).* Ark. Game Fish Comm. St. Fish Hatchery Centerton. 16 p.

C293   Curtis, B. and J. C. Fraser, 1948. Kokanee in California. Calif. Fish Game, 34(3): 111-14.

C294   Cope, O. B., 1957. *Salmo clarki.* Data for Handbook of Biological Data. 11 p.

C295   Clemens, W. A., 1928. The food of trout from the streams of Oneida County, New York State. Trans. Amer. Fish. Soc., 58: 183-97.

C297 Carbine, W. F., 1953. Further data on the development of a nonmigratory strain of rainbow trout. Progr. Fish Cult., 15(2): 83-4.

C298 Crossman, E. J., 1962. Predator-prey relationships in pikes (Esocidae). J. Fish. Res. Bd. Can., 19(5): 979-80.

C299 ____, 1962. The grass pickerel, *Esox americanus vermiculatus* LeSueur, in Canada. Roy. Ont. Mus. Contr., 55: 29 p.

C300 Cooper, E. L., 1959. Trout stocking as an aid to fish management. Penn. Agric. Exp. Sta. Bull., 663: 21 p.

C303 Christie, W. J., 1963. Effects of artificial propagation and the weather on recruitment in the Lake Ontario whitefish fishery. J. Fish. Res. Bd. Can., 20(3): 597-646.

C306 Clemmer, G. H. and F. J. Schwartz, 1964. Age, growth and weight relationships of the striped killifish, *Fundulus majalis,* near Solomons, Maryland. Trans. Amer. Fish. Soc., 93(2): 197-98.

C310 Clark, C. F., 1958. Northern pike, *Esox lucius* Linnaeus. Data for Handbook of Biological Data. 10 p.

C311 ____, 1950. Observations on the spawning habits of the northern pike, *Esox lucius,* in northwestern Ohio. Copeia, 1950(4): 285-88.

C312 Cross, F. B. and C. E. Hastings, 1956. Ages and sizes of thirty-nine flathead cat. Trans. Kans. Acad. Sci., 59(1): 85-6.

C313 Clemens, W. A. and J. A. Munro, 1934. Food of squawfish. Progr. Rep. Fish. Res. Bd. Can. Pacif. Biol. Sta., 19: 3-4.

C314 Coker, R. E., 1927. Black-nosed dace in North Carolina. Copeia, 1927 (162): 4.

C315 Cross, F. B., 1958. Suckermouth minnow, *Phenecobius mirabilis* (Girard); Plains minnow, *Hybognathus placita* Girard; Red shiner, *Notropis lutrensis* (Baird and Girard); Plains killifish, *Fundulus kansae* Garman. Data for Handbook of Biological Data. 14 p.

C316 Coyle, E. E., 1930. The algal food of *Pimephales promelas* (fathead minnow). Ohio J. Sci., 30(1): 23-35.

C317 Cozart, C. E., 1959. The culture of carp and goldfish to feed largemouth black bass broodstock. U. S. Fish Cult. Sta. Tishomingo, Okla. 8 p. mimeo.

C318 Chamberlain, E. B., Jr., 1948. Ecological factors influencing the growth and management of certain waterfowl food plants on Back Bay National Wildlife Refuge. Trans. N. Amer. Wildl. Conf., 13: 347-56.

C319 Cahn, A. R., 1929. The effect of carp on a small lake: the carp as a dominant. Ecology, 10(3): 271-74.

C320 Cahoon, W. G., 1953. Commercial carp removal at Lake Mattamuskat. J. Wildl. Mgmt., 17(3): 312-17.

C321 Cross, F. B., 1954. Fishes of Cedar Creek and the south fork of the Cottonwood River, Chase County, Kansas. Trans. Kans. Acad. Sci., 57(3): 303-14.

C322 Calhoun, A. J., 1940. Note on a hybrid minnow, *Apocope* x *Richardsonius.* Copeia, 1940(2): 142-43.

C324 Cross, F. B. and W. L. Minckley, 1960. Five natural hybrid combinations in minnows (Cyprinidae). Univ. Kans. Mus. Nat. Hist. Publ., 13(1): 1-18.

C326 Canfield, H. L., 1922. Care and feeding of buffalo fish in ponds. U. S. Bur. Fish. Econ. Circ., 56: 1-3.

C327 Cook, S. F., Jr., R. L. Moore, and J. D. Conners, 1966. The status of the native fishes of Clear Lake, Lake County, California. Wasmann J. Biol. 24(1): 141-60.

C328   Coots, M., 1955. The Pacific lamprey, *Entospheus tridentatus*, above
        Copco Dam, Siskiyou County, California. Calif. Fish Game, 41(1): 118-
        19.

C329   Coventry, A. F., 1922. Breeding habits of the land locked sea lamprey
        (*Petromyzon marinus* var. *dorsatus* Wilder). Univ. Toronto Stud. Biol.
        Ser., 9: 129-36.

C330   Cook, F. A., 1952. Occurrence of the lamprey *Lampetra aepyptera* in
        the Tombigbee and Pascagoula drainages, Mississippi. Copeia, 1952(4):
        268.

C331   Cheek, R. P., 1965. Pugheadedness in an American shad. Trans.
        Amer. Fish. Soc., 94(1): 97-8.

C332   Clemens, W. A. and N. K. Bigelow, 1922. The food of ciscoes *(Leuci-
        chthys)* in Lake Erie. Univ. Toronto Stud. Biol. Ser., 20: 41-53.

C333   Cucin, D. and H. A. Regier, 1965. Dynamics and exploitation of lake
        whitefish in southern Georgian Bay. J. Fish. Res. Bd. Can., 23(2): 221-
        74.

C334   Chapman, D. W. and J. D. Fortune, Jr., 1963. Ecology of kokanee
        salmon. Ore. St. Game Comm. Res. Div. Rep., 1963: 11-42.

C335   Colburn, L. G., 1966. The limnology and cutthroat trout fishery of
        Trapper's Lake, Colorado. Colo. Coop. Fish. Res. Unit Spec. Rep., 9:
        26 p.

C336   Campbell, H. J. and D. J. Hansen, 1963. Coastal cutthroat trout re-
        search. Ore. St. Game Comm. Res. Div. Rep., 1963: 3-7.

C337   Cope, O. B., 1964. Revised bibliography on the cutthroat trout. U. S.
        Fish Wildl. Serv. Res. Rep., 65: 43 p.

C338   Crisp, D. T., 1963. A preliminary survey of brown trout (*Salmo trutta*
        L.) and bullheads (*Cottus gobio* L.) in high altitude becks. Salm. Trout
        Mag., 1963(Jan.): 45-59.

C339   Campbell, R. N., 1963. Some effects of impoundment on the environ-
        ment and growth of brown trout (*Salmo trutta* L.) in Loch Garry
        (Inverness-shire). Freshw. Salm. Fish. Res., 30: 27 p.

C340   Crossman, E. J., 1966. A taxonomic study of *Esox americanus* and its
        subspecies in eastern North America. Copeia, 1966(1): 1-20.

C341   Clemens, H. T., 1965. A goldfish bibliography. Aquarium Publ. Co.,
        Norristown, Penn. 41 p.

C342   Clemens, H. P., S. Bhinyoying, and N. Youngstead, 1966. An evaluation
        of dylox medication on growth and reproduction of the goldfish and
        guppy. Progr. Fish Cult., 28(3): 159-61.

C343   Chiba, K., 1961. [The basic study on the production of fish seedling
        under possible control. I. The effect of food in quality and quantity on
        the survival and growth of common carp fry.] (Japanese with English
        summary.) Bull. Freshw. Fish. Res. Lab., 11(1): 105-28.

C344   _____, 1965. [A study on the influence of oxygen concentration on the
        growth of juvenile common carp.] (Japanese with English summary.)
        Bull. Freshw. Fish. Res. Lab., 15(1): 35-47.

C345   Carlson, D. R., 1967. Fathead minnow, *Pimephales promelas* Rafines-
        que, in the Des Moines River, Boone County, Iowa, and the Skunk River
        drainage, Hamilton and Story Counties, Iowa. Iowa St. J. Sci., 41(3):
        363-74.

C346   Corthell, R. A., 1961. Coos-Coquille District. Annu. Rep. Ore. St.
        Game Comm. Fish. Div., 1961: 273-87.

C347   Conder, J. R. and R. Hoffarth, 1965. Growth of channel catfish, *Ictalu-
        rus punctatus*, and blue catfish, *Ictalurus furcatus*, in the Kentucky

Lake portion of the Tennessee River in Tennessee. Proc. S. E. Assoc. Game Fish Comm., 16: 348-54.

C348 Carlander, K. D., 1966. Relationship of limnological features to growth of fishes in lakes. Verh. Int. Ver. Limnol., 16: 1172-75.

D1 Dannevig, A. and P. Host, 1931. Sources of error in computing l_1-l_1 etc. from scales taken from different parts of the fish. J. Cons. Int. Explor. Mer., 6: 64-93.

D2 Davis, H. S., 1934. Growth and heredity in trout. Trans. Amer. Fish. Soc., 64: 197-202.

D3 _____ and M. C. James, 1924. Some experiments on the addition of vitamins to trout foods. Trans. Amer. Fish. Soc., 54: 77-91.

D4 _____ and R. F. Lord., Jr., 1930. Experiments with meat and meat substitutes as trout food. Rep. U. S. Comm. Fish., 1930(Append.VII): 123-47.

D6 Dean, H. D., 1913. Grayling. Trans. Amer. Fish. Soc., 42: 139-44.

D8 Deason, H. J. and R. Hile, 1947. Age and growth of the kiyi, *Leucichthys kiyi* Koelz, in Lake Michigan. Trans. Amer. Fish. Soc., 74(1944): 88-142.

D9 De Bellesme, J., 1897. New method of pond culture. Trans. Amer. Fish. Soc., 25: 69-94.

D10 Dence, W. A., 1948. Life history, ecology and habits of the dwarf sucker, *Catostomus commersonnii utawana* Mather, at the Huntington Wildlife Station. Roos. Wildl. Bull., 8(4): 82-150.

D11 Detwiler, J. D., 1930. Feeding experiments with brook trout fingerlings. Trans. Amer. Fish. Soc., 60: 146-57.

D12 Dickenson, F. B., 1898. The protection of fish and a closed season. Trans. Amer. Fish. Soc., 27: 32-46.

D13 Dill, W. A., 1944. The fishery of the lower Colorado River. Calif. Fish Game, 30(3): 109-211.

D15 Dinsmore, A. H., 1934. Effect of heredity on the growth of brook trout. Trans. Amer. Fish. Soc., 64: 203-4.

D16 Dixon, B., 1931. The age and growth of salmon caught in the Polish Baltic. J. Cons. Int. Explor. Mer., 6(3): 438-48.

D17 _____, 1931. Age and growth rate of the sea-trout *(Salmo trutta)* of the Rivers Reda and Dunajec. J. Cons. Int. Explor. Mer., 6(3): 449-57.

D18 _____, 1934. The age and growth of salmon caught in the Polish Baltic in the years 1931-1933. J. Cons. Int. Explor. Mer., 9(1): 66-78.

D22 Doan, K. H., 1944. The winter fishery in western Lake Erie, with a census of the 1942 catch. Ohio J. Sci., 44(2): 69-74.

D24 Donaldson, L. R. and F. J. Foster, 1938. Comparative feeding values of some of the meals fed to fish. Progr. Fish Cult., 40: 20-3.

D25 Downing, S. W., 1910. A plan for promoting the whitefish production of the Great Lakes. Proc. 4th Int. Cong. Part 1, Bull. U. S. Bur. Fish., 28: 637.

D26 Doze, J. B., 1925. The barbed trout of Kansas. Trans. Amer. Fish. Soc., 55: 167-83.

D27 _____, 1925. Barbed trout of Kansas. Kans. Fish Game Dept. Bull., 8: 1-22.

D29 Dunstan, 1928. (Note in a Symposium on fish propagation.) Trans. Amer. Fish. Soc., 58: 199-200.

D32 Dymond, J. R., 1928. Some factors affecting the production of lake trout *(Cristivomer namaycush)* in Lake Ontario. Univ. Toronto Stud. Biol. Ser., 31: 27-41.

D33   Dymond, J. R., 1933. Biological and oceanographic conditions in Hudson Bay. 8. Coregonine fishes of Hudson and James Bays. Contr. Canad. Biol. Fish., 8(1): 3-12.

D34   _____, 1943. The Coregonine fishes of northwestern Canada. Trans. Roy. Canad. Inst., 24(2): 172-233.

D35   Doan, K. H., 1948. Speckled trout in the Lower Nelson River Region, Manitoba. Fish. Res. Bd. Can. Bull., 79: 1-12.

D36   Dean, B. and F. B. Sumner, 1898. Notes on the spawning habits of the brook lamprey *(Petromyzon wilderi)*. Trans. N. Y. Acad. Sci., 16:321-24.

D40   Dobie, J., 1952. A method of calculating fish production in rearing ponds. Progr. Fish Cult., 14(1): 22-6.

D41   _____, O. L. Meehean, and G. N. Washburn, 1948. Propagation of minnows and other bait species. U. S. Fish Wildl. Serv. Circ., 12: 1-113.

D42   Douglas, P. A., 1952. Notes on the spawning of the humpback sucker, *Xyrauchen texanus* (Abbott). Calif. Fish Game, 38(2): 149-56.

D45   Dunbar, M. J. and H. H. Hildebrand, 1952. Contribution to the study of the fishes of Ungava Bay. J. Fish. Res. Bd. Can., 9(2): 83-128.

D46   Dence, W. A., 1952. Establishment of white perch, *Morone americana*, in central New York. Copeia, 1952(3): 200-201.

D48   Dahl, K., 1915. Salmon and trout: a handbook. Part I. Salmon and salmon fisheries. Chapter I. The life history of the salmon. Salm. Trout Mag., 1(10): 21-55.

D49   _____, 1916. Salmon and trout: a handbook. Part III. Trout and trout fisheries. Chapter III. The sea trout. Salm. Trout Mag., 13(13): 9-29.

D50   _____, 1917. Salmon and trout: a handbook. Salm. Trout Mag., 15: 18-34.

D51   _____, 1919. Studies of trout and trout-waters in Norway. Salm. Trout Mag., 17: 58-79; 18: 16-33.

D52   _____, 1928. The dwarf salmon of Lake Byglandsfjord. Salm. Trout Mag., 51: 108-12.

D53   _____, 1928. Secondary male characters developed in an old hen salmon. Salm. Trout Mag., 52: 242-43.

D54   _____, 1933. Are brown trout and sea trout interchangeable? Salm. Trout Mag., 71: 132-38.

D56   Darnell, R. M., 1958. Food habits of fishes and larger invertebrates of Lake Pontchartrain, Louisiana, an estuarine community. Inst. Mar. Sci., 5: 353-416.

D58   Davis, W. S., 1957. Ova production of American shad in Atlantic Coast rivers. U. S. Fish Wildl. Serv. Res. Rep., 49: 5 p.

D59   Day, L. A., 1932. The introduction of trout in Natal. Salm. Trout Mag., 69: 345-52.

D60   DeJean, J. A., 1952. Some factors affecting the reproduction of the chain pickerel, *Esox niger* LeSueur, in ponds. [Abstract.] Ala. Polytech. Inst. Grad. Sch. Bull., 4: 112-13.

D61   DeLacy, A. C., 1941. Contributions to the life histories of two Alaskan chars *Salvelinus malma* Walbaum and *Salvelinus alpinus* Linnaeus. PhD thesis Univ. Wash. 114 p.

D63   Dendy, J. S. and D. C. Scott, 1953. Distribution, life history and morphological variations of the southern brook lamprey, *Ichthyomyzon gagei*. Copeia, 1953(3): 152-62.

D64   DeRoche, S. E., 1958. Lake trout, *Cristivomer namaycush* (Walbaum). Fishes of Maine, 2nd ed., ed. W. H. Everhart: 42-5.

D65   DeRoche, S. E. and L. H. Bond, 1957. The lake trout of Cold Stream
        Pond, Enfield, Maine. Trans. Amer. Fish. Soc., 85: 257-70.

D66   DeWitt, J. W., Jr., 1954. A survey of the coast cutthroat trout, *Salmo
        clarki clarki* Richardson, in California. Calif. Fish Game, 40(3): 329-
        35.

D75   Dimick, J. B., 1957. Lake rehabilitation review. Ore. Game Comm.
        Bull., 12(3): 3, 6.

D81   Dobie, J., O. L. Meehean, S. F. Snieszko, and G. N. Washburn, 1956.
        Raising bait fishes. U. S. Fish Wildl. Serv. Circ., 35: 124 p.

D82   _____ and J. B. Moyle, 1956. Methods used for investigating produc-
        tivity of fish-rearing ponds in Minnesota. Minn. Fish. Res. Unit Spec.
        Publ., 5: 54 p.

D83   Donaldson, L. R., D. D. Hansler, and T. N. Buckridge, 1957. Inter-
        racial hybridization of cutthroat trout, *Salmo clarkii,* and its use in
        fisheries management. Trans. Amer. Fish. Soc., 86: 350-60.

D84   _____ and P. R. Olson, 1957. Development of rainbow trout brood stock
        by selective breeding. Trans. Amer. Fish. Soc., 85: 93-101.

D85   Druschba, L. J., 1959. Are the carp moving north? Wis. Conserv.
        Bull., 24(11): 22-5.

D89   Dunham, D. K., 1956. How old is that fish? Wis. Conserv. Bull., 21(7):
        11-13.

D90   Dussart, B., 1952. L'Omble-chevalier du Leman (*Salvelinus alpinus,*
        Linné, 1758). Ann. Sta. Cent. Hydrobiol. Appl., 4: 355-78.

D91   _____, 1954. L'Omble-chevalier en France. Biometrie et statistique.
        Ann. Sta. Cent. Hydrobiol. Appl., 5: 129-58.

D92   Derback, B., 1947. The adverse effect of cold weather upon the suc-
        cessful reproduction of pickerel, *Stizostedion vitreum,* at Heming Lake,
        Manitoba in 1947. Canad. Fish Cult., 2(1): 22-3.

D94   Desrochers, R., 1953. Déplacements de dorés *(Stizostedion vitreum)*
        libérés a Chambly Bassin au printemps, 1952. Rev. Canad. Biol., 11(5):
        502-5.

D97   Davis, J., 1959. Management of channel catfish in Kansas. Univ. Kans.
        Mus. Nat. Hist. Misc. Publ., 21: 1-56.

D99   Davis, J. T. and L. E. Posey, Jr., 1959. Length at maturity of channel
        catfish in Louisiana. Proc. S. E. Assoc. Game Fish Comm., 12: 72-4.

D100   Donaldson, I., 1958. White or "Oregon" sturgeon of the Columbia
        River. Data for Handbook of Biological Data. 11 p.

D101   Danforth, C. H., 1911. A 74-millimeter *Polyodon.* Biol. Bull. Wood's
        Hole, 20: 201-4.

D107   Dowell, V. E. and C. D. Riggs, 1955. Further observations on *Astyanax
        fasciatus* and *Menidia audens* in Lake Texoma. Proc. Okla. Acad. Sci.,
        36: 52-3.

D108   Donley, B. F., 1948. A study of survival and growth of fingerling east-
        ern brook trout, *Salvelinus fontinalis fontinalis* (Mitchill), in two cen-
        tral Pennsylvania streams. MS thesis Penn. St. Coll. 65 p.

D109   Desmarais, Y., 1959. Fécondite et Croissance de la Truite *(Salvelinus
        fontinalis)* dans trois lacs du Parc de Laurentides. Le Natur. Canad.,
        86(2): 33-45.

D110   _____, 1958. Croissance et alimentation comparee de la Truite du Parc
        des Laurentides et de Rimouski. Le Natur. Canad., 85(4): 73-8.

D161   Davis, J. T., 1960. Fish populations and aquatic conditions in polluted
        waters in Louisiana. La. Wildl. Fish Comm. Bull., 1(1960): 121 p.

D162   _____, 1955. Contributions to the ecology of fishes of Perche Creek,

Missouri. Part I. Movement of fishes. Part II. Age and growth of the northern river carpsucker, *Carpiodes c. carpio* (Rafinesque). MA thesis Univ. Mo. 75 p.

D163 Dendy, J. S., 1946. Food of several species of fish, Norris Reservoir, Tennessee. J. Tenn. Acad. Sci., 21(1): 105-27.

D165 Dumas, R. F., 1961. Effect of light, diet, and age of spawning brown trout upon certain characteristics of their eggs and fry. N. J. Fish Game J., 8(1): 49-56.

D166 Deacon, J. E., 1961. Fish populations, following a drought, in the Neosho and Marais des Cygnes Rivers of Kansas. Univ. Kans. Publ. Mus. Nat. Hist., 13(9): 359-427.

D168 Dobie, J., 1962. Role of the tiger salamander in natural ponds used in Minnesota for rearing suckers. Progr. Fish Cult., 24(2): 85-7.

D169 Dalziel, J. A. and K. G. Shillington, 1962. Development of a fast-growing strain of Atlantic salmon *(Salmo salar)*. Canad. Fish Cult., 30: 57-9.

D170 Darnell, R. M. and R. R. Meierotto, 1962. Determination of feeding chronology in fishes. Trans. Amer. Fish. Soc., 91(3): 313-20.

D171 Dymond, J. R., 1928. The trout of British Columbia. Trans. Amer. Fish. Soc., 58: 71-7.

D173 DeRoche, S. E., 1963. Slowed growth of lake trout following tagging. Trans. Amer. Fish. Soc., 92(2): 185-86.

D174 Dryer, W. R., 1963. Age and growth of the whitefish in Lake Superior. Fish. Bull. U. S., 63(1): 77-95.

D175 Daly, R., V. A. Hacker, and L. Wiegert, 1962. The lake trout, its life history, ecology and management. Wis. Cons. Dept. Publ., 233: 15 p.

D176 Dymond, J. R., 1926. The fishes of Lake Nipigon. Publ. Ont. Fish. Res. Lab., 27: 3-108.

D177 Douglas, P. A., 1957. Humpback sucker, *Xyrauchen texanus* (Abbott). Data for Handbook of Biological Data. 5 p.

D179 Dietz, E. M. C. and K. C. Jurgens, 1963. An evaluation of selective shad control at Medina Lake, Texas. Tex. Parks Wildl. Dept. IF Rep., 5: 1-32.

D180 Doan, K. H., 1938. Observations on dogfish *(Amia calva)* and their young. Copeia, 1938(4): 204.

D181 Davenport, D. and M. Warmuth, 1965. Notes on the relationship between the freshwater mussel, *Anodonta implicata* Say and the alewife, *Pomolobus pseudoharengus* (Wilson). Limnol. Oceanogr., 10(Suppl.): R74-R78.

D182 Dryer, W. R. and J. Beil, 1964. Life history of lake herring in Lake Superior. Fish. Bull. U. S., 63(3): 493-530.

D183 Dobie, J., 1966. Food and feeding habits of the walleye, *Stizostedion v. vitreum,* and associated game and forage fishes in Lake Vermilion, Minnesota, with special reference to the tullibee, *Coregonus (Leucichthys) artedi.* Minn. Fish. Invest., 4: 39-71.

D184 Dryer, W. R., 1964. Movements, growth and rate of recapture of whitefish tagged in the Apostle Islands area of Lake Superior. Fish. Bull. U. S., 63(3): 611-18.

D185 Drummond, R. A., 1966. Reproduction and harvest of cutthroat trout at Trapper's Lake, Colorado. Colo. Fish. Res. Div. Spec. Rep., 10: 26 p.

D186 _____ and T. D. McKinney, 1965. Predicting the recruitment of cutthroat trout fry in Trapper's Lake, Colorado. Trans. Amer. Fish. Soc., 94(4): 389-93.

D187 Dollar, A. M., M. Katz, M. F. Tripple, and R. C. Simon, 1963. Trout hepatoma. Univ. Wash. Fish. Res. Inst. Contr., 147: 33-5.

D188 Dymond, J. R., 1963. Family Salmonidae. Sears Found. Mar. Res. Mem., 1(3): 457-546.

D189 Dumas, R. F., 1966. Observations on yolk sac constriction in land-locked Atlantic salmon fry. Progr. Fish Cult., 28(2): 73-5.

D190 Domrose, R. J., 1963. Age and growth of brook trout *(Salvelinus fontinalis)* in Montana. Proc. Mont. Acad. Sci., 23: 47-62.

D191 DeLacy, A. C. and R. L. Dryfoos, 1964. The longfin smelt in Lake Washington. Univ. Wash. Coll. Fish. Contr., 166: 41-2.

D192 _____ and _____, 1965. The longfin smelt in Lake Washington. Univ. Wash. Coll. Fish. Contr., 184: 37.

D193 Domanevskii, L. N., 1963. Characteristics of the growth of pike *Esox lucius* (L.). [Russian with English summary.] Zool. Zhur., 42(10): 1539-45.

D194 Davis, B. J. and R. J. Miller, 1967. Brain patterns in minnows of the genus *Hybopsis* in relation to feeding habits and habitat. Copeia, 1967 (1): 1-39.

D195 Dinsmore, J. J., 1962. Life history of the creek chub, with emphasis on growth. Proc. Iowa Acad. Sci., 69: 296-301.

D196 Dawe, C. J., M. F. Stanton, and F. J. Schwartz, 1964. Hepatic neoplasms in native bottom-feeding fish of Deep Creek Lake, Maryland. Cancer Res., 24(7): 1194-1201.

D197 Darnell, R. M. and R. R. Meierotto, 1965. Diurnal periodicity in the black bullhead, *Ictalurus melas* (Rafinesque). Trans. Amer. Fish. Soc., 94(1): 1-8.

D198 DeRoth, G. C., 1965. Age and growth studies of channel catfish in western Lake Erie. J. Wildl. Mgmt., 29(2): 280-86.

D199 Deyoe, C. W. and O. W. Tiemeier, 1966. Channel catfish feeding research. Feedstuffs, 38(8): 56.

E2 Eddy, S. and K. Carlander, 1939. Growth of Minnesota fishes. Minn. Conserv., 69: 8-10.

E3 _____ and _____, 1942. Growth rates studies of Minnesota fish. Minn. Dept. Conserv. Fish. Res. Invest. Rep., 28: 64 p. mimeo.

E4 _____ and T. Surber, 1947. Northern fishes. Ref. ed. Univ. Minn. Press. 276 p.

E7 Elson, P. F., 1941. Rearing maskinonge in a protected area. Trans. Amer. Fish. Soc., 70: 421-29.

E8 Embody, G. C., 1914. Fish meal as a food for trout. Trans. Amer. Fish. Soc., 44(1): 57-60.

E9 _____, 1915. The farm fishpond. Cornell Reading Courses, Country Life Ser., 3: 213-52.

E10 _____, 1918. Artificial hybrids between pike and pickerel. J. Hered., 9(6): 253-56.

E11 _____, 1921. Concerning high water temperatures and trout. Trans. Amer. Fish. Soc., 51: 58-64.

E13 _____, 1928. Stocking policy for the streams, smaller lakes and ponds of the Oswego watershed. Annu. Rep. N. Y. Conserv. Dept., 17(Suppl.): 17-39.

E14 _____ and C. O. Hayford, 1925. The advantage of rearing brook trout from selected breeders. Trans. Amer. Fish. Soc., 55: 135-48.

E15 Eschmeyer, R. W., 1937. The Michigan creel census. Trans. N. Amer. Wildl. Conf., 2: 625-34.

E19   Eschmeyer, R. W., 1938. Experimental management of a group of
       small Michigan lakes. Trans. Amer. Fish. Soc., 67: 120-29.

E23   _____, 1942. The catch, abundance, and migration of game fishes in
       Norris Reservoir, Tennessee, 1940. J. Tenn. Acad. Sci., 17(1): 90-114.

E28   _____, R. H. Stroud, and A. M. Jones, 1944. Studies of the fish popula-
       tion on the shoal area of a TVA main-stream reservoir. J. Tenn.
       Acad. Sci., 19(1): 70-122.

E31   Evermann, B. W. and H. C. Bryant, 1919. California trout. Calif. Fish
       Game, 5(3): 105-35.

E32   _____ and H. W. Clark, 1920. Lake Maxinkuckee. Ind. Dept. Conserv.,
       1: 1-660.

E46   English, T. S., 1952. Growth studies of the carp, Cyprinus carpio Lin-
       neaus, in Clear Lake, Iowa. Iowa St. J. Sci., 24(4): 527-40.

E47   _____, 1952. Method of sectioning carp spines for growth studies.
       Progr. Fish Cult., 14(1): 36.

E53   Everhart, W. H., 1950. The blueback trout, Salvelinus oquassa (Gi-
       rard), in Maine. Copeia, 1950(3): 242.

E54   Evermann, B. W., 1902. Description of a new species of shad (Alosa
       ohiensis) with notes on other food-fishes of the Ohio River. Rep. U. S.
       Comm. Fish Fish., 27(1901): 273-88.

E55   _____ and E. L. Goldsborough, 1902. Notes on the fishes and mollusks
       of Lake Chautauqua, New York. Rep. Comm. (Fish) for year ending
       June 30, 1901. 169-75.

E56   Eddy, S., 1940. Do muskellunge and pickerel interbreed? Progr. Fish
       Cult., 48: 25-7.

E57   _____, 1941. Muskellunge and muskie hybrids. Conserv. Bull. Minn.,
       3(14): 41-4.

E58   Echo, J. B., 1955. Some ecological relationships between yellow perch
       and cutthroat trout in Thompson Lakes, Montana. Trans. Amer. Fish.
       Soc., 84: 239-48.

E61   Elkin, R. E., 1954. The fish population of two cut-off pools in Salt
       Creek, Osage County, Oklahoma. Proc. Okla. Acad. Sci., 35: 25-9.

E62   Ellis, R. J. and H. Gowing, 1957. Relationship between food supply and
       condition of wild brown trout, Salmo trutta Linnaeus, in a Michigan
       stream. Limnol. Oceanogr., 2(4): 299-308.

E77   Elson, P. F., 1957. The importance of sex in the change from parr to
       smolt in Atlantic salmon. Canad. Fish Cult., 21: 1-6.

E78   _____ and C. J. Kerswill, 1955. Studies on Canadian Atlantic salmon.
       Trans. N. Amer. Wildl. Conf., 20: 415-25.

E83   Embody, G. C., 1914. The horned dace. Nature Stud. Rev., 10: 168-74.

E84   Erickson, J. G. and W. M. Zarbock, 1954. A preliminary evaluation of
       the effects of the removal of rough fish upon crappies in Lake St.
       Mary's, Ohio. Midwest Wildl. Conf., 16: 19 p. mimeo.

E85   Eschmeyer, P. H., 1955. The reproduction of lake trout in southern
       Lake Superior. Trans. Amer. Fish. Soc., 84: 47-74.

E86   _____, 1956. The early life history of the lake trout in Lake Superior.
       Mich. Dept. Conserv. Inst. Fish. Res. Misc. Publ., 10: 31 p.

E87   _____, 1957. The near extinction of lake trout in Lake Michigan.
       Trans. Amer. Fish. Soc., 85: 102-19.

E88   _____, 1958. Michigan. Upp. G. Lakes Fish. Comm. Annu. Meet. 42-
       52.

E89   _____, 1959. Survival and retention of tags, and growth of tagged lake
       trout in a rearing pond. Progr. Fish Cult., 21(1): 17-22.

E90   Eschmeyer, P. H. and R. M. Bailey, 1955. The pygmy whitefish, *Coregonus coulteri*, in Lake Superior. Trans. Amer. Fish. Soc., 84: 161-99.

E93   Everhart, W. H., 1958. Fishes of Maine, 2nd ed. Maine Dept. Inl. Fish. Game. 5-94.

E94   Eddy, S. and K. D. Carlander, 1940. The effect of environmental factors upon the growth rates of Minnesota fishes. Proc. Minn. Acad. Sci., 8: 14-19.

E99   Elsey, C. A., 1946. An ecological study of the competitor fish in Pyramid Lake, Jasper, with special reference to the northern sucker. MA thesis Univ. Sask. 61 p.

E100   Edsall, T. A., 1960. Age and growth of the whitefish, *Coregonus clupeaformis*, of Munising Bay, Lake Superior. Trans. Amer. Fish. Soc., 89(4): 323-32.

E102   Embody, G. C., 1910. The ecology, habits and growth of the pike, *Esox lucius*. PhD thesis Cornell Univ. 88 p.

E103   Eipper, A. W., 1959. Trout for farm ponds. N. Y. State Conserv., 13 (5): 20-22.

E104   _____, 1961. Vital statistics of trout populations in New York farm ponds. Canad. Fish Cult., 29: 13-14.

E105   Elser, H. J., 1961. Record Maryland fish. Md. Conserv., 38(2): 15-17.

E106   Erickson, J. G., 1961. Muskellunge stocking evaluation in Deer Creek Reservoir, Stark County, Ohio. Ohio Dept. Nat. Resources Tech. Rep., W-314: 6 p.

E107   Elson, P. F., 1962. Predator-prey relationship between fish-eating birds and Atlantic salmon. Fish. Res. Bd. Can. Bull., 133: 87 p.

E108   _____, 1942. Effect of temperature on activity of *Salvelinus fontinalis*. J. Fish. Res. Bd. Can., 5(5): 461-70.

E109   Evropeizeva, N. W., 1960. Experimental analysis of the young salmon (*Salmo salar* L.) in the stage of transition to life in the sea. Rapp. Cons. Int. Explor. Mer., 148: 29-39.

E110   Embody, G. C., 1934. Relation of temperature to the incubation periods of eggs of four species of trout. Trans. Amer. Fish. Soc., 64: 281-92.

E111   Erkkila, L. F., 1936. Notes on feeding the freshwater shrimp, *Gammarus*, to rainbow trout. Progr. Fish Cult., 20: 10-12.

E112   Eipper, A. W. and H. A. Regier, 1962. Fish management in New York farm ponds. Cornell Extens. Bull., 1089: 40 p.

E113   _____, 1963. Effect of hatchery rearing conditions on stream survival of brown trout. Trans. Amer. Fish. Soc., 92(2): 132-39.

E114   Eschmeyer, P. H., 1957. Life history and ecology of the lake trout of the Great Lakes. For Handbook of Biological Data. 6 p.

E115   _____, 1957. Life history and ecology of the siscowet of the Great Lakes. For Handbook of Biological Data. 2 p.

E116   _____, R. Daly, and L. F. Erkkila, 1953. The movement of tagged lake trout in Lake Superior, 1950-1952. Trans. Amer. Fish. Soc., 92: 68-77.

E117   _____, 1957. The lake trout *(Salvelinus namaycush)*. Fish. Leafl. Wash., 441: 11 p.

E118   Edsall, T. A., 1964. Feeding by three species of fishes on the eggs of spawning alewives. Copeia, 1964(1): 226-27.

E119   Eddy, S., 1957. How to know the freshwater fishes. Wm. C. Brown Co., Dubuque, Iowa. 253 p.

E120   Ewers, L. A., 1934. Summary report of crustacea used as food by the fishes of the western end of Lake Erie. Trans. Amer. Fish. Soc., 63: 379-90.

E121 Ewers, L. A. and M. W. Boesel, 1935. The food of some Buckeye Lake fishes. Trans. Amer. Fish. Soc., 65: 57-70.

E122 Essing, L., 1898. Dreissigjahriger Goldfische. Jahresb. Westphal. Prov. Ver., 26: 41 not seen.

E123 Evans, H. E., 1952. The correlation of brain pattern and feeding habits in four species of cyprinid fishes. J. Compar. Neur., 97(1): 133-42.

E124 Evans, G., 1936. The relation between vitamins and the growth and survival of goldfishes in hemotypically conditioned water. J. Exp. Zool., 74(3): 449-76.

E126 Elkin, R. E., 1955. An estimate of the fish population of a 16-acre lake based on recovery during draining. Proc. Okla. Acad. Sci., 36: 53-9.

E127 Eddy, S. and P. H. Simer, 1929. Notes on the food of paddlefish and the plankton of its habitat. Trans. Ill. St. Acad. Sci., 21: 59-68.

E128 Elliott, J. M., 1966. Downstream movements of trout fry *(Salmo trutta)* in a Dartmoor stream. J. Fish. Res. Bd. Can., 23(1): 157-59.

E129 Everhart, W. H. and C. A. Waters, 1965. Life history of the blueback trout (Arctic char, *Salvelinus alpinus* (Linnaeus)) in Maine. Trans. Amer. Fish. Soc., 94(4): 393-97.

E130 Eipper, A. W., 1964. Growth, mortality rates, and standing crops of trout in New York farm ponds. Cornell Univ. Agric. Exp. Sta. Mem., 388: 67 p.

E131 Eschmeyer, P. H. and A. M. Phillips, Jr., 1965. Fat content of the flesh of siscowets and lake trout from Lake Superior. Trans. Amer. Fish. Soc., 94(1): 62-74.

E132 _____, 1964. The lake trout *(Salvelinus namaycush)*. Fish. Leafl. Wash., 555: 8 p.

E133 Ehrlich, S., 1964. Studies on the influence of nutria on carp growth. Hydrobiologica, 23: 196-210.

E134 Emig, J. W., 1966. Brown bullhead. Freshwater fisheries management, ed. A. Calhoun, Calif. Dept. Fish Game. 463-75.

F3 Fish, C., 1929. Preliminary report on the cooperative survey of Lake Erie, season of 1928. Bull. Buffalo Soc. Nat. Sci., 14(3): 1-220.

F4 Fish, M. P., 1930. Contributions to the natural history of the burbot, *Lota maculosa* (LeSueur). Bull. Buffalo Soc. Nat. Sci., 15(1): 1-20.

F5 _____, 1932. Contributions to the early life histories of sixty-two species of fishes from Lake Erie and its tributary waters. Bull. U. S. Bur. Fish., 47(10): 293-398.

F6 Flower, S. S., 1925. Contributions to our knowledge of the duration of life in vertebrate animals. I. Fishes. Proc. Zool. Soc. Lond., 1925: 247-68.

F7 _____, 1935. Further notes on duration of life in animals. I. Fishes determined by otolith and scale readings and direct observation on live individuals. Proc. Zool. Soc. Lond., 1935(2): 265-304.

F8 Fontaine, P. A., 1944. Notes on the spawning of the shovelhead catfish, *Pilodicti olivaris* (Rafinesque). Copeia, 1944(1): 50-51.

F9 Foote, L. E. and B. P. Blake, 1945. Life history of the eastern pickerel in Babcock Pond, Connecticut. J. Wildl. Mgmt., 9(2): 89-96.

F10 Forbes, S. A. and R. E. Richardson, 1909. The fishes of Illinois. Nat. Hist. Surv. Ill., 3: 1-357.

F12 Frey, D. G., 1942. Studies on Wisconsin carp. I. Influence of age, size, and sex on time of annulus formation by 1936 year class. Copeia, 1942(4): 214-23.

F13   Frey, D. G. and H. Pedracine, 1938. Growth of the buffalo in Wisconsin lakes and streams. Trans. Wis. Acad. Sci. Arts Lett., 31: 513-25.

F16   Frost, N., 1940. A preliminary study of Newfoundland trout. Res. Bull. Dept. Nat. Resources Newf., 9: 1-30.

F17   Frost, W. E., 1945. River Liffey Survey. VI. Discussion on the results obtained from investigations on the food and growth of brown trout (*Salmo trutta* L.) in alkaline and acid waters. Proc. R. Irish Acad., 50B: 321-42.

F18   _____ and A. E. J. Went, 1940. River Liffey Survey. III. The growth and food of young salmon. Proc. R. Irish Acad., 46B(4): 53-80.

F19   Fry, D. H., Jr., 1936. Life history of *Hesperoleucas venustus* Sayder. Calif. Fish Game, 22(2): 65-98.

F20   Fry, F. E. J., 1937. The summer migration of the cisco, *Leucichthys artedi*, in Lake Nipissing, Ontario. Univ. Toronto Stud. Biol. Ser., 44: 1-91.

F21   _____, 1940. A comparative study of the lake trout fisheries in Algonquin Park, Ontario. Ont. Fish. Res. Lab. Publ., 58: 1-69.

F23   _____ and W. A. Kennedy, 1937. Report on 1936 lake trout investigations, Lake Opeongo, Ontario. Univ. Toronto Stud. Biol. Ser., 42:1-20.

F26   Fuller, J. L. and G. P. Cooper, 1946. A biological survey of the lakes and ponds of Mount Desert Island, and the Union and Lower Penobscot River drainage systems. Maine Dept. Inl. Fish Game Fish Surv. Rep., 7: 1-221.

F28   Fuqua, C. L., 1939. Feeding of Montana grayling at the Bozeman, Montana Station. Progr. Fish Cult., 43: 12-17.

F29   Fry, F. E. J., 1949. Lake trout in our inland waters. Sylva, 5(3): 3-13.

F30   _____, 1949. Statistics of a lake trout fishery. Biometrics, 5(1): 27-67.

F31   Fessler, F. R., 1949. A survey of fish populations in small ponds by two methods of analysis. MS thesis Iowa St. Coll. Lib. 47 p.

F32   Fisheries Service Bulletin, 1915. Rearing buffalofish in ponds. U. S. Fish. Serv. Bull., 7: 4-5.

F33   _____, 1916. Growth of channel catfish in ponds. U. S. Fish. Serv. Bull., 16: 4.

F34   _____, 1919. Rearing paddlefish in confinement. U. S. Fish. Serv. Bull., 48: 2.

F35   _____, 1921. A noteworthy production of buffalofish. U. S. Fish. Serv. Bull., 78: 4.

F38   _____, 1924. Notes from the Division of Fish Culture. U. S. Fish. Serv. Bull., 109: 2-3.

F39   _____, 1926. Remarkable yield of wild brook trout eggs. U. S. Fish. Serv. Bull., 139: 3.

F40   _____, 1926. Notes from the Division of Fish Culture. U. S. Fish. Serv. Bull., 131: 4-5.

F41   _____, 1927. Great Lakes trout introduced into an alpine lake. U. S. Fish. Serv. Bull., 141: 6-7.

F42   _____, 1928. Large rainbow trout. U. S. Fish. Serv. Bull., 152: 5.

F44   _____, 1929. Precocious male brook trout. U. S. Fish. Serv. Bull., 167: 3.

F46   _____, 1939. Correction in fish-measurement table. U. S. Fish. Serv. Bull., 294: 7.

F47   Fry, F., J. P. Cuerrier, and G. Prefontaine, 1941. Premiere croissance du maskinonge dans le lac Saint-Louis en 1941. Rapp. Sta. Biol. Montreal et Sta. Biol. du Parc Laurentides, 1941: 171-75.

F50 Fleener, G. C., 1952. Life history of the cutthroat trout *Salmo clarki* Richardson, in Logan River, Utah. Trans. Amer. Fish. Soc., 81: 235-48.

F53 Fautin, R. W., 1951. Age and growth studies of eastern brook trout, *Salvelinus f. fontinalis,* in the Snowy Range of Wyoming. (Abstract.) J. Colo.-Wyo. Acad. Sci., 4(3): 85.

F54 Fenderson, C. N., 1954. The brown trout in Maine. Maine Fish Bull., 2: 16 p.

F55 _____, 1958. Brown trout, *Salmo trutta* Linnaeus. Fishes of Maine, 2nd ed., ed. W. H. Everhart. 34-7.

F62 Finnell, J. C., 1954. Comparison of growth-rates of fishes in Stringtown Sub-prison Lake prior to, and three years after draining and restocking. Proc. Okla. Acad. Sci., 35: 30-36.

F63 _____ and R. M. Jenkins, 1954. Growth of channel catfish in Oklahoma waters: 1954 revision. Okla. Fish. Res. Lab. Rep., 41: 37 p.

F64 _____, _____, and G. E. Hall, 1956. The fishery resources of the Little River system, McCurtain County, Oklahoma. Okla. Fish. Res. Lab. Rep., 55: 82 p. mimeo.

F65 Forney, J. L., 1955. Life history of the black bullhead, *Ameiurus melas* (Rafinesque), of Clear Lake, Iowa. Iowa St. J. Sci., 30(1): 145-62.

F66 _____, 1957. Bait fish production in New York ponds. N. Y. Fish Game J., 4(2): 150-94.

F67 _____, 1957. Raising bait fish and crayfish in New York ponds. Cornell Exten. Bull., 986: 3-30.

F68 Foster, R. F., L. R. Donaldson, A. D. Welander, K. Bonham, and A. H. Seymour, 1949. The effect on embryos and young of rainbow trout from exposing the parent fish to X-rays. Growth, 13: 119-42.

F69 Foye, R. E., 1956. Reclamation of potential trout ponds in Maine. J. Wildl. Mgmt., 20(4): 389-99.

F71 Frank, S., 1955. Prispevek k. biologii sumecka americkeho (*Ameiurus nebulosus* LeSueur 1819). [A contribution to the biology of the common bullhead *Ameiurus nebulosus* LeSueur 1819.] Acta Soc. Zool. Bohemoslovenicae, 19(1): 62-81.

F72 _____, 1959. Zăvislost růstu některých druhů ryb na potravních podmínkăch v polabskě tůni Poltruba. Věstník Československé Zool. Spolecnosti, 23(3): 247-53.

F74 Freeman, B. O. and M. T. Huish, 1953. A summary of a fish population control investigation conducted in two Florida lakes. Fla. Game Freshw. Fish Comm. 109 p. mimeo.

F75 French, R. R. and R. J. Wahle, 1959. Biology of chinook and blueback salmon and steelhead in the Wenatchee River System. U. S. Fish Wildl. Serv. Spec. Sci. Rep. Fish, 304: 1-17.

F76 Frost, W. E., 1940. Rainbows in acid water. Salm. Trout Mag., 100: 234-40.

F77 _____ and C. Kipling, 1959. The determination of the age and growth of pike (*Esox lucius* L.) from scales and opercular bones. J. Cons. Int. Explor. Mer., 24(2): 314-42.

F78 _____ and W. J. P. Smyly, 1952. The brown trout of a moorland fishpond. J. Anim. Ecol., 21(1): 62-86.

F80 Fry, F. E. J., 1953. The 1944 year class of lake trout in South Bay, Lake Huron. Trans. Amer. Fish. Soc., 82: 178-92.

F81   Fry, F. E. J. and J. C. Budd, 1958. The survival of yearling lake trout
       planted in South Bay, Lake Huron. Canad. Fish Cult., 23: 13-23.
F83   Fuller, W. A. L., 1955. The inconnu *(Stenodus leucichthys mackenziei)*
       in Great Slave Lake and adjoining waters. J. Fish. Res. Bd. Can., 12(5):
       768-80.
F86   Fetterolf, C. M., Jr., 1957. Stocking as a management tool in Tennes-
       see reservoirs. Proc. S. E. Assoc. Game Fish Comm., 10: 275-84.
F87   Finnell, J. C., 1955. Growth of fishes in cutoff lakes and streams of
       the Little River System, McCurtain County, Oklahoma. Proc. Okla.
       Acad. Sci., 35: 61-6.
F89   Franklin, D. R. and L. L. Smith, Jr., 1960. Note on development of
       scale patterns in the northern pike, *Esox lucius* L. Trans. Amer.
       Fish. Soc., 89(1): 83.
F90   Fry, F. E. J., D. Cucin, J. C. Kennedy, and A. Papson, 1960. The use
       of lead versenate to place a time mark on fish scales. Trans. Amer.
       Fish. Soc., 89(2): 149-53.
F92   _____, J. R. Brett, and G. H. Clawson, 1942. Lethal limits of tempera-
       ture for young goldfish. Rev. Canad. Biol., 1(1): 50-56.
F93   _____, J. S. Hart, and K. F. Walker, 1946. Lethal temperature rela-
       tions for a sample of young speckled trout, *Salvelinus fontinalis.* Publ.
       Ont. Fish. Res. Lab., 66: 9-35.
F94   Franklin, D. R. and L. L. Smith, Jr., 1960. Notes on the early growth
       and allometry of the northern pike, *Esox lucius* L. Copeia, 1960(2):
       143-44.
F95   Fuller, W. A. L., 1947. The inconnu *(Stenodus leucichthys mackenziei)*
       in Great Slave Lake and adjoining waters. MA thesis Univ. Sask. 96 p.
F97   Fetterrolf, C. M., Jr., 1952. A population study of the fishes of Win-
       tergreen Lake, Kalamazoo County, Michigan; with note on movement
       and effect of netting on condition. MS thesis Mich. St. Coll. 127 p.
F98   Frey, D. G., 1940. Growth and ecology of the carp *Cyprinus carpio*
       Linnaeus in four lakes of the Madison Region, Wisconsin. PhD thesis
       Univ. Wis. 248 p.
F100   Fogle, N. E., 1961. Report of fisheries investigations during the third
       year of impoundment of Oahe Reservoir, South Dakota, 1960. S. D.
       Dept. Game Fish Parks D-J Proj., F-1-R-10(Jobs 9-12): 57 p. mimeo.
F101   Freshwater Biological Association, 1961. Twenty-ninth annual report,
       for the year ended 31 March, 1961. Freshw. Biol. Assoc., Far Sawrey,
       Ambleside, Westmoreland. 88 p.
F104   Fry, F. E. J., 1960. Requirements for the aquatic habitat. Pulp Paper
       Mag. Can., 1960(Feb.): 8 p.
F105   Fogle, N. E., 1961. Report of fisheries investigations during the sec-
       ond year of impoundment of Oahe Reservoir, South Dakota, 1959. S. D.
       Dept. Game Fish Parks D-J Proj., F-1-R-9(Jobs 12-14): 43 p.
F106   Frost, W. E. and C. Kipling, 1961. Some observations on the growth of
       pike, *Esox lucius,* in Windermere. Verh. Int. Ver. Limnol., 14: 776-81.
F107   Flick, W. A. and D. A. Webster, 1962. Problems in sampling wild and
       domestic stocks of brook trout *(Salvelinus fontinalis).* Trans. Amer.
       Fish. Soc., 91(2): 140-44.
F108   Forbes, S. A., 1883. The first food of the common white-fish *(Corego-
       nus clupeaformis* Mitch.). Bull. Ill. St. Lab. Nat. Hist., 1(6): 87-99.
F109   Fraser, J. C. and A. F. Pollitt, 1951. The introduction of Kokanee red
       salmon *(Oncorhynchus nerka kennerlyi)* into Lake Tahoe, California and
       Nevada. Calif. Fish Game, 37(2): 125-27.

F110   Foerster, R. E., 1947. Experiment to develop sea-run from landlocked sockeye salmon *(Oncorhynchus nerka kennerlyi)*. J. Fish. Res.
Bd. Can., 7: 88-91.

F113   Flick, W. A. and D. A. Webster, 1961. Brandon Park. Fish management report, 1961. Cornell Univ. Dept. Conserv. 47 p. mimeo.

F114   Franklin, D. R. and L. L. Smith, Jr., 1963. Early life history of the
northern pike, *Esox lucius* L., with special reference to the factors influencing the numerical strength of year classes. Trans. Amer. Fish.
Soc., 92(2): 91-110.

F115   Frost, W. E., 1956. The growth of brown trout *(Salmo trutta* L.) in
Haweswater before and after the raising of the level of the lake. Salm.
Trout Mag., 1956(Sept.): 267-75.

F116   ____, 1957. Brown trout, char, pike, eel and minnow. Data for Handbook of Biological Data.

F117   Fabricius, E., 1953. Aquarium observations on the spawning behaviour
of the char, *Salmo alpinus*. Rep. Inst. Freshw. Res. Drottning., 34: 14-
48.

F118   ____ and K. J. Gustafson, 1954. Further aquarium observations on the
spawning behaviour of the char, *Salmo alpinus* L. Rep. Inst. Freshw.
Res. Drottning., 35: 58-104.

F120   Flick, W. A. and D. A. Webster, 1964. Comparative first year survival
and production in wild and domestic strains of brook trout, *Salvelinus
fontinalis*. Trans. Amer. Fish. Soc., 93(1): 58-69.

F121   Fenderson, O. C., 1964. Evidence of subpopulations of lake whitefish,
Coregonus clupeaformis, involving a dwarfed form. Trans. Amer.
Fish. Soc., 93(1): 77-94.

F123   Fabricius, E. and K. J. Gustafson, 1955. Observations on the spawning
behaviour of the grayling, *Thymallus thymallus* (L.). Rep. Inst.
Freshw. Res. Drottning., 36: 75-103.

F124   Fogle, N. E., 1963. Report of fisheries investigations during the fourth
year of impoundment of Oahe Reservoir, South Dakota, 1961. S.D.
Dingell-Johnson Proj., F-1-R-11(Jobs 10-12): 43 p.

F125   ____, 1963. Report of fisheries investigations during the fourth year
of impoundment of Oahe Reservoir, South Dakota, 1962. S. D. D-J
Proj., F-1-R-12(Jobs 10-12): 43 p. (The title is "fourth year," but the
report apparently refers to the fifth year of impoundment.)

F126   Foerster, R. E. and W. E. Ricker, 1941. The effect of reduction of
predaceous fish on the survival of young sockeye salmon at Cultus
Lake. J. Fish. Res. Bd. of Can., 5(4): 315-36.

F127   Frank, S., 1959. Rust lina obecneho a okouna ricniho ve Slapske udolni
nadrzi. Sbornik Ceskoslovenske Akad. Zemedelskych Ved Zivocisna
Vyroba, 4(32): 893-96.

F128   ____, 1960. Über das Wachstum der Schleie und des Flussbarsches in
der Talspeire von Slape (Böhemen). Vestnik Ceskoslov. Zool. Spol.
Acta Soc. Zool. Bohemoslov., 24(3): 258-70.

F129   Fridriksson, A., 1939. Um murtuna i Thingvallavatni med hlidsjön af
odrum silung i vatninu. Natturufraedingurinn, 9: 1-30.

F130   Frost, W. E., 1955. An historical account of char in Windermere.
Salm. Trout Mag., 143: 15-24.

F131   Flick, W. A. and D. A. Webster, 1966. Brandon Park. Fish management report, 1965. Cornell Univ. Dept. Conserv. 37 p.

F132   Fitz, R. B., 1966. Unusual food of a paddlefish *(Polyodon spathula)* in
Tennessee. Copeia, 1966(2): 356.

F133   Forbes, S. A., 1888. Studies of the food of fresh water fishes. Bull.
        Ill. St. Lab. Nat. Hist., 2: 433-73.
F134   Flick, W. A., D. A. Webster, and J. Nagel, 1961. Brandon Park. Fish
        management report, 1960. Cornell Univ. Dept. Conserv. 40 p.
F135   _____ and _____, 1962. Brandon Park. Fish management report, 1961.
        Cornell Univ. Dept. Conserv. 47 p.
F136   _____ and _____, 1964. Brandon Park. Fish management report, 1963.
        Cornell Univ. Dept. Conserv. 36 p.
F137   Finnell, L. M., 1966. Granby Reservoir studies. Colo. Dept. Game
        Fish Parks, Ft. Collins, Fish. Res. Rev., 3: 4-6.
F138   Fox, A. C., 1962. Parasite incidence in relation to size and condition
        of trout from two Montana lakes. Trans. Amer. Midl. Nat., 81(2): 179-
        84.
F139   Fish, G. R., 1963. Limnological conditions and growth of trout in three
        lakes near Rotorua. N. Z. Ecol. Soc. Proc., 10: 7 p.
F140   Frank, S., 1966. A contribution to the growth of sea-trout (*Salmo trutta
        trutta* Linnaeus 1759) in Poland. Acta Soc. Zool. Bohemoslov., 30(1):
        14-21.
F141   Frost, W. E., 1950. The growth and food of young salmon *(Salmo salar)*
        and trout *(S. trutta)* in the River Forss, Caithness. J. Anim. Ecol., 19:
        147-58.
F142   _____ and M. E. Brown, 1967. The trout. New Naturalist Series, Col-
        lins, London.
F143   Flick, W. A. and D. A. Webster, 1962. Brandon Park. Fish manage-
        ment report, 1962. Cornell Univ. Dept. Conserv. 39 p.
F144   _____ and _____, 1964. Brandon Park. Fish management report, 1964.
        Cornell Univ. Dept. Conserv. 40 p.
F145   _____ and _____, 1965. Brandon Park. Fish management report, 1965.
        Cornell Univ. Dept. Conserv. 37 p.
F146   Fry, F. E. J., 1964. Changes in age composition in the smelt, *Osmerus
        mordax,* in South Bay, Lake Huron. [Abstract.] Univ. Mich. Gr. Lakes
        Res. Div. Publ., 11: 141.
F147   Ferguson, R. G., 1965. Bathymetric distribution of American smelt,
        Osmerus mordax, in Lake Erie. Univ. Mich. Gr. Lakes Res. Div.
        Publ., 13: 47-60.
F148   Flemer, D. A. and W. S. Woolcott, 1966. Food habits and distribution
        of the fishes of Tackahoe Creek, Virginia, with special emphasis on the
        bluegill, *Lepomis m. macrochirus* Rafinesque. Chesapeake Sci., 7(2):
        75-89.
F149   Frost, W. E. and C. Kipling, 1965. Some observations on the age and
        growth of pike (*Esox lucius* L.) in Windermere. Salm. Trout Mag.,
        1965(Jan.): 21-7.
F150   _____, 1963. The pike—its age, growth and predatory habits. Proc.
        Brit. Coarse Fish Conf. Univ. Liverpool, 1: 35-9.
G1   Gage, S. H., 1893. The lake and brook lampreys of New York, espe-
        cially those of Cayuga and Seneca lakes. Wilder Quarter Century Book:
        421-93.
G2   _____, 1928. Lampreys of New York State. Life history and economics.
        Annu. Rep. N. Y. Conserv. Dept., 1927(Suppl.17): 158-91.
G3   Gardiner, A. C., 1942. More about trout growth. Some observations
        and suggestions. Salm. Trout Mag., 106: 231-45.
G5   Gerrish, C. S., 1935. What is the deciding factor in trout growth? An
        experiment on the Lambourn. Salm. Trout Mag., 78: 13-21.

G6   Gerrish, C. S., 1935. Drought and trout growth. Salm. Trout Mag., 80: 201-10.

G7   ____, 1936. Scales from Avon trout and grayling. Rep. Avon Biol. Res. S'hampton, 1934-35: 81-95.

G8   ____, 1937. Scales of Avon trout in 1936. Rep. Avon Biol. Res. S'hampton, 4: 44-58.

G9   ____, 1938. Scales of Avon trout and grayling. Rep. Avon Biol. Res. S'hampton, 1936-37: 70-78.

G10   ____, 1939. Progress report No. IX. Scales of Avon trout and grayling. Rep. Avon Biol. Res. S'hampton, 6: 54-62.

G11   ____, 1939. A trout marking experiment. Salm. Trout Mag., 96: 224-32.

G12   Gill, T., 1905. The family of cyprinids and carp as its type. Smithson. Misc. Coll., 48: 195-217.

G13   Gilmore, R. J., 1937. Fish propagation in mountain regions. Trans. N. Amer. Wildl. Conf., 2: 635-38.

G14   Glass, J. T., 1935. Growth of fingerling trout in natural lakes in Wyoming. Univ. Wyo. Publ., 2(3): 17-30.

G15   Godby, M. H., 1934. Atlantic salmon in New Zealand. The evidence of the Lake Coleridge fish. Salm. Trout Mag., 75: 173-76.

G16   ____, 1947. Divided winter band in brown trout scales from the Rakaia River. Trans. and Proc. Roy. Soc. N. Z., 76: 516.

G17   Gray, J. and S. B. Setna, 1931. The growth of fish—4. Effects of food supply on scales of *Salmo irrideus*. Brit. J. Exp. Biol., 8: 55-62.

G18   Greeley, J. R., 1933. The growth rate of rainbow trout from some Michigan waters. Trans. Amer. Fish. Soc., 63: 361-78.

G19   ____, 1934. Fishes of the Raquette watershed. Annu. Rep. N. Y. Conserv. Dept., 23(Suppl.): 53-108.

G20   ____, 1935. Fishes of the Mohawk-Hudson watershed. Annu. Rep. N. Y. Conserv. Dept., 24(Suppl.): 63-101.

G21   ____, 1936. Fishes of the area with annotated list. (Delaware and Susquehanna watersheds.) Annu. Rep. N. Y. Conserv. Dept., 25(Suppl.): 45-88.

G22   ____, 1937. Fishes of the area with annotated list. (Lower Hudson watershed.) Annu. Rep. N. Y. Conserv. Dept., 26(Suppl.): 45-104.

G23   ____, 1938. Fishes of the area with annotated list. (Allegheny and Chemung watersheds.) Annu. Rep. N. Y. Conserv. Dept., 27(Suppl.): 48-73.

G24   ____, 1939. Fresh-water fishes of Long Island and Staten Island with annotated list. Annu. Rep. N. Y. Conserv. Dept., 28(Suppl.): 29-44.

G25   ____, 1940. Fishes of the Lake Ontario watershed with annotated list. Annu. Rep. N. Y. Conserv. Dept., 29(Suppl.): 42-81.

G26   ____, 1948. Four years of landlocked salmon study. N. Y. Fish Wildl. Info. Bull., 1948(2): 1-16.

G28   ____ and C. W. Greene, 1931. Fishes of the area. (St. Lawrence watershed.) Annu. Rep. N. Y. Conserv. Dept., 20(Suppl.): 44-94.

G29   Greene, C. W., 1930. The smelts of Lake Champlain. Annu. Rep. N. Y. Conserv. Dept., 19(Suppl.): 105-30.

G31   Griffiths, F. P. and E. D. Yeoman, 1940. A comparative study of Oregon coastal lakes from a fish-management standpoint. Proc. 6th Pacif. Sci. Congr., 3: 323-33.

G32   Grimm, W. W. and R. V. Bangham, n.d. Growth of Buckeye Lake fishes in 1930—six common species compared. Ohio Div. Conserv. Bull., 71: 1 p. mimeo.

G33   Gudger, E. W., 1942. Giant fishes of North America. Nat. Hist., 49: 115-21.

G34   Gutsell, J. S. and S. F. Snieszko, 1949. Response of brook, rainbow, and brown trout to various doses of sulfamerazine. Trans. Amer. Fish. Soc., 77: 93-101.

G35   Greenbank, J., 1941. Selective poisoning of fish. Trans. Amer. Fish. Soc., 70: 80-86.

G36   Gabrielson, I. N. and F. LaMonte, 1950. The fisherman's encyclopedia. Stackpole and Hech, N. Y. 698 p.

G39   Gerking, S. D., 1950. A carp removal experiment at Oliver Lake, Indiana. Invest. Ind. Lakes Streams, 3(10): 373-88.

G40   Gowanloch, J. N., 1951. Lake management. La. Conserv., 3(7): 10-13, 20-22.

G42   Greenbank, J., 1950. The length-weight relationship of some upper Mississippi River fishes. Upp. Miss. R. Conserv. Commit. 12 p. ms.

G43   _____ and M. A. Monson, 1947. Size and maturity of channel catfish. Upp. Miss. R. Conserv. Comm. Tech. Commit. Fish., Progr. Rep., 3: 28-31.

G45   Greene, C. W., 1952. Results from stocking brook trout of wild and hatchery strains at Stillwater Pond. Trans. Amer. Fish. Soc., 81: 43-52.

G47   Geagan, D., 1955. Cane River poisoning project. La. Conserv., 8(1): 4-6.

G51   Gerlach, A. H., 1956. Lincoln District. Annu. Rep. Ore. St. Game Comm. Fish Div., 1955: 201-9.

G53   _____, 1958. Klamath District. Annu. Rep. Ore. St. Game Comm. Fish Div., 1957: 160-73.

G54   _____, 1959. Klamath District. Annu. Rep. Ore. St. Game Comm. Fish Div., 1958: 164-80.

G56   Gerrish, C. S., 1935. Hatchery stock and the trout streams. Salm. Trout Mag., 81: 331-44.

G57   _____, 1938. A trout-marking experiment. Salm. Trout Mag., 91: 128-33.

G60   Godby, M. H., 1920. Scale reading and growth of trout in New Zealand. Review. Salm. Trout Mag., 22: 92.

G62   Godfrey, H., 1955. On the ecology of Skeena River whitefishes, *Coregonus* and *Prosopium*. J. Fish. Res. Bd. Can., 12(4): 499-542.

G63   _____, 1957. Feeding of eels in four New Brunswick salmon streams. Fish. Res. Bd. Can. Progr. Rep. Atlant. Coast Sta., 67: 19-22.

G64   Graham, J. J., 1956. Observations on the alewife, *Pomolobus pseudoharengus* (Wilson), in fresh water. Univ. Toronto Stud. Biol. Ser., 62: 43 p.

G65   Grainger, E. H., 1953. On the age, growth, migration, reproductive potential and feeding habits of the arctic char *(Salvelinus alpinus)* of Frobisher Bay, Baffin Island. J. Fish. Res. Bd. Can., 10(6): 326-70.

G66   Great Lakes Fishery Investigations, 1958. (Report.) Upp. Gr. Lakes Fish. Commit. Annu. Mtg. Minutes, 1958: 5-31.

G67   Greeley, J. R., 1930. A contribution to the biology of the horned dace, *Semotilus atromaculatus* (Mitchill). PhD thesis Cornell Univ. 114 p. + append., figs., lit. cited.

G71   Greenbank, J. and P. R. Nelson, 1959. Life history of the threespine stickleback, *Gasterosteus aculeatus* Linnaeus, in Karluk and Bare Lakes, Kodiak Island, Alaska. Fish. Bull. U. S., 59(153): 537-59.

G73   Greene, A. F. C., 1955. Will stunted brook trout grow? Progr. Fish Cult., 17(2): 91.

G74   Greer, J. K. and F. B. Cross, 1956. Fishes of El Dorado City Lake, Butler County, Kansas. Trans. Kans. Acad. Sci., 59(3): 358-63.

G75   Grenfell, R. A., 1956. Warm-water game fish. Ore. St. Game Comm. Fish Div. Annu. Rep., 1955: 221-34.

G79   Gudger, E. W., 1939. Cuthbert, the two-headed trout. Salm. Trout Mag., 96: 220-23.

G81   Gunning, G. E. and W. M. Lewis, 1955. The fish population of a spring-fed swamp in the Mississippi bottoms of southern Illinois. Ecology, 36(4): 552-57.

G82   _____ and _____, 1956. Age and growth of two important bait species in a coldwater stream in southern Illinois. Amer. Midl. Nat., 55(1): 118-20.

G83   Grice, F., 1959. Elasticity of growth of yellow perch, chain pickerel, and largemouth bass in some reclaimed Massachusetts waters. Trans. Amer. Fish. Soc., 88(4): 332-35.

G84   Garside, E. T., 1959. Some effects of oxygen in relation to temperature on the development of lake trout embryos. Canad. J. Zool., 37: 689-98.

G86   Gunter, G., 1945. Studies on marine fishes of Texas. Publ. Inst. Mar. Sci. Univ. Tex., 1(1): 1-190.

G87   Grosslein, M. D. and L. L. Smith, Jr., 1959. The goldeye, *Amphiodon alosoides* (Rafinesque), in the commercial fishery of the Red Lakes, Minnesota. Fish. Bull. U. S., 60(157): 33-41.

G88   Gurzeda, A., 1957. [An analysis of the growth of carp in ponds on the basis of control catches.] (Eng. summary.) Rocz. Nauk Rolniczych, 72-B-2: 183-97.

G89   Gard, R., 1960. The survival of a brook trout with two mutilated gill arches. Progr. Fish Cult., 22(3): 108.

G91   Gilbert, C. R., 1953. Age and growth of the yellow stone catfish, *Noturus flavus* (Rafinesque). MS thesis Ohio St. Univ. 67 p.

G92   Govindan, P., 1950. The age and growth of the Great Lakes cisco, *Leucichthys artedi* (LeSueur), from Saginaw Bay, Green Bay and Grand Traverse Bay. PhD thesis Mich. St. Coll. 76 p. + appendix.

G93   Gudjonsson, T. V., 1946. Age and body length at the time of seaward migration of immature steelhead trout (*Salmo gairdnerii* Richardson) in Minter Creek, Washington. MS thesis Univ. Wash. 52 p.

G95   Gilmour, W. M., 1950. A study of the Lower Bow River trout with special reference to taxonomy. MS thesis Univ. Alta. 59 p.

G97   Gross, R. W., 1959. A study of the alewife, *Alosa pseudoharengus* (Wilson), in some New Jersey lakes, with special reference to Lake Hopatcong. MS thesis Rutgers. 52 p.

G98   Gerlach, A. H., 1960. Klamath District. Annu. Rep. Ore. St. Game Comm. Fish Div., 1959: 175-94.

G100   Gray, J. W., 1942. Studies of *Notropis atherinoides atherinoides* Rafinesque, in the Bass Islands region of Lake Erie. MS thesis Ohio St. Univ. 29 p.

G102   Gray, J. L., 1958. Effects of five diets on the growth and mortality of fry to fingerling channel catfish (*Ictalurus punctatus*). Trans. Kans. Acad. Sci., 61(3): 288-98.

G103 Gard, R., 1961. Creation of trout habitat by constructing small dams. J. Wildl. Mgmt., 52(4): 384-90.

G104 Gerlach, A. H., 1960. Klamath District. Annu. Rep. Ore. St. Game Comm. Fish Div., 1960: 184-211.

G105 Gordon, W. G., 1961. Food of the American smelt in Saginaw Bay, Lake Huron. Trans. Amer. Fish. Soc., 9(4): 439-43.

G106 Gard, R., 1961. Effects of beaver on trout in Sagehen Creek, California. J. Wildl. Mgmt., 25(3): 221-42.

G108 Graham, R. J., 1961. Biology of the Utah chub in Hebgen Lake, Montana. Trans. Amer. Fish. Soc., 90(3): 269-76.

G109 Geibel, G. E. and P. J. Murray, 1961. Channel catfish culture in California. Progr. Fish Cult., 23(3): 99-105.

G110 Groebner, J. F., 1960. Appraisal of the sport fishery catch in a bass-panfish lake of southern Minnesota. Lake Francis, LeSueur County, 1952-1957. Minn. Dept. Conserv. Invest. Rep., 225: 17 p.

G111 Gunning, G. E. and C. R. Shoop, 1962. Restricted movement of the American eel, *Anguilla rostrata* (LeSueur), in freshwater streams with comments on growth rate. Tulane Stud. Zool., 9(5): 265-72.

G112 Grudniewski, C., 1961. [The development of some morphological features during the larvae stage of Wdzydze Lake trout (*Salmo trutta morpha lacustris* L.).] (Eng. summary.) Rocz. Nauk Rolniczych, 93: 595-626.

G113 _____, 1961. [An attempt to determine the critical temperature and oxygen contents for fry of Wdzydze Lake trout (*Salmo trutta morpha lacustris* L.).] (Eng. summary.) Rocz. Nauk Rolniczych, 93: 627-48.

G114 Galligan, J. P., 1962. Depth distribution of lake trout and associated species in Cayuga Lake, New York. N. Y. Fish Game J., 9(1): 44-68.

G116 Greeley, J. R., 1932. The spawning habits of brook, brown and rainbow trout, and the problem of egg predators. Trans. Amer. Fish. Soc., 62: 239-48.

G117 Gibbs, E. D., 1956. A bisexual steelhead. Calif. Fish Game, 42(3): 229-30.

G118 Griffith, D. deG., 1961. A preliminary investigation into the diet of brown trout in Lough Feeagh, July and September, 1961. Rep. Salm. Res. Trust Ire. Inc. Dublin, 1961: 42-6.

G119 Gammon, J. R., 1963. Conversion of food in young muskellunge. Trans. Amer. Fish. Soc., 92(2): 183-84.

G120 Grainger, E. H., 1957. Tabulated data on arctic char for Handbook of Biological Data.

G121 Graham, J. M., 1949. Some effects of temperature and oxygen pressure on the metabolism and activity of the speckled trout, *Salvelinus fontinalis*. Canad. J. Res., 27(D): 270-88.

G124 Gee, J. H. and T. G. Northcote, 1963. Comparative ecology of two sympatric species of dace *(Rhinichthys)* in the Fraser River system, British Columbia. J. Fish. Res. Bd. Can., 20(1): 105-18.

G125 Gurzeda, A. and P. Wolny, 1962. [Food and growth of carp fry and the dynamics of food animals.] (In Polish with English summary.) Rocz. Nauk Rolniczych, 81B(2): 151-69.

G126 Gottberg, G., 1917. Om gäddens tillvaxt i Ålands skärgård. Finlands Fisherier, 4: 223-43 not seen, quoted in W130.

G128 Geyer, F. and H. Mann, 1939. Limnologische und fischereibiologische Untersuchungen am Ungarischen Teil des Fertő. Arb. Ungarischen Biol. Forsuch., 11: 64-193 not seen, quoted in F98.

G129   Gilbert, C. R., 1961. Hybridization versus intergradation: an inquiry
        into the relationship of two cyprinid fishes. Copeia, 1961(2): 181-92.

G130   Griswold, B. L., 1963. Food and growth of spottail shiners and other
        forage fishes of Clear Lake, Iowa. Proc. Ia. Acad. Sci., 70: 215-23.

G131   Gray, R. A., 1942. Rough fish history. Wis. Conserv. Bull., 7(4): 8-11.

G132   _____, 1943. Carp control and utilization during the war. Trans. N.
        Amer. Wildl. Conf., 8: 263-67.

G133   Gill, T., 1906. Parental care among fresh-water fishes. Annu. Rep.
        Smithson. Inst., 1905: 403-531.

G135   Greenbank, J., 1957. Length-weight relationship of northern redhorse
        (Moxostoma aureolum) in Upper Mississippi River (St. Paul-Dubuque).
        Data for Handbook of Biological Data. 2 p.

G137   Ganssle, D., 1966. Fishes and decapods of San Pablo and Suisun Bays.
        Calif. Fish. Bull., 133: 64-94.

G140   Gennings, R. M., 1965. Seining of young of the year fishes. Okla. Fish.
        Res. Lab. Semi-annu. Rep., 1965(Jan.-June): 79-107.

G141   Green, O. L., 1966. Observations on the culture of the bowfin. Progr.
        Fish Cult., 28(3): 179.

G142   Gerdes, J. H. and W. J. McConnell, 1963. Food habits and spawning of
        the threadfin shad in a small, desert impoundment. J. Ariz. Acad. Sci.,
        2(3): 113-16.

G143   Gerlach, A. H., 1963. Klamath District. Annu. Rep. Ore. St. Game
        Comm. Fish Div., 1963: 220-39.

G144   Garside, E. T., 1966. Developmental rate and vertebral number in
        salmonids. J. Fish. Res. Bd. Can., 23(10): 1537-51.

G145   Graham, T. R. and J. W. Jones, 1962. The biology of Llyn Tegid trout,
        1960. Proc. Zool. Soc. Lond., 139(4): 657-83.

G146   Garofalo, T. R., 1964. A study of the population dynamics and trophic
        ecology of the brook trout in two subalpine lakes, in southeastern Wyo-
        ming. Univ. Wyo. Coop. Res. Proj., 2(Part VI): 96 p.

G147   Gerlach, A. H., 1962. Klamath District. Annu. Rep. Ore. St. Game
        Comm. Fish Div., 1962: 22-43.

G148   Grimas, U. and N. A. Nilsson, 1962. Nährungsfauna und Kanadische
        Seefarelle in Berner Gebirgsseen. Schweiz. Z. Hydrol., 24(1): 49-75.

G149   Gould, W. R., 1966. Cutthroat trout (Salmo clarkii Richardson) x golden
        trout (Salmo aquabonita Jordan) hybrids. Copeia, 1966(3): 599-600.

G150   Gibson, R. J., 1966. Some factors influencing the distributions of brook
        trout and young Atlantic salmon. J. Fish. Res. Bd. Can., 23(12): 1977-
        80.

G151   Grimas, U. and N. A. Nilsson, 1965. On the food chain in some north
        Swedish river reservoirs. Rep. Inst. Freshw. Res. Drottning., 46:31-
        48.

G152   Gammon, J. R. and A. D. Hasler, 1965. Predation by introduced mus-
        kellunge on perch and bass, 1: years 1-5. Trans. Wis. Acad. Sci. Arts
        Lett., 54: 249-72.

G153   Gerald, J. W., 1966. Food habits of the longnose dace, Rhinichthys
        cataractae. Copeia, 1966(3): 478-85.

G154   Geen, G. H., T. G. Northcote, G. F. Hartman, and C. C. Lindsey, 1966.
        Life histories of two species of catostomid fishes in Sixteenmile Lake,
        British Columbia, with particular reference to inlet stream spawning.
        J. Fish. Res. Bd. Can., 23(11): 1761-88.

G155   Giudice, J. J., 1966. Growth of a blue x channel catfish hybrid as com-
        pared to its parent species. Progr. Fish Cult., 28(3): 142-45.

G156   Gould, W. R., III and W. H. Irwin, 1965. The suitabilities and relative resistances of twelve species of fish as bioassay animals for oil-refinery effluents. Proc. S. E. Assoc. Game Fish Comm., 16: 333-48.

G157   Giudice, J. J., 1965. Investigations on the propagation and survival of flathead catfish in troughs. Proc. S. E. Assoc. Game Fish Comm., 17: 178-80.

G159   Godby, M. H., 1919. A preliminary investigation of the age and manner of growth of brown trout in Canterbury, as shown by a microscopic examination of their scales. Trans. N. Z. Inst., 51: 42-67.

H1   Haakh, T., 1929. Studien über Alter und Wachstum der Bodenseefische. Arch. Hydrobiol., 20: 214-95.

H2   Haig-Brown, R. L., 1947. The western angler. William Morrow and Co., N. Y. 356 p.

H3   Haime, J., 1874. A history of fish culture in Europe from its earlier records to 1854. Rep. U. S. Comm. Fish Fish., 1871-72(Append. D): 463-92.

H4   Hall, C. B., 1929. The culture of fish in ponds. G. Brit. Ministry Agric. Fish. Misc. Publ., 64: 1-24.

H5   Hammer, R. C., 1943. Maryland commercial fish hatchery operations, 1943. Md. Bd. Nat. Resources, Chesapeake Biol. Lab. Publ., 60: 1-16.

H6   _____, 1946. Maryland commercial fish hatchery operations, 1944 and 1945. Md. Dept. Res. Educ., Educ. Ser., 11: 1-20.

H7   Hammett, F. S. and D. W. Hammett, 1939. Proportional length growth of gar (*Lepidosteus platyrhinchus* DeKay). Growth, 3: 197-209.

H8   Hankinson, T. L., 1919. Notes of life-histories of Illinois fish. Trans. Ill. St. Acad. Sci., 12: 132-50.

H11   Hardisty, M. W., 1944. The life-history and growth of the brook lamprey, *(Lampetra planeri)*. J. Anim. Ecol., 13: 110-22.

H13   Harkness, W. J. K., 1923. Rate of growth and food of lake sturgeon *(Acipenser rubicundus)*. Publ. Ont. Fish. Res. Lab., 18: 15-42.

H15   _____, 1945. Rate of growth of game fish. 6 p. mimeo.

H16   Harrington, R. W., Jr., 1947. The early life history of the bridled shiner, *Notropis bifrenatus* (Cope). Copeia, 1947(2): 97-102.

H17   _____, 1948. The life cycle and fertility of the bridled shiner, *Notropis bifrenatus* (Cope). Amer. Midl. Nat., 39(1): 83-92.

H23   Hart, J. L., 1930. The spawning and early life history of the whitefish, *Coregonus clupeaformis*, in the Bay of Quinte, Ontario. Contr. Canad. Biol. Fish., NS6(7): 167-214.

H24   _____, 1931. The growth of the whitefish, *Coregonus clupeaformis*. Contr. Canad. Biol. Fish., NS6(20): 429-44.

H25   _____, 1932. Statistics of the whitefish *(Coregonus clupeaformis)* population of Shakespeare Island Lake, Ontario. Publ. Ont. Fish. Res. Lab., 42: 1-28.

H26   _____, 1940. Growth in lingcod. Fish. Res. Bd. Can. Progr. Rep. Pacif. Biol. Sta., 44: 14-15.

H28   Hartley, G. W., 1937. Salmon caught in the sea—northwest Sutherland, 1936. Salm. Fish. Scot., 1937(3): 1-21.

H29   _____, 1938. Salmon caught in the sea—west Sutherland, 1937. Salm. Fish. Scot., 1938(2): 1-21.

H30   Hartley, P. H. T., 1947. The coarse fishes of Britain, being the final report on the coarse fish investigation. Freshw. Biol. Assoc. Sci. Publ., 12: 1-40.

H31   Hartley, P. H. T., 1947. The natural history of some British freshwater fishes. Proc. Zool. Soc. Lond., 117(1): 129-206.

H32   Hasler, A. D., 1938. Fish biology and limnology of Crater Lake, Oregon. J. Wildl. Mgmt., 2: 94-103.

H35   _____ and D. S. Farner, 1942. Fisheries investigations in Crater Lake,
Oregon, 1937-1940. J. Wildl. Mgmt., 6(4): 319-27.

H36   _____, R. K. Meyer, and H. M. Field, 1940. The use of hormones for
the conservation of muskellunge, *Esox masquinongy immaculatus* Girard. Copeia, 1940(1): 43-46.

H37   Hatton, S. R., 1940. Progress report on the Central Valley fisheries
investigations, 1939. Calif. Fish Game, 26(4): 334-73.

H38   Hayes, F. R. and F. H. Armstrong, 1943. Growth of the salmon embryo. Canad. J. Res., 21D(2): 19-33.

H39   Hayford, C. O., 1920. Trout feeding experiments. Trans. Amer. Fish.
Soc., 50: 251-56.

H41   _____ and G. C. Embody, 1930. Further progress in the selective
breeding of brook trout at the New Jersey State Hatchery. Trans.
Amer. Fish. Soc., 60: 109-13.

H42   Hazzard, A. S., 1932. Some phases of the life history of the eastern
brook trout, *Salvelinus fontinalis* Mitchill. Trans. Amer. Fish. Soc.,
62: 344-50.

H43   _____, 1935. A preliminary study of an exceptionally productive trout
water, Fish Lake, Utah. Trans. Amer. Fish. Soc., 65: 122-28.

H44   _____, 1945. Warm-water fish management. Wis. Conserv. Bull.,
10(10): 12-15.

H47   _____ and D. S. Shetter, 1939. Results from experimental plantings of
legal-sized brook trout *(Salvelinus fontinalis)* and rainbow trout *(Salmo
irideus)*. Trans. Amer. Fish. Soc., 68: 196-210.

H48   Heacox, C., 1946. The Chautauqua Lake muskellunge: research and
management applied to a sport fishery. Trans. N. Amer. Wildl. Conf.,
11: 419-25.

H50   Hein, E. N., 1947. The new Weber Lake. Wis. Conserv. Bull., 12(5):
15-16.

H51   Henshall, J. A., 1898. Some preliminary observations concerning the
artificial culture of the grayling. Trans. Amer. Fish. Soc., 27: 105-11.

H55   Hessel, R., 1874. The salmon of the Danube or the Hucho *(Salmo hucho)*
and its introduction into American waters. Rep. U. S. Comm. Fish
Fish., 1871-72(Append. B): 161-65.

H56   _____, 1878. The carp and its culture in rivers and lakes: and its introduction in America. Rep. U. S. Comm. Fish Fish., 1875-76: 865-97.

H57   Hewitt, E. R., 1931. Better trout streams. Chas. Scribner's, N. Y.
140 p.

H58   Hey, D., 1947. The culture of freshwater fish in South Africa. Inl.
Fish. Dept. Stellenbosch. 124 p.

H60   Hildebrand, S. F., 1917. Notes on the life history of the minnows *Gambusia affinis* and *Cyprinodon variegatus*. Rep. U. S. Comm. Fish.,
1917(Append. 6): 1-15.

H61   _____ and W. C. Schroeder, 1928. Fishes of Chesapeake Bay. Bull.
U. S. Bur. Fish., 43: 1-366.

H62   Hile, R., 1931. Rate of growth of fishes of Indiana. Ind. Dept. Conserv.
Div. Fish Game Publ., 107: 9-55.

H63   _____, 1936. Summary of investigations on the morphometry of the cisco, *Leucichthys artedi* (LeSueur), in the lakes of the northeastern highlands, Wisconsin. Pap. Mich. Acad. Sci. Arts Lett., 21: 619-34.

H64   Hile, R., 1936. Age and growth of the cisco, *Leucichthys artedi* (Le-Sueur), in the lakes of the northeastern highlands, Wisconsin. Bull. U. S. Bur. Fish., 48(19): 211-317.

H68   _____, 1948. Standardization of methods of expressing lengths and weights of fish. Trans. Amer. Fish. Soc., 75: 157-64.

H69   _____ and H. J. Deason, 1934. Growth of the whitefish, *Coregonus clupeaformis* (Mitchill), in Trout Lake, northeastern highlands, Wisconsin. Trans. Amer. Fish. Soc., 64: 231-37.

H73   Ho, H. J., 1936. The growth of the godfish *(Carassius auratus)*. China J., 24: 101-5.

H74   Hoar, W. S., 1939. The weight-length relationship of the Atlantic salmon. J. Fish. Res. Bd. Can., 4(5): 441-60.

H75   Hoffbauer, C., 1898. Die Alterbestimmung des Karpfen an seiner Schuppe. Allgem. Fisch. Ztg. 23: 341-43.

H76   Hoffman, C. H. and E. W. Surber, 1948. Effects of an aerial application of wettable DDT on fish and fish-food organisms in Back Creek, West Virginia. Trans. Amer. Fish. Soc., 75: 48-58.

H77   Hogan, J., 1936. Are young carp of any value as a forage fish? Progr. Fish Cult., 20: 22-3.

H80   Hooper, F. F., 1949. Age analysis of a population of the ameiurid fish, *Schilbeodes mollis* (Hermann). Copeia, 1949(1): 34-8.

H81   Hoover, E. E., 1936. Contributions to the life history of the chinook and landlocked salmon in New Hampshire. Copeia, 1936(4): 193-98.

H82   _____, 1936. The spawning activities of fresh water smelt, with special reference to the sex ratio. Copeia, 1936(2): 85-91.

H83   _____, 1938. Fish populations of primitive brook trout streams of northern New Hampshire. Trans. N. Amer. Wildl. Conf., 3: 486-96.

H84   _____, 1939. Age and growth of brook trout in northern breeder streams. J. Wildl. Mgmt., 3(2): 81-91.

H86   Hubbs, C. L., 1921. An ecological study of the life-history of the freshwater atherine fish, *Labidesthes sicculus*. Ecology, 2(4): 262-76.

H87   _____, 1922. Variation in the number of vertebrae and other meristic characters of fishes correlated with the temperature of water during development. Amer. Nat., 56: 360-72.

H88   _____, 1923. Seasonal variation in the number of vertebrae of fishes. Pap. Mich. Acad. Sci. Arts Lett., 2: 207-14.

H89   _____, 1925. The life-cycle and growth of lampreys. Pap. Mich. Acad. Sci. Arts Lett., 4: 587-703.

H90   _____ and G. P. Cooper, 1935. Age and growth of the long-eared and the green sunfishes in Michigan. Pap. Mich. Acad. Sci. Arts Lett., 20: 669-96.

H91   _____ and _____, 1936. Minnows of Michigan. Cranbrook Inst. Sci. Bull., 8: 1-95.

H92   _____ and C. W. Creaser, 1924. On the growth of young suckers and the propagation of trout. Ecology, 5(4): 372-78.

H95   _____ and S. C. Whitlock, 1929. Diverse types of young in a single species of fish, the gizzard shad. Pap. Mich. Acad. Sci. Arts Lett., 10: 461-82.

H98   Hueske, E. E., 1941. An experiment in producing golden shiners. Wildl. in N. C., 13(5): 17,20.

H99   Hutton, J. A., 1920. Wye salmon. Results of scale reading, 1919.
       Salm. Trout Mag., 21: 8-31.

H100   ____, 1921. Wye salmon. Results of scale reading, 1920. Salm.
       Trout Mag., 25: 118-45.

H101   ____, 1922. Wye salmon. Results of scale reading, 1921. Salm.
       Trout Mag., 29: 96-128.

H102   ____, 1923. The salmon of the Derwent. Salm. Trout Mag., 32: 152-58.

H103   ____, 1923. Something about grayling scales. Salm. Trout Mag., 31:
       59-64.

H104   ____, 1923. Salmon scales from the Aberdeenshire Dee. Salm. Trout
       Mag., 32: 141-51.

H105   ____, 1923. Wye salmon. Results of scale reading, 1922. Salm.
       Trout Mag., 31: 4-31.

H106   ____, 1924. Salmon scales from the Aberdeenshire Dee, 1923. Salm.
       Trout Mag., 36: 109-19.

H107   ____, 1924. The salmon of the Derwent, 1923. Salm. Trout Mag., 36:
       120-25.

H108   ____, 1924. The life-history of the salmon. Publ. Aberdeen Nat. Hist.
       Antiqu. Soc., 5: 56 p.

H109   ____, 1924. Wye salmon. Results of scale reading, 1923. Salm.
       Trout Mag., 35: 12-44.

H110   ____, 1925. The salmon of the Derwent, 1924. Salm. Trout Mag., 40:
       213-18.

H111   ____, 1925. Salmon scales from the Aberdeenshire Dee. Salm. Trout
       Mag., 39: 165-72.

H112   ____, 1925. Wye salmon, 1924. Salm. Trout Mag., 39: 127-64.

H113   ____, 1927. Wye salmon. Results of scale reading. Salm. Trout
       Mag., 47: 167-92.

H114   ____, 1927. Wye salmon. Results of scale reading, 1927. Salm.
       Trout Mag., 51: 159-89.

H115   ____, 1937. Wye salmon. Results of scale reading, 1936. Salm.
       Trout Mag., 86: 42-66.

H116   ____, 1940. Wye salmon. Results of scale reading. Salm. Trout
       Mag., 98: 52-68.

H127   Hall, G. E., 1951. Preimpoundment fish populations of the Wister Res-
       ervoir area in the Poteau River Basin, Oklahoma. Trans. N. Amer.
       Wildl. Conf., 16: 266-83.

H131   Haskell, D. C., 1952. Comparison of the growth of lake trout finger-
       lings from eggs taken at Seneca, Saranac, and Raquette Lakes. Progr.
       Fish Cult., 14(1): 15-18.

H132   ____, R. G. Zilliox, and W. M. Lawrence, 1952. Survival and growth
       of stocked lake trout yearlings from Seneca and Raquette Lake
       breeders. Progr. Fish Cult., 14(2): 71-3.

H137   Hooper, F. F., 1951. Limnological features of a Minnesota seepage
       lake. Amer. Midl. Nat., 46(2): 462-81.

H138   Hourston, A. S., 1952. The food and growth of the maskinonge (*Esox
       masquinongy* Mitchill) in Canadian waters. J. Fish. Res. Bd. Can.,
       8(5): 347-68.

H140   Hall, G. E. and R. M. Jenkins, 1952. The rate of growth of channel cat-
       fish in Oklahoma waters. Okla. Fish. Res. Lab. Rep., 27: 15 p.
       mimeo.

H141   Hart, J. S., 1952. Geographic variations of some physiological and morphological characters in certain freshwater fish. Publ. Ont. Fish. Res. Lab., 72: 1-79.

H143   Hendricks, L. J., 1952. Erythrocyte counts and hemoglobin determinations for two species of suckers, genus *Catostomus,* from Colorado. Copeia, 1952(4): 265-66.

H144   Hacker, V. A., 1957. Biology and management of lake trout in Green Lake, Wisconsin. Trans. Amer. Fish. Soc., 86: 71-83.

H150   Hall, G. E. and R. M. Jenkins, 1953. Continued fisheries investigation of Tenkiller Reservoir, Oklahoma, during its first year of impoundment, 1953. Okla. Fish. Res. Lab. Rep., 33: 1-54.

H151   _____ and _____, 1954. Notes on the age and growth of the pirateperch, *Aphredoderus sayanus,* in Oklahoma. Copeia, 1954(1): 69.

H152   _____ and _____, 1954. The rate of growth of channel catfish, *Ictalurus punctatus,* in Oklahoma waters. Proc. Okla. Acad. Sci., 33: 121-29.

H154   _____ and G. A. Moore, 1954. Oklahoma lampreys; their characterization and distribution. Copeia, 1954(2): 127-35.

H155   Halver, J. E. and J. A. Coates, 1957. A vitamin test diet for long-term feeding studies. Progr. Fish Cult., 19(3): 112-18.

H156   Hancock, H. M., 1955. Age and growth of some of the principal fishes in Canton Reservoir, Oklahoma, 1951, with particular emphasis on the white crappie. Okla. Fish Game Coun. Proj. Rep., Part 2: 110 p. mimeo.

H157   Harle, T. G. A. and R. Brooks, 1942. Practical mechanics of trout-scale reading. Salm. Trout Mag., 104: 59-63.

H158   Harmic, J. L., 1952. Fresh water fisheries survey. Del. Bd. Game Fish Fish. Publ., 1: 154 p.

H160   Harrington, R. W., Jr. and L. R. Rivas, 1958. The discovery in Florida of the Cyprinodont fish, *Rivulus marmoratus,* with a redescription and ecological notes. Copeia, 1958(2): 125-30.

H161   Harrison, A. C., 1948. The condition factor. Piscator, 2(7): 90-91.

H163   _____, 1954. The story of the Paarde Vlei. Part II. Introduction of non-indigenous fish. Piscator, 8(29): 29-32.

H164   _____, 1954. The story of Paarde Vlei. Part III. The change in water conditions. Piscator, 30: 56-64.

H166   _____, 1956. Steenbras Reservoir records. Piscator, 36: 13-21.

H169   Harrison, H. M., 1954. Returns from tagged channel catfish in the Des Moines River, Iowa. Proc. Ia. Acad. Sci., 60: 636-44.

H170   _____, 1955. Further studies on the catfish in the Humboldt area of the Des Moines River. Ia. Conserv. Comm. Quart. Biol. Rep., 6(4): 24-32.

H172   _____, 1956. Preliminary study of age and growth of channel catfish, Des Moines River, 1955. Ia. Conserv. Comm. Quart. Biol. Rep., 8(1): 38-41.

H174   _____, 1957. Growth of the channel catfish, *Ictalurus punctatus* (Rafinesque), in some Iowa waters. Proc. Ia. Acad. Sci., 64: 257-66.

H176   _____, 1958. Increased growth of channel catfish and black bullheads in the Humboldt area of the Des Moines River following the removal of a vast population of rough fish by chemical treatment. Ia. Conserv. Comm. Quart. Biol. Rep., 10(1): 23-8.

H177   _____, 1959. Progress report of fish populations, Humboldt study area. Ia. Conserv. Comm. Quart. Biol. Rep., 11(2): 24-9.

H179   Hartman, W. L., 1957. Finger Lake rainbows. III. A chronicle of their progress from egg to adult. N. Y. State Conserv., 11(6): 120-22.

H182 Haskell, D. C. and R. Griffiths, 1956. Growth in relation to sex among brook trout. N. Y. Fish Game J., 3(1): 93-107.

H183 _____, L. E. Wolf, and L. G. Bouchard, 1956. The effect of temperature on the growth of brook trout. N. Y. Fish Game J., 3(1): 108-13.

H185 Hassler, W. W., 1956. The influence of certain environmental factors on the growth of Norris Reservoir sauger, *Stizostedion canadense canadense* (Smith). Proc. S. E. Assoc. Game Fish Comm., 1955: 111-19.

H187 Havey, K. A., 1958. Eastern brook trout, *Salvelinus fontinalis* (Mitchill). Fishes of Maine, 2nd ed., ed. W. H. Everhart: 46-8.

H188 _____, 1959. Validity of the scale method for aging hatchery-reared Atlantic salmon. Trans. Amer. Fish. Soc., 88(3): 193-96.

H191 Hayes, M. L., 1955. Sucker life histories in Shadow Mountain Reservoir, Colorado. Colo. Fish. Res. Unit Quart. Rep., 2(1+2): 1-10.

H194 Heaton, J. R. and O. E. Orr, 1956. Preliminary surveys of public power and irrigation reservoirs in Nebraska with special reference to the fish populations. Midw. Wildl. Conf., 18th: 23 p.

H195 Hedges, S. B. and R. C. Ball, 1953. Production and harvest of bait fishes in Michigan. Mich. Dept. Conserv. Misc. Publ., 6: 1-30.

H199 Hendricks, L. J., 1956. Growth of the smallmouth buffalo in carp ponds. Progr. Fish Cult., 18(1): 45-46.

H200 Hennemuth, R. C., 1955. Growth of crappies, bluegill, and warmouth in Lake Ahquabi. Ia. St. J. Sci., 30(1): 119-37.

H201 Herald, E. S. and R. R. Strickland, 1949. An annotated list of the fishes of Homosassa Springs, Florida. Quart. J. Fla. Acad. Sci., 11(4): 99-109.

H202 Herke, W. H., 1959. Comparison of the length-weight relationship of several species of fish from two different, but connected, habitats. Proc. S. E. Assoc. Game Fish Comm., 13: 299-313.

H205 Hewson, L. C., 1955. Age, maturity, spawning and food of burbot *(Lota lota)* in Lake Winnipeg. J. Fish. Res. Bd. Can., 12(6): 930-40.

H207 _____, 1959. A seven-year study of the fishery for lake whitefish, *Coregonus clupeaformis*. J. Fish. Res. Bd. Can., 16(1): 107-20.

H208 Higgs, A., 1942. Big trout from big eggs. Salm. Trout Mag., 105: 216-30.

H210 Hobbs, D. F., 1954. Factors affecting the size and growth of young salmonidae in New Zealand. Proc. Pacif. Sci. Congr., 4: 562-75.

H212 Haslbauer, O. F., 1945. III. Relation of the bottom to fish distribution, Norris Reservoir. J. Tenn. Acad. Sci., 20(1): 135-38.

H213 Holloway, A. D., 1954. Notes on the life history and management of the shortnose and longnose gars in Florida waters. J. Wildl. Mgmt., 18(4): 438-48.

H215 Holton, G. D., 1953. A trout population study on a small creek in Gallatin County, Montana. J. Wildl. Mgmt., 17(1): 62-82.

H219 Hourston, A. S., 1955. A study of variations in the maskinonge from three regions in Canada. Contr. Roy. Ont. Mus. Zool. Palaeon., 40: 13 p. + 15 fig.

H220 Houser, A., 1958. A summary of fisheries investigations of Fort Gibson Reservoir, Oklahoma. Okla. Dept. Wildl. Conserv. Dingell-Johnson Job Completion Rep., Proj. F-6-R-1.

H221 _____ and M. G. Bross, 1959. Observations on growth and reproduction of the paddlefish. Trans. Amer. Fish. Soc., 88(1): 50-53.

H222 _____ and W. R. Heard, 1958. A one-year creel census on Fort Gibson Reservoir. Proc. Okla. Acad. Sci., 38: 137-46.

H225   Huet, M., 1948. Esociculture. Bull. Francais de Pisciculture, 148: 121-24.

H227   Huish, M. T., 1957. Gizzard shad removal in Deer Island Lake, Florida. Proc. S. E. Assoc. Game Fish Comm., 11: 312-18.

H230   ____, 1957. Studies of gizzard shad reduction at Lake Beulah, Florida. Proc. S. E. Assoc. Game Fish Comm., 11: 66-71.

H233   Hunt, B. P., 1953. Food relationships between Florida spotted gar and other organisms in the Tamiami Canal, Dade County, Florida. Trans. Amer. Fish. Soc., 82: 13-33.

H234   Huntsman, A. G., 1958. Shubenacadie salmon. J. Fish. Res. Bd. Can., 15(6): 1213-18.

H235   Hutton, J. A., 1914. Wye salmon. Results of scale reading, 1909-1913. Salm. Trout Mag., 1(7): 6-48, 79-104.

H237   ____, 1927. Some salmon scales from a river in Canada. Salm. Trout Mag., 49: 360-65.

H238   ____, 1927. Salmon scales from the Aberdeenshire Dee, 1926. Salm. Trout Mag., 47: 155-66.

H239   ____, 1928. Some scales from big salmon. Salm. Trout Mag., 50: 43-56.

H240   ____, 1931. Atlantic salmon in New Zealand. Salm. Trout Mag., 65: 308-10.

H241   ____, 1941. Wye salmon. Salm. Trout Mag., 102: 159-69.

H243   Hoopes, D. T., 1960. Utilization of mayflies and caddisflies by some Mississippi River fishes. Trans. Amer. Fish. Soc., 89(1): 32-4.

H244   Hunt, B. P., 1960. Digestion rate and food consumption of Florida gar, warmouth, and largemouth bass. Trans. Amer. Fish. Soc., 89(2): 206-11.

H245   Havey, K. A., 1960. Recovery, growth and movement of hatchery-reared lake Atlantic salmon at Long Pond, Maine. Trans. Amer. Fish. Soc., 89(2): 212-17.

H246   Hartman, W. L., 1959. Biology and vital statistics of rainbow trout in the Finger Lakes region, New York. N. Y. Fish Game J., 6(2): 121-78.

H247   Haskell, D. C., 1959. Trout growth in hatcheries. N. Y. Fish Game J., 6(2): 204-37.

H248   Haskell, W. L., 1959. Diet of the Mississippi threadfin shad, Dorosoma petenense atchafalaya, in Arizona. Copeia, 1959(4): 298-302.

H250   Helm, W. T., 1958. Notes on the ecology of panfish in Lake Wingra with special reference to the yellow bass. PhD thesis Univ. Wis.

H252   Huet, M. and J. A. Timmermans, 1958. Esociculture. Production de Brochetons de sept semaines. Sta. de Rech. Eaux Forets Groenendaal Trav. Ser. D, 24: 1-10.

H253   Hart, J. S., 1947. Lethal temperature relations of certain fish of the Toronto region. Trans. Roy. Soc. Can., 41: 57-71.

H254   Huntsman, A. G. and M. I. Sparks, 1924. Limiting factors for marine animals. 3. Relative resistance to high temperatures. Contr. Canad. Biol., 2: 95-114.

H256   Hepworth, W., 1959. A study of the population dynamics of brook trout in two sub-alpine lakes in southeastern Wyoming. Wyo. Game Fish Comm. Coop. Res. Proj., 2: 69-203.

H257   Helm, J. M., 1960. Returns from muskellunge stocking. Wis. Conserv. Bull., 25(6): 9-10.

H258   Hansen, D. W., 1952. Life history studies of the trout of Pathfinder
        Reservoir, Wyoming. MS thesis Ia. St. Coll. Lib. 55 p.

H260   Harkness, W. J. K., 1958. Lake sturgeon, *Acipenser fulvescens* Rafi-
        nesque. Data for Handbook of Biological Data. 4 p.

H261   Hoar, W. S., 1952. Thyroid function in some anadromous and land-
        locked teleosts. Trans. Roy. Soc. Can. Ser. 3, 46(5): 39-53.

H263   Hagen, H. K., 1956. Age, growth and reproduction of the mountain
        whitefish in Phelps Lake, Wyoming. PhD thesis Univ. Wash. 176 p.

H267   Haskell, D. C., R. Davies, and J. Reckahn, 1960. Effect of sorting
        brown trout on sex ratio and growth. N. Y. Fish Game J., 7(1): 39-45.

H268   Hayes, M. L., 1956. Life history studies of two species of suckers in
        Shadow Mountain Reservoir, Grand County, Colorado. MS thesis Colo.
        A. M. Coll. 126 p.

H269   Harris, R. H. D., 1952. A study of the sturgeon sucker in Great Slave
        Lake, 1950-51. MS thesis Univ. Alta. 44 p.

H271   Hale, J. G., 1960. Some aspects of the life history of the smelt *(Os-
        merus mordax)* in western Lake Superior. Minn. Fish Game Invest.
        Fish Ser., 2: 25-41.

H279   Hatch, R. W., 1961. Regular occurrence of false annuli in four brook
        trout populations. Trans. Amer. Fish. Soc., 90(1): 6-12.

H280   Havey, K. A., 1950. The freshwater fisheries of Long Pond and Echo
        Lake, Mount Desert Island, Maine. MS thesis Univ. Maine. 83 p.

H281   Hildebrand, S. F. and I. L. Towers, 1928. Annotated list of fishes col-
        lected in the vicinity of Greenwood, Mississippi, with descriptions of
        three new species. Bull. U. S. Bur. Fish., 43: 105-36.

H285   Hatch, R. W. and D. A. Webster, 1961. Trout production in four central
        Adirondack Mountain lakes. Cornell Univ. Agric. Exp. Sta. Mem., 373:
        82 p.

H287   Hildebrand, S. F. and I. L. Towers, 1927. Food of trout in Fish Lake,
        Utah. Ecology, 8(4): 389-97.

H289   Hallock, R. L., W. F. Van Woert, and L. Shapovalov, 1961. An evalua-
        tion of stocking hatchery-reared steelhead rainbow trout *(Salmo gaird-
        nerii gairdnerii)* in the Sacramento River system. Calif. Fish. Bull.,
        114: 74 p.

H290   Hendricks, L. J., 1961. The threadfin shad, *Dorosoma petenense* (Gun-
        ther). Calif. Fish. Bull., 113: 93-4.

H292   Huet, M., 1961. Reproduction et migrations de la truite commune
        (Salmo trutta fario L.) dans un ruisselet salmonicole de l'Ardenne
        belge. Verh. Int. Ver. Limnol., 14: 757-62.

H294   Hacker, V. A., 1962. A summarization of life history information of
        the lake trout, *Salvelinus namaycush,* obtained in gill netting, finclip-
        ping and tagging studies at Green Lake, Wisconsin, 1956-61. Wis.
        Conserv. Dept. E. Cent. Area Invest. Memo., 3: 24 p. mimeo.

H295   Hewkin, J. A., 1960. John Day drainage. Annu. Rep. Ore. St. Game
        Comm. Fish. Div., 1960: 212-31.

H296   Heckeroth, D. N., 1960. Umatilla fishery district, northeast region.
        Annu. Rep. Ore. St. Game Comm. Fish. Div., 1960: 302-13.

H297   Havey, K. A., 1961. Restoration of anadromous alewives at Long Pond,
        Maine. Trans. Amer. Fish. Soc., 90(3): 281-86.

H298   Hiner, L. E., 1961. Propagation of northern pike. Trans. Amer. Fish.
        Soc., 90(3): 298-302.

H300   Horak, D. L., 1961. The vertical distribution of fishes—Horsetooth

Reservoir. Quart. Rep. Colo. Coop. Fish. Res. Unit, 7(Nov.1960-Sept. 1961): 20-40.

H301   Houser, A., 1960. The effect of homing on channel catfish population estimates in large reservoirs. Proc. Okla. Acad. Sci., 40: 121-33.

H303   ____, 1960. A fishery survey by population estimation techniques in Lake Lawtonka. Rep. Okla. Fish. Res. Lab. Norman, 76: 18 p.

H304   Harrison, H. M., C. O'Farrell, and T. E. Moen, 1961. Progress report, renovation of the Winnebago River, Iowa. Ia. Conserv. Comm. Quart. Biol. Rep., 13(1): 32-7.

H305   Harrington, R. W., Jr. and E. S. Harrington, 1961. Food selection among fishes invading a high subtropical salt marsh: from onset of flooding through the progress of a mosquito brood. Ecology, 42(4): 646-66.

H306   Hartman, G. F., T. G. Northcote, and C. C. Lindsey, 1962. Comparison of inlet and outlet spawning runs of rainbow trout in Loon Lake, British Columbia. J. Fish. Res. Bd. Can., 19(2): 173-200.

H307   Harris, R. H. D., 1962. Growth and reproduction of the longnose sucker, *Catostomus catostomus* (Forster), in Great Slave Lake. J. Fish. Res. Bd. Can., 19(1): 113-26.

H308   Hall, A. R., 1925. Effects of oxygen and carbon dioxide on the development of the whitefish. Ecology, 6(2): 104-16.

H309   Hart, J. L., 1931. The food of the whitefish, *Coregonus clupeaformis* (Mitchill), in Ontario waters, with a note on the parasites. Canad. Biol. Fish., 6(21): 1-10.

H310   Hile, R. and H. J. Deason, 1947. Distribution, abundance, and spawning season and grounds of the kiyi, *Leucichthys kiyi* Koelz, in Lake Michigan. Trans. Amer. Fish. Soc., 74: 143-65.

H311   Harrison, A. C., 1961. Tiger trout (*Salmo trutta* female x *Salvelinus fontinalis* male). Piscator, 50: 85-93.

H312   Hansen, M. J. and D. W. Hayne, 1962. Sea lamprey larvae in Ogontz Bay and Ogontz River, Michigan. J. Wildl. Mgmt., 26(3): 237-47.

H313   Heard, W. R., 1962. The use and selectivity of small-meshed gill nets at Brooks Lake, Alaska. Trans. Amer. Fish. Soc., 91(3): 263-68.

H314   Huntsman, A. G., 1942. Death of salmon and trout with high temperature. J. Fish. Res. Bd. Can., 5(5): 485-501.

H315   Hubbs, C. L. and B. W. Walker, 1942. Habitat and breeding behavior of the American cyprinid fish, *Notropis longirostris*. Copeia, 1942(2): 101-4.

H317   Hatch, R. W., 1957. Success of natural spawning of rainbow trout in the Finger Lakes region of New York. N. Y. Fish Game J., 4(1): 69-87.

H318   ____, 1957. Finger Lakes rainbows—spawning habits. N. Y. State Conserv., 11(4): 20-22.

H319   Hume, L. C., 1955. Rainbow trout spawn twice a year. Calif. Fish Game, 41(1): 117.

H320   Huntsman, A. G., 1962. Method in ecology—ectology. Ecology, 43(3): 552-56.

H321   Hida, T. S. and D. A. Thomson, 1962. Introduction of the threadfin shad to Hawaii. Progr. Fish Cult., 24(4): 159-63.

H322   Hunt, R. L., O. M. Brynildson, and J. T. McFadden, 1962. Effects of angling regulations on a wild brook trout fishery. Wis. Conserv. Dept. Tech. Bull., 26: 56 p.

H324   Hartman, G. F., 1958. Mouth size and food size in young rainbow trout, *Salmo gairdneri*. Copeia, 1958(3): 233-34.

H325   Hutton, J. A., 1915. Wye salmon. Results of scale-reading, 1908-14.
        Salm. Trout Mag., 1(10): 27-52.

H327   Huntsman, A. G., 1937. The cause of periodic scarcity in Atlantic
        salmon. Trans. Roy. Soc. Can., III 31(5): 17-28.

H328   _____, 1941. Cyclic abundance and birds versus salmon. J. Fish. Res.
        Bd. Can., 5: 227-35.

H329   _____, 1931. The maritime salmon of Canada. Biol. Bd. Can. Bull.,
        21: 1-99.

H330   _____, 1931. Big catch of Miramichi salmon every three years.
        Atlant. Biol. Sta. Progr. Rep., 2: 18-19.

H331   _____, 1935. The sucker (Catostomus commersonii) in relation to
        salmon and trout. Trans. Amer. Fish. Soc., 65: 152-56.

H332   Hastings, C. E. and F. B. Cross, 1962. Farm ponds in Douglas County,
        Kansas. Univ. Kans. Mus. Nat. Hist. Misc. Pub., 29: 1-21.

H333   Hubley, R. C., Jr., 1963. Movements of tagged channel catfish in the
        Upper Mississippi River. Trans. Amer. Fish. Soc., 92(2): 165-68.

H334   Hartman, G. F., 1963. Observations on behavior of juvenile brown
        trout in a stream aquarium during winter and spring. J. Fish. Res.
        Bd. Can., 20(3): 769-87.

H335   Henderson, N. E., 1963. Influence of light and temperature on the re-
        productive cycle of the eastern brook trout, Salvelinus fontinalis
        (Mitchill). J. Fish. Res. Bd. Can., 20(4): 859-97.

H336   _____, 1963. Extent of atresia in maturing ovaries of the eastern brook
        trout, Salvelinus fontinalis (Mitchill). J. Fish. Res. Bd. Can., 20(4):
        899-908.

H337   Hill, C. W., Jr., 1962. Observations on the life histories of the pea-
        mouth (Mylocheilus caurinus) and the northern squawfish (Ptycho-
        cheilus oregonensis) in Montana. Proc. Mont. Acad. Sci., 22: 27-44.

H338   Horak, D. L. and H. A. Tanner, 1964. The use of vertical gill nets in
        studying fish depth distribution, Horsetooth Reservoir, Colorado.
        Trans. Amer. Fish. Soc., 93(2): 137-45.

H339   Hunt, R. L. and O. M. Brynildson, 1964. A five-year study of a head-
        waters trout refuge. Trans. Amer. Fish. Soc., 93(2): 194-97.

H340   Henshall, J. A., 1907. Culture of the Montana grayling. U. S. Bur.
        Fish. Doc., 628: 1-7.

H342   Harlan, J. R. and E. B. Speaker, 1956. Iowa fish and fishing. 3d ed.
        State of Iowa. 377 p.

H343   Hunt, B. P. and W. F. Carbine, 1950. Food of young pike, Esox lucius
        L., and associated fishes in Peterson's ditches, Houghton Lake, Michi-
        gan. Trans. Amer. Fish. Soc., 80: 67-83.

H345   Hoffbauer, C., 1901. Weitere Beiträge zur Bestimmung des Alters und
        Wachstumsverlaufes an die Struktur der Fisch Schuppe. Jahresbericht
        der teichwirtschaftliche Versuch Station zur Trochenberg. 50 p.

H347   Hankinson, T. L., 1908. A biological survey of Walnut Lake, Michigan.
        Mich. St. Bd. Geol. Surv. Rep., 1907: 157-288 not seen, cited in H8.

H348   Hubbs, C., 1954. Corrected distributional records for Texas fresh-
        water fishes. Tex. J. Sci., 6(3): 277-91.

H349   _____, R. A. Kuehne, and J. C. Ball, 1953. The fishes of the upper
        Guadalupe River, Texas. Tex. J. Sci., 5(2): 216-44.

H350   Hubbs, C. L. and R. R. Miller, 1943. Mass hybridization between two
        genera of cyprinid fishes in the Mohave Desert, California. Pap. Mich.
        Acad. Sci. Arts Lett., 28: 343-78.

H351   Harry, R. R., 1951. The embryonic and early larval stages of the tui chub, *Siphateles bicolor* (Girard), from Eagle Lake, California. Calif. Fish Game, 37(2): 129-32.

H352   Harrington, R. W., Jr., 1957. Life history table for the bridled shiner, *Notropis bifrenatus* (Cope). Data for Handbook of Biological Data. 5 p.

H353   _____, 1947. The breeding behavior of the bridled shiner, *Notropis bifrenatus*. Copeia, 1947(3): 186-92.

H354   _____, 1951. Notes on spawning in an aquarium by the bridled shiner, *Notropis bifrenatus*, with counts of the eggs deposited. Copeia, 1951(1): 85-6.

H355   _____, 1948. The food of the bridled shiner, *Notropis bifrenatus* (Cope). Amer. Midl. Nat., 40(2): 353-61.

H356   _____, 1950. Preseasonal breeding by the bridled shiner, *Notropis bifrenatus*, induced under light-temperature control. Copeia, 1950(4): 304-11.

H357   _____, 1957. Sexual photoperiodicity of the cyprinid fish, *Notropis bifrenatus* (Cope), in relation to the phases of its annual reproductive cycle. J. Exp. Zool., 135(3): 1-27.

H358   Hasler, A. D., H. P. Thomsen, and J. C. Neess, 1946. Facts and comments on raising two common bait minnows. Wis. Conserv. Dept. Bull., 210A 46: 14 p.

H359   Hankinson, T. L., 1920. Report on investigations of the fish of the Galein River, Berrien County, Michigan. Occ. Pap. Mus. Zool. Univ. Mich., 89: 1-14.

H360   _____, 1932. Observations on the breeding behavior and habits of fishes in southern Michigan. Pap. Mich. Acad. Sci. Arts Lett., 15: 411-25.

H361   _____, 1930. Breeding behavior of the silverfin minnow, *Notropis whipplii spilopterus* (Cope). Copeia, 1930(3): 73-4.

H362   Hubbs, C. L., 1941. A systematic study of two Carolinian minnows, *Notropis scepticus* and *Notropis altipinnis*. Copeia, 1941(3): 165-74.

H363   Hickling, C. F., 1961. Tropical inland fisheries. Longman's, London, and J. Wiley and Sons, N. Y. 287 p.

H364   _____, 1962. Fish culture. Faber and Faber, London. 295 p.

H365   Hendricks, L. J., 1955. The effect of carp *(Cyprinus carpio)* on the limnology of central Missouri farm ponds. Disser. Abstr., 15(1): 1984.

H366   Hubbs, C., 1951. Observations on the breeding of *Dionda episcopa serena* in the Nueces River, Texas. Tex. J. Sci., 3: 490-92.

H368   _____, 1956. Relative variability of hybrids between the minnows, *Notropis lepidus* and *N. proserpinus*. Tex. J. Sci., 8(4): 463-69.

H369   _____ and V. G. Springer, 1957. A revision of the *Gambusia nobilis* species group, with descriptions of three new species, and notes on their variation, ecology, and evolution. Tex. J. Sci., 9(3): 279-327.

H370   Hubbs, C. L. and R. R. Miller, 1948. Two new, relict genera of cyprinid fishes from Nevada. Occ. Pap. Mus. Zool. Univ. Mich., 507: 1-30.

H371   _____ and K. F. Lagler, 1949. Fishes of Isle Royale, Lake Superior, Michigan. Pap. Mich. Acad. Sci. Arts Lett., 33: 73-133.

H372   _____ and E. C. Raney, 1951. Status, subspecies, and variations of *Notropis cummingsae*, a cyprinid fish of the southeastern United States. Occ. Pap. Mus. Zool. Univ. Mich., 535: 1-25.

H373   _____, 1951. *Notropis amnis*, a new cyprinid fish of the Mississippi fauna, with two subspecies. Occ. Pap. Mus. Zool. Univ. Mich., 530: 1-30.

H374   _____ and E. C. Raney, 1948. Subspecies of *Notropis altipinnis*, a cyprinid fish of the eastern United States. Occ. Pap. Mus. Zool. Univ. Mich., 506: 20 p.

H375   Hubbs, C. L. and E. C. Raney, 1947. *Notropis alborus,* a new cyprinid fish from North Carolina and Virginia. Occ. Pap. Mus. Zool. Univ. Mich., 498: 1-17.

H376   _____ and _____, 1944. Systematic notes on North American siluroid fishes of the genus *Schilbeodes.* Occ. Pap. Mus. Zool. Univ. Mich., 487: 1-36.

H377   _____, 1942. Sexual dimorphism in the cyprinid fishes, *Margariscus* and *Couesius,* and alleged hybridization between these genera. Occ. Pap. Mus. Zool. Univ. Mich., 468: 1-6.

H378   Haempel, O., 1924. Studien am Seesaibling mehrerer österreichischer Alpenseen. Verh. Int. Ver. Limnol., 2: 129-35.

H379   Huitfeldt-Kaas, H., 1927. Studiei over aldersforholde og veksttyper hos norske fervannsfisker. Oslo. 358 p.

H380   Hubbs, C. L., 1951. An American cyprinid fish, *Notropis germanus* Hay, interpreted as an intergeneric hybrid. Amer. Midl. Nat., 45(2): 446-54.

H381   _____ and R. M. Bailey, 1952. Identification of *Oxygeneum pulverulentium* Forbes, from Illinois, as a hybrid cyprinid. Pap. Mich. Acad. Sci. Arts Lett., 37: 143-52.

H382   _____ and K. F. Lagler, 1964. Fishes of the Great Lakes region, with a new preface. Univ. Mich. Press, Ann Arbor. 213 p.

H383   Hunter, G. W., III and J. S. Rankin, Jr., 1939. The food of pickerel. Copeia, 1939(4): 194-99.

H384   Hubbs, C., 1955. On a Texas record of *Notropis boops* based on an apparent hybrid between *N. venustus* and *N. fumeus.* Tex. J. Sci., 7(3): 346-48.

H385   _____ and K. Strawn, 1956. Interfertility between two sympatric fishes, *Notropis lutiensis* and *N. venustus.* Evolution, 10(4): 341-44.

H390   Heuts, M. J., 1958. Three-spined stickleback, *Gasterosteus aculeatus* L. Material for Handbook of Biological Data. 9 p.

H391   Houser, A., 1962. A trout fishery in Oklahoma. Proc. Okla. Acad. Sci., 42: 272-74.

H394   Hall, J. D., 1962. Production of the chestnut lamprey and its effect on a trout population in the Manistee River, Michigan. Abstr. Amer. Fish. Soc. Meet. 92, Jackson, Wyo.: 2 p.

H395   Heard, W. R., 1966. Observations on lampreys in the Naknek River system of southwest Alaska. Copeia, 1966(2): 332-39.

H396   Howell, J. H., 1966. The life cycle of the sea lamprey and a toxicological approach to its control. Phylogeny of Immunity, Univ. Fla. Press: 263-70.

H397   Hubbs, C. L. and T. E. B. Pope, 1937. The spread of the sea lamprey through the Great Lakes. Trans. Amer. Fish. Soc., 66: 172-76.

H398   Helms, D. R., 1966. 1965 annual survey of the Coralville Reservoir fish population. Ia. Conserv. Comm. Quart. Biol. Rep., 18(2): 27-32.

H399   Houser, A., 1965. Growth of paddlefish in Fort Gibson Reservoir, Oklahoma. Trans. Amer. Fish. Soc., 94(1): 91-3.

H400   Hussakof, L., 1914. Fishes swallowed by gar pike. Copeia, 11: 2.

H401   Hildebrand, S. F., 1963. Family Clupeidae. Fishes of the western north Atlantic. Mem. Sears Found. Mar. Res., 1(3): 257-442.

H402   Heard, W. R. and W. L. Hartman, 1966. Pygmy whitefish *Prosopium coulteri* in the Naknek River system of southwest Alaska. Fish. Bull. U. S., 65(3): 555-79.

H403   Halver, J. E., 1964. Western fish nutrition laboratory. U. S. Fish Wildl. Serv. Circ., 178: 32-66.

H404   Horak, D. L., 1966. Evaluation of hatchery-reared rainbow trout. Colo. Game Fish Parks, Fish. Res. Rev., 3: 18-21.

H405   Hunt, R. L., 1965. Surface-drift insects as trout food in the Brule River. Wis. Acad. Sci. Arts Lett., 54: 51-61.

H406   Huntsman, A. G., 1965. The ectology of Margaree salmon. Limnol. Oceanogr., 10(Suppl.): R137-47.

H408   Hardy, C. J., 1963. An examination of eleven stranded redds of brown trout *(Salmo trutta)* excavated in the Selwyn River during July and August, 1960. N. Z. J. Sci., 6(1): 107-19.

H409   Hunter, J. G., 1966. The Arctic char. Fish. Can., 19(3): 17-19.

H410   Hunt, R. L., 1966. Evaluation of fly-fishing-only at Lawrence Creek. (A three-year progress report.) Wis. Conserv. Dept. Misc. Res. Rep., 10: 15 p.

H411   ____, 1966. Production and angler harvest of wild brook trout in Lawrence Creek, Wisconsin. Wis. Conserv. Dept. Tech. Bull., 35: 52 p.

H412   Heiser, D. W., 1966. Age and growth of anadromous Dolly Varden char, *Salvelinus malma* (Walbaum), in Eva Creek, Baranof Island, southeastern Alaska. Alaska Dept. Fish Game Res. Rep., 5: 29 p.

H413   Hacker, V. A., 1964. Addition to ECA investigational memorandum number 3, relating to continuing studies of lake trout at Green Lake, Wisconsin, 1962-1964. Wis. Conserv. Dept. 4 p.

H414   Hart, J. L. and R. G. Ferguson, 1966. The American smelt. Trade News, Dept. Fish. Can., 18(9): 22-3.

H415   Hunt, R. L., 1965. Food of northern pike in a Wisconsin trout stream. Trans. Amer. Fish. Soc., 94(1): 95-7.

H416   Hart, J. L. and J. L. McHugh, 1944. The smelts (Osmeridae) of British Columbia. Bull. Fish. Res. Bd. Can., 54: 27 p.

H417   Hacker, V. A., 1966. An analysis of the muskellunge fishery of Little Green Lake, Green Lake County, Wisconsin, 1957-65. Wis. Conserv. Dept. Fish Mgmt. Rep., 4: 17 p.

H418   Huet, M. and J. A. Timmermans, 1966. La Production de Cyprins et de Voraces de repeuplement a la Pisciculture de Bokrijk de 1958 a 1963. Sta. Rech. Eaux Forets Groenendaal Trav. Ser. D, 38: 68 p. + 30 fig.

H419   Houser, A. and M. G. Bross, 1963. Average growth rates and length-weight relationships for fifteen species of fish in Oklahoma waters. Okla. Res. Lab. Rep., 85: 75 p.

H420   Hargis, H. L., 1966. Development of improved fishing methods for use in southeastern and south-central reservoirs. Tenn. Game Fish Comm. Dingell-Johnson Job Completion Rep., 4-5-R-1: 34 p.

H421   Houser, A. and C. Collins, 1962. Growth of black bullhead catfish in Oklahoma. Okla. Fish. Res. Lab. Rep., 79: 18 p.

H422   Harrison, H. M., 1962. The status of the fish population, Humboldt study area—spring, 1962. Ia. Conserv. Comm. Quart. Biol. Rep., 14(2): 1-6.

H423   Helms, D. R., 1965. Age and growth of the channel catfish in the Coralville Reservoir. Ia. Conserv. Comm. Quart. Biol. Rep., 17(2): 6-12.

H424   Houser, A. and R. Grinstead, 1961. The effect of black bullhead catfish and bluegill removals on the fish populations of a small lake. Proc. S. E. Assoc. Game Fish Comm., 15: 193-200.

H425   Hastings, W. H., 1964. Fish feed processing research. Feedstuffs, 36(21).

H426   _____, 1964. Catering for channel catfish. Feedstuffs, 36(23).

H427   Henderson, H., 1965. Observation on the propagation of flathead catfish in the San Marcos State Fish Hatchery, Texas. Proc. S. E. Assoc. Game Fish Comm., 17: 173-77.

H428   Hackney, P. A., 1966. Predator-prey relationships of the flathead catfish in ponds under selected forage fish conditions. Proc. S. E. Assoc. Game Fish Comm., 19: 217-22.

H429   Hubbs, C. L. and A. I. Ortenburger, 1929. Fishes collected in Arkansas in 1927. Publ. Univ. Okla. Biol. Surv., 1: 45-112.

I1   Ivanova-Berg, M. M., 1931. Über die Lebensdauer der Larve von *Lampetra planeri* auf dem Gebiete des Finnischen Busens. Zool. Anz. Leipzig, 96: 330-34.

I2   Indian Council of Agricultural Research, 1951. Madras rural piscicultural scheme. Progress Report. Govt. Press, Madras. 75 p.

I4   Iowa Conservation Commission, 1955. Report of State Conservation Commission for Biennium ending June 30, 1954. 244 p.

I5   Iowa Conservationist, 1955. A record carp? Ia. Conserv., 14(7): 148.

I6   Irving, R. B., 1955. Ecology of the cutthroat trout in Henry's Lake, Idaho. Trans. Amer. Fish. Soc., 84: 275-96.

I7   _____ and P. E. Cuplin, 1956. The effect of hydroelectric developments on the fishery resources of Snake River. Idaho Fish Game Dept., Dingell-Johnson Proj. F-8-R.

I8   _____, 1953. Ecology of the cutthroat trout, *Salmo clarki* Richardson, in Henry's Lake, Idaho. MS thesis Utah St. Agric. Coll. 101 p.

I9   Iontaobhas Taighde Brodon na h'Eureann Ioncorpartha, 1961. Report and statement of accounts for the year ended 31st December 1960. 13 p.

I10   Iowa Conservation Commission, Biology Section, 1961. Calculated total lengths at each annulus for various species of fish taken from De Soto Bend, September 25-29, 1961. 13 p. mimeo.

I11   Idyll, C. P., 1942. Food of rainbow, cutthroat and brown trout in the Cowichan River system, B. C. J. Fish. Res. Bd. Can., 5(5): 448-58.

I12   Ivlev, V. S. and G. G. Galkin, 1960. Survival and growth of Baltic salmon of different ages in natural condition. Rapp. Cons. Int. Explor. Mer., 148: 48-52.

I13   Iowa Conservationist, 1964. Two large muskies taken by netting crews. Ia. Conserv., 23(5): 40.

I14   Isaacson, P. A. and R. L. Poole, 1965. The threadfin shad, *Dorosoma petenense* in northern California ocean waters. Calif. Fish Game, 51(1): 56-7.

I15   Isaak, D., 1961. The ecological life history of the fathead minnow, *Pimephales promelas* (Rafinesque). PhD thesis Univ. Minn. Lib. 150 p. Microfilm 61-4598, Ann Arbor, Mich.

J1   Jacobson, H., 1883. Stocking the Stettiner Haff with carp. Bull. U. S. Fish Comm., 1882: 1-8.

J2   Jarvi, T. H., 1948. On the periodicity of salmon reproduction in the northern Baltic area and its causes. Rapp. Cons. Int. Explor. Mer., 119: 1-131.

J3   Jenkinson, J. W., 1912. Growth, variability, and correlation in young trout. Biometrika, 8: 444-55.

J5   Jobes, F. W., 1943. The age, growth, and bathymetric distribution of

Reighard's chub, *Leucichthys reighardi* Koelz, in Lake Michigan.
Trans. Amer. Fish. Soc., 72: 108-35.

J6   Jobes, F. W., 1949. The age, growth, and bathymetric distribution of
        the bloater, *Leucichthys hoyi* (Gill), in Lake Michigan. Pap. Mich.
        Acad. Sci. Arts Lett., 33: 135-72.

J7   _____, 1949. The age, growth, and distribution of the longjaw cisco,
        Leucichthys alpenae Koelz, in Lake Michigan. Trans. Amer. Fish.
        Soc., 76: 215-47.

J8   Johnson, S. M., 1883. Growth of German carp sent to Savoy, Texas, by
        U. S. Fish Commission. Bull. U. S. Fish Comm., 2: 14.

J11   Jones, J. W., 1939. Salmon of the Cheshire Dee, 1937 and 1938. Proc.
        Liverpool Biol. Soc., 52: 19-80.

J12   _____, 1948. Salmon and trout hybrids. Proc. Zool. Soc. Lond., 117(4):
        708-15.

J13   Jordan, D. S. and B. W. Evermann, 1908. American food and game
        fishes. Doubleday, Page and Co. 571 p.

J15   Juday, C. and C. L. Schloemer, 1938. Growth of game fish in Wiscon-
        sin waters—fifth report. Notes Limnol. Lab. Wis. Geol. Nat. Hist.
        Surv. 26 p. mimeo.

J23   Jones, J. W., 1951. Salmon studies, 1950. G. Brit. Minist. Agric.
        Fish. Fish Invest. Ser., 1, 5(5): 1-11.

J24   Jenkins, R. M., 1954. Growth of the flathead catfish, *Pilodictis olivaris*,
        in Grand Lake (Lake O' the Cherokees), Oklahoma. Proc. Okla. Acad.
        Sci., 33: 11-20.

J25   _____, E. M. Leonard, and G. E. Hall, 1952. An investigation of the
        fisheries resources of the Illinois River and pre-impoundment study of
        Tenkiller Reservoir, Oklahoma. Okla. Fish. Res. Lab. Rep., 26: 136 p.
        mimeo.

J26   _____, 1951. A fish population study of Claremore City Lake. Proc.
        Okla. Acad. Sci., 30: 84-93.

J27   Jackson, S. W., Jr., 1954. Rotenone survey of Black Hollow on Lower
        Spavinaw Lake, November, 1953. Proc. Okla. Acad. Sci., 35: 10-14.

J28   _____, 1957. Comparison of the age and growth of four fishes from
        lower and upper Spavinaw Lakes, Oklahoma. Proc. S. E. Assoc. Game
        Fish Comm., 11: 232-49.

J31   Jenkins, R. M., 1953. Growth histories of the principal fishes in Grand
        Lake (O' the Cherokees), Oklahoma, through thirteen years of impound-
        ment. Okla. Fish. Res. Lab. Rep., 34: 1-87.

J34   _____, 1955. A summary of fish population studies conducted during
        1954 at Ardmore City Lake, Stringtown Sub-Prison Lake, Fairfax City
        Lake, and Pawhuska City Lake. Okla. Fish. Res. Lab. Rep., 48: 31 p.

J36   _____, 1958. Growth of blue catfish *(Ictalurus furcatus)* in Lake Tex-
        oma. S. W. Nat., 1(4): 166-73.

J37   _____, 1957. The effect of gizzard shad on the fish population of a
        small Oklahoma lake. Trans. Amer. Fish. Soc., 85: 58-74.

J41   _____, 1953. A pre-impoundment survey of Fort Gibson Reservoir,
        Oklahoma (Summer, 1952). Okla. Fish. Res. Lab. Rep., 29: 1-53.

J42   _____, 1953. A report on the growth of fishes in Fort Gibson Reser-
        voir collected in July and October, 1953—the first year of complete im-
        poundment. Okla. Fish. Res. Lab. Rep., 32: 10 p. mimeo.

J45   _____ and J. C. Finnell, 1957. The fishery resources of the Verdigris
        River in Oklahoma. Okla. Fish. Res. Lab. Rep., 59: 1-46.

J47   _____ and E. M. Leonard, 1954. Initial effects of impoundment on the growth-rate of channel catfish in two Oklahoma reservoirs. Proc. Okla. Acad. Sci., 33: 79-86.

J48   Jensen, C. C., 1956. Upper Willamette. Annu. Rep. Ore. St. Game Comm. Fish. Div., 1955: 37-50.

J49   ____, 1957. Upper Willamette. Annu. Rep. Ore. St. Game Comm. Fish. Div., 1956: 37-52.

J52   Jeppson, P., 1957. The control of squawfish by use of dynamite, spot treatment and reduction of lake levels. Progr. Fish Cult., 19(4): 168-71.

J53   ____ and W. S. Platts, 1959. Ecology and control of the Columbia River squawfish in northern Idaho lakes. Trans. Amer. Fish. Soc., 88(3): 197-202.

J55   John, K. R., 1959. Ecology of the chub, *Gila atraria*, with special emphasis on vertebral curvatures, in Two Ocean Lake, Teton National Park, Wyoming. Ecology, 40(4): 564-71.

J56   Johnson, F. H., 1955. Rainbow trout mortality and carrying capacity of a rehabilitated trout lake in Minnesota. Progr. Fish Cult., 17(3): 129-31.

J57   ____ and A. R. Peterson, 1955. Comparative harvest of northern pike by summer angling and winter darkhouse spearing from Ball Club Lake, Itasca County, Minnesota. Minn. Div. Game Fish Invest. Rep., 164: 1-11.

J59   Johnson, L. D., 1954. Muskellunge culture in ponds. Midw. Wildl. Conf., 16: 4 p. mimeo.

J61   ____, 1958. Muskellunge growth in Wisconsin. Abstr. Midw. Wildl. Conf., 20: 2-3.

J62   ____, 1958. Pond culture of muskellunge in Wisconsin. Wis. Conserv. Dept. Tech. Bull., 17: 1-54.

J63   ____, 1959. Story of a thousand stomachs. Wis. Conserv. Bull., 24(3): 7-10.

J64   Johnson, M. C., 1954. Preliminary experiment on fish culture in brackish water ponds. Progr. Fish Cult., 16(3): 131-33.

J65   ____, 1959. Food-fish farming in the Mississippi Delta. Progr. Fish Cult., 21(4): 154-60.

J66   Johnson, W. E. and A. D. Hasler, 1954. Rainbow trout production in dystrophic lakes. J. Wildl. Mgmt., 18(1): 113-34.

J68   Jokiel, J., 1958. Losos (*Salmo salar* L.) rzeki wisly. Rocz. Nauk Rolniczych, 73-B-2: 159-213.

J69   Jonas, H. C., 1921. Taw and Torridge salmon. Salm. Trout Mag., 24: 146-56.

J70   ____, 1922. Taw and Torridge salmon—season 1921. Salm. Trout Mag., 30: 221-31.

J71   ____, 1923. Taw and Torridge salmon. Season 1922. Salm. Trout Mag., 32: 182-88.

J72   ____, 1924. Taw and Torridge salmon. Salm. Trout Mag., 36: 126-32.

J73   ____, 1926. Taw and Torridge salmon. Salm. Trout Mag., 42: 90-95.

J74   Jones, J. W., 1953. Salmon studies, 1951. G. Brit. Fish. Invest. Ser. I, 5(6): 1-16.

J75   ____, 1953. Age and growth of the trout *(Salmo trutta)*, grayling *(Thymallus thymallus)*, perch *(Perca fluviatilis)*, and roach *(Rutilus rutilus)* of Llyn Tegid (Bala) and the roach of the River Birket. G. Brit. Fish. Invest. Ser. I, 5(7): 7-18.

J76   Jones, J. W. and G. M. King, 1939. Dee salmon parr. Salm. Trout
        Mag., 95: 164-74.

J78   Johnson, L. D., 1960. Let's compare muskies. Wis. Conserv. Bull.,
        25(7): 13-16.

J80   Jones, J. W. and H. B. N. Hynes, 1950. The age and growth of *Gasteros-
        teus aculeatus, Pygosteus pungitius* and *Spinachia vulgaris,* as shown
        by their otoliths. J. Anim. Ecol., 19(1): 59-73.

J81   John, K. R., 1954. An ecological study of the cisco, *Leucichthys artedi*
        (LeSueur), in Lake Mendota, Wisconsin. PhD thesis Univ. Wis. 121 p.

J82   Jhingran, V. G., 1948. A contribution to the biology of the Klamath
        black dace, *Rhinichthys osculus klamathensis* (Evermann and Meek).
        PhD disser. Stanford Univ. 99 p.

J84   Johannes, R. E. and P. A. Larkin, 1961. Competition for food between
        redside shiners *(Richardsonius balteatus)* and rainbow trout *(Salmo
        gairdneri)* in two British Columbia lakes. J. Fish. Res. Bd. Can.,
        18(2): 203-20.

J86   Judy, M. H., 1961. Validity of age determination from scales of
        marked American shad. U. S. Fish Wildl. Serv. Fish. Bull., 185: 161-
        70.

J87   Jones, J. W. and H. Evans, 1962. Salmon rearing in mountain tarns—
        a preliminary report. Proc. Zool. Soc. Lond., 138(4): 499-515.

J88   ____, 1959. The salmon. Collins, London. 192 p.

J89   Johnson, R. L., 1962. The yield and standing crop of fish in Dailey
        Lake, Montana. MS thesis Mont. St. Coll. 25 p.

J90   Johnson, R. P., 1963. Studies on the life history and ecology of the big-
        mouth buffalo, *Ictiobus cyprinellus* (Valenciennes). J. Fish. Res. Bd.
        Can., 20(6): 1397-1429.

J91   Johnson, L. D., J. H. Klingbiel, C. A. Wistrom, and A. A. Oehmcke,
        1957. Life history and ecology of the northern muskellunge (*Esox
        masquinongy immaculatus* Garrard). Data for Handbook of Biological
        Data. 6 p.

J92   John, K. R., 1957. Comparative rates of survival of normal and de-
        formed chub, *Gila atraria* Girard, in Two Ocean Lake, Teton County,
        Wyoming. Proc. Penn. Acad. Sci., 31: 77-82.

J93   Jessen, R. L. and J. H. Kuehn, 1920. A preliminary report on the ef-
        fects of the elimination of carp on submerged vegetation. Minn. Dept.
        Conserv. Fish Ser., 2: 1-12.

J96   Jenkins, R. M., 1959. Some results of the partial fish population re-
        moval techniques in lake management. Proc. Okla. Acad. Sci., 36: 164-
        73.

J97   Johnson, M. G., 1964. Production of brook trout in eight Ontario farm
        ponds. Progr. Fish Cult., 26(4): 147-54.

J98   Johnson, L., 1966. Experimental determination of food consumption of
        pike, *Esox lucius,* for growth and maintenance. J. Fish. Res. Bd. Can.,
        23(10): 1495-1505.

J99   ____, 1966. Consumption of food by the resident population of pike,
        Esox lucius, in Lake Windermere. J. Fish. Res. Bd. Can., 23(10):
        1523-35.

J100   Jennings, T. L. and F. Fronk, 1965. Results of an experimental
        muskellunge-silver northern pike cross. Ia. Conserv. Comm. Quart.
        Biol. Rep., 17(4): 1-3.

J101   Jackson, S. W., Jr., 1966. Summary of fishery management activities on Lakes Eucha and Spavinaw, Oklahoma, 1951-1964. Proc. S. E. Assoc. Game Fish Comm., 19: 315-43.

J102   Jacobs, D. L., 1948. Nesting of the brook stickleback (abstract). Proc. Minn. Acad. Sci., 16: 33-4.

K1   Katoh, G., 1932. On the sex ratio and the growth of body in *Carassius auratus* and its variety "the iron fish." Sci. Rep. Tohoku Imper. Univ. 4th Ser. Biol., 7(3): 365-87.

K2   Keil, W. M., 1921. The domestication of landlocked salmon breeders. Trans. Amer. Fish. Soc., 51: 48-57.

K3   _____, 1922. The biological significance of the smolt period in certain salmonoids. Trans. Amer. Fish. Soc., 52: 178-83.

K4   Kendall, W. C., 1917. The pikes; their geographical distribution, habits, culture, and commercial importance. Rep. U. S. Comm. Fish., 1917(Append.5): 5-45.

K5   _____, 1917. An unusual catch of the young of Maine whitefish *(Coregonus labradorisus)*. Copeia, 44: 45.

K6   _____, 1918. The Rangely Lakes, Maine: with special reference to the habits of the fishes, fish culture, and angling. Fish. Bull. U. S., 35: 487-594.

K7   _____, 1919. Scotch sea trout in Maine. Copeia, 75: 85-87.

K8   _____, 1921. Fresh water crustacea as food for young fishes. Trans. Amer. Fish. Soc., 51: 70-75.

K9   _____, 1927. The smelts. Fish. Bull. U. S., 42: 217-375.

K10   _____ and W. A. Dence, 1927. A trout survey of the Allegheny State Park in 1922. Roosevelt Wildl. Bull., 4(3): 291-482.

K11   _____ and _____, 1929. The fishes of the Cranberry Lake region. Roosevelt Wildl. Bull., 5(2): 219-310.

K13   Kennedy, W. A., 1943. The whitefish, *Coregonus clupeaformis* (Mitchill), of Lake Opeongo, Algonquin Park, Ontario. Publ. Ont. Fish. Res. Lab., 62: 21-66.

K15   King, W., 1942. Trout management studies at Great Smoky Mountains National Park. J. Wildl. Mgmt., 6(2): 147-61.

K16   Klak, G. E., 1941. The condition of brook trout and rainbow trout from four eastern streams. Trans. Amer. Fish. Soc., 70: 282-89.

K17   Kopec, J. A., 1949. Ecology, breeding habits and young stages of *Crenichthys baileyi*, a cyprinodont fish of Nevada. Copeia, 1949(1): 56-61.

K18   Koster, W. J., 1939. Some phases of the life history and relationships of the cyprinid, *Clinostomus elongatus* (Kirtland). Copeia, 1939(4): 201-8.

K22   Krumholz, L. A., 1948. Reproduction in the western mosquitofish, *Gambusia affinis affinis* (Baird and Girard), and its use in mosquito control. Ecol. Monogr., 18: 1-43.

K27   Kuntz, A., 1914. Notes on the habits, morphology of the reproductive organs and embryology of the viviparous fish, *Gambusia affinis*. Fish. Bull. U. S., 33: 177-90.

K28   _____, 1916. Notes on the embryology and larval development of five species of teleostean fishes. Fish. Bull. U. S., 34: 407-29.

K29   Krumholz, L. A., 1949. Length-weight relationship of the muskellunge, *Esox m. masquinongy*, in Lake St. Clair. Trans. Amer. Fish. Soc., 77: 42-8.

K32   Kennedy, W. A., 1949. Some observations on the coregonine fish of Great Bear Lake, N. W. T. Fish. Res. Bd. Can. Bull., 82: 1-10.

K33   Kuehn, J. H., 1949. Statewide average total length in inches at each year. Minn. Fish. Res. Lab. Invest. Rep., 51(Suppl., 2nd rev.).

K34   Kathrein, J. W., 1951. Growth rate of four species of fish in a section of the Missouri River between Holster Dam and Cascade, Montana. Trans. Amer. Fish. Soc., 80: 93-8.

K37   Keleher, J. J., 1952. Growth and *Triaenophorus* parasitism in relation to taxonomy of Lake Winnipeg ciscoes *(Leucichthys)*. J. Fish. Res. Bd. Can., 8(7): 469-78.

K41   Krumholz, L. A., 1950. New fish stocking policies for Indiana ponds. Trans. N. Amer. Wildl. Conf., 15: 251-69.

K44   Kuehn, J. H., 1949. A study of a population of longnose dace *(Rhinichthys c. cataractae)*. Proc. Minn. Acad. Sci., 17: 81-7.

K49   Katz, M. and W. C. Howard, 1955. The length and growth of 0-year class creek chubs in relation to domestic pollution. Trans. Amer. Fish. Soc., 84: 228-38.

K51   Kawamoto, N. Y., Y. Inouye, and S. Nakanishi, 1957. Studies on effects by the pond-areas and the densities of fish in the water upon the growth rate of carp (*Cyprinus carpio* L.). Rep. Faculty Fish. Prefect. Univ. Mie., 2(3): 437-47.

K55   Kennedy, W. A., 1953. Growth, maturity, fecundity and mortality in the relatively unexploited whitefish, *Coregonus clupeaformis*, of Great Slave Lake. J. Fish. Res. Bd. Can., 10(7): 413-41.

K56   _____, 1954. Tagging returns, age studies and fluctuations in abundance of Lake Winnipeg whitefish, 1931-1951. J. Fish. Res. Bd. Can., 11(3): 284-309.

K57   _____, 1954. Growth, maturity and mortality in the relatively unexploited lake trout, *Cristivomer namaycush*, of Great Slave Lake. J. Fish. Res. Bd. Can., 11(6): 827-52.

K58   Kimsey, J. B., 1954. The life history of the Tui Chub, *Siphateles bicolor* (Girard), from Eagle Lake, California. Calif. Fish Game, 40(4): 395-410.

K61   _____ et al., 1956. A survey of the fish population of Pardee Reservoir, Amador/Calaveras Counties. Calif. Inl. Fish. Admin. Rep., 56-18.

K62   _____ and R. R. Bell, 1956. Notes on the status of the pumpkinseed sunfish, *Lepomis gibbosus*, in the lower Susan River, Lassen County, California. Calif. Inl. Fish. Admin. Rep., 56-1: 1-19.

K63   _____ and _____, 1956. Observations on the ecology of the largemouth black bass and the tui chub in Big Sage Reservoir, Modoc County. Calif. Inl. Fish. Admin. Rep., 55-15: 1-17.

K64   _____, R. H. Hagy, and G. W. McCammon, 1957. Progress report on the Mississippi threadfin shad, *Dorosoma petenensis atchaflayae*, in the Colorado River for 1956. Calif. Inl. Fish. Admin. Rep., 57-23: 48 p.

K65   King, J. E., 1954. Three years of partial fish population removal at Lake Hiwasse, Oklahoma. Proc. Okla. Acad. Sci., 35: 21-24.

K66   _____, 1955. Growth rates of fishes of Lake Hiwasse, Oklahoma, after two years of attempted population control. Proc. Okla. Acad. Sci., 34: 53-6.

K69   Klingbiel, J. H. and L. E. Morehouse, 1954. Does musky stocking pay? Wis. Conserv. Bull., 19(10): 17-19.

K71   Krumholz, L. A., 1956. Observations on the fish population of a lake contaminated by radioactive wastes. Bull. Amer. Mus. Nat. Hist., 110(4): 281-367.

K72   Kruse, T. E., 1959. Grayling of Grebe Lake, Yellowstone National
       Park, Wyoming. Fish. Bull. U. S., 59(149): 307-51.

K73   Kutkuhn, J. H., 1955. Food and feeding habits of some fishes in a
       dredged Iowa lake. Proc. Ia. Acad. Sci., 62: 576-88.

K74   _____, 1958. Utilization of gizzard shad by game fishes. Proc. Ia.
       Acad. Sci., 65: 571-79.

K75   Kramer, R. H. and L. L. Smith, Jr., 1960. Utilization of nests of
       largemouth bass, *Micropterus salmoides,* by golden shiners, *Notemi-
       gonus crysoleucas.* Copeia, 1960(1): 73-4.

K76   Kimsey, J. B., 1960. Observations on the spawning of Sacramento hitch
       in a lacustrine environment. Calif. Fish Game, 46(2): 211-15.

K77   _____, 1960. Note on spring food habits of the lake trout, *Salvelinus
       namaycush.* Calif. Fish Game, 46(2): 229-30.

K79   Krogh, A., 1939. Osmotic regulation in aquatic animals. Cambridge
       Univ. Press, London.

K80   Klavano, W. C., 1958. Age and growth of fish from Oregon farm ponds.
       MS thesis Ore. St. Coll. 41 p.

K81   Karvelis, E. G., 1952. Growth characteristics of a bluegill population
       in a Michigan trout lake. MS thesis Mich. St. Coll. 72 p.

K85   Kinney, E. C., Jr., 1950. The life history of the trout perch, *Percopsis
       omiscomaycus* (Walbaum), in western Lake Erie. MS thesis Ohio St.
       Univ. 75 p.

K86   _____, 1954. A life history study of the silver chub, *Hybopsis storeri-
       ana* (Kirtland), in western Lake Erie with notes on associated species.
       PhD thesis Ohio St. Univ. 99 p.

K87   Kutkuhn, J. H., 1958. Utilization of plankton by juvenile gizzard shad in
       a shallow prairie lake. Trans. Amer. Fish. Soc., 87: 80-103.

K88   Koelz, W., 1929. Coregonid fishes of the Great Lakes. Fish. Bull.
       U. S., 43: 297-643.

K90   Kerr, R. B., 1961. Scale to length ratio, age and growth of Atlantic
       salmon in Miramichi fisheries. J. Fish. Res. Bd. Can., 18(1): 117-24.

K91   Kawamoto, N. Y., 1961. The influence of excretory substances of
       fishes on their own growth. Progr. Fish Cult., 23(2): 70-75.

K92   Kinne, O., 1960. Growth, food intake, and food conversion in a eury-
       plastic fish exposed to different temperatures and salinities. Physiol.
       Zool., 33(4): 288-317.

K93   Keleher, J. J., 1961. Comparison of largest Great Slave Lake fish with
       North American records. J. Fish. Res. Bd. Can., 18(3): 417-21.

K97   Krueger, W. H., 1961. Meristic variation in the fourspine stickleback,
       Apeltes quadracus. Copeia, 1961(4): 442-50.

K98   Keleher, J. J., 1961. Largest fish from Great Slave Lake. Fish. Res.
       Bd. Can. Biol. Sta. Tech. Unit Circ., 3: 12-16.

K99   Kessler, S., G. Wohlfarth, M. Lachman, and R. Moav, 1961. Monosex
       culture of carp. Bamidgeh, 13(3/4): 57-60.

K100   Kimsey, J. B., 1955. Post-spawning behavior of the kokanee, *Oncor-
       hynchus nerka kennerlyi,* in Donner Lake, California. Copeia, 1: 51-2.

K101   _____, 1951. Notes on kokanee spawning in Donner Lake, California,
       1949. Calif. Fish Game, 37(3): 273-80.

K102   Kipling, C., 1962. The use of the scales of the brown trout (*Salmo
       trutta* L.) for the back-calculation of growth. J. Cons. Int. Explor.
       Mer., 27(3): 304-15.

K103   Knight, A. E., 1963. The embryonic and larval development of the rain-
       bow trout. Trans. Amer. Fish. Soc., 92: 344-55.

K104   Kirka, A., 1962. [Age and growth of *Salmo trutta* m. *fario, Salmo gairdneri irideus, Salvelinus fontinalis* and *Thymallus thymallus* in the brook of Vrica near Klastor pod Znievom.] (Czech. with Eng. summary.) Prace Lab. Rybarstva, 1: 153-61.

K107   Kraatz, W. C., 1923. A study of the food of the minnow, *Campostoma anomalum*. Ohio J. Sci., 23: 265-83.

K108   Kuehn, J. H., 1957. Life history and ecology of the long-nose dace. Data for Handbook of Biological Data. 2 p.

K109   Koster, W. J., 1957. Guide to the fishes of New Mexico. Univ. N. M. Press, Albuquerque.

K110   Kimsey, J. B., 1957. Sacramento squawfish, *Ptychocheilus grandis* Data for Handbook of Biological Data. 5 p.

K111   ____, 1957. Greaser blackfish, *Orthodon microlepidotus;* Sacramento hitch, *Lavinia e. exilacauda*. Data for Handbook of Biological Data. 9 p.

K112   Kraatz, W. C., 1928. Study of the food of the blunt-nosed minnow, *Pimephales notatus*. Ohio J. Sci., 28(2): 86-98.

K113   Kostomarov, B. and A. Pulankova, 1942. Untersuchugen über die Beziehung der Körpergrobe die Schleie (*Tinca tinca* L.) zu ihrer Darmbrobe und über den Einfluss des Geschlechtes auf diese Beziehungen. Z. Fisch., 40: 157-70.

K114   Kobayasi, H., 1951. On the scales of the hybrid of *Cyprinus carpio* and *Carassius carassius*. Japanese J. Ichthyol., 1(5): 300-303.

K116   Kemp, R. J. and C. Hubbs, 1954. First record of the cypress minnow (*Hybognathus hayi* Jordan) in Texas. Tex. J. Sci., 6(1): 113-14.

K117   Koster, W. J., 1948. Notes on the spawning activities and young stages of *Plancterus kansae* (Garman). Copeia, 1948(1): 25-33.

K118   Katz, M., 1954. Reproduction of fish. Data for Handbook of Biological Data. 22 p.

K119   Kinney, E. C., 1957. Trout-perch, *Percopsis omiscomaycus* (Walbaum). Data for Handbook of Biological Data. 3 p.

K120   Knapp, F. T., 1951. Additional records of lampreys from Texas. Copeia, 1951(1): 87.

K121   Kimsey, J. B., 1958. Introduction of the redeye black bass and the threadfin shad into California. Calif. Fish Game, 40(2): 203-4.

K122   Kennedy, W. A., 1963. Growth and mortality of whitefish in three unexploited lakes in northern Canada. J. Fish. Res. Bd. Can., 20(2): 265-72.

K123   Kramer, R. H. and L. L. Smith, Jr., 1965. Effects of suspended wood fiber on brown and rainbow trout eggs and alevins. Trans. Amer. Fish. Soc., 94(3): 252-58.

K124   Kleinert, S. J. and D. Mraz, 1966. Life history of the grass pickerel, *(Esox americanus vermiculatus)* in southeastern Wisconsin. Wis. Conserv. Dept. Tech. Bull., 37: 39 p.

K125   Keast, A. and D. Webb, 1966. Mouth and body form relative to feeding ecology in the fish fauna of a small lake, Lake Opinicon, Ontario. J. Fish. Res. Bd. Can., 23(12): 1845-74.

K126   Karvelis, E. G., 1964. The true pikes. Fish. Leafl. Wash., 569: 11 p.

K127   Koura, R. and A. R. El-Bolock, 1960. Acclimatization and growth of mirror carp in Egyptian ponds. Inst. Freshw. Biol., Gizira, Cairo, U.A.R., Notes and Mem., 51: 15 p.

K128   King, D. R. and G. S. Hunt, 1967. Effect of carp on vegetation in a Lake Erie marsh. J. Wildl. Mgmt., 31(1): 181-88.

K129   Keast, A., 1965. Resource subdivision amongst cohabiting fish species in a bay, Lake Opinicon, Ontario. Univ. Mich. Gr. Lakes Res. Div. Publ., 13: 106-32.

K130   Koehn, R. K. and W. L. Minckley, 1965. Changes with growth in selected body proportions of the cyprinid fish, *Notropis lutrensis*. S. W. Nat., 10(3): 151-55.

K131   Keeton, D., 1963. Growth of fishes in the Des Moines River, Iowa, with particular reference to water levels. PhD disser. Ia. St. Univ. 208 p.

K132   Knapp, F. T., 1950. Survey of systems used in measuring lengths and weights of fishes and of systems proposed as standard in the Southwest. Progr. Fish Cult., 12(4): 207-8.

K133   Kelley, J. R., Jr. and D. C. Carver, 1966. Age and growth of blue catfish, *Ictalurus furcatus* (LeSueur) in the recent delta of the Mississippi River. Proc. S. E. Assoc. Game Fish Comm., 19: 296-9.

K134   Kipling, C., 1957. The effect of gill-net selection on the estimation of weight-length relationships. J. Cons. Int. Explor. Mer., 23(1): 51-63.

K136   Krisnandhi, S. and W. Shell, 1966. Utilization of casein and soybean protein by channel catfish, *Ictalurus punctatus* (Rafinesque). Proc. S. E. Assoc. Game Fish Comm., 19: 205-9.

L1   Lagler, K. F., 1949. Studies in freshwater fishery biology. Edwards Bros., Ann Arbor, Mich. 231 p.

L2   _____ and V. C. Applegate, 1942. Age and growth of the gizzard shad, *Dorosoma cepedianum* (LeSueur), with a discussion of its value as a buffer and as forage of game fishes. Invest. Ind. Lakes Streams, 2: 99-110.

L3   _____ and C. Hubbs, 1943. Fall spawning of the mud pickerel, *Esox vermiculatus* LeSueur. Copeia, 1943(2): 131.

L6   Langlois, T. H., n.d. Bait culturist's guide. Ohio Dept. Conserv. Mimeo.

L7   _____, 1929. Breeding habits of the northern dace. Ecology, 10(1): 161-63.

L10   _____, 1935. Notes on the spawning habits of the Atlantic smelt. Copeia, 1935(3): 141-42.

L12   _____, 1936. Length-weight relationships in Ohio state fish ponds. Copeia, 1936(2): 120.

L13   _____, 1939. Ohio fish management progress report. Ohio Conserv. Bull., 3(1): 16-19.

L14   _____, 1945. Ohio's fish program. Ohio Div. Conserv. Nat. Resources. 40 p.

L15   Leach, G. C., 1919. Culture of rainbow trout and brook trout in ponds. U. S. Bur. Fish. Econ. Circ., 41: 1-19.

L16   _____, 1919. The artificial propagation of carp. U. S. Bur. Fish. Econ. Circ., 39: 1-19.

L17   _____, 1923. Artificial propagation of whitefish, grayling and lake trout. Rep. U. S. Comm. Fish., 1923(Append.III): 1-32.

L18   _____, 1923. Artificial propagation of brook trout and rainbow trout, with notes on three other species. Rep. U. S. Comm. Fish., 1923 (Append.VI): 1-74.

L19   _____, 1925. Artificial propagation of shad. Rep. U. S. Comm. Fish., 1924(Append.VIII): 459-86.

L20   _____, 1927. Artificial propagation of pike perch, yellow perch and pike. Rep. U. S. Bur. Fish., 1927(Append.I): 1-27.

L21   Leach, W. J., 1940. Occurrence and life history of the northern brook lamprey, *Ichthyomyzon fossor*, in Indiana. Copeia, 1949(1): 21-34.

L23   Leim, A. H., 1924. Life history of the shad *(Alosa sapidissima* (Wilson)) with special reference to the factors limiting its abundance. Contr. Canad. Biol. NS 2(11): 161-284.

L24   Leonard, A. K., 1927. The rate of growth and the food of the horned dace, *(Semotilus atromaculatus)* in Quebec, with some data on the food of the common shiner *(Notropis cornutus)* and of the brook trout *(Salvelinus fontinalis)* from the same region. Publ. Ont. Fish. Res. Lab., 30: 35-44.

L25   Leonard, J. W., 1938. Feeding habits of brook trout fry in natural waters. Pap. Mich. Acad. Sci. Arts Lett., 23: 645-6.

L26   Lewis, R. C., 1944. Selective breeding of rainbow trout at Hot Creek Hatchery. Calif. Fish Game, 30(2): 95-7.

L27   Lewis, W. M., 1949. The use of vertebrae as indicators of the age of the northern black bullhead, *Ameiurus m. melas* (Rafinesque). Ia. St. J. Sci., 23(2): 209-18.

L29   ____, 1950. Fisheries investigations on two artificial lakes in southern Iowa. II. Fish populations. Ia. St. J. Sci., 24(3): 287-324.

L30   ____ and K. D. Carlander, 1948. Growth of the yellow bass, *Morone interrupta* Gill, in Clear Lake, Iowa. Ia. St. J. Sci., 22(2): 185-95.

L33   Lindeborg, R. G., 1941. Records of fishes from the Quetico Provincial Park of Ontario, with comments on the growth of the yellow pike-perch. Copeia, 1941(3): 159-61.

L34   Locke, F., 1947. Trout fishery on Diamond Lake. Ore. St. Game Comm. Bull., 2(5): 1,4,6.

L35   Lord, R. F., Jr., 1927. Notes on the use of the blackhead minnow, *Pimephales promelas*, as a forage fish. Trans. Amer. Fish. Soc., 57: 92-9.

L36   ____, 1932. Notes on Montana graylings at the Pittsford, Vermont, experimental trout hatchery. Trans. Amer. Fish. Soc., 62: 171-8.

L38   Luce, W. M., 1933. A survey of the fishery of the Kaskaskia River. Ill. Nat. Hist. Surv. Bull., 20(2): 71-123.

L39   Lyall, W. A., 1941. Problems of fish management in Grand Mesa National Forest in Colorado. Trans. N. Amer. Wildl. Conf., 6: 252-5.

L41   Lydell, D., 1926. Raising trout in bass ponds and perch in blue gill ponds in the same season. Trans. Amer. Fish. Soc., 56: 47-9.

L44   Lord, R. F., Jr., 1935. The 1935 trout harvest from Furnace Brook, Vermont's "test stream." Trans. Amer. Fish. Soc., 65: 224-33.

L51   Laakso, M., 1951. Food habits of the Yellowstone whitefish, *Prosopium williamsoni cismontanus* (Jordan). Trans. Amer. Fish. Soc., 80: 99-109.

L52   Lachner, E. A., 1946. Studies of the biology of the chubs (genus *Nocomis*, family Cyprinidae) of northeastern United States. Cornell Univ. Abstr. Theses, 1946: 207-10.

L53   ____, 1950. The comparative food habits of the cyprinid fishes *Nocomis biguttatus* and *Nocomis micropogon* in western New York. J. Wash. Acad. Sci., 40(7): 229-36.

L57   Lagler, K. F. and H. Van Meter, 1951. Abundance and growth of gizzard shad, *Dorosoma cepedianum* (LeSueur), in a small Illinois lake. J. Wildl. Mgmt., 15(4): 357-60.

L59   Larimore, R. W., 1950. Gametogenesis of *Polyodon spathula* (Walbaum): a basis for regulation of the fishery. Copeia, 1950(2): 116-24.

L61   Larkin, P. A., 1951. The effects on fisheries of proposed West

Kootenay water-storage project at Trout Lake. B. C. Game Dept. Mgmt. Publ., 1: 1-25.

L62   Larkin, P. A., G. C. Anderson, W. A. Clemens, and D. C. G. MacKay, 1950. The production of kamloops trout (*Salmo gairdnerii kamloops*, Jordan) in Paul Lake, British Columbia. Univ. B. C. 37 p.

L65   Lin, S. Y., 1952. Fish culture project in Haiti. Proc. Gulf Cari. Fish. Inst., 4: 110-18.

L66   Louisiana Conservationist, 1951. (Picture of large gar.) La. Conserv., 4(1,2): 24.

L69   Laakso, M. and O. B. Cope, 1956. Age determination in Yellowstone cutthroat trout by the scale method. J. Wildl. Mgmt., 20(2): 138-53.

L70   Lachner, E. A., 1952. Studies of the biology of the cyprinid fishes of the chub genus *Nocomis* of northeastern United States. Amer. Midl. Nat., 48(2): 433-66.

L71   Lagler, K. F., 1956. The pike, *Esox lucius* Linneaus, in relation to waterfowl on the Seney National Wildlife Refuge, Michigan. J. Wildl. Mgmt., 20(2): 114-24.

L80   Langlois, T. H., 1954. The western end of Lake Erie and its ecology. J. W. Edwards, Ann Arbor, Mich. 479 p.

L82   LaPointe, D. F., 1958. Age and growth of the American shad, from three Atlantic coast rivers. Trans. Amer. Fish. Soc., 87: 139-50.

L83   Lapworth, E. D., 1956. The effect of fry plantings on whitefish production in eastern Lake Ontario. J. Fish. Res. Bd. Can., 13(4): 547-58.

L87   Larkin, P. A., 1950. Report on the preliminary survey of the steelhead of the lower Fraser River. Rep. B. C. Game Comm., 1948: 12 p.

L90   _____ and S. B. Smith, 1954. Some effects of introduction of the redside shiner on the kamloops trout in Paul Lake, British Columbia. Trans. Amer. Fish. Soc., 83: 161-75.

L91   _____, J. G. Terpenning, and R. R. Parker, 1957. Size as a determinant of growth rate in rainbow trout, *Salmo gairdneri*. Trans. Amer. Fish. Soc., 86: 84-96.

L92   Larsen, K., 1955. Fish population analyses in some small Danish trout streams by means of D. C. electro-fishing, with special reference to the populations of trout (*Salmo trutta L.*). Medd. Komm. Havundersøg., Kbh., 1(10): 1-69.

L93   Larson, R. W. and J. M. Ward, 1955. Management of steelhead trout in the state of Washington. Trans. Amer. Fish. Soc., 84: 261-74.

L96   Lawler, G. H., 1954. Observations on the trout-perch, *Percopsis omiscomaycus* (Walbaum), at Heming Lake, Manitoba. J. Fish. Res. Bd. Can., 11(1): 1-4.

L97   Lawrence, J. M., 1957. Estimated sizes of various forage fishes largemouth bass can swallow. Proc. S. E. Assoc. Game Fish Comm., 11: 220-26.

L98   LeCren, E. D., 1958. Preliminary observations on population of *Salmo trutta* in becks in northern England. Verh. Int. Ver. Limnol., 13: 754-7.

L100 Lehman, B. A., 1953. Fecundity of Hudson River shad. Res. Rep. U. S. Fish Serv., 33: 1-8.

L101 Leitritz, E., 1959. Trout and salmon culture (hatchery methods). Fish Bull. Calif. Dept. Fish Game, 107: 1-169.

L103 Lennon, R. E., 1955. Artificial propagation of the sea lamprey, *Petromyzon marinus*. Copeia, 1955(3): 235-6.

L104 _____, 1957. The stoneroller in the Smokies. Progr. Fish Cult., 19(1): 25.

L105   Lennon, R. E. and P. S. Parker, 1959. The reclamation of Indian and
       Abrams Creeks, Great Smoky Mountains National Park. Spec. Sci. Rep.
       U. S. Fish Wildl. Serv., 306: 1-22.

L107   Lewis, W. M., 1953. Analysis of the gizzard shad population of Crab
       Orchard Lake, Illinois. Ill. Acad. Sci. Trans., 46: 231-4.

L108   _____, 1957. The fish population of a spring-fed stream system in
       southern Illinois. Ill. Acad. Sci. Trans., 50: 23-9.

L110   _____ and D. Elder, 1953. The fish population of the headwaters of a
       spotted bass stream in southern Illinois. Trans. Amer. Fish. Soc., 82:
       193-202.

L115   Lin, S. Y., 1955. Chinese systems of pond stocking. Proc. Indo-Pacif.
       Fish. Coun., 5(II,III): 113-25.

L118   Loeb, H. A., 1958. Comparison of estimates of fish populations in
       lakes. N. Y. Fish Game J., 5(1): 66-76.

L119   Loftus, K. H., 1958. Studies on river-spawning populations of lake
       trout in eastern Lake Superior. Trans. Amer. Fish. Soc., 87: 259-77.

L126   Lynch, T. M., P. A. Buscemi, and D. G. Lemons, 1953. Limnological
       and fishery conditions of Two Buttes Reservoir, Colorado, 1950 and
       1951. Colo. Game Fish Dept. Rep., 92 p.

L127   Lux, F. E. and L. L. Smith, Jr., 1960. Some factors influencing sea-
       sonal changes in angler catch in a Minnesota lake. Trans. Amer. Fish.
       Soc., 89(1): 67-79.

L129   Lopinot, A., 1958. How fast do Illinois fish grow? Outdoors in Ill.,
       5(4): 8-10.

L132   Lux, F. E., 1960. Notes on first-year growth of several species of
       Minnesota fish. Progr. Fish Cult., 22(2): 81-2.

L133   Lennon, R. E. and P. S. Parker, 1960. The stoneroller, *Campostoma
       anomalum* (Rafinesque), in Great Smoky Mountains National Park.
       Trans. Amer. Fish. Soc., 89(3): 263-70.

L134   Lewis, W. M. and G. E. Gunning, 1959. Notes on the life history of the
       steel color shiner, *Notropis whipplei* (Girard). Ill. Acad. Sci. Trans.,
       52: 59-64.

L135   Lagler, K. F., 1956. Freshwater fishery biology, Second edition. Wm.
       C. Brown Co., Dubuque, Ia. 421 p.

L137   Lambou, V. W., 1952. Food and habitat of gar fish in the tide water of
       southeastern Louisiana. MS thesis La. St. Univ. 54 p.

L140   _____, 1961. Utilization of macrocrustaceans for food by freshwater
       fishes in Louisiana and its effects on the determination of predator-
       prey relations. Progr. Fish Cult., 23(1): 18-25.

L141   Lindroth, A., 1960. Body/scale relationship in Atlantic salmon. Int.
       Coun. Explor. Sea, Salm. Trout Comm., 104: 11 p. + 8 fig.

L143   Lawler, G. H. and G. P. McRae, 1961. A method for preparing glycerin-
       stored otoliths for age determination. J. Fish. Res. Bd. Can., 18(1): 47-
       50.

L145   _____, 1961. Abnormalities in Lake Erie whitefish. J. Fish. Res. Bd.
       Can., 18(2): 283-4.

L146   _____, 1961. Heming Lake experiment. Progr. Rep. Biol. Sta. Tech.
       Unit, London, Ont., Fish. Res. Bd. Can., 2: 48-50.

L147   Larsen, K., 1961. The fish population of a peat pit as determined by
       rotenone poisoning. Medd. Komm. Havundersøg., Kbh., 3(5): 117-32.

L149   Lawrence, J. M., 1960. Estimated sizes of various forage fishes chain
       pickerel can swallow. Proc. S. E. Assoc. Game Fish Comm., 14:
       257-8.

L150 Larkin, P. A. and K. V. Ajyangar, 1960. Applications of the Parker equation to growth of aquatic organisms. Proc. Alaska Sci. Congr., 2: 103-24.

L155 Lagler, K. F. and A. T. Wright, 1962. Predation of the dolly varden, *Salvelinus malma*, on young salmons, *Oncorhynchus* spp., in an estuary of southeastern Alaska. Trans. Amer. Fish. Soc., 91(1): 90-93.

L157 Louder, D. E., 1961. Coastal plain lakes of southeastern North Carolina. N. C. Wildl. Res. Comm. D-J Job Completion Rep., Proj. F5R + F6R, 1: 9-55.

L158 Lopukhina, A. M., 1961. Effect of *Triaenophorus nodulosus* Pallas (Cestoda, Pseudophyllidae) on yearling rainbow trout. Doklady Biol. Sci. Sect., Transl. by AIBS, 137: 302-4.

L159 Lennon, R. E., 1961. The trout fishery in Shenandoah National Park. Spec. Sci. Rep. Fish Wildl. Serv., 395: 16 p.

L160 _____ and P. S. Parker, 1960. The fishing-for-fun program on trout streams in Great Smoky Mountains National Park. Proc. Soc. Amer. Foresters, 1960: 106-11.

L161 Larsen, K., 1961. Fish populations in small Danish streams. Verh. Int. Ver. Limnol., 14: 769-72.

L162 Lord, R. F., Jr., 1930. Rearing a brood stock of blackspotted trout. Trans. Amer. Fish. Soc., 60: 164-66.

L163 Long, J. B. and L. E. Griffin, 1937. Spawning and migratory habits of the Columbia River steelhead trout as determined by scale studies. Copeia, 1937(1): 62.

L164 Leonard, J. W., 1938. Feeding habits of trout in waters carrying a heavy population of naturally hatched fry. Copeia, 1938(3): 144.

L165 _____ and F. A. Leonard, 1949. An analysis of the feeding habits of rainbow trout and lake trout in Birch Lake, Cass County, Michigan. Trans. Amer. Fish. Soc., 76: 301-14.

L166 Lagunov, I. I. and V. V. Azbelev, 1960. On the efficiency of natural propagation in Atlantic salmon (*Salmo salar* L.). Rapp. Cons. Int. Explor. Mer., 148: 45-7.

L167 Lishev, M. N., 1960. Some peculiarities of the population dynamics of the salmon stock of the eastern Baltic. Rapp. Cons. Int. Explor. Mer., 148: 72-5.

L168 Lord, R. F., Jr., 1934. Hatchery trout as foragers and game fish. Trans. Amer. Fish. Soc., 64: 339-45.

L169 Latta, W. C., 1962. Periodicity of mortality of brook trout during the first summer of life. Trans. Amer. Fish. Soc., 91(4): 408-11.

L170 _____, 1963. Semiannual estimates of natural mortality of hatchery brook trout in lakes. Trans. Amer. Fish. Soc., 92(1): 53-9.

L171 Logan, S. M., 1963. Winter observations on bottom organisms and trout in Bridge Creek, Montana. Trans. Amer. Fish. Soc., 92(2): 140-45.

L172 Larsen, K., 1960. The effect of the liberation of sea-trout fry in Gudena Area, as shown by the trout catch in the Lower River and the Randers Fjord. Rapp. Cons. Int. Explor. Mer., 148: 26-8.

L175 Lindsey, C. C., 1963. Sympatric occurrence of two species of humpback whitefish in Squanga Lake, Yukon Territory. J. Fish. Res. Bd. Can., 20(3): 749-67.

L176 _____ and T. G. Northcote, 1963. Life history of redside shiners, *Richardsonius balteatus*, with particular reference to movements in and out of Sixteenmile Lake streams. J. Fish. Res. Bd. Can., 20(4): 1001-30.

L177   Lindroth, A., 1963. The body/scale relationship in Atlantic salmon
        (*Salmo salar* L.). A preliminary report. J. Cons. Int. Explor. Mer.,
        28(1): 137-52.

L179   Leonard, J. W., 1939. Feeding habits of the Montana grayling (*Thy-
        mallus montanus* Milner) in Ford Lake, Michigan. Trans. Amer. Fish.
        Soc., 68: 188-95.

L180   _____, 1940. Further observations on the feeding habits of the Montana
        grayling *(Thymallus montanus)* and the bluegill *(Lepomis macrochirus)*
        in Ford Lake, Michigan. Trans. Amer. Fish. Soc., 69: 244-56.

L181   Lawler, G. H., 1964. Further evidence of hardiness of "silver" pike.
        J. Fish. Res. Bd. Can., 21(3): 651-2.

L182   Lindsey, C. C., 1953. Variation in anal fin ray count of the redside
        shiner, *Richardsonius balteatus* (Richardson). Canad. J. Zool., 31: 211-
        25.

L183   Lindquist, A. W., C. C. Deonier, and J. E. Hancy, 1943. The relation-
        ship of fish to the Clear Lake gnat in Clear Lake, California. Calif.
        Fish Game, 29(4): 196-202.

L184   Loeb, H. A., 1954. Experimental carp control. N. Y. State Conserv.,
        9(1): 10-11.

L185   _____, 1955. An electrical surface device for carp control and fish
        collection in lakes. N. Y. Fish Game J., 2(2): 220-31.

L186   _____, 1960. Reactions of aquarium carp to food and flavors. N. Y.
        Fish Game J., 7(1): 60-71.

L187   Lamby, K., 1941. Zur Fishereibiologie des Myvatn, Nord-Island.
        Cons. Int. Explor. Mer. Publ. Circ., 53: 7-174.

L188   Lagler, K. F. and F. V. Hubbs, 1940. Food of the long-nosed gar *(Le-
        pisosteus osseus oxyurus)* and the bowfin *(Amia calva* in southern
        Michigan. Copeia, 1940(4): 239-41.

L191   Laakso, M., 1955. Variability in scales of cutthroat trout in mountain
        lakes. Proc. Utah Acad. Sci. Arts Lett., 32: 81-7.

L192   _____, 1956. Body-scale regressions in juvenile cutthroat from Yellow-
        stone Lake. Proc. Utah Acad. Sci. Arts Lett., 33: 107-11.

L193   LaRivers, I., 1962. Fishes and fisheries of Nevada. Nev. St. Fish
        Game Comm. 781 p.

L194   Lennon, R. E., 1954. Feeding mechanism of the sea lamprey and its
        effect on host fishes. Fish. Bull. U. S., 56(98): 247-93.

L195   Linton, T. L., 1961. A study of fishes of the Arkansas and Cimarron
        Rivers in the area of the proposed Keystone Reservoir. Okla. Fish.
        Res. Lab. Rep., 81: 30 p.

L196   Lagler, K. F., C. B. Obrecht, and G. V. Harry, 1942. The food and
        habits of gars (*Lepisosteus* spp.) considered in relation to fish manage-
        ment. Invest. Ind. Lakes Streams, 2: 117-35.

L197   _____ and V. C. Applegate, 1942. Further studies on the food of the
        bowfin *(Amia calva)* in southern Michigan with notes on the inadvisabil-
        ity of using trapped fish in food analyses. Copeia, 1942: 190-91.

L198   Langford, R. R., 1938. The food of the Lake Nipissing cisco, *Leuci-
        chthys artedi* (LeSueur) with special reference to the utilization of the
        limnetic crustacea. Ont. Fish. Res. Lab., 45: 145-90.

L199   Lawler, G. H., 1961. Egg counts of Lake Erie whitefish. J. Fish. Res.
        Bd. Can., 18(2): 293-4.

L200   Lowry, G. R., 1966. Production and food of cutthroat trout in three
        Oregon coastal streams. J. Wildl. Mgmt., 30(4): 754-67.

L201   _____, 1965. Movement of cutthroat trout, *Salmo clarkii clarkii*

(Richardson) in three Oregon coastal streams. Trans. Amer. Fish. Soc., 94(4): 334-8.

L202   LeBrasseur, R. J., 1966. Stomach contents of salmon and steelhead trout in the northeastern Pacific Ocean. J. Fish. Res. Bd. Can., 23(1): 85-100.

L203   Lindroth, A., 1965. First winter mortality of Atlantic salmon parr in the hatchery. Canad. Fish Cult., 36: 23-26.

L204   _____, 1963. Salmon conservation in Sweden. Trans. Amer. Fish. Soc., 92(3): 286-91.

L205   Lane, E. D., 1964. Brown trout *(Salmo trutta)* in the Hinds River. Proc. N. Z. Ecol. Soc., 11: 10-16.

L206   Lake, J. S., 1957. Trout population and habitats in New South Wales. Aust. J. Mar. Freshw. Res., 8: 414-50.

L207   Latta, W. C., 1965. Relationship of young-of-the-year trout to mature trout and ground water. Trans. Amer. Fish. Soc., 94(1): 32-9.

L208   Lynch, D. D., 1966. Inland Fisheries Commission Report for year ending 30th June, 1966. Hobart, Tasmania, Australia. 19 p.

L209   Lindström, T., 1965. Char and whitefish recruitment in north Swedish lake reservoirs. Rep. Inst. Freshw. Res. Drottning., 46: 124-40.

L210   Lindsey, C. C., 1964. Problems in zoogeography of the lake trout, *Salvelinus namaycush*. J. Fish Res. Bd. Can., 21(5): 977-94.

L211   Lawler, G. H., 1966. Pugheadedness in perch, *Perca flavescens*, and pike, *Esox lucius*, of Heming Lake, Manitoba. J. Fish. Res. Bd. Can., 23(11): 1807-8.

L212   Laventer, H. and Z. Perah, 1966. Preliminary observations on late-spawnings of carp. Bamidgeh, 18(2): 31-6.

L213   LeCren, E. D., 1951. The length-weight relationship and seasonal cycle in gonad weight and condition in the perch *(Perca fluviatilis)*. J. Anim. Ecol., 20(2): 201-19.

L214   LaFaunce, D. A., 1965. Long-term retention of tags by some freshwater fish. Calif. Fish Game, 51(1): 52-3.

L215   Loeb, H. A., 1964. Submergence of brown bullheads in bottom sediment. N. Y. Fish Game J., 11(2): 119-24.

L216   Lambou, V. W., 1961. Efficiency and selectivity of flag gillnets fished in Lake Bistineau, Louisiana. Proc. S. E. Assoc. Game Fish Comm., 15: 319-59.

L217   Lewis, W. M., M. Anthony, and D. R. Helms, 1965. Selection of animal forage to be used in the culture of channel catfish. Proc. S. E. Assoc. Game Fish Comm., 17: 364-7.

L219   Langemeier, R. N., 1964. Effects of channelization on growth and food habits of the flathead catfish in the Missouri River. Midw. Wildl. Conf. Program Abstr., 26: 52-3.

L220   Lindsey, C. C., 1962. Observations on meristic variation in ninespine sticklebacks, *Pungitius pungitius*, reared at different temperatures. Canad. J. Zool., 40: 1237-47.

L221   Laurent, P-J., 1965. Que deviennent les truitelles arc-en-ciel lâchées dans le Léman? Pêcheur et Chasseur suisses, 1965(Sept.). Reprint.

M1   McCabe, B. C., 1942. Section 3. Fishes. Mass. Dept. Conserv. Fish. Surv. Rep., 1942: 30-68.

M2   McCay, C. M. and staff Cortland Hatchery, 1940. Experimental studies of the growth of brook trout. Progr. Fish Cult., 49: 27-30.

M3   _____, F. C. Bing, and W. S. Dilley, 1927. The effect of variations in vitamins, protein, fat and mineral matter in the diet upon the growth and mortality of eastern brook trout. Trans. Amer. Fish. Soc., 57: 240-49.

M4   McCay, C. M. and W. E. Dilley, 1927. Factor H in the nutrition of trout. Trans. Amer. Fish. Soc., 57: 250-60.

M5   _____, L. A. Maynard, J. W. Titcomb, and M. F. Crowell, 1930. Influence of water temperature upon the growth and reproduction of brook trout. Ecology, 11(1): 30-34.

M6   _____ and A. V. Tunison, 1937. Cortland (N.Y.) Hatchery Rep. No. 5, for the year 1936. 3-18.

M7   McDonald, M., 1891. Report of the Commissioner. U. S. Fish Comm. Rep. 1887, 15: I-LXIII.

M8   _____, 1892. Report of the U. S. Commission of Fish and Fisheries for the fiscal year ending June 30, 1889. Rep. U. S. Fish Comm., 16: ix-xxxix.

M9   _____, 1893. Report of the U. S. Commission of Fish and Fisheries for the fiscal year 1889-90 and 1890-91. Rep. U. S. Fish Comm., 17: 1-96.

M10   _____, 1894. Report of the U. S. Commission of Fish and Fisheries for the fiscal year ending June 30, 1892. Rep. U. S. Fish Comm., 18: VII-LXXXVII.

M11   Macfarlane, P. R. C., 1928. Salmon *(Salmo salar)* of the River Moisie (eastern Canada), 1926-27. Proc. Roy. Soc. Edinb., 48(2): 134-39.

M12   _____, 1931. Salmon of the River Dee (Kirkcudbright), 1928 and 1929. Salm. Fish. Scot., 1931(2): 1-23.

M13   _____, 1938. Salmon of the Upper Solway district, 1934. Salm. Fish. Scot., 1938(3): 1-18.

M15   McHugh, J. L., 1939. The whitefishes, *Coregonus clupeaformis* (Mitchill) and *Prosopium williamsoni* (Girard), of the lakes of the Okanagan Valley, British Columbia. Bull. Fish. Res. Bd. Can., 56: 39-50.

M16   _____, 1941. Growth of the Rocky Mountain whitefish. J. Fish. Res. Bd. Can., 5(4): 337-43.

M18   MacKay, H. H. and W. H. R. Werner, 1934. Some observations on the culture of maskinonge. Trans. Amer. Fish. Soc., 64: 313-17.

M20   MacKenthun, K. M., 1948. Age-length and length-weight relationship of southern area lake fishes. Wis. Conserv. Dept. Fish. Biol. Invest. Rep., 586(Rev.): 1-7.

M22   McKenzie, R. A., 1947. The effect of crowding of smelt eggs on the production of larvae. Fish. Res. Bd. Can. Progr. Rep. Atlant. Res. Sta., 39: 11-13.

M23   Markus, H. C., 1934. Life history of blackheaded minnow *(Pimephales promelas)*. Copeia, 1934(3): 116-22.

M24   _____, 1934. The fate of our forage fish. Trans. Amer. Fish. Soc., 64: 93-6.

M25   _____, 1939. Propagation of bait and forage fish. Fish. Circ. U. S., 28: 1-19.

M26   Marshall, N., 1939. Annulus formation in scales of the common shiner, *Notropis cornutus chrysocephalus* (Rafinesque). Copeia, 1939(3): 148-54.

M27   _____, 1947. Studies on the life history and ecology of *Notropis chalybaeus* (Cope). Quart. J. Fla. Acad. Sci., 9(3-4).

M28   Martin, W. R., 1941. Rate of growth of the ling, *Lota lota maculosa* (LeSueur). Trans. Amer. Fish. Soc., 70: 77-9.

M29   Mather, F., 1889. Brown trout in America. Bull. U. S. Fish Comm., 7: 21-2.

M30 Meehan, W. E., 1907. The shad work on the Delaware River in 1907 and its lessons. Trans. Amer. Fish. Soc., 36: 105-18.

M35 Meigs, R. C. and C. F. Pautzke, 1940. Studies on the life history of the Puget Sound steelhead (*Salmo gairdnerii*). Wash. Dept. Game Biol. Bull., 3: 1-24.

M36 _____ and _____, 1941. Additional notes on the life history of the Puget Sound steelhead (*Salmo gairdnerii*). Wash. Dept. Game Biol. Bull., 5: 1-13.

M37 Mellen, I. M., 1923. Whitefishes reared in the New York Aquarium. Zoologica, 2(17): 375-80.

M38 Menzel, R. W., 1945. The catfish fishery of Virginia. Trans. Amer. Fish. Soc., 73: 364-72.

M39 Menzies, W. J. M., 1921. Notes on the salmon of Thurso Bay: May-September, 1920. Salm. Fish. Scot., 1921(2): 13-22.

M40 _____, 1922. Salmon investigations in Scotland, 1921. I. Salmon of the River Dee (Aberdeenshire). II. Salmon of the River Spey. Salm. Fish. Scot., 1921(I and II): 48 and 57 p.

M41 _____, 1923. Salmon investigations in Scotland, 1921. I. Salmon of the River Don. II. Salmon of the River Findhorn. III. Salmon of the River Forth. IV. Salmon of the River Tweed. Salm. Fish. Scot., 1921(3): 14 p.

M42 _____, 1923. Salmon investigations in Scotland, 1921. IV. Summary of results. Salm. Fish. Scot., 1921(4): 18 p.

M43 _____, 1925. The salmon: its life story. Wm. Blackwood and Sons, Edinb. and Lond. 211 p.

M44 _____, 1926. Salmon (*Salmo salar*) of the River Moisie (eastern Canada). Proc. Roy. Soc. Edinb., 45: 334-45.

M46 _____, 1928. Salmon of the River Conon, 1927. Salm. Fish. Scot., 1928(8): 1-16.

M47 _____, 1927. Some aspects of the growth of salmon in river and sea. Salm. Fish. Scot., 1927(1): 1-23.

M48 _____, 1927. Salmon of the Grimersta District, Lewis, 1925. Salm. Fish. Scot., 1926(6): 1-14.

M49 _____ and P. R. C. Macfarlane, 1926. Salmon of the River Dee (Aberdeenshire). Salm. Fish. Scot., 1926(4): 1-46.

M50 _____ and _____, 1926. Salmon of the River Spey, 1923. Salm. Fish. Scot., 1926(5): 1-36.

M51 _____ and _____, 1927. Salmon of the River Dee (Aberdeenshire), 1924. Salm. Fish. Scot., 1927(3): 1-30.

M52 _____ and _____, 1928. Salmon of the River Spey, 1925. Salm. Fish. Scot., 1928(3): 1-18.

M53 _____ and _____, 1928. Salmon of the River Spey, 1924. Salm. Fish. Scot., 1928(1): 1-24.

M54 _____ and _____, 1928. Some further notes on the salmon (*Salmo salar*) of the Moisie River. Proc. Roy. Soc. Edinb., 47(4): 359-65.

M55 _____ and _____, 1932. Salmon of the River Dee (Aberdeenshire). Salm. Fish. Scot., 1931(4): 1-22.

M56 Merriman, D., 1935. The effect of temperature on the development of the eggs and larvae of the cut-throat trout (*Salmo clarkii clarkii* Richardson). J. Exp. Biol., 12(4): 297-305.

M57 Miller, R. B., 1945. Effect of Triaenophorus on growth of two fishes. J. Fish. Res. Bd. Can., 6(4): 334-7.

M58   Miller, R. B., 1946. Notes on the Arctic grayling, *Thymallus signifer* Richardson, from Great Bear Lake. Copeia, 1946(4): 227-36.

M59   _____, 1946. The effects of ammonium nitrate fertilizer in trout rearing ponds. Canad. Fish Cult., 1(1): 13-17.

M60   _____, 1947. Great Bear Lake. Bull. Fish. Res. Bd. Can., 72: 31-44.

M61   _____, 1947. The effects of different intensities of fishing on the whitefish populations of two Alberta lakes. J. Wildl. Mgmt., 11(4): 289-301.

M62   _____ and W. A. Kennedy, 1948. Pike *(Esox lucius)* from four northern Canadian lakes. J. Fish. Res. Bd. Can., 7(4): 190-99.

M63   _____ and _____, 1948. Observations on the lake trout of Great Bear Lake. J. Fish. Res. Bd. Can., 7(4): 176-89.

M64   Miller, S., 1948. Do fish die of old age? Wis. Conserv. Bull., 13(11): 22.

M65   Milner, J. W., 1874. Report on the fisheries of the Great Lakes. The results of inquiries prosecuted in 1871 and 1872. Rep. U. S. Comm. Fish Fish., 1871-72(Append.A): 1-75.

M66   Moen, T. E., 1946. Why rough fish removal? Ia. Conserv., 6(8): 153, 156-7.

M67   Moffett, J. W., 1942. A fishery survey of the Colorado River below Boulder Dam. Calif. Fish Game, 28(2): 76-86.

M69   Moore, E., 1926. Culture of the maskinonge ("Muskellunge"). Annu. Rep. N. Y. St. Conserv. Comm., 1925: 131-7.

M71   _____, J. R. Greeley, C. W. Greene, H. M. Faigenbaum, F. R. Nevin, and H. K. Townes, 1934. A problem in trout stream management. Trans. Amer. Fish. Soc., 64: 68-80.

M72   Moore, G. A., 1944. Notes on the early life history of *Notropis girardi.* Copeia, 1944(4): 209-14.

M73   Moore, H. F., 1917. The burbot. U. S. Bur. Fish. Econ. Circ., 25: 1-4.

M74   Morris, A. G., 1939. Propagation of channel catfish. Progr. Fish Cult., 44: 23-7.

M76   Mottley, C. McC., 1932. The propagation of trout in the Kamloops district, British Columbia. Trans. Amer. Fish. Soc., 62: 144-51.

M77   _____, 1938. Loss of weight by rainbow trout at spawning time. Trans. Amer. Fish. Soc., 67: 207-10.

M78   _____, 1941. The effect of increasing the stock in a lake on the size and condition of rainbow trout. Trans. Amer. Fish. Soc., 70: 413-20.

M79   _____, 1947. The Kamloops rainbow trout. Fish. Leafl. Wash., 235: 1-3.

M86   Murphy, G. I., 1948. Notes on the biology of the Sacramento hitch *(Lavinia e. exilicauda)* of Clear Lake, Lake County, California. Calif. Fish Game, 34(3): 101-10.

M87   McCabe, B. C., 1946. Fisheries report for lakes of central Massachusetts, 1944-1945. Mass. Dept. Conserv. 254 p.

M88   Miller, R. B., 1949. Problems of the optimum catch in small whitefish lakes. Biometrics, 5(1): 14-26.

M89   Massachusetts Fish Commissioners, 1870. Pickerel. Report of the Commissioners of Fisheries for the year ending January 1, 1870. 36-37.

M90   Mottley, C. McC., 1940. The production of rainbow trout at Paul Lake, British Columbia. Trans. Amer. Fish. Soc., 69: 187-91.

M98   Martin, W. R., 1949. The mechanics of environmental control of body form in fishes. Publ. Ont. Fish. Res. Lab., 70: 1-91.

M99   McClane, A. J., (ed.), 1951. The Wise fishermen's encyclopedia. Wm. H. Wise and Co., N. Y. 1336 p.

M101   McHugh, J. L. and W. E. Barraclough, 1951. An abnormal carp, *Cyprinus carpio*, from California waters. Calif. Fish Game, 37(4): 391-3.

M103   McRae, R., 1921. Noteworthy growth of rainbow trout. U. S. Fish. Serv. Bull., 76: 4.

M108   Mayer, R., 1950. Trout and minnows. Piscator, 4(13): 2-6.

M109   Merriman, D. and Y. Jean, 1949. The capture of an Atlantic salmon *(Salmo salar salar)* in the Connecticut River. Copeia, 1949(3): 220-21.

M110   Millenbach, C., 1950. Rainbow brood-stock selection and observations on its application to fishery management. Progr. Fish Cult., 12(3): 151-2.

M112   Miller, N. J., 1952. Carp; control and utilization. Wis. Conserv. Bull., 17(5): 3-7.

M113   Miller, R. B., 1949. Preliminary biological surveys of Alberta watersheds, 1947-1949. Alta. Dept. Lands Forests. 139 p.

M114   Mississippi Game and Fish, 1951. Huge gar taken at Moon Lake. Miss. Game Fish, 14(8): 10.

M115   Missouri Conservationist, 1939. Scale study reveals rate of growth of Lake of Ozarks fish. Mo. Conserv., 2(1): 6.

M122   Missouri River Basins Studies, 1950. Reservoir fishery investigations for summer, 1949, Deerfield Reservoir, South Dakota. U. S. Fish Wildl. Serv., Billings, Mont. 24 p. mimeo.

M130   Moen, T. E., 1950. Notes on the growth of Lost Island Lake bullheads. Biol. Seminar Ia. Conserv. Comm., 1950(Oct. 10). 11 p. mimeo.

M132   Monson, M. A. and J. Greenbank, 1947. Size and maturity of hackleback sturgeon. Upp. Miss. R. Conserv. Comm. Tech. Commit. Fish. Progr. Rep., 3: 42-5.

M133   Moore, G. A. and F. B. Cross, 1950. Additional Oklahoma fishes with validation of *Poecilichthys parvipinnis* (Gilbert and Swain). Copeia, 1950(2): 139-48.

M136   Moyle, J. B., 1952. Age and growth of fishes. Conserv. Bull. Minn., 15(86): 14-17.

M139   Murphy, G. I., 1950. The life history of the greaser blackfish *(Orthodon microlepidotus)* of Clear Lake, Lake County, California. Calif. Fish Game, 36(2): 119-33.

M140   _____, 1951. The fishery of Clear Lake, Lake County, California. Calif. Fish Game, 37(4): 439-84.

M141   McConnell, W. J., 1952. The opercular bone as an indicator of age and growth of the carp, *Cyprinus carpio* Linnaeus. Trans. Amer. Fish. Soc., 81: 138-49.

M142   Martin, N. V., 1952. A study of the lake trout, *Salvelinus namaycush*, in two Algonquin Park, Ontario, lakes. Trans. Amer. Fish. Soc., 81: 111-37.

M145   Miller, R. B., 1952. Survival of hatchery-reared cutthroat trout in an Alberta stream. Trans. Amer. Fish. Soc., 81: 35-42.

M146   McCabe, B. C., 1953. Fisheries report for lakes and ponds of north central Massachusetts (1950). Mass. Div. Fish Game. 122 p.

M147   _____, 1958. Tabular treatment of the life history and ecology of the chain pickerel. Springfield College, Mass. 45 p. mimeo.

M149   McCarraher, D. B., 1957. The natural propagation of northern pike in small drainable ponds. Progr. Fish Cult., 19(4): 185-7.

M150   McCarraher, D. B., 1959. The northern pike-bluegill combination in north-central Nebraska farm ponds. Progr. Fish Cult., 21(4): 188-9.

M151   McConnell, W. J., 1953. Why age carp? Utah Farm and Home Sci., 14(1): 6-7, 23.

M152   ____, W. J. Clark, and W. F. Sigler, 1957. Bear Lake—its fish and fishing. Utah Dept. Fish Game Publ. 76 p.

M153   McCoy, H. A., 1955. The rate of growth of flathead catfish in twenty-one Oklahoma lakes. Proc. Okla. Acad. Sci., 34: 47-52.

M154   McCrimmon, H. R., 1954. Stream studies on planted Atlantic salmon. J. Fish. Res. Bd. Can., 11(4): 362-403.

M155   ____, 1958. Observations on the spawning of lake trout, *Salvelinus namaycush*, and the post-spawning movement of adult trout in Lake Simcoe. Canad. Fish Cult., 23: 3-13.

M156   ____, 1959. Observations on spawning of burbot in Lake Simcoe, Ontario. J. Wildl. Mgmt., 23(4): 447-9.

M157   ____ and O. E. Devitt, 1954. Winter studies on the burbot *(Lota lota lacustris)* of Lake Simcoe. Canad. Fish Cult., 16: 35-41.

M161   McFadden, J. T., 1959. Relationship of size and age to time of annulus formation in brook trout. Trans. Amer. Fish. Soc., 88(3): 176-7.

M162   McKenzie, R. A., 1958. Age and growth of smelt, *Osmerus mordax* (Mitchill), of the Miramichi River, New Brunswick. J. Fish. Res. Bd. Can., 15(6): 1313-27.

M165   McMynn, R. G. and P. A. Larkin, 1953. The effects on fisheries of present and future water utilization in the Campbell River drainage area. B. C. Game Comm. Mgmt. Publ., 2: 61 p.

M166   MacOnie, J., 1938. Dwarf rainbow trout. Salm. Trout Mag., 92: 236-40.

M167   Maher, F. B. and P. A. Larkin, 1955. Life history of the steelhead trout of the Chilliwach River, B. C. Trans. Amer. Fish. Soc., 84: 27-38.

M173   Mansueti, R. J. and H. Kolb, 1953. A historical review of the shad fisheries of North America. Chesapeake Biol. Lab. Publ., 97: 293 p.

M175   Martin, M., 1954. Age and growth of the goldeye, *Hiodon alosoides* (Rafinesque), of Lake Texoma, Oklahoma. Proc. Okla. Acad. Sci., 33: 37-49.

M176   Martin, N. V., 1955. Limnological and biological observations in the region of the Ungava or Chubb Crater, Province of Quebec. J. Fish. Res. Bd. Can., 12(4): 487-98.

M179   Marvich, E. S., A. H. McRea, R. J. Simon, and W. J. Cahill, 1953. Sport fish. Annu. Rep. Alaska Fish. Bd. and Alaska Dept. Fish. 1953, 5: 55-66.

M180   ____, ____, and ____, 1955. Sport fish. Alaska Dept. Fish. Rep., 6: 58-70.

M181   ____, ____, and ____, 1955. Sport fish. Annu. Rep. Alaska Fish. Bd. and Alaska Dept. Fish., 7: 107-23.

M183   Marzolf, R. C., 1955. Use of pectoral spines and vertebrae for determining age and rate of growth of the channel catfish. J. Wildl. Mgmt., 19(2): 243-8.

M185   Maxfield, G. H., 1953. The food habits of hatchery-produced pond-cultured shad *(Alosa sapidissima)* reared to a length of two inches. Chesapeake Biol. Lab. Publ., 98.

M186   May, O. D., Jr., n.d. (1955?). Experiments with commercial fishing traps in the Oconee, Ocmulgee and Altamaha Rivers of Georgia. Ga. Game Fish Comm. D.-J. Proj., F-1-R-3: 11 p.

M187   May, O. D., Jr., n.d. (1956?). An evaluation of two years commercial trapping in the Oconee, Ocmulgee and Altamaha Rivers of Georgia. Ga. Game Fish Comm. D.-J. Proj., F-4-R-2: 16 p.

M189   Mayhew, J., 1957. The occurrence of blindness in the black bullhead, *Ictalurus melas* (Rafinesque), of East Okoboji Lake, Iowa. Proc. Ia. Acad. Sci., 64: 654-6.

M195   ____, 1957. Population studies—fish population of a southern Iowa artificial lake. Ia. Conserv. Comm. Quart. Biol. Rep., 9(1): 1-5.

M196   ____, 1957. Notes on the observations of the reproductive habits of the gizzard shad in two Iowa lakes. Ia. Conserv. Comm. Quart. Biol. Rep., 10(1): 20-22.

M197   ____, 1958. The fish population of a southern Iowa artificial lake. Proc. Ia. Acad. Sci., 65: 565-70.

M204   Menzies, W. J. M., 1929. Some Newfoundland salmon. Salm. Trout Mag., 54: 64-7.

M205   Meredith, W. G. and F. J. Schwartz, 1959. Summer food of the minnow, *Notropis a. ardens*, in the Roanoke River, Virginia. (Abstract) Proc. W. Va. Acad. Sci., 29 + 30: 23.

M207   Miller, R. B., 1954. Comparative survival of wild and hatchery-reared cutthroat trout in a stream. Trans. Amer. Fish. Soc., 83: 120-30.

M209   ____, 1956. The collapse and recovery of a small whitefish fishery. J. Fish. Res. Bd. Can., 13(1): 135-46.

M210   ____ and R. C. Thomas, 1957. Alberta's "Pothole" trout fisheries. Trans. Amer. Fish. Soc., 86: 261-8.

M212   Milne, J. A., 1921. A curious salmon. Salm. Trout Mag., 24(24): 240-41.

M213   ____, 1924. Age and growth of some brown and rainbow trout in Blagdon Lake. Salm. Trout Mag., 38: 17-25.

M214   ____, 1925. Some Sebago salmon scales from the Argentine. Salm. Trout Mag., 40: 268-73.

M215   Minckley, W. L., 1959. Fishes of the Big Blue River Basin, Kansas. Univ. Kans. Publ. Mus. Nat. Hist., 11(7): 401-42.

M225   Moen, T. E., 1953. Summary of hatchery studies at Spirit and Clear Lake hatcheries, spring of 1953. Ia. St. Conserv. Comm. Quart. Biol. Rep., 5(2): 70-76.

M228   ____, 1954. Food habits of the carp in northwest Iowa lakes. Proc. Ia. Acad. Sci., 60: 665-86.

M229   ____, 1955. Summary of hatchery studies, Spring, 1955. Ia. St. Conserv. Comm. Quart. Biol. Rep., 6(4): 15-17.

M230   ____, 1956. Age and growth of gizzard shad, *Dorosoma cepedianum* (LeSueur), in Black Hawk Lake. Ia. St. Conserv. Comm. Biol. Quart. Rep., 8(1): 42-4.

M232   ____, 1958. Notes on the fish population of Center Lake. Ia. St. Conserv. Comm. Quart. Biol. Rep., 5(4): 15-18.

M234   ____, 1959. Sexing channel catfish. Trans. Amer. Fish. Soc., 88(2): 149-50.

M236   ____ and M. Lindquist, 1954. The northern pike hatch at the Clear Lake (Iowa) hatchery. Progr. Fish Cult., 16(2): 89-90.

M238   Montgomery, M. L., 1957. River basin investigation. Annu. Rep. Ore. St. Game Comm. Fish. Div., 1956: 120-33.

M242   Moody, H. L., 1957. A fishery study of Lake Panasoffkee, Florida. Quart. J. Fla. Acad. Sci., 20(1): 21-88.

M243   Moore, E., 1922. The primary sources of food of certain food and game, and bait fishes, of Lake George. A biological survey of Lake George, New York. N. Y. St. Conserv. Comm. 52-63.

M246   Moorman, R. B., 1957. Reproduction and growth of fishes in Marion County, Iowa, farm ponds. Ia. St. J. Sci., 32(1): 71-88.

M249   Morrow, J. E., 1957. Fish records from Long Island Sound. Copeia, 1957(3): 240-41.

M251   Moss, D. D., 1956. The effects of the slider turtle on production of fish in farm ponds. Proc. S. E. Assoc. Game Fish Comm., 1955: 97-100.

M252   Mottram, J. C., 1936. Experiments on the starvation of trout fry. Salm. Trout Mag., 83: 264-67.

M258   Moyle, J. B. and C. R. Burrows, 1954. Manual of instructions for lake survey. Minn. Bur. Fish. Fish Res. Unit Spec. Publ., 1: 70 p.

M259   _____ and W. D. Clothier, 1959. Effects of management and winter oxygen levels on the fish population of a prairie lake. Trans. Amer. Fish. Soc., 88(3): 178-85.

M265   Mraz, D. and E. L. Cooper, 1957. Natural reproduction and survival of carp in small ponds. J. Wildl. Mgmt., 21(1): 66-9.

M266   _____ and _____, 1957. Reproduction of carp, largemouth bass, blue-gills and black crappies in small rearing ponds. J. Wildl. Mgmt., 21(2): 127-33.

M270   Mullan, J. W., 1958. The sea-run for "salter" brook trout *(Salvelinus fontinalis)* fishery of the coastal streams of Cape Cod, Massachusetts. Mass. Div. Fish. Game Bull., 17: 1-25.

M273   _____, 1959. Observations on three reclaimed trout ponds in Massachusetts. Progr. Fish Cult., 21(3): 121-31.

M275   Müller, K., 1953. Die Schuppenmessbildungen bei der Forelle, *Salmo trutta* L. und eine Deutung dieser Erscheinung. Rep. Inst. Freshw. Res. Drottning., 34: 78-89.

M276   _____, 1954. Produktionsbiologische Untersuchungen in Nordschwedischen Fliessgewassern. Teil 2. Untersuchungen über Verbreitung, Bestanddichte Wachstum und Ernährung der Fische der Nordschwedischen Waldregion. Rep. Inst. Freshw. Res. Drottning., 35: 149-83.

M277   Muller, W., 1958. Das Wachstum der Quappe (*Lota lota* L.) im Oderhaff und in deutschen Gewassern. Verh. Int. Ver. Limnol., 13(Part2): 743-8.

M279   Muncy, R. J., 1959. Age and growth of channel catfish from the Des Moines River, Boone County, Iowa, 1955 and 1956. Ia. St. J. Sci., 34(2): 127-37.

M281   Munro, W. R., 1957. The pike of Loch Choin. Scot. Freshw. Salm. Fish. Res., 16: 3-16.

M283   Murray, A. R., 1958. Preliminary biology of Atlantic salmon of the Little Codroy River, Newfoundland. Fish. Res. Bd. Can. Progr. Rep. Pacif. Coast Sta., 68: 20-27.

M287   McCann, J. A., 1960. Estimates of the fish populations of Clear Lake, Iowa. PhD thesis Ia. St. Univ. 163 p. typewritten.

M290   _____, 1959. Life history studies of the spotted shiner of Clear Lake, Iowa, with particular reference to some sampling problems. Trans. Amer. Fish. Soc., 88(4): 336-43.

M291 Minckley, W. L. and J. E. Deacon, 1959. Biology of the flathead catfish in Kansas. Trans. Amer. Fish. Soc., 88(4): 344-55.

M292 McCraig, R. S. and J. W. Mullan, 1960. Growth of eight species of fishes in Quabbin Reservoir, Massachusetts, in relation to age of reservoir and introduction of smelt. Trans. Amer. Fish. Soc., 89(1): 27-31.

M293 McCarraher, D. B., 1960. Pike hybrids (*Esox lucius* x *E. vermiculatus*) in a sandhill lake, Nebraska. Trans. Amer. Fish. Soc., 89(1): 82-3.

M294 McCombie, A. M. and F. E. J. Fry, 1960. Selectivity of gill nets for lake whitefish, *Coregonus clupeaformis*. Trans. Amer. Fish. Soc., 89(2): 176-84.

M295 Moen, T. E., 1959. Notes on the growth of bullheads. Ia. St. Conserv. Comm. Quart. Biol. Rep., 11(3): 29-31.

M296 McCraig, R. S., J. W. Mullan, and C. O. Dodge, 1960. Five-year report on the development of the fishery of a 25,000 acre domestic water supply reservoir in Massachusetts. Progr. Fish Cult., 22(1): 15-23.

M299 Micklus, R. C. and J. G. Hale, 1959. Fishing harvest and population estimate for brook trout, Duck Lake, Lake County. Minn. Dept. Conserv. Invest. Rep., 223: 6 p. mimeo.

M300 Miller, R. B., 1950. Observations on mortality rates in fishes and unfished cisco populations. Trans. Amer. Fish. Soc., 79: 180-86.

M302 Moen, T. E., 1960. Length-weight tables for fishes from northwest Iowa lakes. Typewritten ms.

M303 MacPhee, C., 1960. Postlarval development and diet of the largescale sucker, *Catostomus macrocheilus*, in Idaho. Copeia, 1960(2): 119-25.

M305 Meyer, F. P., 1960. Life history of *Marsipometra hastata* and the biology of its host, *Polyodon spathula*. PhD thesis Ia. St. Univ. Ames. 145 p.

M312 McReynolds, H., 1952. An age and growth study of the golden shiner *(Notemigonus crysoleucas auratus)* from Lower Loch Alpine, Michigan. MS thesis Univ. Mich. 11 p.

M313 Matveeva, R. R., 1955. The nutrition of the young pike in the spawning-cultivation establishment in 1953. (Pitanie molodi sudaka v nerostovo-vyrostnom khozyaistne v 1953.) Vop. Ikhtiol., 5: 61-70; Biol. Abstr., 32(7): 22146.

M314 McClellan, W. G., 1954. A study of the southern spotted channel catfish, *Ictalurus punctatus* (Rafinesque). MS thesis N. Tex. St. Coll. 69 p.

M315 Martin, N. V. and N. S. Baldwin, 1960. Observations on the life history of the hybrid between eastern brook trout and lake trout in Algonquin Park, Ontario. J. Fish. Res. Bd. Can., 17(4): 541-51.

M316 Matsushima, M., 1956. Spawning activity of wild gold-fish in Lake Sagami Reservoir, observed through the eggs deposited on "fish-nest" and the ovarian eggs. (Japanese, with Eng. summary.) Bull. Freshw. Fish. Res. Lab. Tokyo, 6(1): 1-20.

M319 Maryland Tidewater News, 1957. Alligator gar. Md. Tidewater News, 4(3): 6.

M324 Magnuson, J. J., 1958. Some phases of the life history of troutperch, *Percopsis omisomaycus* (Walbaum), in Lower Red Lake, Minnesota. MS thesis Univ. Minn. 104 p.

M325 Miller, R. G., 1951. The natural history of Lake Tahoe fishes. PhD thesis Stanford Univ. 160 p.

M326 Miller, R. R., 1960. Systematics and biology of the gizzard shad

(Dorosoma cepedianum) and related fishes. Fish. Bull. U. S., 60(173): 371-92.

M327　Moffett, J. W., 1957. Recent changes in the deep-water fish populations of Lake Michigan. Trans. Amer. Fish. Soc., 86: 393-408.

M328　McCammon, G. W. and D. A. LaFaunce, 1961. Mortality rates and movement in the channel catfish population of the Sacramento Valley. Calif. Fish Game, 47(1): 5-26.

M329　Muir, B. S., 1960. Comparison of growth rates for native and hatchery stocked populations of *Esox masquinongy* in Nogies Creek, Ontario. J. Fish. Res. Bd. Can., 17(6): 919-27.

M330　McCrimmon, H. R., 1960. Observations on the standing trout populations and experimental plantings in two Ontario streams. Canad. Fish Cult., 28: 45-55.

M334　_____ and A. H. Berst, 1961. An analysis of sixty-five years fishing in a trout pond unit. J. Wildl. Mgmt., 25(2): 168-78.

M335　McCammon, G. W. and C. M. Seeley, 1961. Survival, mortality, and movements of white catfish and brown bullheads in Clear Lake, California. Calif. Fish Game, 47(3): 237-55.

M337　Merrell, T. R., 1961. Unusual white sturgeon diet. Res. Briefs Ore. Fish. Comm., 8(1): 77.

M340　McDonald, D. B. and P. A. Dotson, 1960. Fishery investigations of the Glen Canyon and Flaming Gorge impoundment areas. Utah St. Dept. Fish Game Info. Bull., 60-3: 70 p.

M348　Mallet, J., 1961. Middle Fork of Salmon River trout fisheries investigations. Idaho Dept. Fish Game D.-J. Completion Rep., Proj. F-37-R-2: 66 p.

M350　Montgomery, M. L., 1960. Bend District. Annu. Rep. Ore. St. Game Comm. Fish. Div., 1960: 148-83.

M351　Moss, D. D. and D. C. Scott, 1961. Dissolved-oxygen requirements of three species of fish. Trans. Amer. Fish. Soc., 90(4): 377-93.

M353　McCammon, G. W. and D. A. LaFaunce, 1961. Mortality rates and movement in the channel catfish population of the Sacramento Valley. Calif. Fish Game, 47(1): 5-26.

M354　Miller, R. V., 1961. The food habits and some aspects of the biology of the threadfin shad, *Dorosoma petenense* (Gunther). MS thesis Univ. Ark. 46 p.

M356　McCombie, A. M., 1961. Gill-net selectivity of lake whitefish from Goderich-Bayfield area, Lake Huron. Trans. Amer. Fish. Soc., 90(3): 337-40.

M360　McFadden, J. T. and E. L. Cooper, 1962. An ecological comparison of six populations of brown trout *(Salmo trutta)*. Trans. Amer. Fish. Soc., 91(1): 53-62.

M362　_____, _____, and J. K. Andersen, 1962. Sex ratios in wild populations of brown trout. Trans. Amer. Fish. Soc., 91(1): 94-5.

M363　Minckley, W. L., 1962. Spring foods of juvenile blue catfish from the Ohio River. Trans. Amer. Fish. Soc., 91(1): 95.

M364　Muncy, R. J., 1960. A study of the comparative efficiency between nylon and linen gillnets. Chesapeake Sci., 1(2): 96-102.

M367　Mills, D. H., 1959. I. (2) Salmon smolt tagging experiments at Invergarry salmon hatchery. Rapp. Cons. Int. Explor. Mer., 148: 8-10.

M370　McCormick, J. H., 1960. A preliminary study of hematology as a diagnostic tool in fisheries medicine. Quart. Rep. Colo. Coop. Fish. Res. Unit, 6: 53-60.

M371 McCrimmon, H. R. and A. H. Berst, 1961. The native fish population and trout harvests in an Ontario farm pond. Progr. Fish Cult., 23(3): 106-13.

M375 McFadden, J. T., 1961. A population study of the brook trout, *Salvelinus fontinalis*. Wildl. Monogr., 7: 73 p.

M377 Meyer, F. P. and J. H. Stevenson, 1962. Studies on the artificial propagation of the paddlefish. Progr. Fish Cult., 24(2): 65-7.

M378 Morton, K. E., 1962. Experimental heating of pond water to start rainbow trout fry on a dry diet. Progr. Fish Cult., 24(2): 94-6.

M379 Munro, W. R., 1961. The effect of mineral fertilizers on the growth of trout in some Scottish lochs. Verh. Int. Ver. Limnol., 14: 718-21.

M380 Meister, A. L., 1962. Atlantic salmon production in Cover Brook, Maine. Trans. Amer. Fish. Soc., 91(2): 208-12.

M381 Miller, R. B., 1952. Relative strengths of whitefish year classes as affected by egg plantings and weather. J. Wildl. Mgmt., 16(1): 39-49.

M382 Mansueti, R. J., 1962. Distribution of small, newly metamorphosed sea lampreys, *Petromyzon marinus*, and their parasitism on menhaden, *Brevoortia tyrannus*, in Mid-Chesapeake Bay during winter months. Chesapeake Sci., 3(2): 137-9.

M383 Meyers, C. D. and R. J. Muncy, 1962. Summer food and growth of chain pickerel, *Esox niger*, in brackish waters of the Severn River, Maryland. Chesapeake Sci., 3(2): 125-8.

M384 Moore, R. L., 1953. Kokanee in Colorado. Colo. Conserv., 2(3): 19-23.

M385 McHugh, J. L., 1940. Food of the Rocky Mountain whitefish, *Prosopium williamsoni* (Girard). J. Fish. Res. Bd. Can., 5(2): 131-7.

M386 Miller, R. J., 1962. Reproductive behavior of the stoneroller minnow, *Campostoma anomalum pullum*. Copeia, 1962(2): 407-17.

M387 Miller, J. G., 1962. Occurrence of ripe chain pickerel in the fall. Trans. Amer. Fish. Soc., 91(3): 323.

M388 McCarraher, D. B., 1962. Northern pike, *Esox lucius*, in alkaline lakes of Nebraska. Trans. Amer. Fish. Soc., 91(3): 326-9.

M390 Metzelaar, J., 1928. The food of the rainbow trout in Michigan. Trans. Amer. Fish. Soc., 58: 178-82.

M391 Moore, G. A., 1937. The germ cells of the trout (*Salmo irideus* Gibbons). Trans. Amer. Microscop. Soc., 56(1): 105-12.

M392 Maciolek, J. A. and P. R. Needham, 1952. Ecological effects of winter conditions on trout and trout foods in Convict Creek, California, 1951. Trans. Amer. Fish. Soc., 81: 202-17.

M393 Mansueti, R. J., 1962. Eggs, larvae, and young of the hickory shad, *Alosa mediocris*, with comments on its ecology in the estuary. Chesapeake Sci., 3(3): 173-205.

M395 Meyer, W. H., 1962. Life history of three species of redhorse (*Moxostoma*) in the Des Moines River, Iowa. Trans. Amer. Fish. Soc., 91(4): 412-19.

M396 Markus, H. C., 1962. Hatchery-reared Atlantic salmon smolts in ten months. Progr. Fish Cult., 24(3): 127-30.

M397 Mill, C. K., 1961. Report for the year ended 31st December, 1961. Salm. Res. Trust Ire., Inc. 3-11.

M398 McFarland, W. L., 1925. Salmon of the Atlantic. Parke, Austin and Lipscomb, N. Y. 156 p.

M400 Määr, A., 1949. Fertility of char (*Salmo alpinus* L.) in the Faxälven water system, Sweden. Rep. Inst. Freshw. Res. Drottning., 29: 57-70.

M401   Määr, A., 1950. A supplement to the fertility of char (*Salmo alpinus* L.) in Faxälven water system, Sweden. Rep. Inst. Freshw. Res. Drottning., 31: 125-36.

M402   McFadden, J. T., J. Brasch, and S. Kmiotek, 1957. Life history of eastern brook trout. Data for Handbook of Biological Data. 16 p.

M403   Mansueti, A. J., 1963. Some changes in morphology during ontogeny in the pirateperch, *Aphredoderus s. sayanus.* Copeia, 1963(3): 546-57.

M404   Minckley, W. L., 1963. The ecology of a spring stream, Doe Run, Meade County, Kentucky. Wildl. Monogr., 11: 1-124.

M405   McCormack, J. C., 1962. The food of young trout *(Salmo trutta)* in two different becks. J. Anim. Ecol., 31: 305-16.

M406   Mraz, D., 1964. Age and growth of round whitefish in Lake Michigan. Trans. Amer. Fish. Soc., 93(1): 46-52.

M407   Minckley, W. L., R. H. Goodyear, and J. E. Craddock, 1964. Incidence of aberrant scalation in Catostomid fishes from Doe Run, Meade County, Kentucky. Trans. Amer. Fish. Soc., 93(2): 202-3.

M408   Massmann, W. H., 1963. Summer food of juvenile American shad in Virginia waters. Chesapeake Sci., 4(4): 167-71.

M410   MacKay, H. H., 1931. The maskinonge and its conservation. Fish Cult. Bull. Ont. Dept. Game Fish, 1: 1-11.

M411   Magnin, E., 1964. Croissance en longueur de trois esturgeons d' Amerique du Nord: *Acipenser oxyrhynchus* Mitchill, *Acipenser fulvescens* Rafinesque, et *Acipenser brevirostris* LeSueur. Verh. Int. Ver. Limnol., 15: 968-74.

M412   Moore, E., 1932. Certain minnows showing adaptability to conditions in impounded waters. Trans. Amer. Fish. Soc., 62: 290-91.

M413   Miller, R. R. and C. L. Hubbs, 1960. The spiny-rayed cyprinid fishes (Plagopterini) of the Colorado River system. Misc. Publ. Mus. Zool. Univ. Mich., 115: 39 p.

M414   Mottley, C. McC., 1938. Carp control studies with special reference to Chautauqua Lake. Annu. Rep. N. Y. Conserv. Dept., 27(Suppl.): 226-34.

M415   Moyle, J. B., 1949. Fish population concepts and management of Minnesota lakes for sport fishing. Trans. N. Amer. Wildl. Conf., 14: 283-94.

M417   Marshall, N., 1947. The spring run and cave habitats of *Erimystax harperi* (Fowler). Ecology, 28(1): 68-75.

M418   Minckley, W. L. and F. B. Cross, 1959. Distribution, habitat and abundance of the Topeka shiner *Notropis topeka* (Gilbert) in Kansas. Amer. Midl. Nat., 61(1): 210-17.

M419   Markley, M. H., 1940. Notes on the food habits and parasites of the stickleback, *Gasterosteus aculeatus* (Linnaeus), in the Sacramento River, California. Copeia, 1940(4): 223-5.

M420   Moore, G. A., 1957. Fishes. Vertebrates of the United States, Blair, et al., McGraw Hill Book Co., N. Y.: 31-210.

M421   Muir, B. S. and J. G. Sweet, 1964. The survival, growth and movement of *Esox masquinongy* transplanted from Nogies Creek Sanctuary to public fishing waters. Canad. Fish Cult., 32: 31-44.

M422   McConnell, W. J. and J. H. Gerdes, 1964. Threadfin shad, *Dorosoma petenense,* as food of yearling centrarchids. Calif. Fish Game, 50(3): 170-75.

M423   Magnin, E., 1962. Recherches sur la Systematique et la Biologie des Acipensendes, *Acipenser sturio* L., *Acipenser oxyrhynchus* Mitchill et *Acipenser fulvescens* Raf. Ann. Sta. Cent. Hydrobiol. Appl., 9: 7-242.

M425   Moen, T. E., 1954. Food of the bigmouth buffalo, *Ictiobus cyprinellus* (Valenciennes), in northwest Iowa lakes. Proc. Ia. Acad. Sci., 61: 561-9.

M426   Migdalski, E. C., 1955. Reproduction and classification of some of the better known fishes of North America. Data for Handbook of Biological Data. 42 p.

M430   McFadden, J. T., E. L. Cooper, and J. K. Andersen, 1965. Some effects of environment on egg production in brown trout *(Salmo trutta)*. Limnol. Oceanogr., 10(1): 88-95.

M431   Magnin, E., 1964. Premier inventaire ichthyologique du Lac et de la Riviere Waswanipi. Naturaliste Canadien, 91(11): 273-308.

M432   Miller, R. J., 1964. Behavior and ecology of some North American cyprinid fishes. Amer. Midl. Nat., 72: 313-57.

M433   Morris, L. A., 1965. Age and growth of the river carpsucker, *Carpiodes carpio*, in the Missouri River. Amer. Midl. Nat., 73: 423-9.

M434   Mattson, C. R., 1949. The lamprey fishery at Willamette Fall, Oregon. Ore. Fish. Comm. Res. Briefs, 2(2): 23-7.

M435   Ming, A., 1965. Life history of the grass pickerel and chain pickerel. Okla. Fish. Res. Lab. Semi-Annu. Rep., 1965(Jan.-June): 12-15.

M436   Mense, J. B., 1965. Mississippi silversides investigations. Okla. Fish. Res. Lab. Semi-Annu. Rep., 1965(Jan.-June): 108-9.

M437   Mason, J. W., O. M. Brynildson, and P. E. Degurse, 1966. Survival of trout fed dry and meat-supplemented dry diets. Progr. Fish Cult., 28(4): 187-92.

M438   Magnin, E., 1963. Notes sur la repartition la biologie et particuliérement la croissance de l'*Acipenser brevirostris* LeSueur 1817. Naturaliste Canadien, 80: 87-96.

M440   _____, 1966. Croissance de l'esturgeon *Acipenser fulvescens* Raf. vivant dans le bassin hydrographique de la riviere Nottaway, tributaire de la baie James. Naturaliste Canadien, 92: 193-204.

M441   Meehan, W. E., 1910. Experiments in sturgeon culture. Trans. Amer. Fish. Soc., 39: 85-91.

M442   Mansueti, R. J., 1956. Alewife herring eggs and larvae reared sucessfully in lab. Md. Tidewater News, 13(1): 2-3.

M443   Miller, R. R., 1963. Genus *Dorosoma* Rafinesque 1820. Gizzard shads, Threadfin shads. Fishes of the western north Atlantic. Mem. Sears Found. Mar. Res., 1(Part 3): 443-51.

M444   Mansueti, R. J., 1955. Natural history of the American shad in Maryland waters. Md. Tidewater News, 11(11, Supp.4).

M445   Massmann, W. H., 1952. Characteristics of spawning areas of shad, *Alosa sapidissima* (Wilson), in some Virginia streams. Trans. Amer. Fish. Soc., 81: 78-93.

M446   Miller, R. R., 1957. Origin and dispersal of the alewife, *Alosa pseudoharengus,* and the gizzard shad, *Dorosoma cepedianum,* in the Great Lakes. Trans. Amer. Fish. Soc., 86: 97-111.

M447   Moss, D. D., 1946. Preliminary studies of the shad *(Alosa sapidissima)* catch in the lower Connecticut River, 1944. N. Amer. Wildl. Conf. Trans., 11: 230-39.

M448   Mayhew, J., 1965. Pre-impoundment studies of the Chariton River in the vicinity of Rathbun dam and reservoir. Ia. Conserv. Comm. Quart. Biol. Rep., 17(4): 4-10.

M449   Minckley, W. L. and L. A. Krumholz, 1960. Natural hybridization between the Clupeid genera *Dorosoma* and *Signalosa,* with a report on the distribution of *S. petenense.* Zoologica, 44: 171-80.

M450   Miller, R. V., 1964. The morphology and function of the pharyngeal organs in the Clupeid, *Dorosoma petenense* (Gunther). Chesapeake Sci., 5(4): 194-9.

M451   Mraz, D., 1964. Age, growth, sex ratio, and maturity of the whitefish in central Green Bay and adjacent waters of Lake Michigan. Fish. Bull. U. S., 63(3): 619-34.

M452   McCart, P. J., 1963. Growth and morphometry of the pygmy whitefish *(Prosopium coulteri)* in British Columbia. MS thesis Univ. B. C. 97 p. not seen.

M453   Montana Fish and Game Department, 1964. Age and growth studies. D.-J. Completion Rep., F-23-R-7 (Jobs I, II): 6 p. mimeo.

M454   McKnight, I. M., 1966. A hematological study on the mountain whitefish, *Prosopium williamsoni.* J. Fish. Res. Bd. Can., 23(1): 45-64.

M455   Miller, R. R., 1950. Notes on the cutthroat and rainbow trouts with the description of a new species from the Gila River, New Mexico. Occ. Pap. Univ. Mich. Mus. Zool., 529: 42 p.

M456   Mastin, H. E., 1962. Lake County district. Annu. Rep. Ore. St. Game Comm. Fish. Div., 1962: 164-72.

M457   Mills, L. E., 1966. Environmental factors and egg mortality of cutthroat trout, *Salmo clarkii,* in three tributaries of Yellowstone Lake. Colo. Dept. Game Fish Park, Fish. Res. Rev., 3: 24-5.

M458   McAfee, W. R., 1966. Mountain whitefish. Calif. Dept. Fish Game, Inland Fisheries Management, ed. A. Calhoun: 299-303.

M459   _____, 1966. Lahontan cutthroat trout. Calif. Dept. Fish Game, Inland Fisheries Management, ed. A. Calhoun: 225-31.

M460   _____, 1966. Piute cutthroat trout. Calif. Dept. Fish Game, Inland Fisheries Management, ed. A. Calhoun: 231-3.

M461   _____, 1966. Golden trout. Calif. Dept. Fish Game, Inland Fisheries Management, ed. A. Calhoun: 216-21.

M462   _____, 1966. Rainbow trout. Calif. Dept. Fish Game, Inland Fisheries Management, ed A. Calhoun: 192-216.

M463   _____, 1966. Eagle Lake rainbow trout. Calif. Dept. Fish Game, Inland Fisheries Management, ed. A. Calhoun: 221-5.

M464   McCrimmon, H. R. and A. H. Berst, 1966. A water recirculation unit for use in fishery laboratories. Progr. Fish Cult., 28(3): 165-70.

M465   Mills, D. H., 1964. The ecology of the young stages of the Atlantic salmon in the River Bran, Ross-shire. Freshw. Salm. Fish. Res. Scot., 32: 58 p.

M466   McCormack, J. C., 1962. The food of young trout *(Salmo trutta)* in two different becks. J. Anim. Ecol., 31: 305-16.

M467   Miller, W. H., 1966. Distribution, abundance and growth of the fishes of Little Beaver Creek, Colorado. Colo. Fish. Res. Rev., 3: 22-3.

M468   McFadden, J. T. and E. L. Cooper, 1964. Population dynamics of brown trout in different environments. Physiol. Zool., 37(4): 355-63.

M489   Munro, W. R. and K. H. Balmain, 1956. Observations on the spawning runs of brown trout in the South Queich, Loch Leven. Freshw. Salm. Fish. Res. Scot., 13: 17 p.

M490   Macan, T. T., 1965. The effect of the introduction of *Salmo trutta* into a moorland fishpond. Mitt. Int. Ver. Limnol., 13: 194-9.

M491   McAfee, W. R., 1966. Eastern brook trout. Calif. Dept. Fish Game, Inland Fisheries Management, ed. A. Calhoun: 242-60.

M492   McAfee, W. R., 1966. Dolly varden trout. Calif. Dept. Fish Game, Inland Fisheries Management, ed. A. Calhoun: 271-4.

M493   ____, 1966. Lake trout. Calif. Dept. Fish Game, Inland Fisheries Management, ed. A. Calhoun: 260-71.

M494   Martin, N. V., 1966. The significance of food habits in the biology, exploitation and management of Algonquin Park, Ontario, lake trout. Trans. Amer. Fish. Soc., 95(4): 415-22.

M495   Magnin, E. and G. Beaulieu, 1965. Quelques données sur la biologie de l'eperlan, *Osmerus eperlanus mordax* (Mitchill) du Saint-Laurent. Naturaliste Canadien, 92: 81-105.

M496   McKenzie, R. A., 1964. Smelt life history and fishery in the Miramichi River, New Brunswick. Fish. Res. Bd. Can. Bull., 144: 77 p.

M497   Marcotte, A. and J. L. Tremblay, 1948. Notes sur la biologie de l'eperlan de la province de Quebec. Sta. Biol. St.-Laurent Contr., 18: 107 p.

M498   Marr, D. H. A., 1966. Factors affecting the growth of salmon alevins and their survival and growth during the fry stage. Freshw. Fish. Lab. Pitlochry. 9 p. Reprint.

M499   Muench, B., 1963. Length-weight relationship of eighteen species of fish in northeastern Illinois. N. E. Area Fish. HQ, Rt. 2, Box 51, Marengo, Ill. Typed ms.

M500   Muir, B. S., 1963. Vital statistics of *Esox masquinongy* in Nogies Creek, Ontario. I. Tag loss, mortality due to tagging, and the estimate of exploitation. J. Fish. Res. Bd. Can., 20(5): 1213-30.

M501   Mills, H. B., W. C. Starrett, and F. C. Bellrose, 1966. Man's effect on the fish and wildlife of the Illinois River. Ill. Nat. Hist. Surv. Biol. Notes, 57: 24 p.

M502   Mark, M., 1966. Carp breeding in drainage water. Bamidgeh, 18(2): 51-54.

M503   Mayhew, J., 1964. Coralville Reservoir fisheries investigation, 1963. Part II: Limnology and fish populations. Ia. Conserv. Comm. Quart. Rep., 16(1): 25-31.

M504   McKechnie, R. J., 1966. Golden shiner. Calif. Dept. Fish Game, Inland Fisheries Management, ed. A. Calhoun: 488-92.

M505   Martin, R. E., S. I. Auerbach, and D. J. Nelson, 1964. Growth and movement of smallmouth buffalo, *Ictiobus bubalus* (Rafinesque), in Watts Bar Reservoir, Tennessee. PhD thesis Univ. Tenn.; Oak Ridge Nat. Lab., 03530: 98 p.

M506   McComish, T. S., 1967. Food habits of bigmouth and smallmouth buffalo in Lewis and Clark Lake and in the Missouri River. Trans. Amer. Fish. Soc., 96(1): 70-74.

M507   Mitzner, L., 1966. Age and growth of bigmouth buffalo in Coralville Reservoir. Ia. Conserv. Comm. Quart. Biol. Rep., 18(4): 66-74.

M508   Miller, E. E., 1966. White catfish. Calif. Dept. Fish Game, Inland Fisheries Management, ed. A. Calhoun: 430-40.

M509   ____, 1966. Black bullhead. Calif. Dept. Fish Game, Inland Fisheries Management, ed. A. Calhoun: 476-9.

M510   ____, 1966. Channel catfish. Calif. Dept. Fish Game, Inland Fisheries Management, ed. A. Calhoun: 440-63.

M511   Muncy, R. J., 1957. Distribution and movements of channel and flathead catfish in the Des Moines River, Boone County, Iowa. PhD disser. Ia. St. Univ. Lib. 118 p.

M512 Magnuson, J. J. and L. L. Smith, Jr., 1963. Some phases of the life history of the trout-perch. Ecology, 44(1): 83-95.

M513 Magnin, E. and G. Beaulieu, 1966. Divers aspects de la biologie et de l'ecologie de la barbue *Ictalurus punctatus* (Rafinesque) du fleuve Saint-Laurent d'apres les données du marquage. Naturaliste Canadien, 92: 277-91.

M514 ____, 1966. Quelques données biologiques sur la reproduction des esturgeons *Acipenser fulvescens* Raf. de la Rivière Nottaway, tributaire de La Baie James. Canad. J. Zool., 44: 257-63.

M515 ____ and G. Beaulieu, 1960. Deplacements des esturgeons *(Acipenser fulvescens* et *Acipenser oxyrhynchus)* du Fleuve Saint-Laurent d-apres les données du marquage. Naturaliste Canadien, 87(11): 237-52.

M516 ____, 1966. Recherches sur les cycles de reproduction des esturgeons, *Acipenser fulvescens* Raf., de la Rivière Nottaway tributaire de la baie James. Verh. Internat. Ver. Limnol., 16: 1018-24.

N1 Nall, G. H., 1925. Report on a collection of sea trout scales from the River Hope and Loch Hope in Sutherland. Salm. Fish. Scot., 1925(1): 1-22.

N2 ____, 1926. The sea trout of the River Ewe and Loch Maree. Salm. Fish. Scot., 1926(1): 1-42.

N3 ____, 1926. Sea trout of the River Ailort and Loch Eilt. Salm. Fish. Scot., 1926(3): 1-24.

N4 ____, 1927. Report on a collection of salmon scales from the River Hope and Loch Hope in Scotland. Salm. Fish. Scot., 1926(7): 1-8.

N5 ____, 1928. Sea trout of the River Ailort and Loch Eilt. Salm. Fish. Scot., 1928(9): 1-38.

N6 ____, 1928. Sea trout of South Uist, the Howmore, Kildonan and Loch a Bharp districts. With a note on the Howmore and Loch a Bharp salmon. Salm. Fish. Scot., 1928(7): 1-47.

N7 ____, 1928. Sea trout from the Broom of Moy waters of the Findhorn and from the tidal waters of the Ugie. Salm. Fish. Scot., 1928(6): 1-24.

N8 ____, 1928. Report on a collection of sea trout scales from the River Carron and Loch Dhughaill (Doule), western Ross-shire. Salm. Fish. Scot., 1928(4): 1-16.

N9 ____, 1928. The sea trout of the River Ewe and Loch Maree, Part 2, 1926-1927. Salm. Fish. Scot., 1928(2): 1-16.

N10 ____, 1929. Sea trout from the Beauly Firth and from the tidal waters of the Beauly and Ness rivers. Salm. Fish. Scot., 1929(3): 1-40.

N11 ____, 1929. Sea trout of the River Tweed. Salm. Fish. Scot., 1929(5): 1-59.

N12 ____, 1929. Sea trout of South Uist. Salm. Fish. Scot., 1929(4): 1-31.

N13 ____, 1930. The life of sea trout. Seeley, Service and Co., Lond. 335 p.

N14 ____, 1931. Sea trout of the River Tay. Salm. Fish. Scot., 1931(1): 1-24.

N15 ____, 1931. Irish sea-trout. Notes on collections of scales from the west coast of Ireland. Proc. R. Irish Acad., 40: 1-36.

N16 ____, 1932. Sea trout of the Laerdal. Salm. Trout Mag., 69: 353-64.

N17 ____, 1932. Sea trout of the Solway rivers. Salm. Fish. Scot., 1932 (3): 1-72.

N18 ____, 1932. Notes on collections of sea trout scales from Lewis and Harris and from North Uist. Salm. Fish. Scot., 1932(1): 1-37.

N19   Nall, G. H., 1934. Sea trout of Lewis and Harris. Further notes on
      collections of scales from this district. Salm. Fish. Scot., 1934(4): 1-
      71.

N20   _____, 1938. Sea trout of the River Conon. Salm. Fish. Scot., 1937(4):
      1-31.

N21   _____, 1939. Notes on scales from Avon salmon smolts in 1938. Rep.
      Avon Biol. Res. S'hampton, 6: 16-26.

N22   _____ and W. J. M. Menzies, 1932. Difficulties of age determinations
      and length calculations from the scales of sea trout (Salmo trutta).
      Salm. Fish. Scot., 1931(5): 1-12.

N23   Neave, F. and A. Bajkov, 1929. Reports of the Jasper Park lakes in-
      vestigations, 1926-28. V. Food and growth of Jasper Park fishes.
      Contr. Canad. Biol. Fish., 4: 199-217.

N24   _____ and G. C. Carl, 1940. The brown trout on Vancouver Island.
      Proc. 6th Pacif. Sci. Cong., 3: 341-3.

N25   Needham, J. G., 1920. Clean waters for New York State. Cornell
      Rural School Leafl., 13: 153-82.

N26   Needham, P. R., 1936. The Hot Creek rearing ponds. Calif. Fish
      Game, 22(2): 118-25.

N27   _____, 1937. A biological survey of Lake Arrowhead, California.
      Calif. Fish Game, 23(4): 311-28.

N28   _____, 1938. Trout streams. Comstock Publ. Co., Ithaca. 233 p.

N29   _____, 1938. Notes on the introduction of Salmo nelsoni Evermann into
      California from Mexico. Trans. Amer. Fish. Soc., 67: 139-46.

N31   _____ and R. C. Lewis, 1937. Experiments in pond rearing of trout on
      natural foods. Progr. Fish Cult., 29: 18-24.

N32   _____, J. W. Moffett, and D. W. Slater, 1945. Fluctuations in wild
      brown trout populations in Convict Creek, California. J. Wildl. Mgmt.,
      9(1): 9-25.

N33   _____ and _____, 1944. Survival of hatchery-reared brown and rainbow
      trout as affected by wild trout populations. J. Wildl. Mgmt., 8(1): 22-
      36.

N34   _____ and _____, 1945. Seasonal changes in growth, mortality, and
      condition of rainbow trout following planting. Trans. Amer. Fish. Soc.,
      73: 117-24.

N35   _____ and F. K. Sumner, 1942. Fish management problems of high
      western lakes with returns from marked trout planted in Upper Angora
      Lake, California. Trans. Amer. Fish. Soc., 71: 249-69.

N36   _____ and E. H. Vestal, 1938. Notes on growth of golden trout (Salmo
      aqua-bonita) in two High Sierra lakes. Calif. Fish Game, 24(3): 273-9.

N37   Nelson, M. N. and A. D. Hasler, 1942. The growth, food, distribution
      and relative abundance of the fishes of Lake Geneva, Wisconsin, in
      1941. Trans. Wis. Acad. Sci. Arts Lett., 34: 137-48.

N39   Nevin, J., 1898. Artificial propagation versus a closed season for the
      Great Lakes. Trans. Amer. Fish. Soc., 27: 17-27.

N40   _____, 1901. The propagation of muskellunge in Wisconsin. Trans.
      Amer. Fish. Soc., 30: 90-93.

N41   _____, 1911. Reminiscences of forty-one years' work in fish culture.
      Trans. Amer. Fish. Soc., 40: 313-18.

N43   New York Conservation Department, 1942. Thirty-first annual report
      for the year 1941. Legis. Doc., 1942(32): 402 p.

N45   _____, 1946. Thirty-fourth—Thirty-fifth annual reports for the years
      1944-1945. Legis. Doc., 1946(61): 322 p.

N46 New York State Hatcheries, n.d. Constants for trout in New York State hatcheries. Mimeo. table.

N47 Norris, T., 1883. The Michigan grayling. Sport with gun and rod in American woods and waters, Century Co., N. Y.: 493-506.

N49 Neave, F., 1949. Game fish populations of the Cowichan River. Fish. Res. Bd. Can. Bull., 84: 1-32.

N50 New Jersey Fisheries Survey, 1950. Lakes and ponds. N. J. Dept. Conserv. Rep., 1: 1-189.

N52 Noland, W. E., 1951. The hydrography, fish, and turtle population of Lake Wingra. Trans. Wis. Acad. Sci. Arts Lett., 40(2): 5-58.

N54 Nall, G. H., 1924. A salmon with four spawning marks. Salm. Trout Mag., 38: 48-69.

N55 _____, 1925. The sea trout of Loch Maree. Salm. Trout Mag., 40: 280-97.

N56 _____, 1927. Salmon of the River Ewe and Loch Maree. Salm. Trout Mag., 48: 296-315.

N57 _____, 1928. Spring runs of large sea trout in the Waterville River. Salm. Trout Mag., 50: 69-71.

N58 _____, 1928. The fish of some west coast Highland rivers. Salm. Trout Mag., 52: 299-308.

N59 _____, 1928. The fish of some west coast Highland rivers. Salm. Trout Mag., 53: 373-86.

N60 _____, 1929. Irish Sea trout. Salm. Trout Mag., 55: 182-6.

N62 _____, 1933. The sea trout of the Dovey. I. The river and its recent history. Salm. Trout Mag., 71: 169-86.

N63 _____ and M. H. G. Fell, 1935. Sea trout of some lake district rivers. Salm. Trout Mag., 79: 157-73.

N64 _____, 1938. A report on the sea trout of the Cumberland and Lancashire Rivers. Salm. Trout Mag., 91: 164-8.

N67 Needham, P. R. and J. P. Welsh, 1953. Rainbow trout in the Hawaiian Islands. J. Wildl. Mgmt., 17(3): 233-55.

N68 Neess, J. C., W. T. Helm, and C. W. Threinen, 1957. Some vital statistics in a heavily exploited population of carp. J. Wildl. Mgmt., 21(3): 279-92.

N69 Nelson, B. A., 1957. Propagation of channel catfish in Arkansas. Proc. S. E. Assoc. Game Fish Comm., 10: 165-8.

N70 Nelson, P. H., 1954. The American grayling in Montana. J. Wildl. Mgmt., 18(3): 324-42.

N71 Neuhold, J. M., 1957. Age and growth of the Utah chub, *Gila atraria* (Girard) in Panguitch Lake and Navajo Lake, Utah, from scales and opercular bones. Trans. Amer. Fish. Soc., 85: 217-33.

N73 Newell, A. E., 1956. Trout stream management investigations in Swift River, Albany, New Hampshire. N. H. Fish Game Dept. Dingell-Johnson Proj., F-5-R.

N74 _____, 1958. The life history and ecology of the sunapee trout, *Salvelinus aureolus* (Bean). N. H. Fish Game Dept. 17 p. ms.

N75 _____, 1958. Trout stream management investigations of the Swift River watershed in Albany, New Hampshire. N. H. Fish Game Dept. Surv. Rep., 7: 40 p.

N78 Nielson, R. S., N. Reimers, and H. D. Kennedy, 1957. A six-year study of the survival and vitality of hatchery-reared rainbow trout of catchable size in Convict Creek, California. Calif. Fish Game, 43(1): 5-42.

N80   Nilsson, N-A., 1955. Studies on the feeding habits of trout and char in north Swedish lakes. Rep. Inst. Freshw. Res. Drottning., 36: 163-225.

N82   Northcote, T. G. and G. F. Hartman, 1959. A case of "schooling" behavior in the prickly sculpin, *Cottus asper* Richardson. Copeia, 1959 (2): 156-8.

N83   Nichols, P. R., 1959. Extreme loss in body weight of an American shad *(Alosa sapidissima).* Copeia, 1959(4): 343-4.

N84   Nielson, R. S., 1960. California-Nevada sport fishery investigations. U. S. Fish Wildl. Serv. Circ., 81: 18-23.

N85   Nelson, B. A., 1959. Progress report on golden channel catfish. Proc. S. E. Assoc. Game Fish Comm., 12: 75-8.

N86   Nakamura, K., M. Shimadate, H. Koyama, and H. Okubo, 1954. Fish production in seven farm ponds in Shioda Plain, Nagano Prefecture, with reference to natural limnological environment and artificial treatment. (Japanese with English summary.) Bull. Freshw. Fish. Lab. Tokyo, 3(1): 27-79.

N89   New York State Conservationist, 1958. Cayuga record. N. Y. State Conserv., 13(1): 39.

N91   Neth, P. C., 1955. Assessment of a control program for common whitefish *(Coregonus clupeaformis)* and round whitefish *(Prosopium quadrilaterale)* in Little Moose Lake, New York. MS thesis Cornell Univ. 101 p.

N92   Nilsson, N-A., 1961. The effect of water-level fluctuations on the feeding habits of trout and char in the Lakes Blåsjön and Jormsjön, North Sweden. Rep. Inst. Freshw. Res. Drottning., 42: 238-61.

N93   Nelson, W. R., 1961. Report of fisheries investigations during the eighth year of impoundment of Fort Randall Reservoir, South Dakota, 1960. S. D. Dept. Game Fish Parks. 30 p. mimeo.

N94   _____, 1961. Report of fisheries investigations during the sixth year of impoundment of Gavins Point Reservoir, South Dakota, 1960. S. D. Dingell-Johnson Proj., F-1-R-10(Jobs2-4): 59 p.

N95   Newell, A. E., 1960. Biological survey of the lakes and ponds in Coos, Grafton and Carroll Counties. N. H. Fish Game Dept. Surv. Rep., 8a: 297 p.

N96   Needham, P. R., 1961. Observations on the natural spawning of eastern brook trout. Calif. Fish Game, 47(1): 27-40.

N98   Northcote, T. G., 1962. Migratory behaviour of juvenile rainbow trout, *Salmo gairdneri,* in outlet and inlet streams of Loon Lake, British Columbia. J. Fish. Res. Bd. Can., 19(2): 201-70.

N101 Needham, P. R. and R. Gard, 1959. Rainbow trout in Mexico and California with notes on the cutthroat series. Univ. Cal. Publ. Zool., 67(1): 1-124.

N102 Netsch, N. F. and A. Witt, Jr., 1962. Contributions to the life history of the longnose gar *(Lepisosteus osseus)* in Missouri. Trans. Amer. Fish. Soc., 91(3): 251-62.

N104 Nurnberger, P. K., 1928. A list of the plant and animal food of some fishes of Jay Cooke Park. Trans. Amer. Fish. Soc., 58: 175-7.

N105 Needham, P. R. and A. C. Taft, 1934. Observations on the spawning of steelhead trout. Trans. Amer. Fish. Soc., 64: 332-8.

N106 Newell, A. E., 1957. Two-year study of movements of stocked brook trout and rainbow trout in a mountain trout stream. Progr. Fish Cult., 19(2): 76-81.

N107 Northcote, T. G. and R. J. Paterson, 1960. Relationship between number of pyloric caeca and length of juvenile rainbow trout. Copeia, 1960(3): 249-50.

N108   Nicholls, A. G., 1957. The Tasmanian trout fishery. 1. Sources of information and treatment of data. Aust. J. Mar. Freshw. Res., 8: 451-75.

N109   Nielson, R. S., 1963. California-Nevada sport fishery investigations. U. S. Fish Wildl. Serv. Circ., 160: 56-62.

N110   _____, 1953. Should we stock brown trout? Progr. Fish Cult., 15(3): 125-6.

N111   Needham, P. R., 1949. Survival of trout in streams. Trans. Amer. Fish. Soc., 77: 26-31.

N113   Nichols, P. R. and W. H. Massmann, 1963. Abundance, age and fecundity of shad, York River, Virginia, 1953-59. Fish. Bull. U. S., 63(1): 179-87.

N114   Nilsson, N-A., 1964. Effects of impoundment on the feeding habits of brown trout and char in Lake Ransaren (Swedish Lappland). Verh. Int. Ver. Limnol., 15: 444-52.

N115   Nelson, W. R., 1962. Report of fisheries investigations during the seventh year of impoundment of Gavins Point Reservoir, South Dakota, 1961. S. D. Dingell-Johnson Proj., F-1-R-11(Jobs1-3,7): 40 p.

N116   Nurnberger, P. K., 1931. Observations on the spawning temperature of *Luxilus cornutus*. Trans. Amer. Fish. Soc., 61: 215.

N117   _____, 1930. The plant and animal food of the fishes of Big Sandy Lake. Trans. Amer. Fish. Soc., 60: 253-9.

N118   Nordeng, H., 1961. On the biology of char (*Salmo alpinus* L.) in Salangen, North Norway. I. Age and spawning frequency determined from scales and otoliths. Nytt Mag. Zool., 10: 67-123.

N120   Nichols, P. R., 1960. Homing tendency of American shad, *Alosa sapidissima*, in the York River, Virginia. Chesapeake Sci., 1(3-4): 200-201.

N121   Northcote, T. G. and H. W. Lorz, 1966. Seasonal and diet changes in food of adult kokanee *(Oncorhynchus nerka)* in Nicola Lake, British Columbia. J. Fish. Res. Bd. Can., 23(8): 1259-63.

N122   Nielson, R. S., 1964. California-Nevada sport fishery investigations. U. S. Fish Wildl. Serv. Circ., 178: 67-78.

N123   Nose, T., 1963. Determination of nutritive value of food protein on fish —II. Effect of amino and composition of high protein diets on growth and protein utilization of the rainbow trout. Bull. Freshw. Fish. Res. Lab. Tokyo, 13(1): 41-50.

N124   Nicholls, A. G., 1958. The egg yield from brown and rainbow trout in Tasmania. Aust. J. Mar. Freshw. Res., 9(4): 526-36.

N125   _____, 1958. The Tasmanian trout fishery. II. The fishery of the north-west region. Aust. J. Mar. Freshw. Res., 9: 19-59.

N126   _____, 1958. The Tasmanian trout fishery. III. The rivers of the North and East. Aust. J. Mar. Freshw. Res., 9(2): 167-90.

N127   _____, 1958. The population of a trout stream and the survival of released fish. Aust. J. Mar. Freshw. Res., 9(3): 319-50.

N128   _____, 1961. The Tasmanian trout fishery. IV. The rivers of the South and South-east. Aust. J. Mar. Freshw. Res., 12(1): 17-53.

N129   Nall, G. H., 1955. Movements of salmon and sea trout, chiefly kelts, and of brown trout tagged in the Tweed between January and May, 1937 and 1938. Freshw. Salm. Fish. Res. Scot., 10: 19 p.

N130   Nilsson, N-A., 1965. Food segregation between salmonoid species in north Sweden. Rep. Inst. Freshw. Res. Drottning., 46: 58-78.

N131   Nilsson, N-A., 1963. Interaction between trout and char in Scandinavia. Trans. Amer. Fish. Soc., 92(3): 276-85.

N132   Nagata, T. H., 1967. Artificial spawning of anadromous Dolly Varden. Progr. Fish Cult., 29(1): 26.

N133   Nikol'skii, G. V., 1961. Special ichthyology. Trans. from Russian, Nat. Sci. Found. and Smithson. Inst.: 538 p.

N134   Nail, M. L., 1965. The protein requirement of channel catfish, *Ictalurus punctatus* (Rafinesque). Proc. S. E. Assoc. Game Fish Comm., 16: 307-16.

O1   Odell, T. T., 1934. The life history and ecological relationships of the alewife (*Pomolobus pseudoharengus* (Wilson)) in Seneca Lake, New York. Trans. Amer. Fish. Soc., 64: 118-24.

O2   Odenwall, E., 1927. Fiskfaunan i Lappajarvi sjo. Acta Soc. Fauna Flora Fennica, 56(13): 1-48.

O5   Oehmcke, A. A., 1949. Muskellunge fingerling culture. Progr. Fish Cult., 11(1): 3-18.

O7   Ohio Bureau of Scientific Research, 1934. Length-weight relationship of several Ohio food and game fishes. Ohio Bur. Sci. Res. Bull., 70: 2 p. mimeo.

O8   _____, n.d. Age and rate of growth of several game fishes in the inland fishing district of Ohio. Ohio Bur. Sci. Res. Bull., 9: 1 p. mimeo.

O11   Okkelberg, P., 1921. The early history of the germ cells in the brook lamprey, *Entosphenus wilderi* (Gage), up to and including the period of sex differentiation. J. Morph., 35: 1-151.

O12   _____, 1922. Notes on the life-history of the brook lamprey, *Ichthyomyzon unicolor*. Occ. Pap. Mus. Zool. Univ. Mich., 125: 1-14.

O13   Otterstrom, C. V., 1933. Reif Lachse *(Salmo salar)* in Teichen. J. Cons. Int. Explor. Mer., 8: 83-9.

O15   Oregon State Game Commission, 1949. How old is a sturgeon? Ore. St. Game Comm. Bull., 4(11): 4.

O19   O'Driscoll, D., 1950. Salmon of the Drumcliffe River. Sci. Proc. Roy. Dublin Soc., NS 25(10): 117-30.

O20   Oehmcke, A. A., 1951. Muskellunge yearling culture and its application to lake management. Progr. Fish Cult., 13(2): 63-70.

O21   Oklahoma City Times, 1951. Record weight spoonbill catfish. Okla. Game Fish News, 7(2): 14.

O22   Oregon State Game Commission, 1950. Annual report, Fishery Division, 1949. 205 p.

O23   _____, 1951. Annual report, Fishery Division, 1950. 159 p.

O24   _____, 1952. Annual report, Fishery Division, 1951. 238 p.

O26   Oehmcke, A. A., L. D. Johnson, J. H. Klingbiel, and C. A. Wistrom, 1958. The Wisconsin muskellunge, its life history, ecology, and management. Wis. Conserv. Dept. Publ., 225: 1-11.

O28   Oliva, O., 1955. Contribution to the biology and growth of the carp in back-waters of the River Elbe region. Univ. Carolina, Biologica, 1(3): 225-73.

O30   _____, 1956. K biologii Stiky (*Esox lucius* L.) [Biology of the pike.] Acta Soc. Zool. Bohemoslov., 20(3): 208-23.

O33   Oregon State Game Commission, 1953. Annual report, Fishery Division, 1952. 308 p.

O34   _____, 1954. Annual report, Fishery Division, 1953. 239 p.

O35   Orton, J. H., J. W. Jones, and G. M. King, 1938. The male sexual stage in salmon parr (*Salmo salar* L. juv.). Proc. Roy. Soc. Ser. B, 125: 103-14.

O37   Orr, O. E., 1958. The populations of fishes and limnological conditions of Heyburn Reservoir with reference to productivity. PhD thesis Okla. St. Univ. 68 p.

O38   Olund, L. J. and F. B. Cross, 1961. Geographic variation in the North American cyprinid fish, *Hybopsis gracilis*. Univ. Kans. Publ. Mus. Nat. Hist., 13(7): 323-48.

O39   Onodera, K. and T. Ueno, 1961. On the survival of trout fingerlings stocked in a mountain brook. II. Survival rate measured and scouring effect of flood as a cause of mortality. Bull. Jap. Soc. Sci. Fish., 27 (6): 530-57.

O40   O'Donnell, D. J. and W. S. Churchill, 1954. Certain physical, chemical and biological aspects of the Brule River, Douglas County, Wisconsin. Wis. Acad. Sci. Arts Lett., 43: 201-55.

O41   Olson, H. F., 1957. Age and growth of the Utah chub, *Gila atraria* (Girard) in Fish Lake, Utah. Proc. Utah Acad. Sci., 34: 83-5.

O42   _____, 1959. The biology of the Utah chub, *Gila atraria* (Girard), of Scofield Reservoir, Utah. MS thesis Utah St. Univ. 34 p. not seen, quoted from S463.

O43   Ostdiek, J. L. and R. M. Nardone, 1959. Studies on the Alaskan blackfish *Dallia pectoralis*. I. Habitat, size and stomach analysis. Amer. Midl. Nat., 61(1): 218-29.

O44   Outten, L. M., 1957. A study of the life history of the cyprinid fish *Notropis coccogenis*. J. Elisha Mitchell Sci. Soc., 73(1,May): 68-84.

O45   _____, 1958. Studies of the life history of the cyprinid fishes *Notropis galacturus* and *rubricoceus*. J. Elisha Mitchell Sci. Soc., 74(2): 122-34.

O46   Oppenheimer, J. M., 1937. The normal stages of *Fundulus heteroclitus*. Anat. Rec., 68(1): 1-15.

O47   _____, 1947. Organization of the teleost blastoderm. Quart. Rev. Biol., 22(2): 105-18.

P1   Page, W. F., 1895. Feeding and rearing fishes, particularly trout, under domestication. Bull. U. S. Fish Comm., 14: 289-314.

P2   Parrott, A. W., 1932. Age and growth of trout from Eilden Weui, Victoria, Australia. N. Z. J. Sci. Technol., 14(2): 101-10.

P4   _____, 1932. The age and growth of trout in New Zealand. Fish. Bull. N. Z., 4: 1-46.

P5   _____, 1932. Age and growth of the Te Anau salmon: some new data on Atlantic salmon growth in New Zealand. Salm. Trout Mag., 66: 86-94.

P6   Pautzke, C. F. and R. C. Meigs, 1941. Studies on the life history of the Puget Sound steelhead trout *(Salmo gairdnerii)*. Trans. Amer. Fish. Soc., 70: 209-20.

P9   Pentelow, F. T. K., 1939. The relation between growth and food consumption in the brown trout *(Salmo trutta)*. J. Exp. Biol. (Lond.), 16(4): 446-73.

P10   _____, B. A. Southgate, and R. Bassindale, 1933. The relation between the size, age and time of migration of salmon and sea trout smolts in the River Tees. Fish. Invest. Lond. Ser. 1, 3(4): 1-14.

P11   Phillips, A. M., Jr., D. R. Brockway, and E. O. Rodgers, 1949. Use of cod-liver oil in the diet of hatchery-reared trout. Progr. Fish Cult., 11(2): 109-12.

P13   Potter, G. E., 1926. Ecological studies of the short-nosed gar-pike *(Lepidosteus platystomus)*. Univ. Ia. Stud. Nat. Hist., 11(9): 17-27.

680 CITATIONS

P19   Pritchard, A. L., 1929. The alewife *(Pomolobus pseudoharengus)* in Lake Ontario. Univ. Toronto Stud. Biol. Ser., 33: 39-54.

P20   _____, 1930. Spawning habits and fry of the cisco *(Leucichthys artedi)* in Lake Ontario. Contr. Canad. Biol. Fish., NS 6(9): 227-40.

P21   _____, 1931. Taxonomic and life history studies of the ciscoes of Lake Ontario. Publ. Ont. Fish. Res. Lab., 41: 1-78.

P22   Privolnev, T., 1934. Zur Biologie von *Salmo trutta*. Trav. Inst. Biol. Peterhof, 12: 311-23.

P23   Percival, E., 1932. On the depreciation of trout-fishing in the Oreti (or New River), Southland. N. Z. Mar. Dept. Fish. Bull., 5: 1-48.

P24   Pincher, C., 1948. A study of fish. Duell, Sloan and Pearce, N. Y. 343 p.

P31   Patriarche, M. H., 1951. The grass pickerel. Mo. Conserv., 12(6): 16.

P35   Pearson, J. C., 1952. Rearing young shad in ponds. Progr. Fish Cult., 14(1): 33-6.

P37   Phenicie, C. K., 1949. Progress report of the Fort Peck Reservoir fishery survey. Mont. Game Fish. Bull., 3: 1-19 mimeo.

P39   Piscator, 1950. Secretarial notes. Piscator, 4(13): 14-15; (15): 67-8; (16): 98-9.

P40   _____, 1950. Rapid growth of rainbow trout in a lake at Maclear. Piscator, 4(14): 35.

P41   _____, 1951. Secretarial notes. Piscator, 5(17): 17-19.

P42   _____, 1951. Big fish records and notable fish. Piscator, 5(17): 3-5; (19): 72-74.

P43   _____, 1951. Liesbeek notes. Piscator, 5(18): 35-6; (19): 67; (20): 99.

P44   _____, 1951. Record carp for the British Isles. Piscator, 5(20): 113.

P45   _____, 1951. City trout. Piscator, 5(20): 98-9.

P46   _____, 1951. Lower Eerste River rainbows. Piscator, 5(20): 100-101.

P47   _____, 1951. American eastern brook trout. Piscator, 5(18): 34-5.

P49   _____, 1951. Further remarkable growth of rainbow trout in the lake at Maclear. Piscator, 5(18): 37-8.

P50   _____, 1951. More large rainbow trout from Matatiele. Piscator, 5(18): 43.

P51   _____, 1951. Large rainbow trout from Mabele River, Matatiele. Piscator, 5(17): 16.

P52   _____, 1951. Rapid growth of American eastern brook trout at the Jonkenshoek hatchery. Piscator, 5(17): 16.

P53   _____, 1952. Big fish records and notable fish. Piscator, 6(21): 7.

P54   _____, 1952. Letters to the editor, and Liesbeek notes and Lower Eerste River notes. Piscator, 6(22): 37,38,42,49,55; 6(21): 4-7; 6(23): 68-70.

P55   Purkett, C. A., Jr., 1951. Growth rate of trout in relation to elevation and temperature. Trans. Amer. Fish. Soc., 80: 251-9.

P56   Phillips, A. M., Jr., 1940. The development of anemia in trout fed a synthetic diet and its cure by the feeding of fresh beef liver. Progr. Fish Cult., 48: 11-13.

P57   _____ *et al.*, 1946. The nutrition of trout. N. Y. Fish. Res. Bull., 9: 1-21.

P58   _____ _____, 1947. The nutrition of trout. N. Y. Fish. Res. Bull., 10: 1-35.

P59   _____ _____, 1949. The nutrition of trout. N. Y. Fish. Res. Bull., 12: 1-31.

P60   _____ and D. R. Brockway, 1947. This question of anemia. Progr. Fish Cult., 9(3): 151-4.

P63   Paloumpis, A. A., 1958. Responses of some minnows to flood and drought conditions in an intermittent stream. Ia. St. J. Sci., 32(4): 547-62.

P64   Parker, P. S. and R. E. Lennon, 1956. Biology of the sea lamprey in its parasitic phase. Res. Rep. U. S. Fish Serv., 44: 32 p.

P66   Parsons, J. W., 1954. The fish species composition and chemical and physical features of a seventy-two acre reservoir in Tennessee. J. Tenn. Acad. Sci., 29(1): 55-65.

P68   _____, 1957. The trout fishery of the tailwater below Dale Hollow Reservoir. Trans. Amer. Fish. Soc., 85: 75-92.

P69   _____, 1959. Muskellunge in Tennessee streams. Trans. Amer. Fish. Soc., 88(2): 136-41.

P71   Patriarche, M. H. and R. S. Campbell, 1958. The development of the fish population in a new flood-control reservoir in Missouri, 1948 to 1954. Trans. Amer. Fish. Soc., 87: 240-58.

P73   _____, 1953. The fishery in Lake Wappapello, a flood-control reservoir on the St. Francis River, Missouri. Trans. Amer. Fish. Soc., 82: 242-54.

P74   _____ and E. M. Lowry, 1953. Age and rate of growth of five species of fish in Black River, Missouri. Univ. Mo. Stud., 26(2): 86-109.

P76   Peart, A. R., 1922. Trout scales and the size limit. Salm. Trout Mag., 28: 52-60.

P77   _____, 1936. How farm fish are really bred. Salm. Trout Mag., 82: 19-24.

P78   Pechacek, L. S., 1956. The effects of tags on rate of growth and condition of several species of cold-water fish in Wyoming. Progr. Fish Cult., 18(3): 120-25.

P80   Pelgen, D. E. and G. W. McCammon, 1955. Second progress report on the tagging of white catfish (Ictalurus catus) in the Sacramento-San Joaquin Delta. Calif. Fish Game, 41(4): 261-9.

P85   Pfeiffer, R. A., 1955. Studies on the life history of the rosyface shiner, Notropis rubellus. Copeia, 1955(2): 95-103.

P86   Phillips, A. M., Jr., H. A. Podoliak, D. R. Brockway, and R. R. Vaughn, 1958. The nutrition of trout. N. Y. Fish. Res. Bull., 21: 93 p.

P89   Phillips, R. W., 1958. Lincoln district. Annu. Rep. Ore. St. Game Comm. Fish. Div., 1957: 211-24.

P92   Piscator, 1952. (Various notes.) Piscator, 6(24): 98-100, 103.

P93   _____, 1953. Secretarial notes. Piscator, 7(25): 30-32.

P95   _____, 1953. Piscator, 7(28): 99,100,103,107,125.

P96   _____, 1954. Piscator, 8(29): 4,8,32.

P99   _____, 1955-61. Piscator, numbers 33-50.

P103   Pitt, T. K., E. T. Garside, and R. L. Hepburn, 1956. Temperature selection of the carp (Cyprinus carpio Linn.). Canad. J. Zool., 34(1956): 555-7.

P104   Platts, W. S., 1959. Homing, movements, and mortality of wild cutthroat trout (Salmo clarki Richardson) spawned artificially. Progr. Fish Cult., 21(1): 36-9.

P106   Power, G., 1958. The evolution of the freshwater races of the Atlantic salmon (Salmo salar L.) in eastern North America. Arctic, 11(2): 86-92.

P107   Prather, E. E., 1957. Experiments on the commercial production of golden shiners. Proc. S. E. Assoc. Game Fish Comm., 10: 150-55.

P108   _____, 1957. Preliminary experiments on winter feeding small fathead minnows. Proc. S. E. Assoc. Game Fish Comm., 11: 249-53.

P109   Prather, E. E., J. R. Fielding, M. C. Johnson, and H. S. Swingle, 1953.
        Production of bait minnows in the Southeast. Ala. Poly. Inst. Agric.
        Exp. Sta. Circ., 112: 71 p.

P110   Probst, R. T., 1954. Why study sturgeon? Wis. Conserv. Bull., 19(3):
        3-5.

P111   _____ and E. L. Cooper, 1955. Age, growth, and production of the lake
        sturgeon (Acipenser fulvescens) in the Lake Winnebago region, Wiscon-
        sin. Trans. Amer. Fish. Soc., 84: 207-27.

P112   Purkett, C. A., Jr., 1958. Growth of the fishes in the Salt River, Mis-
        souri. Trans. Amer. Fish. Soc., 87: 116-31.

P113   _____, 1958. Growth rates of Missouri stream fishes. Missouri
        Dingell-Johnson Ser., 1: 46 p.

P115   Pycha, R. L., 1956. Progress report on white sturgeon studies. Calif.
        Fish Game, 42(1): 23-6.

P119   Platts, W. S., 1958. Age and growth of the cutthroat trout in Strawberry
        Reservoir, Utah. Utah Acad. Sci. Proc., 35: 101-3.

P120   Parker, P. S., 1960. Appalachian sport fishery investigations. U. S.
        Fish Wildl. Serv. Circ., 81: 3-6.

P121   Progressive Fish Culturist, 1960. Note on lake trout. Progr. Fish
        Cult., 22(2): 84.

P122   Parker, R. A., 1958. Some effects of thinning on a population of fishes.
        Ecology, 39(2): 304-17.

P125   Pratt, H. P., 1938. Ecology of the trout of the Gunnison River, Colo-
        rado. PhD thesis Univ. Colo. 197 p.

P126   Prather, E. E., 1959. The use of channel catfish as sport fish. Proc.
        S. E. Assoc. Game Fish Comm., 13: 331-5.

P127   _____ and H. S. Swingle, 1960. Preliminary results on the production
        and spawning of white catfish in ponds. Proc. S. E. Assoc. Game Fish
        Comm., 14: 143-5.

P128   Peckham, R. S., 1955. Ecology and life history of the central mudmin-
        now, Umbra limi (Kirtland). PhD thesis Univ. Notre Dame. 71 p.

P130   Purkett, C. A., Jr., 1961. Reproduction and early development of the
        paddlefish. Trans. Amer. Fish. Soc., 90(2): 125-9.

P132   Palmer, E. L. and A. H. Wright, 1920. A biological reconnaissance of
        the Okefenokee Swamp in Georgia: the fishes. Proc. Ia. Acad. Sci.,
        27: 353-77.

P133   Potter, G. E., 1923. The food of the short-nosed garpike (Lepidosteus
        platostomus) in Lake Okoboji. Proc. Ia. Acad. Sci., 30: 167-70.

P134   Price, J. W., 1961. Food habits of some Lake Erie fish. [Abstract.]
        Proc. Conf. Gr. Lakes Res. Univ. Mich., 4: 160.

P136   Pope, J. A., D. H. Mills, and W. M. Shearer, 1961. The fecundity of
        Atlantic salmon (Salmo salar Linn.). Rep. Freshw. Salm. Fish. Res.
        Scot., 26: 12 p.

P137   Pyle, E. A., G. L. Hammer, and A. M. Phillips, Jr., 1961. The effect
        of grading on the total weight gained by brook trout. Progr. Fish Cult.,
        23(4): 162-8.

P139   Phillips, A. M., Jr., H. A. Podoliak, D. L. Livingston, R. F. Dumas,
        and G. L. Hammer, 1961. Cortland hatchery report 29 for the year
        1960. N. Y. Conserv. Dept. Fish. Res. Bull., 24: 76 p.

P140   Price, J. W., 1940. Time-temperature relations in the incubation of
        the whitefish, Coregonus clupeaformis (Mitchill). J. Gen. Physiol.,
        23(4): 449-68.

P141   Pruginin, Y. and J. Shechter, 1960. The elimination of uncontrolled spawns in carp fattening ponds. Bamidgeh, 12(4): 100-102.

P142   Piavis, G. W., 1961. Embryological stages in the sea lamprey and effects of temperature on development. Fish. Bull. U. S., 182(61): 111-43.

P144   Piper, R. G. and R. F. Stephens, 1962. A comparative study of the blood of wild and hatchery reared lake trout. Progr. Fish Cult., 24(2): 81-4.

P145   Priegel, G. R., 1962. Plentiful but unknown. Wis. Conserv. Bull., 27(3): 13.

P146   Patrick, N. D. and P. Graf, 1962. The effect of temperature on the artificial culture of aurora trout. Canad. Fish Cult., 30: 49-55.

P150   Piggins, D. J., 1961. The age and growth of sea trout of the Burrishoole River. Rep. Salm. Res. Trust Ire., Inc., 1961: 21-27.

P151   ____, 1961. Salmon and sea trout kelts. Rep. Salm. Res. Trust Ire., Inc., 1961: 12-20.

P152   ____, 1961. Fish population and fish food, October, 1960. The Gleenamong River survey. Rep. Salm. Res. Trust Ire., Inc., 1961: 35-8.

P153   Pyefinch, K. A., 1955. A review of the literature on the biology of the Atlantic salmon (*Salmo salar* Linn.). Freshw. Salm. Fish. Res. Scot., 9: 1-24.

P154   Phillips, A. M., Jr., D. L. Livingston, H. A. Poston, and H. A. Booke, 1963. The effect of diet mixture and calorie source on growth, mortality, conversion, and chemical composition of brook trout. Progr. Fish Cult., 25(1): 8-14.

P155   Purkett, C. A., Jr., 1963. Artificial propagation of paddlefish. Progr. Fish Cult., 25(1): 31-3.

P156   Posewitz, J. A., 1962. Observations on the fish population of Willow Creek Reservoir, Montana. Proc. Mont. Acad. Sci., 21: 49-69.

P157   Pycha, R. L., 1962. The relative efficiency of nylon and cotton gill nets for taking lake trout in Lake Superior. J. Fish. Res. Bd. Can., 19(6): 1085-94.

P159   Montana Fish and Game Dept. Fisheries Division, ed. Peters, J. C., 1964. Summary of calculated growth data on Montana fishes, 1948-61. D.-J. Job Completion Rep., F-23-R-6(Jobs I-II): 76 p. mimeo.

P162   Penn, G. H., 1950. Utilization of crayfishes by cold-blooded vertebrates in the eastern United States. Amer. Midl. Nat., 44(3): 643-58.

P163   Pozalujena, E. V., 1928. Beiträge zur Kenntnis des Alters und Wachstumstempo des Karpfen aus dem Aralsee. Rep. sci. Inst. Fisch. Mosk., 3(2): 17-34 original not seen, quoted in F98.

P164   Pate, V. S. L., 1933. Studies on fish food in selected areas. Annu. Rep. N. Y. State Conserv. Dept., 22(Suppl.): 130-56.

P165   Pearse, A. S., 1918. The food of the shore fishes of certain Wisconsin lakes. Bull. U. S. Bur. Fish., 35: 249-92.

P166   Pruginin, Y., 1956. Improvements in the construction of spawning nests for carp. Bamidgeh, 8(5): 88-90.

P167   Pendleton, R. C. and E. W. Smart, 1954. A study of the food relations of the least chub, *Iotichthys phlegethontis* (Cope), using radioactive phosphorus. J. Wildl. Mgmt., 18(2): 226-8.

P168   Patten, B. G., 1960. A high incidence of the hybrid *Acrocheilus alutaceum* x *Ptychocheilus oregonense*. Copeia, 1960(1): 71-3.

P169   Poulson, T. L., 1963. Cave adaptation in Amblyopsid fishes. Amer. Midl. Nat., 70(2): 257-90.

P170   Pike, G. C., 1951. Lamprey marks on whales. J. Fish. Res. Bd. Can., 8(4): 275-80.

P171   Purkett, C. A., Jr., 1963. The paddlefish fishery of the Osage River and the Lake of the Ozarks, Missouri. Trans. Amer. Fish. Soc., 92(3): 239-44.

P173   Pahl, G. and J. Maurer, 1962. Food habits of the young gizzard shad. Proc. Minn. Acad. Sci., 39(1): 30(1): 20-21.

P174   _____ and G. Willfahrt, 1962. Growth study of young gizzard shad. Proc. Minn. Acad. Sci., 30(1): 22-5.

P175   Parsons, J. W. and J. B. Kimsey, 1954. A report on the Mississippi threadfin shad. Progr. Fish Cult., 16(4): 179-81.

P176   Percival, E. and A. M. R. Burnet, 1963. A study of the Lake Lyndon rainbow trout (Salmo gairdnerii). N. Z. J. Sci., 6(2): 273-303.

P177   Phillips, A. M., Jr., A. V. Tunison, and G. C. Balzer, 1963. Trout feeds and feeding. U. S. Fish Wildl. Serv. Circ., 159: 38 p.

P178   Pyefinch, K. A., 1963. Salmon and freshwater fisheries research. Fish. Scot. Rep., 1963: 19 p. extract.

P179   _____, 1964. Salmon and freshwater fisheries research. Fish. Scot. Rep., 1964: 25 p. extract.

P180   Pyle, E. A., 1966. A 42-week study on the growth in length of brook trout. Fish. Res. Bull. Cortland (N. Y.), 29: 34-9.

P181   _____, 1966. Growth rates of large and small brook trout from a graded population. Fish. Res. Bull., Cortland (N.Y), 29: 30-33.

P182   Phillips, A. M., Jr., D. L. Livingston, and H. A. Poston, 1966. Use of calorie sources by brook trout. Progr. Fish Cult., 28(2): 67-72.

P183   _____, 1964. Eastern fish nutrition laboratory. U. S. Fish Wildl. Serv. Circ., 178: 24-31.

P184   Piggins, D. J., 1964. Salmon and sea trout hybrids. Salm. Res. Trust Ire., Inc., 1964: 27-37.

P185   Peckham, R. S. and C. F. Dineen, 1957. Ecology of the central mud-minnow, Umbra limi (Kirtland). Amer. Midl. Nat., 58(1): 222-31.

P186   Peer, D. L., 1966. Relationship between size and maturity in the spot-tail shiner, Notropis hudsonius. J. Fish. Res. Bd. Can., 23(3): 455-7.

P187   Parker, H. L., 1963. Natural history and characteristics of Pimephales vigilax (Baird and Girard). Disser. Abstr., 24(4): 1764.

P188   Priegel, G. R., 1966. Age-length and length-weight relationship of bullheads from Little Lake Butte des Mortes, 1959. Wis. Conserv. Dept. Res. Rep. Fish, 17: 6 p.

P190   Prather, E. E., 1961. A comparison of production of albino and normal channel catfish. Proc. S. E. Assoc. Game Fish Comm., 15: 302-3.

Q1   Qadri, S. U., 1955. The whitefish population of Lac LaRonge. MS thesis Univ. Sask. 136 p.

Q3   _____, 1961. Food and distribution of lake whitefish in Lac LaRonge, Saskatchewan. Trans. Amer. Fish. Soc., 90(3): 303-7.

Q4   Quigley, J. J., 1962. Landlocked arctic char in Newfoundland. Trade News, 15(3): 11-12.

R2   Raney, E. C., 1939. The breeding habits of Ichthyomyzon greeleyi Hubbs and Trautman. Copeia, 1939(2): 111-12.

R3   _____, 1939. The breeding habits of the silvery minnow, Hybognathus regius Girard. Amer. Midl. Nat., 21(3): 674-80.

R4   _____, 1942. The summer food and habits of the chain pickerel (Esox niger) of a small New York pond. J. Wildl. Mgmt., 6(1): 58-66.

R5   Raney, E. C., 1942. Propagation of the silvery minnow *(Hybognathus muchalis regius* Girard) in ponds. Trans. Amer. Fish. Soc., 71: 215-18.

R6   ____, 1947. Subspecies and breeding behavior of the cyprinid fish *Notropis procne* (Cope). Copeia, 1947(2): 103-9.

R8   ____ and E. A. Lachner, 1942. Autumn food of recently planted young brown trout in small streams of central New York. Trans. Amer. Fish. Soc., 71: 106-11.

R11   ____ and ____, 1946. Age, growth, and habits of the hog sucker, *Hypentelium nigricans* (Le Sueur) in New York. Amer. Midl. Nat., 36(1): 76-86.

R12   ____ and ____, 1946. Age and growth of the rustyside sucker, *Thoburnia rhothoeca* (Thoburn). Amer. Midl. Nat., 36(3): 675-81.

R13   ____ and D. A. Webster, 1940. The food and growth of the young of the common bullhead, *Ameiurus nebulosus nebulosus* (LeSueur) in Cayuga Lake, New York. Trans. Amer. Fish. Soc., 69: 205-09.

R14   ____ and ____, 1942. The spring migration of the common white sucker, *Catostomus c. commersonnii* (Lacépède), in Skaneateles Lake Inlet, New York. Copeia, 1942(3): 139-48.

R15   Rawson, D. S., 1932. The pike of Waskesiu Lake, Saskatchewan. Trans. Amer. Fish. Soc., 62: 323-30.

R17   ____, 1941. The eastern brook trout in the Maligne River system, Jasper National Park. Trans. Amer. Fish. Soc., 70: 221-35.

R18   ____, 1946. Successful introduction of fish in a large saline lake. Canad. Fish Cult., 1(1): 5-8.

R19   ____, 1947. Great Slave Lake. Bull. Fish. Res. Bd. Can., 72: 45-8.

R20   ____, 1947. Lake Athabaska. Bull. Fish. Res. Bd. Can., 57: 69-85.

R21   ____ and J. E. Moore, 1944. The saline lakes of Saskatchewan. Canad. J. Res., 22(D): 141-201.

R23   Reighard, J. E., 1914. A plea for the preservation of records concerning fish. Trans. Amer. Fish. Soc., 43: 106-10.

R24   ____, 1915. An ecological reconnaissance of the fishes of Douglas Lake, Cheboygan County, Michigan, in midsummer. Bull. U. S. Comm. Fish., 33: 215-49.

R26   Richardson, R. E., 1913. Observations on the breeding of the European carp in the vicinity of Havana, Illinois. Bull. Ill. St. Lab. Nat. Hist., 9(7): 387-404.

R27   ____, 1913. Observations on the breeding habits of fishes at Havana, Illinois, 1910 and 1911. Bull. Ill. St. Lab. Nat. Hist., 9: 405-16.

R28   Riepe, E., 1906. Einige Daten über das Alter der Goldfische und deren Abarter. Wochenschr. Aquar. Terrar-Kunde, 3: 347-8.

R29   Ricker, W. E., 1932. Studies of speckled trout *(Salvelinus fontinalis)* in Ontario. Publ. Ont. Fish Res. Lab., 44: 68-110.

R31   ____, 1942. Fish populations of two artificial lakes. Invest. Ind. Lakes Streams, 2(13): 255-65.

R34   ____, 1945. Abundance, exploitation and mortality of the fishes in two lakes. Invest. Ind. Lakes Streams, 2(17): 345-448.

R38   ____, 1948. Hybrid sunfish for stocking small ponds. Trans. Amer. Fish. Soc., 75: 84-96.

R41   ____ and D. Merriman, 1945. On the methods of measuring fish. Copeia, 1945(4): 184-91.

R45   Roach, L. S., 1947. Muskellunge. Ohio Conserv. Bull., 11(12): 13.

R46   ____, 1947. Channel catfish. Ohio Conserv. Bull., 11(10): 13.

R51   Roach, L. S., 1948. Common sucker. Ohio Conserv. Bull., 12(5): 13.

R52   _____, 1948. The grass pike. Ohio Conserv. Bull., 12(11): 13.

R55   _____, 1948. Golden mullet. Ohio Conserv. Bull., 12(2): 13.

R56   _____, 1948. In fishing circles. Ohio Conserv. Bull., 12(12): 12-14.

R58   _____, 1948. In fishing circles. Ohio Conserv. Bull., 12(6): 12-13.

R60   _____, 1948. Minnows. Ohio Conserv. Bull., 12(4): 13.

R63   _____, 1949. Long-nosed gar, *Lepisosteus osseus*. Ohio Conserv. Bull., 13(3): 13.

R64   _____, 1949. In fishing circles. Ohio Conserv. Bull., 13(1): 12-13, 25, 16.

R65   Robertson, O. H., 1947. An ecological study of two high mountain trout lakes in the Wind River Range, Wyoming. Ecology, 28(2): 87-112.

R66   _____, 1948. The occurrence of increased activity of the thyroid gland in rainbow trout at the time of transformation from parr to silvery smolt. Physiol. Zool., 21(3): 282-95.

R67   Rodd, J. A., 1930. Unproductive water areas made productive. Trans. Amer. Fish. Soc., 60: 116-18.

R68   _____, 1936. Fish management, Paul Lake, British Columbia. Proc. N. Amer. Wildl. Conf., 1: 324-6.

R72   Roszman, F. D., 1940. Dogfish data. Ohio Conserv. Bull., 4(5): 14.

R74   Royce, W. F., 1942. Standard length versus total length. Trans. Amer. Fish. Soc., 71: 270-74.

R75   Ryder, J. A., 1890. The sturgeons and sturgeon industries of the eastern coast of the United States, with an account of experiments bearing upon sturgeon culture. Bull. U. S. Fish Comm., 8: 231-328.

R76   Rawson, D. S., 1949. Estimating the fish production of Great Slave Lake. Trans. Amer. Fish. Soc., 77: 81-92.

R82   Roach, L. S. and J. Z. Pelton, 1947. Lake management reports, 1. Lake Alma. Ohio Div. Conserv. 27 p. mimeo.

R84   Rayner, H. J., 1949. Rainbow trout. Ore. St. Game Comm. Bull., 4(1): 1, 6-8.

R87   Ricker, W. E., 1946. Production and utilization of fish populations. Ecol. Monogr., 16: 373-91.

R89   Rawson, D. S., 1950. The grayling *(Thymallus signifer)* in northern Saskatchewan. Canad. Fish Cult., 6: 3-10.

R90   Ricker, W. E., 1934. An ecological classification of certain Ontario streams. Univ. Toronto Stud. Biol. Ser., 37: 1-114.

R91   Raney, E. C. and E. A. Lachner, 1947. *Hypentelium roanokense*, a new catostomid fish from the Roanoke River in Virginia. Amer. Mus. Novitates, 1333: 1-15.

R97   Rawson, D. S., 1951. Studies of the fish of Great Slave Lake. J. Fish. Res. Bd. Can., 8(4): 207-40.

R98   _____ and C. A. Elsey, 1950. Reduction in the longnose sucker population of Pyramid Lake, Alberta, in an attempt to improve angling. Trans. Amer. Fish. Soc., 78: 13-31.

R99   Reed, G. B. and J. R. Dymond, 1951. Annual report of the Fisheries Research Board of Canada, 1950. Ottawa, Can. 138 p.

R100   Roach, L. S., 1949. In fishing circles. Ohio Conserv. Bull., 13(12): 12-13.

R105   Rose, E. T., 1950. A fish population study of Storm Lake. Proc. Ia. Acad. Sci., 56: 385-95.

R115   _____ and T. E. Moen, 1951. Results of increased fish harvest in Lost Island Lake. Trans. Amer. Fish. Soc., 80: 50-55.

R120 Rounsefell, G. A. and L. H. Bond, 1950. Growth-control charts applied to Atlantic salmon. Trans. Amer. Fish. Soc., 78: 189-91.

R121 _____, _____, and G. K. White, 1950. Diet experiments on Atlantic salmon in Maine. Progr. Fish Cult., 12(3): 169-72.

R122 Royce, W. F., 1950. The effect of lamprey attacks upon lake trout in Seneca Lake, New York. Trans. Amer. Fish. Soc., 79: 71-6.

R123 Runnström, S., 1949. Control of trout migration by a fish ladder. Rep. Inst. Freshw. Res. Drottning., 29: 85-8.

R124 _____, 1950. Director's report for the year 1949. Rep. Inst. Freshw. Res. Drottning., 31: 5-18.

R125 Ruhr, C. E., 1952. Fish population of a mining pit lake, Marion County, Iowa. Ia. St. J. Sci., 27(3): 55-77.

R128 Rounsefell, G. A. and L. D. Stringer, 1945. Restoration and management of the New England alewife fisheries with special reference to Maine. Trans. Amer. Fish. Soc., 73: 394-424.

R129 Raney, E. C., 1955. Natural hybrids between two species of pickerel (Esox) in Stearns Pond, Massachusetts. Mass. Fish. Annu. Rep., 1951-52(Suppl.): 406-19.

R134 Rawson, D. S., 1953. The limnology of Amethyst Lake, a high alpine type near Jasper, Alberta. Canad. J. Zool., 31: 193-210.

R136 _____, 1957. Limnology and fisheries of five lakes in the Upper Churchill drainage, Saskatchewan. Sask. Dept. Nat. Res. Fish. Rep., 3: 61 p.

R137 _____, 1957. The life history and ecology of the yellow walleye, Stizostedion vitreum, in Lac LaRonge, Saskatchewan. Trans. Amer. Fish. Soc., 86: 15-37.

R138 _____, 1959. Limnology and fisheries of Cree and Wollaston Lakes in northern Saskatchewan. Sask. Dept. Nat. Res. Fish. Rep., 4: 5-73.

R139 _____ and F. M. Atton, 1953. Biological investigation and fisheries management at Lac LaRonge, Saskatchewan. Sask. Dept. Nat. Res. Fish. Branch. 39 p.

R140 Raymond, C. J., 1936. The value of hatchery trout for stocking rivers. Salm. Trout Mag., 83: 135-44.

R141 _____, 1937. Hebridean lochs. Salm. Trout Mag., 87: 134-9.

R142 _____, 1938. Brown-trout geology. Salm. Trout Mag., 93: 329-38.

R143 Reed, R. J., 1957. Phases of the life history of the rosyface shiner, Notropis rubellus, in northwestern Pennsylvania. Copeia, 1957(4): 286-90.

R144 _____, 1958. The early life history of two cyprinids, Notropis rubellus and Campostoma anomalum pullum. Copeia, 1958(4): 325-7.

R145 _____, 1959. Age, growth, and food of the longnose dace, Rhinichthys cataractae, in northwestern Pennsylvania. Copeia, 1959(2): 160-62.

R146 Regan, C. T., 1927. Atlantic salmon in New Zealand. Tasmanian and New Zealand salmon at the Natural History Museum. With a note on the scales by J. A. Hutton. Salm. Trout Mag., 48: 234-9.

R148 Rehder, D. D., 1959. Some aspects of the life history of the carp, Cyprinus carpio, in the Des Moines River, Boone County, Iowa. Ia. St. J. Sci., 34(1): 11-27.

R149 Reimers, N., 1957. Some aspects of the relation between stream foods and trout survival. Calif. Fish Game, 43(1): 43-69.

R150 _____, 1958. Conditions of existence, growth, and longevity of brook trout in a small, high-altitude lake of the eastern Sierra Nevada. Calif. Fish Game, 44(4): 319-34.

R151 Reimers, N., J. A. Maciolek, and E. P. Pister, 1955. Limnological study of the lakes in Convict Creek Basin, Mono County, California. U. S. Fish. Bull., 56(103): 437-503.

R152 Ricker, W. E., 1955. Fish and fishing in Spear Lake, Indiana. Invest. Ind. Lakes Streams, 4: 117-62.

R153 Ridenhour, R. L., 1957. Northern pike, *Esox lucius* L., population of Clear Lake, Iowa. Ia. St. J. Sci., 32(1): 1-18.

R155 Rivers, C. M. and H. E. Mastin, 1956. Rogue River and south coastal streams. Annu. Rep. Ore. St. Game Comm. Fish. Div., 1955: 21-36.

R158 ____ and ____, 1959. Rogue River and south coastal streams. Annu. Rep. Ore. St. Game Comm. Fish. Div., 1958: 27-48.

R159 Roberts, R. C., 1932. The pathology of four rainbow kelts. Salm. Trout Mag., 68: 268-73.

R160 Robins, C. R. and E. E. Deubler, Jr., 1955. The life history and systematic status of the burbot, *Lota lota lacustris* (Walbaum), in the Susquehanna River system. N. Y. Mus. Sci. Serv. Circ., 39. 49 p.

R162 Roelofs, E. W., 1958. Age and growth of whitefish, *Coregonus clupeaformis* (Mitchill) in Big Bay De Noc and northern Lake Michigan. Trans. Amer. Fish. Soc., 87: 190-99.

R179 Rounsefell, G. A., 1957. Fecundity of North American salmonidae. U. S. Fish. Bull., 57(122): 451-68.

R180 Roussow, G., 1957. Some considerations concerning sturgeon spawning periodicity. J. Fish. Res. Bd. Can., 14(4): 553-72.

R181 Rowley, W., 1955. 1954 creel census, North Fork Feather River. Calif. Inl. Fish. Admin. Rep., 55-9. 18 p. mimeo.

R182 Ruhr, C. E., 1956. Nature page (paddlefish). Tenn. Conserv., 22(4): 23-4.

R183 Runnström, S., 1952. The population of trout, *Salmo trutta* Linne, in regulated lakes. Inst. Freshw. Res. Drottning. Bull., 33: 179-98.

R184 ____, 1955. Changes in fish production in impounded lakes. Proc. Int. Assoc. Theor. Appl. Limnol., 12: 176-82.

R185 Rupp, R. S., 1955. Studies of the eastern brook trout population and fishery in Sunkhaze Stream, Maine. J. Wildl. Mgmt., 19(3): 336-45.

R186 ____, 1958. Smelt family (Osmeridae), American smelt, *Osmerus mordax* (Mitchill). Fishes of Maine, 2nd ed., ed. W. H. Everhart: 55-6.

R187 Rushton, W., 1941. Biological notes—growth rate of trout. Salm. Trout Mag., 103: 249-51.

R189 Ryhanen, R., 1959. Summary on observations concerning the trout in the Isojoki (Finland). Madt. Kala. Tutk., 6(1): 1-9.

R192 Rupp, R. S., 1959. Variation in the life history of the American smelt in inland waters of Maine. Trans. Amer. Fish. Soc., 88(4): 241-52.

R193 Roos, J. F., 1959. Feeding habits of the Dolly Varden, *Salvelinus malma* (Walbaum), at Chignik, Alaska. Trans. Amer. Fish. Soc., 88(4): 253-60.

R196 Ricker, W. E., 1949. Mortality rates in some little-exploited populations of fresh-water fishes. Trans. Amer. Fish. Soc., 66: 114-28.

R199 Roussow, G., 1958. *Acipenser fulvescens* Rafinesque. Data for Handbook of Biological Data. 28 p.

R202 Rabe, F. W., III, 1957. Brook trout populations in Colorado beaver ponds. MS thesis Colo. St. Univ. 104 p.

R203 Rawson, D. S., 1960. Five lakes on the Churchill River near Stanley, Saskatchewan. Sask. Dept. Nat. Res. Fish. Rep., 5: 38 p.

R206   Riggs, C. D. and G. A. Moore, 1958. The occurrence of *Signalosa pete-nensis* in Lake Texoma. Proc. Okla. Acad. Sci., 38: 64-7.

R208   Rothschild, B. J., 1961. Production and survival of eggs of the American smelt, *Osmerus mordax* (Mitchill), in Maine. Trans. Amer. Fish. Soc., 90(1): 42-8.

R210   Ryer, R., III, 1938. Contributions to the life history of *Notropis cornutus cornutus* (Mitchill). MS thesis Cornell Univ. 41 p.

R211   Rawson, D. S., 1961. The lake trout of Lac La Ronge, Saskatchewan. J. Fish. Res. Bd. Can., 18(3): 423-62.

R214   Ryder, R. A., 1961. First Ontario record of the Arctic char, *Salvelinus alpinus*. Copeia, 1961(3): 359-60.

R215   Ricker, W. E., 1962. Connecticut River shad. Trans. Amer. Fish. Soc., 91(1): 98-9.

R216   Royer, L. M., 1960. A study of the ecology of the brook trout in Little Brooklyn Lake and Towner Lake, Medicine Bow Forest, Wyoming. MS thesis Univ. Wyo.; Wyo. Game Fish Comm. Univ. Wyo. Coop. Res. Proj., 2: 1-92.

R220   Rozanska, Z., 1961. [The food of lake trout fry (*Salmo trutta morpha lacustris* L.) and other fish species in Trzebiocha Stream.] (English summary.) Rocz. Nauk Rolniczych, 93: 387-422.

R222   Robel, R. J., 1962. Weight increases of carp in populations of different densities. Trans. Amer. Fish. Soc., 91(2): 234-6.

R223   Rogers, D. E., 1962. Reproduction of three spine stickleback. Univ. Wash. Fish. Res. Inst. Contr., 139: 9-10.

R224   Ricker, W. E., 1959. Additional observations concerning residual sockeye and kokanee *(Oncorhynchus nerka)*. J. Fish. Res. Bd. Can., 16(6): 897-902.

R225   _____, 1938. "Residual" and kokanee salmon in Cultus Lake. J. Fish. Res. Bd. Can., 4(3): 192-218.

R226   _____, 1937. The food and food supply of sockeye salmon (*Oncorhynchus nerka* Walbaum) in Cultus Lake, British Columbia. J. Biol. Bd. Can., 3(5): 450-68.

R227   _____, 1940. On the origin of kokanee, a freshwater type of sockeye salmon. Trans. Roy. Soc. Can. Ser. III, 34(5): 121-36.

R228   Raney, E. C., 1942. Alligator gar feeds upon birds in Texas. Copeia, 1942(1): 50.

R229   Rayner, H. J., 1942. The spawning migration of rainbow trout at Skaneateles Lake, New York. Trans. Amer. Fish. Soc., 71: 180-83.

R230   Reed, E. B., 1962. Limnology and fisheries of the Saskatchewan River in Saskatchewan. Sask. Fish. Rep., 6: 1-48.

R232   Robertson, O. H., 1949. Production of the silvery smolt stage in rainbow trout by intramuscular injection of mammalian thyroid extract and thyrotropic hormone. J. Exp. Zool., 110(3): 337-55.

R233   Reimers, N., 1963. Body condition, water temperature, and overwinter survival of hatchery-reared trout in Convict Creek, California. Trans. Amer. Fish. Soc., 92(1): 39-46.

R234   Ricker, W. E., 1930. Feeding habits of speckled trout in Ontario waters. Trans. Amer. Fish. Soc., 60: 64-72.

R235   Rothschild, B. J., 1963. A critique of the scale method for determining the age of the alewife, *Alosa pseudoharengus* (Wilson). Trans. Amer. Fish. Soc., 92: 409-13.

R236   Royce, W. F., 1951. Breeding habits of lake trout in New York. Fish. Bull. U. S., 52(59): 59-76.

R238   Rawstrom, R. R., 1964. Spawning of threadfin shad, *Dorosoma pete-nense,* at low temperatures. Calif. Fish Game, 50(1): 59.

R239   Regier, H. A., 1963. Ecology and management of largemouth bass and golden shiners in farm ponds in New York. N. Y. Fish Game J., 10(2): 139-69.

R240   _____, 1963. Ecology and management of channel catfish in farm ponds in New York. N. Y. Fish Game J., 10(2): 170-85.

R241   Ross, D. A., 1940. Jackfish investigations, 1940, Athabaska Delta, Alberta. Publ. Ducks Unlimited, Winnipeg.

R242   Runnström, S., 1964. Effects of impoundment on the growth of *Salmo trutta* and *Salvelinus alpinus* in Lake Ransaren (Swedish Lappland). Verh. Int. Ver. Limnol., 15: 453-61.

R243   Raney, E. C., 1940. Comparison of the breeding habits of two subspecies of black-nosed dace, *Rhinichthys atratulus* (Hermann). Amer. Midl. Nat., 23(2): 399-403 not seen, quoted in S441.

R244   Ricker, W. E., 1933. Destruction of sockeye salmon by predatory fish. Fish. Res. Bd. Can. Progr. Rep. Pacif. Biol. Sta., 18: 3-4.

R245   _____, 1941. The consumption of young sockeye salmon by predaceous fish. J. Fish. Res. Bd. Can., 5: 293-313.

R246   Raney, E. C., 1947. *Nocomis* nests used by other breeding cyprinid fishes in Virginia. Zoologica, 32(3): 125-32.

R247   Radcliffe, L., 1931. Propagation of minnows. Trans. Amer. Fish. Soc., 61: 131-8.

R248   Reighard, J. E., 1910. Methods of studying the habits of fishes with an account of the breeding habits of the horned dace. Bull. U. S. Bur. Fish., 28: 1111-36.

R249   Rimsky-Korsakoff, V. N., 1930. The food of certain fishes of the Lake Champlain watershed. Annu. Rep. N. Y. State Conserv. Dept., 19 (Suppl.): 88-104.

R250   Raney, E. C., 1940. Reproductive activities of a hybrid minnow, *Notropis cornutus* x *Notropis rubellus.* Zoologica, 25(3): 361-8.

R251   _____, 1940. *Rhinichthys bowersi* from West Virginia, a hybrid, *Rhinichthys cataractae* x *Nocomis micropogon.* Copeia, 1940(4): 270-71.

R252   _____, 1940. The breeding behavior of the common shiner, *Notropis cornutus* (Mitchill). Zoologica, 25(1): 1-14.

R253   Rose, E. T. and T. E. Moen, 1953. The increase in game fish populations in East Okoboji Lake, Iowa, following intensive removal of rough fish. Trans. Amer. Fish. Soc., 82: 104-14.

R254   Ramaswami, L. S. and A. D. Hasler, 1955. Hormones and secondary sex characters in the minnow, *Hyborhynchus.* Physiol. Zool., 28(1): 62-8.

R255   Reisinger, E., 1953. Zum Saiblingsproblem. Corinthia II(2): 74-102.

R256   Reed, R. J., 1954. Hermaphroditism in the rosyface shiner, *Notropis rubellus.* Copeia, 1954(4): 293-94.

R257   _____, 1957. The prolonged spawning of the rosyface shiner, *Notropis rubellus* (Agassiz), in northwestern Pennsylvania. Copeia, 1957(3): 250.

R258   Raney, E. C., 1939. Observations on the nesting habits of *Parexoglossum laurae* Hubbs and Trautman. Copeia, 1939(2): 112-13.

R259   Richardson, L. R., 1939. The spawning behaviour of *Fundulus diaphanus* (Le Sueur). Copeia, 1939(3): 165-7.

R260   Reighard, J. E., 1920. Breeding behavior of suckers and minnows. I. Suckers. Biol. Bull., Wood's Hole, 38: 1-3.

R262   Richards, S. W. and A. M. McBean, 1966. Comparison of postlarval and juveniles of *Fundulus heteroclitus* and *Fundulus majalis* (Pisces: Cyprinodontidae). Trans. Amer. Fish. Soc., 95(2): 218-26.

R263   Riggs, C. D. and G. A. Moore, 1960. Growth of young gar *(Lepisosteus)* in aquaria. Proc. Okla. Acad. Sci., 40: 44-6.

R264   Raney, E. C., 1952. A new lamprey, *Ichthyomyzon hubbsi* from the Upper Tennessee River system. Copeia, 1952(2): 93-9.

R265   Reighard, J. E. and H. Cummins, 1916. Description of a new species of lamprey of the genus *Ichthyomyzon*. Occ. Pap. Mus. Zool. Univ. Mich., 31: 1-12.

R266   Rodman, D. T., 1963. Anesthetizing and air-transporting young white sturgeons. Progr. Fish Cult., 25(2): 71-8.

R267   Rothschild, B. J., 1965. Aspects of the population dynamics of the alewife, *Alosa pseudoharengus* (Wilson), in Cayuga Lake, New York. Amer. Midl. Nat., 74(2): 479-96.

R268   Regier, H. A. and D. S. Robson, 1966. Selectivity of gill nets, especially to lake whitefish. J. Fish. Res. Bd. Can., 23(3): 423-54.

R269   Robbins, O., Jr., 1966. Flathead Lake (Montana) fishery investigations, 1961-64. Tech. Pap. U. S. Bur. Sport Fish. Wildl., 4: 26 p. + 19 fig.

R270   Rogers, D. E., 1964. Some morphological and life history data on pygmy whitefish from Wood River. Univ. Wash. Coll. Fish. Contr., 166: 24-5.

R271   Regan, D. M., 1966. Ecology of Gila trout in Main Diamond Creek in New Mexico. Tech. Pap. U. S. Bur. Sport Fish. Wildl., 5: 24 p.

R272   Rogers, D. E., M. O. Nelson, J. J. Pella, and R. L. Burgner, 1963. Relative abundance and distribution of fish species in Lake Aleknagik. Univ. Wash. Coll. Fish. Contr., 17: 14-5.

R273   Rabe, F. W., III, and V. E. Dyer. 1964. Age and growth study of brook trout from three cirque lakes in the Uinta Mountains. Proc. Utah Acad. Sci. Arts Lett., 41(2): 243-54.

R274   ____, 1967. The transplantation of brook trout in an alpine lake. Progr. Fish Cult., 29(1): 53-5.

R275   Rahrer, J. F., 1963. Age and growth of four species of fish from Flathead Lake, Montana. Proc. Mont. Acad. Sci., 23: 144-56.

R276   Reed, R. J., 1967. Observations of fishes associated with spawning salmon. Trans. Amer. Fish. Soc., 96(1): 62-7.

R277   Rahrer, J. F., 1965. Age, growth, maturity and fecundity of "humper" lake trout, Isle Royale, Lake Superior. Trans. Amer. Fish. Soc., 94(1): 75-83.

R278   Rupp, R. S. and M. A. Redmond, 1966. Transfer studies of ecologic and genetic variation in the American smelt. Ecology, 47(2): 253-9.

R279   Reeves, J. E., 1964. Age and growth of hardhead minnow, *Mylopharadon conocephalus* (Baird and Girard) in the American River basin of California with notes on its ecology. MS thesis Univ. Calif. Berkeley. 90 p. not seen, quoted from B340.

R280   Rawstrom, R. R., 1967. Harvest, mortality, and movement of selected warmwater fishes in Folsom Lake, California. Calif. Fish Game, 53(1): 40-48.

R281   Russell, T. R., 1964. Age, growth and food habits of channel catfish in channelized and unchannelized portions of the Missouri River, Nebraska (abstract). Midw. Wildl. Conf. Program Abstr., 26: 53-4.

R282   Rounsefell, G. A. and W. H. Everhart, 1953. Fishery science, its methods and applications. J. Wiley and Sons, N. Y. 444 p.

R283   Ricker, W. E., 1958. Handbook of computations for biological statistics of fish populations. Fish. Res. Bd. Can. Bull., 119: 300 p.

R284   Riffenburgh, R. H., 1966. On growth parameter estimation for early life stages. Biometrics, 22(1): 162-78.

S1   Schaffner, D. C., 1902. Notes on the occurrence of ammocoetes, the larval form of *Lampetra wilderi*, near Ann Arbor. Rep. Mich. Acad. Sci., 3: 71-2.

S2   Schloemer, C. L., 1936. The growth of the muskellunge, *Esox masquinongy immaculatus* (Garrard), in various lakes and drainage areas of northern Wisconsin. Copeia, 1936(4): 185-93.

S3   _____, 1938. A second report on the growth of the muskellunge, *Esox masquinongy immaculatus* (Garrard), in Wisconsin waters. Trans. Wis. Acad. Sci. Arts Lett., 31: 507-12.

S5   Schiemenz, F., 1930. Pisces. Tabulae Biologicae, 6: 582-97.

S7   Schneberger, E., 1937. The biological and economic importance of the smelt in Green Bay. Trans. Amer. Fish. Soc., 66: 139-42.

S8   _____, 1937. The food of small dogfish, *Amia calva*. Copeia, 1937(1): 61.

S9   _____ and L. A. Woodbury, 1944. The lake sturgeon, *Acipenser fulvescens* Rafinesque, in Lake Winnebago, Wisconsin. Trans. Wis. Acad. Sci. Arts Lett., 36: 131-40.

S10   Schneider, P. W. and F. P. Griffiths, 1943. Production of trout in a small artificial pond in western Oregon. J. Wildl. Mgmt., 7(2): 148-54.

S15   Schoffman, R. J., 1942. Age and growth of the carp in Reelfoot Lake. J. Tenn. Acad. Sci., 17(1): 68-77.

S16   _____, 1943. Age and growth of the gourdhead buffalo in Reelfoot Lake. J. Tenn. Acad. Sci., 18(1): 36-46.

S17   _____, 1944. Age and growth of the smallmouth buffalo in Reelfoot Lake. J. Tenn. Acad. Sci., 29(1): 3-9.

S20   Schuck, H. A., 1942. The effect of jaw-tagging upon the condition of trout. Copeia, 1942(1): 33-9.

S21   _____, 1945. Survival, population density, growth, and movement of the wild brown trout in Crystal Creek. Trans. Amer. Fish. Soc., 73: 209-30.

S22   _____, 1948. Survival of hatchery trout in streams and possible methods of improving the quality of hatchery trout. Progr. Fish Cult., 10(1): 3-14.

S23   _____ and O. R. Kingsbury, 1948. Survival and growth of fingerling brown trout *(Salmo fario)* reared under different hatchery conditions and planted in fast and slow water. Trans. Amer. Fish. Soc., 75: 147-56.

S24   Schultz, L. P., 1930. The life history of *Lampetra planeri* Block, with a statistical analysis of the rate of growth of larvae from western Washington. Occ. Pap. Mus. Zool. Univ. Mich., 221: 1-35.

S27   Segerstrale, C., 1937. Studies rorande havsforellen (*Salmo trutta* L.) i Sodra Finland, speciellt pa Karelska naset och i Nyland. Acta Soc. Fauna Flora Fenn., 60: 696-750.

S28   Shapovalov, L., 1937. Experiments in hatching steelhead eggs in gravel. Calif. Fish Game, 23(3): 208-14.

S30   Shebley, W. H., 1931. Trout lives 19 years. Calif. Fish Game, 17(4): 441.

S33   Shetter, D. S., 1939. Success of plantings of fingerling trout in Michigan waters as demonstrated by marking experiments and creel census. Trans. N. Amer. Wildl. Conf., 4: 318-25.

S34 Shetter, D. S., 1944. Anglers' catches from portions of certain Michigan trout streams in 1939 and 1940, with discussion of indices to angling quality. Pap. Mich. Acad. Sci. Arts Lett., 29: 305-14.

S37 _____ and A. S. Hazzard, 1939. Species composition by age groups and stability of fish populations in sections of three Michigan trout streams during the summer of 1937. Trans. Amer. Fish. Soc., 68: 281-302.

S39 _____ and J. W. Leonard, 1943. A population study of a limited area in a Michigan trout stream, September, 1940. Trans. Amer. Fish. Soc., 72: 35-51.

S40 Shillington, K. G., 1934. Selective breeding of speckled trout. Trans. Amer. Fish. Soc., 64: 274-5.

S41 Shira, A. F., 1917. Notes on the rearing, growth, and food of the channel catfish, *Ictalurus punctatus*. Trans. Amer. Fish. Soc., 46: 77-88.

S49 Smallwood, W. M. and M. L. Smallwood, 1929. The German carp, an invited immigrant. Sci. Monthly, 29: 394-401.

S50 _____ and P. H. Struthers, 1928. Carp control studies in Oneida Lake. Biol. Surv. of Oswego River System. Annu. Rep., N. Y. Conserv. Dept., 17(Suppl.): 67-83.

S55 Smith, H. M., 1896. A review of the history and results of the attempts to acclimatize fish and other water animals in the Pacific states. Bull. U. S. Fish Comm., 1895, 15: 379-472.

S56 _____, 1899. Report on the inquiry respecting food fishes and the fishing grounds. Rep. U. S. Fish Comm., 1898, 24: CXXIII-CXLVI.

S57 _____, 1907. The fishes of North Carolina. N. C. Geol. Econ. Surv., II: 1-453.

S58 _____, 1912. The prolificness of *Gambusia*. Science, NS 36: 224.

S59 _____ and L. G. Harron, 1903. Breeding habits of the yellow cat-fish. Bull. U. S. Fish Comm., 22: 149-54.

S60 Smith, L. L., Jr., 1941. The results of planting brook trout of legal length in the Salmon Trout River, northern Michigan. Trans. Amer. Fish. Soc., 70: 249-59.

S62 _____ and N. L. Moe (compilers), 1944. Minnesota fish facts. Minn. Dept. Conserv. Bull., 7: 1-31.

S65 Smith, M. W., 1939. Fish population of Lake Jesse, Nova Scotia. Proc. Nova Scotian Inst. Sci., 19(4): 389-427.

S67 _____, 1943. Atlantic salmon in Lake Jesse, Nova Scotia. Copeia, 1943(4): 257.

S70 _____, 1947. Food of killifish and white perch in relation to supply. J. Fish. Res. Bd. Can., 7(1): 22-34.

S71 _____, 1947. Yield of speckled trout to anglers from a Prince Edward Island pond. Progr. Rep. Atlant. Coast Sta. Fish. Res. Bd. Can., 38: 3-5.

S76 Smith, O. H. and J. Van Oosten, 1940. Tagging experiments with lake trout, whitefish, and other species of fish from Lake Michigan. Trans. Amer. Fish. Soc., 69: 63-84.

S77 Smith, O. R., 1947. Returns from natural spawning of cutthroat trout and eastern brook trout. Trans. Amer. Fish. Soc., 74: 281-96.

S78 Snieszko, S. F., 1941. Pond fish farming in Poland. A Symposium on Hydrobiology, Univ. Wis. Press: 227-40.

S79 Snyder, C. O., 1938. A study of the trout (*Salmo irideus* Gibbons) from Waddell Creek, California. Calif. Fish Game, 24(4): 354-75.

S80   Snyder, J. O., 1925. The half-pounder of Eel River, a steelhead trout.
       Calif. Fish Game, 11: 49-55.

S81   _____, 1926. The trout of the Sierra San Pedro Martir, Lower Califor-
       nia. Univ. Calif. Publ. Zool., 21(17): 419-26.

S82   _____, 1933. California trout. Calif. Fish Game, 19: 81-112.

S83   Solman, V. E. F., 1945. The ecological relations of pike, *Esox lucius*
       L., and waterfowl. Ecology, 26(2): 157-70.

S84   Southern, R., 1928. Salmon of the River Shannon—1924, 1925, and 1926.
       Proc. Roy. Irish Acad., 28(13): 38-64.

S85   _____, 1932. The food and growth of brown trout. Salm. Trout Mag.,
       69: 168-76, 243-58, 339-44.

S88   Speaker, E. B., 1946. Oscar the sturgeon. Ia. Conserv., 5(8): 60.

S89   Spoor, W. A., 1938. Age and growth of the sucker, *Catostomus com-
       mersonnii* (Lacépède), in Muskellunge Lake, Vilas County, Wisconsin.
       Trans. Wis. Acad. Sci. Arts Lett., 31: 457-505.

S90   Starrett, W. C., 1948. An ecological study of the minnows of the Des
       Moines River, Boone County, Iowa. PhD thesis Ia. St. Coll. Lib. 161 p.

S91   Stewart, N. H., 1926. Development, growth, and food habits of the white
       sucker *(Catostomus commersonii)*. Fish. Bull. U. S., 42: 147-84.

S92   Stokell, G., 1934. New light on New Zealand salmon. Salm. Trout
       Mag., 76: 260-76.

S93   Stone, U. B., 1938. Growth, habits, and fecundity of the ciscoes of
       Irondequoit Bay, New York. Trans. Amer. Fish. Soc., 67: 234-45.

S94   _____, 1947. A study of the deep-water cisco fishery of Lake Ontario
       with particular reference to the bloater, *Leucichthys hoyi* (Gill).
       Trans. Amer. Fish. Soc., 74(1944): 230-49.

S100   Stroud, R. H., 1949. Rate of growth and condition of game and pan fish
       in Cherokee and Douglas Reservoirs, Tennessee, and Hiwassee Reser-
       voir, North Carolina. J. Tenn. Acad. Sci., 24(1): 60-74.

S101   Struthers, P. H., 1929. Carp control studies in the Erie canal. A bio-
       logical survey of the Erie-Niagara system. Annu. Rep. N. Y. Conserv.
       Dept., 18(Suppl.): 208-19.

S102   _____, 1930. Carp control studies in the Cayuga and Owasco Lake
       Basin. A biological survey of Champlain watershed. Annu. Rep. N. Y.
       Conserv. Dept., 19(Suppl.): 261-80.

S103   Sumner, F. H., 1948. Age and growth of steelhead trout, *Salmo gaird-
       nerii* Richardson, caught by sport and commercial fishermen in Tilla-
       mook County, Oregon. Trans. Amer. Fish. Soc., 75: 77-83.

S104   Surber, E. W., 1933. A quantitative study of rainbow trout production
       in one mile of stream. Trans. Amer. Fish. Soc., 63: 251-6.

S105   _____, 1937. Rainbow trout and bottom fauna production in one mile of
       stream. Trans. Amer. Fish. Soc., 66: 193-202.

S107   _____, 1940. An appraisal of the results of planting fingerling trout in
       St. Mary River, Virginia. Progr. Fish Cult., 49: 1-13.

S112   Surber, T., 1933. Rearing lake trout to maturity. Trans. Amer. Fish.
       Soc., 63: 64-5.

S113   _____, 1940. Propagation of minnows. Minn. Dept. Conserv. 22 p.

S114   Swartz, A. H., 1942. (A fisheries survey of some lakes and ponds in the
       Connecticut River drainage system.) Mass. Dept. Conserv. Fish. Surv.
       Rep., 1942: 1-180.

S117   Swingle, H. S., 1949. Experiments with combinations of largemouth
       black bass, bluegills, and minnows in ponds. Trans. Amer. Fish. Soc.,
       76: 46-62.

CITATIONS   695

S122 Schaeperclaus, W., 1933. Lehrbuch der Teichwirtschaft. Book Pub-
      lishing House Paul Parey, Berlin, 289 p. Transl. Frederick Hund as
      Textbook of Pond Culture. U. S. Fish Wildl. Serv. Leafl., 311: 260 p.
S123 Smith, E., 1902. The home aquarium. N. Y. 213 p.
S125 Snedecor, G. W., 1956. Statistical methods. 5th edition. Ia. St. Univ.
      Press, Ames. 534 p.
S126 Sumner, F. H., 1948. The coast cutthroat trout. Ore. St. Game Comm.
      Bull., 3(12): 1,6-8.
S132 Stenton, J. E., 1950. Artificial hybridization of eastern brook trout and
      lake trout. Canad. Fish. Cult., 3: 20-22.
S134 Svärdson, G., 1949. Competition between trout and charr (*Salmo trutta*
      and *S. alpinus*). Rep. Inst. Freshw. Res. Drottning., 1948: 108-11.
S137 Scott, W. B., 1951. Fluctuations in abundance of the Lake Erie cisco,
      (*Leucichthys artedi*) population. Contr. Roy. Ont. Mus. Zool., 32: 1-41.
S139 Seaman, E. A., 1951. Snaring through the ice for nongame fish in West
      Virginia. Progr. Fish Cult., 13(2): 86-90.
S140 Shetter, D. S., 1950. Results from plantings of marked fingerling brook
      trout (*Salvelinus f. fontinalis* Mitchill) in Hunt Creek, Montmorency
      County, Michigan. Trans. Amer. Fish. Soc., 79: 77-93.
S142 _____, 1951. The effect of fin removal on fingerling lake trout (*Cris-
      tivomer namaycush*). Trans. Amer. Fish. Soc., 80: 260-77.
S144 Sigler, W. F., 1951. The life history and management of the Rocky
      Mountain whitefish, *Prosopium williamsoni* (Girard) in the Logan
      River, Utah. Utah St. Agric. Coll. Bull., 347: 1-21.
S146 Simon, J. R., 1946. Wyoming fishes. Wyo. Game Fish Dept. Bull., 4:
      1-129.
S147 Simpson, J. C., 1951. Fishes of Idaho, No. 13. Northern brown bull-
      head. Idaho Wildl. Rev., 1951(Apr.-May): 7.
S148 _____, 1952. Idaho fish hatchery production. Idaho Wildl. Rev., 1952
      (May-June): 4-5.
S149 Simson, J. and J. Culbreath, 1950. Sagebrush trout. Colo. Conserv.,
      1950(Dec.): 3-4,22.
S151 Smith, C. G., W. M. Lewis, and H. M. Kaplan, 1952. A comparative
      morphologic and physiologic study of fish blood. Progr. Fish Cult.,
      14(4): 169-72.
S155 Smith, M. W., 1952. The lake whitefish in Kerr Lake, New Brunswick.
      J. Fish. Res. Bd. Can., 8(5): 340-46.
S156 _____, 1952. Limnology and trout angling in Charlotte County lakes,
      New Brunswick. J. Fish. Res. Bd. Can., 8(6): 383-452.
S159 Sneed, K. E., 1951. A method for calculating the growth of channel
      catfish, *Ictalurus lacustris punctatus*. Trans. Amer. Fish. Soc., 80:
      174-83.
S161 Solman, V. E. F., 1950. Limnological investigation of Fundy (New
      Brunswick) National Park, 1948. Wildl. Mgmt. Bull. Canad. Wildl.
      Serv., 3(2): 1-48.
S162 _____, 1951. Limnological investigations in Cape Breton Highlands
      National Park, Nova Scotia, 1947. Wildl. Mgmt. Bull. Canad. Wildl.
      Serv., 3(3): 1-52.
S164 Speirs, J. M., 1952. Nomenclature of the channel catfish and the burbot
      of North America. Copeia, 1952(2): 99-103.
S165 Sprules, W. M., 1952. The Arctic char of the west coast of Hudson
      Bay. J. Fish. Res. Bd. Can., 9(1): 1-15.

S168   Stokell, G., 1951. The American lake charr *(Cristivomer namaycush)*. Trans. Roy. Soc. N. Z., 79(2): 213-17.

S178   Shetter, D. S., 1952. The mortality and growth of marked and unmarked lake trout fingerlings in the presence of predators. Trans. Amer. Fish. Soc., 81: 17-34.

S179   Sigler, W. F., 1952. Age and growth of the brown trout, *Salmo trutta fario* Linnaeus, in Logan River, Utah. Trans. Amer. Fish. Soc., 81: 171-8.

S185   Saila, S. B., 1952. Report on farm fish ponds for New York State Conservation Council. 3 p. mimeo.

S186   _____, 1956. Estimates of the minimum size-limit for maximum yield and production of chain pickerel, *Esox niger* LeSueur, in Rhode Island. Limnol. Oceanogr., 1(3): 195-201.

S188   _____ and D. Horton, 1957. Fisheries investigations and management in Rhode Island lakes and ponds. R. I. Div. Fish Game, Fish. Publ., 3: 1-134.

S190   Salmon and Trout Magazine, 1921. The Hampshire Avon. Past, present and future. Salm. Trout Mag., 24: 263-87.

S192   Sanderson, A. E., Jr., 1958. Smallmouth bass management in the Potomac River basin. Trans. N. Amer. Wildl. Conf., 23: 248-62.

S194   Saunders, J. W. and M. W. Smith, 1955. Standing crops of trout in a small Prince Edward Island stream. Canad. Fish. Cult., 17: 32-9.

S200   Schoffman, R. J., 1954. Age and rate of growth of the channel catfish in Reelfoot Lake, Tennessee. J. Tenn. Acad. Sci., 29(1): 2-8.

S201   _____, 1955. Age and rate of growth of the yellow bullhead in Reelfoot Lake, Tennessee. J. Tenn. Acad. Sci., 30(1): 4-7.

S203   _____, 1957. Age and rate of growth of the carp in Reelfoot Lake, Tennessee, for 1941 and 1956. J. Tenn. Acad. Sci., 32(1): 3-8.

S207   Schoonover, R. and W. H. Thompson, 1954. A post-impoundment study of the fisheries resources of Fall River Reservoir, Kansas. Trans. Kans. Acad. Sci., 57(2): 172-9.

S208   Schultz, F. H., 1955. Investigation of the spawning of northern pike in Prince Albert National Park, Saskatchewan, 1953. Canad. Wildl. Serv. Wildl. Mgmt. Bull., 3(4): 21 p.

S209   Schumacher, R. E., 1955. Growth of brown trout fingerlings on a diet fortified with aureomycin and thiamine hydrochloride. Progr. Fish Cult., 17(3): 123-5.

S211   _____, 1958. Dusche Creek creel census and population study. Minn. Fish Game Invest. Fish Serv., 1: 1-25.

S212   Schwartz, F. J. and J. Norvell, 1958. Food, growth and sexual dimorphism of the redside dace *Clinostomus alongatus* (Kirtland) in Linesville Creek, Crawford County, Pennsylvania. Ohio J. Sci., 58(5): 311-16.

S213   Scidmore, W. J., 1955. Notes on the fish population structure of a typical rough-fish-crappie lake of southern Minnesota. Minn. Dept. Conserv. Invest. Rep., 162: 1-11.

S214   _____ and A. W. Glass, 1953. Use of pectoral fin rays to determine age of the white sucker. Progr. Fish Cult., 15(3): 114-15.

S218   Sequin, L. R., 1957. Scientific fish culture in Quebec since 1945. Trans. Amer. Fish. Soc., 86: 136-43.

S219   Seversmith, H. F., 1953. Distribution morphology and life history of *Lampetra aepyptera*, a brook lamprey, in Maryland. Copeia, 1953(4): 225-31.

S220   Shapovalov, L. and A. C. Taft, 1954. Life histories of the steelhead
        rainbow trout and silver salmon. Calif. Fish. Bull., 98: 375 p.

S221   Shetter, D. S. and L. N. Allison, 1955. Comparison of mortality be-
        tween fly-hooked and worm-hooked trout in Michigan streams. Mich.
        Dept. Conserv. Misc. Publ., 9.

S223   _____, M. J. Whalls, and O. M. Corbett, 1954. The effect of changed
        angling restrictions on a trout population of the Au Sable River. Trans.
        N. Amer. Wildl. Conf., 19: 222-38.

S224   Shields, J. T., 1955. Carp control through water drawdowns, Fort
        Randall Reservoir, South Dakota. Midw. Wildl. Conf., 17: 10 p. mimeo.

S225   _____, 1955. Report of fisheries investigations during the second year
        of impoundment of Fort Randall Reservoir, South Dakota, 1954. S. D.
        Dept. Game Fish Parks. 100 p. mimeo.

S226   _____, 1956. Report of fisheries investigations during the third year of
        impoundment of Fort Randall Reservoir, South Dakota, 1955. S. D.
        Dept. Game Fish Parks Dingell-Johnson Proj., F-1-R-5: 91 p.

S227   _____, 1957. Report of fisheries investigations during the second year
        of impoundment of Gavins Point Reservoir, South Dakota, 1956. S. D.
        Dept. Game Fish Parks Dingell-Johnson Proj., F-1-R-6: 34 p. mimeo.

S228   _____, 1957. Report of fisheries investigations during the fourth year
        of impoundment of Fort Randall Reservoir, South Dakota, 1956. S.D.
        Dept. Game Fish Parks Dingell-Johnson Proj., F-1-R-6: 1-60.

S229   _____, 1958. Report of fisheries investigations during the third year of
        impoundment of Gavins Point Reservoir, South Dakota, 1957. S. D.
        Dept. Game Fish Parks Dingell-Johnson Proj., F-1-R-7: 1-48.

S230   _____, 1958. Report of fisheries investigations during the fifth year of
        impoundment of Fort Randall Reservoir, South Dakota, 1957. S. D.
        Dept. Game Fish Parks Dingell-Johnson Proj., F-1-R-7: 1-27.

S231   _____, 1958. Experimental control of carp reproduction through water
        drawdowns in Fort Randall Reservoir, South Dakota. Trans. Amer.
        Fish. Soc., 87: 23-33.

S234   Sigler, W. F., 1953. The rainbow trout in relation to the other fish in
        Fish Lake. Utah State Agric. Exp. Sta. Bull., 358: 1-26.

S235   _____, 1958. The white fishes of Bear Lake. Utah Fish Game, 14(12):
        20-22.

S236   _____, 1958. The ecology and use of carp in Utah. Utah St. Univ.
        Agric. Exp. Sta. Bull., 405: 1-63.

S237   _____ and J. B. Low, 1952. Age composition and growth of fish and
        fisherman success in Utah's high Uinta lakes. Proc. Utah Acad. Sci.
        Arts Lett., 27: 32-6.

S239   Skinner, J. E., 1957. Status of the striped bass-sturgeon study and
        suggestions for its future. Calif. Inl. Fish. Admin. Rep., 57-11: 1-16.

S240   Sklower, A., 1951. Carp breeding in Palestine. Arch. Fisch Wiss.,
        3(1/2): 42-54.

S242   Slastenenko, E. P., 1954. The relative growth of hybrid char (*Salve-
        linus fontinalis* x *Cristivomer namaycush*). J. Fish. Res. Bd. Can.,
        11(5): 652-9.

S246   Smith, H. H., 1916. Trout in the Punjab. Salm. Trout Mag., 12: 37-40.

S250   Smith, M. W., 1952. Fertilization and predator control to improve
        trout production in Crecy Lake, New Brunswick. Canad. Fish Cult.,
        13: 33-9.

S251   _____, 1954. Annual crops of speckled trout from a Prince Edward Is-
        land pond. Fish. Res. Bd. Can. Progr. Rep. Atlant. Coast Sta., 58: 21-2.

S252   Smith, M. W., 1954. Planting hatchery stocks of speckled trout in improved waters. Canad. Fish Cult., 16: 1-5.

S253   _____, 1955. Fertilization and predator control to improve trout angling in natural lakes. J. Fish. Res. Bd. Can., 12(2): 210-37.

S254   _____, 1956. Further improvement in trout angling at Crecy Lake, New Brunswick, with predator control extended to large trout. Canad. Fish Cult., 19: 13-16.

S255   _____, 1957. Comparative survival and growth of tagged and untagged brook trout. Canad. Fish Cult., 20: 1-6.

S256   _____ and J. W. Saunders, 1955. The American eel in certain fresh waters of the Maritime Provinces of Canada. J. Fish. Res. Bd. Can., 12(2): 238-69.

S257   _____ and _____, 1958. Movements of brook trout, *Salvelinus fontinalis* (Mitchill), between and within fresh and salt water. J. Fish. Res. Bd. Can., 15(6): 1403-49.

S259   Smith, R. F. and R. W. Gross, 1955. An evaluation of the 15-inch minimum size limit on pickerel in three New Jersey lakes. N. J. Fish. Lab. Misc. Rep., 15.

S262   Smith, S. B., 1955. The relation between scale diameter and body length of Kamloops trout, *Salmo gairdneri kamloops*. J. Fish. Res. Bd. Can., 12(5): 742-53.

S263   _____, 1957. Survival and growth of wild and hatchery rainbow trout *(Salmo gairdnerii)* in Corbett Lake, British Columbia. Canad. Fish Cult., 20: 7-12.

S265   Smith, S. H., 1956. Life history of lake herring of Green Bay, Lake Michigan. Fish. Bull. U. S., 57: 87-138.

S267   Snow, H. E., 1958. Northern pike of Murphy Flowage. Wis. Conserv. Bull., 23(2): 15-18.

S271   Snow, J. R., 1959. Notes on the propagation of the flathead catfish, *Pilodictis olivaris* (Rafinesque). Progr. Fish Cult., 21(1): 75-81.

S272   Southern, R., 1928. Notes on the scales of spawning salmon and the changes which they undergo. Salm. Trout Mag., 50: 75-7.

S274   Sowards, C. L., 1959. Experiments in hybridizing several species of trout. Progr. Fish Cult., 21(4): 147-50.

S275   Spindler, J. C. and J. E. Bailey, 1955. Comparison of survival, growth, and condition of hatchery rainbow trout and wild trout in Flint Creek. Mont. D.-J. Completion Rep., Proj. F-13-R-1(Job 2): 22 p.

S276   Sprague, J. W., 1959. Report of fisheries investigations during the fourth year of impoundment of Gavins Point Reservoir, South Dakota, 1958. S. D. Dept. Game Fish Parks. 42 p. mimeo.

S279   Starrett, W. C., 1958. Fishery values of a restored Illinois river bottomland lake. Trans. Ill. St. Acad. Sci., 50: 41-8.

S280   Stenton, J. E., 1952. Additional information on eastern brook trout x lake trout hybrids. Canad. Fish Cult., 13: 15-21.

S282   Stevens, R. I. and J. R. Truog, 1956. A comparison of two dried yeasts, *Candida utilis* and *Saccharomyces cerevisiae* as supplements in a diet for rainbow trout. Wis. Conserv. Dept. 23 p. mimeo.

S283   _____ and _____, 1957. Comparison of dried yeasts, *Candida utilis*, and *Saccaromyces cerevisiae*, as supplements in a diet for rainbow trout. Trans. Amer. Fish. Soc., 86: 161-8.

S285   Stone, U. B. and W. L. Hartman, 1957. Finger Lakes rainbows—tagging studies and future management. N. Y. State Conserv., 11(5): 6-8.

S287	Storrow, B., 1928. The salmon as a sea fish. Salm. Trout Mag., 51: 128-40.

S289	Stroud, R. H., 1955. Fisheries report for some central, eastern and western Massachusetts lakes, ponds, and reservoirs, 1951-1952. Mass. Div. Fish Game. 447 p.

S291	Stuart, T. A., 1957. The migrations and homing behavior of brown trout (*Salmo trutta* L.). Freshw. Salm. Fish. Res. Scot., 18: 3-27.

S292	Stube, M., 1958. The fauna of a regulated lake. Rep. Inst. Freshw. Res. Drottning., 39: 162-224.

S300	Swan, M. A., 1954. When is a smolt not a smolt? Salm. Trout Mag., 141: 454-6.

S305	Schwartz, F. J., 1958. The breeding behavior of the southern black-nose dace, *Rhinichthys atratulus obtusus* Agassiz. Copeia, 1958(2): 141-3.

S306	Swift, D. R., 1955. Seasonal variations in the growth rate, thyroid gland activity and food reserves of brown trout. J. Exp. Biol., 32(4): 751-64.

S309	Swingle, H. S., 1957. Preliminary results on the commercial production of channel catfish in ponds. Proc. S. E. Assoc. Game Fish Comm., 10: 160-62.

S310	_____, 1957. Commercial production of red cats (speckled bullheads) in ponds. Proc. S. E. Assoc. Game Fish Comm., 10: 156-60.

S311	_____, 1957. Revised procedures for commercial production of big-mouth buffalo fish in ponds in the southeast. Proc. S. E. Assoc. Game Fish Comm., 10: 162-5.

S313	Swynnerton, G. H. and E. B. Worthington, 1939. Brown-trout growth in the lake district. Salm. Trout Mag., 97: 337-55.

S324	Seamans, R. G., Jr., 1959. Trout stream management investigations of the Saco River watershed. N. H. Fish Game Dept. Surv. Rep., 9: 1-71.

S325	Swingle, H. S., 1960. Comparative evaluation of two tilapias as pond-fishes in Alabama. Trans. Amer. Fish. Soc., 89(2): 142-8.

S326	Smith, D. C. W., 1959. The biology of the rainbow trout (*Salmo gaird-neri*) in the lakes of the Rotorua District, North Island. N. Z. J. Sci., 2(3): 275-312.

S328	Stickney, A. P., 1960. Atlantic salmon investigations. U. S. Fish Wildl. Serv. Circ., 81: 6-17.

S330	Sprague, J. W., 1959. Report of fisheries investigations during the sixth year of impoundment of Fort Randall Reservoir, South Dakota. S. D. Dept. Game Fish Parks, Dingell-Johnson Proj., F-1-R-8: 32 p.

S331	Schumacher, R. E., 1958. A test of dry corn fermentation solubles as a substitute for corn distillers dried solubles in pelleted trout food. Minn. Dept. Conserv. Invest. Rep., 191: 4 p. mimeo.

S333	Strawn, K., 1958. Optimum and extreme temperatures for growth and survival: various fishes. For Handbook of Biological Data. 1 p. table.

S334	Swingle, H. S., 1952. Pounds of fish per acre in central Alabama power reservoirs. Pounds of fish per acre in Alabama rivers. Temperatures of surface water of ponds at Auburn, Alabama, when the first young fish hatch in the spring. Tables for Handbook of Biological Data.

S335	Stewart, R. K., 1959. A study of the scale markings and the age determination of certain brook trout populations in the Rocky Mountain region of southeastern Wyoming. Wyo. Game Fish Comm. Coop. Res. Proj., 2: 1-68.

S336   Salmon Research Trust of Ireland, Inc., 1960. Report for the year ending 31st December, 1959. 16 p.

S337   Sneed, K. E. and H. P. Clemens, 1960. Use of fish pituitary to induce spawning in channel catfish. Spec. Sci. Rep. U. S. Fish Wildl. Serv., 329: 12 p.

S338   Sawyer, P. J., 1960. A new geographic record for the American brook lamprey, *Lampetra lamottei*. Copeia, 1960(2): 136-7.

S340   Schoonover, R., 1958. The rehabilitation of warm-water lakes for fishery improvement. Central Mts. Plains Sect. Wildl. Soc. Proc. Annu. Conf., 3: 12 p. mimeo.

S342   Smith, W. A., Jr., 1959. Shad management in reservoirs. Proc. S. E. Assoc. Game Fish Comm., 12: 143-7.

S345   Swingle, H. S., 1959. Experiments on growing fingerling channel catfish to marketable size in ponds. Proc. S. E. Assoc. Game Fish Comm., 12: 63-72.

S347   Skinner, J. E., 1956. White sturgeon, *Acipenser transmontanus* (Richardson), with particular reference to the life history and ecology of the California population. Data for Handbook of Biological Data. 2 p.

S349   Sprague, J. W., 1958. Age and growth determination of the black bullhead from soft fin rays. MS thesis S. D. St. Coll. 38 p.

S352   Smith, M. W., 1960. An attempt to improve trout fishing in a New Brunswick lake by deferment of opening of angling season. J. Fish Res. Bd. Can., 17(5): 677-85.

S353   Smith, S. B., 1960. A note on two stocks of steelhead trout *(Salmo gairdneri)* in Capilano River, British Columbia. J. Fish. Res. Bd. Can., 17(5): 739-41.

S354   Saunders, J. W., 1960. The effect of impoundment on the population and movement of Atlantic salmon on Ellerslie Brook, Prince Edward Island. J. Fish. Res. Bd. Can., 17(4): 453-73.

S356   Shimadate, M., K. Nakamura, H. Koyama, T. Ito, and J. Toi, 1957. Effect of fertilization and significance of artificial feeding to fish production in farm pond, Shioda Plain, Nagano Prefecture. (Japanese with English summary.) Bull. Freshw. Fish. Res. Lab. Tokyo, 7(1): 1-32.

S357   Scidmore, W. J. and D. E. Woods, 1960. Some observations on competition between several species of fish for summer foods in four southern Minnesota lakes in 1955, 1956 and 1957. Minn. Fish Game Invest. Fish. Ser., 2: 13-24.

S359   Scott, R. N., 1950. The time of spawning, age, and secondary sex characters of adult chubs *(Mylocheilus caurinum)* from Flathead Lake, Montana, as correlated with histological changes in their testes. MA thesis Mont. St. Univ. 56 p.

S361   Sanderson, A. E., Jr., 1950. An ecological survey of the fishes of the Severn River with reference to the eastern chain pickerel, *Esox niger* LeSueur, and the yellow perch, *Perca flavescens* (Mitchill). MS thesis Univ. Md. 48 p.

S363   Sayre, R. C., 1960. Northeastern Oregon. Annu. Rep. Ore. St. Game Comm. Fish. Div., 1959: 86-108.

S364   Sumner, F. H., 1960. North coast—Tillamook District. Annu. Rep. Ore. St. Game Comm. Fish. Div., 1959: 226-42.

S365   Stevens, R. E., 1959. The white and channel catfishes of the Santee-Cooper Reservoir and tailrace sanctuary. Proc. S. E. Assoc. Game Fish Comm., 13: 203-19.

S367   Snow, H. E., 1960. Murphy Flowage. Res. in Wis., 1959: 14-17.

S368 Snyder, G. R. and H. A. Tanner, 1960. Cutthroat trout reproduction in the inlets to Trappers Lake. Colo. Dept. Game Fish Tech. Bull., 7: 85 p.

S369 Smith, S. H., 1957. An outline of the life history of *Leucichthys artedi*. Data for Handbook of Biological Data. 11 p.

S370 Stone, U. B., 1940. Studies on the biology of the satinfin minnows, *Notropis analostanus* and *Notropis spilopterus*. PhD thesis Cornell Univ. 98 p. + vii + 14 pl.

S372 Swenson, E. A., Jr., 1954. Analysis of the fish populations of two farm ponds in central New York. MS thesis St. Univ. N. Y. Coll. Forestry. 100 p.

S373 Smith, M. W., 1961. A limnological reconnaissance of a Nova Scotian brown-water lake. J. Fish. Res. Bd. Can., 18(3): 463-78.

S377 Saksena, V. P., K. Yamamoto, and C. D. Riggs, 1961. Early development of channel catfish. Progr. Fish Cult., 23(4): 156-61.

S378 Spindler, J. C. and J. E. Bailey, 1956. Year's travel by tagged rainbow trout. Progr. Fish Cult., 18(3): 119.

S380 Sayre, R. C., 1960. Northeastern Oregon. Annu. Rep. Ore. St. Game Comm. Fish. Div., 1960: 89-106.

S382 Shetter, D. S., 1961. Survival of brook trout from egg to fingerling stage in two Michigan trout streams. Trans. Amer. Fish. Soc., 90(3): 252-8.

S385 Schumacher, R. E., 1961. Some effects of increased angling pressure on lake trout populations in four northeastern Minnesota lakes. Minn. Fish Game Invest. Fish. Ser., 3: 20-42.

S386 Schwartz, F. J., 1961. Record Maryland fish: salt and brackish species. Md. Conserv., 38(3).

S387 Sumner, F. H., 1962. Migration and growth of the coastal cutthroat trout in Tillamook County, Oregon. Trans. Amer. Fish. Soc., 91(1): 77-83.

S388 Schreiber, M. R., 1962. Observations on the food habits of juvenile white sturgeon. Calif. Fish Game, 48(1): 79-80.

S390 Swingle, H. S., 1956. A repressive factor controlling reproduction in fishes. Oceanog. Zool. Proc. Eighth Pacif. Sci. Cong., IIIA (1953): 865-71.

S391 Sandoz, O., 1960. A pre-impoundment study of Arbuckle Reservoir, Rock Creek, Murray County, Oklahoma. Okla. Fish. Res. Lab. Rep., 77: 28 p.

S393 Sprague, J. W., 1961. Report of fisheries investigations during the seventh year of impoundment of Fort Randall Reservoir, South Dakota, 1959. S.D. Dept. Game Fish Parks, Dingell-Johnson Proj., F-1-R-9 (Jobs 5-8): 49 p.

S394 Schoffman, R. J., 1961. Age and rate of growth of the channel catfish in Reelfoot Lake, Tennessee, for 1953 and 1960. J. Tenn. Acad. Sci., 36(1): 5-11.

S396 Sharpe, F. P., 1960. Investigation of Shadow Mountain trout fishery. Quart. Rep. Colo. Coop. Fish. Res. Unit, 6: 36-49.

S397 _____, 1962. Some observations of the feeding habits of brown trout. Progr. Fish Cult., 24(2): 60-64.

S398 Schwartz, F. J. and H. J. Elser, 1962. Additions to Maryland list: new record fish. Md. Conserv., 39(2): 26-7.

S399 Sakowicz, S., 1961. [The migration of trout (*Salmo trutta morpha lacustris* L.) from Wdzydze Lake] (English summary). Rocz. Nauk Rolniczych, 93: 703-34.

S400   Sakowicz, S. and A. Stegman, 1961. [Rearing of Wdzydze lake trout
        (*Salmo trutta morpha lacustris* L.) in ponds] (English summary). Rocz.
        Nauk Rolniczych, 93: 751-70.

S402   Smith, M. W., 1961. Bottom fauna in a fertilized natural lake and its
        utilization by trout *(Salvelinus fontinalis)* as food. Verh. Int. Ver.
        Limnol., 14: 722-6.

S403   Saunders, J. W. and M. W. Smith, 1962. Physical alteration of stream
        habitat to improve brook trout production. Trans. Amer. Fish. Soc.,
        91(2): 185-8.

S404   Schwartz, F. J., 1962. Artificial pike hybrids, *Esox americanus ver-
        miculatus* x *E. lucius*. Trans. Amer. Fish. Soc., 91(2): 229-30.

S405   Scott, W. B., 1961. Summaries of current information on kiyi, bloater,
        and inconnu. Ont. Dept. Lands Forests Res. Info. Pap. Fish., 10: 25 p.

S406   _____, 1960. Summaries of current information on shortjaw cisco,
        shortnose cisco, and blackfin cisco. Ont. Dept. Lands Forests Res.
        Info. Pap. Fish., 9: 24 p.

S407   _____, 1960. Summaries of current information on round whitefish and
        mountain whitefish. Ont. Dept. Lands Forests Res. Info. Pap. Fish., 8:
        19 p.

S408   Scattergood, L. W., 1949. Notes on the kokanee *(Oncorhynchus nerka
        keenerlyi)*. Copeia, 1949(4): 297-8.

S410   Schultz, L. P. and students, 1935. The breeding activities of the little
        redfish, a landlocked form of the sockeye salmon, *Oncorhynchus nerka*.
        J. Pan-Pacif. Res. Inst., 10(1): 68-77.

S411   _____, 1937. The breeding habits of salmon and trout. Annu. Rep.
        Smithson. Inst., 1937: 365-76.

S412   Snyder, J. O., 1918. The fishes of the Lahontan system of Nevada and
        northeastern California. Bull. U. S. Bur. Fish., 35: 31-86.

S413   Shapovalov, L., 1951. A remarkable sea journey by a rainbow trout
        (Salmo gairdnerii) of "interior stock." Calif. Fish Game, 37(4): 489-90.

S415   Seppovaara, O., 1962. Zur Systematik und Okologie des Lachses und
        der Forellen in den Binnengewässern Finnlands. Ann. Zool. Soc.
        Vanamo, 24(1): 1-86.

S416   Sinn, J. A., 1961. A study of the brook trout in Libby Lake, Medicine
        Bow National Forest, in southeastern Wyoming. Wyo. Game Fish.
        Comm. Coop. Res. Rep., 2: 54 p.

S417   Stauffer, T. M., 1962. Duration of larval life of sea lampreys in Carp
        Lake River, Michigan. Trans. Amer. Fish. Soc., 91(4): 422-3.

S418   Scott, D. P., 1962. Effect of food quantity on fecundity of rainbow trout,
        Salmo gairdneri. J. Fish. Res. Bd. Can., 19(4): 715-31.

S420   Snow, J. R., 1962. A comparison of rearing methods for channel cat-
        fish fingerlings. Progr. Fish Cult., 24(3): 112-18.

S421   Swift, D. R., 1962. Evidence for the absence of an endogenous growth-
        rate rhythm in brown trout *(Salmo trutta* Linn.). Compar. Biochem.
        Physiol., 6: 91-3.

S422   _____, 1961. The annual growth-rate cycle in brown trout *(Salmo trutta*
        Linn.) and its cause. J. Exp. Biol., 38: 595-604.

S423   Shearer, W. M., 1958. Very small Atlantic salmon. Salm. Trout Mag.,
        1958(Sept.): 164-8.

S425   Scott, W. B. and S. H. Smith, 1962. The occurrence of the longjaw
        cisco, *Leucichthys alpenae*, in Lake Erie. J. Fish. Res. Bd. Can.,
        19(6): 1013-23.

S426   Schwartz, F. J. and B. W. Dutcher, 1962. Tooth replacement and food of the cyprinid, *Notropis cerasinus*, from the Roanoke River, Virginia. Amer. Midl. Nat., 68(2): 369-75.

S427 Stevens, E. D. and O. W. Tiemeier, 1961. Daily movement of channel catfish, *Ictalurus punctatus* (Rafinesque), in a farm pond. Trans. Kans. Acad. Sci., 64(3): 218-24.

S428 Schindler, O. and E. Wagler, 1936. Zur Biologie der Seeforelle (*Trutta lacustris* L.). Int. Rev. Ges. Hydrobiol. Hydrogr., 33: 327-56.

S429 Stewart, R. K., 1959. A study of the scale markings and the age determination of certain brook trout populations in southeastern Wyoming. MS thesis Univ. Wyo. 69 p. not seen, quoted in R216.

S430 Sullivan, C. M., 1954. Temperature reception and responses in fish. J. Fish. Res. Bd. Can., 11(2): 107-29.

S431 Shell, E. W., 1963. Effects of changed diets on the growth of channel catfish. Trans. Amer. Fish. Soc., 92: 432-4.

S432 Swift, D. R., 1962. Activity cycles in the brown trout (*Salmo trutta* Linn.). I. Fish feeding naturally. Hydrobiol., 20(3): 241-7.

S433 Smith, L. L., Jr., and R. H. Kramer, 1964. The spottail shiner in Lower Red Lake, Minnesota. Trans. Amer. Fish. Soc., 93(1): 35-45.

S434 Smith, S. H., 1964. Status of the deepwater cisco population of Lake Michigan. Trans. Amer. Fish. Soc., 93(2): 155-63.

S435 Scott, W. B., 1958. Inconnu, *Stenodus leucichthys* (Guldenstadt). Tabulation for Handbook of Biological Data. 6 p.

S436 Svetovidov, A. N., 1936. Graylings, genus *Thymallus* Cuvier, of Europe and Asia. [Russian, English summary.] Trav. de l' Institut Zool. de l' Acad. des Sci. de l'URSS, 3: 183-301 not seen, quoted in B76.

S437 Sprules, W. M., 1958. Goldeye, *Hiodon alosoides* (Rafinesque). Data for Handbook of Biological Data. 12 p.

S438 Smith, L. L., Jr., D. R. Franklin, and R. H. Kramer, 1958. Determination of factors influencing year class strength in northern pike and largemouth bass. Minn. Dingell-Johnson Completion Rep., F-12-R (Jobs II, III): 200 p. + 116 p. Append.

S439 Svärdson, G., 1948. Note on spawning habits of *Esox lucius* L. Rep. Inst. Freshw. Res. Drottning., 29: 102-7.

S440 Scott, W. B., 1963. A review of the changes in the fish fauna of Ontario. Trans. Roy. Canad. Inst., 34(2): 111-25.

S441 Schultz, L. P., 1935. The spawning habits of the chub, *Myocheilus courinus* —a forage fish of some value. Trans. Amer. Fish. Soc., 65: 143-7.

S442 _____, 1941. Fishes of Glacier National Park, Montana. U. S. Dept. Interior Conserv. Bull., 22: 1-42 not seen, quoted in G56.

S443 _____ and M. B. Schaefer, 1936. Descriptions of new intergeneric hybrids between certain cyprinid fishes of northwestern United States. Proc. Biol. Soc. Wash., 49: 1-10.

S444 Schmulbach, J. C., 1957. The life history of the central stoneroller, *Campostoma anomalum pullum* (Agassiz). MS thesis, South. Ill. Univ. Lib., Carbondale.

S445 Starrett, W. C., 1950. Food relationships of the minnows of the Des Moines River, Iowa. Ecology, 31(2): 216-33.

S446 _____, 1951. Some factors affecting the abundance of minnows in the Des Moines River, Iowa. Ecology, 32(1): 13-27.

S447 Sibley, C. K. and V. N. Rimsky-Korsakoff, 1931. Food of certain fishes in the watershed. Annu. Rep. N. Y. State Conserv. Dept., 20 (Suppl.): 102-20.

S448   Sibley, C. K., 1932. Fish food studies. Annu. Rep. N. Y. State Conserv. Dept., 21(Suppl.): 120-32.

S449   Smith, O. R., 1935. The breeding habits of the stone roller minnow (*Campostoma anomalum* Rafinesque). Trans. Amer. Fish. Soc., 65: 148-51.

S450   Sarbahi, D. S., 1951. Studies of the digestive tracts and digestive enzymes of the goldfish, *Carassius auratus* (Linnaeus) and the largemouth black bass, *Micropterus salmoides* (Lacépède). Biol. Bull., Wood's Hole, 100(3): 244-57.

S451   Stromsten, F. A., 1931. The development of the gonads in the gold fish, *Carassius auratus* (L.). Univ. Ia. Stud. Nat. Hist., 13(7): 1-45.

S454   Struthers, P. H., 1932. A review of the carp control problem in New York waters. Annu. Rep. N. Y. Conserv. Dept., 21(Suppl.): 272-89.

S455   Sharp, R., 1942. Some studies of the distribution and ecology of the German carp in Minnesota with suggested control measures. Minn. Bur. Fish. Res. Invest. Rep., 45: 25 p. mimeo.

S456   Sigler, W. F., 1955. An ecological approach to understanding Utah's carp population. Proc. Utah Acad. Sci. Arts Lett., 32: 95-104.

S457   Suttkus, R. D. and E. C. Raney, 1955. *Notropis asperifrons*, a new cyprinid fish from the Mobile Bay drainage of Alabama and Georgia, with studies of related species. Tulane Stud. Zool., 3(1): 1-33.

S458   _____ and _____, 1955. *Notropis baileyi*, a new cyprinid fish from the Pascagoula and Mobile Bay drainages of Mississippi and Alabama. Tulane Stud. Zool., 2(5): 71-85.

S459   _____ and _____, 1955. *Notropis hypsilepis*, a new cyprinid fish from the Apalachicola River system of Georgia and Alabama. Tulane Stud. Zool., 2(7): 161-70.

S460   Starrett, W. C., 1950. Distribution of the fishes of Boone County, Iowa, with special reference to the minnows and darters. Amer. Midl. Nat., 43(1): 112-27.

S461   Schwartz, F. J., 1963. The fresh-water minnows of Maryland. Md. Conserv., 40(2): 19-29.

S462   Smith, B. G., 1908. The spawning habits of *Chrosomus erythrogaster* Rafinesque. Biol. Bull. Wood's Hole, 15: 9-18.

S463   Sigler, W. F. and R. R. Miller, 1963. Fishes of Utah. Utah St. Dept. Fish Game, Salt Lake City. 203 p.

S464   Seaburg, K. G. and J. B. Moyle, 1964. Feeding habits, digestive rates, and growth of some Minnesota warmwater fishes. Trans. Amer. Fish. Soc., 93(3): 269-85.

S465   Svetovidov, A. N., 1963. Fauna of USSR fishes. Volume II, No. 1, Clupeidae. Transl. pub. by Nat. Sci. Found. and Smithson. Inst. 428 p.

S466   Saunders, J. W. and M. W. Smith, 1964. Planting brook trout (*Salvelinus fontinalis* (Mitchill)) in estuarial waters. Canad. Fish Cult., 32: 25-30.

S470   Sneed, K. E. and E. M. Leonard, 1959. Age and growth of the channel catfish, *Ictalurus punctatus*, in Lake Texoma. Proc. Okla. Acad. Sci., 37: 73-8.

S471   Slastenenko, E. P., 1958. The freshwater fishes of Canada. Kiev Printers, Toronto, Ont. 385 p.

S472   Swingle, W. E., 1965. Length-weight relationships of Alabama fishes. Auburn Univ. Agric. Exp. Sta. Zool.-Ent. Ser. Fish., 3: 87 p.

S473   Swingle, H. A., 1965. Growth rates of paddlefish receiving supplemental feeding in fertilized ponds. Progr. Fish Cult., 27(4): 220.

S474   Suttkus, R. D., 1963. Order Lepisostei. Fishes of western North Atlantic, Part 3. Mem. Sears Found. Mar. Res., 1(3): 61-88.
S475   Scott, W., 1938. The food of *Amia* and *Lepidosteus*. Invest. Ind. Lakes Streams, 1: 112-15.
S476   Sykes, J. E., 1956. Shad fishery of the Ogeechee River, Georgia, in 1954. Spec. Sci. Rep. U. S. Fish Wildl. Serv., 191: 11 p.
S477   Swanson, G. A., 1932. A mid-winter migration of gizzard shad. Copeia, 1932: 34.
S478   Svärdson, G., 1951. The coregonid problem. III. Whitefish from the Baltic successfully introduced into fresh waters in the north of Sweden. Rep. Inst. Freshw. Res. Drottning., 32: 5-145.
S479   _____, 1952. The coregonid problem. IV. The significance of scales and gillrakers. Rep. Inst. Freshw. Res. Drottning., 33: 204-32.
S480   Seeley, C. M. and G. W. McCammon, 1966. Kokanee. Calif. Dept. Fish Game, Inland Fisheries Management, ed. A. Calhoun: 274-94.
S481   Stocek, R. and H. R. McCrimmon, 1965. The co-existence of rainbow trout (*Salmo gairdneri* Richardson) and largemouth bass (*Micropterus salmoides* Lacépède) in a small Ontario lake. Canad. Fish Cult., 35: 37-58.
S482   Seamans, R. G., Jr., and H. C. Nowell, Jr., 1962. Trout stream management investigations of the Ammonoosuc River watershed, 1961. N. H. Dingell-Johnson Proj., F-5-R-10(Job II): 25 p.
S483   Swan, R. L., 1963. Upper Willamette district. Annu. Rep. Ore. St. Game Comm. Fish Div., 1963: 41-55.
S484   Salmon Research Trust of Ireland, 1964. Report and statement of accounts for the year ended 31st December, 1964. 67 p.
S485   Saunders, J. W., 1966. Estuarine spawning of Atlantic salmon. J. Fish. Res. Bd. Can., 23(1): 1803-4.
S486   Saunders, R. L. and J. H. Gee, 1964. Movements of young Atlantic salmon in a small stream. J. Fish. Res. Bd. Can., 21(1): 27-36.
S487   Staley, J., 1966. Brown trout. Inland Fisheries Management, ed. A. Calhoun, Calif. Dept. Fish Game: 233-42.
S488   Spence, J. A. and A. B. West, 1964. The Cottage River experiment. Salm. Res. Trust Ire., Inc., Rep., 1964: 40-53.
S489   Schoumacher, R., 1965. Some observations on trout populations in two Iowa streams. Ia. St. Conserv. Comm. Quart. Biol. Rep., 17(4): 11-13.
S490   Swift, D. R., 1964. Activity cycles in the brown trout (*Salmo trutta* L.) 2. Fish artificially fed. J. Fish. Res. Bd. Can., 21(1): 133-8.
S491   Stoklosowa, S., 1966. Sexual dimorphism in the skin of sea-trout, *Salmo trutta*. Copeia, 1966(3): 613-14.
S492   _____, 1964. The effect of temperature and oxygen on the growth rate of the Windermere char *(Salvelinus alpinus willoghbii)*. Compar. Biochem. Physiol., 12: 179-83.
S493   Schofield, C. L., Jr., 1965. Water quality in relation to survival of brook trout, *Salvelinus fontinalis* (Mitchill). Trans. Amer. Fish. Soc., 94(3): 227-35.
S494   Stangenberg, M., 1965. Coarse fish research in Poland. Proc. Brit. Coarse Fish Conf., Univ. Liverpool, 2: 72-86.
S495   Spanovskaya, V. D., 1963. Food of pike fingerlings, *Esox lucius*. (Russian with English summary.) Zool. Zhur., 42(7): 1071-9.
S496   Swift, D. R., 1965. Effect of temperature on mortality and rate of development of the eggs of the pike (*Esox lucius* L.) and the perch (*Perca fluviatilis* L.). Nature, 206: 528.

S497 Sorenson, L., K. Buss, and A. D. Bradford, 1966. The artificial propagation of esocid fishes in Pennsylvania. Progr. Fish Cult., 28(2): 133-41.

S498 Smith, W. E. and R. W. Saalfeld, 1955. Studies on Columbia River smelt *Thalichthys pacificus* (Richardson). Wash. Dept. Fish., Fish. Res. Pap., 1(3): 3-26.

S499 Schmitz, W. R. and R. E. Hetfeld, 1965. Predation by introduced muskellunge on perch and bass, 11: years 8-9. Trans. Wis. Acad. Sci. Arts Lett., 54: 273-82.

S500 Scott, D. P., 1964. Thermal resistance of pike (*Esox lucius* L.), muskellunge *(E. masquinongy)* and their F_1 hybrid. J. Fish. Res. Bd. Can., 21(5): 1043-9.

S501 Summerfelt, R. C., W. M. Lewis, and M. G. Ulrich, 1967. Measurement of some hematological characteristics of the goldfish. Progr. Fish Cult., 29(1): 13-20.

S502 Stevenson, J. H., 1965. Observations on grass carp in Arkansas. Progr. Fish Cult., 27(4): 203-6.

S503 Swee, U. B. and H. R. McCrimmon, 1966. Reproductive biology of the carp, *Cyprinus carpio* L., in Lake St. Lawrence, Ontario. Trans. Amer. Fish. Soc., 95(4): 372-80.

S504 Stevenson, J. H., 1964. Fish farming experimental station. U. S. Fish Wildl. Serv. Circ., 178: 79-100.

S505 Shaheen, A. H., A. E. Imam, and M. T. Hashem, 1959. Fish culture in Egyptian rice fields. Hydrobiol. Dept. UAR Notes and Mem., 55: 15 p.

S506 Schontz, C. J., 1963. The effects of altitude and latitude on the morphometry, meristics, growth and fecundity of the eastern blacknose dace, *Rhinichthys atratulus atratulus* (Hermann). Disser. Abst., 24(3): 1303.

S507 Smith, G. R., 1966. Distribution and evolution of the North American catostomid fishes of the subgenus *Pantosteus*, Genus *Catostomus*. Univ. Mich. Mus. Zool. Misc. Publ., 129: 133 p.

S508 Shoemaker, H. H., 1951. Can we establish a standard system of measuring lengths and weights of fishes? Progr. Fish Cult., 13(3): 157-8.

S509 Schwartz, F. J. and R. Jachowski, 1965. The age, growth, and length-weight relationship of the Patuxent River, Maryland, ictalurid white catfish, *Ictalurus catus*. Chesapeake Sci., 6(4): 226-9.

S510 Simco, B. A. and F. B. Cross, 1966. Factors affecting growth and production of channel catfish, *Ictalurus punctatus*. Univ. Kans. Publ. Mus. Nat. Hist., 17(4): 191-256.

S511 Schoumacher, R., 1964. A brief preliminary report on commercial channel catfish catch studies in the Mississippi River in 1963. Ia. Conserv. Comm. Quart. Biol. Rep., 16(1): 10-15.

S512 _____, 1964. Mississippi River catfish investigations in 1963. Part II. Small channel catfish and flathead catfish studies. Ia. Conserv. Comm. Quart. Biol. Rep., 16(2): 42-7.

S513 _____, 1965. Commercial channel catfish catch studies in the Mississippi River in 1964. Ia. Conserv. Comm. Quart. Biol. Rep., 17(2): 14-16.

S514 Schoffman, R. J., 1966. Age and rate of growth of the channel catfish in Reelfoot Lake for 1960 and 1966. J. Tenn. Acad. Sci., 41: 12-14.

S515 Strawn, K., 1965. Resistance of threadfin shad to low temperatures. Proc. S. E. Assoc. Game Fish Comm., 17: 290-93.

S517 Sneed, K. E., 1964. Southeastern Fish Cultural Laboratory. U. S. Fish Wildl. Serv. Circ., 178: 111-18.

S518 ____, H. K. Dupree, and O. L. Green, 1961. Observations on the culture of flathead catfish *(Pylodictis olivaris)* fry and fingerlings in troughs. Proc. S. E. Assoc. Game Fish Comm., 15: 298-302.

S519 Smith, M. W., 1966. Amount of organic matter lost to a lake by migration of eels. J. Fish. Res. Bd. Can., 23(11): 1799-1801.

S520 Scott, W., 1912. The regenerated scales of *Fundulus heteroclitus* L. with a preliminary note on their formation. Proc. Ind. Acad. Sci., 1911: 439-44.

S521 Solberg, A. N., 1938. The development of a bony fish. Progr. Fish Cult., 40: 1-19.

S522 Swarup, H., 1958. States in the development of the stickleback, *Gasterosteus aculeatus* (1). J. Embryol. Exper. Morphol., 6(3): 373-83.

S523 Svärdson, G., 1949. Natural selection in egg number in fish. Rep. Inst. Freshw. Res. Drottning., 29: 115-22.

S524 Sequin, R. L., 1957. Management, fishing results and growth of speckled trout *(Salvelinus fontinalis)* in Baldwin Pond, Stanstead County, Quebec. Canad. Fish Cult., 20: 29-37.

T2 Tarzwell, C. M., 1938. An evaluation of the methods and results of stream improvement in the Southwest. Trans. N. Amer. Wildl. Conf., 3: 339-64.

T3 ____, 1938. Factors influencing fish food and fish production in southwestern streams. Trans. Amer. Fish. Soc., 67: 246-55.

T5 ____, 1941. The fish population of a small pond in northern Alabama. Trans. N. Amer. Wildl. Conf., 5: 245-51.

T10 Thompson, D. H., 1928. The "knothead" carp of the Illinois River. Bull. Ill. Nat. Hist. Surv., 17: 285-320.

T11 ____, 1933. The finding of very young *Polyodon*. Copeia, 1933(1): 31-3.

T16 ____ and G. W. Bennett, 1938. Lake management reports. I. Horseshoe Lake near Cairo, Illinois. Ill. Nat. Hist. Surv. Biol. Notes, 8: 1-6.

T23 Thorpe, L. M., 1942. Application of fishery survey data to heavily fished lakes. Trans. N. Amer. Wildl. Conf., 7: 436-42.

T24 Titcomb, J. W., 1897. Wild trout spawn: methods of collection and utility. Trans. Amer. Fish. Soc., 26: 73-86.

T25 ____, 1904. Report on the propagation and distribution of food fishes. Rep. U. S. Fish Comm. 1902, 28: 22-110.

T27 ____, 1920. Some fish-cultural notes. Trans. Amer. Fish. Soc., 50: 200-211.

T28 ____, 1921. Growth of fish and location of hatcheries. Trans. Amer. Fish. Soc., 51: 65-9.

T29 ____ and E. W. Cobb, 1928. The nutritional requirements and growth rates of brook trout. Trans. Amer. Fish. Soc., 58: 205-31.

T30 ____, ____, M. F. Crowell, and C. W. McCay, 1929. The relative value of plant and animal by-products as feeds for brook trout and the basic nutritional requirements of brook trout in terms of proteins, carbohydrates, vitamins, inorganic elements, and roughage. Trans. Amer. Fish. Soc., 59: 126-45.

T32 Traver, J. R., 1929. The habits of the black-nosed dace, *Rhinichthys atronasus* (Mitchill). J. Elisha Mitchell Sci. Soc., 45(1): 101-29.

T33 Trembley, F. J., 1930. The gar-pike of Lake Champlain. "A biological survey of the Champlain watershed," Annu. Rep. N. Y. Conserv. Dept., 19(Suppl.): 139-45.

T34   Tulian, E. A., 1911. Five years' progress in fish culture in Argentina. Trans. Amer. Fish. Soc., 40: 413-22.

T39   Taft, A. C. and G. I. Murphy, 1950. The life history of the Sacramento squawfish (Ptychocheilus grandis). Calif. Fish Game, 36(2): 147-64.

T44   Thompson, W. H., 1950. Investigation of the fisheries resources of Grand Lake. Okla. Game Fish Dept. Fish Mgmt. Rep., 18: 1-46.

T49   Threinen, C. W., 1949. An analysis and appraisal of the rough fish problem of Wisconsin. Wis. Div. Fish Mgmt. Invest. Rep., 715: 19 p. mimeo.

T53   _____ and A. A. Oehmcke, 1950. The northern invades the musky's domain. Wis. Conserv. Bull., 15(9): 10-12.

T56   Toole, M., 1951. Channel catfish culture in Texas. Progr. Fish Cult., 13(1): 3-10.

T57   Tunison, A. V. et al., 1943. The nutrition of trout. Cortland Hatchery Report No. 12. N. Y. Fish. Res. Bull., 5: 1-26.

T61   Taylor, G. T. and R. J. LeBrasseur, 1957. Distribution, age and food of steelhead trout Salmo gairdneri caught in the northeast Pacific. Fish. Res. Bd. Can. Progr. Rep. Atlant. Coast Sta., 109: 9-11.

T62   Taylor, W. R., 1954. Records of fishes in the John N. Lowe Collection from the Upper Peninsula of Michigan. Univ. Mich. Mus. Zool. Misc. Publ., 87: 50 p.

T70   Thompson, R. B., 1959. Food of the squawfish Ptychocheilus orego nensis (Richardson) of the lower Columbia River. Fish. Bull. U.S., 158(60): 43-58.

T75   Threinen, C. W., 1958 (?). Life history, ecology, and management of the alewife. Wis. Conserv. Dept. Publ., 223: 1-7.

T76   _____ and W. T. Helm, 1952. Composition of the fish population and carrying capacity of Spaulding's Pond, Rock County, as determined by rotenone treatment. Wis. Invest. Rep., 656: 19 p.

T78   Tiemeier, O. W., 1957. Notes on stunting and recovery of the channel catfish. Trans. Kans. Acad. Sci., 60(3): 294-6.

T79   _____ and J. B. Elder, 1957. Limnology of Flint Hills farm ponds for 1956 and preliminary report on growth studies of fishes. Trans. Kans. Acad. Sci., 60(4): 379-92.

T81   Tremblay, J. L. and A. Marcotte, 1954. Le Saumon. Rapp. Ann. Sta. Biol. Mar. Que., 1953(Append.IV): 60-67.

T82   Trout, C., 1937. Sex and condition. Salm. Trout Mag., 87: 140-47.

T84   Turner, W. R., 1953. The age and growth of the gizzard shad, Dorosoma cepedianum (LeSueur), in Herrington Lake, Kentucky. Ky. Div. Fish. Bull., 13: 14 p.

T88   Thompson, R. B., 1959. Fecundity of the Arctic char, Salvelinus alpinus, of the Wood River Lakes, Bristol Bay, Alaska. Copeia, 1959(4): 345-6.

T93   Tomkins, F. T., 1951. The life history and reproduction of Georgian Bay lake trout, with some notes on the commercial fishery. MA thesis Univ. Toronto. 65 p.

T95   Tiemeier, O. W. and J. B. Elder, 1960. Growth of stunted channel catfish. Progr. Fish Cult., 22(4): 172-6.

T96   Toth, S. J. and R. F. Smith, 1960. Soil over which water flows affects ability to grow fish. N. J. Agric., 42(6): 5-11.

T97   Thomas, M. L. H., 1961. Ammocoete biology. Progr. Rep. Biol. Sta. and Technol. Unit, London, Ont., Fish. Res. Bd. Can., 2: 14-17.

T98   Trade News, 1961. Athabaska lake trout. Trade News, 14(3): 10.

T100 Timmermans, J. A., 1960. Observations concernant les Populations de Truite commune (*Salmo trutta fario* L.) dans les Eaux courantes. Trav. Sta. Rech. Groenendaal Ser. D, 28: 36 p.

T101 Tennant, D. L. and G. Billy, 1963. Artificial hybridization of the muskellunge and grass pickerel in Ohio. Progr. Fish Cult., 25(2): 68-70.

T102 Trembley, F. J., 1960. Research project on effect of condenser discharge water on aquatic life. Inst. Res. Lehigh Univ. Progr. Rep., 1956-1959.

T105 Tiemeier, O. W., 1962. Supplemental feeding of fingerling channel catfish. Progr. Fish Cult., 24(2): 88-90.

T106 Tesch, F. W. and M. L. Albrecht, 1961. Über den Einfluss verschiedener Umweltfaktoren auf Wachstum und Bestand der Bachfarelle (*Salmo trutta fario* L.) in Mittelgebirgsgewässern. Verh. int. Ver. Limnol., 14: 763-8.

T107 Thurston, C. E., 1962. Physical characteristics and chemical composition of two subspecies of lake trout. J. Fish. Res. Bd. Can., 19(1): 39-44.

T108 Taft, A. C. and L. Shapovalov, 1938. Homing instinct and straying among steelhead trout *(Salmo gairdnerii)* and silver salmon *(Oncorhynchus kisutch)*. Calif. Fish Game, 24(2): 118-25.

T109 Turner, C. L., 1946. A case of hermaphroditism in the cut-throat trout. Chicago Acad. Sci. Nat. Hist. Misc., 1: 1-2.

T110 Trembley, G. L., 1945. Results from plantings of tagged trout in Spring Creek, Pennsylvania. Trans. Amer. Fish. Soc., 73: 158-72.

T111 Tiemeier, O. W., 1962. Increasing size of fingerling channel catfish by supplemental feeding. Trans. Kans. Acad. Sci., 65(2): 144-53.

T112 Taub, S. H., 1963. Recovery of a marked lake trout after 21 years. Trans. Amer. Fish. Soc., 92: 432.

T113 Trautman, M. B., 1957. The fishes of Ohio. Ohio St. Univ. Press. 683 p.

T114 Taft, A. C., 1957. Sacramento squawfish, *Ptychocheilus grandis* (Ayres). Data for Handbook of Biological Data. 3 p.

T115 Threinen, C. W., 1952. Carp in the ecology of southern Wisconsin lakes. Wis. Conserv. Bull., 17(7): 14-15.

T116 _____ and W. T. Helm, 1954. Carp destruction of aquatic vegetation. J. Wildl. Mgmt., 18(2): 247-50.

T117 Tryon, C. A., Jr., 1954. Effect of carp exclosures on aquatic vegetation. J. Wildl. Mgmt., 18(2): 251-4.

T118 Timmermans, J. A., 1960. L'Elevage de la Carpe au Liban. Trav. Sta. Rech. Groenendaal Ser. D, 29: 1-23.

T119 Talbot, G. B., 1961. The American shad. Fish. Leafl. Wash., 504: 7 p.

T120 Tiffany, L. H., 1920. Algal food of the young gizzard shad. Ohio J. Sci., 21: 113-22.

T121 _____, 1921. The gizzard shad in relation to plants and game fishes. Trans. Amer. Fish. Soc., 50: 381-6.

T122 Tebo, L. B., Jr., and W. W. Hassler, 1963. Food of brook, brown, and rainbow trout in streams in western North Carolina. J. Elisha Mitchell Sci. Soc., 79(1): 44-53.

T123 Thomas, J. D., 1962. The food and growth of brown trout (*Salmo trutta* L.) and its feeding relationships with the salmon parr (*Salmo salar* L.)

and the eel (*Anguilla anguilla* (L.)) in the River Teify, West Wales. J. Anim. Ecol., 31: 175-205.

T124   Thomas, J. D., 1964. Studies on the growth of trout, *Salmo trutta,* from four contrasting habitats. Proc. Zool. Soc. Lond., 142(3): 459-509.

T125   Threinen, C. W., C. A. Wistrom, B. Apelgren, and H. E. Snow, 1966. The northern pike: its life history, ecology, and management. Wis. Conserv. Dept. Publ., 235: 16 p.

T126   Timmermans, J. A., 1962. Influence de la Temperature sur la Production Piscicole en Etang. Bull. Francais Pisciculture, 207: 67-71.

T127   Tennessee Valley Authority, 1964. Average growth rates for East Tennessee Valley fishes, 1964. 6 p. mimeo.

T128   Tomlinson, P. K. and N. J. Abramson, 1961. Fitting a von Bertalanffy growth curve by least squares. Calif. Fish Bull., 116: 69 p.

U1   Ulrey, L., C. Risk, and W. Scott, 1938. The number of eggs produced by some of our common fresh-water fishes. Invest. Ind. Lakes Streams, 1(6): 73-8.

U2   Underhill, A. H., 1939. Cross between *Esox niger* and *E. lucius.* Copeia, 1939(4): 237.

U3   _____, 1941. Estimation of a breeding population of chub suckers. Trans. N. Amer. Wildl. Conf., 5: 251-6.

U4   Upper Mississippi River Conservation Committee, 1946. Second progress report of the technical committee for fisheries. 27 p. mimeo.

U5   _____, 1948. Fifth progress report of the technical committee for fisheries. 23 p. mimeo.

U6   Underhill, A. H., 1949. Studies on the development, growth and maturity of the chain pickerel, *Esox niger* LeSueur. J. Wildl. Mgmt., 13(4): 377-91.

U8   Upper Mississippi River Conservation Committee, 1959. Supplemental report, fish technical sub-committee, proceedings of thirteenth annual meeting. 147 p. mimeo.

U10   Underhill, J. C. and D. J. Merrell, 1959. Intra-specific variation in the bigmouth shiner *(Notropis dorsalis).* Amer. Midl. Nat., 61: 133-47.

U11   U. S. Bureau of Commercial Fisheries, Mobridge, S. D., 1965. Missouri River reservoir commercial fishing, investigations. A documentation of 1963-64 activities and findings. 74 p. mimeo.

V1   Van Cleave, H. J. and H. C. Markus, 1929. Studies in the life-history of the blunt-nosed minnow. Amer. Nat., 63: 530-39.

V3   Van Engel, W. A., 1940. The rate of growth of the northern pike, *Esox lucius* Linnaeus, in Wisconsin waters. Copeia, 1940(3): 177-88.

V4   Van Oosten, J., 1923. The whitefishes *(Coregonus clupeaformis).* A study of the scales of whitefishes of known ages. Zoologica (N. Y.), 2: 381-412.

V5   _____, 1929. Life history of the lake herring (*Leucichthys artedi* LeSueur) of Lake Huron as revealed by its scales, with a critique of the scale method. Bull. U. S. Bur. Fish., 44: 265-428.

V6   _____, 1935. First record of the alewife, *Pomolobus pseudoharengus,* for the State of Michigan. Copeia, 1935(4): 194-5.

V7   _____, 1937. First records of the smelt, *Osmerus mordax,* in Lake Erie. Copeia, 1937(1): 64-5.

V8   _____, 1937. The age, growth, and sex ratio of the Lake Superior longjaw, *Leucichthys zenithicus* (Jordan and Evermann). Pap. Mich. Acad. Sci. Arts Lett., 22: 691-711.

V10 Van Oosten, J., 1939. The age, growth, sexual maturity, and sex ratio of the common whitefish, *Coregonus clupeaformis* (Mitchill) of Lake Huron. Pap. Mich. Acad. Sci. Arts Lett., 24: 195-221.

V13 ____, 1944. Lake trout. Fish. Leafl. Wash., 15: 1-8.

V15 ____, 1946. The pikes. Fish. Leafl. Wash., 166: 1-6.

V16 ____, 1947. Mortality of smelt, *Osmerus mordax* (Mitchill), in Lakes Huron and Michigan during the fall and winter of 1942-1943. Trans. Amer. Fish. Soc., 74: 310-37.

V17 ____, 1948. Turbidity as a factor in the decline of Great Lakes fishes with special reference to Lake Erie. Trans. Amer. Fish. Soc. 1945, 75: 281-322.

V18 ____ and H. J. Deason, 1938. The food of the lake trout, *(Cristivomer namaycush namaycush)* and of the lawyer *(Lota maculosa)* of Lake Michigan. Trans. Amer. Fish. Soc., 67: 155-77.

V19 ____ and ____, 1939. The age, growth, and feeding habits of the whitefish, *Coregonus clupeaformis* (Mitchill) of Lake Champlain. Trans. Amer. Fish. Soc., 68: 152-62.

V20 ____ and ____, 1939. Age, growth, and condition of the walleyed pike, yellow perch, and goldeye of Lower Red Lake, Beltrami and Clearwater Counties, Minnesota. Pap. given Midw. Wildl. Conf., 1939. Typewritten.

V21 Vessel, M. F. and S. Eddy, 1941. A preliminary study of the egg production of certain Minnesota fishes. Minn. Bur. Fish. Res. Invest. Rep., 26: 26 p.

V22 Vestal, E. H., 1943. Creel returns from hatchery trout in June Lake, California. Calif. Fish Game, 29(2): 51-63.

V23 Viosca, P., Jr., 1931. The bullhead, *Ameiurus melas catulus,* as a dominant in small ponds. Copeia, 1931(1): 17-19.

V28 Vladykov, V. D. and V. Legendre, 1940. The determination of the number of eggs in ovaries of brook trout, *(Salvelinus fontinalis)*. Copeia, 1940(4): 218-20.

V29 ____, ____, and J. P. Cuerrier, 1941. Variations regionales du nombre des oeufs chez la truite mouchetee *(Salvelinus fontinalis)* du parc des Laurentides. Ann. L'Acfas Montreal, 7: 121-2.

V30 Van Oosten, J. and R. Hile, 1949. Age and growth of the lake whitefish, *Coregonus clupeaformis* (Mitchill), in Lake Erie. Trans. Amer. Fish. Soc., 77: 178-249.

V31 ____, 1940. The smelt, *Osmerus mordax* (Mitchill). U. S. Bur. Fish. Great Lakes Invest. 13 p. mimeo.

V33 Vrat, V., 1949. Reproductive behavior and development of eggs of the three-spined stickleback *(Gasterosteus aculeatus)* of California. Copeia, 1949(4): 252-60.

V39 Van Oosten, J., 1950. Progress report on the study of Great Lakes trout. The Fisherman, 18(5): 5,8-10; (6): 5,8.

V41 Vibert, R., 1950. Recherches sur le Saumon de L'Adour *(Salmo salar* Linné) (ages, croissance, cycle genetique, races) 1942-1948. (English summary.) Ann. Sta. Cent.Hydrobiol. Appl., 3: 27-149.

V43 Vladykov, V. D., 1951. Fecundity of Quebec lampreys. Canad. Fish Cult., 10: 1-14.

V44 Viosca, P., Jr., 1952. Eleventh report to International Paper Company, August 5, 1952. 21 p. typewritten. (Most of data also given in a paper at Amer. Fish. Soc., Dallas, Tex., Sept., 1952.)

V51   Van Oosten, J. and P. H. Eschmeyer, 1956. Biology of young lake trout
       (Salvelinus namaycush) in Lake Michigan. Res. Rep. U. S. Fish. Serv.,
       42: 88 p.

V52   van Someren, V. D., 1939. Research at Wonersh. Salm. Trout Mag.,
       95: 156-63.

V53   _____, 1950. The biology of trout in Kenya Colony. Game Dept., Kenya
       Colony. 114 p.

V54   Van Oosten, J., 1937. Artificial propagation of commercial fish of the
       Great Lakes. Trans. N. Amer. Wildl. Conf., 2: 605-12.

V55   Van Woert, W. F., 1957. Time pattern of migration of adult salmon and
       steelhead into the upper Sacramento River during the 1956-57 season,
       and length-weight relationship of Sacramento River steelhead. Calif.
       Inl. Fish. Admin. Rep., 57-19: 1-5.

V57   Vernon, E. H. and R. G. McMynn, 1957. Scale characteristics of year-
       ling coastal cutthroat and steelhead trout. J. Fish. Res. Bd. Can.,
       14(2): 203-12.

V60   Vladykov, V. D., 1955. Fishes of Quebec—Sturgeons. Que. Dept. Fish.
       Album, 5: 11 p.

V61   _____, 1956. Fecundity of wild speckled trout *(Salvelinus fontinalis)* in
       Quebec lakes. J. Fish. Res. Bd. Can., 13(6): 799-841.

V62   _____ and W. I. Follett, 1958. Redescription of *Lampetra ayresii*
       (Gunther) of western North America, a species of lamprey *(Petromy-
       zontidae)* district from *Lampetra fluviatilis* (Linnaeus) of Europe. J.
       Fish. Res. Bd. Can., 15(1): 47-77.

V63   Vincent, R. E., 1960. Some influences of domestication upon three
       stocks of brook trout (*Salvelinus fontinalis* Mitchill). Trans. Amer.
       Fish. Soc., 89(1): 35-52.

V66   Vladykov, V. D., 1945. Trois poissons nouveaux pour la Province de
       Quebec. Canad. Nat., 72: 27-39.

V67   Van Oosten, J., 1961. Records, ages, and growth of the mooneye, *Hio-
       don tergisus*, of the Great Lakes. Trans. Amer. Fish. Soc., 90(2): 170-
       74.

V68   _____, 1956. Tabular data on whitefish submitted to National Research
       Council, 1956(Dec.4): 3 p.

V69   _____, 1961. Formation of an accessory annulus on the scales of
       starved whitefish. Progr. Fish Cult., 23(3): 135.

V70   Vanicek, D., 1962. Life history of the quillback and highfin carp-
       suckers in the Des Moines River. Proc. Ia. Acad. Sci., 68: 238-46.

V71   Van Oosten, J., 1942. Relationship between the plantings of fry and
       production of whitefish in Lake Erie. Trans. Amer. Fish. Soc., 71:
       118-21.

V72   Vernon, E. H., 1957. Morphometric comparison of three races of
       kokanee *(Oncorhynchus nerka)* within a large British Columbia lake.
       J. Fish. Res. Bd. Can., 14(4): 573-98.

V73   Vladimirskaya, M. I., 1960. Young salmon in the upper part of the
       river Petchora. Rapp. Cons. Int. Explor. Mer., 148: 53-7.

V75   Vaas-van Oven, A., 1958. Experiments on different stocking rates of
       the common carp (*Cyprinus carpio* L.) in nursing ponds. Proc. Indo.-
       Pacif. Fish. Coun., 7(II): 13-47.

V76   Van Duzer, E. M., 1939. Observations on the breeding habits of the
       cut-lips minnow, *Exoglossum maxillingua.* Copeia, 1939(2): 65-75.

V77   Vladykov, V. D., 1949. Quebec lampreys. Dept. Fish. Que. Contr.,
       26: 1-67.

V78 Vladykov, V. D. and J.-M. Roy, 1948. Biologie de la lamproie d'eau douce *(Ichthyomyzon unicuspis)* apres la metamorphose (Résumé). Rev. Canad. Biol., 7(3): 483-5.

V79 _____ and J. R. Greeley, 1963. Order Acipenseroidei. Soft-rayed bony fishes. Fishes of Western North Atlantic, Mem. Sears Found. Mar. Res., 1(3): 24-60.

V80 Vincent, R. E., 1960. Experimental introductions of fresh-water alewives. Progr. Fish Cult., 22(1): 38-42.

V81 Von Geldern, C. E., Jr., 1965. Evidence of American shad reproduction in a landlocked environment. Calif. Fish Game, 51(3): 212-13.

V82 Vladykov, V. D., 1950. Movements of Quebec shad, *Alosa sapidissima,* as demonstrated by tagging. Canad. Nat., 77(5-6): 121-35.

V83 Velasquez, G. T., 1939. On the viability of algae obtained from the digestive tract of the gizzard shad, *Dorosoma cepedianum* (LeSueur). Amer. Midl. Nat., 22: 376-405.

V84 Voix, C. A., 1967. Bisexual cutthroat trout. Progr. Fish Cult., 29(1): 51-2.

V85 Vandermeer, J. H., 1966. Statistical analysis of geographic variation of fathead minnow, *Pimephales promelas.* Copeia, 1966(3): 457-66.

W1 Wales, J. H., 1946. Castle Lake trout investigations. First phase: interrelationships of four species. Calif. Fish Game, 32(3): 109-43.

W2 _____, 1947. Castle Lake trout investigations: 1946 catch and chemical removal of all fish. Calif. Fish Game, 33(4): 267-86.

W3 _____, 1947. Growth rate and fin regeneration in trout. Progr. Fish Cult., 9(2): 86-9.

W4 _____ and R. C. Lewis, 1935. Progress report of trout feeding experiments. Calif. Fish Game, 21(2): 110-24.

W5 _____ and _____, 1936. Progress report of trout feeding experiments, 1935. Calif. Fish Game, 22(2): 105-17.

W7 Ward, H. C., 1949. A study of fish populations, with special reference to the white bass, *Lepibema chrysops* (Rafinesque), in Lake Duncan, Oklahoma. Master's thesis Univ. Okla. 44 p. typewritten.

W8 Warfel, H. E., T. P. Frost, and W. H. Jones, 1943. The smelt, *Osmerus mordax,* in Great Bay, New Hampshire. Trans. Amer. Fish. Soc., 72: 257-62.

W9 Wascko, H., 1939. Notes on bait propagation. Ohio Div. Conserv. Nat. Resour. Leafl., 156.

W10 Washburn, G. N., 1948. Propagation of the creek chub in ponds with artificial raceways. Trans. Amer. Fish. Soc., 75: 336-50.

W12 Webster, B. O., 1930. Propagation of muskellunge. Trans. Amer. Fish. Soc., 59: 202-3.

W13 Webster, D. A., 1942. The life histories of some Connecticut fishes. A fishery survey of important Connecticut lakes. Bull. Geol. Nat. Hist. Surv. Conn., 63: 122-227.

W15 Went, A. E. J., 1938. Salmon of the River Shannon. Proc. Roy. Irish Acad., 44B(11): 261-322.

W17 _____, 1941. Salmon of the Owenduff (Ballycroy) River. Proc. Roy. Irish Acad., 47B(6): 161-78.

W18 _____, 1941. Salmon of the Ballisodare River. II. Age and growth. Sci. Proc. Roy. Dublin Soc., NS 22: 327-44.

W19 _____, 1942. Salmon of the River Erne. Results of the examination of a small collection of scales and data. Sci. Proc. Roy. Dublin Soc., NS 22: 471-80.

W20 Went, A. E. J., 1943. Salmon of the River Shannon. Proc. Roy. Irish Acad., 49B: 151-75.

W21 ____, 1943. Salmon of the River Corrib, together with notes on the growth of brown trout in the Corrib system. Proc. Roy. Irish Acad., 48B: 269-98.

W22 ____, 1946. River Liffey Survey. VII. Salmon of the River Liffey. Proc. Roy. Irish Acad., 51B: 9-26.

W23 ____, 1948. Salmon and sea trout of the River Inny. Sci. Proc. Roy. Dublin Soc., NS 24(29): 335-47.

W24 ____ and W. E. Frost, 1942. The growth of brown trout (*Salmo trutta* L.) in alkaline and acid waters. Proc. Roy. Irish Acad., 48B: 67-84.

W26 Westman, J. R., 1938. Studies on reproduction and growth of blunt-nosed minnow, *Hyborhynchus notatus*. Copeia, 1938: 57-61.

W27 Weyer, A. E., 1940. The Lake of the Ozarks—a problem in fishery management. Progr. Fish Cult., 51: 1-10.

W28 Whitaker, H., 1886. The Michigan grayling. Trans. Amer. Fish Cult. Assoc. 1885, 15: 59-67.

W29 White, H. C., 1924. A quantitative determination of the number of sur-vivors from planting 5000 trout fry in each of two streams. Contr. Canad. Biol., NS 2(9): 137-52.

W30 ____, 1930. Trout fry planting experiments in Forbes Brook, P. E. I., in 1928. Contr. Canad. Biol., NS 5: 203-11.

W31 ____, 1940. Life history of sea-running brook trout *(Salvelinus fonti-nalis)*. J. Fish. Res. Bd. Can., 5(2): 176-86.

W34 Wiesner, E. R., 1934. Investigations on the artificial insemination and hatching of brook trout eggs from different groups of brood fish and their further development in ponds to the age of one year. Z. Fisch., 33(1): 69-151.

W35 Wilkinson, J. T., 1939. Notes on the use of supplements for fresh meat in the propagation of brook, rainbow and brown trout in Michigan. Trans. Amer. Fish. Soc., 68: 96-117.

W37 Williams, J. E., 1948. The muskellunge in Michigan. Mich. Conserv., 17(10): 10-11,15.

W38 Williamson, L. O., 1940. Muskellunge tagging—Progress Report No. 1. Wis. Conserv. Bull., 5(6): 51-3.

W39 ____, 1940. Length-weight relationship of fish. Wis. Conserv. Bull., 5(9): 37-9.

W40 ____, 1942. Spawning habits of muskellunge, northern pike. Wis. Conserv. Bull., 7(5): 10-11.

W41 Wilson, C. H., 1913. The whitefish, minimum size limits. The scales vs. the yard stick. Trans. Amer. Fish. Soc., 42: 89-100.

W42 Wingfield, C. A., 1940. The effect of certain environmental factors on the growth of brown trout (*Salmo trutta* L.). J. Exp. Biol. (Lond.), 17: 435-48.

W43 Worth, S. G., 1893. Observations on the spawning habits of the shad. Bull. U. S. Fish Comm., 11: 201-6.

W44 Worthington, E. B. and G. H. Swynnerton, 1939. The growth of brown trout. Rep. Freshw. Biol. Assoc. Plymouth, 7: 66-71.

W47 Wunder, W., 1939. Die "Hungerform" und die "Mastform" das Karpfens (*Cyprinus carpio* L.). Z. Morph. Ökol. Tiere, 35: 594-614.

W51 Walker, M. C. and P. T. Frank, 1952. The propagation of buffalo. Progr. Fish Cult., 14(3): 129-30.

W52   Wallis, O. L., 1951. The status of the fish fauna of the Lake Mead National Recreational Area, Arizona-Nevada. Trans. Amer. Fish. Soc., 80: 84-92.

W53   Ward, H. C., 1951. A study of fish populations, with special reference to the white bass, *Lepibema chrysops* (Rafinesque), in Lake Duncan, Oklahoma. Proc. Okla. Acad. Sci., 30: 69-84.

W56   Weisel, G. F. and H. W. Newman, 1951. Breeding habits, development and early life history of *Richardsonius balteatus*, a northwestern minnow. Copeia, 1951(3): 187-94.

W57   Weiss, G., 1952. Armored gars. Mo. Conserv., 13(4): 16.

W58   Went, A. E. J., 1944. Sea trout of the Waterville (Currane) River. Sci. Proc. Roy. Dublin Soc., NS 23(20): 201-13.

W59   _____, 1947. Salmon of the Kerry Blackwater. Sci. Proc. Roy. Dublin Soc., NS 24(20): 179-87.

W60   _____, 1949. Sea trout of the Owengowla (Gowla River). Sci. Proc. Roy. Dublin Soc., NS 25(5): 55-64.

W64   West Virginia Conservation, 1951. Two fish of record size for West Virginia lakes. W. Va. Conserv., 15(8): 18.

W66   Wilde, C. W. and V. B. Romeo, 1951. Practical gill net sizes for sampling average Connecticut lakes. Progr. Fish Cult., 13(3): 153-6.

W67   Williams, J. E., 1951. The lake sturgeon. Mich. Conserv., 20(6): 15-18.

W71   Wohlschlag, D. E. and C. A. Woodhull, 1953. The fish populations of Salt Springs Valley Reservoir, Calaveras County, California. Calif. Fish Game, 39(1): 5-14.

W72   Wolf, L. E., 1952. Experiments with antibiotics and vitamin B_{12} in the diets of brown trout fingerlings. Progr. Fish Cult., 14(4): 148-53.

W74   Wales, J. H. and M. Moore, 1938. Progress of trout feeding experiments, 1937. Calif. Fish Game, 24(2): 126-32.

W75   Wilder, D. G., 1952. A comparative study of anadromous and freshwater populations of brook trout (*Salvelinus fontinalis* (Mitchill)). J. Fish. Res. Bd. Can., 9(4): 169-203.

W76   Went, A. E. J., 1950. Salmon of the River Shannon in 1946 and 1947. J. Cons. Int. Explor. Mer., 16(3): 341-57.

W78   Wynne-Edwards, V. C., 1952. Freshwater vertebrates of the Arctic and subarctic. Fish. Res. Bd. Can. Bull, 94: 1-28.

W79   Walburg, C. H., 1957. Observations on the food and growth of juvenile American shad, *Alosa sapidissima*. Trans. Amer. Fish. Soc., 86: 302-6.

W81   Wales, J. H., 1957. Castle Lake investigation—progress report for 1956. Calif. Inl. Fish. Admin. Rep., 57-6: 1-23.

W82   _____ and E. R. German, 1956. Castle Lake investigations—Second phase: eastern brook trout. Calif. Fish Game, 42(2): 93-108.

W84   Warner, K., 1958. Landlocked salmon *Salmo salar* Linnaeus. Fishes of Maine, 2nd ed., ed. W. H. Everhart: 30-32.

W85   _____, 1959. Migration of landlocked salmon in the Fish River Lakes, Maine. J. Wildl. Mgmt., 23(1): 17-27.

W85a   _____, 1959. Migration of landlocked salmon in the Fish River chain of lakes. Trans. N. E. Wildl. Conf., 10: 72-3.

W86   Waters, J., 1953. There's a tiger in our waters. Ohio Conserv. Bull., 1953(Oct.).

W90   Weatherley, A. H. and A. G. Nicholls, 1956. The effects of artificial enrichment of a lake. Aust. J. Mar. Freshw. Res., 6(3): 443-68.

W91   Webster, D. A., 1954. A survival experiment and an example of selective sampling of brook trout *(Salvelinus fontinalis)* by angling and rotenone in an Adirondack pond. N. Y. Fish Game J., 1(2): 214-19.

W93   _____, W. G. Bentley, and J. P. Galligan, 1959. Management of the lake trout fishery of Cayuga Lake, New York, with special reference to the role of hatchery fish. Cornell Univ. Agric. Exp. Sta. Mem., 357: 5-83.

W94   Weise, J. G., 1957. The spring cave-fish, *Chologaster papilliferus*, in Illinois. Ecology, 38(2): 195-204.

W95   Weisel, G. F. and J. B. Dillon, 1954. Observations on the pygmy whitefish, *Prosopium coulteri*, from Bull Lake, Montana. Copeia, 1954(2): 124-7.

W96   Went, A. E. J., 1939. Salmon in Eire. Salm. Trout Mag., 94: 44-55.

W97   _____, 1954. Sea trout of the River Mattock (Boyne). Salm. Trout Mag., 141: 417-23.

W98   Wenzel, L. E. and T. H. Leik, 1956. A fish management study of Guernsey Reservoir and the adjacent portion of North Platte River. Wyo. Game Fish Comm. Fish. Tech. Rep., 5: 41 p.

W100   Westerman, F. A., 1954. Fish Division. Mich. Dept. Conserv. Bienn. Rep., 17: 77-120.

W101   Wetherbee, J. J., 1956. Central Willamette. Annu. Rep. Ore. St. Game Comm. Fish. Div., 1955: 51-61.

W105   Whitacre, M. A., 1952. The fishes of Crab Orchard Lake, Illinois. Midw. Wildl. Conf., 14: 41 p. mimeo.

W108   Wich, K. F. and J. W. Mullan, 1958. A compendium of the life history and ecology of the chain pickerel, *Esox niger* (LeSueur). Mass. Fish. Bull., 22: 27 p.

W109   Wigley, R. L., 1959. Life history of the sea lamprey of Cayuga Lake, New York. Fish. Bull. U. S., 59(154): 561-617.

W112   Wirszubski, A. and P. Ivri, 1954. Breeding of mother carp in 1951-52. Bamidgeh, 6(1): p. iii.

W113   Wirth, T. L., 1954. Fishes of Lake Winnebago. Midw. Wildl. Conf., 16: 10 p. mimeo.

W115   Wisconsin Conservation Bulletin, 1953. [Picture of rock sturgeon.] Wis. Conserv. Bull., 18(4): 32.

W117   Withler, I. L., 1956. A limnological survey of Atlin and Southern Tagish Lake. B. C. Game Comm. Mgmt. Publ., 5: 36 p.

W120   Wohlschlag, D. E., 1954. Growth peculiarities of the cisco, *Coregonus sardinella* (Valenciennes), in the vicinity of Point Barrow, Alaska. Stanford Ichthyol. Bull., 4(3): 189-209.

W121   _____, 1954. Mortality rates of whitefish in an arctic lake. Ecology, 35(3): 388-96.

W122   _____, 1956. Information from studies of marked fishes in the Alaskan Arctic. Copeia, 1956(4): 237-41.

W124   Worthington, E. B., 1940. Rainbows. Salm. Trout Mag., 100: 241-62.

W125   _____, 1941. Rainbows. Salm. Trout Mag., 101: 62-99.

W128   Webster, D. A., W. A. Lund, Jr., R. W. Wahl, and W. D. Youngs, 1960. Observed and calculated lengths of lake trout *(Salvelinus namaycush)* in Cayuga Lake, New York. Trans. Amer. Fish. Soc., 89(3): 274-9.

W129   Wirth, T. L., 1958. Data on lake sturgeon. Data for Handbook of Biological Data. 6 p.

W130   Williams, J. E., 1955. Determination of age from the scales of northern pike (*Esox lucius* L.). PhD thesis Univ. Mich. 185 p.

W132   Ward, C. M., 1960. A survey of the fishes of Nolichacky River. Tenn. Game Fish Comm. 30 p. mimeo.

W133   Ward, J. C., 1951. The biology of the arctic grayling in the southern Athabaska Drainage. MS thesis Univ. Alta. 71 p.

W134   Wallis, O. L., 1948. Trout studies and a stream survey of Crater Lake National Park, Oregon. MS thesis Ore. St. Coll. 124 p.

W137   Wardle, W. D., 1953. An ecological study of a farm fish pond. MS thesis Univ. Utah. 80 p.

W139   Wilder, D. G., 1944. A comparative study of anadromous and freshwater populations of the eastern speckled trout *(Salvelinus fontinalis)*. PhD thesis Univ. Toronto. 115 p. + append.

W141   Walburg, C. H., 1960. Abundance and life history of shad, St. Johns River, Florida. Fish. Bull. U. S., 60(177): 487-501.

W142   Walker, G. W., 1951. A fish population study of an artificial lake in southern Illinois. MS thesis South. Ill. Univ. 33 p.

W143   Webster, D. A., 1958. Cayuga lake trout, Part 3: Their food, growth, survival and management. N. Y. St. Conserv., 13(2): 11-13.

W144   Westman, J. R., 1941. A consideration of population life-history studies in their relation to the problems of fish management research, with special reference to the small-mouthed bass, *Micropterus dolomieu* Lacépède, the lake trout, *Cristivomer namaycush* (Walbaum), and the mud minnow, *Umbra limi* (Kirtland). PhD thesis Cornell Univ. 182 p.

W145   Welsh, J. P., 1952. A population study of Yellowstone blackspotted trout (*Salmo clarkii lewisi* Girard). PhD thesis Stanford Univ. 180 p.

W146   Watling, H. and C. J. D. Brown, 1955. The embryological development of the American grayling *(Thymallus signifer tricolor)* from fertilization to hatching. Trans. Amer. Microscop. Soc., 74(1): 85-93.

W147   Walker, B. W., R. R. Whitney, and G. W. Barlow, 1961. The fishes of the Salton Sea. Calif. Fish. Bull., 113: 77-91.

W150   Wales, J. H. and D. P. Borgeson, 1961. Castle Lake investigation—third phase: rainbow trout. Calif. Fish Game, 47(4): 399-414.

W151   Wohlfarth, G., R. Moav, and M. Lahman, 1961. Genetic improvement of carp. III. Progeny tests for differences in growth rate, 1959-60. Bamidgeh, 13(2): 40-54.

W152   Warner, K. and K. A. Havey, 1961. Body-scale relationships in landlocked salmon, *Salmo salar*. Trans. Amer. Fish. Soc., 90(4): 457-61.

W153   _____, 1961. A new longevity record for the landlocked population of *Salmo salar*. Copeia, 1961(4): 483-4.

W154   _____, 1962. Contribution of hatchery-reared salmon to the fishery of the Fish River Lakes, Maine. Trans. Amer. Fish. Soc., 91(1): 99-102.

W156   Wojno, T., 1961. [The growth of Wdzydze Lake trout (*Salmo trutta morpha lacustris* L.)] (English summary.) Rocz. Nauk Rolniczych, 93: 649-80.

W157   _____, 1961. [The feeding of trout (*Salmo trutta morpha lacustris* L.) from Wdzydze Lake.] (English summary.) Rocz. Nauk Rolniczych, 93: 681-702.

W158   Warner, K., 1962. The landlocked salmon spawning run at Cross Lake Thoroughfare, Maine. Copeia, 1962(1): 131-8.

W159   _____ and O. C. Fenderson, 1962. Effect of DDT spraying for forest insects on Maine trout streams. J. Wildl. Mgmt., 26(1): 86-93.

W160   Webster, D. A., 1962. Artificial spawning facilities for brook trout, *Salvelinus fontinalis*. Trans. Amer. Fish. Soc., 91(2): 168-74.

W161   Wickliff, E. L., 1933. The practical value of determining the fertility of whitefish eggs. Trans. Amer. Fish. Soc., 63: 144-50.

W163   Webster, D. A. and W. A. Flick, 1960. Results of planting kokanee salmon in two Adirondack Mountain lakes, New York. Progr. Fish Cult., 22(2): 59-63.

W165 White, H. C., 1941. Migrating behaviour of sea-running *Salvelinus fontinalis*. J. Fish. Res. Bd. Can., 5(3): 258-64.

W166 _____, 1942. Sea life of the brook trout *(Salvelinus fontinalis)*. J. Fish. Res. Bd. Can., 5(5): 471-3.

W167 Wood, R., R. H. Macomber, and R. K. Franz, 1956. Trends in fishing pressure and catch, Allatoona Reservoir, Georgia, 1950-53. J. Tenn. Acad. Sci., 31(3): 215-23.

W168 Wilkens, L. P., 1956. Trout management investigations. Tenn. Dingell-Johnson Proj., F-6-R, 1956(Sept.16): 9 p. mimeo.

W169 _____, 1957. Trout management investigations. Tenn. Dingell-Johnson Proj., F-6-R, 1957(Sept.16).

W170 White, H. C., 1936. Age determinations of salmon parr by effect of rate of growth on body proportions. J. Biol. Bd. Can., 2(4): 379-82.

W171 _____, 1937. Local feeding of kingfishers and mergansers. J. Biol. Bd. Can., 3: 323-38.

W172 _____, 1939. Bird control to increase the Margaree River salmon. Bull. Fish. Res. Bd. Can., 58: 1-30.

W174 Woynarovich, E., 1962. Hatching of carp eggs in "zugar" glasses and breeding of carp larvae until an age of 10 days. Bamidgeh, 14(2): 38-46.

W175 Warner, K., 1963. Natural spawning success of landlocked salmon, *Salmo salar*. Trans. Amer. Fish. Soc., 92(2): 161-4.

W176 Walburg, C. H., 1963. Parent-progeny relation and estimate of optimum yield for American shad in the Connecticut River. Trans. Amer. Fish. Soc., 92: 436-9.

W178 Wolny, P., 1962. [The influence of increasing the density of stocked fish population on the growth and survival of carp fry.] (Polish with English summary.) Rocz. Nauk Rolniczych, 81B(2): 171-88.

W180 Wojcik, F. J., 1955. Life history and management of the grayling in interior Alaska. MS thesis Univ. Alaska. (Not seen, quoted by K72.)

W181 Wistrom, C. A., B. Apelgren, and C. W. Threinen, 1957. Northern pike *Esox lucius*. Data for Handbook of Biological Data. 7 p.

W182 Walter, E., 1934. Grundlagen der allgemeinen fischereilichen Produktionslehre. Handb. Binnenfisch. Mitteleurop., IV.

W183 Weisel, G. F., 1954. A rediscovered cyprinid hybrid from western Montana, *Mylocheilus caurinum* x *Richardsonius balteatus balteatus*. Copeia, 1954(4): 278-82.

W184 _____, 1955. Three new intergeneric hybrids of cyprinid fishes from western Montana. Amer. Midl. Nat., 53(2): 396-411.

W185 _____, 1957. Redside shiner, *Richardsonius balteatus balteatus* (Richardson). Data for Handbook of Biological Data. 8 p.

W186 Wynne-Edwards, V. C., 1932. The breeding habits of the black-headed minnow *(Pimephales promelas* Raf.). Trans. Amer. Fish. Soc., 62: 382-3.

W187 Wiebe, A. H., 1935. The pond culture of black bass. Tex. Game Fish Oyster Comm. Bull., 8: 1-58.

W188 Wascko, H. and C. F. Clark, 1948. Pond propagation of bluntnose and blackhead minnows. Ohio Wildl. Conserv. Bull., 4: 1-16.

W189 Westman, J. R. and W. E. Fahy, 1940. Carp problem of the area. Survey of the Lake Ontario watershed. Annu. Rep. N. Y. Conserv. Dept., 29(Suppl.): 226-31.

W190 Wunder, W., 1949. Fortschrittliche Karpfenteichwirtschaft. E. Schweizerbartsche Verlags., Stuttgart. 385 p.

W191 Wales, J. H., 1942. Carp control work in Lake Almanor, 1941. Calif. Fish Game, 28(1): 28-33.

W192 Walburg, C. H., 1964. Fish population studies, Lewis and Clark Lake, Missouri River, 1956 to 1962. Spec. Sci. Rep., U. S. Fish Wildl. Serv., 482: 1-27.

W193 Webster, D. A., 1958. Adirondack League Club fishery management report for 1957. 26 p. mimeo.

W194 Watson, N. H. F., 1963. Summer food of lake whitefish, *Coregonus clupeaformis* Mitchill, from Heming Lake, Manitoba. J. Fish. Res. Bd. Can., 20(2): 279-86.

W195 Wells, L., 1966. Seasonal and depth distribution of larval bloaters *(Coregonus hoyi)* in southeastern Lake Michigan. Trans. Amer. Fish. Soc., 95(4): 388-96.

W196 Whitt, C. R., 1958. Age and growth characteristics of Lake Pend Oreille kokanee, 1956. Idaho Dept. Fish Game D.-J. Annu. Progr. Rep., F-3-R-6: 34 p. not seen, quoted in S480.

W197 Webster, D. A., 1955. Adirondack League Club fishery management report for 1954. 26 p.

W198 _____, 1956. Adirondack League Club fishery management report for 1955. 26 p.

W199 _____, 1960. Adirondack League Club fishery management report for 1959. 33 p.

W200 _____, 1963. Adirondack League Club fishery management report for 1962. 31 p.

W201 Withler, I. L., 1966. Variability in life history characteristics of steelhead trout *(Salmo gairdneri)* along the Pacific coast of North America. J. Fish. Res. Bd. Can., 23(3): 365-93.

W202 Wagner, H. H., 1967. A summary of investigations of the use of hatchery-reared steelhead in the management of a sport fishery. Ore. St. Game Comm. Fish. Dept., 5: 62 p.

W203 Walker, J. D., 1963. Trout movement and harvest in the Mad River, Ohio. Ohio Div. Wildl. Publ., W-65: 3 p.

W204 Wendt, C., 1965. Liver and muscle glycogen and blood lactate in hatchery-reared *Salmo salar* L. following exercise in winter and summer. Rep. Inst. Freshw. Res. Drottning., 46: 167-84.

W205 Weatherley, A. H., 1958. Growth, production and survival of brown trout in a large farm dam. Aust. J. Mar. Freshw. Res., 9: 159-66.

W206 Wydoski, R. S. and E. L. Cooper, 1966. Maturation and fecundity of brook trout from infertile streams. J. Fish. Res. Bd. Can., 23(5): 623-49.

W207 Webster, D. A., 1961. Adirondack League Club fishery management report for 1961. 29 p.

W208 _____, 1960. Adirondack League Club fishery management report for 1960. 43 p.

W210 Wesloh, M. L. and D. E. Olson, 1962. The growth and harvest of stocked yearling northern pike, *Esox lucius* Linnaeus, in a Minnesota walleye lake. Invest. Rep. Minn. Dept. Conserv., 242: 9 p.

W211 Wohlfarth, G., M. Lahman, and R. Moav, 1963. Genetic improvement of carp. IV. Leather and line carp in fish ponds of Israel. Bamidgeh, 15(1): 3-8.

W212 Wenke, T. L., 1960. Probable maximum age of native Michigan minnows, family Cyprinidae. MS thesis Univ. Mich. Lib. (not seen).

W213 Wyatt, H. N. and H. A. Zeller, 1965. Fish population dynamics following a selective shad kill. Proc. S. E. Assoc. Game Fish Comm., 16: 411-18.

W214 Ward, C. M. and W. H. Irwin, 1961. The relative resistance of thirteen species of fishes to petroleum refinery effluent. Proc. S. E. Assoc. Game Fish Comm., 15: 255-76.

W215   Wunder, W., 1930. Experimentelle Untersuchungen an dreistachligen Stichling (*Gasterosteus aculeatus* L.) während der Laichzeit. Z. Morph. Ökol. Tiere, 16: 453-98.

W216   _____, 1928. Experimentelle Untersuchungen an Stichlingen (Kampfe, Nestbau, Laichen, Brutpflege). Zool. Anz., 3(Suppl.) [Verh. Deutsch. Zool. Ges., 32]: 115-27.

W217   Winn, H. E., 1960. Biology of the brook stickleback, *Eucalia inconstans* (Kirtland). Amer. Midl. Nat., 63(2): 424-38.

W218   Walford, L. A., 1946. A new graphic method of describing the growth of animals. Biol. Bull. Wood's Hole, 90(2): 141-7.

Y1   Yarrow, H. C., 1874. Notes on the shad as observed at Beaufort Harbor, North Carolina, and vicinity. Rep. U. S. Comm. Fish Fish., 1871-72(Append.C): 452-6.

Y2   Yashouv, A., 1954. The value of natural food in fish breeding. Bamidgeh 6(3): 103-8.

Y3   _____, 1955. The "Punten" carp and its attributes. Bamidgeh, 7(3): 46-54.

Y4   _____, 1955. Barrages lakes in Israel and their utilization in fish culture. Bamidgeh, 7(2): 19-34.

Y5   _____, 1958. Winter culture of carps at the fish culture research station Dor. Bamidgeh, 10(4): 85-90.

Y6   _____, 1958. The excreta of carp as a growth limiting factor. Bamidgeh, 10(4): 90-95.

Y7   _____, 1958. Report on the growth of the buffalo fish *(Megastomatobus cyprinella)*. Bamidgeh, 10(4): 81-4.

Y9   Yoshihara, T., 1952. Effect of population density and pond-area on the growth of fish. J. Tokyo Univ. Fish., 39(1): 47-61.

Y10   _____, 1952. A table and alinement chart to facilitate the calculation of the condition factor. J. Tokyo Univ. Fish., 39(1): 63-7.

Y13   Youngs, W. D., 1958. Effect of the mandible ring tag on growth and condition of fish. N. Y. Fish Game J., 5(2): 184-205.

Y14   Yashouv, A., 1959. Studies on the productivity of fish ponds. I. Carrying capacity. Proc. Gen. Fish. Coun. Medit., 5: 409-19.

Y18   Yamada, J., 1961. Studies on the structure and growth of the scales in the goldfish. Mem. Faculty Fish. Hokkaido Univ., 9(2): 181-226.

Y19   Youn, P. M. D., 1962. Comparisons of some carp populations. MS thesis Ia. St. Univ. 48 p.

Y20   Yessipov, W. K., 1935. Materials on the life history and fishery of the char in Novaya Zemlya (*Salvelinus alpinus* L.). Trans. Arct. Inst. Leningr., 17: 5-70.

Z2   Zilliox, R. G. and M. Pfeiffer, 1956. Restoration of brook trout fishing in a chain of connected waters. N. Y. Fish Game J., 3(2): 167-90.

Z3   Zilliox, R. G. and W. D. Youngs, 1958. Further studies on the smelt of Lake Champlain. N. Y. Fish Game J., 5(2): 164-75.

Z4   _____ and M. Pfeiffer, 1960. The use of rotenone for management of New York trout waters. Canad. Fish Cult., 28: 3-12.

Z5   Zawisza, J. and B. Antosiak, 1961. The rate of growth of tench (*Tinca tinca* L.) in lakes of Wegorzewo District. Polish Agric. Annu., 77B(2): 524-5.

Z6   Zimmer, P. D., 1967. A note about squawfish. Progr. Fish Cult., 29(1): 35.

Author Index

Subject Index

739

SUBJECT INDEX